19 20/N EE
P9-CKX-139

Reverse Osmosis

REVERSE OSMOSIS

S. SOURIRAJAN

Division of Applied Chemistry
National Research Council of Canada
Ottawa

Logos Press Limited

Published by
LOGOS PRESS LIMITED
in association with
ELEK BOOKS LIMITED
2 All Saints Street, London, N.1.

Library of Congress Catalog Card Number 77-80935

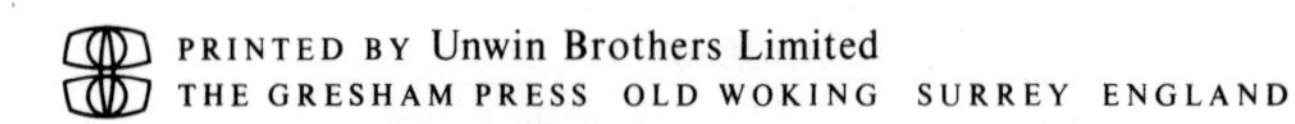
PRINTED BY Unwin Brothers Limited
THE GRESHAM PRESS OLD WOKING SURREY ENGLAND

Produced by 'Uneoprint'

A member of the Staples Printing Group (UO6096)

DEDICATED TO MY PAST, PRESENT, AND FUTURE ASSOCIATES

Contents

Preface

The Reverse Osmosis Process is a general separation technique. In principle, the process is applicable for the separation, concentration, and fractionation of inorganic or organic substances in aqueous or nonaqueous solutions in the liquid or the gaseous phase, and hence it opens a new and versatile field of separation technology in chemical process engineering. To present this point of view is the purpose of this monograph.

By far the major current research effort in Reverse Osmosis Process is in the field of water treatment in general, and saline water conversion in particular. This application is, and will always be, one of tremendous engineering, economic, and social significance. This monograph is particularly dedicated to the advancement of this application.

Even more significant is the possibility of developing the technique as a general process for the separation of any substance in solution. It is the object of this monograph to contribute to the development of the science, technology, and engineering involved in the Reverse Osmosis Membrane Separation Process in all its applications.

Ottawa
1969

S. Sourirajan

Note

In this book, an extended summary is given at the end of each chapter. Readers who wish to concentrate on certain parts of the text will find that the summaries enable them to appreciate the essence of the rest.

For the solute transport parameter, the expression $(D_{AM}/K\delta)$ is used throughout this book, although it may appear more convenient to use a single symbol in its place. The intention of this practice is to reinforce the fact that $(D_{AM}/K\delta)$ is not a single quantity and is not a simple proportionality constant; it is a parameter of much physical and practical significance, involving several quantities.

Pertinent references to the application of irreversible thermodynamics to reverse osmosis are included in the literature list. This book does not use the approach of irreversible thermodynamics to reverse osmosis. The book provides a more practical and effective approach to the engineering development of the subject based on experimental reverse osmosis data.

Chapters 3, 4, 5, 6, and 8 include extensive mathematical analysis; a list of symbols used appears at the end of each of these chapters.

As in most current literature in industrial and engineering chemistry, both metric and imperial units of mass, length, area, and volume are used in this book.

The figures given in Chapter 8 for the cost of certain reverse osmosis operations are not intended as a real illustration of actual costs; the object of this discussion is simply to illustrate the analytical technique which can be used for such studies.

Acknowledgments

The author expresses his deep gratitude to (late) Professor S. T. Yuster who initiated him in the field of Reverse Osmosis research in 1956, and to (late) Professor L. M. K. Boelter, and Professors W. C. Hurty, Myron Tribus, J. W. McCutchan, and G. C. Kennedy for their generous support to his work at the University of California, Los Angeles, during 1956-61, and to Dr. I. E. Puddington, Dr. G. L. Osberg, and Mr. W. S. Peterson for their generous and continuing support to his work at the National Research Council of Canada, Ottawa, since 1961.

Most of the material presented in this book formed part of the author's course on Reverse Osmosis given at the Thayer School of Engineering, Dartmouth College, Hanover, N.H., U.S.A., during 8-19 July 1968; the author is particularly grateful to Professor Myron Tribus (formerly Dean of the Thayer School of Engineering) and every one of the participants in the course, for their valuable contributions to the final form of this book.

The author gratefully acknowledges the permission granted to him by the following publishers and journals to include in this book many of their copyrighted materials, and figures and tables indicated below:

American Association for the Advancement of Science, Washington, D.C., U.S.A.
Table: 1-10

American Chemical Society, Washington, D. C., U. S. A.
Figures: 1-1, 1-2, 1-4, 1-9, 1-16, 1-17, 1-18, 1-19, 1-20, 1-21, 1-22, 1-23, 1-24, 1-26, 1-29, 1-33, 2-4, 2-6, 2-8, 2-12, 2-30, 2-31, 3-2, 3-3, 3-4, 3-7, 3-8, 3-9, 3-10, 3-11, 3-12, 3-13, 3-14, 3-15, 3-16, 3-17, 3-18, 3-19, 3-20, 3-21, 3-22, 3-23, 3-24, 3-26, 3-27, 3-28, 3-31, 3-32, 3-33a, 3-38, 3-39a, 3-39b, 3-40, 3-42, 3-43, 3-45, 3-46, 3-48, 3-50, 3-51, 3-52, 3-53, 3-54, 3-55, 3-56, 3-57, 3-58, 3-59, 3-60, 3-61, 3-62, 3-63, 3-64, 3-65, 3-66, 3-67, 3-68, 3-69, 3-70, 3-71, 3-72, 3-73, 3-74, 3-75, 3-76, 3-77, 3-78, 3-79, 3-80, 3-81, 3-82, 3-83, 4-2, 4-3, 4-4, 4-5, 4-6, 4-7, 4-8, 4-9, 4-10, 4-11, 4-12, 4-13, 4-14, 4-16, 4-17, 4-18, 4-19, 4-20, 4-21, 4-22, 4-23, 4-24, 4-25, 4-26, 4-29, 4-30, 4-31, 4-34, 4-35, 4-36, 4-49, 4-50, 4-51, 4-52, 4-53, 4-54, 5-1, 5-2, 5-3, 5-4, 5-5, 5-7, 5-8, 5-9, 5-10, 5-12, 6-1, 6-2, 6-3, 7-5, 7-6, 7-7, 7-8, 7-9, 7-10, 7-11, 7-12, 7-13, 8-1
Tables: 1-5, 1-6, 1-7, 2-7, 2-8, 2-9, 2-11, 2-12, 2-13, 2-38, 2-44, 3-2, 3-3, 3-4, 3-5, 3-6, 3-7, 3-8, 3-9, 3-10, 3-11, 4-1, 4-2, 6-4, 6-5, 6-6, 6-7, 6-8, 7-2, 7-3, 7-4, 7-5, 7-6, 8-1, 8-2, 8-3, 8-4, 8-5, 8-6, 8-7, 8-8, 8-10

American Institute of Chemical Engineers, New York, U.S.A.
Figures: 3-30, 3-47, 3-49, 3-50, 4-1, 4-37, 4-38, 4-39, 4-40, 4-41, 4-42, 4-43, 4-44, 4-45, 4-46, 4-47, 4-48, 8-9, 8-52, 8-53
Tables: 1-11, 1-12, 8-13

Dechema Deutsche Gesellschaft fur Chemisches Apparatwesen, Frankfurt, W. Germany
Figures: 2-8, 2-9, 2-10, 8-6, 8-7, 8-8

Elsevier Publishing Co., Amsterdam, The Netherlands
Figures: 2-16, 2-17, 2-18, 2-19, 2-20, 2-21, 2-22, 2-23, 2-24, 8-10, 8-39, 8-40, 8-41, 8-42, 8-43, 8-44, 8-45, 8-46, 8-47, 8-48
Tables: 2-39, 2-40, 2-41, 2-42, 2-43, 8-25, 8-26, 8-30, 8-31, 8-32, 8-33

Gulf General Atomic Inc., San Diego, California, U.S.A.
Figures: 8-22, 8-31, 8-32, 8-33
Tables: 8-19, 8-23, 8-24

Havens Industries, San Diego, California, U.S.A.
Figures: 8-11, 8-12, 8-13
Plates: 8-2, 8-3
Table: 8-17

John Wiley & Sons, Inc., New York, U.S.A.
Figures: 2-11, 2-13, 2-29, 2-31, 2-32, 2-33, 2-34, 2-35, 2-36, 2-37, 2-38, 3-61
Tables: 2-24, 2-57, 2-58

Marcel Dekker, Inc., New York, U.S.A.
Figure: 2-19

National Research Council of Canada, Ottawa—Canadian Journals of Research
Figures: 7-2, 7-3, 7-4

Nature, London, England
Tables: 1-8, 1-9, 7-1

Office of Saline Water, U.S. Department of the Interior, Washington, D.C., U.S.A.
Figures: 2-14, 2-15, 2-25, 2-26, 2-27, 3-5, 3-6, 8-15, 8-16, 8-17, 8-18, 8-19, 8-20, 8-21, 8-23, 8-24, 8-25, 8-26, 8-27, 8-28, 8-29, 8-30, 8-38
Plate: 8-5
Tables: 1-2, 1-3, 2-25, 2-26, 2-27, 2-28, 2-29, 2-30, 2-31, 2-32, 2-33, 2-34, 2-35, 2-36, 2-37, 2-44, 2-45, 2-46, 2-47, 2-48, 2-49, 2-50, 2-51, 2-54, 2-55, 8-18, 8-20, 8-21, 8-22, 8-28, 8-29

Pergamon Press Ltd., Oxford, England
Table: 8-9

Pulp Manufacturers Research League, Inc., Appleton, Wisconsin, U.S.A.
Tables: 8-11, 8-12

Society of Chemical Industry, London, England
Figures: 1-25, 1-27, 1-28, 1-30, 1-31, 1-32, 1-34
Tables: 2-14, 2-15, 2-16, 2-17, 2-18, 2-19, 2-20, 2-21, 2-22, 2-23

Southern Research Institute, Birmingham, Alabama, U.S.A.
Table: 8-27

The M.I.T. Press, Cambridge, Massachusetts, U.S.A.
Figure: 8-2

Universal Water Corporation, Del Mar, California, U.S.A.
Figure: 8-14
Plate: 8-4
Tables: 2-52, 2-53

University of California, Los Angeles, California, U.S.A.
Figures: 1-1, 1-2, 1-3, 1-4, 1-5, 1-6, 1-7, 1-8, 1-9, 1-10, 1-11, 1-12, 1-13, 1-14, 1-15, 2-1, 2-2, 2-3, 2-4, 2-5, 2-6, 2-7, 8-3, 8-4, 8-5
Plate: 1-1
Tables: 1-1, 1-4, 2-1, 2-2, 2-3, 2-4, 2-5, 2-6, 2-10, 6-2, 8-14, 8-15, 8-16

University of Toronto Press, Toronto, Canada
Figures: 1-1, 1-2, 1-16, 1-17, 2-25, 3-5, 3-6, 3-8, 3-11, 3-16, 3-47, 3-49
Tables: 2-55, 8-10

Reference to the source of information of all the materials used is given at the appropriate places in the text, and in every figure and table, and detailed literature references are given in the Literature section of the book.

The author gratefully acknowledges the valuable assistance of Mr. Edward Selover at the University of California, Los Angeles, during 1956-60, and the valuable and continuing assistance of Messrs. A. G. Baxter, A. E. McIlhinney, L. Pageau, N. Scheel, W. L. Thayer and several others in the administrative, technical and secretarial staff of the National Research Council of Canada, to his work since 1961.

CHAPTER 1

General Applicability of the Technique

The 'Reverse Osmosis Membrane Separation Process' (also called 'Ultrafiltration' or 'Hyperfiltration' in the literature) is a general and widely applicable technique for the separation, concentration or fractionation of substances in fluid solutions. It consists in letting the fluid mixture flow under pressure through an appropriate porous membrane, and withdrawing the membrane-permeated product generally at atmospheric pressure and surrounding temperature; the product is enriched in one or more constituents of the mixture, leaving a concentrated solution of other constituents on the upstream side of the membrane. No heating of the membrane, and no phase change in product recovery are involved in this separation process.

This process is an exciting new development in the field of solute-solvent separation. Its most significant application is in the field of saline water conversion, to which problem it owes its origin. In spite of the rapid advances which are being made with respect to this application, the process is still at its very early stages of development. The basic principles involved are still controversial, and no currently available theory on the mechanism of the process is beyond question. However, whatever is known about the process is useful, and what is yet to be known appears full of promise, and that makes the process especially interesting. Several approaches to this subject are in vogue. The purpose of this monograph is to present a review of the fundamental and practical aspects of this separation process from the point of view of the general applicability of the technique.

1. Osmosis and Reverse Osmosis

The term 'Osmosis' is familiarly used to describe the spontaneous flow of pure water into an aqueous solution, or from a less to a more concentrated aqueous solution, when separated by a suitable membrane. In order to obtain potable water from saline water in a similar process, the direction of flow of the pure water must be reversed, i.e. pure water must flow from a more to a less concentrated solution. Consequently the latter process has been conveniently termed 'Reverse Osmosis'. This term has now gained such wide popular usage that it seems necessary to point out that the process is not restricted to the passage of water from aqueous solutions, and that it is not restricted to 100 per cent solute separation, and that neither 'osmosis' nor 'reverse osmosis' is an explanation of the mechanism of the process involved, and hence it is misleading to explain 'reverse osmosis' as the reverse of 'osmosis'. Under isothermal conditions, in both 'osmosis' and 'reverse osmosis', the preferential transport of material through the membrane is always in the direction of lower chemical potential. This is a thermodynamic requirement, and it does not, and cannot, specify which component, if any, of a solution will be preferentially transported through a given membrane, and the mechanism by which such transport takes place. In the literature, the 'osmosis' process is always associated with the existence of a 'semipermeable' membrane; the cause of such semipermeability however continues to remain obscure. Hence it must be understood that the mechanism of both 'osmosis' and 'reverse osmosis' is still an open question, and the distinction between the two terms is entirely one of arbitrary convention and popular usage.

2. The Preferential Sorption-Capillary Flow Mechanism

The Gibbs equation relating interfacial tension (σ) of a solution, and the adsorption (Γ) of the solute at an interface can be given in the form (Harkins and McLaughlin, 1925; McBain, 1950)

$$\Gamma = -\frac{1}{RT}\left[\frac{d\sigma}{d\ln \underline{a}}\right] \tag{1-1}$$

where R is gas constant, T is absolute temperature and $\underline{a}$ is activity of the solute. Equation 1-1 predicts the possible existence of a steep concentration gradient at interfaces. For aqueous sodium chloride solutions, there is a negative adsorption of the solute resulting in a monomolecular layer of pure water at the air-solution interface (Goard, 1925; Harkins and McLaughlin, 1925; Langmuir, 1917; McBain and Dubois, 1929). From Equation 1-1, the thickness, t, of the fresh water layer at the air-solution interface can be derived to be (Harkins and McLaughlin, 1925)

$$t = -\frac{1000\alpha}{2RT}\left[\frac{d\sigma}{d(\alpha m)}\right] \tag{1-2}$$

where α is the activity coefficient of the salt in solution, and m is the molality of solution. Yuster (1956) first suggested the possibility of developing a practical desalination process based on the concept of skimming this surface layer of pure water. This suggestion was the starting point of the author's investigations on the subject at the University of California, Los Angeles (UCLA) in September 1956.

The problems involved in trying to skim a monomolecular layer of pure water are obviously formidable. But it is clear from the Gibbs equation that the thickness of the pure water layer is a function of the nature of the interface. If the latter is such that there exists at the interface a multimolecular layer of pure water, then, from an engineering standpoint, the problem would seem different.

The above considerations led to the conceptual model given in Figure 1-1 for recovering fresh water from aqueous salt solutions (Sourirajan, 1963a, 1967b, 1967c; Yuster et al., 1958). Here the solution is in contact, not with air, but with a solid material in the form of a porous membrane. If only the surface of a porous membrane in contact with the solution is of such a chemical nature that it has a preferential sorption for water or preferential repulsion for the solute, then a multimolecular layer of preferentially sorbed pure water could exist at the membrane-solution interface. A continuous removal of this interfacial water can then be effected by letting it flow under pressure through the membrane capillaries.

This model also gives rise to the concept of a critical pore diameter for maximum separation and permeability. This is obviously twice the thickness, t, of the interfacial pure water layer (Figure 1-2). If the pore diameter is bigger, permeability will be higher but solute separation will be lower since the effective feed solution (concentrated boundary solution) will also flow through the pores; if the diameter is smaller, the separation could be maximum, but permeability will be reduced. When there are no pores at all on the film surface, then there can be no flow of the interfacial fluid through the film material by the above mechanism, and the fluid permeability due to diffusion through the nonporous film surface can only be insignificantly small.

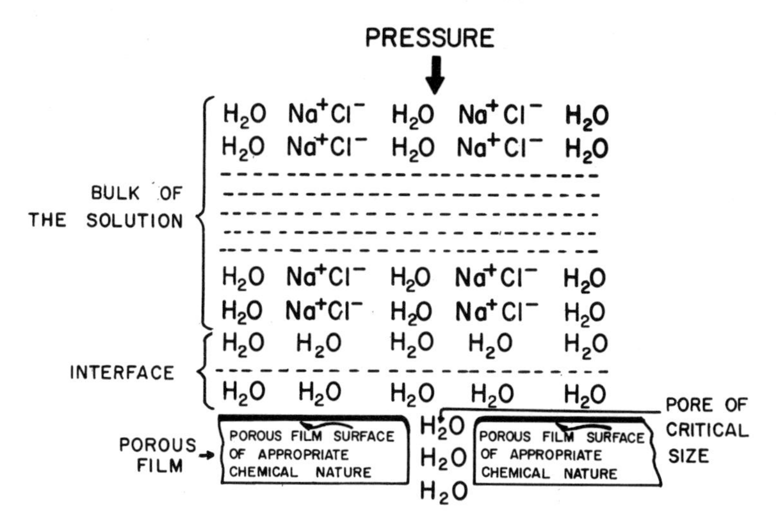

Figure 1-1. Schematic representation of the preferential sorption-capillary flow mechanism (Sourirajan, 1963a).

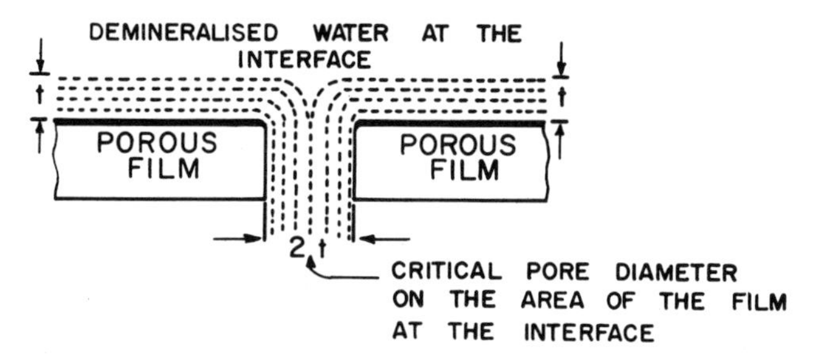

Figure 1-2. Critical pore diameter for maximum separation and permeability (Sourirajan, 1963a).

For maximum separation and permeability, it is necessary to maintain the pore size equal to 2t only on the area of the film at the interface, i.e. on an infinitesimally thin layer of the film at the interface. The connecting pores in the interior bulk of the film material, away from the film-solution interface, could be, and should be, bigger. These requirements are essential from a practical point of view, because then the total resistance to fluid flow will be less, and consequently the operating pressure needed to maintain the continuous flow of the interfacial water at the desired rate will also be less.

This model thus sets the fundamental approach for the practical development of the reverse osmosis process for sea water conversion. Accordingly, the chemical nature of the film surface must be such that it has a preferential sorption for water and/or preferential repulsion for the solutes; and the porous film must have the largest number of pores at size 2t or less on the area of the film (i.e. on the thinnest layer of the film) at the film-solution interface.

The model also illustrates the generality of this separation technique. In principle, this technique is applicable for the separation, concentration or fractionation of inorganic or organic substances in aqueous or nonaqueous solutions in the liquid or the gaseous state, and hence it opens a new and versatile field of separation technology in chemical process engineering. On the basis of the preferential sorption-capillary flow mechanism, an appropriate chemical nature of the film surface in contact with the solution, as well as the existence of pores of appropriate size on the area of the porous film at the interface, constitute the indispensable twin requirement for the success of this separation process.

From an industrial standpoint, the application of the reverse osmosis technique for a given separation problem involves essentially the problem of (a) choosing the appropriate film material whose chemical nature is such that it has a preferential sorption and/or preferential repulsion for one or more constituents of the fluid mixture, and (b) developing methods for making films containing the largest number of pores of the required size on the area of the film (i.e. on the thinnest layer of the film) at the film-solution interface, with comparatively big interconnecting pores in the interior of the film material. This approach formed the basis of the successful development of the Loeb-Sourirajan type porous cellulose acetate membranes for saline water conversion and other applications (Loeb and Sourirajan, 1961, 1963, 1964, 1965; Loeb et al., 1964; Sourirajan and Govindan, 1965; UCLA, 1960).

The Gibbs equation. Equation 1-1 simply gives the thermodynamic condition under which a negatively or positively adsorbed interfacial fluid layer can exist; it says nothing about the possible means of establishing such a condition in any given system. In particular, the ordinary surface tension data at the air-solution interface do not offer a valid means of predicting the preferential sorption characteristics of a membrane. Thus, a positive Gibbs slope at the air-solution interface does not necessarily predict solute rejection by a membrane; also, a negative Gibbs slope at the air-solution interface does not necessarily predict solvent rejection by a membrane. Precise interfacial tension data at solid-solution interfaces are difficult to obtain; in the absence of such data the Gibbs equation predicts nothing about the preferential sorption characteristics of any membrane. Consequently, from a practical point of view, Equation 1-1 by itself is not a very helpful one, but it does point to a new approach for the development of the reverse osmosis separation process. Precise physicochemical criteria of preferential sorption or preferential repulsion based on intermolecular forces at solid-solution interfaces are needed for the full scientific and industrial development of this separation process.

Some consequences of the preferential sorption-capillary flow mechanism. At the present state of development of the subject, the physicochemical criteria of preferential sorption or preferential repulsion are not precisely known, and hence there is no way of determining the thickness t of the preferentially sorbed layer on the membrane surface. Still, this concept has several consequences of practical interest from the point of view of the development of the reverse osmosis membrane separation process in all its applications. The most important one is that the process is governed by a surface phenomenon; hence the chemical nature of the membrane surface, relative to that of the solution in contact with it, is a controlling parameter in reverse osmosis separation. The magnitude of t can be different for different membrane material-solution systems. Therefore an infinite variety of membrane-solution combinations can give rise to different levels of solute separation by the reverse osmosis technique. Also, t is not necessarily a function of the size of the solute and solvent molecules. Hence the critical pore diameter on a membrane surface could be several times bigger than the size of the solute and solvent molecules, and yet solute separation is possible by the reverse osmosis technique. Consequently, this separation process is not simple sieve separation or ultrafiltration.

The higher the value of t, the higher is the permissible pore size on the membrane surface for a given level of solute separation. The size of the pores in the interior bulk of the membrane material is independent of t. Both the porous structure and the chemical nature of the membrane surface together determine the solute and solvent flux through the membrane, which is a function of the magnitude of t, the effective thickness of the membrane, the size, number and distribution of pores on the membrane surface, and the operating pressure and flow conditions in the apparatus. Solute separation and membrane flux are two independent parameters with respect to a given membrane material-solution system.

The solute can be organic or inorganic, or ionic or nonionic. The solution can be aqueous or nonaqueous, and it can be in the liquid or the gaseous state. With respect to a given solution containing components A and B, the membrane-permeated product can be enriched either in A or in B, depending upon the chemical nature of the membrane surface in contact with the solution. This again means that this separation process is not ultrafiltration. With respect to a given membrane material-solution system, in which the membrane surface has a preferential sorption for the solvent, practically any degree of solute separation is possible depending upon the porous structure of the membrane and other operating conditions; when the latter are fixed, the separation characteristics of a membrane can be changed by changing its surface pore structure. Since t can be different for different membrane material-solution systems, a given membrane can give different levels of solute separation for different solution systems under otherwise identical experimental conditions. Some degree of solute separation is possible under any operating pressure which establishes flow of the fluid through the membrane, provided, of course, that the membrane material has a preferential sorption for the solvent. All these consequences have been demonstrated experimentally (Govindan and Sourirajan, 1966; Ironside and Sourirajan, 1967; Kopeček and Sourirajan, 1969d; Sourirajan, 1963a, 1963b, 1964a, 1964b, 1964c, 1965, 1967a: Sourirajan and Govindan, 1965; Sourirajan and Sirianni, 1966). This fact, in itself, is one of tremendous significance. This opens up a new field of science, technology, and engineering whose potential industrial applications are unlimited.

3. Some Early UCLA Experiments in Reverse Osmosis Desalination

The first successful results in the UCLA program were obtained by Sourirajan during October-December 1956. A summary of these results is given in Table 1-1. These experiments were carried out under static conditions at the laboratory temperature* in a crude apparatus shown in Figure 1-3. A stainless steel pressure chamber (Plate 1-1) was connected to a transfer cylinder which served as the salt water reservoir. The transfer cylinder was in turn connected to a Carver Laboratory Press through which oil pressure was applied to the salt water in the cylinder, and hence to that in the pressure chamber.

Commercial cellophane and cellulose ester films were chosen for investigation since cellulosic structure, because of its ability for hydrogen-bonding, was particularly appropriate for the preferential sorption of water from aqueous salt solutions. These films were manufactured primarily for packaging purposes and hence had very little porosity; their porous structure was entirely fortuitous and nonuniform. Still, they were sufficiently useful for the purpose of these preliminary experiments.

---

* Temperatures given are in degree Centigrade unless otherwise stated.

Table 1-1. Summary of some early UCLA experiments in reverse osmosis desalination

Data of Sourirajan (Yuster et al., 1958)

Nature of film and film treatment	Wt. % NaCl in feed	Operating pressure (psig)	Solute separation (%)	Product rate ($cm^3/hr/ft^2$)	Remarks on desalination efficiency of the film
Cellophane film simply treated with either (i) a 10% solution of Dri-Film 9987 in toluene, or (ii) a 10% solution of Linde-Y-1001 silicone thermosetting resin in acetone or (iii) Linde Y-1045 silicone elastomer solution	3.570	1250	≈10.0	53.9	Decreased rapidly with time
Cellophane film simply treated with Dow Corning Antifoam 'A' emulsion	3.510	1250	17.7	29.2	Decreased rapidly with time
Cellophane film simply treated with Dow Corning Antifoam 'B' emulsion	3.591	1250	18.4	27.2	Decreased rapidly with time
Cellophane film pressure treated with Dow Corning Antifoam 'B' emulsion	3.510	1250	32.2	19.3	Decreased slowly with time
Untreated cellulose acetate film	3.720	1250	94.4	4.8	Decreased rapidly after the first two days continuous service
Cellulose acetate film simply treated with Dow Corning 200 fluid viscosity grade 12, 500 at 25°	3.720	1500	94.7	4.3	Remained constant for a period of 7 days of continuous service, and then decreased slowly
Cellulose acetate film, pressure treated with Dow Corning Antifoam 'B' emulsion	3.720	2000	96.4	2.2	Remained constant throughout the experimental period of 30 days of continuous service
Untreated cellulose acetate-butyrate film	3.590	4000	>99.9	1.2	Remained constant throughout the experimental period of 30 days of continuous service

The cellophane and the cellulose acetate films were supplied by DuPont Co., and the cellulose acetate butyrate films were supplied by Eastman Kodak Co.

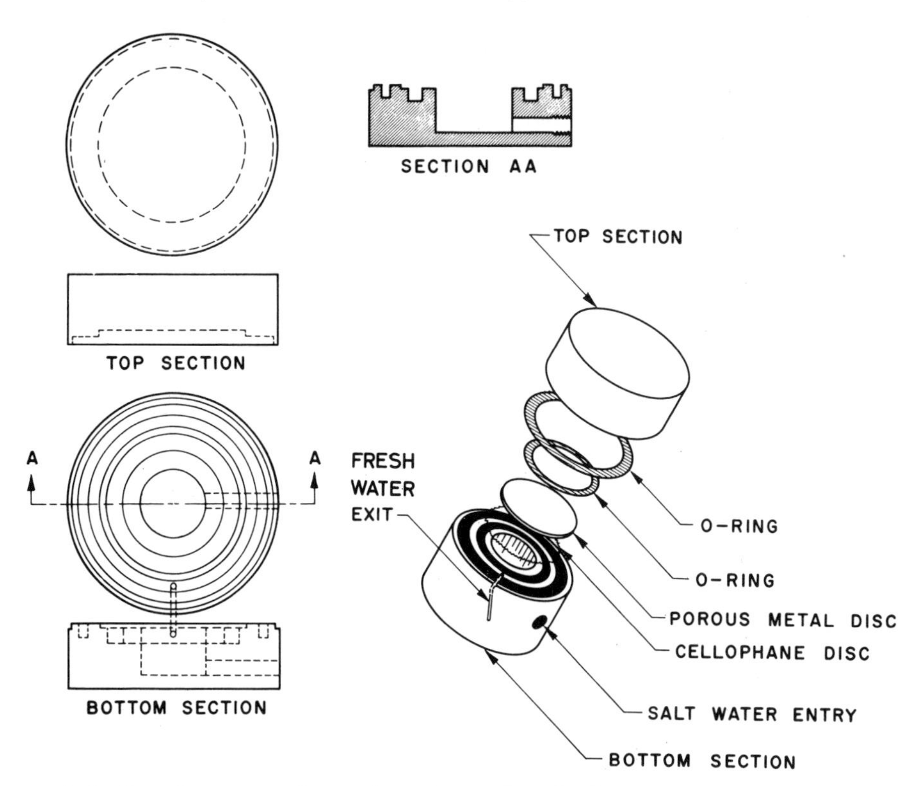

Figure 1-3. Desalination cell used in the early UCLA static experiments (Yuster et al., 1958).

Aqueous solutions, containing about 3.5 per cent by weight of sodium chloride, were used as the feed. The porous structure of the cellophane films used was such that their use in the reverse osmosis experiments did not result in any significant solute separation. But the same films, when simply (i.e. superficially) surface-treated with different kinds of silicones and then used in the reverse osmosis experiments with the hydrophobic surface facing the feed solution, did give some significant solute separation.

Among the early film-treating solutions investigated, Dri-Film 9987 and Linde silicones Y-1001 and Y-1045 appeared promising. The use of cellophane films simply surface-treated with any of the above silicones resulted in an initial reduction of salt content in the membrane-permeated product to the extent of about 10 per cent under an operating pressure of 1250 psig, and the product rate was 53.9 $cm^3/hr/ft^2$ of film area.

The Dow Corning Antifoam emulsions were found to possess superior properties as film treating agents. Experiments making use of cellophane films simply surface treated with the Dow Corning Antifoam 'A' emulsion or the Dow Corning Antifoam 'B' emulsion showed that the extent of solute separation and the product rate were 17.7 per cent and 29.2 $cm^3/hr/ft^2$, and 18.4 per cent and 27.2 $cm^3/hr/ft^2$ of film surface respectively under an operating pressure of 1250 psig. In both cases, however, the separation efficiency of the films decreased rapidly with time on continued experimentation.

A pressure treatment on the cellophane film with the Dow Corning Antifoam emulsion appeared to induce still superior and more stable surface properties to the film. The use of cellophane films treated with Dow Corning Antifoam 'B' emulsion under a pressure of 1500 psig for a period of two hours for the permeation of salt water under a pressure of 1250 psig resulted in a solute separation of 32.2 per cent and a product rate of 19.3 $cm^3/hr/ft^2$ of film surface. The separation efficiency of the film decreased slowly with time on continued use.

These results confirmed the basic postulate that the mechanism of solute separation is governed by a surface phenomenon. The results showed that a silicone-treated hydrophobic cellophane surface had a greater preferential sorption for water or preferential repulsion for the solute than the untreated hydrophillic cellophane surface. Consequently, the next step in these investigations was to use cellulose ester films which, by themselves, were more hydrophobic than the cellophane film.

Using cellulose acetate film for the permeation of salt water, an initial solute separation as high as 94.4 per cent was obtained at 1250 psig, and the product rate was 4.8 $cm^3/hr/ft^2$ of film surface. The experiment with the cellulose acetate film was carried out continuously for a period of 4 days. After the first two days, the desalination efficiency of the film became progressively less.

The effect of different surface treatments on the cellulose acetate film on its desalination efficiency was also investigated. Experiments using cellulose acetate film superficially treated with Dow Corning 200 silicone fluid gave 94.7 per cent solute separation at 1500 psig, with a product rate of 4.3 $cm^3/hr/ft^2$ of film surface; the level of solute separation remained constant for a period of 7 days of continuous operation, and then decreased slowly. Further, the use of the cellulose acetate film, initially pressure treated with Dow Corning Antifoam 'B' emulsion, resulted in 96.4 per cent solute separation at 2000 psig, yielding a product rate of 2.2 $cm^3/hr/ft^2$ of film surface; and the separation efficiency of the film remained constant throughout the experimental period of 30 days of continuous operation.

The decrease in desalination efficiency of the films observed in these experiments is at least partly attributable to the build-up in solute concentration on the high pressure side of the membrane.

Cellulose acetate butyrate, which is even more hydrophobic than cellulose acetate film, was the next material investigated for the reverse osmosis separation of aqueous salt solutions. The permeability of the film was low and an operating pressure of 4000 psig was necessary to obtain some significant product rate. But the results obtained were truly remarkable. The film was found to give essentially complete solute separation in a single pass. The experiment was carried out continuously for a period of 30 days, during which period the results obtained in 20 different product samples were recorded. Solute separation was >99.9 per cent in all cases. The average product rate was only 1.2 $cm^3/hr/ft^2$ of film surface at 4000 psig. Except for the very low product rates, the results seemed to offer a simple and complete solution to the problem of saline water conversion.

These experiments not only established the technical feasibility of demineralising aqueous sodium chloride solution by simply letting it flow under pressure through appropriate porous membranes, but they also showed that cellulose ester membranes were appropriate for the purpose, if only they could be made sufficiently porous with appropriate pore size on the membrane surface. In particular, cellulose acetate and cellulose acetate butyrate materials seemed to have the required chemical nature for the preferential sorption for water and/or preferential repulsion for the solute from aqueous salt solutions.

Therefore, from a practical stand point, the problem was only to develop a means of making membranes of appropriate porosity using cellulose acetate or cellulose acetate butyrate as the membrane material.

4. Work of Reid and Breton

The United States Department of Interior, Office of Saline Water (OSW) Research and Development Progress Report No. 16 and related papers (Breton, 1957; Breton and Reid, 1959; Reid and Breton, 1959) are widely referred documents in connection with the development of the reverse osmosis desalination process. These reports give the details of the work of Reid and Breton on the subject at the University of Florida. In order to find a 'semipermeable' membrane for strong electrolytes, they used a trial and error approach, although they felt that the desired membrane should contain some hydrophillic groups. The results of their screening programme are given in Tables 1-2 and 1-3. They discovered that cellulose acetate membrane was capable of rejecting electrolytes from aqueous solutions, and the semipermeability of cellulose acetate far exceeded that of any other film tested. They however concluded that the cellulose acetate membranes tested by them could not be used as 'ultrafilters' for purifying saline waters because their rate of filtration was too slow and their life was limited to only a few weeks, and consequently some other type of semipermeable membrane would probably be developed for the purpose.

According to Reid and Breton, the semipermeability of cellulose acetate was caused by regions of bound water within the membrane, and the transfer of water and ions through the membrane was governed by two different mechanisms. Those ions and molecules that could associate with the membrane through hydrogen bonding actually combined with the membrane and were transported through it by alignment-type diffusion; those which could not enter into hydrogen bonding with the membrane were transported by hole-type diffusion.

The work of Reid and Breton and the author's programme at UCLA. The author first came to know of the work of Reid and Breton only in May 1958. Thus the author's prior work at UCLA was completely independent of the work of Reid and Breton at the University of Florida. Although one could see some broad similarities between the concepts of bound water region and hydrogen bonding on the one hand, and the concept of preferential sorption on the other, the two approaches were fundamentally different, and, in at least one important respect, were diametrically opposed to each other. While the presence of pores of appropriate size on the area of the film at the interface was necessary for the practical success of this separation process from the author's point of view, the presence of such pores could only contribute to the leakage of the salt through the membrane and hence must be avoided according to the approach of Reid and Breton; further, the concept of a critical pore size characteristic of the membrane-solution system had no place in the approach of Reid and Breton. Consequently, in the author's subsequent programme at UCLA (which was concerned with the study and production of porous membranes, and which ultimately did produce the Loeb-Sourirajan type porous cellulose acetate membranes), the work of Reid and Breton (Breton, 1957) played no part whatsoever. This is a historical fact, and not a commentary on the mechanism of Reid and Breton.

Table 1-2. Breton's data on the semipermeability of various membranes to aqueous sodium chloride solutions

(Breton, 1957)

Membrane tested	Pressure on membrane ($lb/in.^2$)	% Rejection of NaCl by membrane	Remarks
Polyvinyl alcohol (Du Pont Elvanol 71-24)	100	None	Effluent pH = 11
Polyvinyl alcohol (baked at 145°)	600	22	No pH change
Polyvinyl alcohol (baked at 145°)	780	26	No pH change
50% Polyvinyl alcohol, 50% phenol formaldehyde	100	None	Effluent pH = 12
75% Polyvinyl alcohol, 25% polytetraallylammonium bromide	100	19	Effluent pH = 12
Polytetraallylammonium bromide	70	20-35	
Amberplex A-1 anion membrane (Rohm and Haas)	700	None	
Amberplex C-1 cation membrane (Rohm and Haas)	700	None	
Ethyl cellulose	800	None	
Nylon (Plastex Process Co.)	800	None	
Cellophane (Du Pont PT 300)	700	6	No pH change
Rubber hydrochloride	700	—	No flow
Polystyrene (Dow Chemical Co.)	800	None	
Saran (Dow Chemical Co.)	800	None	Very slow flow rate
Polyethylene	800	None	
Cellulose acetate butyrate	800	None	
Cellulose acetate (Celanese S-604)	800	—	No flow
Cellulose acetate (Du Pont 88 CA-43)	850	96	
Cellulose acetate (Du Pont 88 CA-43)	400	97.4	

Table 1-3. Breton's data on the semipermeability of cellulose acetate to aqueous solutions of various electrolytes

(Breton, 1957)

Pressure on membrane (lb/in^2)	Electrolyte	Concentration	% Rejection
850	NaCl	0.11M	96
715	$MgCl_2$	0.031M	99
800	$CaCl_2$	0.0054M	98
500	Na_2SO_4	0.018M	97
850	NaBr	0.0012M	93
840	NaF	0.0012M	86
800	H_3BO_3	0.0011M	46
800	NH_3	0.04M	30
830	NaCl	0.17M	90*
800	ocean water	4%	96

* Same film as used for NH_3 experiment

Membrane: Cellulose acetate (Du Pont CA-48), 0.88 mil thick.

5. UCLA Investigations During the Period June 1958-August 1960

This is the most significant period in the history of the development of the Reverse Osmosis Membrane Separation Process. During this period Loeb and Sourirajan tested various commercially available plastic films, then discovered the desalination characteristics of the preshrunk ultrafine superdense cellulose acetate filter membranes supplied by Schleicher and Schuell Co. (Keene, New Hampshire, U.S.A.), and finally developed their own technique for making porous cellulose acetate membranes capable of giving both high flow rates and high levels of solute separation from aqueous sodium chloride solutions. The development of the Loeb-Sourirajan technique was announced by UCLA on 23 August 1960 (Appendix I). It is this particular development which has led to the current world-wide interest and activity in the reverse osmosis process.

Further UCLA Tests With Some Commercial Films For Reverse Osmosis Desalination. A summary of the results of these tests is given in Table 1-4 (Loeb and Sourirajan, 1958; Sourirajan and Leob, 1958) where the term 'product rate' refers to the membrane-permeated product solution. The polystyrene film gave negligible flow rate at pressures up to 3500 psig, and hence it was not possible to obtain any separation data. Polyethylene terephthalate, rubber hydrochloride, cellulose triacetate, ethyl cellulose and compressed teflon films gave significant desalting, but the product rates were too low. Tests were also conducted with neoprene, Buna-N, soft natural rubber, and latex films (Loeb and Sourirajan, 1959a); no product was obtained with any of these films. A fibreglass-reinforced silicone film gave reasonable product rate but no measurable desalination (Loeb and Sourirajan, 1959a). From these results it was concluded that the materials tested above were practically unsuitable for reverse osmosis desalination.

Table 1-4. Summary of some further UCLA tests of commercial films for reverse osmosis desalination

(Loeb and Sourirajan, 1958; Sourirajan and Loeb, 1958)

Film type	Nominal film thickness	Wt. % NaCl in feed	Operating pressure (psig)	Solute separation (%)	Product rate (lb/hr/ft²)	Remarks
Polystyrene (Trycite supplied by Dow Chemical Co.)	0.001″	3.5	1800 to 3500	—	negligible	—
Polyethelene terephthalate (Mylar supplied by Du Pont Co.)	0.005″	3.5	350 to 2500	48 to 79	2.9×10^{-3} at start	Product rate dropped to zero in 24 hours
Rubber hydrochloride (Pliofilm FM-1 supplied by Goodyear Co.)	0.0008″	3.5	15 to 1500	30 to 82	7.0×10^{-3} at start	Product rate dropped to zero in 24 hours
Cellulose triacetate (Kodapack IV supplied by Eastman Kodak Co.)	0.0008″	3.5	3800 4600 5000	65 80 96	2.8×10^{-3} 3.0×10^{-3} 3.5×10^{-3}	
Ethyl cellulose	0.001″	3.5	1900 4000	91 93	1.0×10^{-3} 2.0×10^{-3}	
Ethyl cellulose	0.0005″	3.5	3100 8500	95 91	1.0×10^{-3} 4.0×10^{-3}	Film backed with 0.0014″ thick cellophane disc
Porous Teflon (initial density: 1.7 g/cm³ initial size of pores: 10 microns compression pressure: 4800 lb/in² heating temperature; 625°-630°F Final density: 2.2 g/cm³	0.01″	3.5	7000-7500	17 to 22	5.0×10^{-3}	

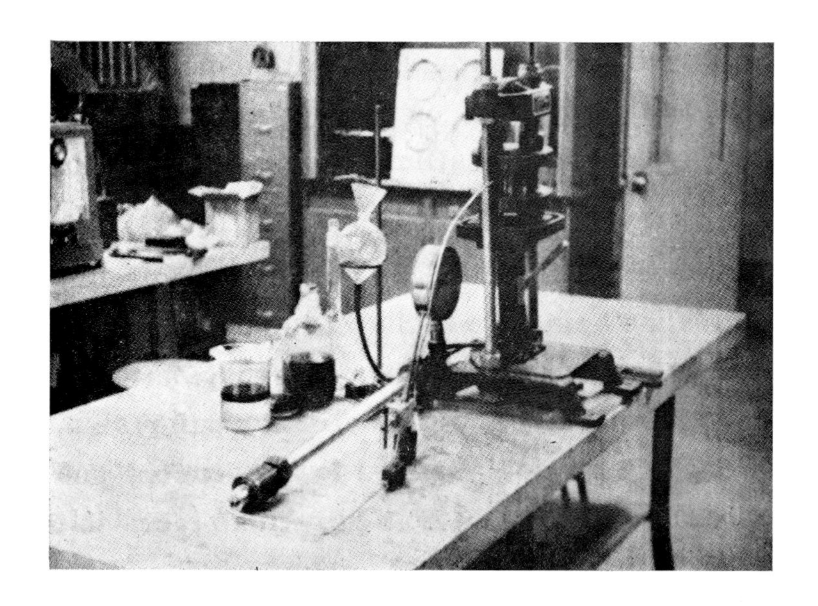

Plate 1–1. Desalination cell-pressure chamber assembly used in the early UCLA static experiments (Yuster et al., 1958).

6. UCLA Experiments With Schleicher and Schuell Type Membranes

Schleicher and Schuell (S & S) type UA-Ultrafine Superdense Cellulose acetate membranes were of particular interest for reverse osmosis desalination because they seemed to satisfy the twin requirement stated earlier for the success of this separation process. They were made of cellulose acetate which had already been found to have a preferential sorption for water from aqueous salt solutions; and they were specified to contain pores of size 50 Å or less, which was interesting enough for the purpose. Consequently detailed studies on their characteristics were undertaken by Loeb and Sourirajan (1959a, 1959b, 1960a, 1960b). A summary of the results is given below.

As received from the vendor, the membranes were 0.003 to 0.005 inch thick, and had a pure water permeability of about 700 gallons* per day per square foot at 15 psig. They were always stored under water or dilute alcohol solution to preserve their porous structure. If allowed to dry in air, they shrivelled, and their porous structure changed irreversibly. In the as-received condition, the membranes gave no measurable desalination in the reverse osmosis experiments.

In accordance with the preferential sorption-capillary flow mechanism, the size of the membrane pores was obviously too big to give any significant desalination, and it was hence necessary to reduce the pore size. This was accomplished easily. By simply heating the membrane under water between two glass plates, the membranes were found to shrink in all directions, and the preshrunk membranes did exhibit excellent desalination properties with aqueous sodium chloride feed solutions. The higher the temperature of hot water treatment, the higher was the membrane shrinkage, and consequently, the membrane-permeated product rate was lower and solute separation was higher at a given set of experimental conditions. These results were both remarkable and entirely consistent with the preferential sorption-capillary flow mechanism.

Early in this work, it was also found that the two sides of the same membrane behaved differently under the reverse osmosis experimental conditions. One could easily feel that one side of the membrane was 'rough' compared to the other side which was 'smooth'. The separation technique worked only when the rough side was facing the feed solution. If the smooth side was facing the feed solution under the reverse osmosis experimental conditions, the product rate increased and the solute separation decreased, and it appeared as if the feed solution was piercing through the membrane damaging its pore structure permanently. On the basis of extensive experimental results, it was recognised that the rough side of the membrane had the fine porous structure responsible for desalination; this structure extended only to a very small thickness of the membrane material, and it offered almost the entire resistance to fluid flow. The remainder of the film material was a spongy mass containing relatively very large pores (Loeb and Sourirajan, 1960a).

The test cell and the flow diagram of the apparatus used in these early experiments are illustrated in Figures 1-4 and 1-5. The cell had a volume of about 12 cm^3, and the feed was always a 3.5 per cent aqueous sodium chloride solution. An air-actuated single-stroke reciprocating sprague pump was used to maintain a constant pressure in the cell. The solution from the cell was

---

* U.S. gallons are used throughout this book, abbreviated as gal. At some points, flow rates are indicated in units of 1000 gal, abbreviated as K (1K = 1000 gal).

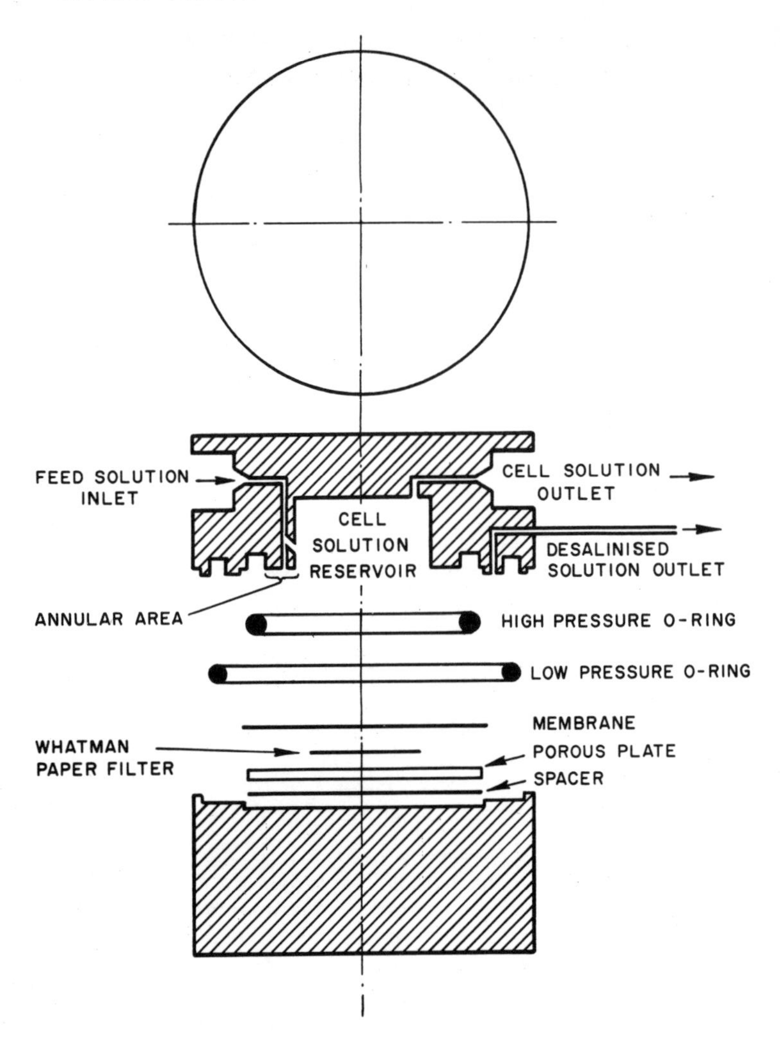

Figure 1-4. Desalination cell used in the early UCLA experiments with S & S membranes (Loeb and Sourirajan, 1959b).

purged either continuously (at the rate of 4 to 30 cm^3 per hour) or intermittently (at the rate of 15 to 30 cm^3 at a time). Under these semiflow conditions, the experimental results obtained, and summarised below, were partly controlled by the concentration build-up in the test cell due to the withdrawal of the water enriched product solution.

Effect of Temperature on Membrane Shrinkage and Porosity. The diameter of the membrane was measured at three equidistant diametric locations both before and after heating. The effects of temperature on membrane thickness and porosity were also determined. The porosity (= pore volume/ total volume) measurements were only approximate; they were calculated from the difference between the wet and dry weights of the membrane. The results obtained are given in Figures 1-6 and 1-7. The membrane contracted upon heating, the extent of shrinkage being a function of the time and temperature of heating. The decrease in membrane thickness was of the same order of mag-

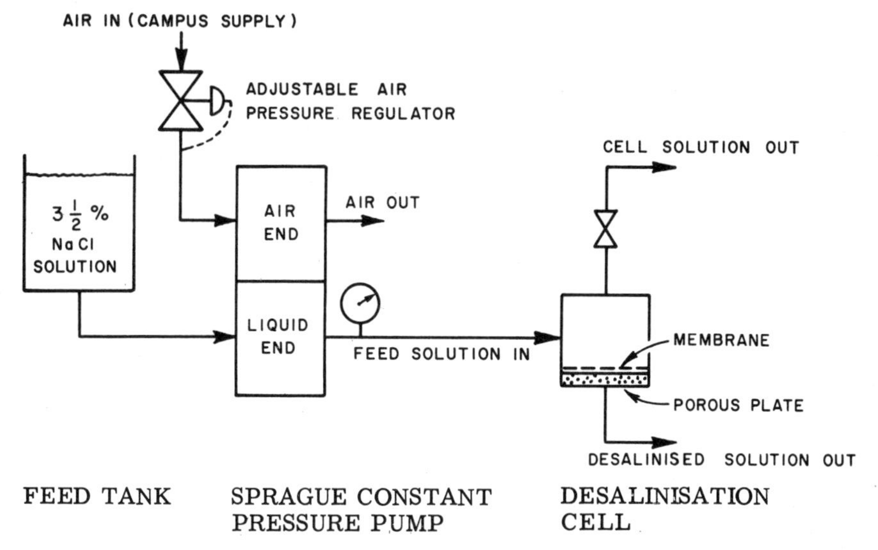

Figure 1-5. Flow diagram of apparatus used in the early UCLA experiments with S & S membranes (Loeb and Sourirajan, 1959b).

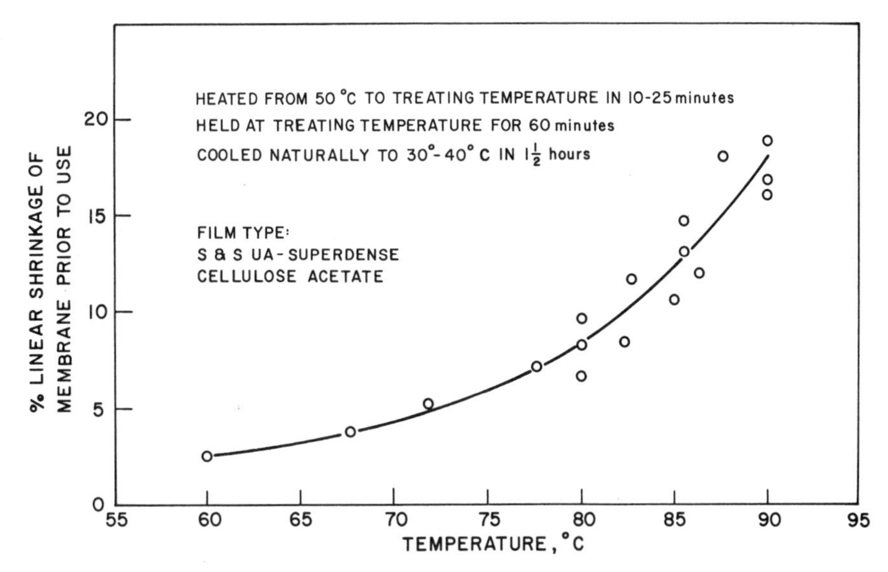

Figure 1-6. Effect of temperature on the shrinkage of S &S porous cellulose acetate membranes. Data of Loeb and Sourirajan (1959b).

nitude as the linear shrinkage. Porosity was inversely proportional to linear shrinkage. The membrane, as received from the vendor (unshrunk), had a porosity of about 65 per cent; Figure 1-7 indicates that with about 28 per cent linear shrinkage, the membrane became practically nonporous.

Desalination and Permeability Characteristics. These are illustrated in Figures 1-8 and 1-9 as functions of operating pressure and membrane shrinkage. These results showed that by increasing membrane shrinkage, higher levels of solute separation were obtainable; and the product rate be-

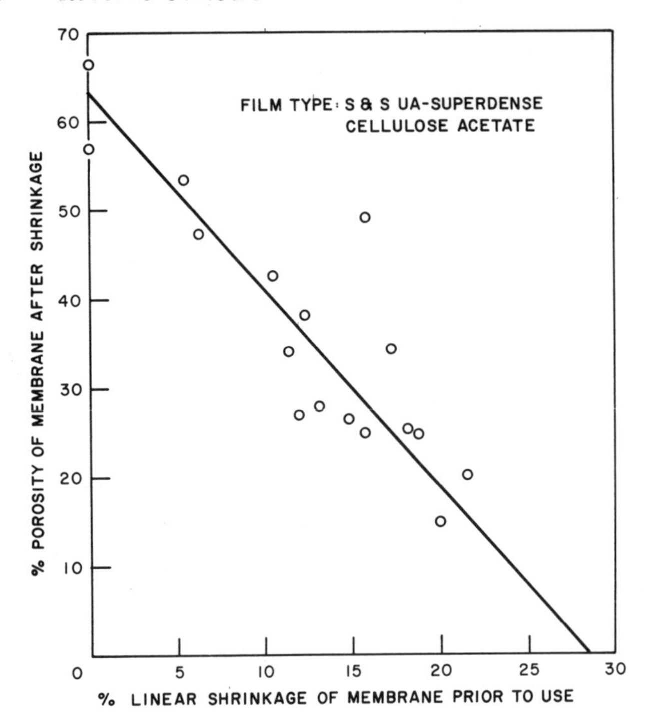

Figure 1-7. Porosity of preshrunk S & S porous cellulose acetate membranes. Data of Loeb and Sourirajan (1959b).

came practically insignificant with about 20 to 25 per cent membrane shrinkage. Figure 1-8 shows that solute separation was essentially unaffected by the operating pressure; this is later shown to be not true. Obviously, under the experimental conditions used, other factors (such as change in pore structure of the membrane) must have interfered with the effect of operating pressure on solute separation.

Membrane Stability. The stability of the membrane was affected by the pH of the feed solution. This is illustrated in Figure 1-10. Both the membranes used in these experiments were cut from a larger film shrunk to 18.1 per cent. Film 4A, which was tested with a solution having a pH of 10.6, deteriorated seriously after only 2 days of use. On the other hand, Film 4E, which was tested with solutions having a pH of about 8, maintained its performance for at least one week of continuous service. The data indicated that higher pH affected the chemical nature of the membrane surface.

Effect of Time. The variations of solute separation and product rate as a function of time are shown in Figures 1-11 and 1-12. Under conditions of continuous purge, the membrane appeared to maintain practically constant desalination capacity and product flow rate for at least 17 days. Further, at higher operating pressures, it was observed that product flow rates tended to decrease from an initial high value indicating the possible compaction of the membrane material and the possible closure of some of the pores on the membrane surface.

Differential Heating of the Two Sides of the Membranes. This was accomplished by keeping the spongy smooth side cool with wet filter paper,

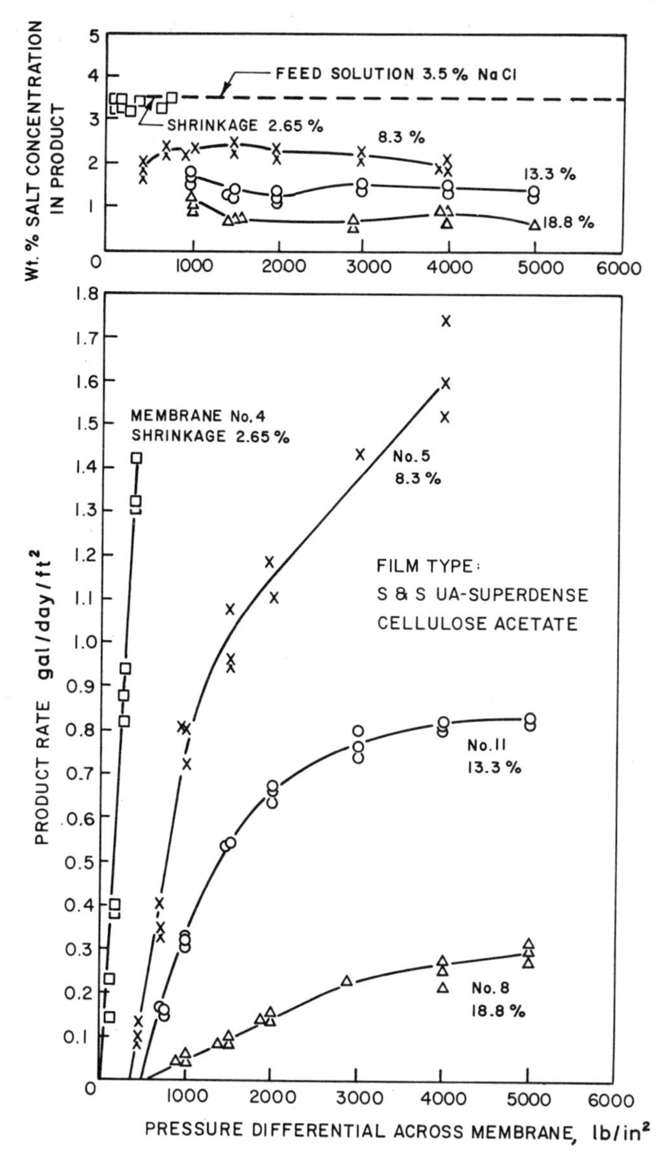

Figure 1-8. Effect of film shrinkage and operating pressure on product rate and solute separation using S & S porous cellulose acetate membranes. Data of Loeb and Sourirajan (1959b).

and heating the rough side with boiling water or steam. It was found (Loeb and Sourirajan, 1960a) that film shrinkage occurred as usual, but desalination was only fair, and there was no significant gain in the product rate. In another experiment, the smooth side of a membrane was heated and used in contact with the feed solution in the reverse osmosis experiment; there was practically no desalination.

Shrinkage of Films by Drying. Since the films shrunk on drying, several film drying techniques were investigated. These included (a) complete drying in air followed by heating to 105°; (b) complete drying without

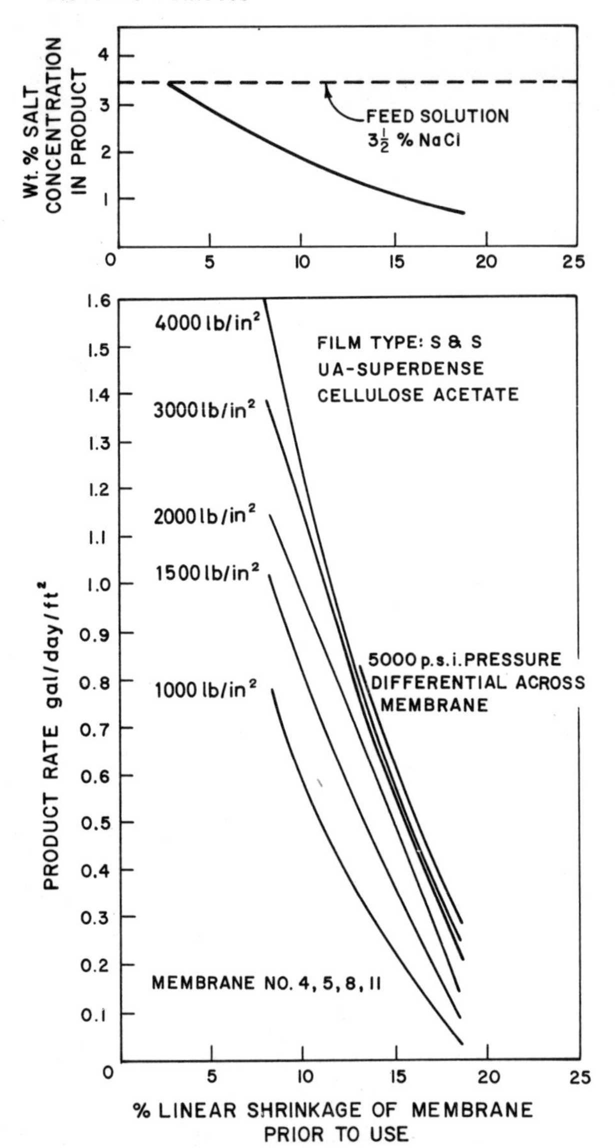

Figure 1-9. Effect of film shrinkage and operating pressure on product rate and solute separation using S & S porous cellulose acetate membranes. Data of Loeb and Sourirajan (1959b).

subsequent heating; (c) partial drying of both sides of the film; and (d) drying one side of the film keeping the other side wet. Films subjected to the above treatments gave desalination results generally inferior to those subjected to the simple hot water treatment. It was considered that the above drying treatments caused rupture of some of the fine pores on the 'rough' side of the film (Loeb and Sourirajan, 1960a).

Other Film Treatments. In view of the earlier results, an effort was made to improve the desalination characteristics of the films by treating them with Dow Corning DC-200-10 silicone oil and Antifoam 'B' emulsion. Attempts

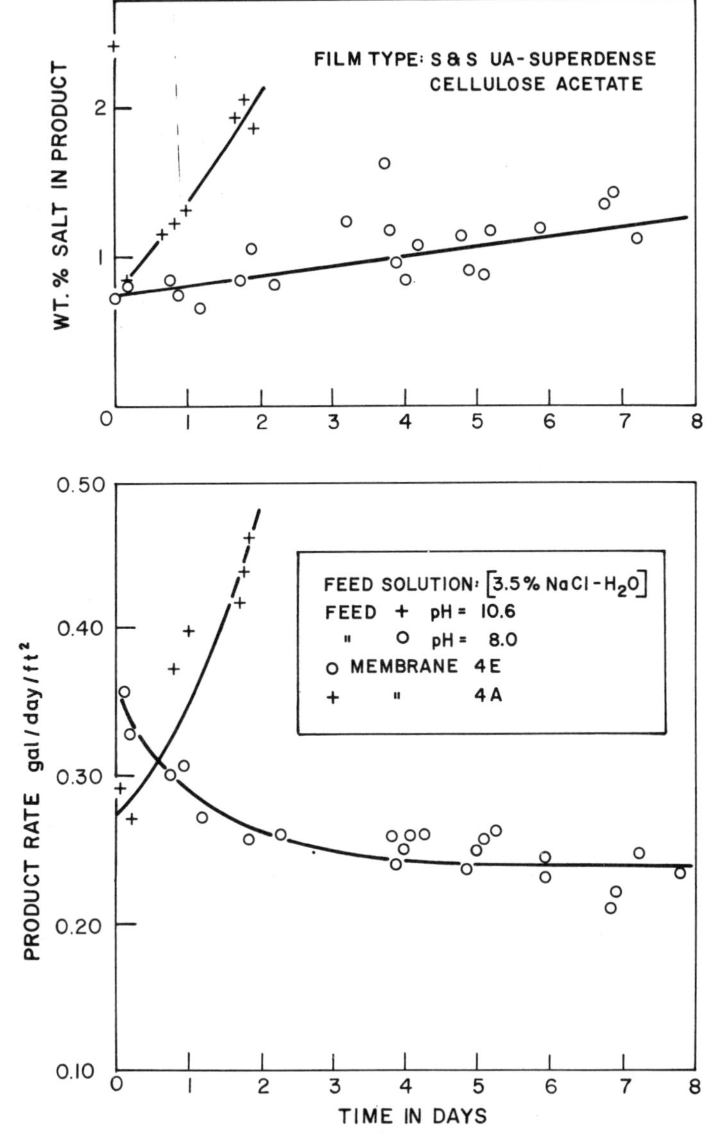

Figure 1-10. Effect of pH of feed solution on the performance of a S & S porous cellulose acetate membrane. Data of Loeb and Sourirajan (1959b).

were also made to completely acetylate the film material before use in the reverse osmosis experiments. No improvement in film performance was achieved by the above treatments (Loeb and Sourirajan, 1960a).

Effect of pH of the Heating Solution. Tests were made with films shrunk in hot water adjusted to various pH values. A set of results of such tests is shown in Figure 1-13. At lower pH values, the membrane seemed to shrink more at a given temperature (Loeb and Sourirajan, 1960a); no particular advantage however seemed to result from such shrinkage treatment.

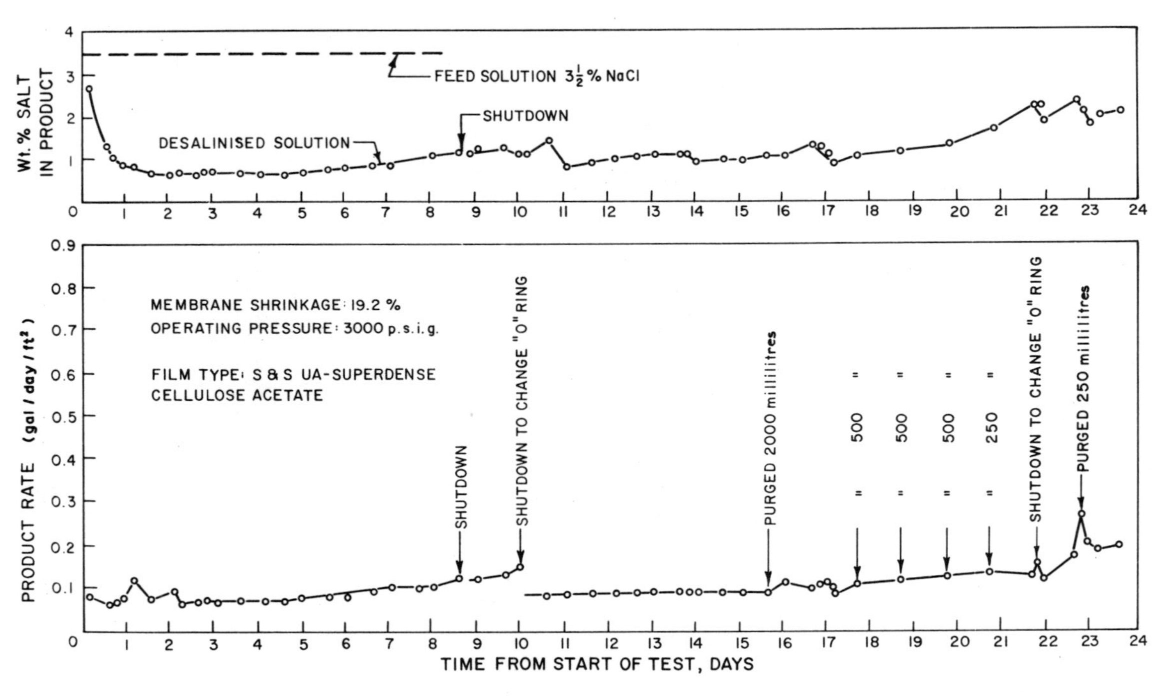

Figure 1-11. Effect of time on the performance of a S & S porous cellulose acetate membrane. Data of Loeb and Sourirajan (1959b).

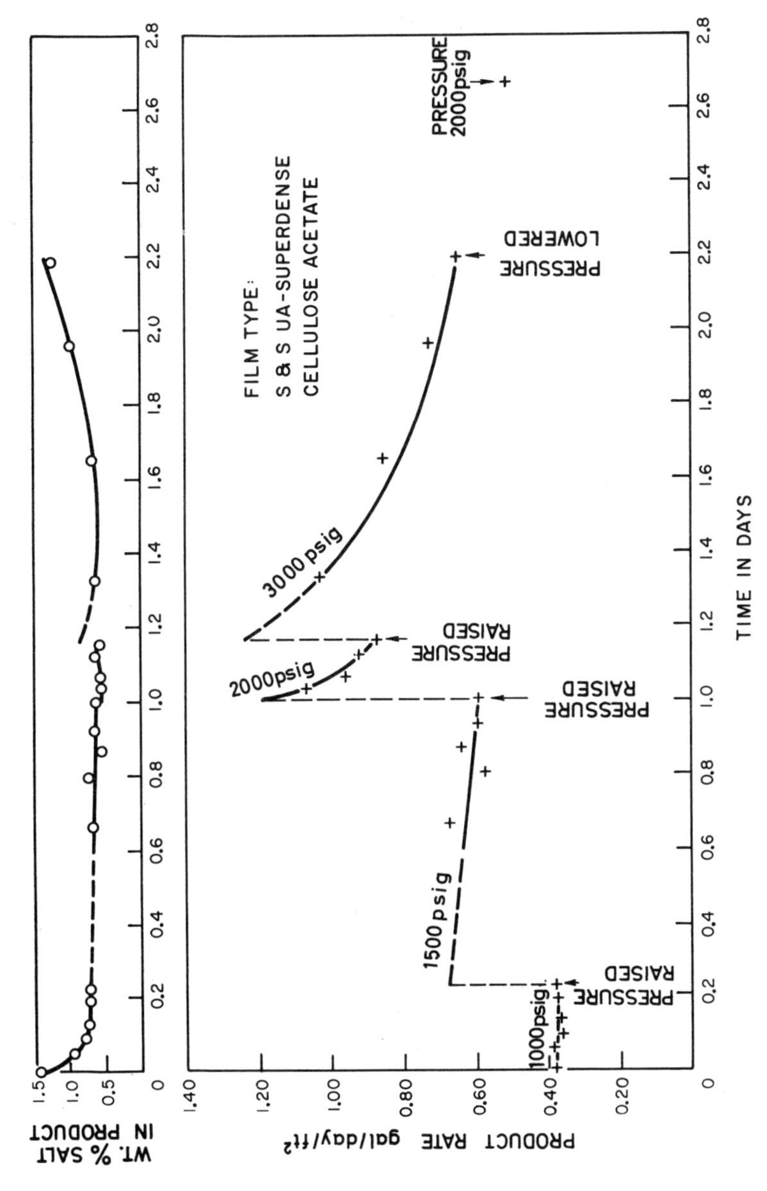

Figure 1-12. Effect of time, and operating pressure on the performance of a S & S porous cellulose acetate membrane. Data of Loeb and Sourirajan (1959b).

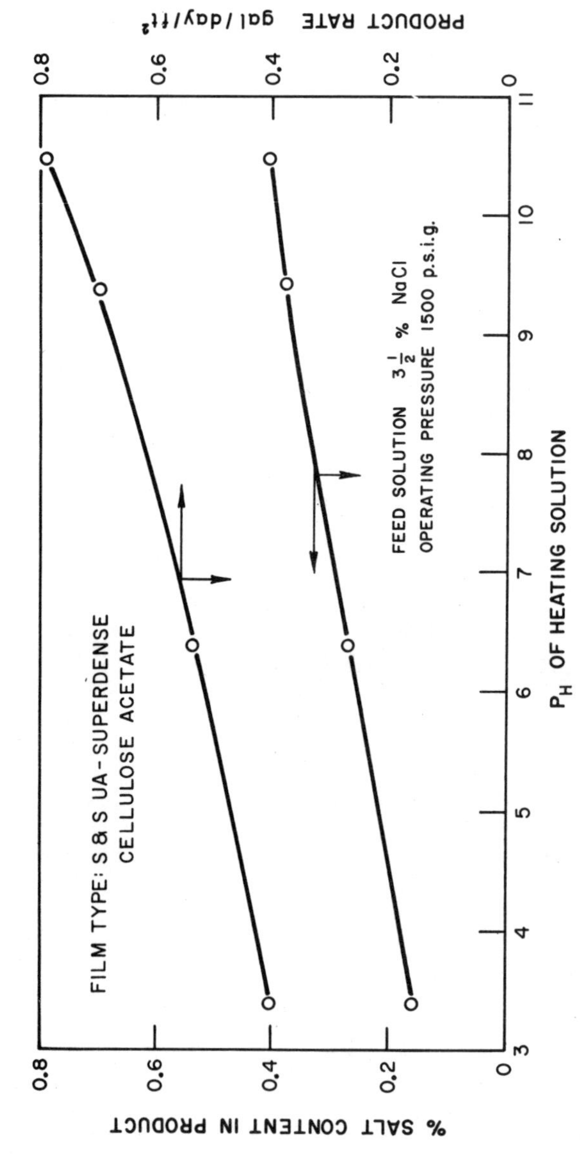

Figure 1-13. Effect of pH of heating solution for shrinking S & S porous cellulose acetate membranes. Data of Loeb and Sourirajan (1960a).

Tests with Other S & S Films. Tests were made with the nitrocellulose films (Type LSG, pore size > 50 Å), and 'very dense' acetyl cellulose films (pore size 50 to 100 Å), both supplied by Schleicher and Schuell Co. No significant desalination was obtained with the nitrocellulose films even when shrunk up to 11 per cent. The very dense acetyl cellulose films, preshrunk to 11 per cent, gave 65 per cent solute separation with a product flow rate of 1.3 gal/day/ft^2 at an operating pressure of 1500 psig.

Results Obtained with the S & S UA-Superdense Cellulose Acetate Films. The best results are illustrated in Figures 1-14 and 1-15. Using a 3.5 per cent NaCl feed solution, a solute separation of about 97 per cent was obtained with a product rate of about 0.8 gal/day/ft^2 at 1500 psig; using a 1.25 per cent NaCl feed solution, potable water containing less than 500 ppm of salt was obtained with a product rate of 2.3 gal/day/ft^2 at 1500 psig. An extended test run for a continuous period of 48 days was also conducted with a film shrunk at 83° (linear shrinkage = 7.2 per cent) using a 0.8 per cent NaCl solution as the feed in an apparatus with a flow system. The feed rate was maintained at 12 cm^3/min. The product flow rate was initially 2.7 gal/day/ft^2, and it dropped to 1.7 gallons in the first 5 days. During the remainder of the test period the product rate varied between 1.7 and 1.5 gal/day/ft^2. The salt content of the product was initially 350 ppm, and finally 560 ppm.

Significance of results. The discovery of the desalination characteristics of the S & S porous cellulose acetate membranes described above, was of extraordinary significance at this stage of the development of the reverse osmosis process. These results were entirely consistent with the twin-requirement (i.e. appropriate chemical nature and porosity of membrane surface) concept for the practical success of the reverse osmosis process. Even though the S & S porous cellulose acetate membranes did give excellent results for the laboratory demonstration of the technical feasibility of the reverse osmosis desalination process, their permeability characteristics were not considered good enough for the economic conversion of saline waters on an industrial scale. In other words, the 'rough' side of the S & S cellulose acetate membranes had the appropriate chemical nature and pore size, but the number of desirable pores, and the overall porosity of the membrane were inadequate for their industrial application for the economic conversion of saline waters. Consequently investigations were next undertaken by Loeb and Sourirajan to produce more porous cellulose acetate membranes capable of giving both high levels of solute separation and high rates of desalinised water from aqueous salt solutions (Loeb and Sourirajan, 1960b, 1961). The details of these investigations, along with the more recent developments in the field, are given in Chapter 2.

7. Further Studies on Schleicher and Schuell Porous Cellulose Acetate Membranes

The preshrunk S & S UA-Superdense cellulose acetate filter membranes tested as described above had sufficient surface porosity and permeability for use in the laboratory experiments to demonstrate the general applicability of this separation technique, and to make a preliminary survey of the process; hence they were studied further by Sourirajan (1963a, 1963b, 1964a, 1964b, 1964c, 1965). A summary of the results obtained is given below.

Experimental details. The apparatus and the experimental procedure employed in the above and subsequent studies of Sourirajan and co-workers are described below in detail in view of their simplicity and general suitability for the laboratory investigation of the reverse osmosis process.

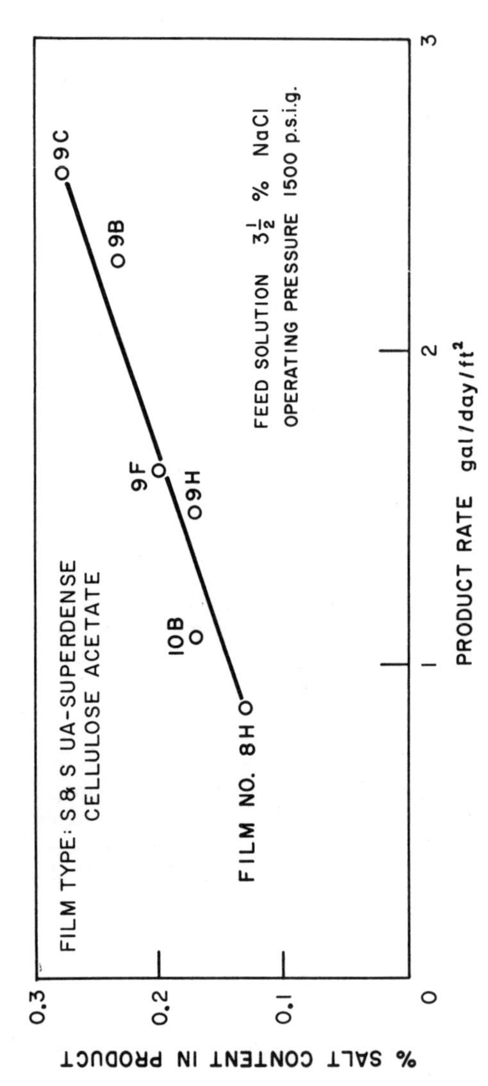

Figure 1-14. Performance of preshrunk S & S porous cellulose acetate membranes for water desalination. Data of Loeb and Sourirajan (1960a).

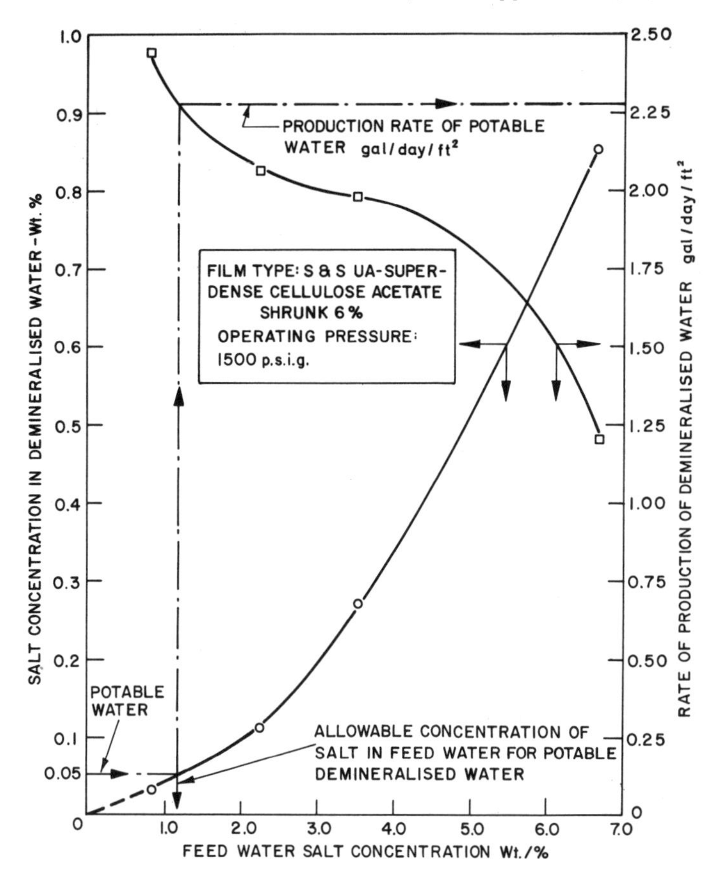

Figure 1-15. Performance of a typical preshrunk S & S porous cellulose acetate membrane for water desalination. Data of Loeb and Sourirajan (1961).

The reverse osmosis cell used, shown in Figure 1-16, was simple in design and construction and made of 310 Stainless Steel. It consisted of two detachable parts. The upper part was a high pressure chamber provided with inlet and outlet openings for the flow of the feed solution under pressure. The lower part was the membrane stand provided with an outlet opening for the withdrawal of the membrane-permeated product solution. The outside diameter of the cell was $2^{13}/_{16}$ in., and the inside diameter of the pressure chamber was $1^{1}/_{8}$ in. The effective area of the membrane in the cell was 7.6 cm^2. The wet preshrunk membrane was mounted on a stainless steel porous plate (2 in. in diameter × $^{1}/_{16}$ in. thick) embedded in the lower part of the cell. A wet Whatman filter paper was placed between the membrane and the porous plate to protect the membrane from abrasion and also aid the flow of the product solution through the porous plate. Under the operating conditions, the porous plate and the Whatman filter paper offered practically no resistance to fluid flow. The upper and lower parts of the cell were set in proper alignment with rubber O-ring contacts between the high pressure chamber and the wet membrane. A pressure-tight joint was obtained by clamping the cell tightly between two thick end-plates.

A flow diagram of the experimental arrangement is shown in Figure 1-17. A motor-driven controlled-volume Milton Roy (or similar) duplex pump was

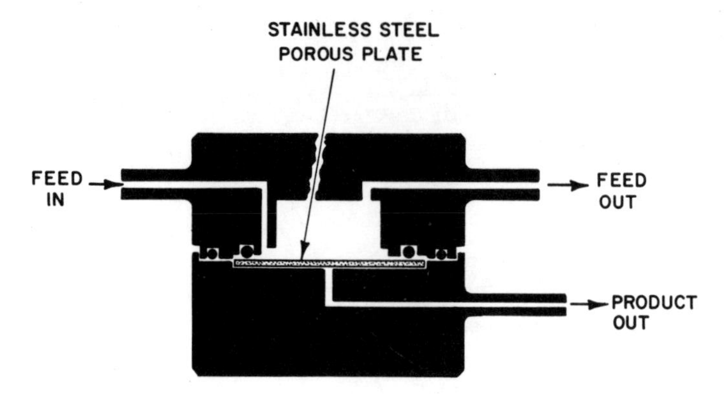

Figure 1-16. Desalting cell (Sourirajan, 1964a).

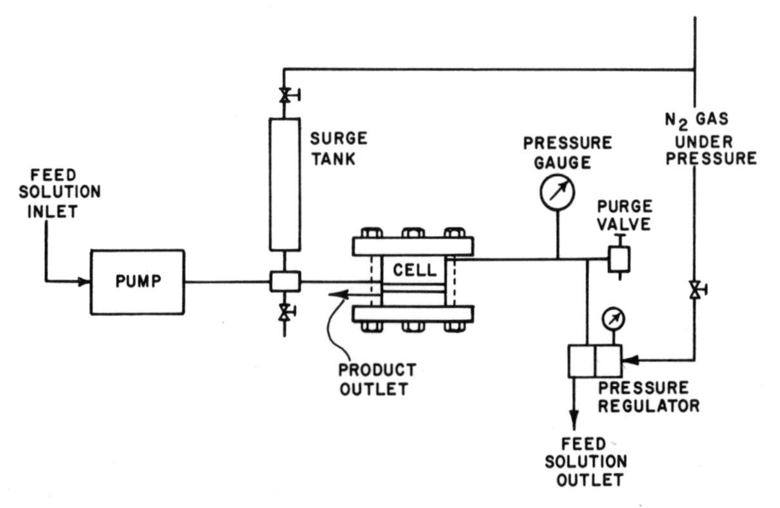

Figure 1-17. Flow diagram of apparatus for reverse osmosis experiments (Sourirajan, 1964a).

used to pump the feed solution under pressure through the cell. All parts of the pump coming into contact with the feed liquid were made of either Hastalloy or stainless steel. The surge tank, a stainless steel high pressure cylinder (2 in. o.d. × $1\frac{1}{2}$ in. i.d. × 14 in. high), was used to minimise the pressure fluctuations in the cell. A porous stainless steel plate specified to have pores of average size equal to 5 microns was mounted between the pump and the cell to act as a filter for dust particles which might otherwise clog the pores on the membrane surface. Under the operating conditions the fluid pressure in the cell was indicated by a liquid-sealed Ashcroft pressure gauge. The purge valve was used to drain the system whenever necessary. A stainless steel Grove pressure regulator was used to maintain a constant operating pressure in the cell. Nitrogen gas under pressure from commercial gas cylinders was used to load the dome of the Grove regulator. Monel metal high pressure tubings and HIP high pressure fittings made of 316 Stainless Steel were used throughout the system.

Six cells were used in series (only one cell is illustrated in Figure 1-17) so that six different membranes could be tested at the same time. Samples of the feed solution entering each cell could be drawn separately, if necessary.

Membranes shrunk at different temperatures were used to give different levels of solute separation at a given set of operating conditions. Unless otherwise stated, the experiments were of the short run type, each lasting for about 2 hours; they were carried out at the laboratory temperature; the effective area of the film used was 7.6 cm^2; and the reported product rates were those corrected to 25° using the relative viscosity and density data for pure water.

Occasionally experiments were also carried out in the temperature range 5° to 35°. For such experiments, the cells and solutions were placed in an insulated box which was cooled or warmed as needed. An automatic recorder was used to record the temperature of the solution flowing through each of the cells to which thermocouples were attached through conax fittings.

Each film was subjected to a temperature and pressure treatment before use in the reverse osmosis experiments. The temperature treatment consisted of heating the film gradually in water from the laboratory temperature to the required temperature where it was kept for 10 minutes and subsequently cooled rapidly. The pressure treatment consisted of pumping distilled water over the surface of the film mounted in the desalination cell at a pressure about 15 per cent higher than the maximum intended operating pressure for at least an hour.

In each experiment, the fraction solute separation, f, defined as

$$f = \frac{\text{molality of feed } (m_1) - \text{molality of product } (m_3)}{\text{molality of feed } (m_1)},$$

the product rate [PR], and the pure water permeability [PWP] in grams per hour per 7.6 cm^2 of effective film area, were determined at the preset operating conditions of pressure, feed molality and feed rate. In all cases, the terms 'product' and 'product rate' refer to the membrane-permeated solutions. Unless otherwise stated, the concentrations of the solute in the feed and product solutions were determined by refractive index measurements using a Bausch and Lomb refractometer or by specific resistance measurements using a conductivity cell. The accuracy of the separation data was within 1 per cent in most cases, and that of [PR] and [PWP] was within 3 per cent in all cases.

Inorganic Salts in Aqueous Solution. The general applicability of the reverse osmosis separation technique was tested with several commonly available inorganic salts in aqueous solution using different samples of preshrunk S & S cellulose acetate membranes. A set of experimental data given in Table 1-5 shows that the technique was successful in all the cases tested. The extent of solute separation obtainable depends on the pore structure and chemical nature of the membrane with respect to that of the solution; hence the data similar to those given in Table 1-5 do not represent the limits of solute separation obtainable with the different systems tested.

The ability of a given cellulose acetate membrane to separate the various inorganic ions in aqueous solution was found to be in the order: citrate > tartrate = sulfate > acetate > chloride > bromide > nitrate > iodide > thiocyanate, and Mg, Ba, Sr, Ca > Li, Na, K—the same as the lyotropic series with respect to both anions and cations (Glasstone 1951b).

The above order of solute separation is, in some respects, different from the one given by Ambard and Trautmann (1960a) from their ultrafiltration studies using cellophane membranes. Their results show that the extent of separation

Table 1-5. Separation of some inorganic salts in aqueous solution using preshrunk S & S porous cellulose acetate membranes

Data of Sourirajan (1964a)

Film no.	Solute	Feed rate cm^3/min	Concn. of feed solution		Solute separation (%)	Product rate (g/hr at 20°)
			Molality	Wt.%		
M	Sodium citrate	30	0.5	11.43	90.7	5.5
M	Sodium tartrate	30	0.5	8.85	75.9	6.9
M	Sodium sulphate	30	0.5	6.63	75.1	7.4
M	Sodium acetate	30	0.5	3.94	34.0	9.2
M	Sodium chloride	30	0.5	2.84	25.4	10.8
J	Lithium sulphate	30	0.5	5.21	86.6	3.6
J	Sodium sulphate	30	0.5	6.63	87.1	4.4
J	Potassium sulphate	30	0.5	8.02	84.9	4.7
31	Magnesium chloride	30	0.5	4.54	66.3	9.3
31	Calcium chloride	30	0.5	5.26	61.4	9.3
31	Strontium chloride	30	0.5	7.34	65.2	9.8
31	Barium chloride	30	0.5	9.43	66.5	9.5
30	Magnesium nitrate	30	0.5	6.91	68.0	8.0
30	Calcium nitrate	30	0.5	7.58	65.2	8.4
30	Strontium nitrate	30	0.5	9.57	70.5	8.8
30	Barium nitrate	30	0.25	6.13	78.6	10.8
B	Aluminum nitrate	17	0.25	5.06	93.0	2.1
B	Chromic nitrate	16	0.25	5.62	87.5	2.0
B	Manganese nitrate	17	0.25	4.29	83.5	2.8
B	Ferrous chloride	18	0.25	3.07	90.0	2.7
B	Ferric chloride	18	0.25	3.90	85.4	2.3
B	Ferric nitrate	18	0.25	5.70	85.0	2.5
B	Cobaltous nitrate	17	0.25	4.38	84.5	2.5
B	Nickel sulphate	17	0.25	3.73	96.5	2.8
B	Nickel chloride	17	0.25	3.14	86.5	2.5
B	Nickel nitrate	18	0.25	4.37	85.0	2.8
B	Cupric sulphate	18	0.25	3.84	97.5	2.9
B	Cupric chloride	17	0.25	3.25	87.5	2.4
B	Cupric nitrate	16	0.25	4.48	80.0	2.6
B	Zinc nitrate	18	0.25	4.52	82.5	2.8
B	Cadmium nitrate	17	0.25	5.58	81.0	2.8
B	Lead nitrate	18	0.25	7.65	77.0	3.0
B	Thorium nitrate	17	0.25	10.72	91.5	2.6

Operating pressure, 1500 psig. Effective area of film, 7.6 cm^2

of LiCl, NaCl, and KCl are identical, and the salts with bivalent cations and monovalent anions such as $MgCl_2$ and $CaCl_2$ are hardly separated at all compared to salts with monovalent cations and anions such as NaCl and KCl; according to them, the smaller the volume of the hydrated anion, the greater is the filterability of electrolytes, and the cation influences the electrolyte filterability only in so far as it dehydrates the anion. Van Oss (1963) has pointed out that the differences in hydration of ions can explain neither the difference in the influence of multi-valency of anions and cations nor the enhanced salt retention at higher pressures in the reverse osmosis process.

The lyotropic series of ions is of general significance in many fields of physical chemistry. The correspondence of the order of separation of the ions with the lyotropic series simply emphasises the fundamental physicochemical nature of the mechanism involved in this separation technique. Hydration of ions also affects the relative mobility of the ions and hence their separation. Consequently there should be consistency between the hydration of the ions and their relative order of separation. Such is the case. Even though the estimated hydration numbers for the various ions differ widely in their actual values (Bell, 1958; Bockris, 1949; Remy, 1915), they are also in the order of the lyotropic series. Exceptions to the lyotropic series are not uncommon, and they are found in the order of separation too. For example, the lyotropic order with respect to bivalent cations is Mg $>$ Ca $>$ Sr $>$ Ba; the corresponding separation order was found to be Mg $=$ Ba $>$ Sr $>$ Ca at certain levels of solute separation.

Related Solution Systems. Under identical experimental conditions of constant feed molality, feed flow rate, and operating pressure and temperature, the separation and permeability characteristics of the S & S porous cellulose acetate membranes were found to be uniquely related for all solution systems containing (monovalent or bivalent) cations and anions of the same respective valency. These relations could be expressed by a system of experimentally determined characteristic curves such as those shown in Figures 1-18, 1-19, and 1-20 which could then be used to predict the performance of any similar membrane. The general form of these curves representing the relative separation data for aqueous solutions containing ions of the same respective valency gave rise to the concept of related solution systems (Sourirajan, 1964a, 1964b). The experimental criteria of such systems may be stated explicitly as follows. When the results of solute separation and product rate in a related system are plotted in the y-axis against the corresponding values for the reference system plotted in the x-axis, (i) all points fall on a unique line characteristic of the systems under comparison, and (ii) the characteristic relative separation line extrapolates uniquely to 0 per cent and 100 per cent respectively at the points representing 0 per cent and 100 per cent solute separation for the reference system (Figure 1-18, 1-19, and 1-20). Thus, only the above experimental criteria determine whether or not any two solutions can be classified as related systems. In terms of the preferential sorption-capillary flow mechanism, Figures 1-18 and 1-19 would seem to indicate that the thickness of the preferentially sorbed interfacial water layer is the same for all related systems; this probably is a sufficient, but not a necessary, condition to satisfy the above experimental criteria of related solution systems, which seem to be better explained on the basis of the relative flows of the preferentially sorbed pure water and feed solution through the membrane for the systems under comparison.

Several consequences of the concept of related solutions are of practical interest. For instance, the separation and permeability characteristics of a membrane are uniquely fixed for all related solution systems; hence they are

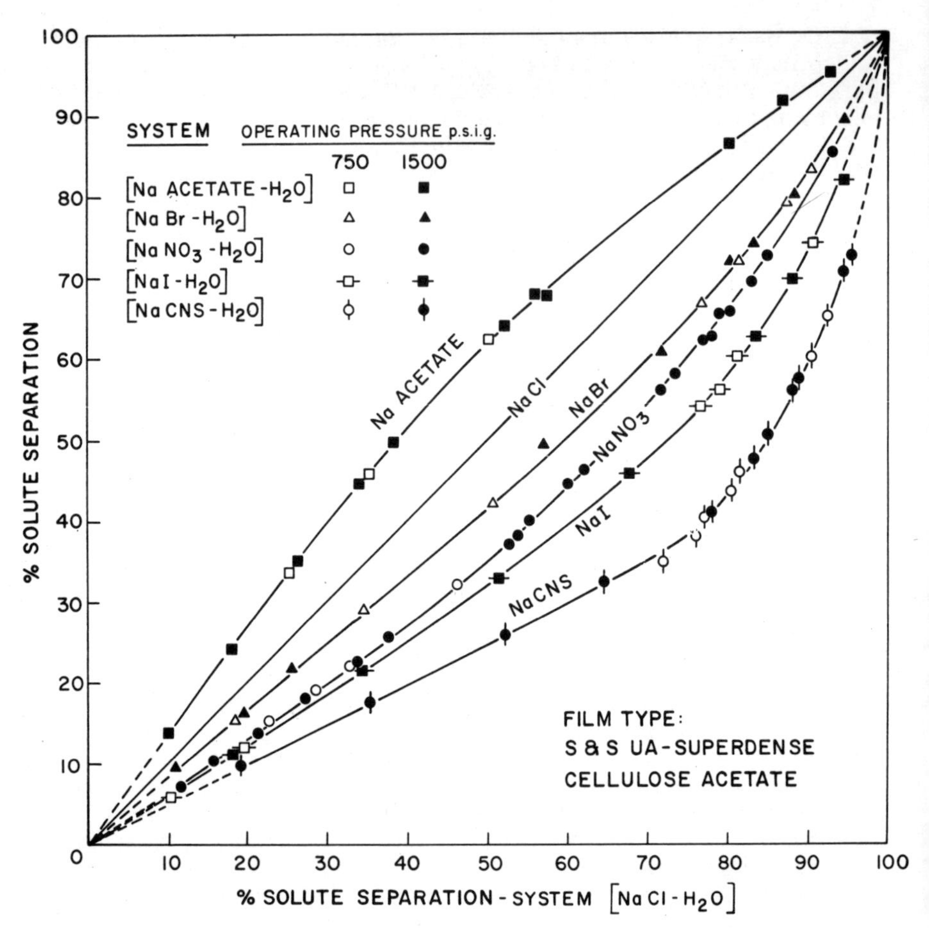

Figure 1-18. Solute separation characteristics of S & S porous cellulose acetate membranes for some aqueous related solution systems. Data of Sourirajan (1964a).

Feed solution molality: 0.5M. Feed rate: 30 cm^3/min.

predictable for such systems on the basis of the above criteria alone. Several groups of related solution systems can exist or can be synthesised. The latter possibility has not yet been explored, and it may offer a useful means of establishing procedures for expressing the separation and permeability characteristics of membranes for a variety of solution systems in the reverse osmosis process. These considerations also indicate that the relative number of pores of size equal to or less than the characteristic critical diameter in a given heteroporous film will be different for different groups of related solution systems.

Organic Solutes in Aqueous Solution. The relative separation data for a number of commonly available organic solutes in aqueous solution using S & S porous cellulose acetate membranes are given in Figures 1-21 and 1-22 using 0.5M ethyl alcohol-water as a reference system. The available data cover the range of 8.3 to 33 per cent EtOH separation for the system 0.5M ethyl alcohol-water; the corresponding NaCl separation data cover the range 34 to 94 per cent for the system 0.5M sodium chloride-water. The following

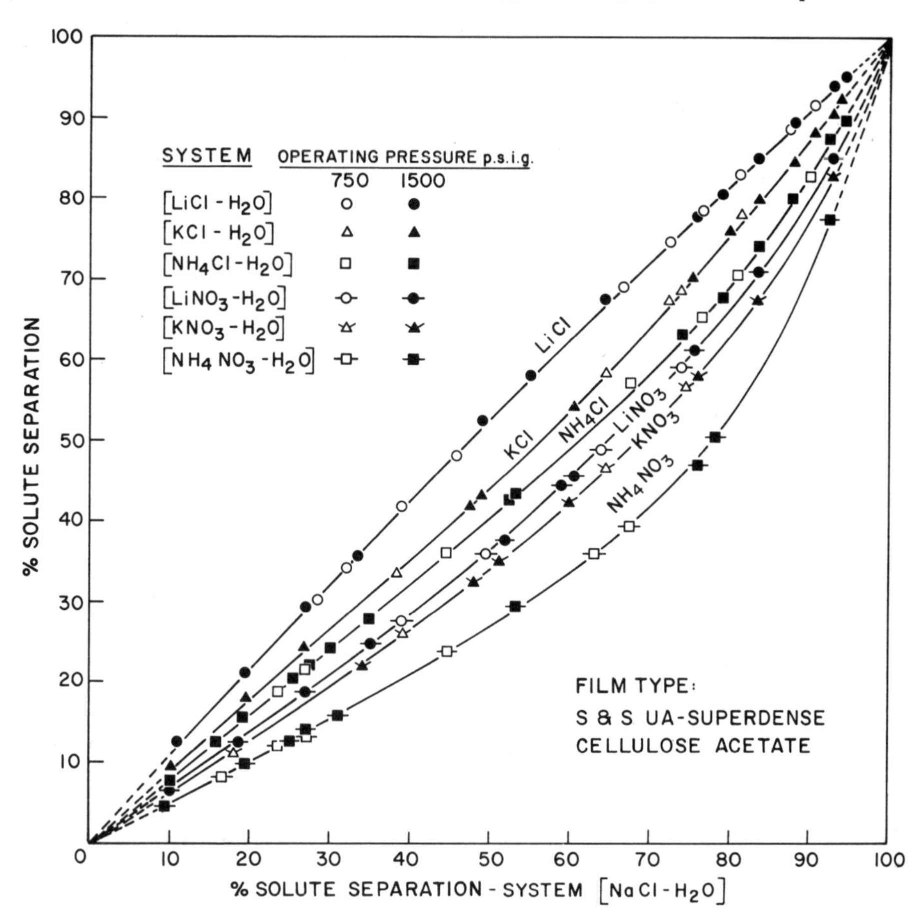

Figure 1-19. Solute separation characteristics of S & S porous cellulose acetate membranes for some aqueous related solution systems. Data of Sourirajan (1964a).

Feed solution molality: 0.5M. Feed rate: 30 cm^3/min.

orders of separation exist consistently under the experimental conditions specified in Figures 1-21 and 1-22:

n-PrOH > EtOH > n-BuOH
Iso-PrOH > n-PrOH
tert-BuOH > sec-BuOH > iso-BuOH > n-BuOH
glycerol > ethylene glycol > n-PrOH
acetaldehyde > ethyl alcohol > acetic acid
propionic acid > acetic acid
NaCl > any of the above organic solutes

The above orders may be related to the possible differences in the thickness of the interfacial pure water layers, depending on the chemical nature and molecular structure of organic solutes. From the experimental data plotted in Figures 1-21 and 1-22, it does not seem likely that any simple criteria of related systems could emerge for organic solutes in aqueous solution, similar

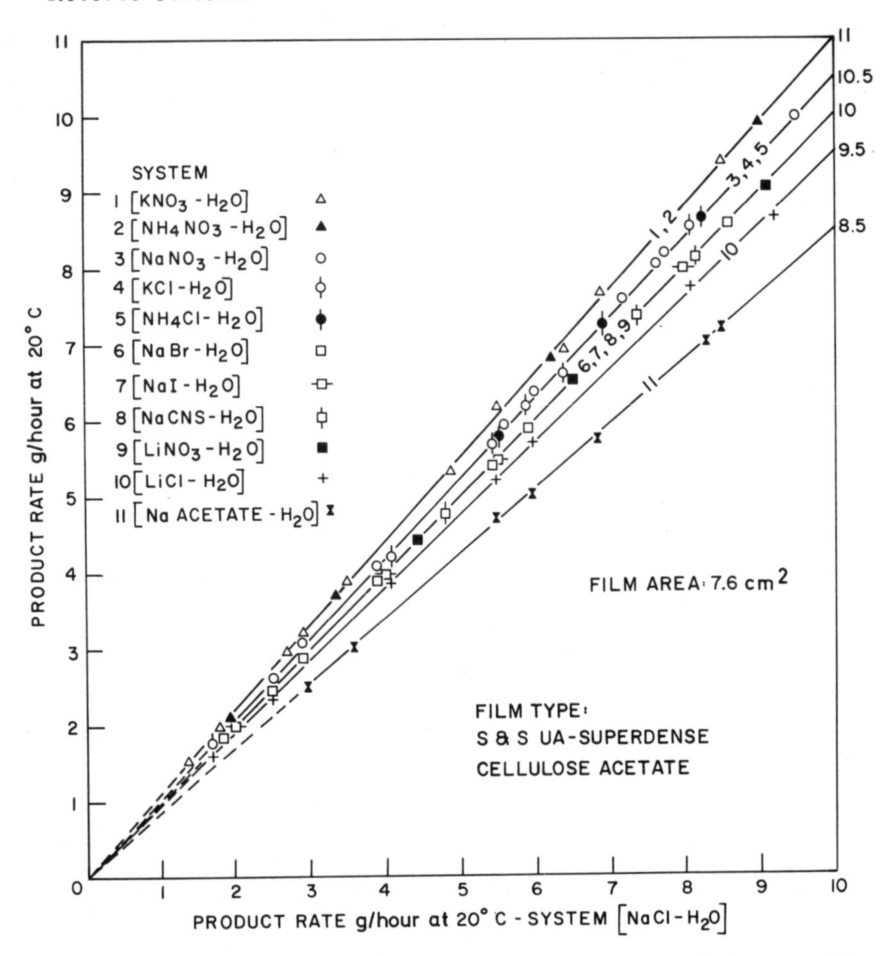

Figure 1-20. Product rate characteristics of S & S porous cellulose acetate membranes for some aqueous related solution systems. Data of Sourirajan (1964a).

Feed solution molality: 0. 5M. Feed rate: 30 cm^3/min. Operating pressure: 1500 psig.

to the criteria of ionic valencies for the relative separation of inorganic solutes in aqueous solution.

When organic solutes of various molecular sizes are involved, more than one mechanism could control the separation process by the reverse osmosis technique. The solute molecule might be retained on the film surface purely by ultrafiltration—i.e. by virtue of the fact that its molecular size is bigger than the size of the pores on the film surface—or by the negative adsorption of the solute at the membrane-aqueous solution interface, together with the capillary flow of the preferentially sorbed interfacial fluid through the membrane pores. In the latter case, the pores on the membrane surface need not be smaller, and could be bigger, than the molecules involved. While both the mechanisms could be operative simultaneously to different extents with reference to any given membrane-solution system, the preferential sorption-capillary flow mechanism is probably the predominant one here, since the

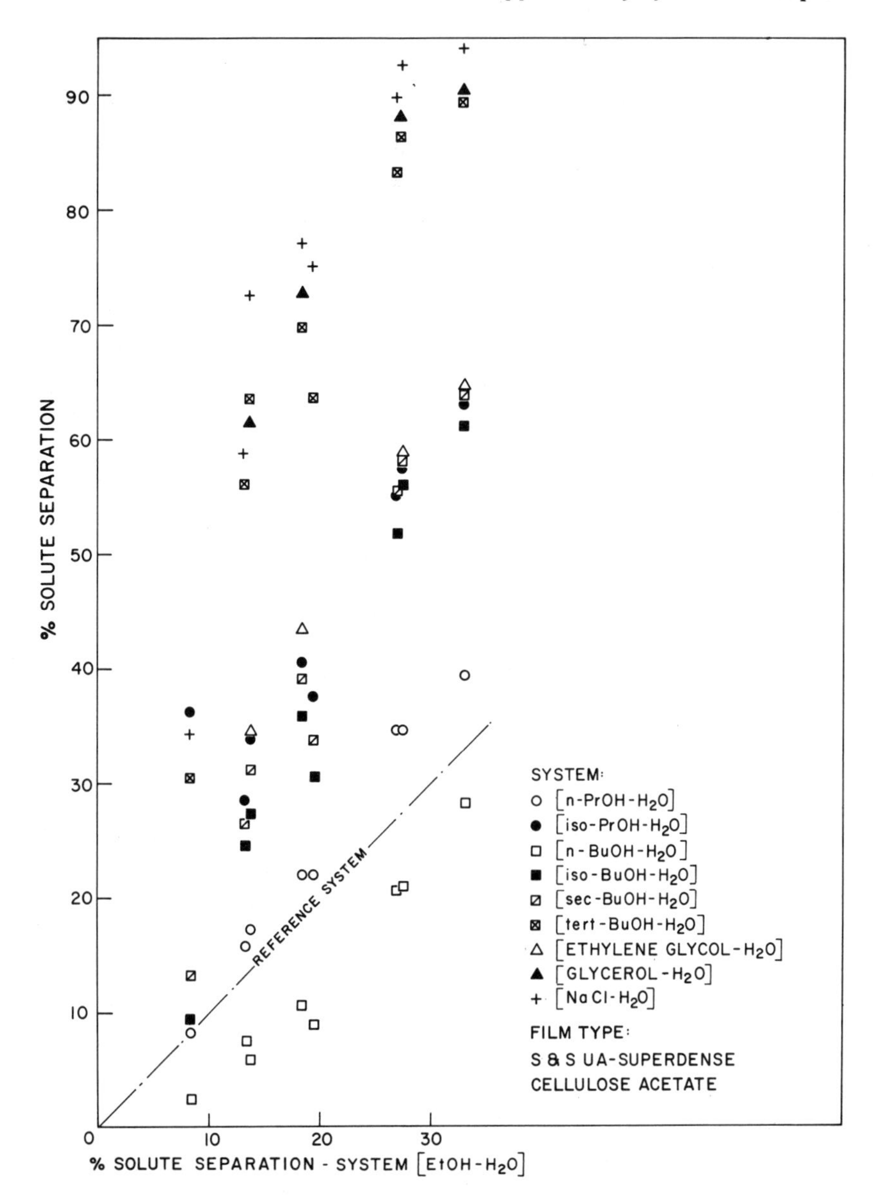

Figure 1-21. Relative separation of some organic solutes in aqueous solution using S & S porous cellulose acetate membranes. Data of Sourirajan (1965).

Feed solution molality: 0.5M. Feed rate: 30 cm^3/min. Operating pressure: 1500 psig.

average size of the pores on the membrane surface is very likely to be several times bigger than the size of the molecules involved in the solution systems studied in this work.

The permeability of the cellulose acetate membrane seems affected by its contact with aqueous organic solutions. This is illustrated by a typical set of

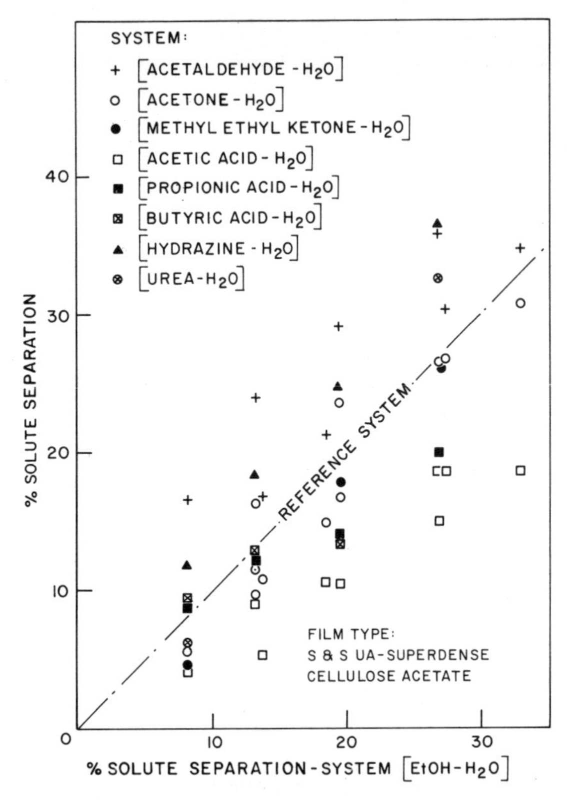

Figure 1-22. Relative separation of some organic solutes in aqueous solution using S & S porous cellulose acetate membranes. Data of Sourirajan (1965).

Feed solution molality: 0.5M. Feed rate: 30 cm^3/min. Operating pressure: 1500 psig.

data given in Table 1-6, in which the experiments are listed in the order in which they were performed. The experiment with the system 0.5M ethyl alcohol-water was repeated frequently to serve as a reference. The pure water permeability, [PWP], of the film at 20° decreased from 7.35 to 2.5 grams per hour in the course of 25 experiments involving several aqueous organic solutions. This decrease in permeability, however, did not affect solute separation significantly. This is evident from the fact that EtOH separation remained essentially constant within the experimental error of ±2 per cent applicable for this system. It is probable that the various aqueous organic solutions used affected primarily the spongy bulk structure of the film.

Tritons in Aqueous Solution. Table 1-7 gives a set of illustrative data on the separation and permeability characteristics of an S & S-UA-Medium type membrane using several polyoxyethylated nonionic surface active agents (Tritons, supplied by Rohm and Hass Co.) in aqueous feed solutions. This membrane was specified to have an average pore size of 200 to 350 Å. A consideration of the available data in the literature (Becher, 1961; Schick et al., 1962; Shinoda et al., 1963) indicates that the micellar size of the Tritons, even in their hydrated state, whether spherical, rod, or disk type, is probably of the order of 60 to 70 Å. This size is considerably smaller than

Table 1-6. Separation and permeability characteristics of an S & S porous cellulose acetate membrane for some organic solutes in aqueous solutions

Data of Sourirajan (1965)

Solute	Operating temp. (°)	Solute separation (%)	[PR] (g/hr at 20°)	[PWP] (g/hr at 20°)
Sodium chloride	24.5	75.0	5.67	7.35
Ethyl alcohol	23.5	18.0	6.28	7.32
Ethyl alcohol	25.0	18.4	6.54	7.03
Isopropyl alcohol	26.0	37.4	5.75	6.56
n-Propyl alcohol	24.5	20.6	5.35	6.66
Ethyl alcohol	26.0	18.4	6.18	6.09
tert-Butyl alcohol	25.5	63.6	4.91	6.42
sec-Butyl alcohol	25.5	33.8	4.16	5.74
Ethyl alcohol	25.5	18.4	5.88	5.94
Isobutyl alcohol	25.5	30.4	3.38	5.98
n-Butyl alcohol	25.5	8.9	2.48	5.00
n-Propyl alcohol	25.5	23.1	3.67	4.51
Ethyl alcohol	25.5	18.4	4.52	4.62
Acetone	25.5	16.7	3.97	4.41
Methyl ethyl ketone	26.0	17.7	2.82	4.44
Ethyl alcohol	27.0	20.4	3.89	3.70
Acetic acid	25.0	10.5	3.58	3.96
Urea	26.0	23.5	4.00	3.73
Ethyl alcohol	23.0	20.0	3.85	3.84
Hydrazine	26.0	24.7	4.01	3.76
Ethyl alcohol	26.0	20.0	3.91	3.89
Acetaldehyde	24.5	28.4	3.77	3.93
Propionic acid	25.5	13.9	2.87	3.89
Ethyl alcohol	24.5	18.4	3.67	3.89
Butyric acid	20.5	13.1	0.99	3.85
Ethyl alcohol	24.0	20.4	2.35	2.59

Film no. 147
Feed solution molality: 0.5M
Feed rate: 30 cm^3/min
Operating pressure: 1500 psig
Effective area of film: 7.6 cm^2

Table 1-7. Separation of Tritons in aqueous solutions using an S & S-UA-medium type porous cellulose acetate membrane

Data of Sourirajan and Sirianni (1966)

Solute	Solute separation (%)	Product rate (g/hr)
Triton n-128	90.0	38.7
Triton x-165	71.4	41.1
Triton x-205	69.2	41.4
Triton x-305	71.8	36.5

Feed concentration: 2 g Triton/100g water
Feed flow rate: 120 cm^3/min
Operating pressure: 1000 psig
Effective area of film: 7.6 cm^2

the average size of the membrane pores. The fact that 70 to 90 per cent of the Tritons was separated by the use of the above membrane is consistent with the concept that the mechanism of separation involved here is not primarily ultrafiltration.

Hydrocarbon Liquid Mixtures. Table 1-8 gives a set of experimental reverse osmosis separation and permeability data for the systems xylene-ethyl alcohol, xylene-n-heptane and n-heptane-ethyl alcohol using S & S porous cellulose acetate membranes (Sourirajan, 1964c). These experiments were carried out at the laboratory temperature at 1000 psig using a static cell shown in Figures 1-23 and 1-24. A different sample of the film was used for

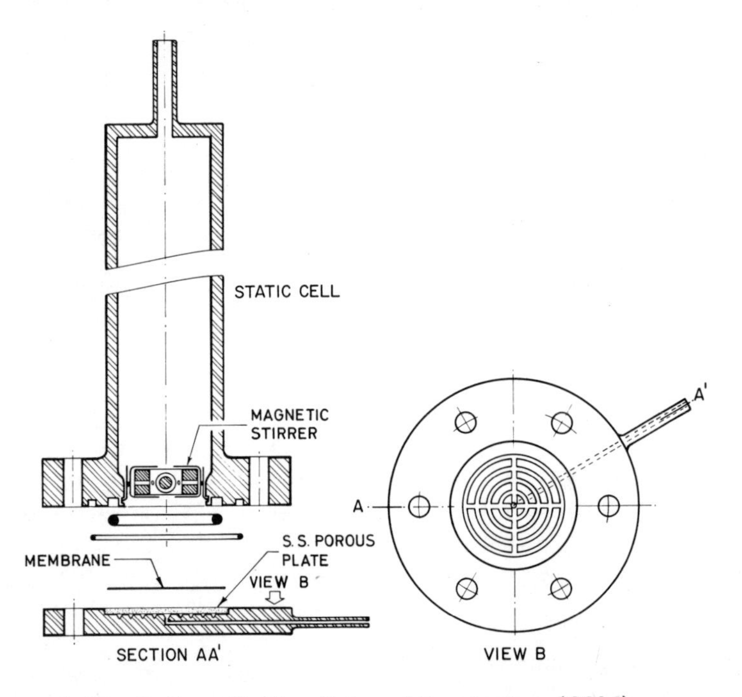

Figure 1-23. Static cell (Kopeček and Sourirajan, 1969d).

Table 1-8. Separation of some hydrocarbon liquid mixtures using S & S porous cellulose acetate membranes

Data of Sourirajan (1964c)

Expt. no.	Film type	Components in mixture	Composition (mole %) Feed	Composition (mole %) Product	Product rate (g/hr)
1	Preshrunk S & S-UA-Superdense	Xylene EtOH	19.5 80.5	16.5 83.5	Not determined
2	Above type film surface sprayed with 'Fluoroglide'	Xylene EtOH	19.5 80.5	23.0 77.0	Not determined
3	Preshrunk S & S-UA-Superdense	Xylene EtOH	19.5 80.5	12.5 87.5	Not determined
4	Above type film surface sprayed with 'Fluoroglide'	Xylene EtOH	14.0 86.0	15.5 84.5	Not determined
5	Preshrunk S & S-UA-Superdense	Xylene EtOH	50.5 49.5	38.0 62.0	1.23
6	Preshrunk S & S-UA-Superdense	Xylene EtOH	51.0 49.0	41.0 59.0	1.04
7	Preshrunk S & S-UA-Superdense	Xylene n-Heptane	50.0 50.0	57.5 42.5	1.85
8	Preshrunk S & S-UA-Superdense	Xylene n-Heptane	50.0 50.0	60.5 39.5	0.44
9	Preshrunk S & S-UA-Superdense	n-Heptane EtOH	33.0 }* 67.0	9.5 90.5	2.90
10	Preshrunk S & S-UA-Superdense	n-Heptane EtOH	33.0 }* 67.0	6.0 94.0	0.51

* Azeotropic mixture.

Area of film surface: 9.6 cm^2
Operating pressure: 1000 psig

each experiment. Just before use in the experiment, the film was superficially dried between filter papers and quickly immersed in ethyl alcohol several times to remove the water from the membrane pores; it was again dried superficially between filter papers, and quickly immersed in the feed solution several times; finally, it was again dried superficially between filter papers and quickly mounted in the cell. The cell was a stainless steel pressure chamber consisting of two detachable parts. The film was mounted on a stainless steel porous plate embedded in the lower part of the cell through which the membrane-permeated liquid was withdrawn at the atmospheric pressure. The upper part of the cell contained the feed solution under pressure in contact with the membrane. The two parts of the cell were clamped and sealed tight using rubber O-rings. Compressed nitrogen gas was used to pressurize the system. About 250 cm^3 of feed solution were used each time. The feed solution was kept well stirred during the experiment by means of a magne-

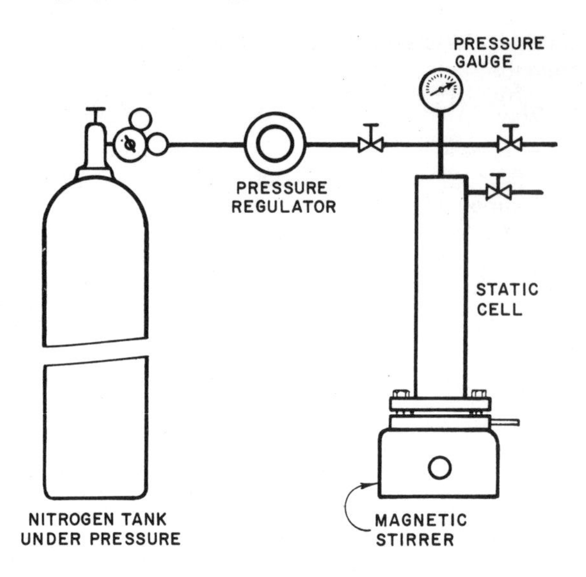

Figure 1-24. Static cell flow diagram (Kopeček and Sourirajan, 1969d).

tic stirrer fitted in the cell about one quarter of an inch above the membrane surface. The quantity of liquid removed by membrane permeation was small compared to the amount of feed solution in the pressure chamber. The compositions of the feed and the membrane-permeated liquids were analysed by means of a Bausch and Lomb precision refractometer.

The experimental results given in Table 1-8 illustrate that the separation technique is applicable for hydrocarbon systems, that the cellulose acetate material has a preferential sorption for EtOH for xylene-EtOH and n-heptane-EtOH feed systems and it has a preferential sorption for xylene for xylene-n-heptane feed system, and that azeotropic composition is no limitation to this separation process. When the surface of the cellulose acetate film was sprayed uniformly with 'Fluoro-glide' (a fluorocarbon polymer supplied by Chemplast, Inc.) and used in contact with the xylene-EtOH feed system, the direction of separation was reversed. This result again shows that the mechanism of the process is governed by a surface phenomenon.

Separation of Gases. The separation of gases by permeation through membranes has been reviewed by several authors (Barrer, 1951, 1965; Brown and Tuwiner, 1962; Choo, 1962; Kammermeyer, 1956). Since the observed separations do not always follow Graham's law, the mechanism is explained on the basis of solubility, diffusion and re-evaporation in the case of flow through 'non-porous' plastic films, and sorbed-flow through porous media.

There is sufficient evidence in the literature to show that the physicochemical nature of the film surface affects the degree of separation of different gases (Brubaker and Kammermeyer, 1953a, 1953b; Agrawal and Sourirajan, 1969a). Oxygen permeates faster than nitrogen in rubber and plastic films (Brubaker and Kammermeyer, 1954; Van Amerongen, 1950; Weller and Steiner, 1950), and the reverse is true of flow through porous glass (Brubaker and Kammermeyer, 1953); this is understandable since preferential sorption is a surface property.

Flow through microporous media has been discussed by Rutz and Kammermeyer (1958, 1959). From the point of view of the preferential sorption-capillary flow mechanism, the reverse osmosis technique should be applicable for gas separation also. A few experiments with air were carried out to test this approach using the preshrunk S & S porous cellulose acetate membranes in the static apparatus shown in Figure 1-24 (Sourirajan, 1963b). Compressed air at the laboratory temperature was used in the experiments. The compositions of the feed and the product samples were analysed using the mass spectrometer. The product gas was collected over water and its rate determined. A set of typical results is given in Table 1-9.

Table 1-9. Separation of gases using S & S porous cellulose acetate membranes

Data of Sourirajan (1963b)

Film no.	Operating pressure (psig)	Product rate cm^3/min	Composition of product gas* (%)		
			N_2	O_2	A
1	1000	1.0	76.7	22.4	0.9
2	250	0.19	74.2	24.7	1.1
2	500	0.42	72.0	26.8	1.2
3	1000	0.15	73.4	25.6	1.0
3	1500	0.22	73.7	25.3	1.0
4	1000	(not determined)	71.0	27.2	1.8

* Feed air composition (per cent): nitrogen, 78.1; oxygen, 21.0; argon, 0.9.

The feed air contained about 0.06 per cent carbon dioxide. The product gas showed an enrichment in carbon dioxide also. Since the analysis for carbon dioxide is not considered accurate in this investigation, the gas compositions here are given on a carbon dioxide-free basis.

Effective area of film: 9.6 cm^2

The membrane-permeated air was found to be enriched in oxygen, argon and carbon dioxide. The higher the temperature of film shrinkage, the higher was the degree of separation. Increase of operating pressure increased the rate of production of the membrane-permeated gas almost proportionately. Even though the experimental conditions used are far from ideal for the given separation problem, yet the results obtained are significant with respect to both the degree of separation and product rate; they illustrate the essential validity of the reverse osmosis approach to gas separation.

The reverse osmosis technique should be applicable for the separation of vapours also (Kammermeyer and Wyrick, 1958); in this case the relative ease of condensibility of the vapour components under the experimental conditions becomes an important additional factor affecting their separation and flow characteristics.

Operating Pressure. Figures 1-25 and 1-26 illustrate the effect of operating pressure on the separation and product rate characteristics of the S & S porous cellulose acetate membranes for the systems [$NaCl$-H_2O] and [iso-PrOH-H_2O] respectively. The data plotted are typical of those obtained for several of the systems studied (Sourirajan, 1964b, 1965). An in-

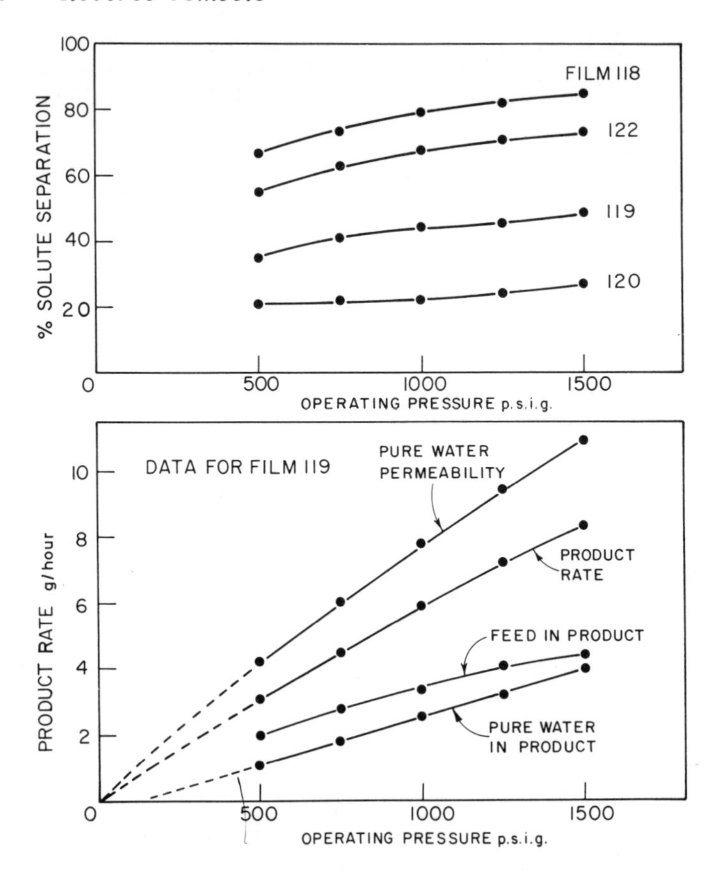

Figure 1-25. Effect of operating pressure on solute separation and product rate for the system [NaCl—H_2O] using S & S porous cellulose acetate membranes. Data of Sourirajan (1964b).

Film type: S & S UA-Superdense cellulose acetate. Feed solution molality: 0.5M. Feed rate: 30 cm^3/min. Film 119 has film area 7.6 cm^2.

crease in operating pressure increases both solute separation and product rate. The latter effect is understandable on the basis of increased driving pressure for fluid flow; the increase in solute separation with increase in operating pressure might be due to a decrease in the average pore size on the membrane surface, and/or increase in the preferential sorption of the membrane for pure water at higher pressures. The films used in this work were initially subjected to a pure water pressure of 1700 psig for about 2 hours. For films shrunk at high temperatures, the plot of [PWP] data v. operating pressure is essentially a straight line; but for films shrunk at low temperatures, the above plot is a curve indicating the increasing effect of compaction of the spongy bulk layer of the film under pressure. If the product is considered as a mixture of (i) the feed solution and (ii) the preferentially sorbed pure water, the effect of operating pressure on the separation and permeability characteristics of the film depends on the relative extents to which the flow of (i) and (ii) above are individually affected by the change in the operating pressure and the bulk structure of the film.

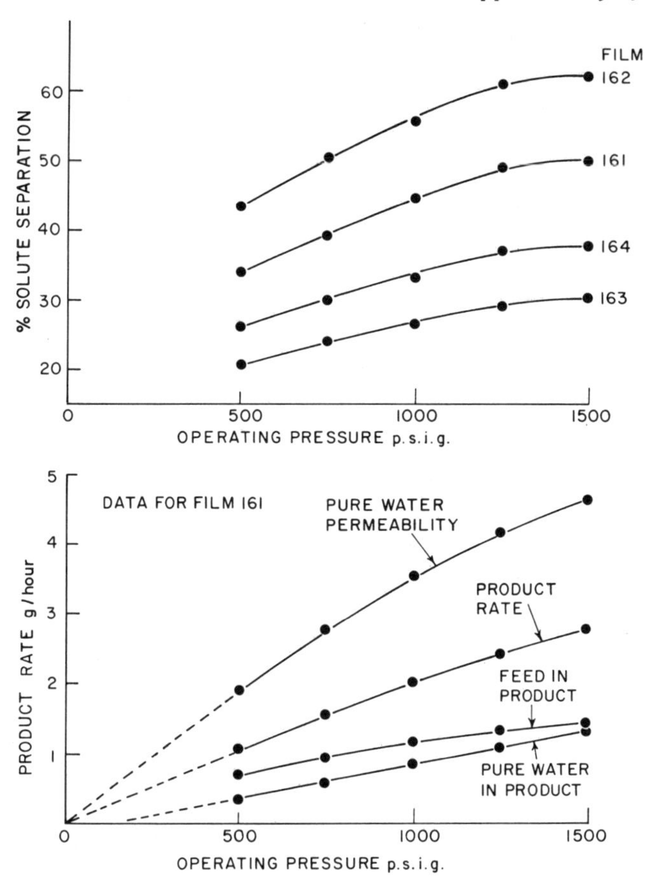

Figure 1-26. Effect of operating pressure on solute separation and product rate for the system [iso—PrOH—H_2O] using S & S porous cellulose acetate membranes. Data of Sourirajan (1965).

Film type: S & S UA-Superdense cellulose acetate. Feed solution molality: 0.5M. Feed rate: 30 cm^3/min. Film 161 has film area 7.6 cm^2.

Feed Concentration. Figure 1-27 illustrates the effect of feed concentration on the separation and permeability characteristics of S & S porous cellulose acetate membranes for the system [NaCl-H_2O]. At a given operating pressure, both solute separation and product rate decrease with increase in feed concentration in the range 0.25M to 5.0M. The latter effect is simply due to the increase in osmotic pressure, and hence a decrease in effective driving pressure for fluid flow, with increase in feed concentration. The data also show that with a given solute, feed solutions of different concentrations can be considered as related systems. Further, the slopes of the curves representing the separation and product rate data v. feed molality depend not only on the level of feed concentration but also on the nature of the solute; this is illustrated by the data plotted in Figure 1-28 for the systems [NaCl-H_2O] and [$NaNO_3$-H_2O]. With feed concentrations lower than about 2.9M, the separation of NaCl is higher than that of $NaNO_3$; with higher feed concentrations, the relative order of separation is reversed. These data indicate that the order of solute separation is governed not only by the physicochemical nature of the

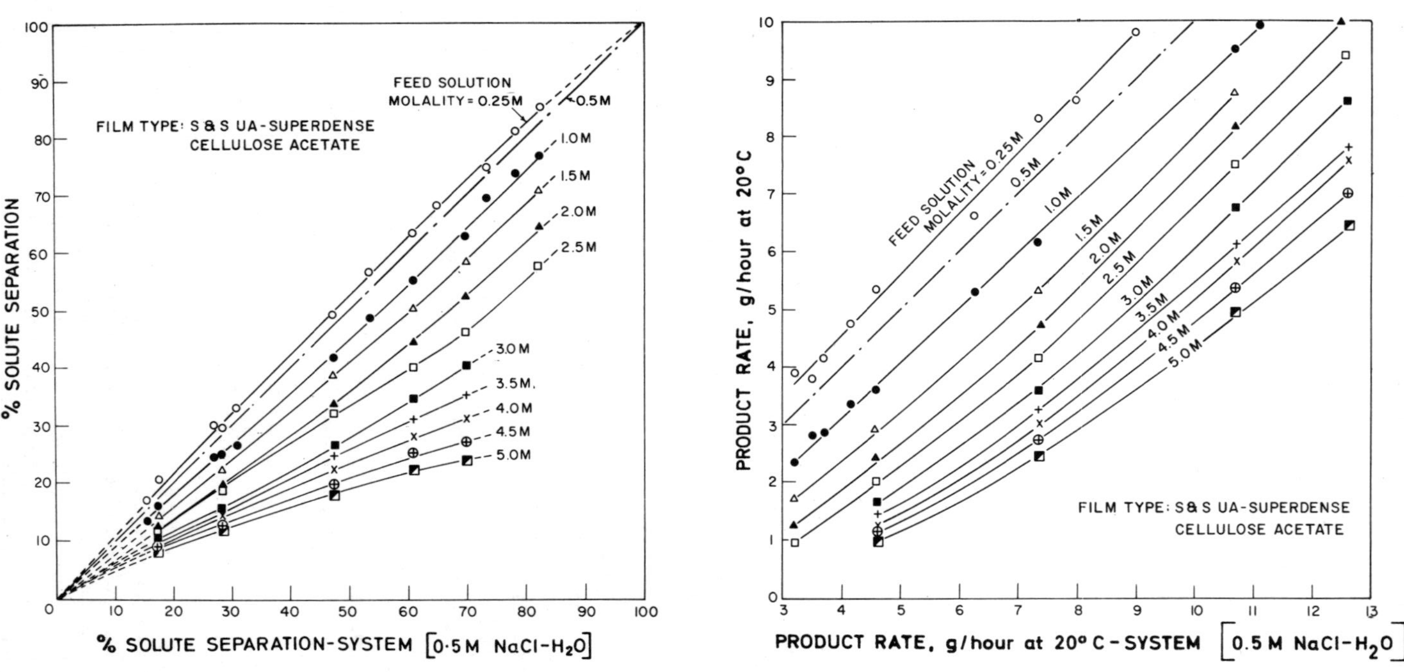

Figure 1-27. Effect of feed concentration on solute separation and product rate for the system [NaCl—H_2O] using S & S porous cellulose acetate membranes. Data of Sourirajan (1964b).

Feed rate: 30 cm^3/min. Operating pressure: 1500 psig. Film area: 7.6 cm^2.

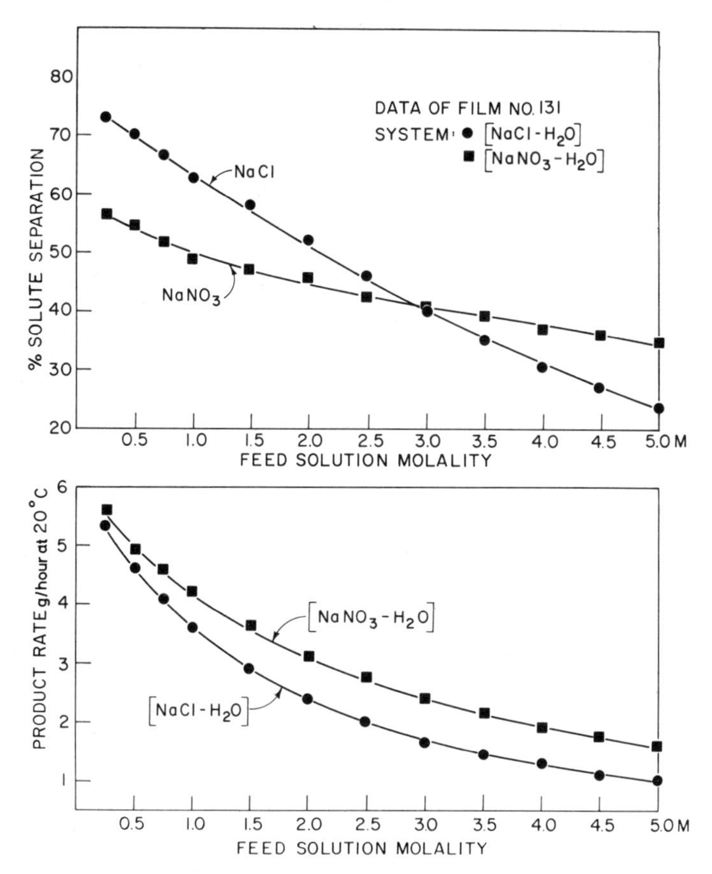

Figure 1-28. Effect of feed concentration on solute separation and product rate characteristics of S & S porous cellulose acetate membranes for the systems [NaCl—H_2O] and [$NaNO_3$—H_2O]. Data of Sourirajan (1964b).

Film type: S & S UA-Superdense cellulose acetate. Feed rate: 30 cm^3/min. Operating pressure: 1500 psig. Film area: 7.6 cm^2.

membrane-solution system, but also by the factors governing material transport in the reverse osmosis process.

Figure 1-29 illustrates the effect of feed concentration on the reverse osmosis separation of some organic solutes in aqueous solution using S & S cellulose acetate membranes. Solute separation decreases, but only slowly, with increase in feed concentration, while the corresponding product rate decreases rapidly. Only a part of the latter change could be attributed to the effect of the organic solute on membrane permeability referred to above (Table 1-6). According to Ambard and Trautmann (1960b), the nonelectrolyte molecules in aqueous solution do not mutually dehydrate each other and their volumes in the hydrated state remain constant at all concentrations; consequently, the solute concentration has no effect on the filterability of the nonelectrolyte through an ultrafilter membrane. The chemical nature of the membrane surface has no controlling influence in the ultrafiltration mechanism. On the basis of the preferential sorption-capillary flow mechanism, however, the separation and product rate characteristics of a membrane depend on the extent of the negative adsorption of the solute and the membrane-solution interface, osmotic

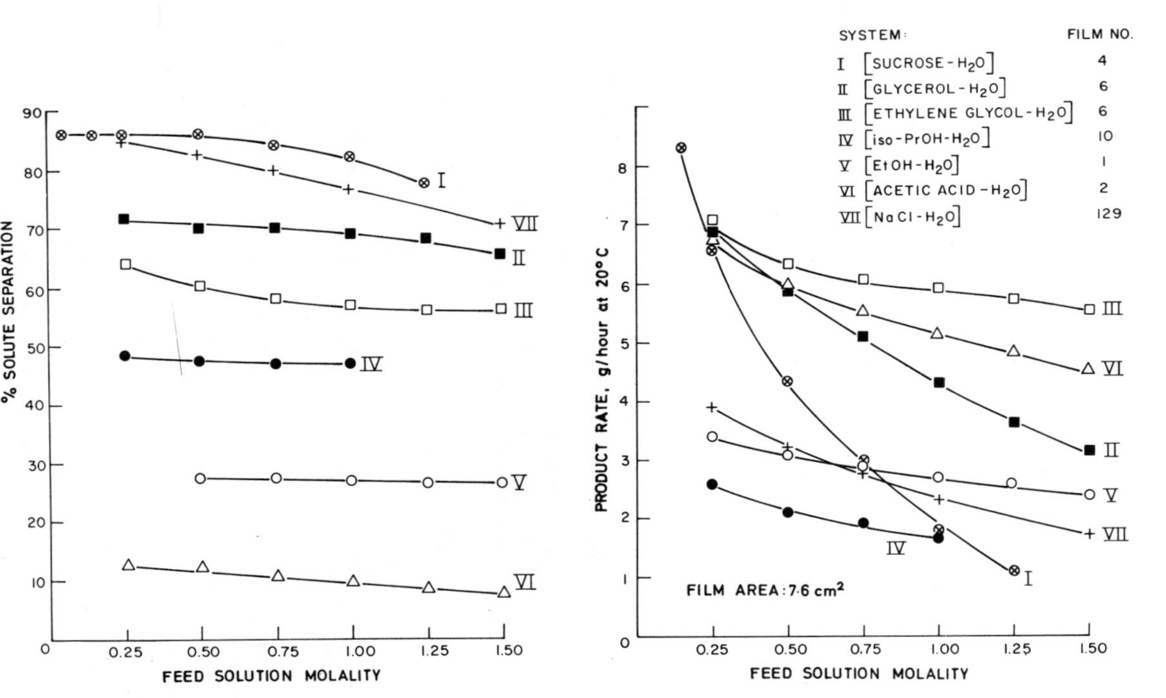

Figure 1-29. Effect of feed concentration on solute separation and product rate characteristics of S & S porous cellulose acetate membranes for some aqueous organic systems. Data of Sourirajan (1965).

Film type: S & S UA-Superdense cellulose acetate. Feed rate: 30 cm^3/min. Operating pressure: 750 psig.

pressure of the feed solution and other factors governing the transport of material through the reverse osmosis membrane. As Levich (1962) has pointed out, the parameter $d\sigma/dc$ (where σ is the surface tension and c is the concentration of the surface active agent) is an important variable affecting mass transfer across interfaces especially in the viscous flow regime.

Feed Flow Rate. It is necessary to impart a high degree of turbulence in the desalting cell to prevent the concentration build up in the solution at the vicinity of the membrane surface. A high feed flow rate combined with proper geometry of the flow system can accomplish this purpose. A flow rate of 30 cm^3 per minute was arbitrarily chosen in this work with the S & S membranes. The actual effect of variation of feed flow rate on the separation and permeability characteristics of a given type of membranes will depend upon the design of the apparatus and the experimental conditions employed. For the experimental conditions used in this work, Figure 1-30 illustrates the effect of feed flow rate in the range 16 to 270 cm^3/min on the characteristics of six membranes of different surface porosities using [0.5M $NaCl-H_2O$] as the feed solution. Both solute separation and product rate increased with increase in feed flow rate; the change was greatest for films giving lower separations especially in the feed rate range 16 to 100 cm^3/min.

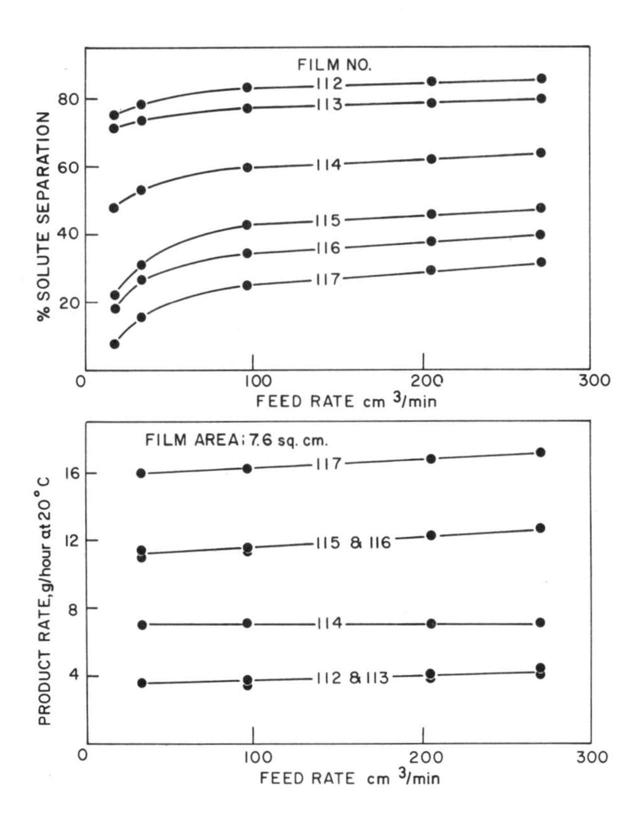

Figure 1-30. Effect of feed flow rate on solute separation and product rate characteristics of S & S porous cellulose acetate membranes for the system [$NaCl-H_2O$]. Data of Sourirajan (1964b).

Film type: S & S UA-Superdense cellulose acetate. Feed solution molality: 0.5M. Operating pressure: 1500 psig.

The increase in product rate was not more than 10 per cent in the entire range of feed rate studied. A variation in feed flow rate had, however, no effect on the characteristic separation and product rate curves obtained for related solution systems so long as the results referred to conditions of constant feed rate. This is illustrated in Figure 1-31 for the related systems [0. 5M $NaCl-H_2O$] and [0. 5M $NaNO_3-H_2O$] at feed flow rates in the range 16 to 270 cc. per minute.

Operating Temperature. With a [0. 5M $NaCl-H_2O$] feed solution, a change in operating temperature from 10° to 25° had no effect on the extent of solute separation. The product rates, however, increased with increase in operating temperature such that the product rate multiplied by its viscosity at the operating temperature was essentially constant. This is illustrated in Figure 1-32 with the data obtained with four different S & S films, numbers 32, 33, 34, and 35 giving 78. 3, 62. 0, 47. 1 and 41. 0 per cent solute separation respectively.

With a [2. 0M $NaCl-H_2O$] feed solution, solute separation at 10° was significantly higher than that at 25°, and the product rate results were similar to those observed for the [0. 5M $NaCl-H_2O$] feed solution.

A change in operating temperature changes (i) the densities and viscosities of both the feed solution and the preferentially sorbed pure water, and (ii) the osmotic pressure of the system. Consequently the relative permeation of the pure water and the concentrated boundary solution through the membrane changes with temperature. An increase of temperature from 10° to 25° increases the osmotic pressure of aqueous sodium chloride solutions (Tribus et al., 1960), resulting in a decrease in the effective pressure (i.e. the operating pressure minus the osmotic pressure) for the flow of the preferentially sorbed water. Thus while change (i) above increases the relative flow of the pure water through the membrane with increase in temperature, the change in the osmotic pressure has the opposite effect. The experimental results obtained seem at least qualitatively understandable on the basis of the above considerations. Probably the preferential sorption characteristics of the membrane also change with temperature.

The effect of operating temperature, in the range 10° to 30°, on the separation and product rate characteristics of the S & S membranes for the systems $EtOH-H_2O$, $n-PrOH-H_2O$, ethylene glycol-water, glycerol-water, and hydrazine-water is illustrated in Figure 1-33. In all the above cases, solute separation decreased and the product rate increased with increase in operating temperature. The product rate results were again essentially similar to those observed with the sodium chloride-water feed solutions.

Separation of Mixed Salts in Aqueous Solution. Equimolal solutions of [$NaCl-H_2O$] and [$NaNO_3-H_2O$] behaved as related systems under identical experimental conditions, and their relative separation values were essentially on the same characteristic curve for the feed solution concentration range 0. 25 to 1. 0 molality. A few experiments were carried out on the separation characteristics of the membrane with feed solutions containing equimolal mixtures of sodium chloride and sodium nitrate in aqueous solutions corresponding to the total molality of 0. 25M, 0. 5M, and 1. 0M in the feed. The results were compared with those obtained for the systems sodium chloride-water and sodium nitrate-water corresponding to the feed solution at the same molalities. Thus both the single solute system and the mixed solute system under comparison contained the same solute/solvent ratio. Three general observations emerged (Sourirajan, 1964b).

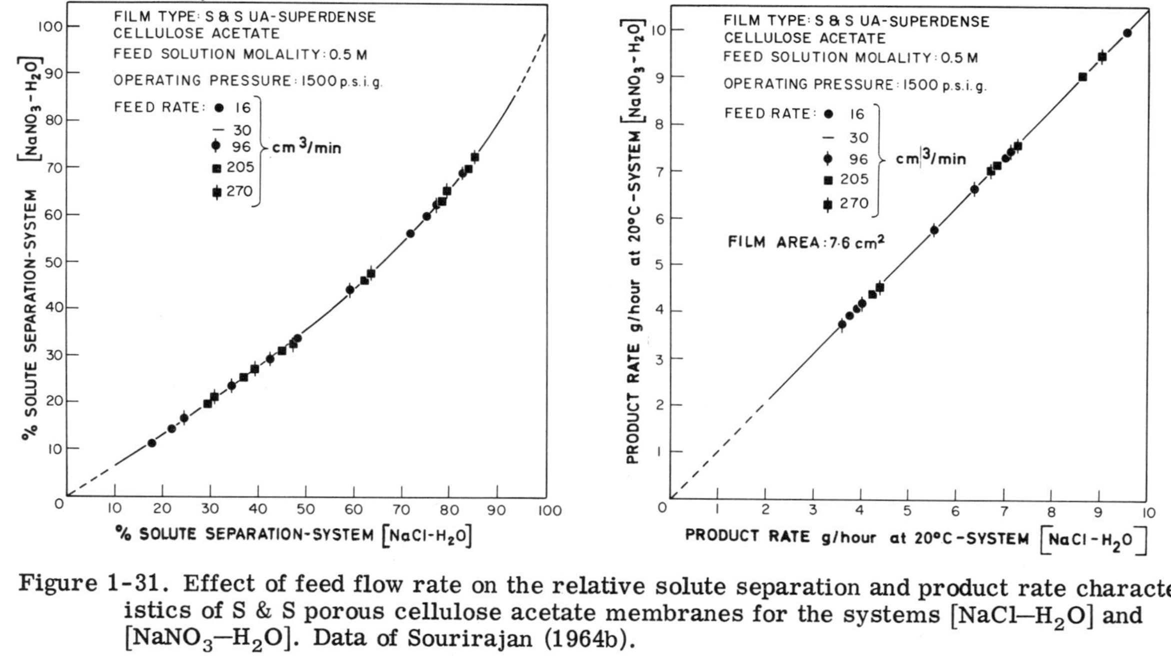

Figure 1-31. Effect of feed flow rate on the relative solute separation and product rate characteristics of S & S porous cellulose acetate membranes for the systems [$NaCl—H_2O$] and [$NaNO_3—H_2O$]. Data of Sourirajan (1964b).

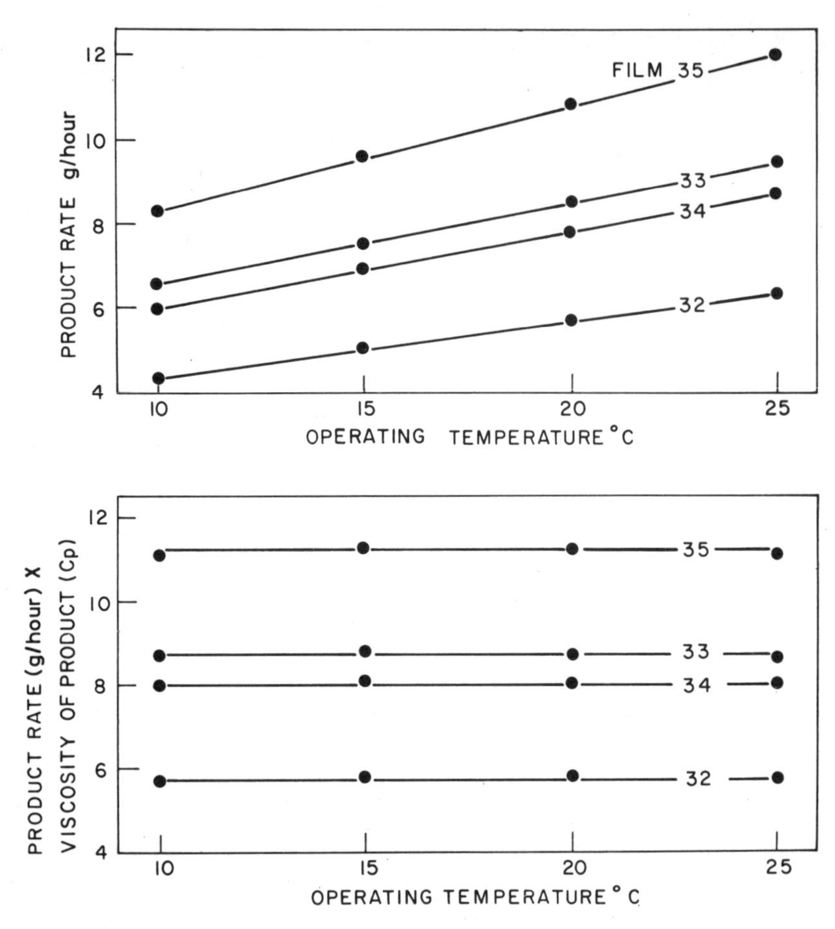

Figure 1-32. Effect of operating temperature on the product rate characteristics of S & S porous cellulose acetate membranes for the system [NaCl–H_2O]. Data of Sourirajan (1964b).

Film type: S & S UA-Superdense cellulose acetate. Feed solution molality: 0.5M. Feed rate: 30 cm^3/min. Operating pressure: 1500 psig. Film area: 7.6 cm^2.

There was no significant change in the extent of sodium chloride separation in the two types of feed systems. The extent of sodium nitrate separation however was always less in the mixed solute system. The following is a set of illustrative results. For six different films, the NaCl separations in the feed system [0.5M NaCl-H_2O] were 83.7, 71.4, 62.0, 48.4, 28.3 and 18.0; the corresponding data for the feed system [0.25M NaCl-0.25M $NaNO_3$-H_2O] were 83.8, 71.2, 61.6, 48.4, 28.8, and 18.8 respectively. On the other hand, the results for $NaNO_3$ separation for the system [0.5M $NaNO_3$-H_2O] for the above films were 69.0, 55.2, 45.2, 33.3, 19.0, and 12.0; the corresponding data for the feed system [0.25M NaCl-0.25M $NaNO_3$-H_2O] were 65.5, 51.3, 43.3, 30.6, 15.2 and 9.1 respectively. Similar results were obtained for the other concentrations of feed solutions studied.

The relative separation data for NaCl and $NaNO_3$ in the mixed solute feed systems also seem to fall on a characteristic line. This is illustrated in

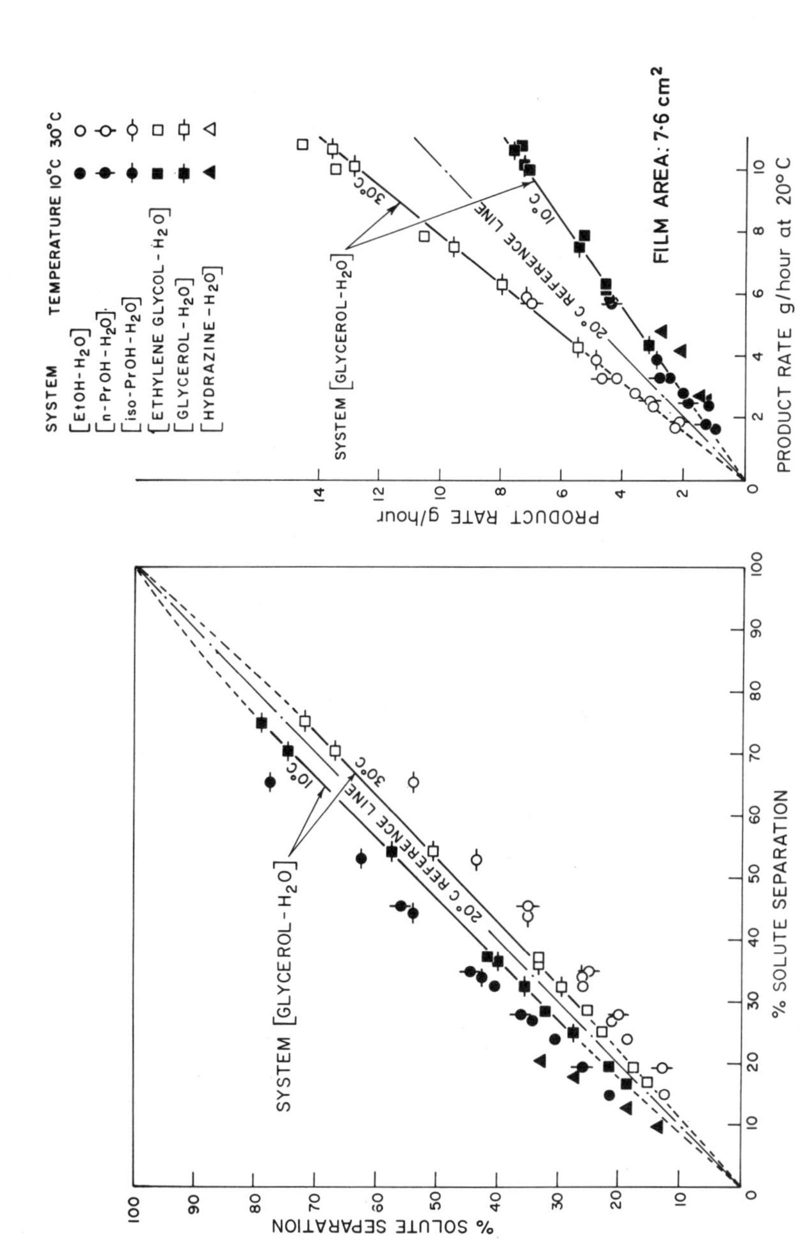

Figure 1-33. Effect of operating temperature on solute separation and product rate characteristics of S & S porous cellulose acetate membranes for some aqueous organic systems. Data of Sourirajan (1965).

Film type: S & S UA-Superdense cellulose acetate. Feed solution molality: 1.0M. Feed rate: 30 cm^3/min. Operating pressure: 1500 psig.

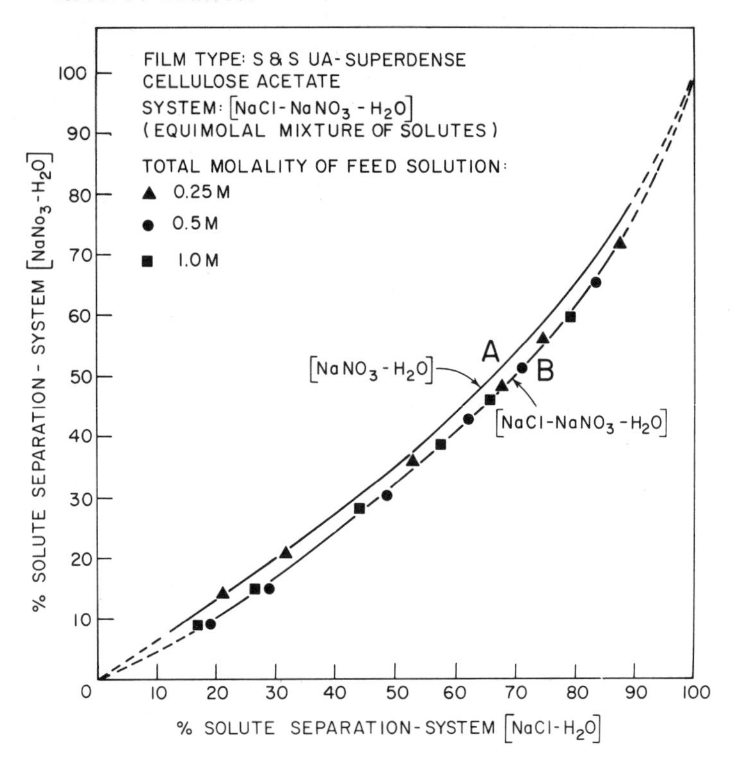

Figure 1-34. Separation of mixed solutes in aqueous solution using S & S porous cellulose acetate membranes. Data of Sourirajan (1964b).

Operating pressure: 1500 psig. Feed rate: 30 cm^3/min.

Figure 1-34, where line A is the characteristic solute separation curve for the system [$NaNO_3$-H_2O] with reference to [NaCl-H_2O], and curve B is another characteristic $NaNO_3$ separation curve with reference to NaCl separation in the mixed solute system [NaCl-$NaNO_3$-H_2O]. The curve B appears to be valid for feed solutions of total molality 0.5M and 1.0M.

The product rate data for the mixed solute system were in general about the same or intermediate between those obtained for the corresponding single solute systems.

8. Experiments With Porous Glass Membranes

Inorganic Salts in Aqueous Solution. Kraus et al. (1966a) investigated the characteristics of porous Vycor glass membranes for the reverse osmosis separation of some inorganic salts in aqueous solutions. Ninety-five per cent of the pores in the membrane had diameters between 36 and 56 Å. Feed solutions were circulated under pressure past the membrane surface. A summary of their results is given in Table 1-10. They found that Vycor glass separated NaCl and Na_2SO_4 to a small but appreciable extent from neutral solutions. The separation of Na_2SO_4 was relatively higher than that of NaCl.

Since the isoelectric point of silica is about 2 (Parks, 1965), Vycor glass was expected to have cation exchange properties when in contact with solutions of neutral pH. Two methods of modifying pore-surface charge, and hence the

Table 1-10. Separation of some inorganic salts in aqueous solution using porous Vycor glass membranes at 25°

Data of Kraus et al. (1966a)

Solute	Feed molality	Additive*	Feed pressure (atm)	Product rate $\left(\frac{cm^3}{hr\ atm\ cm^2}\right)$	Solute separation (%)
NaCl	0.03		25 to 55	0.003 to 0.01	5 to 10
NaCl	0.03	pH 10, silicate	120	0.009	35
NaCl	0.01	pH 9.4	80	0.008	35
NaCl	0.01	pH 10.2	100	0.009	55
NaCl	0.03	Th (IV)	50	0.003	55
NaCl	0.10	Th (IV)	50	0.003	30
Na_2SO_4	0.015		50	0.009	20
Na_2SO_4	0.005	pH 10, silicate	100	0.004	70
$MgCl_2$	0.015		50	0.01	−10
$MgCl_2$	0.02	Th (IV)	40	0.005	45
$MgSO_4$	0.015		55	0.008	5
$CaCl_2$	0.015		50	0.01	−5
$CaCl_2$	0.03	Th (IV)	35	0.006	40

* Where mentioned, pH was adjusted with $NaHCO_3$ or Na_2CO_3; Th (IV) concentration, 10^{-3}M; silicate concentration, 0.007M.

Membrane thickness: 0.5 to 1.0 mm

separation characteristics, were investigated. In one, the pH was changed; in the other, polyvalent ions or polyelectrolytes of charge opposite to that of silica were adsorbed.

Solute separation increased when the pH was raised; with 0.03M solution, NaCl separation increased from about 10 to 35 per cent with increase in pH from 6 to 10; and, with 0.01M solution, NaCl separation increased from 35 to 55 per cent between pH 9.4 and 10.2. Product rates increased with time particularly with high pH feed solutions; this was attributed to the possible dissolution of the glass. The other technique of treating the glass with polyvalent cations, such as Fe(III), Zr(IV), Bi(III), Al(III), did not result in any increase in solute separation in the reverse osmosis experiments.

Solute separation was significantly modified by addition of thorium salts to the feed solution. For example, a membrane which rejected 6 per cent of solute from a 0.03M NaCl solution rejected 63 per cent of total chloride during pretreatment with a solution 0.03M in NaCl and 0.05M in $ThCl_4$; it then rejected 55 per cent of the salt from a solution 0.03M in NaCl and 0.001M in $ThCl_4$, and 31 per cent from a solution 0.1M in NaCl and 0.001M in $ThCl_4$. With a feed of 0.015M $MgCl_2$ or $CaCl_2$ alone, the product was 5 to 10 per cent more concentrated than the feed solution. When $ThCl_4$ was added to bring the feed to 0.001M in Th(IV), solute rejection (based on total chloride) from 0.02 to 0.03M $MgCl_2$ or $CaCl_2$ solutions was 40 to 50 per cent. Similar results were reported by Kuppers et al. (1966) using pretreated cellophane membranes.

These results again illustrate the relevance of the porous structure and chemical nature of the membrane surface with respect to that of the feed solution, to the success of the reverse osmosis separation process.

Hydrocarbon Liquid Mixtures. Kammermeyer and Hagerbaumer (1955) accomplished the separation of ethyl acetate-carbon tetrachloride, cyclohexane-ethyl alcohol, benzene-methanol, and benzene-ethyl alcohol mixtures by the reverse osmosis technique using porous Vycor glass membranes containing approximately 1.34 × 10^{12} pores per cm^2 with an average pore diameter of 40 Å. Some of their data are given in Table 1-11. In all the above cases, the membrane-permeated product was found to be richer in the more polar constituent of the mixture for which the surface of glass may be expected to possess a preferential sorption.

Table 1-11. Separation of some hydrocarbon liquid mixtures using porous Vycor glass membranes

Data of Kammermeyer and Hagerbaumer (1955)

Operating pressure (psig)	System A-B	Wt.% A		Product rate ($g/cm^2/min$) × 10^5
		Feed	Product	
60	EtAc*-CCl_4	43.30	44.31	5.16
125	EtAc*-CCl_4	43.64	45.36	10.62
200	EtAc*-CCl_4	42.95	45.06	16.15
60	Cyclohexane-EtOH	69.0	63.16	1.606
125	Cyclohexane-EtOH	69.0	62.24	4.00
200	Cyclohexane-EtOH	69.0	62.40	6.57
60	Benzene-MeOH	60.40	59.25	3.87
125	Benzene-MeOH	60.56	57.95	7.74
200	Benzene-MeOH	60.11	56.28	12.57
60	Benzene-EtOH	67.4	67.0	3.15
125	Benzene-EtOH	67.4	67.0	6.97
200	Benzene-EtOH	67.4	66.6	11.29

* EtAc = Ethyl Acetate

9. Experiments With Charged Membranes

Cellulose and cellulose ester membranes are not generally considered to be charged membranes (Prideaux, 1942; Reid and Kuppers, 1959). Ion exchange membranes have long been known to have the property of excluding electrolytes; the literature on the subject is extensive (Erschler, 1934; Helfferich, 1962; Lorimer et al., 1956: McBain and Kistler, 1928; McBain and Stuewer, 1936; Schmid and Schwarz, 1952; Sollner, 1930, 1958; Sollner et al., 1954; Spiegler, 1958; Staverman, 1952; Trautman and Ambard, 1952). The above property is the basis of the ion exclusion process (Wheaton and Bauman, 1953) and of the electrodialysis method (Wilson, 1960) for desalination. McKelvey et al. (1959) investigated the separation of NaCl, Na_2SO_4 and $CaCl_2$ from aqueous solutions by the reverse osmosis technique using some synthetic ion exchange membranes. Some of their results are given in Table 1-12. They obtained good solute separation but very low product rates at the operating pressure of 1000 psig. It has been suggested (Dresner and Kraus, 1963; Dresner, 1965) that interesting salt rejection and permeability properties could be expected from porous ion exchange membranes having pores as large as 100 Å. Some successful results have also been reported by Baldwin and Holcomb (1965). The above results again indicate the potential extensive applicability of the reverse osmosis separation technique.

Table 1-12. Characteristics of some synthetic ion exchange membranes for the separation of inorganic salts in aqueous solution

Data of McKelvey et al. (1959)

Membrane**	Membrane thickness (in.)	Solute	Solution normality	Operating pressure (psig)	Solute separation (%)	Hydraulic permeability* $cm^3/in^2/day$
Permaplex C-10	0.03	NaCl	0.1	1200	19	17.2
Nalfilm 1	0.004	NaCl	0.1	1400	61	2.1
Nalfilm 2	0.004	NaCl	0.1	1000	90	0.41
Nalfilm 2	0.004	NaCl	1.0	1000	46	0.33
Nalfilm 2	0.004	NaCl	0.01	1000	91	0.31
Nalfilm 2	0.004	NaCl	1.0	1000	42	0.41
Nalfilm 2	0.004	$CaCl_2$	1.0	1000	72	0.27
AMF	0.006	$CaCl_2$	1.0	1000	94	0.51
AMF	0.006	Na_2SO_4	1.0	1000	75	1.13
AMF	0.006	NaCl	1.0	1000	73	0.58

* Rates are at 1000 psig

** Permaplex C-10 and Nalfilm 1 are sulphonic cation exchange membranes; Nalfilm 2 and AMF are quaternary ammonium anion exchange membranes.

10. Summary

In the light of the preferential sorption-capillary flow mechanism, the reverse osmosis process is a general separation technique. In principle, the process is applicable for the separation, concentration, and fractionation of inorganic or organic (ionic or nonionic) substances in aqueous or nonaqueous solutions in the liquid or the gaseous phase, and hence it opens a new and versatile field of separation technology in chemical process engineering. On the basis of the preferential sorption-capillary flow mechanism, the process is governed by a surface phenomenon; an appropriate chemical nature of the film surface in contact with the solution, as well as the existence of pores of appropriate size on the area of the porous film at the interface, constitutes the indispensable twin-requirement for the practical success of this separation process. This approach to the subject led to the independent discovery of the suitability of cellulose acetate and other cellulose ester membranes for water desalination, and then to the discovery of the desalination characteristics of the porous S & S cellulose acetate filter membranes, and subsequently to the development of superior porous cellulose acetate membranes of industrial interest for the economic desalination of saline waters. Detailed studies on the characteristics of the porous S & S cellulose acetate membranes together with the results of earlier experiments with porous glass and ion exchange membranes, have established beyond doubt the general applicability of the reverse osmosis separation technique.

CHAPTER 2

Preparation and Performance of Porous Cellulose Acetate and Other Membranes for Reverse Osmosis Application

1. Development of Loeb-Sourirajan Type Membranes at UCLA for Saline Water Conversion

The success of the desalination experiments with the preshrunk Schleicher and Schuell cellulose acetate filter membranes led Loeb and Sourirajan to a programme of investigations during the period June 1959-August 1960, which resulted in the development of their own technique for making superior porous cellulose acetate membranes capable of giving high levels of both solute separation and product rate in the reverse osmosis process for saline water conversion and related applications. The details of this technique are now well known (UCLA, 1960; Loeb and Sourirajan, 1961, 1963, 1964).

Basic approach. The purpose of these investigations was to produce more porous cellulose acetate membranes, to obtain pores of the smallest possible size on the surface layer of the membrane, with big interconnecting pores in the interior mass of the membrane material, and to make the above surface layer as thin as possible. Since the preliminary experiments with membranes cast from an acetone solution of cellulose acetate were not particularly encouraging, the basic approach was to incorporate a suitable water-soluble additive in the film casting solution, leach out the additive from the membrane subsequently with water thus creating the required porosity in the membrane, and adjust the film casting conditions so as to create a very thin microporous surface layer embedded on a macroporous spongy mass.

Choice of the additive material. The work of Dobry (1936) and Biget (1950) guided the choice of the appropriate water soluble additive material for use in the film casting solution. Dobry (1936) reported that cellulose acetate was soluble at room temperature in saturated solutions of Ca, Mg, Cu, and Zn perchlorates, and membranes could be formed from solutions of cellulose acetate in saturated magnesium perchlorate. Biget (1950) reported that cellulose acetate gels could be prepared by coagulating acetone solutions of cellulose acetate with aqueous solutions of magnesium perchlorate, and it was possible to prepare, reproducibly, cellulose acetate membranes with porosities from 0.023 to 0.190 micron sizes by the above technique. On the basis of the above results, magnesium perchlorate became the natural choice for test as the additive material in the film casting solutions of Loeb and Sourirajan.

Performance of early laboratory-cast porous cellulose acetate membranes.* The first set of membranes was made from casting solutions containing cellulose acetate, acetone, and water. The object was to find out whether water itself was adequate as the additive material. The films were cast on a glass plate at the laboratory temperature. Acetone was allowed to evaporate from the surface of the film for a few minutes, after which the glass plate was immersed in water at the laboratory temperature for the film to set as a gel.

---

* Unless otherwise stated, all performance data are at laboratory temperature (23-25°)

After about an hour, the film was peeled off easily from the glass plate and tested in the reverse osmosis experiments. The results obtained, given in Table 2-1, were not satisfactory.

Table 2-1. Performance of Membranes Cast From Solutions Containing Cellulose Acetate, Acetone, and Water

Data of Loeb and Sourirajan (1960b)

Casting solution composition (wt. %)					
Cellulose acetate	Acetone	Water	Wt. % NaCl in product	Product rate (gal/day/ft^2)	Thermal treatment
23.1	68.0	8.9	1.6	1.9	None
9.8	82.1	8.1	3.5	10.0	None
8.0	72.0	20.0	2.7	5.0	None
8.0	72.0	20.0	3.5	7.0	Shrunk at 100° for 10 min

Feed concentration: 3.5 wt. % NaCl
Operating pressure: 1500 psig

Following the work of Dobry (1936), casting solutions containing 4, 8, and 10 per cent cellulose acetate in saturated magnesium perchlorate were prepared. Films made from such casting solutions, even when preshrunk, did not give satisfactory solute separation and product rate with aqueous sodium chloride feed solutions. The apparent failure of the above films was attributed largely to the fact that the ratio of magnesium perchlorate solution to cellulose acetate was too high, i.e. the concentration of cellulose acetate was too low; but the concentration of cellulose acetate could not be increased above 10 per cent since the solution became too viscous to handle. This problem was solved by adding acetone to the mix as already indicated by the work of Biget. It was found that with an acetone concentration of about 70 per cent, any ratio of magnesium perchlorate solution to cellulose acetate could be incorporated in the casting solution. Hence in the subsequent work, films were made from casting solutions containing cellulose acetate, acetone, and an aqueous solution of magnesium perchlorate.

In a set of preliminary experiments with films cast from the above casting solutions, the effects of saturated magnesium perchlorate/cellulose acetate ratio, casting temperature, time between casting and immersion in cold water, temperature of water used for gelation, film-shrinkage temperature, operating pressure, and feed solution concentration, on the performance of the membranes in reverse osmosis experiments were investigated. Some of the results obtained are given in Figures 2-1, 2-2, and 2-3*. The results revealed the general success of the film making technique, and indicated that every step in the film making process affected the porous structure of the membrane. In

* The cellulose acetate material used in these studies was the type C-215 supplied by Fisher Scientific Co.

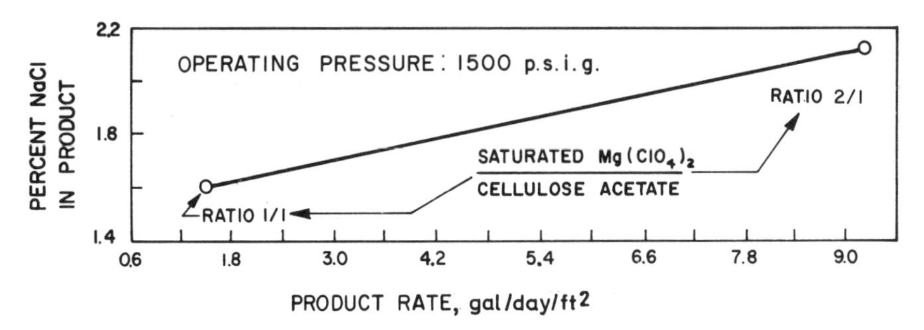

Figure 2-1. Effect of saturated magnesium perchlorate/cellulose acetate ratio on the performance of laboratory-made porous cellulose acetate membranes. Data of Loeb and Sourirajan (1960b).

Casting temperature, 72°F; time between casting and immersion, ½ hour; Acetone/cellulose acetate = 9/1; Immersion water temperature, 65°F; thermal treatment 80°C, one hour; feed solution 3.5% NaCl.

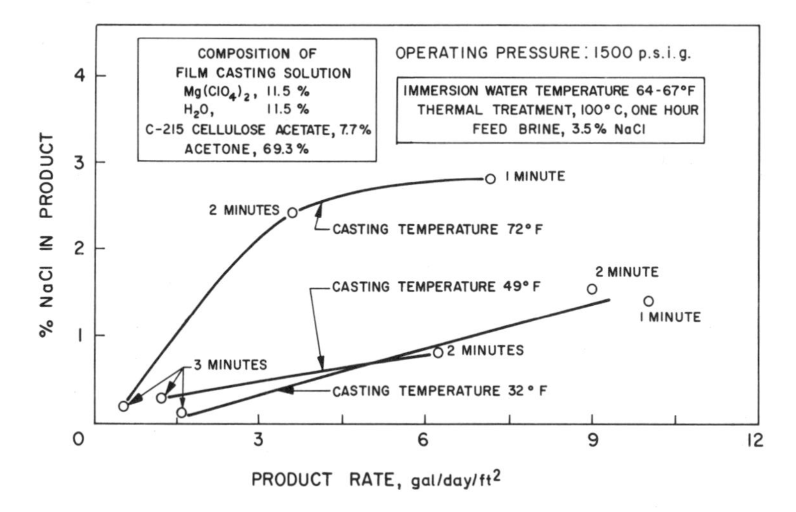

Figure 2-2. Effect of film casting conditions on the performance of laboratory-made porous cellulose acetate membranes. Data of Loeb and Sourirajan (1960b).

particular it was apparent that lower temperatures for film casting and gelation gave better and more reproducible results. The performance data given in Figure 2-3 are particularly interesting. Using a 3.5 wt.% NaCl feed solution, the particular membrane used gave a product containing less than 1000 ppm of salt at a rate of over 3 gal/day/ft^2 of membrane area at an operating pressure of 2000 psig. These results (obtained in December 1959) were better than the best ever obtained before, and thus they represented a significant advance in the direction of producing useful reverse osmosis porous cellulose acetate membranes for saline water conversion and related applications. The above results were followed by a period of tremendous activity in the Loeb-Sourirajan reverse osmosis programme at UCLA.

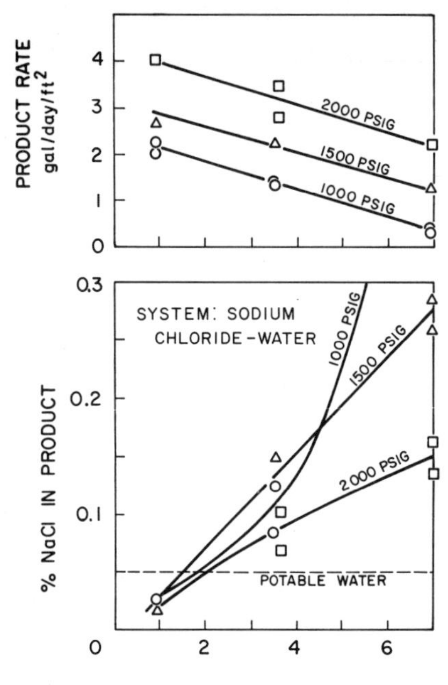

Figure 2-3. Effect of operating pressure and feed concentration on the performance of a laboratory-made porous cellulose acetate membrane (composition of film casting solution same as in Figure 2-2). Data of Loeb and Sourirajan (1960b).

2. The First Successful UCLA Film for Saline Water Conversion

As a result of extensive experimental investigations, Loeb and Sourirajan produced the first successful UCLA film for saline water conversion. The details of their film making technique are summarised below.

The composition (weight per cent) of the film casting solution was as follows (Figure 2-4):

Cellulose acetate *(acetyl content 39.8%):	22.2
Acetone	66.7
Water	10.0
Magnesium perchlorate	1.1

The film was cast in a cold box at 0° to —10° on a glass plate with 0.010 inch side runners to give this thickness to the as-cast film. Uniformity of film thickness was obtained by passing an inclined knife (or thick glass rod) across the top of the plate, the knife (or the glass rod) resting on the side runners only.

The acetone was then allowed to evaporate at the cold box temperature from the surface of the film on the glass plate for a period of 3 to 4 minutes (Figure 2-5). This step in the film casting technique is generally referred as the 'evaporation' step. During this step, the film also partially hardened to a semisolid mass.

---

* The above cellulose acetate material was supplied by Eastman Chemical Co.

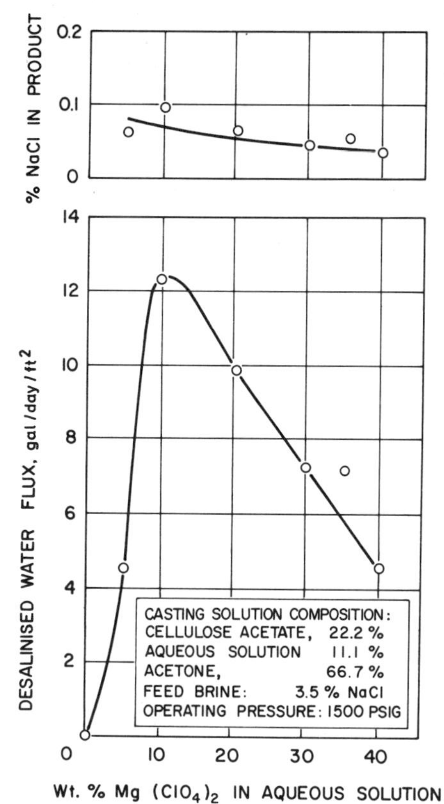

Figure 2-4. Effect of variation of magnesium perchlorate concentration in the casting solution on the performance of laboratory-made porous cellulose acetate membranes. Data of Loeb and Sourirajan (1961).

The glass plate containing the partially hardened film was then carefully immersed into ice-cold water in a tray where it was left at that temperature for at least an hour. This step in the film casting technique is generally referred as the 'gelation' step. During this step, the film finally set to a hard porous gel from which acetone and magnesium perchlorate were leached out practically completely. On standing in ice-cold water for about an hour, the film peeled itself off, or could be removed by hand easily, from the surface of the glass plate.

The above film, in the as-cast condition, was too porous to give any significant solute separation in reverse osmosis involving aqueous sodium chloride solutions. Hence it was subjected to an initial temperature and pressure treatment as described in Chapter 1 for the Schleicher and Schuell membranes. In the initial stages of the investigations with the above film at UCLA, the temperature treatment consisted in shrinking the membrane held freely floating between glass plates immersed in water which was heated gradually from the laboratory temperature to the required temperature (75 to 82°) and held at the latter temperature for about an hour, after which the water was cooled rapidly. The pressure treatment consisted in increasing the pressure on the film in stages at the start of the reverse osmosis operation; for example, if the preset operating pressure was 1500 psig, a pressure of 1000 psig was applied on the film for about ½ hour, after which the pressure was raised to 1500 psig.

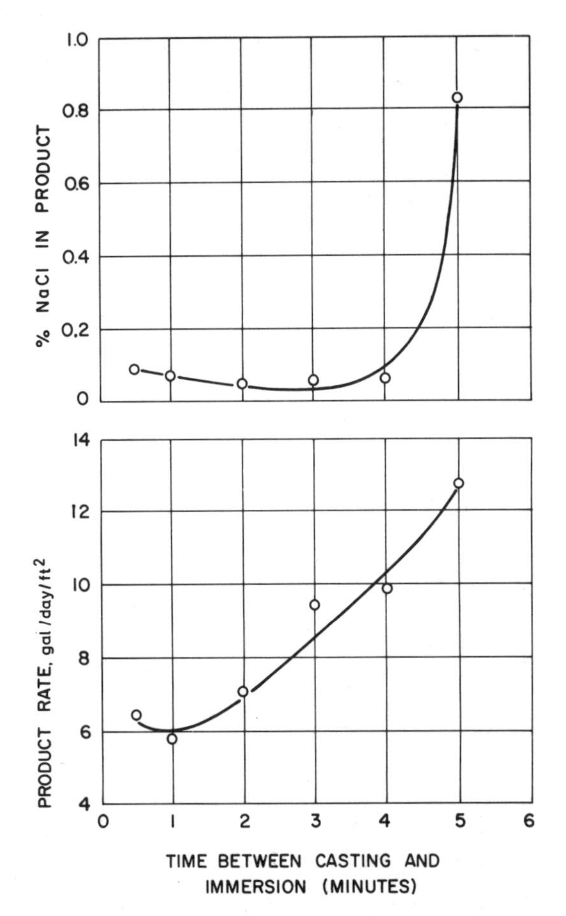

Figure 2-5. Effect of variation of evaporation period on the performance of laboratory-made (Batch 25) porous cellulose acetate membranes. Data of Loeb and Sourirajan (1961).

The films made by the above technique had similar general characteristics as those of the Schleicher and Schuell cellulose acetate membranes, and, in addition, they exhibited superior performance, in terms of solute separation and product rate, for saline water conversion.

The membranes, in the as-cast condition, were too porous; but they could be shrunk to different extents by simply immersing them in hot water at different temperatures. For any particular film, the higher was the temperature of shrinkage, the higher was solute separation and, of course, the lower was product rate.

These membranes also had to be kept at all times under water or dilute alcohol solution or in very humid atmosphere to preserve their porous structure; when allowed to dry in air, they shrivelled, and often their porous structure changed irreversibly.

The two sides of each membrane exhibited different performance characteristics in reverse osmosis as in the case of Schleicher and Schuell membranes. The surface of the film which was away from that of the glass plate, and exposed to the air during casting, appeared to be the one which had the surface layer containing the appropriate size pores for reverse osmosis desalination.

This side was always held facing the feed solution during the reverse osmosis experiments; when the other side was so held in desalination experiments operating at 1500 psig, the product rate increased several fold, and the solute separation decreased practically to zero.

It was found that product rate was not affected significantly by changing the overall film thickness at least within the range of film thickness tested, namely 0.005 to 0.020 inch, before use in the experiment.

The above observations showed that the porous cellulose acetate membrane made by the Loeb-Sourirajan technique had an asymmetric structure consisting of a very thin microporous surface layer embedded on to a relatively thick macroporous spongy mass underneath. The microporous surface layer contained the desired pore structure for reverse osmosis desalination, and it offered most of the resistance to fluid flow. The macroporous spongy mass underneath the surface layer, while serving as a built-in base for the thin surface layer, offered comparatively little resistance to fluid flow (Loeb and Sourirajan, 1961).

The following are some of the typical performance data obtained with the above porous cellulose acetate membranes. These data were obtained in experiments carried out under flow or semi-flow conditions involving continuous or intermitant purging of the concentrated brine on the high pressure side of the membrane. Using 3.5 per cent NaCl feed solution, product waters containing less than 500 ppm of NaCl were obtained in one step with product rates of 8 to 10 gal/day/ft^2 of film area at an operating pressure of 1500 psig. The data obtained with sea water feed solutions, shown in Figures 2-6 and 2-7, were par-

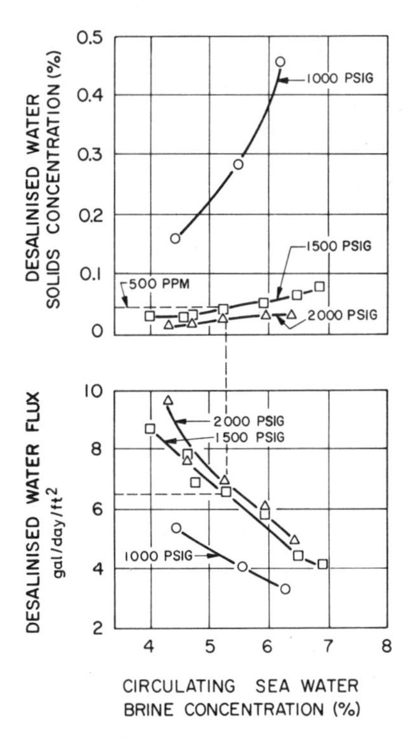

Figure 2-6. Performance of a laboratory-made (Batch 25) porous cellulose acetate membrane for sea water conversion. Data of Loeb and Sourirajan (1961).

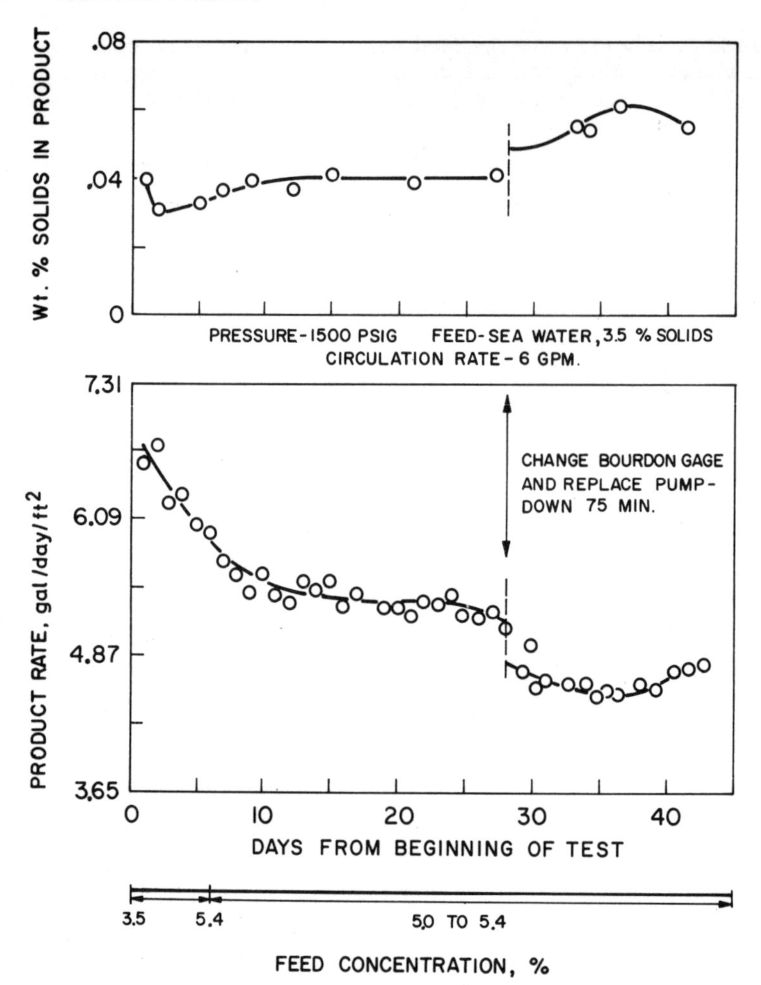

Figure 2-7. Performance of a laboratory-made (Batch 25) porous cellulose acetate membrane for sea water conversion in a 45-day continuous test run. Data of Loeb and Sourirajan (1961).

ticularly significant. Using sea water as the feed solution, product waters containing less than 500 ppm total solids were obtained in one step with product rates of 5 to 11 gal/day/ft^2 of membrane area at operating pressures of 1500 to 2000 psig (Figure 2-6). Further, using sea water as the feed solution, at an operating pressure of 1500 psig, product waters containing 400 to 600 ppm total solids with product rates of 4 to 5 gallons per day per sq. ft. were obtained in continuous reverse osmosis experiments extending to a period of 45 days (Figure 2-7). In the above experiments, the purge rate was controlled such that the concentration of the brine on the high pressure side of the membrane remained at about 5$\frac{1}{4}$ per cent solids.

The performance data given above were far better than any obtained at any time before in such reverse osmosis experiments. They were considered significant enough for engineering evaluation of the reverse osmosis process for the economic conversion of saline waters. In the author's view, these results not only represented the remarkable practical success of the basic approach which governed the reverse osmosis research programme at UCLA during 1956-60, they also represented a good example of applied research. This research programme, in its entirety, not only opened a new water desalting process of tremendous engineering, economic, and social significance, it also opened a new field of science and technology whose potential practical applications are just unlimited.

The Office of Public Information, University of California, Los Angeles, called attention to the above new development at UCLA through a press release dated 23 August 1960* (Appendix I) which, in turn, generated world-wide interest and subsequent activity in the field of reverse osmosis.

Film Nomenclature. Following the Loeb-Sourirajan technique and their general approach discussed in the preceding section, an infinite number of types of porous cellulose acetate and other membranes can be developed for a wide variety of reverse osmosis and other applications because of the many variables involved in the film casting conditions. The composition of the film casting solution, the film casting conditions used in the original development, and the resulting films have since been referred to as 'standard' solution, 'standard' technique and 'standard' films respectively by Loeb and co-workers in their subsequent work on the subject; the same films are referred to in the author's laboratory, and elsewhere in this book, as 'CA-NRC-25' or simply 'Batch 25' type films. Such nomenclature, it must be understood, simply identifies the type of membranes, and does not constitute a specification for any particular membrane.

3. Further UCLA Studies on Membrane Development

Following the success of the Loeb-Sourirajan technique for making porous cellulose acetate membranes as described above, extensive studies were continued at UCLA by Loeb and co-workers on the effects of the different variables involved in the film casting technique. In this programme, a very large number of films were made, and their performance determined for the desalination of aqueous sodium chloride solutions, and of natural sea and brackish waters (Loeb, 1961, 1962, 1963; Loeb and Manijikian, 1963, 1965; Loeb and McCutchan, 1965; Loeb and Milstein, 1962; Loeb and Nagaraj, 1965). A summary of their results is given below. Unless otherwise stated, the 'standard' film casting conditions were employed using different compositions for the film casting solution, the operating pressure was 1500 psig, and the feed solution contained $5\frac{1}{4}$ per cent NaCl.

Effect of degree of acetylation in cellulose acetate material. Membranes were cast from solutions containing the following cellulose acetate materials:

Material	percentage acetylation
XGL-70 (Celanese)	73
E-398-3 (Eastman) (Used in standard solution)	81
A-432-200 (Eastman)	94

* On 31 August 1960, the author's appointment in the Reverse Osmosis project at UCLA terminated.

With the lower acetylated material (XGL-70), the flux corresponding to 0.08 per cent NaCl in product was only 1/3 of that obtained with the standard film. The higher acetylated material, A-432-200 (cellulose triacetate), did not dissolve in acetone; other solvents were employed, but the resulting membranes gave desalination characteristics inferior to those obtained with the standard films.

Other Cellulose Derivatives as Membrane Materials. Ethyl cellulose, cellulose propionate, and cellulose acetate butyrate materials were tested as possible substitutes for cellulose acetate. Films made from ethyl cellulose or cellulose propionate by the standard technique gave poor solute separation and product rates. A representative set of data obtained on the performance of cellulose propionate films is given in Table 2-2.

Table 2-2. Performance of cellulose propionate films

Data of Loeb (1962)

Evaporation time (min)	Product rate* (gal/day/ft^2.)	NaCl* in product (wt. %)
10	1.18	0.41
10	0.90	0.99
5	0.96	0.65
3	1.70	0.54
2	2.06	1.16
1	12.88	3.05

* Feed concentration: 5¼%NaCl
Operating pressure: 1500 psig

Composition of film casting solution (wt. %):

Cellulose propionate (a):	13.95
Acetone :	79.07
Magnesium Perchlorate:	3.49
Water :	3.49

(a) PLFS-70, supplied by Celanese Corp.

Tests were made with membranes made from cellulose acetate butyrate material containing 17.0 per cent combined butyryl (EAB-171-2 supplied by Eastman). These membranes, without any prior thermal treatment, gave fluxes of the order of 7 gal/day/ft^2 with 0.2 per cent NaCl in product water. The membranes lacked physical strength and were quite brittle. In order to increase their physical strength, the gelation step was carried out at —5° in 15 per cent NaCl solution. The resulting membranes gave better desalination (salt content in product was only 0.01 per cent), but the flux was only about 2 gal/day/ft^2.

Tests were then made with a membrane made from cellulose acetate butyrate material containing 26 per cent combined butyryl (EAB-272-3 supplied by Eastman). This membrane was made using 1:1 ratio of 50 per cent aqueous magnesium perchlorate solution: cellulose acetate butyrate, evaporation period

½ minute, and gelation in 15 per cent NaCl-water at —5° (Loeb, 1961). It gave a flux of 4 gal/day/ft² with 0.01 per cent NaCl in the product water.

Solvents. The use of solvents other than acetone for the film casting solution was investigated. They included methyl acetate, methyl ethyl ketone, methanol, ethanol, acetic acid and formic acid. They were used in the casting solution either alone or combined in part with acetone. Some results are illustrated in Table 2-3. In all cases tested, the performance of the membrane was inferior to that of the standard membrane.

Table 2-3. Performance of porous cellulose acetate membranes from casting solutions using different solvents

Data of Loeb (1963)

Casting solution composition (wt. %)	Film shrinkage temp. (°)	Evaporation time (min)	Product rate (gal/day/ft²)	NaCl in product (wt. %)
Cellulose acetate 16.5 Acetic acid 75.3 $Mg(ClO_4)_2$ 1.6 Water 6.6	82.0	3	12.6	0.14
	83.9	3	13.5	0.10
	84.0	3	14.3	0.13
	84.0	10	10.4	0.09
Cellulose acetate 19.9 Acetic acid 70.2 $Mg(ClO_4)_2$ 2.0 Water 7.9	85.0	3	10.4	0.15
	76.0	6	16.3	0.40
	85.5	10	5.6	0.065
Cellulose acetate 16.57 Acetic acid 75.00 $Mg(ClO_4)_2$ 1.64 Water 6.57 HCl 0.22	85	6	14.8	0.175
Cellulose acetate 18.0 Acetic acid 82.0	Unshrunk	3	7.3	0.24
Cellulose acetate 19.8 Acetic acid 70.3 Water 9.9	Unshrunk	3	11.8	0.75
Cellulose acetate 22.2 Formic acid 66.6 $Mg(ClO_4)_2$ 2.3 Water 8.9	83.8	1.5	9.8	0.24

Feed concentration: 5.25 wt. % NaCl
Operating pressure: 1500 psig
Films cast at room temperature

More Experiments with 'Standard' Membranes. The presence of aluminium ions in the feed brine, in concentrations of only a few parts per million, had a marked effect on the separation characteristics of the standard membrane. By the addition of 6 ppm of aluminium sulphate to a 5¼ per cent NaCl-distilled water feed solution, the salt content in the product water decreased from 0.30 per cent to 0.18 per cent; and, on increasing the aluminium sulphate in the feed brine to 12 ppm, the salt content in the product water further dropped to 0.11 per cent (Loeb and Milstein, 1962; Loeb and Nagaraj, 1965).

A 'life' test was carried out on a standard membrane for a 3-month period using a feed brine of concentrated sea water maintained at $5^1/_4$ per cent solids. The membrane gave good desalination characteristics throughout the test period (Figure 2-8) although there was some evidence of derioration of per-

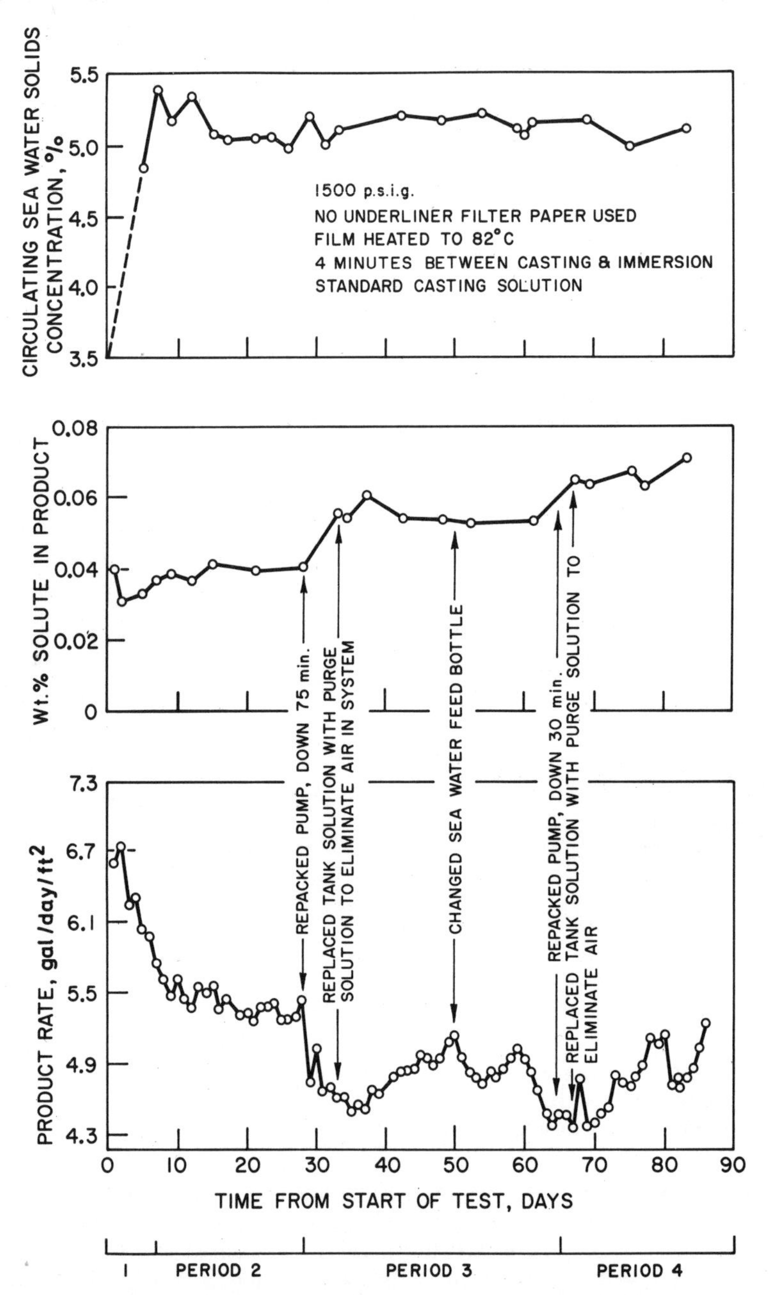

Figure 2-8. Performance of a 'Standard' cellulose acetate membrane in a long-time test run. Data of Loeb and Milstein (1962).

formance which was primarily associated with the shut down periods for pump maintenance.

Studies on Additives

Magnesium perchlorate concentration. A ratio of saturated magnesium perchlorate/cellulose acetate $\approx 1/2$ in the casting solution gave films of best performance (Figure 2-9). Highest water fluxes were obtained when aqueous solutions containing about 10 per cent magnesium perchlorate were used in the casting solution. A membrane was made from a casting solution containing cellulose acetate, acetone and magnesium perchlorate only (i.e. with no water); it gave poor desalination and very low product rate.

Magnesium perchlorate content in the standard film. Analyses of the standard films were made at various stages after casting, and the following results were obtained (Loeb and Milstein, 1962):

	Treatment of Film	Percentage of Magnesium Perchlorate in Dry Film
(a)	Air dried only	7.86
(b)	Dried for 3 minutes in cold box, immersed in ice water for 15 minutes then air dried	0.44
(c)	As in (b), but immersed in ice water for 60 minutes	0.31
(d)	Dried for 3 minutes in cold box, immersed in ice water for 60 minutes then heated in water at 82° for 16 hours, finally air dried	0.26

The above results showed that most of the magnesium perchlorate was leached out during the gelation step. A small amount of the magnesium perchlorate probably remained trapped in the blind pores of the film.

Effect of other perchlorate ions in the casting solution. The magnesium perchlorate additive was replaced by perchlorates of other cations in the 'standard' casting solution. The performance of the resulting films was not too different from each other as illustrated in Table 2-4.

Effect of other additives. Casting solutions were prepared, in which the aqueous magnesium perchlorate solution was replaced by aqueous solutions of Na_2SO_4, Na_2HPO_4, NaH_2PO_4, Na_3PO_4, borax, Na_2CO_3, and K_2CrO_4. The solutions obtained were turbid and inhomogenous, containing solid particles or shreds. They were not found suitable for film casting. Membranes were made in a few cases and they gave very low fluxes.

Since successful results were obtained with perchlorates of all cations tested, and salts with anions other than perchlorate gave poor results, Loeb and Milstein (1962) concluded that perchlorate was the significant ion in the additive. Considering the perchlorate anion in general terms as a halogen atom surrounded by four oxygen atoms, it was postulated that any perhalogenate anion would function as the additive as well as the perchlorate did. Experimental results using $NaIO_4$ instead of $Mg(ClO_4)_2$ in the casting solution confirmed the above postulate. Further a flux of 10 gal/day/ft^2 with 0.09 per cent

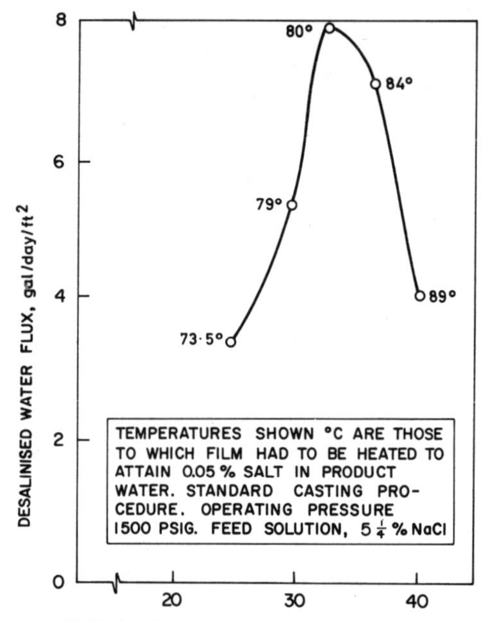

Figure 2-9. Effect of magnesium perchlorate solution/cellulose acetate ratio in the casting solution on membrane performance. Data of Loeb and Milstein (1962).

Table 2-4. Performance of films made from casting solutions containing perchlorate anion and different cations*

Data of Loeb (1962)

Cation	Product rate (gal/day/ft^2)	NaCl in product (wt. %)
Magnesium	8.22	0.05
Sodium	8.49	0.07
Lithium	9.86	0.11
Aluminum	7.59	0.058
Perchloric acid	6.85	0.065
Perchloric acid	4.33	0.038
Ammonium	12.60	0.128

* 'Standard' casting solutions and film casting conditions were used except for the nature of the additive material.

Feed concentration: 5¼% NaCl
Film shrinkage temperature: 77° to 82°
Operating pressure: 1500 psig

salt in the product solution was obtained using membranes cast from standard solutions containing $KMnO_4$ instead of $Mg(ClO_4)_2$.

A casting solution was made in which sodium chlorate was substituted for magnesium perchlorate. Although the resulting casting solution was quite clear, the membranes produced from it, even without any thermal treatment, gave very low fluxes.

Polyethylene glycol, either alone or in admixture with magnesium perchlorate, was also tested as the additive material without much success (Tables 2-5 and 2-6). The other less successful additives tested included K_2CrO_4, NH_4NO_3, $Ca(NO_3)_2$, $Al(NO_3)_3$, H_2SO_4, H_2CrO_7, H_3PO_4, H_2SiF_6, sodium phenolate, sodium benzoate, and zinc nitrate (Table 2-7).

Table 2-5. Performance of films made from casting solutions containing cellulose acetate, acetone, and polyethylene glycol

Data of Loeb (1962)

PEG*/ cellulose acetate	Evaporation time (min)	Film shrinkage Temp(°)	Product rate (gal/day/ft²)	NaCl in product (wt. %)
1/1	3	82	high	5.25
	6	82	high	5.25
	15	Unshrunk	30.14	5.25
	15	82	1.37	0.30
1/2	1	Unshrunk	15.89	4.36
	1	82	22.47	3.0
	3	Unshrunk	7.67	5.0
	3	82	5.75	1.93
	6	Unshrunk	4.38	4.16
	6	82	4.38	1.15
	6	87	2.47	2.0
	15	Unshrunk	2.74	3.13
	15	71	0.16	1.5
	25	Unshrunk	2.47	2.88
	25	82	2.47	1.5
1/4	3	Unshrunk	5.21	3.96
	6	Unshrunk	1.10	3.75
	15	Unshrunk	3.01	3.87

* Polyethylene glycol (E-400) supplied by Dow Chemical Co.

In the above films polyethylene glycol replaced both magnesium perchlorate and water in the 'standard' casting solution.

Feed concentration: 5¼% NaCl
Operating pressure: 1500 psig

Some successful additive materials. It was found that the addition of small amounts of sodium chloride to the casting solution effected an increase of product water flux for the resulting membrane for a given level of solute separation as illustrated in Figure 2-10 (Loeb and Milstein, 1962). The casting solution employed was a mixture of cellulose acetate, acetone, water, magnesium

Table 2-6. Performance of films made from casting solutions containing cellulose acetate, acetone, magnesium perchlorate and polyethylene glycol

Data of Loeb (1962)

Evaporation time (min)	Film shrinkage temp. (°)	Product rate (gal/day/ft²)	NaCl in product (wt. %)
3	Unshrunk	5.75	4.35
3	82	8.49	2.54
6	Unshrunk	0.55	3.25
6	82	0.55	3.66
15	Unshrunk	11.51	4.76
15	82	0.55	0.84

In the above films, polyethylene glycol (E-400 supplied by Dow Chemical Co.) replaced water only in the 'standard' casting solution.

Feed concentration: 5¼% NaCl
Operating pressure: 1500 psig

Table 2-7. Less successful membranes

Data of Loeb and McCutchan (1965)

Casting solution composition (wt. %)						
Cellulose acetate	Acetone	Water	Additive	Film shrinkage temp. (°)	Product rate (gal/day/ft²)	NaCl in product (ppm)
Feed: 5.25% NaCl in Los Angeles Tap Water. Operating pressure: 1500 psig						
22.2	66.7	9.99	1.11 K_2CrO_4	Unshrunk	0.3	9050
22.2	66.7	8.88	2.22 NH_4NO_3	Unshrunk	Very low	—
22.2	66.7	8.88	2.22 $Ca(NO_3)_2$	75	0.8	944
22.2	66.7	7.21	3.89 $Al(NO_3)_3$	Unshrunk	6.0	15,000
22.2	66.7	9.99	1.11 H_2SO_4	Unshrunk	1.6	2940
22.2	66.7	9.99	1.11 $H_2Cr_2O_7$	Unshrunk	Nil	—
22.2	66.7	9.99	1.11 H_3PO_4	Unshrunk	Nil	—
22.2	66.7	7.77	3.33 H_2SiF_6	Unshrunk	1.6	2000
Feed: 0.5% NaCl in Los Angeles Tap Water. Operating pressure: 600 psig						
22.2	66.7	8.32	2.78 Sodium Phenolate	Unshrunk	3.5	1200
22.2	66.7	8.32	2.78 Sodium Benzoate	Unshrunk	Very low	—
22.2	66.7	8.32	2.78 Zinc Nitrate	Unshrunk	3.2	2940

Feed rate: 300 cm³/min

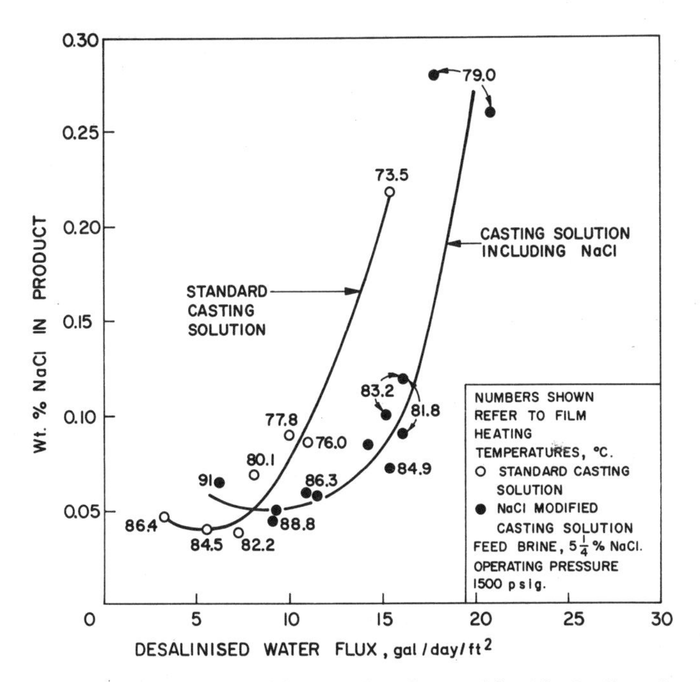

Figure 2-10. Effect of addition of sodium chloride in the standard casting solution on membrane performance. Data of Loeb and Milstein (1962).

perchlorate, and sodium chloride in the weight percentages of 22. 3, 66. 7, 9. 7, 1. 08, and 0. 32 respectively. Membranes obtained from the above casting solution had to be shrunk at a higher temperature than that needed for the standard membranes to obtain a given level of solute separation.

Loeb and McCutchan (1965) reported a large number of successful additive materials. These included $HClO_4$, $LiClO_4$, $NaClO_4$, NH_4ClO_4, NH_4PF_6, $Mg(ClO_4)_2$, $Ca(ClO_4)_2$, $Zn(ClO_4)_2$, $Ba(ClO_4)_2$, H_2PtCl_6, $Al(ClO_4)_3$, $Fe(ClO_4)_3$, $NaIO_4$, $NaReO_4$, $NaBF_4$, NaCNS, KNCS, sodium tetraphenyl boron, sodium salicylate, HF, HCl, HBr, HI, HNO_3, NaCl, NaBr, NaI, KI, CsI, MgI_2, $CdBr_2$, CdI_2, $ZnCl_2$, $ZnBr_2$, ZnI_2, $CuCl_2$, $AlBr_3$ and AlI_3 (Tables 2-8 and 2-9). Loeb and Manjikian found that hydrochloric acid, along with aqueous magnesium perchlorate, were particularly useful as additive materials, especially for membranes meant for brackish water conversion as illustrated by their data given in Table 2-10.

Six-Month Field Test of a Laboratory-Made Porous Cellulose Acetate Membrane for Brackish Water Conversion. Loeb and Manjikian (1965) conducted a 6-month field test of a laboratory-made membrane for the conversion of brackish waters at Coalinga, California, in 1963. The feed water contained about 2500 ppm of dissolved salts, more than half of which consisted of divalent ions. The film was made using the standard film casting conditions with the following composition (weight per cent) for the film casting solution:

Cellulose acetate (Eastman 398-3)	23.2 wt.%
Acetone	69.4 wt.%
Water	5.43 wt.%
Magnesium perchlorate	1.64 wt.%
Hydrochloric acid	0.33 wt.%

Table 2-8. Some successful additive materials

Data of Loeb and McCutchan (1965)

Casting solution composition (wt. %)				Film shrinkage temp. (°)	Product rate (gal/day/ft^2)	NaCl in product (ppm)
Cellulose acetate	Acetone	Water	Additive			
Feed: 5.25% NaCl in Los Angeles Tap Water. Operating pressure: 1500 psig						
22.2	66.7	9.99	1.11 $HClO_4$	80	7.3	700
22.2	66.7	9.99	1.11 NH_4ClO_4	80	13.0	1310
22.2	66.7	9.99	1.11 $LiClO_4$	80	10.0	1090
22.2	66.7	9.99	1.11 $NaClO_4$	80	8.9	700
22.2	66.7	9.99	1.11 $Mg(ClO_4)_2$	82	8.7	525
22.2	66.7	9.99	1.11 $Al(ClO_4)_3$	80	8.0	595
22.2	66.7	9.99	1.11 $NaIO_4$	80	12.0	1500
22.2	66.7	9.99	1.11 $KMnO_4$	81.4	11.0	905
22.2	66.7	8.88	2.22 $NaReO_4$	80	7.0	990
22.2	66.7	8.88	2.22 $NaBF_4$	79.3	15.0	1310
22.2	66.7	8.88	2.22 Sodium tetraphenyl boron	82.5	5.6	1590
16.7	50.0	15.8	17.5 K_2HgI_4	72	6.1	1380
22.2	66.7	8.88	2.22 NaNCS	84.3	11.0	1380
22.2	66.7	8.88	2.22 KNCS	83	8.0	700
22.2	66.7	7.77	3.33 Sodium salicylate	83.3	13.0	1380
Feed: 0.5% NaCl in Los Angeles Tap Water. Operating pressure: 600 psig						
23.2	69.4	5.43	1.64 $Mg(ClO_4)_2$	71.3	27	294
22.2	66.7	9.99	1.11 $Ca(ClO_4)_2$	73	15	538
21.7	65.0	9.87	3.10 $Zn(ClO_4)_2$	76	20	294
22.2	66.7	8.88	2.22 $Ba(ClO_4)_2$	73	23	500
22.2	66.7	8.32	2.78 $Fe(ClO_4)_3$	71.5	12	294
22.2	66.7	8.32	2.78 NH_4PF_6	83	12	200
22.2	66.7	7.21	3.89 H_2PtCl_6	85	14	200
22.2	66.7	8.32	2.78 HNO_3	71	28	333

Feed flow rate: 300 cm^3/min

The membrane was shrunk at 65°. A laboratory test cell was used. The feed brine flowed through a 5-micron Cuno filter prior to entering the cell. An air-operated Sprague pump (simplex, single acting type) was used. To minimise pressure fluctuations caused by this pump, a 6-inch branch line of ½-inch tubing near the cell discharge was used as a pneumatic cushion by filling it with air at operating pressure. The feed water was passed over the membrane surface on a once-through basis at an operating pressure of 600 psig. The feed rate was about 350 cm^3 per minute. Operation was on a 24-hour basis, with effluent and product samples taken once daily. The feed brine filter was changed about every 3 weeks, even though it did not usually appear dirty. The temperature range of the feed brine over the 6-month period was 80° to 90°F. Potable water was produced during the entire period during which the salt content of the desalinised water increased from about 150 to 300 ppm, and the product rate decreased from 48 to 17 gal/day/ft^2. The analyses of the Coalinga feed waters, and those of the corresponding desalinised waters obtained, are

Table 2-9. Performance of membranes made from casting solutions containing halide additives

Data of Loeb and McCutchan (1965)

Casting solution composition (wt. %)				Film shrinkage temp. (°)	Product rate (gal/day/ft^2)	NaCl in product (ppm)
Cellulose acetate	Acetone	Water	Additive			
22.2	66.7	8.88	2.22 HF	Unshrunk	0.28	4160
22.2	66.7	8.32	2.78 HCl	Unshrunk	0.17	2180
22.2	66.7	8.32	2.78 $N(CH_3)_4Cl$	Unshrunk	0.29	3330
22.2	66.7	8.32	2.78 $ZnCl_2$	70.4	14	238
23.3	70.0	5.36	1.34 $CuCl_2$	Unshrunk	0.42	1090
22.2	66.7	5.80	5.30 HBr	73	13	1110
22.2	66.7	8.32	2.78 NaBr	Unshrunk	5.2	3120
22.2	66.7	5.55	5.55 $CdBr_2$	55	6.5	645
22.2	66.7	8.32	2.78 $ZnBr_2$	72	17	238
22.2	66.7	7.77	3.33 $AlBr_3$	Unshrunk	Low	4160
22.2	66.7	5.90	5.20 HI	70	10	385
22.2	66.7	5.55	5.55 NaI	71.5	14	800
22.2	66.7	5.55	5.55 KI	77.5	8.7	510
22.2	66.7	7.21	3.89 CsI	Unshrunk	0.58	4160
22.2	66.7	8.32	2.78 CdI_2	Unshrunk	3.9	590
22.2	66.7	5.55	5.55 MgI_2	83	11	200
22.2	66.7	5.55	5.55 ZnI_2	72.5	28	294
22.2	66.7	5.55	5.55 AlI_3	83	25	600

Feed: 0.5% NaCl in Los Angeles tap water
Operating pressure: 600 psig
Feed rate: 300 cm^3/min

given in Tables 2-11 and 2-12 respectively. One of the main problems encountered was the accumulation of ferric oxide deposits on the membrane surface, which accounted for at least part of the decrease in product rate.

In Table 2-13, the performances obtained from a one-day old membrane, and from a membrane stored in Los Angeles tap water at 70° to 75° F for about 4½ months, are given along with the performance of the membrane in continuous service in the above test run for about 4 months. All the three membrane disks were cut from the same membrane sheet. The results showed that the solute separations obtained with all the three films were comparable, that the membranes could be stored in water without any deterioration for extended periods of time, and that the salts present in the Coalinga feed water did not accelerate deterioration of the desalinising capacity of the membrane.

The above test results are of historic importance. They provided for the first time an experimental basis for the estimated membrane life of six months, which figure is still frequently used in the literature.

4. Work of Banks and Sharples

Following the work of Loeb and Sourirajan (1961), Banks and Sharples (1966a, 1966b, 1966c) made extensive studies of the reverse osmosis process for desalination, and of the relation between the fabrication procedure and performance of the Loeb-Sourirajan type porous cellulose acetate membranes. Their

Table 2-10. Performance of some porous cellulose acetate membranes for brackish water conversion

Data of Loeb and Manjikian (1963)

Casting solution composition (wt. %)					Film shrinkage temp. (°)	Product rate (gal/day/ft^2)	NaCl in product (ppm)
Cellulose acetate	Acetone	$Mg(ClO_4)_2$	Water	HCl			
22.20	66.50	1.67	9.40	0.34	81.8	27	300
22.20	66.50	1.78	8.90	0.53	79.5	26	200
21.80	65.40	1.83	10.60	0.34	81.3	27	270
22.40	67.20	1.53	8.42	0.46	81.8	20	160
21.50	64.40	2.46	10.90	0.81	80.0	22	300
22.50	67.60	1.80	7.75	0.36	75.8	26	220
22.90	68.70	1.53	6.59	0.30	73.8	16	155
23.20	69.70	1.30	5.72	0.26	70.5	19	300
23.15	69.41	1.65	5.46	0.33	71.3	27	230
24.00	72.00	1.78	2.06	0.18	64.2	14	225
23.90	71.80	0.96	3.19	0.19	62.6	16	255
23.30	70.00	1.75	4.67	0.35	70.0	30	340
23.40	70.30	1.76	4.40	0.18	66.5	22	300
24.20	72.40	0.80	2.55	0.08	56.8	10	290
23.10	69.30	1.94	5.27	0.39	71.5	26	270
23.90	71.60	1.77	2.37	0.35	66.0	19	300
21.50	64.60	2.70	10.80	0.36	82.5	21	300
22.30	67.00	2.03	8.34	0.32	78.0	21	300
22.70	68.20	1.75	7.15	0.27	71.7	21	272
22.20	66.70	1.75	8.72	0.74	79.0	26	221
23.10	69.40	1.30	5.90	0.38	70.0	19	300
*21.40	64.50	1.65	10.90	1.43	—	—	—
21.00	63.00	1.75	14.00	0.35	86.5	14	300
20.00	64.00	1.75	18.00	0.35	89.0	11	300

* Poor membrane

Feed concentration: 0.5 wt. % NaCl
Operating pressure: 600 psig

experiments were carried out in a static cell fitted with a magnetically driven stirrer immediately above the membrane surface (Banks and Sharples, 1966a). In their work also, the word 'standard' is used to refer to the film, film casting technique and casting solution composition originally used by Loeb and Sourirajan (1961). Some of their results and conclusions (Banks and Sharples, 1966b) are summarised below.

Effect of methanol in casting solution on membrane performance. Films made from standard casting solutions containing methanol in place of water were tested (the weight of methanol was so chosen to give the same concentration of hydroxyl groups in the standard solution). The properties of one such membrane are given in Table 2-14 along with those of a standard membrane, and one cast from a solution containing no water. The standard membrane and the one cast from the solution containing methanol possessed almost identical properties.

Table 2-11. Coalinga feed water analyses

Data of Loeb and Manjikian (1965)

Description	Dissolved solids (ppm)			
	15 Apr 1963	8 June 1963	30 Aug 1963	5 Oct 1963
Total dissolved solids	2520	2372	2442	2479
Alkalinity (bicarbonate)	—	170.8	170.8	167.8
Chlorides	262.4	297.9	248.2	265.9
Fluorides	—	0.14	—	—
Sulphates	1303.7	1190.0	1253.5	1260.5
Nitrates	5.2	0.7	0.8	4.0
Calcium	116.2	145.3	125.3	118.4
Magnesium	97.2	100.9	88.1	85.6
Sodium	529.0	468.0	524.1	541.0
Potassium	—	15.8	—	—
Iron	—	0.01	—	—
Manganese	—	0.11	—	—
Total hardness ($CaCO_3$)	—	—	659.4	655
pH	7.8	7.3	7.6	7.1

Table 2-12. Coalinga desalinised water analyses at various periods during field test

Data of Loeb and Manjikian (1965)

Description	Dissolved solids (ppm)			
	15 Apr 1963	8 June 1963	30 Aug 1963	5 Oct 1963
Total dissolved solids	180.0	308.0	310.0	302.0
Alkalinity (bicarbonate)	—	36.6	54.9	36.6
Chlorides	56.7	145.4	127.7	131.2
Sulphates	4.1	7.0	12.8	26.3
Nitrates	14.4	4.3	1.1	1.8
Calcium	0.8	4.2	10.2	6.4
Magnesium	0.5	1.5	1.4	3.7
Sodium	52.0	102.5	95.6	98.0
Total hardness ($CaCO_3$)	—	—	31.3	31.0
pH	7.5	7.2	7.8	7.3

Table 2-13. Comparative performance of fresh membrane, stored membrane, and membrane in continuous service at Coalinga

Data of Loeb and Manjikian (1965)

		Desalinised Water	
Test date (1963)	Membrane*	Flux (gal/day/ft^2)	Total dissolved solids (ppm)
Mar. 27	Fresh membrane	40.5	108
Aug. 7	Stored membrane	43.5	274
Aug. 7	Membrane in continuous service since April 9	27.0	234

* Membranes tested with Coalinga water as feed at 600 psig operating pressure.

Table 2-14. Effect of methanol in casting solution on membrane performance

Data of Banks and Sharples (1966b)

		Membrane performance	
Lyophilic solvent	Degree of swelling*	Product rate (cm^3/hr)	Solute separation (%)
None	1.67	35	96
Water	1.95	98	97
Methanol	1.94	102	97

Membrane preparation technique: 'Standard'
Feed concentration: 3.5% NaCl
Operating pressure: 1500 psig
Film area: 80 cm^2

* Ratio of wet weight to dry weight.

Evaporation time and membrane performance. Within the time range investigated (1 to 10 minutes), the properties of the standard membrane were found independent of evaporation time (Table 2-15). The results were explained as follows. As soon as the film was cast, acetone was lost from the surface with the result that a thin 'skin' was formed. Subsequently, acetone from the body of the membrane diffused across this skin and was lost by evaporation. However, this loss was so slight relative to the large amount of acetone present that the main part of the membrane had the same structure, irrespective of whether the evaporation period was 1 or 10 minutes.

Table 2-15. Evaporation time and membrane performance

Data of Banks and Sharples (1966b)

Evaporation time (min)	Membrane performance	
	Product rate (cm^3/hr)	Solute separation (%)
1	105	98
2	100	97
4	105	98
6	98	97
8	95	97
10	100	98

Membrane preparation technique: 'Standard'
Feed concentration: 3.5% NaCl
Operating pressure: 1500 psig
Film area: 80 cm^2

Effect of membrane casting on a filter paper surface. In order to let the water reach both surfaces of the membrane more or less simultaneously during the gelation period, the standard films were cast on a filter paper. It was found (Table 2-16) that such membranes performed just as well as those cast on a glass plate. High desalination was achieved in both cases only when the surface in contact with the air during casting was facing the feed solution in the reverse osmosis experiments. From the above observations it was concluded that the structures of the membranes used were similar, and the thin surface layer was formed during casting and evaporation steps, and not during gelation.

Table 2-16. Effect of membrane casting on a filter paper surface

Data of Banks and Sharples (1966b)

Casting surface	Side of film in contact with feed	Membrane performance	
		Product rate (cm^3/hr)	Solute separation (%)
Glass	Surface layer side	95	98
	Reverse side	680	0
Filter paper	Surface layer side	92	97
	Reverse side	640	0

Membrane preparation technique: 'Standard'
Feed concentration: 3.5% NaCl
Operating pressure: 1500 psig
Film area: 80 cm^2

Effect of different immersion media on membrane performance. The membranes yielded exactly the same results when methanol was used for immersion, as they did with water (Table 2-17). These results again showed that the thin surface layer was not formed during the gelation step.

Table 2-17. Effect of different immersion media on membrane performance

Data of Banks and Sharples (1966b)

		Membrane performance	
Film	Immersion medium	Product rate (cm^3/hr)	Solute separation (%)
a	Water	98	98
a	Methanol	95	98
b	Methanol	100	97

Membrane preparation technique:
(a) 'Standard'
(b) Membrane prepared from casting solution containing methanol in place of water.
Feed concentration: 3.5% NaCl
Operating pressure: 1500 psig
Film area: 80 cm^2

Effect of heat treatment. It was found that the treatment of the cellulose acetate film in hot water led to changes in properties, in addition to solute separation and product rate, as shown in Table 2-18.

The degree of shrinkage progressively increased with increasing temperature of treatment; simultaneously, the product rate decreased (from 700 cm^3/hour originally, to virtually nil after treatment at 100°) and the solute separation also increased (from <10 per cent for the untreated membrane to 95 per cent for one shrunk at 85°). The amount of free water within the membrane was reduced by heat treatment. A reorganisation of the membrane material also occurred as shown by both the change in appearance on drying and on the water uptake after drying. After the membrane had been heated at temperatures in excess of 70°, the surface layer side of the membrane showed decided optical differences from the back; the former presented a glassy, smooth appearance, whereas the latter was matt and rough. The penetration of the as-cast membrane by solutions of Biebrich Scarlet (dye) under pressure was examined. The membrane was found to be heavily stained on its surface layer side. In the heat-treated membrane, the dye was rejected by the surface layer. These observations proved that the membranes were asymmetric prior to heat treatment.

Relation between pressure applied to membrane and its degree of swelling. Table 2-19 gives the degrees of swelling as a function of pressure applied to the as-cast standard membrane. A considerable amount of water was removed irreversibly from the membrane as a result of application of pressure. In nearly all cases, solute separation and product rate of the as-cast membrane

Table 2-18. Effect of heat treatment on the properties of cellulose acetate membranes[a]

Data of Banks and Sharples (1966b)

Temperature and time of heat treatment	Film shrinkage[b] (%)	Degree of swelling (%)	Appearance on drying	Water uptake after drying[c] relative to original water content (%)
No heat treatment	—	2.19	Transparent	12
82° for 15 mins.	7	1.88	Transparent	18
87° for 15 mins.	11	1.91	Slight opacity	30
92° for 15 mins.	15	1.86	More opaque	38
100° for 15 mins.	19	1.72	Opaque	86

(a) 'Standard' technique using Celanese cellulose acetate (acetyl content 41.1%)

(b) Shrinkage in diameter of a 3 cm disc.

(c) Water uptake after drying was determined by immersing the dried material for 24 hours, removing the surface moisture by wiping, and then weighing the sample. The results are quoted as the percentage of the water originally present which has been regained by this treatment.

Table 2-19. Relation between pressure applied to membrane and its degree of swelling[a]

Data of Banks and Sharples (1966b)

Pressure (psig)	Degree of swelling	Pressure (psig)	Degree of swelling
0	2.25	600	1.75
150	2.08	900	1.65
300	1.91	1200	1.60
		1500	1.58

(a) The 'Standard' membrane was brought to the desired pressure in the desalination cell and held at that value for 30 minutes using water as the feed. The pressure was then released, the cell dismantled and the degree of swelling of the membrane measured.

were not affected by the application of pressure; occasionally, however, they were affected as a function of time as illustrated in Table 2-20.

Performance of membranes cast at room temperature. Using the standard film casting composition, membranes were cast at room temperature (~15°), with evaporation period of 1 minute, followed by gelation in ice water and subsequent heat treatment at different temperatures. A set of data on the performance of the above films is given in Table 2-21. The figures for solute separation and product rate for the membrane shrunk at 73° were virtually identical with the corresponding values for a standard membrane.

Table 2-20. Membrane performance during prolonged application of pressure

Data of Banks and Sharples (1966b)

Time (min)	Product rate (cm^3/hr)	Solute separation (%)
15	780	10
30	740	10
60	680	12
120	580	15
180	510	18
240	480	19
300	440	22
420	410	25

Membrane preparation technique: 'Standard'
Feed concentration: 0.1% NaCl
Operating pressure: 1500 psig
Film area: 80 cm^2

Table 2-21. Performance of membrane cast at room temperature

Data of Banks and Sharples (1966b)

Film shrinkage temperature (°)	Product rate (cm^3/hr)	Solute separation (%)
As-cast	780	12
62	180	90
70	120	95.5
73	112	98.5
75	104	98.6
80	64	98.4

Casting solution composition: 'Standard'
Evaporation time: 1 minute
Gelation in ice water
Feed concentration: 3.5% NaCl
Operating pressure: 1500 psig
Film area: 80 cm^2

Using the standard film casting composition, membranes were then cast as above except that both casting and gelation were at room temperature. The performance of such membranes is illustrated by the data given in Table 2-22.

Table 2-22. Performance of membranes cast at room temperature

Data of Banks and Sharples (1966b)

Evaporation time (min)	Film shrinkage temperature	Product rate (cm^3/hr)	Solute separation (%)
1	As-cast	600	34
1	60°	130	94
1	70°	110	95.4
1	80°	56	93
3	As-cast	230	54
3	60°	160	61
3	70°	92	82
3	80°	34	85

Casting solution composition: 'Standard'
Gelation in water at room temperature
Feed concentration: 3.5% NaCl
Operating pressure: 1500 psig
Film area: 80 cm^2

Modifications of the standard casting composition. The casting composition was modified by increasing the amount of magnesium perchlorate relative to the other constituents. The properties of membranes cast at room temperature from the above solutions are illustrated in Table 2-23. The data showed that membranes having solute separations greater than 97 per cent were produced by a variety of procedures.

5. Electron Micrographs of Loeb-Sourirajan Type Membranes

The porous structure of the membranes was examined under the electron microscope by Riley et al. (1964, 1966).

Small cross sections (1cm by 100μ by 20μ) of a membrane were cut with a conventional microtome equipped with a steel knife. The segments were then placed in aqueous osmic acid solution (5 per cent) for 48 hours. After osmic acid fixation, the water within the membrane was replaced with carbon tetrachloride by Soxhlet extraction over a period of several hours. This step did not appear to alter the structure of the membrane significantly. The carbon tetrachloride was then completely replaced by an epoxy resin solution. Finally, the segments in epoxy solution were placed in gelatine capsules and polymerized at 48° for 24 hours. Glass knives were used to section the segments into pieces 100μ by 20μ by 500Å on a Reichert thermal-feed ultramicrotome. A tape-collecting trough was attached to the glass knife and filled with distilled water. The thin sections were collected on the water surface from the cutting edge of the knife. After 30 minutes the sections were removed from the water and placed on standard 200 mesh copper electron microscope grids. The thin

Table 2-23. Effect of variation of magnesium perchlorate in casting solution on the performance of membranes cast at room temperature

Data of Banks and Sharples (1966b)

$Mg(ClO_4)_2$ (wt. %)	Evaporation time (min)	Film shrinkage temp. (°)	Product rate (cm^3/hr)	Solute separation (%)
1.5	1	61	130	92
1.5	1	73	90	97
1.5	1	80	56	95
2.0	1	61	130	91
2.0	1	73	102	97.1
2.0	1	80	80	97.3
2.0	2	61	158	87
2.0	2	73	70	97
2.0	2	80	40	93
3.0	1	61	132	87
3.0	1	73	100	95
3.0	1	80	61	97.7

Casting solution composition (wt. %)
Cellulose acetate: 22.2
Acetone: 66.8
Water
Magnesium perchlorate: (as indicated)
Gelation in water at room temperature
Feed concentration: 3.5% NaCl
Operating pressure: 1500 psig
Film area: 80 cm^2

sections were examined in transmission in a Hitachi HU-11 electron microscope. They confirmed that the membrane consisted of a dense thin surface layer on the side of the film exposed to air during casting, with a highly porous substructure underneath the surface layer. Riley et al. reported that the dense surface layer was devoid of structural characteristics and had a thickness of about 0.25μ compared to the total membrane thickness of 100μ. The remainder of the film appeared to be a spongy mass having a pore size of the order of 0.1μ. The pore structure of the membrane surface, which was in contact with the glass surface during casting, was like that of the interior (Riley et al., 1964).

In a later study using improved techniques, Riley et al. (1966) obtained greater resolutions using a replication method. In this method, the dried membrane was shadowed with palladium metal at an angle of 20° in a vacuum evaporator and a thin (~200Å) carbon film was deposited over the palladium. The membrane was then placed on a 200 mesh copper grid, and subsequently dissolved away with acetone in a reflux column, leaving a palladium-carbon replica. Again a Hitachi HU-11 electron microscope was used to examine the replicas. The results obtained were similar. They reported that the dense surface layer showed no evidence of pores greater than about 100Å, whereas the interior structure of the film was highly porous with pore size of the order of 0.4μ.

6. Hydrolysis Rate of Cellulose Acetate Material Used in Loeb-Sourirajan Type Membranes

As pointed out in Chapter 1, the reverse osmosis process is governed by a surface phenomenon, and hence the chemical nature of the membrane surface relative to that of the solution in contact with it is a controlling parameter in reverse osmosis separation. Since cellulose acetate material is subject to hydrolysis when in contact with aqueous feed solutions of different pH and temperature, it is important to choose practical conditions under which hydrolysis rate is minimum, so that long membrane life and steady performance are obtained. A kinetic study of the hydrolysis of 39.8. wt.% acetyl content cellulose acetate has been made by Vos et al. (1966a) as a function of pH and temperature over the pH range 2.2 to 10, and temperature range 23 to 95°. The hydrolysis reaction was carried out on highly porous membranes under quasi-homogeneous conditions, and the data have been treated as a pseudo-first order reaction in acetyl concentration. The results, given in Figure 2-11, showed that the hydrolysis rate of the cellulose acetate material used was minimum in the

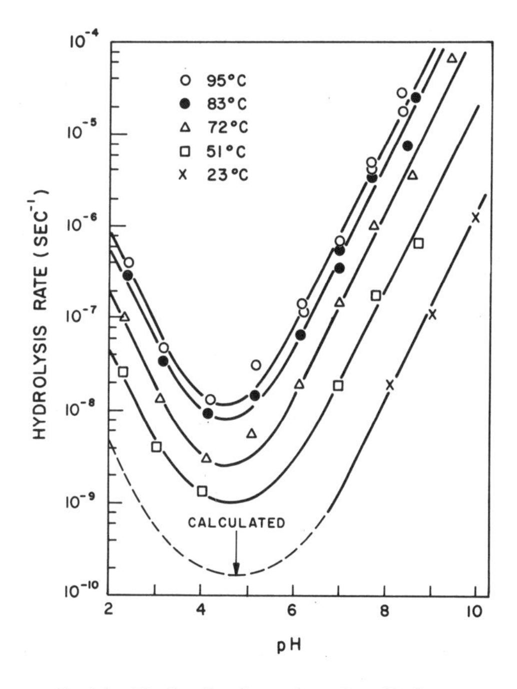

Figure 2-11. Hydrolysis rate of cellulose acetate material (acetyl content 39.8 wt.%) v. pH of feed solution. Data of Vos et al. (1966a).

pH range 4 to 5. Using the hydrolysis rate data given in Figure 2-11, Vos et al. (1966b) calculated that at 23° and a pH of 4.8, the chloride permeation constant would change by a factor of 2 in about 4.3 years because of hydrolysis of the cellulose acetate material when the membrane was used in conjunction with aqueous sodium chloride solutions.

7. Mechanism of Pore Formation in the Process of Casting Loeb-Sourirajan Type Membranes

This subject is one of fundamental importance in the field of porous membrane technology. The subject, however, has not yet progressed far enough to help create better membranes for reverse osmosis applications. The steps involved in the process of casting Loeb-Sourirajan type porous cellulose acetate membranes have been critically studied by Kesting and others (Kesting, 1965; Kesting et al., 1965; Vincent et al., 1965) Their conclusions are briefly as follows.

The Nature of Salt-Polymer Interactions. The role of certain inorganic salts in increasing the flux of water through cellulose acetate membranes is related to the capacity of the component ions to swell the cellulosic substrate. Swelling results in an increase in the total volume of the polymer-liquid system and has as its (sometimes attainable) limit complete solution of the polymer. In order for swelling to occur, the polymer-liquid interaction forces must exceed that of polymer-polymer cohesive energy; in other words, crosslinking sites must be ruptured in the polymer system. Swelling is effected by the formation of metastable complexes involving the highly hydrated cationic fraction of the salt and both the hydroxyl and acetate groups of the cellulose acetate. Since cellulose acetate material swells, and even dissolves, in aqueous solutions of magnesium perchlorate, it is assumed that the hydrated salt ions are capable of association (complex formation) with certain sites on the polymer with the eventual result that polar cross linking sites are ruptured and additional water is incorporated into the gel network. The inclusion of such salts in a casting solution containing polymer and water results in salt-polymer interactions and the formation of a more open network than that which would have occurred in the absence of such salts.

The capacity of a given salt for swelling cellulose acetate has been related to the electrophilicity of the cationic portion of the salt which in turn affects the capacity of this species for both hydration and complex formation with the necleophilic hydroxyl and acetate groups of cellulose acetate. Differences in swelling capacity between a series of salts in which the cation is held constant and the anion varied are attributable to interaction between the two species. Ion-ion interactions might lead to association and result in a decrease of the effective charge density of the cation. Since the dielectric constant of the acetone-water medium is substantially lower than that of pure water, the possibility of ion association is greatly enhanced. For any given cation, the charge density of the anion with which it is associated is the most important factor in determining the extent to which ion association occurs. The relative swelling capacities of a given salt series, in which the cation is held constant and the anion varied, depend upon anionic size (the larger the anion, the more effective the cation). Cation-anion interaction is not limited to ion pairs. The formation of larger ion aggregates decreases cationic hydratability and hence salt swelling capacity (Kesting, 1965).

The Gelation Process. The formation of the Loeb-Sourirajan type porous cellulose acetate membrane is a gelation process involving the coagulation of the polymer solution into a comparatively rigid mass incorporating a large amount of water. Using the concepts of Elford (1940) and Maier and Scheuermann (1960), Kesting et al. (1965) have given the following description of the gelation process.

The size and uniformity of the cells of which a cellulose acetate membrane is composed are functions of the size and uniformity of the pregelation polymer aggregates in the solution and are influenced by such factors as casting solution composition, gelation temperature and uniformity of desolvation. In dilute solutions of secondary cellulose acetate, the polymer molecules are random coils, enclosed in an envelope of solvent which acts as a barrier to contact between macromolecules. As the concentration of polymer is increased, polymer molecules come into contact and aggregation occurs. Coacervation and polymer desolvation (solvent removal) give rise to an open cell pore structure in the resulting membrane.

Coacervation is the coalescence of sol particles whose solvent envelopes diffuse into one another. Coacervating groups of sol particles tend to assume a spherical droplet configuration as the solution becomes more concentrated. Most of the solvent which is associated with the coacervate droplets is merely physically encapsulated in the interior of the spherical polymer aggregates, but probably a small amount is more strongly held to the individual polymer molecules by Van der Waal forces.

Polymer desolvation may occur gradually by solvent evaporation or more suddenly by immersion of the polymer solution into a gelation medium such as water. After desolvation has progressed to a certain degree, coacervate droplets come into contact, deform into polyhedra, and finally bond together through intermingling of polymer molecules from neighbouring polyhedra.

When a film is cast and immediately immersed into a polymer nonsolvent, solvent is rapidly removed from the surface of the droplets (causing contraction of the exterior surfaces) and less rapidly from the interior of the droplets (causing the aggregation of polymer from the droplet interiors at the surfaces). Depletion of the polymer from the interior of the droplet appears to be a more likely reason for the existence of hollow interiors (filled with nonsolvent) than an initial formation of hollow coacervates.

Because of the rapid desolvation which occurs at the droplet surface, stresses are introduced which cause the walls to rupture. The resultant membrane is composed of an open-cell foam or pore system filled with nonsolvent.

Increasing the concentration of water in the casting solution results in corresponding increases in both the thickness and gravimetric swelling ratios of the resultant membranes. Cations of the additive salt participate in formation of complexes with the nucleophilic hydroxyl and acetate groups of cellulose acetate. Therefore, the spherical polymer aggregates possess positive charges resulting in mutual repulsion in a solvent medium of low dielectric constant such as acetone-water. Thus desolvation of the acetone from between the charged droplets will proceed more uniformly, and contact between polymer aggregates will be delayed for a greater length of time than in the absence of salt. A consequence of the uniform rate of desolvation is greater regularity in the size of the individual polymer droplets at the time contact is made between them.

The effects of the simultaneous presence of both water and the additive salt in the casting solution are extremely complex even to the point of being additive in some aspects and opposing in others. An apparently additive effect is the increase in swelling obtained in the resultant membrane. On the other hand, acetone containing large concentrations of water becomes a poorer solvent, and polymer-polymer interaction, and hence the capacity for aggregation, increases. Further, the presence of water in the casting solution increases the dielectric constant of the solvent system, which tends to diminish charge repulsion effects of absorbed salt cations thereby favouring irregular aggregation.

In the complete absence of water in the casting solution, desolvation will be further advanced before droplet formation, coalescence, and gelation occur. As a result, much of the potential capillary space in the centre of the droplets will have disappeared, and the resultant membrane will be quite dense and impermeable. Hence a proper balance between the additive salt and water in the casting solution is essential to control the gelation process and produce more uniform ultragel porous structures.

Pore-Structure of Cellulose Acetate Membranes. Meares (1966) estimates that 99 per cent of the transport of liquid water through the above membranes probably takes place by viscous flow in tortuous channels 6 to 9Å in diameter. These channels occupy about 4 per cent of the total volume of the membrane. About 1 per cent of the water is carried by homogeneous diffusion in the polymer matrix. Discussing the origin of pores in cellulose acetate, Meares (1966) points out that there are several possible explanations for their existence. The pores may be morphological artefacts introduced by the method of membrane preparation or they may be an inherent property of cellulose acetate. They may be a consequence of the natural packing of the molecules and hence exist in the dry and wet membranes, or they may result from a non-uniform swelling on the molecular scale when the membrane imbibes water.

Based on diffusion studies in cellulose acetate gels prepared from cellulose acetate-benzyl alcohol solutions, Klemm and Friedman (1932) concluded that the gel structure was porous and that the magnitude of the average pore radius decreased with increase in cellulose acetate concentration. They estimated that the average pore radius was 180Å at 2 per cent concentration of cellulose acetate and practically 0Å at about 25 per cent concentration (Figure 2-12).

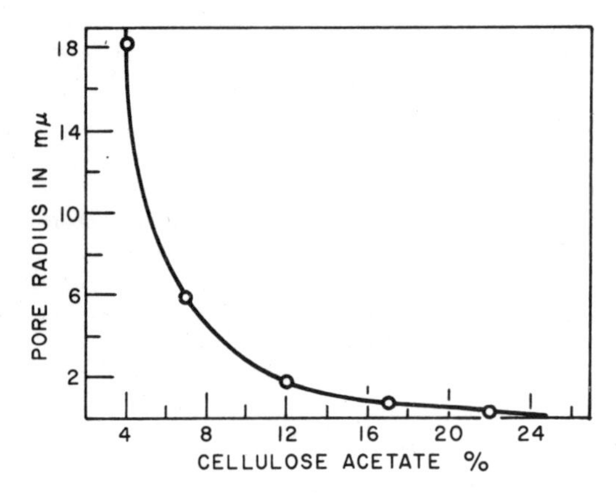

Figure 2-12. Effect of cellulose acetate concentration on pore radius. Data of Klemm and Friedman (1932).

Their results indicated that the number of pore openings also changed with cellulose acetate concentrations. According to Klemm and Friedman (1932), gel formation was the result of incomplete precipitation of the dispersed medium, and the resulting pore openings consisted of minute tubes of varying lengths, chaotically arranged, but each intersecting others to form continuous open passages completely through the gel; the thickness of the pore walls was also undoubtedly a function of the cellulose acetate concentration. Since the cellulose acetate solutions used by Klemm and Friedman did not contain any additive or swelling agent, the cellulose acetate concentration versus pore radius curve for the surface structure of Loeb-Sourirajan type membranes should be expected to be above that given in Figure 2-12.

Structural Changes During Heating of Membranes. Wasilewski (1965) made a microwave investigation of temperature induced transient changes in the Loeb-Sourirajan type porous cellulose acetate membranes. He concluded that parts of the cellulose acetate molecule rotated during the heating period, and a temperature of about 65° (close to the 'glass-transition temperature') had to be reached before rotational motion was set up in the membrane. According to Wasilewski, in the unheated membrane, almost all of the oxygen atoms of the carbonyl groups in the side chains of the cellulose acetate molecule are intramolecularly hydrogen bonded. During the heating process, enough energy is absorbed by such a carbonyl group to break this hydrogen bond and cause the group to rotate about a single bond; in that process two segments from different cellulose acetate molecules are brought close together to cause a physical interaction and form a new intermolecular bond which, being stronger than the intramolecular one, permanently immobilises the segments in question. Thus the formation of an intermolecular bond between two neighbouring cellulose acetate molecules in the membrane during heating brings the segments of these two molecules closer together, resulting in a decrease in the pore size in the surface layer of the membrane.

8. Preparation and Performance of Ultrathin Membranes

Merten et al. (1967) prepared ultrathin cellulose acetate membranes by the Carnell-Cassidy technique (Carnell, 1965; Carnell and Cassidy, 1961) which consisted essentially of slowly drawing a clean glass plate out of a dilute solution of a polymer in a suitable solvent. The details of their film making technique were as follows (Riley et al., 1967).

Dilute acetone solutions of 0.0063, 0.0125, 0.0250, 0.0312, 0.0440 and 0.0525 gram of cellulose acetate (type Eastman 398-10) per cm^3 of spectrograde solvent were prepared by rolling the solution on a mill for several hours. The solutions were stable for several months. Many of the solutions were filtered through 4 to 5.5μ pore-size fritted glass filters before use; several solutions were centrifuged for 1 hour at 17,000 rev/min in an International Model HT centrifuge.

A glass plate, 140 mm long, 37 mm wide and 5 mm thick, was used for the film forming surface. The glass plate was cleaned by vigorous scrubbing of the surfaces with Aquet liquid laboratory detergent. Large volumes of tap water followed by several rinses of distilled water were used to rinse the detergent from the plate. After being rinsed with distilled water, the plate was rinsed with reagent-grade acetone. The glass plates were conditioned in spectro-quality acetone, reagent-grade methyl ethyl ketone, or in some instances with methyl ethyl ketone plus 0.1 per cent tridodecylamine, for one half hour prior to dipping them into the dilute polymer solution. The tridodecylamine was added to the methyl ethyl ketone to assist in the dissipation of any static charge that might have been produced on the glass plate by the rubbing during the cleaning step.

The dilute cellulose acetate solution was placed in a Pyrex tube (460 mm long, 42 mm i.d.) to a depth of 80 mm., requiring approximately 225 cm^3 of solution. The tube was then immersed in a 30° water bath to maintain constant temperature of the dilute polymer solution during the film formation. The glass plate was rapidly taken from the acetone conditioning vessel to prevent drying and immersed into the dilute polymer solution while supported by a fine wire from a variable speed motor which was supported above the pyrex tube. The glass plate remained in the dilute polymer solution for 5 minutes. The plate was then withdrawn from the dilute polymer solution and raised to within 60 mm of the top of the open tube, where it was allowed to dry for 20 minutes. The plate was then removed from the tube, the membrane edges were cut with a sharp blade, the plate was immersed in a clean water trough, and the membranes were floated off each surface of the glass plate on to the surface of the water.

The first step in their study (Riley et al., 1967) was to establish a relationship between the polymer concentration in a given solvent and the thickness of the resulting membrane. The membrane thickness was determined by dimensioning a rectangular membrane section approximately 1 in. by 3 in., weighing the dried membrane with a Cahn microbalance, and using the bulk density (1.3 g/cm^3 for cellulose acetate) to calculate the thickness. In this way, the relationship shown in Figure 2-13 was obtained. The results given in Figure 2-13 apply only to acetone solutions at 30° and the membrane withdrawal rate of 22 cm/sec used in these studies. Membranes as thin as 600Å have been prepared by the above method.

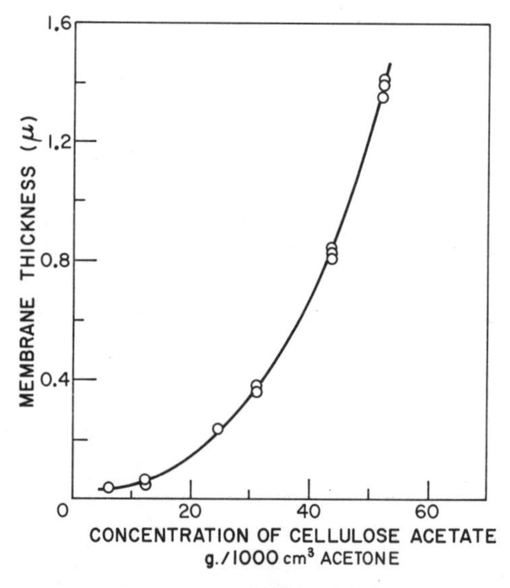

Figure 2-13. Effect of cellulose acetate concentration on the thickness of ultrathin membranes. Data of Riley et al. (1967).

The membranes were removed from the water trough by floating them on to Millipore filters (types VF and VM with mean pore sizes of 100 to 500Å respectively) which were then used to support the thin membranes in reverse osmosis experiments. A set of performance data obtained with such membranes is given in Table 2-24.

Merten et al. (1967) made several such thin membranes from ethyl cellulose, ethyl cellulose-polyacrylic acid, cellulose butyrate, and cellulose nitrate materials; the data on their performance in reverse osmosis experiments are given in Tables 2-25 to 2-29. Francis and Cadotte (1967) made several thin polysaccharide acetate membranes whose performance data in reverse osmosis are given in Table 2-30.

Table 2-24. Reverse osmosis data for six thin cellulose acetate membranes

Data of Riley et al. (1967)

Film thickness (Å)	Operating pressure (psig)	Water flux (gal/day/ft^2)	Solute separation (%)
Acetone Solvent, Acetone Glass Conditioner			
2800	1560	9.0	99.81
2800	1560	9.7	99.81
Methyl Acetate Solvent, Methylethyl Ketone-0.1% Tridodecylamine Glass Conditioner			
3000	840	4.5	99.47
3000	840	4.8	99.10
3000	840	4.5	99.35
3000	840	4.3	99.10

Temperature: 24°
Feed concentration: 0.9 wt.% NaCl
Membranes supported on millipore VF filters

9. Dry Membranes

On drying in air at the laboratory temperature, the porous structure of the Loeb-Sourirajan type cellulose acetate membranes often changes irreversibly. Hence, unless otherwise specified, the membranes are always kept under water, or aqueous alcohol solutions, or just in moist condition. It has been shown by Merten et al. (1967) that the above membranes can be stored dry without change of properties by pretreatment with surface active agents. A summary of their experimental results is given below.

Table 2-31 presents reverse osmosis data on the effects of 8 different surface-active agents which were examined. In each of these tests, several pieces of preshrunk membrane were placed in an aqueous solution of the surface-active agent for 15 minutes. The membranes were then allowed to dry under ambient conditions. After rinsing the membranes in distilled water for about ½ hour, the samples and the controls were tested in reverse osmosis experiments at 800 psig using 1 per cent aqueous sodium chloride feed solutions. Table 2-31 gives the results of the experiments where the uncertainties represent standard deviations for four tests. The control membranes were cut from the same sheets as the test membranes; the controls were never dried. The surface tension data given in Table 2-31 are from the manufacturer's literature as measured by the Du Nouy tensiometer. Of the 8 surface-active agents tested, all except Triton X-114 and FC-170 resulted in membranes whose solute separation and water flux were comparable to the controls. When the membranes were dried with Triton X-100 and rewetted and used, the water flux was about 10 per cent higher than the control.

Tergitol 15S7 (polyoxyethylene, 9 molecules ethylene oxide substituted on C_{11} to C_{15} linear alkyl) was examined in more detail than any of the other surface-

Table 2-25. Reverse osmosis data for thin membranes drawn from dilute solutions of ethyl cellulose

Data of Merten et al. (1967)

Membrane	Solvent	Glass conditioner	Millipore treatment (100Å)	Membrane thickness (Å)	Operating pressure (psig)	Water flux (gal/day/ft^2)	Solute separation (%)
A	Acetone	MEK	wet	1150	800	2.8	27.0
B						6.3	2.5
A	Methyl acetate	Acetone	wet	1800	800	0.5	69.0
B						0.6	69.0
C						0.4	70.0
D						0.6	67.0
A	Chloroform	Acetone	partially dry	2200	870	5.1	10.9
B						1.1	30.0
C						2.8	10.9
D						3.0	15.6
A	Benzene	Acetone	partially dry	2900	860	0.9	10.3
B						2.8	3.4
C						0.2	80.0
D						0.2	78.6

A	MEK	Acetone	dry	2150	860	2.4	11.9
B						0.2	81.2
C						0.3	69.3
D						0.6	38.1
A	p-Dioxane	Acetone	dry	2850	860	1.9	7.1
B						12.1	0
C						2.0	0
D						43.6	0
A	Ethyl alcohol	Acetone	dry	3220	860	0.3	35.5
B						0.2	80.5
C						0.3	47.3
D						0.2	59.4
A	Benzene	Acetone	partially dry	550	800	3.2	38.5
B						1.3	59.0
A	Benzene	Acetone	partially dry	1850	800	0.7	52.3
B						0.5	52.3
C						1.5	44.4
D						0.2	91.5

Operating temperature: 24°
Feed concentration: 0.9 to 1.6 wt.% NaCl

Table 2-26. Reverse osmosis data for ethyl cellulose-polyacrylic acid membranes

Data of Merten et al. (1967)

Ethyl alcohol[(a)] (wt.%)	Second solvent[(b)] (wt.%)	Evapor-ation time (min)	Membrane thickness (μ)	Water flux (gal/day/ft^2)	Solute separation (%)
86	4 DMF	5	90	210.2	0
86	4 DMF	10	66	2.5	50.0
82	8 DMF	5	120	30.1	26.4
82	8 DMF	10	100	18.6	36.8
73	17 DMF	5	130	204.5	0.6
73	17 DMF	15	78	2.3	31.6
82	8 EGMEE	6	100	6.3	19.3
82	8 EGMEE	11	95	9.4	16.1
73	17 EGMEE	6	110	2.0	37.1
73	17 EGMEE	10	86	1.1	45.0
54	36 EGMEE	5	150	8.4	25.0
54	36 EGMEE	10	154	14.9	16.7
54	36 EGMEE	15	120	6.5	16.7

(a) Remaining casting solution constituents: 9.1 wt.% EC, 10 Wt.% PAA based on the weight of ethyl cellulose

(b) DMF: N-N′-dimethylformamide
EGMEE: ethylene glycol monoethyl ether

Operating temperature: 20°-23°
Feed concentration: 0.9-1.9 wt.% NaCl
Operating pressure: 800 psig

active agents. A study was made on the effect of both the immersion time in the solution and the concentration of the surface-active agent in the solution. The experimental reverse osmosis data obtained are given in Table 2-32. The results showed that a concentration of only 1 per cent of the surface-active agent was sufficient to give satisfactory dry membranes and the immersion time in the solution could be as short as 1 minute.

A study was made of different plasticisers which could be usefully incorporated in the Tergitol solution. Membranes were soaked for 15 minutes in these mixed solutions and then dehydrated, rewetted, and tested as before. Reverse osmosis results on membranes treated in this manner are given in Table 2-33. Glycerine, ethylene glycol, and triacetin were found acceptable as plasticisers, but polyethylene glycol 6000 appeared to degrade the membrane. Glycerine and ethylene glycol improved the water flux without decreasing solute separation.

Some samples of membrane were soaked in solutions containing 30 per cent glycerine and decreasing concentrations of Tergitol 15S7 in water, and then dried, rewetted, and used in reverse osmosis experiments, with the results given in Table 2-34. All the membranes except those soaked in the 0.005 per cent Tergitol solution showed very little wrinkling; the membranes soaked in the 0.005 per cent Tergitol solution were somewhat wrinkled and had white and clear spots. All membranes, however, gave good results.

To determine the reason for the increased water flux, 4 membranes were soaked in a 4 per cent Tergitol solution for 15 minutes, washed in distilled water, and tested without drying. Four other membranes were soaked in glycerine without surface active agent for 15 minutes, dried, rewetted and tested.

Table 2-27. Reverse osmosis data for ethyl cellulose-polyacrylic acid membranes

Data of Merten et al. (1967)

Ethyl alcohol[(a)] (wt.%)	Second solvent[(b)] (wt.%)	Evaporation time (min)	Membrane thickness (μ)	Water Flux (gal/day/ft^2)	Solute separation (%)
82	8 EGMEE	0.5	54	0.3	47.2
82	8 EGMEE	1.0	56	10.3	5.6
73.1	12.8 EGMEE[(f)]	0.5	108	10.4	11.4
73.1	12.8 EGMEE	1	90	1.1	25.0
73.1	12.8 EGMEE	2	106	0.7	38.6
73.1	12.8 EGMEE	5	75	2.3	25.0
70.4	17.5 EGMEE[(c)]	0.5	75	48.5	8.6
82	8 THFOH	6	115	4.7	16.7
82	8 THFOH	10	125	8.6	26.6
73	17 THFOH	6	170	7.4	43.3
73	17 THFOH	10	125	29.4	6.7
55	35 THFOH	6	215	152.5	3.3
55	35 THFOH	10	145	52.1	6.7
55	35 THFOH	15	138	24.2	0
74.9	13 Diacetin[(c)]	1	50	145.0	0
74.9	13 Diacetin	1[(d)]	50	115.1	0
74.9	13 Diacetin	1[(e)]	50	31.2	17.1
74.9	13 Diacetin	5	34	39.1	11.4

(a) Remaining casting solution constituents: 9.1 wt.% EC, 10 wt.% PAA based on the weight of EC

(b) EGMEE: ethylene glycol monoethyl ether
THFOH: tetrahydrofurfuryl alcohol

(c) This solution contained 11.0% EC and 10 wt.% PAA based on the weight of ethyl cellulose

(d) This membrane was heat-treated at 80° for 30 min

(e) This membrane was immersed for 24 hours in pH 11.4 solution

(f) This solution contained 12.8% EC, and 10 wt.% PAA based on the weight of EC.

Operating temperature: 20°-23°
Feed concentration: 0.9-1.9 wt.% NaCl
Operating pressure: 800 psig

Table 2-28. Reverse osmosis data for Ethyl Cellulose-Polyacrylic Acid Membranes

Data of Merten et al. (1967)

Evaporation time (min)	Membrane thickness (μ)	Water Flux (gal/day/ft^2)	Solute separation (%)	Evaporation time (min)	Membrane thickness (μ)	Water Flux (gal/day/ft^2)	Solute separation (%)
2	38	154.5	2.2	10	80	35.4	14.3
2	38	29.2	15.6	10	80	41.0	19.0
2	38	36.0	6.8	10	80	41.4	14.3
2	38	43.5	2.2	10	80	28.2	21.4
5	30	83.8	6.8	10*	80	68.5	7.1
5	30	77.3	13.3	10	80	18.6	14.3
5	30	29.8	13.3	10	80	50.3	9.6
				10	80	51.6	19.2
10	25	0.3	56.2	1*	12	88.0	12.8
10	25	0.1	79.0	1	12	66.9	12.8
10	25	0.2	76.1	1	12	70.5	10.3
10	25	0.2	71.4	1	12	44.5	15.4
10*	25	7.1	7.1	2*	8	358.0	0
10	25	5.6	29.5	2	8	59.6	10.3
10	25	1.6	14.3	2	8	59.5	15.4
10	25	0.7	38.1	2	8	26.1	24.1

* Before testing in reverse osmosis, the membranes were immersed in pH 11.4 solution for 16 hr, then immersed in pH 3.0 solution for 1 hr

Operating temperature: 20°-23°
Feed concentration: 0.9 to 1.9 wt.% NaCl
Operating pressure: 800 psig

Composition of membrane casting solution (wt.%):
- Ethyl alcohol: 62.8
- Ethylene glycol monoethyl ether: 26.9
- Ethyl cellulose: 9.0
- Polyacrylic acid: 0.9
- Magnesium perchlorate: 0.4

Table 2-29. Reverse osmosis data for thin cellulose butyrate and cellulose nitrate membranes

Data of Merten et al. (1967)

Membrane thickness (Å)	Operating temperature (°)	Water flux (gal/day/ft^2)	Solute separation (%)
Alcohol Soluble Butyrate, 47.2% Butyryl			
2100	30	0.8	71.0
2100	30	1.9	24.0
Cellulose Nitrate, 11.8 to 12.2% Nitrogen			
700	33	0.8	70.0

Operating pressure: 800 psig
Feed concentration: 1.1 wt.% NaCl

The results, given in Table 2-35, showed that the increase in water flux was primarily due to the plasticising action of the glycerine.

Repeated soaking and drying of the membrane in a Tergitol and glycerine solution did not further improve its performance, as shown by the data in Table 2-36.

Two additional plasticisers were examined very briefly, diacetin and benzyl alcohol. Both of these materials plasticised the membrane, but in the presence of Tergitol 15S7 the membranes curled and had both clear and white spots, indicating that the plasticiser and the surface-active agent were not completely compatible in the membrane. No reverse osmosis experiments were made with these plasticisers.

The effect of the surface-active agent and the plasticiser on the physical properties of the membranes was determined in a number of tensile measurements. These measurements were made on an Instron tensile tester using a 2-in. length and a cross-head speed of 0.05 in./min. All the samples except the control were tested dry. The results of these measurements, given in Table 2-37, showed that Tergitol 15S7 plasticised the membrane, and that glycerine plasticised better, but neither one was as effective as water.

In summary, Merten et al. (1967) showed that the Loeb-Sourirajan type porous cellulose acetate membranes could be held dry after treatment with a surface-active agent such as a mixture of Tergitol 15S7 and glycerol in water.

10. Other Reverse Osmosis Membranes

Graphitic Oxide Membranes. Flowers et al. (1966) have pointed out that graphitic oxide possesses properties which make it particularly interesting for reverse osmosis studies. Graphitic oxide is prepared by oxidising graphite flakes with strong oxidising agents such as nascent chlorine dioxide (from potassium chlorate) in concentrated sulphuric acid or sulphuric acid-nitric acid mixtures. The shape of the graphite particles is retained throughout the oxidation process. A model proposed by Ruess (1947) for a single graphitic

Table 2-30. Reverse osmosis data for thin polysaccharide acetate membranes

Data of Francis and Cadotte (1967)

Membrane description*	Water flux (gal/day/ft^2)	Solute separation (%)
Amylose triacetate, high molecular weight fraction cast from 90/10 dichloromethane/methanol	1.0 to 0.9	99.1
Amylose triacetate, low molecular weight fraction cast from 90/10 dichloromethane/methanol	1.2 to 1.1	89.6
Amylose acetate, 39.5% acetyl membrane air dried, cast from cyclohexanone	2.3 to 1.9	81.0
Guar triacetate high molecular weight fraction cast from cyclohexanone	1.6	97.7
Guar triacetate, low molecular weight fraction, cast from cyclohexanone	1.7 to 1.4	97.4
Guar acetate, 37% acetyl cast from cyclohexanone	5.3 to 5.2	50.5
Locust bean gum triacetate, low molecular weight fraction, cast from cyclohexanone	2.3	98.6
Locust bean gum triacetate, low molecular weight fraction cast from cyclohexanone	1.7	94.6
β-Glucan triacetate, high molecular weight fraction, cast from cyclohexanone	3.6 to 3.5	96.3
β-Glucan triacetate, low molecular weight fraction, cast from cyclohexanone	2.3	97.7
Xylan diacetate not fractionated, cast from 90/10 dichloromethane/methanol	2.9 to 2.8	97.3
Cellulose triacetate cast from 90/10 cyclohexanone/DMF	2.3 to 2.1	99.7
Agar acetate, cast from cyclohexanone	4.4 to 4.2	61.3
Acetylated alginic acid methyl ester, cast from 90/10 cyclohexanone/DMF	20	18.0
Cellulose triacetate/xylan diacetate, 1/1	2.2	99.1

* Membranes were 1500 to 2500Å thick and unshrunk. Support films were cast from cellulose acetate by the method of Loeb and Sourirajan except that the evaporation and shrinking steps were omitted

Operating pressure: 1500 psig
Feed concentration: 3.5 wt.% NaCl

Table 2-31. Reverse osmosis data on cellulose acetate membranes dried with surface-active agents

Data of Merten et al. (1967)

Surface-active agent					
Trade name	Chemical name	Concentration (wt.%)	Surface tension (dynes/cm)	Solute separation (%)	Water flux (gal/day/ft^2)
Triton X-100	Iso-octyl phenoxy polyethoxyl ethanol (10 moles of ethylene oxide)	4	30	97.0 ± 0.3	12.3
	Control experiment			97.1 ± 0.2	11.2
Triton X-114	Iso-octyl phenoxy polyethoxyethanol (8 moles of ethylene oxide)	4	28	94.8 ± 1.2	8.2
	Control experiment			95.0 ± 0.8	12.0
Zonyl A	(Supplied by E.I. du Pont)	4	23	96.6 ± 0.2	9.9
	Control experiment			96.0 ± 0.5	9.8
Duponol C	Na lauryl sulphate	4		95.1 ± 0.7	12.1
	Control experiment			95.0 ± 0.8	12.0
Tween 20	(Supplied by Atlas Chemical Co.)	4		96.6 ± 0.7	11.3
	Control experiment			95.0 ± 0.8	12.0
Tergitol 15S7	Polyoxyethylene (9 moles of ethylene oxide)	4	28	95.3 ± 0.9	12.2
	Control experiment			95.0 ± 0.8	12.0
Aerosol OT-B	Na dioctyl sulphosuccinate 85% Na benzoate 15%	4	26	94.4 ± 0.7	12.4
	Control experiment			95.0 ± 0.8	12.0
FC-170	(Supplied by Minnesota Mining)	0.1	19	69	5.3
	Control experiment			95.0 ± 0.8	12.0

Operating temperature: 20° ± 3°
Operating pressure: 800 psig
Feed concentration: 1 wt.% NaCl

oxide layer is illustrated in Figure 2-14 (Flowers et al., 1966). The oxygen atoms taken into the crystal structure as a result of oxidation are shown here in the form of ether bridges and tertiary hydroxy groups. The intracrystalline space (~12Å) may be pictured as containing a plentiful supply of hydrogen bonding sites. Graphitic oxide is capable of a high degree of intercrystalline swelling on exposure to water or water vapour. On the basis of the results of

Table 2-32. Reverse osmosis data on cellulose acetate membranes dried with Tergitol 15S7

Data of Merten et al. (1967)

Tergitol 15S7[a] concentration (wt.%)	Immersion time (min)	Solute separation (%)	Water flux (gal/day/ft^2)
4	0.25	94.0 ± 0.3	11.9
4	1	94.6 ± 0.4	12.5
4	5	94.9 ± 0.6	12.5
4	30	94.9 ± 0.4	11.2
4	60	95.1 ± 0.5	11.8
4	900	95.5 ± 0.4	12.2
0.5	15	95.7 ± 0.8	10.9
1	15	96.0 ± 0.3	12.0
2	15	95.8 ± 0.6	11.9
4	15	95.3 ± 0.9	12.2
6	15	94.9 ± 0.4	12.2
Non-dried Control		95.0 ± 0.8	12.0

(a) Supplied by Union Carbide Co.

Operating temperature: 20° ± 3°
Operating pressure: 800 psig
Feed concentration: 1 wt.% NaCl

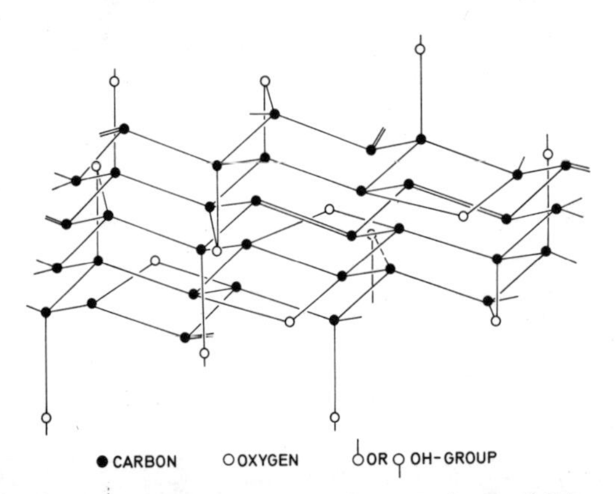

Figure 2-14. Structure of a layer of graphitic oxide with tertiary OH-groups according to Ruess (1947) (Flowers et al., 1966).

Table 2-33. Reverse osmosis data on dried, plasticised cellulose acetate membranes

Data of Merten et al. (1967)

Tergitol 15S7 concentration (wt.%)	Plasticiser	Plasticiser concentration (wt.%)	Solute separation (%)	Water flux (gal/day/ft²)
4	Polyethylene glycol	6	96.3	12.8
	Control experiment		99.1 ± 0.2	9.9
4	Triacetin	1	96.3	8.5
	Control experiment		97.5	9.4
4	Triacetin	1	96.8	11.9
	Control experiment		96.8 ± 0.4	11.4
0.5	Glycerine	20	96.1 ± 0.2	14.2
0.5	Glycerine	30	95.4 ± 0.9	14.6
0.5	Glycerine	40	94.3 ± 0.3	13.4
0.5	Glycerine	60	95.0 ± 0.5	14.4
	Control experiment		95.0 ± 0.8	12.0
4	None		99.1	11.7
4	Glycerine	20	99.3	12.0
4	Glycerine	40	99.1	12.1
4	Glycerine	60	99.1	12.4
	Control experiment		99.1 ± 0.2	9.9
4	Ethylene glycol	2	97.2	12.3
4	Ethylene glycol	5	96.9	12.9
4	Ethylene glycol	10	97.6	12.0
4	Ethylene glycol	20	97.2	13.0
4	Ethylene glycol	60	96.8	14.1
	Control experiment		96.8 ± 0.4	11.4

Operating temperature: 20° ± 3°
Operating pressure: 800 psig
Feed concentration: 1 wt.% NaCl

X-ray and water sorption studies, Clauss et al. (1957) proposed the arrangement of layers and voids (Figure 2-15) which probably consitute the pore system in the structure of graphitic oxide membranes (Flowers et al., 1966).

Flowers et al. (1966) prepared a number of graphitic oxide membranes by depositing a thin graphitic oxide layer on a porous supporting base and applying a water pump vacuum to its lower surface. The uniformity and packing of the graphitic oxide layer was controlled by adjusting the dilution of the suspension and also by the porosity of the base. The thickness of the layer was controlled by the known concentration of graphitic oxide in the dilute suspension used.

Table 2-34. Reverse osmosis data on cellulose acetate membranes soaked in 30% glycerine and Tergitol 15S7

Data of Merten et al. (1967)

Tergitol 15S7 (wt.%)	Solute separation (%)	Water flux ($gal/day/ft^2$)
0.1	96.7 ± 0.2	15.0
0.05	96.7 ± 0.5	14.8
0.01	96.2 ± 0.3	15.3
0.005	96.6 ± 0.2	15.1
Control	97.5 ± 0.3	13.6

Operating temperature: 20° ± 3°
Operating pressure: 800 psig
Feed concentration: 1 wt.% NaCl

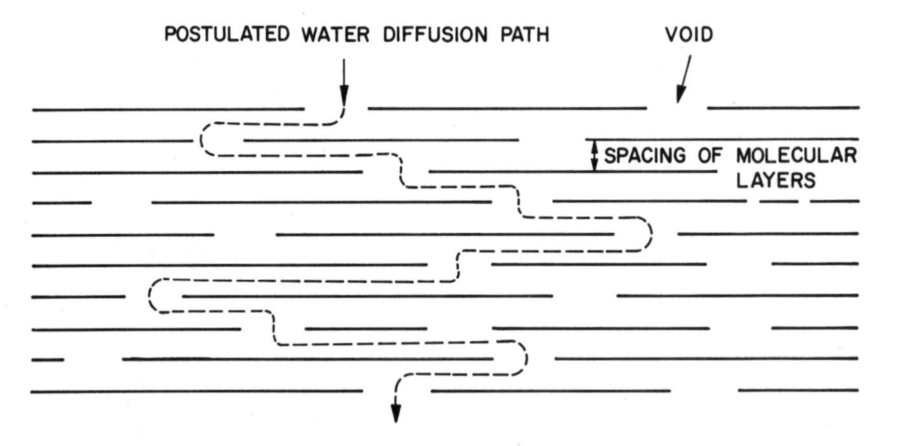

Figure 2-15. Schematic representation of GO membrane composed of lamellar molecule layers (Claus et al., 1957; Flowers et al., 1966).

Best results were obtained when the suspension contained only 0.1 mg graphitic oxide per cm^3 of suspension, and was fed to the 'filtering' porous base in separate portions, one after the other, so that a multi-layer graphitic oxide deposit was formed. The porous supporting base could be inorganic such as glass fibre paper, or organic such as porous polymer filters. In reverse osmosis experiments using aqueous sodium chloride (0.5 per cent NaCl) feed solutions at 600 psig, the water fluxes shown by the good graphitic oxide membranes ranged from 0.7 to 5.8 $gal/day/ft^2$ with solute separations ranging from 65 to 83 per cent. Though such performance data do not appear impressive from the point of view of saline water conversion, there is every reason to believe that better graphitic oxide membranes, showing better performance, will be developed in the future. Further, the graphitic oxide membranes may be found particularly useful in reverse osmosis applications involving the separation of different mixtures of organic liquids.

Table 2-35. Reverse osmosis data on treated cellulose acetate membranes

Data of Merten et al. (1967)

Sample history	Solute separation (%)	Water flux (gal/day/ft^2)
Membrane soaked 15 min in Tergitol 15S7. Not dried	94.6 ± 1.2	12.2
Membrane soaked in 20% glycerine. Dried, rewetted	94.7 ± 0.2	13.6
30% glycerine	96.4 ± 0.3	13.6
40% glycerine	93.5 ± 0.6	13.8
60% glycerine	93.8 ± 0.6	14.5
100% glycerine	95.2 ± 0.1	13.7
Control experiment	95.0 ± 0.8	12.0

Operating temperature: 20° ± 3°
Operating pressure: 800 psig
Feed concentration: 1 wt.% NaCl

Table 2-36. Reverse osmosis data on treated cellulose acetate membranes

Data of Merten et al. (1967)

Number of times soaked in 4% Tergitol 15S7 and 30% glycerine and dried	Solute separation (%)	Water flux (gal/day/ft^2)
1	96.6	13.3
2	96.5	13.8
4	96.5	12.3
Control experiment	96.8 ± 0.4	11.4

Operating temperature: 20° ± 3°
Operating pressure: 800 psig
Feed concentration: 1 wt.% NaCl

Polysalt Complex Membranes. Ionically bonded polymeric network structures synthesised from linear polyelectrolytes possess unusual physical and chemical properties. Michaels and Miekka (1961) reported on the interaction between sodium poly(styrene sulphonate) (NaSS) and poly(vinylbenzyltrimethylammonium) chloride, and on the composition and properties of the resulting precipitate or polysalt. This precipitate contained almost exactly stoichiometric proportions of the component polyions, and furthermore contained none of the counterions initially associated with the individual polymers—i.e. Na^+ and Cl^-. This precipitate was found to be infusible and insoluble in all common solvents. The composition of the polysalt was independent of the relative proportions in which the component polymers were mixed, and of

Table 2-37. Tensile measurements on dried cellulose acetate membranes

Data of Merten et al. (1967)

Membrane	Thickness (10^{-4} cm)	Yield strength (10^7 dynes/cm^2)	Ultimate strength (10^7 dynes/cm^2)	Modulus of elasticity (10^9 dynes/cm^2)	Elongation at failure (%)
Freeze dry	80	7.7	9.4	4.8	3.1
0.5% Tergitol	79	8.8	12.1	6.1	5.0
1.0% Tergitol	80	8.6	10.9	5.3	3.8
2.0% Tergitol	84	8.1	10.4	5.0	3.8
4.0% Tergitol	84	7.6	9.5	4.9	5.0
6.0% Tergitol	85	6.9	9.2	4.8	5.8
40% Glycerine	87	4.9	8.4	4.3	12.1
60% Glycerine	85	5.1	7.9	4.4	8.2
0.5% Tergitol + 20% glycerine	85	5.4	8.6	4.6	8.6
0.5% Tergitol + 20% glycerine	86	5.5	8.6	4.4	8.6
0.5% Tergitol + 20% glycerine	87	5.6	7.9	4.0	5.8
0.5% Tergitol + 20% glycerine	89	5.4	8.0	4.1	9.0
Control-tested wet	88	3.4	4.9	3.2	18.0

the order or rate of addition. The polysalt precipitate could however be dissolved (or, alternatively, the component polyelectrolytes codissolved without reaction) in selected ternary solvent mixtures comprising water, a water-miscible organic solvent such as acetone, and a strongly ionised simple electrolyte such as NaBr. The solubilising activity of these so-called shielding solvents is believed to be caused by loosening of the polyion linkages of the complex by the ions of the electrolyte, and the interaction of the organic solvent component with the organophilic backbone of the polymer chain. There is a small region in the solvent composition field where the component polyions remain in solution to yield a homogeneous, transparent, viscous syrup which can be used to form membranes, because by suitable alteration of the solution composition (by drying and/or washing) a microporous solid structure containing only the polymer and water can be produced. Also, by varying the relative proportions of the component polyelectrolytes in solution, it is possible to produce nonstoichiometric complex structures exhibiting either cation exchange or anion exchange capacity (Michaels, 1965). Several polysalt complex membranes (called 'Diaplex' or 'Diaflo' membranes) for reverse osmosis application have been prepared by Michaels et al. (1965b) as follows.

The NaSS and poly(vinylbenzyltrimethylammonium) chloride polymers used had the structures and molecular weights (MW) given below:

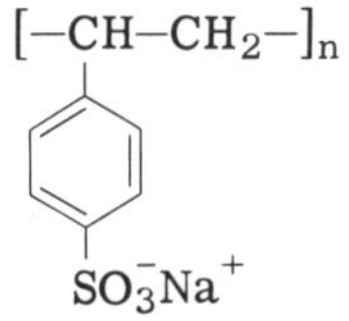

$[-CH-CH_2-]_n$ (with para-phenylene bearing $SO_3^-Na^+$)

NaSS
$n \approx 480$
$MW \approx 10,000$

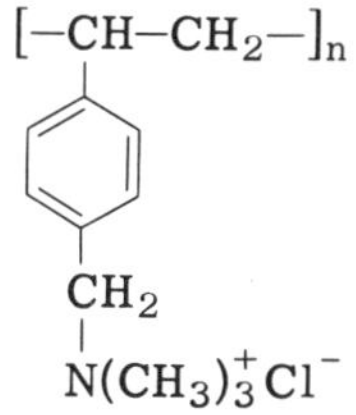

$[-CH-CH_2-]_n$ (with para-phenylene bearing $CH_2-N(CH_3)_3^+Cl^-$)

Poly(vinylbenzyltrimethylammonium) chloride
$n \approx 140\text{-}190$
$MW \approx 30,000\text{-}40,000$

Neutral polysalt complex was precipitated from a solution containing 20 wt.% of the polymers dissolved in a ternary solvent consisting of equal parts of dioxane, calcium nitrate, and water. Precipitation was accomplished by pouring the polymer solution into a large excess of water (1 volume of solution to 10 volumes of water) while the latter was being agitated in a Waring Blender. The precipitate was washed with water and dried in an air oven at 50°. This procedure removed the sodium sulphate (which was present in the NaSS as received from the supplier), as well as the sodium chloride discharged upon reaction of the two polyions.

Neutral casting solutions were prepared by dissolving the powder obtained above into a ternary solvent consisting of equal parts of dioxane and water and one-half part calcium bromide. Non-neutral casting solutions were prepared by adding one polyelectrolyte in a calculated amount of neutral solution. The individual polyelectrolytes were purified and obtained in solid form by precipitation from aqueous solution into acetone, and rinsing the precipitate with acetone.

The following procedure was adopted as the standard method of preparing polysalt films.

(1) Dissolve 20 wt.% of the polyion complex in the calcium bromide shielding solution.

(2) Cast the solution on to a glass plate at room temperature using a drawdown bar set at 8 mil.

(3) Gel in air at 50° for 10 minutes by solvent evaporation.

(4) Wash in water at 25° for 15 minutes to remove calcium bromide and unevaporated dioxane.

(5) Consolidate film by immersion in 25 wt.% $Ca(NO_3)_2$ solution at 50° for 1 hour.

(6) Dry in air at 50° for 25 minutes.

(7) Press at 80° between Mylar sheets for 40 minutes.

(8) Wash in water at 25° for 1 hour.

The film was now ready to be die-cut and placed in the reverse osmosis cell for evaluation.

The polysalt membranes prepared by the above technique did not exhibit the 'one-sidedness' of the Loeb-Sourirajan type porous cellulose acetate membranes. Neutral membranes gave solute separations of about 50 per cent at 1500 psig using aqueous sodium chloride (4 per cent NaCl) feed solutions with product rates of about 2 gal/day/ft^2 for a film 1 mil thick. Improving consolidation of films by varying fabrication conditions increased desalination and decreased product rate, but in no case did desalination significantly exceed 50 per cent. The best non-neutral films were decidedly inferior to the best neutral films for desalination application (Michaels et al., (1965b). Even though the polysalt films do not appear to be useful desalination membranes, they have been suggested as possible effective filtration media for the feed to reverse osmosis systems (Markley et al., 1967). Membranes showing water permeabilities of 80 to 100 gal/day/ft^2 at 100 psig with >90 per cent solute separation from dilute aqueous raffinose feed solutions have been obtained. A set of reverse osmosis performance data for such membranes is shown in Table 2-38 (Michaels, 1965). At the present time, these membranes are being used for the ultrafiltration of biological fluids such as urine and blood, and for the purification of waste and food-processing streams.

Dynamically Formed Membranes. In June 1965, A. E. Marcinkowsky at Oak Ridge National Laboratory observed that by circulating a pressurised salt solution containing a few hundred parts per million of an additive such as $ThCl_4$ past a porous silver filter (pore size ~0.2μ), a high flux salt rejecting layer was formed dynamically on the filter surface. Following this observation, extensive work has been reported by Kraus and co-workers on dynamically formed membranes for reverse osmosis separation of many inorganic and organic substances in aqueous solution (Johnson and Kraus, 1968; Kraus et al., 1966a, 1966b, 1967; Kuppers et al., 1967; Marcinkowsky et al., 1966).

Dynamic formation of membranes was found to be quite general, particularly with materials which might be expected to give ion exchange layers. Hydrous oxides, and salts of hydrolysable metals [such as Zr (IV), Fe (III), Th (IV), and U (VI)], organic polyelectrolytes (such as poly(styrene sulphonates), poly(vinylbenzyltrimethylammonium) salts, polycarboxylic acids and their salts), ground low-cross-linked ion exchangers, humic acid, and polyamines, among others, were found to do so. The chemical nature of the porous body did not seem of primary importance—membranes were formed, for example, on porous metals, ceramics, carbon, and sintered glass. Usable pore sizes varied with the additive, but the materials mentioned above formed membranes on substrates of nominal pore sizes around 0.1μ, and many could be used with larger pore sizes, up to 5μ for example with colloidal hydrous oxide. Some typical additives and

Table 2-38. Performance of Diaplex (polyion complex) membranes in reverse osmosis

Data of Michaels (1965)

Film type	Charge	Water content (g/g resin)	Water permeability*	% Solute separation at 100 psig							
								Fuchsin	Carbowax		
				NaCl	Urea	Sucrose	Raffinose	Red	1500	6000	BSA[a]
I-YLL	Neutral	0.4	3	5	5	50	80	100	—	—	100
I-YL	Neutral	1.3	50	0	0	10	30	98	—	—	100
I-YM	Neutral	2.2	340	0	0	0	0	85	—	—	100
I-YH	Neutral	3.5	810	0	0	0	0	75	—	—	100
I-XL	Anionic	1.3	5.5	0	0	5	20	99	—	—	100
I-XM	Anionic	2.8	110	0	0	0	0	50	—	—	100
I-XH	Anionic	6.5	460	0	0	0	0	18	—	—	100
UM-1	Neutral	8.0	1000	1	1	15	20	—	50	100	100
UM-2	Neutral	8.0	330	5	5	60	90	100	100	100	100
UM-3	Anionic	—	125	25	—	100	100	100	100	100	100

* Water permeability at 100 psig, $(cm^3 mil/cm^2/sec) \times 10^5$

a BSA = Bovine serum albumin

porous supports used for forming dynamic membranes are listed in Tables 2-39 and 2-40 (Kraus et al., 1967). Some data on the performance of such dynamically formed membranes are given below.

Hydrous zirconium oxide membranes (prepared from boiled zirconium oxychloride solution) showed water fluxes of the order of 300 gal/day/ft^2 while giving significant solute separation (40 to 80 per cent at 35 atm. pressure using 0.01 to 0.05N sodium chloride feed solutions. In acidic solutions $MgCl_2$, $BaCl_2$, and $LaCl_3$ were separated better than NaCl by hydrous zirconium oxide films; under these conditions Na_2SO_4 separation was negligible. In alkaline solutions, Na_2SO_4 was separated better than $BaCl_2$ (Table 2-41).

Hydrous Fe_2O_3 films were also pH sensitive. They showed a minimum of solute separation at pH = 8 for 0.08M aqueous sodium chloride feed solutions containing 10^{-4}M Fe(III). In one such experiment at 600 psig, NaCl separation was 56 per cent near pH = 3, negligible near pH = 8, and 20 per cent near pH = 12; the product rates were about 25 gal/day/ft^2.

With polyvinylbenzyltrimethyl ammonium chloride membrane, solute separations ranged from about 25 per cent with 0.1M NaCl to 85 per cent with 0.001M NaCl feed solutions. $MgCl_2$ and $LaCl_3$ were better separated than NaCl at the same chloride concentrations. Water fluxes of about 100 gal/day/ft^2 were obtained at 34 atm pressure with these feed solutions (Figure 2-16).

Salts containing polyvalent counterions (e.g. Mg^{++} with polystyrene sulphonic acid and SO_4^{--} with benzyltrimethylammonium chloride) were poorly, if at all, separated by dynamically formed polyelectrolyte membranes, and their presence in feeds even at low concentrations frequently seriously affected the separation of other salts. For example, 0.001M Na_2SO_4 in 0.02M NaCl dropped NaCl separation from ~60 per cent to 40 per cent, and with 0.01M NaCl-0.005M Na_2SO_4, chloride separation was 20 per cent (Kraus et al., 1966b).

Both humic acid (extract obtained from decaying oak leaves) and mixtures of humic acid and bentonite (a montmorillonite-type clay) in low concentrations

Table 2-39. Typical additives which form membranes capable of solute separation

Data of Kraus et al. (1967)

Hydrous oxides [Al (III), Fe (III), Si (IV), Zr (IV), Th (IV), U (VI)]
Finely ground low cross-linked ion-exchange resis
Clays (Bentonite)
Humic acid
Poly(styrene sulphonic acid)
Poly(vinylbenzyltrimethylammonium chloride)
Cellulose acetate hydrogen phthalate
Cellulose acetate N, N-Diethylamminoacetate
Poly(Methyl vinyl ether/Maleic anhydride) (Gantrez AN)
Poly(4-vinyl pyridine)
Poly(4-vinyl pyridinium butyl chloride)
Poly(vinyl pyrrolidone)

Table 2-40. Typical porous supports for dynamically formed membranes

Data of Kraus et al. (1967)

Material	Nominal pore diameter (microns)	Trade name or supplier
Silver	0.2-5	Selas Corp.
Porcelain	0.5	—
Sintered glass	Ultrafine	—
Carbon	0.2-0.5	Union Carbide Corp.
Polyvinyl chloride	0.45	Metrical VM-6 (Gelman)
Polyvinyl chloride (on nylon)	0.45	Vinyl Acropor VNM-450 (Gelman)
Vinylidene fluoride	0.45	Metricel VF-6 (Gelman)
Cellulose acetate	5	Metricel GA-1 (Gelman)
	0.1-0.5	Millipore
Porous fibreglass tubes	—	American Standard
	—	Havens Industries
Nuclepore	0.5	General Electric

Table 2-41. Solute separation by hydrous zirconium oxide membranes

Data of Kraus et al. (1967)

Salt	Concentration molality	pH	Separation (%)
NaCl	0.05	3-4	70
$MgCl_2$	0.027	3-4	91
$BaCl_2$	0.025	3-4	95
$LaCl_3$	0.029	3-4	96
NaCl	0.05	~11.5	40
Na_2SO_4	0.025	~11.5	60
$BaCl_2$	0.025	~11.5	30

Porous carbon tubes, permeation velocity = 0.05 to 0.4 cm/min*, 400-600 psi, 25°

* 1 cm/min = 355 gal/day/ft^2

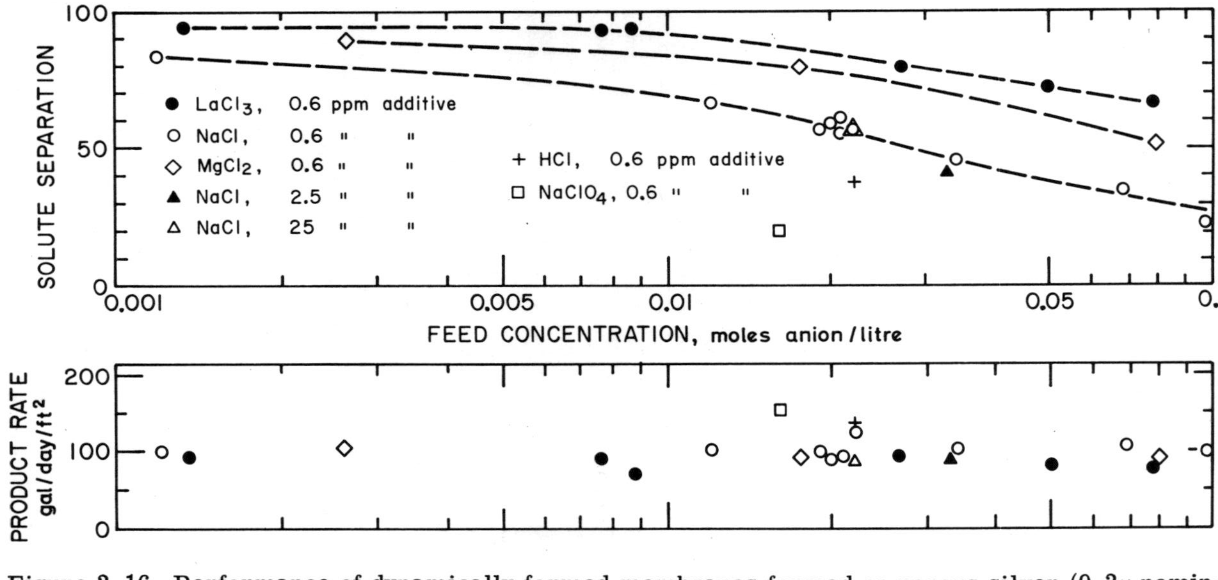

Figure 2-16. Performance of dynamically formed membranes formed on porous silver (0.2μ nominal pore diameter) with polyvinylbenzyltrimethyl ammonium chloride at 34 atm. Data of Kraus et al. (1967).

(25 to 45 ppm) were capable of forming dynamic membranes with separation characteristics for NaCl, Na_2SO_4, and $MgCl_2$ (Figure 2-17).

A set of results on the performance of dynamically formed neutral polyvinylpyridine membrane, 'aged' by several days prior operation, is given in Figure 2-18. The membrane was formed on a flat disk of Metricel VM6 with nominal pore size 0.45μ. Initial concentration of the additive was 100 ppm; during the reverse osmosis experiments, the solutions contained 1 ppm of the additive.

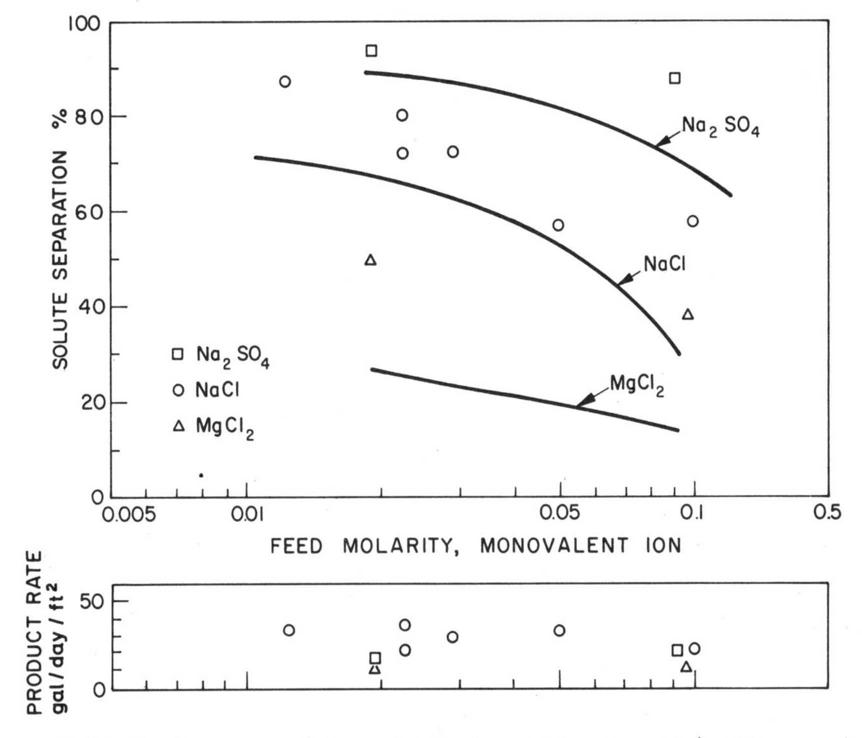

Figure 2-17. Performance of dynamically formed humic acid (solid curves) and bentonite-humic acid (points) membranes (0.8μ Ag filter, 2000 psig, ~25 ppm humic acid, ~45 ppm bentonite). Data of Kraus et al. (1967).

The membrane gave 70 per cent solute separation with 0.05M NaCl feed solutions near pH 3. This rose to 85-90 per cent above pH 4. $MgCl_2$ (from 0.025M feed solution) was separated 85 per cent at pH 3; separation was better at higher pH. Na_2SO_4 (from 0.01M feed solution) separation increased from 43 per cent at pH 3 to better than 80 per cent at high pH. The product rates obtained at high pH range were however low—4 to 7 gal/day/ft^2. The above membrane was also capable of separating salts from sea water; in one 18-hour test using sea water feed, chloride separation was 90 per cent, Mg, Ca, and sulphate separation were 99 per cent, and the product rate was 3 gal/day/ft^2 at 1500 psig.

Kraus et al. (1967) also illustrated the performance of dynamically formed zirconium oxide membranes for the separation of mixed solutes in aqueous solution (Table 2-42), and organic solutes in aqueous solution as shown in Table 2-43 and Figure 2-19 (Kuppers et al., 1967).

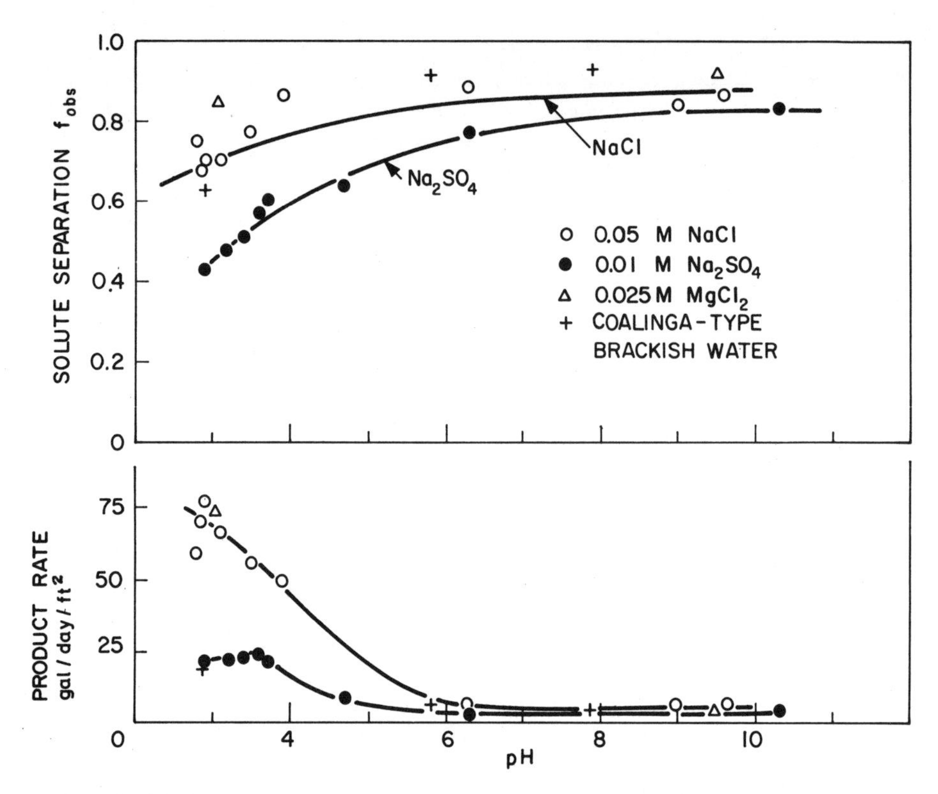

Figure 2-18. Performance of dynamically formed poly (4-vinylpyridine) membrane (25°, 1500 psig, PVC support, 0.45μ). Data of Kraus et al. (1967).

The discovery of the separation characteristics of several dynamically formed membranes by Kraus and co-workers must be considered as a major development in the reverse osmosis process. It is significant to note that all the results illustrated emphasise again the twin-requirement basis of reverse osmosis separation discussed in Chapter 1. From that point of view, the possible number of substances which can form dynamic membranes and exhibit separation characteristics is unlimited. Hence many more such membranes of practical interest for specific separation problems may be expected to be discovered in the future. Under conditions when solute separations and product rates are adequately high, they may be expected to be used industrially for a wide variety of practical separation problems. Even though the current researches seem to be limited to aqueous feed solution systems, future studies may be expected to include many nonaqueous feed solutions also.

11. Recent Developments in Preparation and Pretreatment of Loeb-Sourirajan Type Membranes

While a large number of other membranes have been tested for reverse osmosis separations, the Loeb-Sourirajan type porous cellulose acetate membranes continue to remain the most promising ones for saline water conversion and related water treatment applications. Three major developments have been reported on the preparation and pretreatment of these membranes since 1965, and they are discussed below.

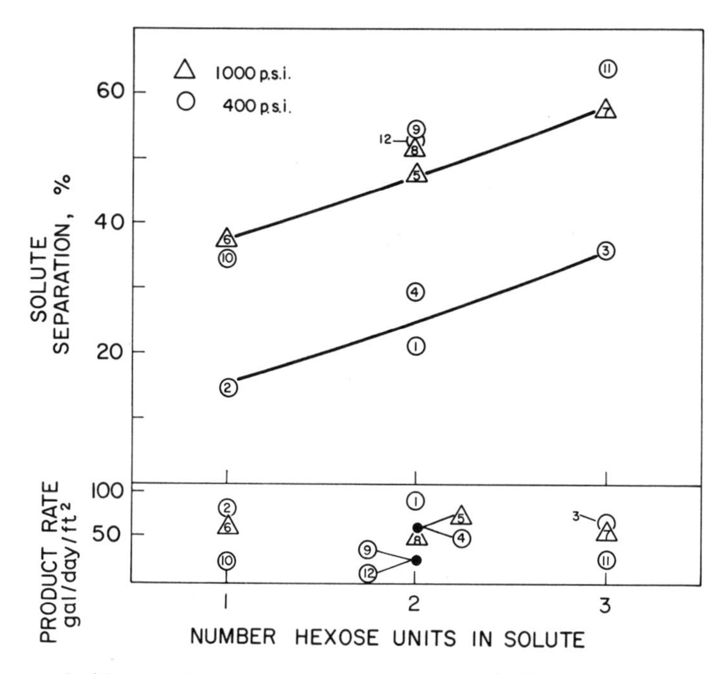

Figure 2-19. Performance of a hydrous Zr (IV) oxide membrane on 0.2μ Ag filter. Feed solutions contained about 50 g/litre of sugar and were 10^{-5}M in Zr (IV). Numbers indicate order of experiments. Data of Kuppers et al. (1967).

Table 2-42. Performance of hydrous zirconium oxide membranes for the separation of mixed solutes in aqueous solution

Data of Kraus et al. (1967)

Solution	Solute separation (%) NaCl	Solute separation (%) MCl	Water flux cm/min*
0.017M $LaCl_3$-0.05M NaCl	70	98	0.19
0.017M $LaCl_3$-0.05M NaCl	35	88	0.15
0.054M $MgCl_2$-0.12M NaCl	52	87	0.19
0.052M $MgCl_2$-0.52M NaCl	27	65	0.19
0.048M $CaCl_2$-0.11M NaCl	49	84	0.20
0.052M $CaCl_2$-0.51M NaCl	24	57	0.20
0.046M $CaCl_2$-1.02M NaCl	14	48	0.21
0.045M $BaCl_2$-0.01M HCl	3 (HCl)	96	0.24

* 1 cm/min = 355 gal/day/ft²

Porous carbon tubes, Dynamic $ZrO_2 \cdot xH_2O$ membrane, 25°, 10^{-3}M HCl, 10^{-4}M Zr(IV), 800 psig

Table 2-43. Separation of some organic solutions using hydrous zirconium oxide membranes

Data of Kraus et al. (1967)

Solute	Concentration (g/l.)	Pressure (atm)	Flux (Gal/day/ft^2)	Separation (%)
Phenol	50	105	230	10
n-Butanol	50	105	150	47
Ethylene glycol	15	35	60	23
Diethylene glycol	25	35	60	26
PEG-300	50	35	50	52
PEG-1000	50	35	45	85
PEG-4000	50	30	40	91
PEG-6000	50	35	40	89

PEG refers to 'polyethylene glycol', polymeric ethers of ethylene glycol.

25°, 0.2 μ-silver frits, ~10^{-4}M Zr(IV)

Further Work of Manjikian, Loeb, and McCutchan at UCLA. These authors made detailed studies on the use of formamide as the additive material in the casting solution replacing magnesium perchlorate and water originally used by Loeb and Sourirajan. The results obtained led to the development of casting solution compositions which could be used to cast cellulose acetate membranes at the laboratory temperature. The performance of some of the resulting membranes were even better than those obtained by the earlier membranes. In these investigations, all the membranes were tested in a flow system at 600 psig, using aqueous feed solutions containing 5000 ppm of NaCl, with a feed rate of 300 cm^3/minute. Unless otherwise stated, the cellulose acetate material used was Eastman grade E-398-3 (acetyl content 39.8 per cent). A summary of their extensive performance data is given in Tables 2-44 to 2-49 (Manjikian et al., 1965; Manjikian, 1967).

The effects of casting solution composition and casting temperature, evaporation time and shrinkage temperature on the performance of membranes cast from cellulose acetate-acetone-formamide solutions are illustrated in Tables 2-44, 2-45, and 2-46 respectively. The results show that a membrane made from a casting solution containing cellulose acetate : acetone : formamide in the proportions 25 : 45 : 30 (parts by weight) and cast at the laboratory temperature (~23°) with an evaporation period of ½ to 1 minute and subsequent gelation in ice water, when shrunk at 71° to 76°, can give initial product rates of about 30 gal/day/ft^2 with solute separations >90 per cent from 0.5 wt.% NaCl-H_2O feed solutions at 600 psig. Such membrane performance is better than any reported earlier. Consequently films obtained by the above procedure are now being extensively used in several laboratories and reverse osmosis units. In the author's laboratory, and elsewhere in this book, cellulose acetate films made by the above procedure are designated as CA-NRC-47, or simply Batch 47, type films.

Tables 2-47 and 2-48 illustrate the effects of molecular weight, and molecular weight distribution of the cellulose acetate material respectively on the performance of the resulting membranes. Changes in molecular weight did not materially affect membrane performance, but the shrinkage-temperature profiles for the films were different for different cellulose acetate materials. The performance of membranes made from cellulose acetate materials with 'narrow' molecular weight distribution appeared to be slightly less sensitive to changes in shrinkage temperature.

The effect of degree of acetylation of cellulose acetate material on subsequent membrane performance was examined by considering two groups of polymers, one having a higher, and the other having a lower degree of acetylation than the standard Eastman E-398-3 cellulose acetate material. The polymers tested from the former group included:

Eastman Designation	Acetyl Content (%)	Degree of Acetylation (%)
602-A	41.6	88
602-B	41.0	86
602-C	40.1	83

Of these materials, only 602-C formed a homogeneous casting solution in proportions used for Batch 47 films. Films made with 602-C materials, when shrunk at 70°, gave 25 gal/day/ft^2 of product water containing 600 ppm of NaCl. Results with polymers having lower degrees of acetylation are given in Table 2-49; the materials tested in this group included:

Eastman Designation	Acetyl Content (%)	Degree of Acetylation (%)
EMP-949-A	38.9	79
EMP-949-B	37.5	75
EMP-949-C	37.1	74

On the basis of the experimental results obtained with the above films, it was concluded that good performance could be obtained with membranes made from cellulose acetate material whose acetyl content was between 37.5 and 40.1 per cent; within this range no specific correlations were obtained between either flux or desalination and acetyl content.

In addition to formamide, several other successful additive materials have been reported by Manjikian et al. (1965). These include (Table 2-50) aqueous solutions of urea, glyoxal, or hydrogen peroxide, dimethyl formamide, dimethyl sulphoxide, tetrahydro-furfuryl-phosphate, triethyl phosphate, acetic acid, and N-methyl-2-pyrrolidone. The performance data of films obtained from binary casting solutions given in Table 2-50 are particularly interesting. They indicate that dimethyl formamide, acetic acid, and triethyl phosphate serve both as solvents and additives simultaneously. Some data on the performance of ethyl cellulose membranes obtained from ethyl cellulose-dioxane-triethyl phosphate casting solutions are given in Table 2-51. These data show that by suitable choice of casting solution composition, better ethyl cellulose membranes can be produced for reverse osmosis applications.

Manjikian (1966) extended the above work at the Universal Water Corporation, and presented performance data for Batch 47 type membranes for sea water conversion (Table 2-52). These tests were run on natural sea water without filtering, pH adjustment, or pretreatment of any kind. The results showed that Batch 47 type films could be used successfully for sea water conversion in one stage at pressures in the range of 800 to 1200 psig. Majikian (1966) also show-

Table 2-44. Performance of membranes cast from various cellulose acetate-acetone-formamide mixtures

Data of Manjikian et al. (1965) and Manjikian (1967)*

Solution number	Casting solution composition (wt. %)			Evaporation time (min)	Film shrinkage temp. (°)	Desalinised water	
	Cellulose acetate	Acetone	Formamide			Flux (gal/day/ft^2)	Salt content (ppm)
Film casting temperature —10°							
1	22.2	66.7	11.1	10	73	12	310
2	21.4	64.4	14.2	10	73	30	610
3	31.1	57.8	11.1	10	Unshrunk	6.5	880
4	25	65	10	10	68.5	4.6	270
5	25	55	20	10	77	17	220
6	25	50	25	6	79.5	27	480
7	25	45	30	6	81.5	22	190
8	25	40	35	3	85.3	14	150

Film casting temperature 10°

*	25	65	10	10	68.5	4.6	272
*	25	55	20	10	77	17	221
*	25	50	25	6	79.5	27	476
*	25	45	30	6	81.5	21.6	187
*	25	40	35	3	85.3	13.8	153

Film casting temperature 23°

9	29.1	54.2	16.7	1	Unshrunk	2.7	310
10	30	55.7	14.3	1	Unshrunk	8.1	410
7	25	45	30	1	74	30	410
6	25	50	25	1	71.5	15	270
5	25	55	20	1	65	15	360
*	25	55	20	1	65	15.1	357
*	25	50	25	1	71.5	15.4	272
*	25	45	30	1	74	30	408
*	25	45	30	½	72	45	600
*	25	45	30	½	76	30	250

Feed concentration: 5000 ppm NaCl
Operating pressure: 600 psig

Table 2-45. Effect of evaporation time on membrane performance

Data of Manjikian et al. (1965)

Evaporation time (min)	Desalinised water: Flux (gal/day/ft^2)	Desalinised water: Salt content (ppm)
0.25	84	1900
0.50	47	650
1.0	26	580
1.5	34	730
2.0	25	1100
2.5	23	1200
3.0	25	1700

Casting solution composition (wt.%)
Cellulose acetate 25
Acetone 45
Formamide 30
Casting temperature: 23°
Film shrinkage temperature: 71.0-71.6°
Operating pressure: 600 psig
Feed concentration: 5000 ppm NaCl

ed that by incorporating pyridine in the cellulose acetate-acetone-formamide casting solution, films with superior mechanical properties and comparable reverse osmosis performance could be obtained (Table 2-53).

Thus the researches of Manjikian, Loeb, and McCutchan have introduced the Batch 47 and related types of porous cellulose acetate membranes for practical reverse osmosis applications.

Fabrication of Tubular Porous Cellulose Acetate Membranes. This is the second major development. The use of reverse osmosis desalination membranes integrally lined within a porous fibreglass-reinforced tube was first disclosed by Havens Industries (1964); however, no details of their technique were available in the open literature. At UCLA, Loeb pioneered the gravity-drop technique for making tubular porous cellulose acetate membranes and their subsequent assembly into perforated tubes for use in reverse osmosis pilot plants (Loeb, 1966a). His technique, described below, is a model of simplicity and ingenuity.

Gravity-drop technique. The details given below are for tubular membranes for subsequent assembly into perforated tubes with a nominal outside diameter of 2.54 cm. However, the method is generally applicable for tubes of different diameters.

The casting solution is a mixture of cellulose acetate (Eastman grade E-398-10) : acetone : formamide in the proportions 25 : 45 : 30 by weight. The membrane

Table 2-46. Effect of shrinkage temperature on membrane performance

Data of Manjikian et al. (1965)

	Desalinised water	
Film shrinkage temp. (°)	Flux (gal/day/ft^2)	Salt content (ppm)
Unshrunk	95	3800
70	45	800
76	30	250
78	17	110
81	12	85

Casting solution composition (wt. %)
- Cellulose acetate 25
- Acetone 45
- Formamide 30

Casting temperature: 23°
Evaporation time: ½ min
Operating pressure: 600 psig
Feed concentration: 5000 ppm NaCl

is cast batchwise at the laboratory temperature in the annulus between a casting bob as shown in Figure 2-20. Both the tube and the bob are made of brass. The casting bob consists of a straight section surmounted by two intersecting conical sections to provide a smooth transition in flow to the casting solution as it moves from the casting bob apex down through the annulus. Figure 2-20(a) shows the casting assembly just prior to casting. The bottom-filled casting solution is sufficiently viscous so that it would flow only very slowly of its own weight through the 0.045 cm annulus between the casting bob and the casting tube. Figure 2-20(b) shows the casting tube falling by gravity during the casting period. The total drop time is 40 to 60 seconds for a 10 ft long film. The casting tube is only roughly centered between the two guides shown, since the casting bob is self centring. The movement of the casting tube downward, past the fixed casting bob, drags the casting solution with it through the annulus, thus forming the membrane tube. After descending 30 cm through air trapped between the casting bob and the surface of ice-water in the immersion tank, the casting tube drops into the latter where gelation takes place. A gelation period of about 1 hour is allowed. During this period the membrane sets to a porous solid gel from which most of the formamide and acetone are leached away. Consequently the membrane tube shrinks and no longer adheres to the casting tube. Owing to this shrinkage, the outside diameter of the tubular membrane is slightly less than the inside diameter of the casting tube. The thickness of the resulting membrane is about 0.020 cm.

Wrapping of tubular membranes. This is illustrated in Figure 2-21. The perforated support tube is of hard-drawn copper having a 2.54 cm outer diameter and a 0.089 cm wall thickness, giving an approximate 2.36 cm inner diameter.

Table 2-47. Effect of cellulose acetate molecular weight on membrane performance

Data of Manjikian et al. (1965)

Cellulose acetate				Desalinised water	
Acetyl content (%)	Viscosity indicator (Eastman)	Molecular weight	Film shrinkage temp. (°)	Flux (gal/day/ft²)	Salt content (ppm)
39.4	30	50,000	79	34	560
(E-394-)	45	51,000	79	32	510
	60	55,000	76.5	33	440
39.8	3	24,000	75.5	30	300
(E-398-)	6	34,000	75.5	30	340
	10	39,000	75.5	29	290

Casting solution composition (wt. %)
- Cellulose acetate 25
- Acetone 45
- Formamide 30

Casting temperature: 23°
Evaporation time: ½ min
Operating pressure: 600 psig
Feed concentration: 5000 ppm NaCl

Thus, for a nominal membrane tube outer diameter of 2.29 cm, there is an annular gap of 0.035 cm between the membrane tube and the support tube. The gap is filled by wrapping the membrane tube with one non-overlapping layer of Whatman filter paper No. 52 followed by one or more layers of Nylon parchment (French Fabrics No. 627T). The filter paper backing is not necessary, and the wrapping may be done entirely by Nylon fabric the purpose of which is only to provide a low resistance path for the desalinised water flowing from the outside of the membrane tube to one of the perforations in the support tube.

Assembly of tubular membranes. Assembly is illustrated in Figure 2-22. A seal is necessary at each end of the assembly. For this purpose, the membrane is flared to conform with the 37° flare at each end of the support tube. It is first necessary to plasticise the end of the membrane tube with a 19 : 1 by volume mixture of n-propyl alcohol: triacetin, after which it is flared by hydraulic expansion of an elastic tube placed within the membrane tube. Plasticising can also be accomplished thermally, but is most conveniently done with a plasticising mixture, as described.

The completion of the seal at each end of the tube is accomplished with standard fittings plus a rubber gasket interposed between the flared part of the membrane tube and the conical metal coupling connecting one tube to another as shown in Figure 2-22. The rubber gasket is a convenient assistance to the completion of the seal between the membrane tube and the conical metal coupling with which it fits. The rubber gasket permits sealing without high torquing, by flowing into any mismatched region. The torque used is about 1.4 mkg.

Table 2-48. Effect of cellulose acetate molecular weight distribution on membrane performance

Data of Manjikian et al. (1965)

Molecular weight distribution	Eastman identification no.*	Film shrinkage temperature (°)	Desalinised water Flux (gal/day/ft^2)	Desalinised water Salt content (ppm)
Narrow	CAE-2196-1	72	32	370
Normal	CAE-2196-2	72	41	480
Wide	CAE-2196-3	72	38	340
Narrow	CAE-2196-1	74.5	27	240
Normal	CAE-2196-2	75.5	24	140
Wide	CAE-2196-3	75.4	27	200
Narrow	CAE-2196-1	77.8	24	120
Normal	CAE-2196-2	77.8	22	100
Wide	CAE-2196-3	77.8	22	100

* Cellulose acetate similar in acetylation and molecular weight to E-398-3.

Casting solution composition (wt. %)
- Cellulose acetate 25
- Acetone 45
- Formamide 30

Casting temperature: 23°
Evaporation time: ½ min.
Operating pressure: 600 psig
Feed concentration: 5000 ppm NaCl

Thermal treatment of tubular membrane assembly. The shrinkage of pores on the inside surface of the membrane is accomplished by the flow of hot water through the tube under a pressure of 7 to 10 psig at the required temperature for about 10 minutes. The above level of pressurisation of hot water is critical for the successful subsequent performance of the membrane.

Performance of tubular membranes. Figures 2-23 and 2-24 illustrate the performance of 10 ft long tubular cellulose acetate membranes for brackish water conversion. The data show that for a desalination factor of 10, the initial product rate is about 38 gal/day/ft^2 under the conditions specified in Figure 2-23. The data given in Figure 2-24 show that the porous structure of the membrane is not uniform in the entire length of the tube. That section of the membrane which first entered the ice water in the immersion tank gives the best desalination. This result is essentially true whether the initially immersed section is positioned at the brine inlet or outlet. However, positioning has some influence as shown by the fact that either terminal immersion section gave better desalination of the brine inlet location. The non-uniformity of the porous structure of the tubular membranes tested above indicates the need for improving the technique of making such membranes.

Table 2-49. Effect of cellulose acetate acetyl content on membrane performance

Data of Manjikian et al. (1965)

			Desalinised water	
Acetyl content (%)	Eastman identification number*	Film shrinkage temp. (°)	Flux (gal/day/ft²)	Salt content (ppm)
38.9	EMP-949-A	69	33	940
37.5	EMP-949-B	69	34	940
37.1	EMP-949-C	69	20	850
38.9	EMP-949-A	72	24	430
37.5	EMP-949-B	72	29	540
37.1	EMP-949-C	72	14	440
38.9	EMP-949-A	75	18	210
37.5	EMP-949-B	75	15	310
37.1	EMP-949-C	75	7.6	250

* Cellulose acetate similar in molecular weight to E-398-3

Casting solution composition (wt.%)

Cellulose acetate	25
Acetone	45
Formamide	30

Casting temperature: 23°
Evaporation time: ½ min
Operating pressure: 600 psig
Feed concentration: 5000 ppm NaCl

Batch 18 Type Membranes. The emergence of CA-NRC-18, or simply, Batch 18 type membranes in the author's laboratory is the third major development. These membranes are cast on flat surfaces by the general method described earlier for Batch 25 films, using the following film casting conditions. Casting solution composition (wt.%):

Cellulose acetate (acetyl content 39.8%, Eastman grade E-398-3)	17
Acetone	68
Magnesium perchlorate	1.5
Water	13.5

Casting temperature: —10°
Evaporation period at —10°: 4 minutes
Gelation in ice-cold water for 1 to 2 hours
Nominal film thickness: 0.004 inch

These films are handled in exactly the same way as Batch 25 and Batch 47 flat films.

Table 2-50. Other casting solution-compositions giving promising membrane performance

Data of Manjikian et al. (1965)

Additive compound	Casting soln. compn. (wt.%)	Cast-ing temp. (°)	Evapn. time (min)	Film shrink-age temp. (°)	Desalinised water	
					Flux (gal/day/ft^2)	Salt content (ppm)
Urea in water	CA 21.0 Acet 62.5 H_2O 13.2 Urea 3.3	–10 –10	6 6	Unshrunk 63.3	9.0 2.6	930 370
Glyoxal in water	CA 21.5 Acet 64.8 H_2O 9.2 Glyoxal 4.5	–10	17	70.3	19	370
H_2O_2 in water	CA 22.0 Acet 65.0 H_2O 10.4 H_2O_2 2.6	–10 –10	10 10	78.5 80.5	20 18	600 150
Dimethyl formamide (DMF)	CA 14.3 Acet 64.3 DMF 21.4	23 23	3½ 3½	Unshrunk 60	19 5.4	550 300
Dimethyl sulphoxide (DMSO)	CA 25.0 Acet 37.5 DMSO 37.5	23	3	62	5	750
Tetrahydro-furfuryl-phosphate (THFP)	CA 20 Acet 60 THFP 20	–10	10	75	18	180
Triethyl-phosphate (TEP)	CA 25 Acet 50 TEP 25	–10	2	79	20	200
Acetic Acid (AA)	CA 22.5 Acet 68.4 AA 9.1	–10	2	77.5	5	750
N-methyl-2-pyrrolidone (NM2P)	CA 22.0 Acet 65.0 NM2P 13.0	–10	16	75	5.7	250
Binary Casting Solutions						
Dimethyl-formamide (DMF)	CA 25 DMF 75	23	8	93	11	140
Acetic Acid (AA)	CA 20 AA 80	23	10	Unshrunk	6.2	650
Triethyl phosphate (TEP)	CA 20 TEP 80	–10 23	1.5 1.0	93 94	8 9	220 240

Operating pressure: 600 psig
Feed concentration: 5000 ppm NaCl
CA denotes Cellulose Acetate
Acet denotes Acetone

Table 2-51. Performance of some ethyl cellulose membranes

Data of Manjikian et al. (1965)

Casting solution compn. (wt. %)					Desalinised water	
Ethyl cellu- lose	P- dioxane	Triethyl phosphate	Evapn. time (min)	Film shrinkage temp. (°)	Flux (gal/day/ft²)	Salt content (ppm)
11	80	9	20	Unshrunk	4.6	600
11	80	9	20	81	0.6	400
13	73	14	20	Unshrunk	4.3	700

Film casting temperature: 23°
Operating pressure: 600 psig
Feed concentration: 5000 ppm NaCl

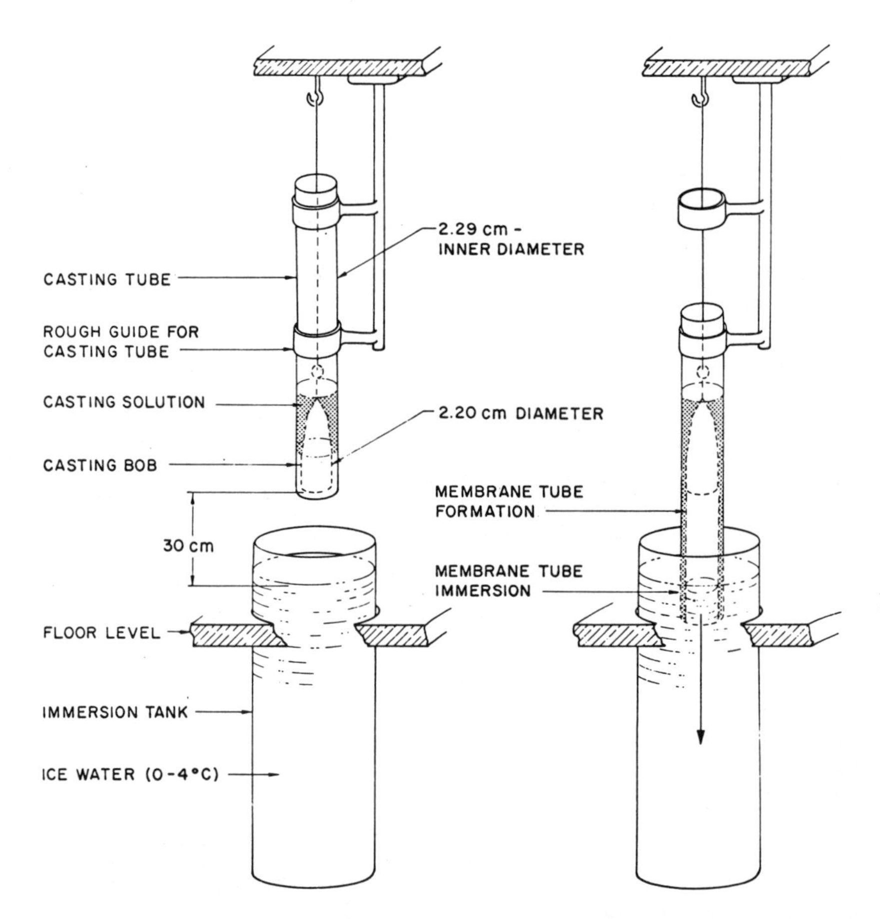

Figure 2-20. Gravity-drop technique for making tubular membranes (Loeb, 1966a).

Table 2-52. Performance of Batch 47 type membranes for sea water conversion

Data of Manjikian (1966)

Film shrinkage temp. (°)	Operating pressure (psig)	Salt in product (ppm)	Product rate (gal/day/ft^2)
Cellulose acetate material E-398-3			
78	600	1950	12.0
78	800	3500	25.0
78	800	2940	23.0
78	800	1820	17.0
78	1000	2310	26.0
78	1000	1190	19.0
78	1200	1080	21.0
81	600	1330	8.0
81	800	1050	15.7
81	800	1330	15.6
81	1000	504	12.8
81	1000	755	17.0
81	1200	532	16.2
81	1200	560	18.0
82	800	730	8.8
82	1000	935	12.0
82	1000	655	11.5
82	1200	595	11.7
83	600	560	6.0
83	600	700	7.0
83	600	840	6.2
83	800	380	9.6
83	800	480	10.9
83	800	540	9.6
83	1200	270	13.0
83	1200	340	18.5
83	1200	390	14.6
84	800	595	8.0
84	1000	504	11.0
84	1200	530	11.0
84	1200	475	11.0
85	600	833	4.3
85	800	525	7.5
85	800	413	8.0
85	1000	336	8.9
85	1000	490	12.9
85	1200	560	10.6
85	1200	343	11.6
85	1200	413	12.6
85	1200	210	10.5
88	600	445	0.17
88	800	280	2.9
88	1000	308	5.1
88	1000	203	4.8
88	1200	196	5.5
Cellulose acetate material E-398-6			
82	800	740	8.8
82	1000	660	11.5
82	1200	590	11.7
84	800	590	8.3
84	1000	500	10.9
84	1200	480	10.9
Cellulose acetate material E-398-10			
82	800	840	9.3
82	1000	630	11.0
82	1200	660	13.5
84	800	650	6.9
84	1000	560	8.3
84	1200	420	9.4
86	800	450	3.4
86	1000	390	5.7
86	1200	350	7.8

Feed: Sea water (single pass)

Table 2-53. Performance of membranes cast from cellulose acetate-acetone-formamide-pyridine solutions

Data of Manjikian (1966)

Casting solution composition (wt.%)				Evaporation time (min)	Shrinkage temp. (°)	Product rate (gal/day/ft^2)	Solute separation (%)
Cellulose acetate	Acetone	Formamide	Pyridine				
25	30	30	15	1/2	78	17.7	97
25	30	30	15	1/2	74	20.8	94
25	30	30	15	1/4	78	19.2	97
25	30	30	15	1/4	74	24.0	93
25	30	30	15	0	74	22.6	95
25	30	25	20	1/2	78	15.1	97
25	30	25	20	1/2	74	18.2	96
25	30	25	20	1/4	78	18.0	97
25	30	25	20	1/4	74	21.8	95
25	30	25	20	0	74	28.0	94
25	30	25	20	1.0	70	28.0	88
25	30	25	20	1.0	74	18.0	93
25	30	25	20	3/4	74	23.4	94

Membrane gelation in ice water
Feed concentration: 5000 ppm NaCl
Operating pressure: 600 psig

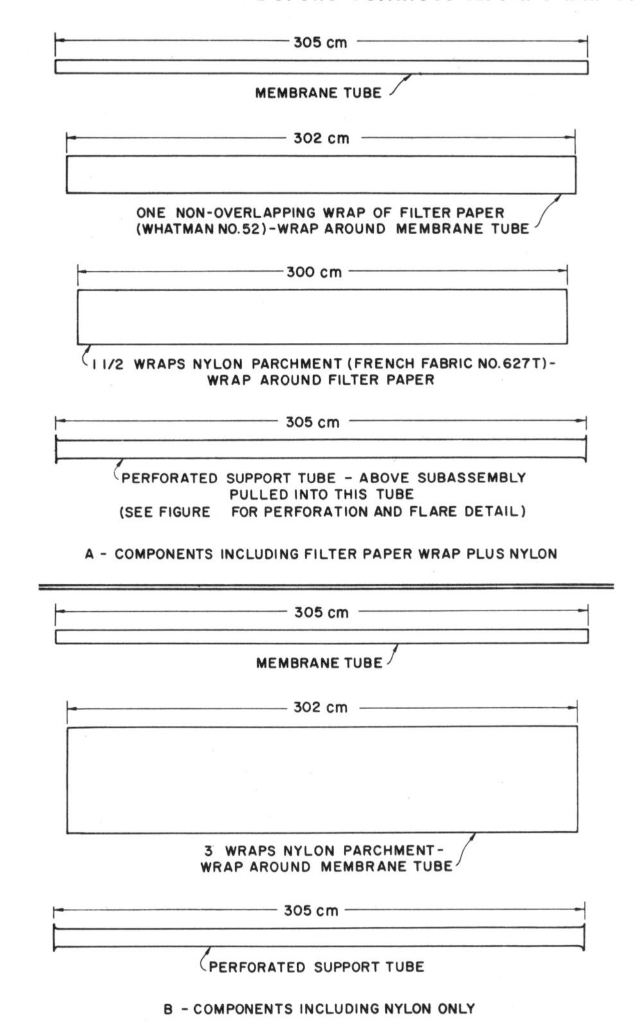

Figure 2-21. Wrapping of tubular membranes (Loeb, 1966a).

In the laboratory, Batch 18 membranes are cast on glass plates and kept under water at all times. Each film is subjected to a temperature and pressure treatment before use in reverse osmosis experiments. The temperature treatment consists in heating the film (held free floating between two glass plates) under water gradually from the laboratory temperature to the required temperature where it is kept for 10 minutes, and subsequently cooling the water rapidly. The pressure treatment consists in pumping distilled water over the surface layer of the film mounted in the reverse osmosis cell at a pressure about 15 per cent higher than the maximum intended operating pressure for at least an hour. The surface layer side of the film (i.e. the surface of the film which was away from that of the glass plate and exposed to the air during casting) is always held facing the feed solution in reverse osmosis experiments.

A typical set of performance data obtained with the above type of films is given in Figures 2-25 to 2-28 and Tables 2-54 and 2-55. The effects of shrinkage temperature, operating pressure and feed concentration on the performance

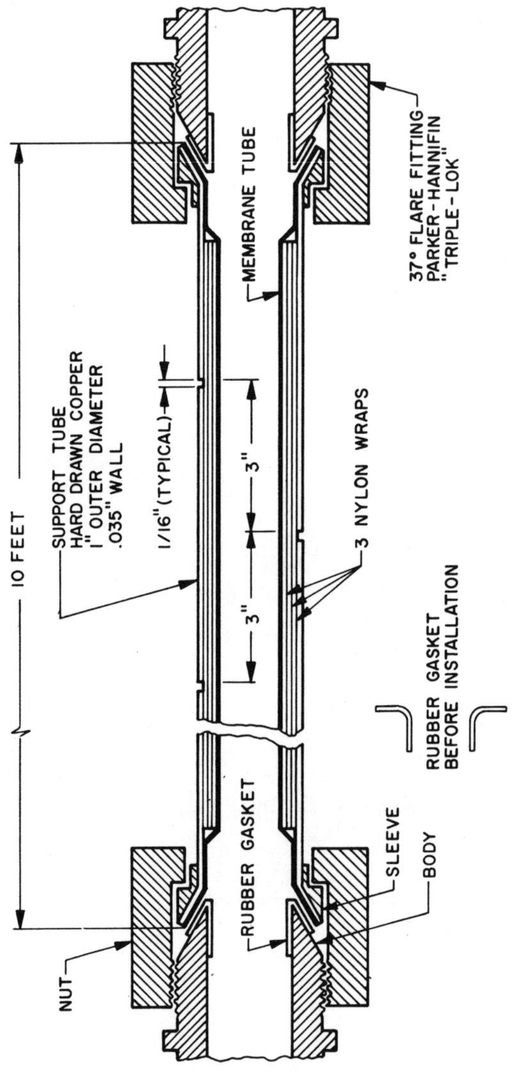

Figure 2-22. Assembly of tubular membranes (Loeb, 1966a).

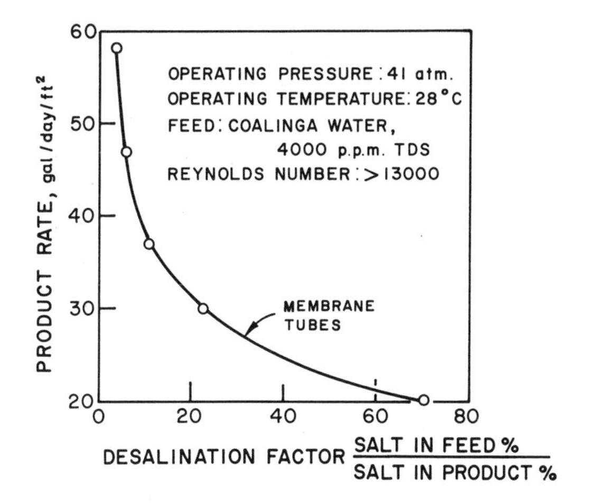

Figure 2-23. Performance of tubular porous cellulose acetate membranes for brackish water conversion. Data of Loeb (1966a).

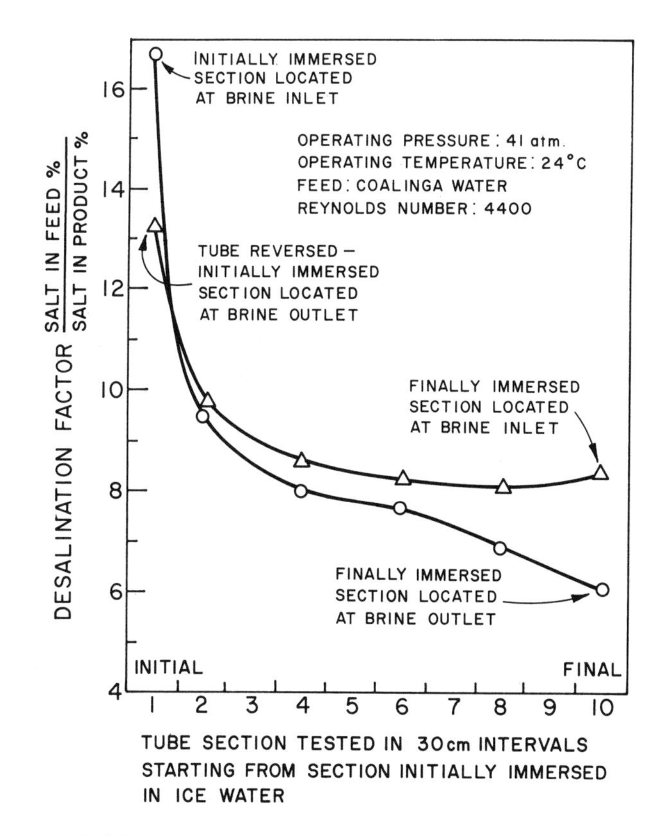

Figure 2-24. Desalination from section-to-section within a given tubular porous cellulose acetate membrane. Data of Loeb (1966a).

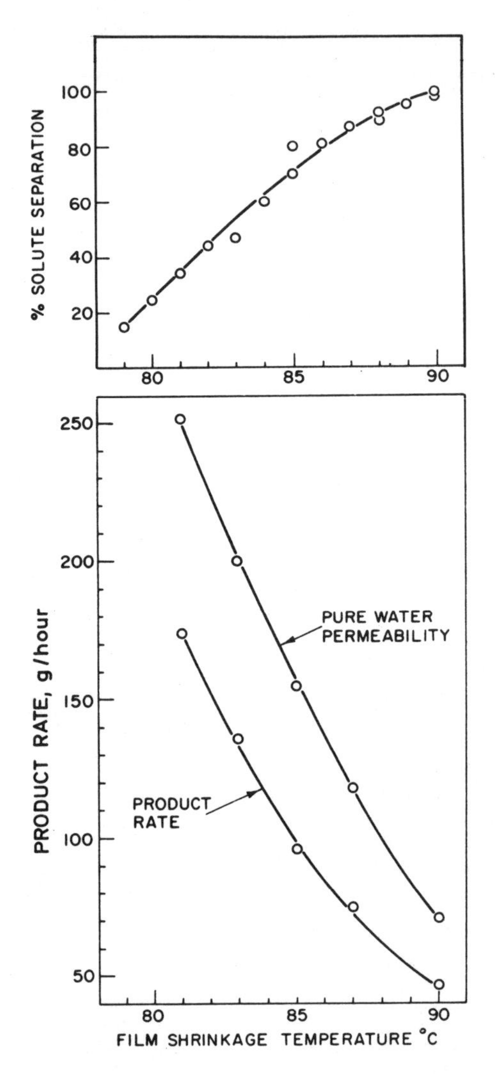

Figure 2-25. Effect of shrinkage temperature on the performance of CA-NRC-18 type films for the system sodium chloride-water. Data of Sourirajan and Govindan (1965).

System: [0.5M NaCl-H_2O]; feed rate: 250 cm^3/min; operating pressure: 1500 psig; film area: 7.6 cm^2.

of the membranes for the system sodium chloride-water are illustrated in Figures 2-25, 2-26, and 2-27 respectively. Table 2-54 shows that the effect of overall film thickness on membrane performance is small, which indicates that the thickness of the surface layer is essentially independent of overall film thickness under otherwise identical film casting conditions. Table 2-55 and Figure 2-28 illustrate the performance of the above film for saline water and sea water conversions respectively. Table 2-55 gives results of short-run experiments while Figure 2-28 gives results obtained in continuous operation of the reverse osmosis experiment for a period of 7 days. The performance data are as good as, or better than, any reported with other types of porous

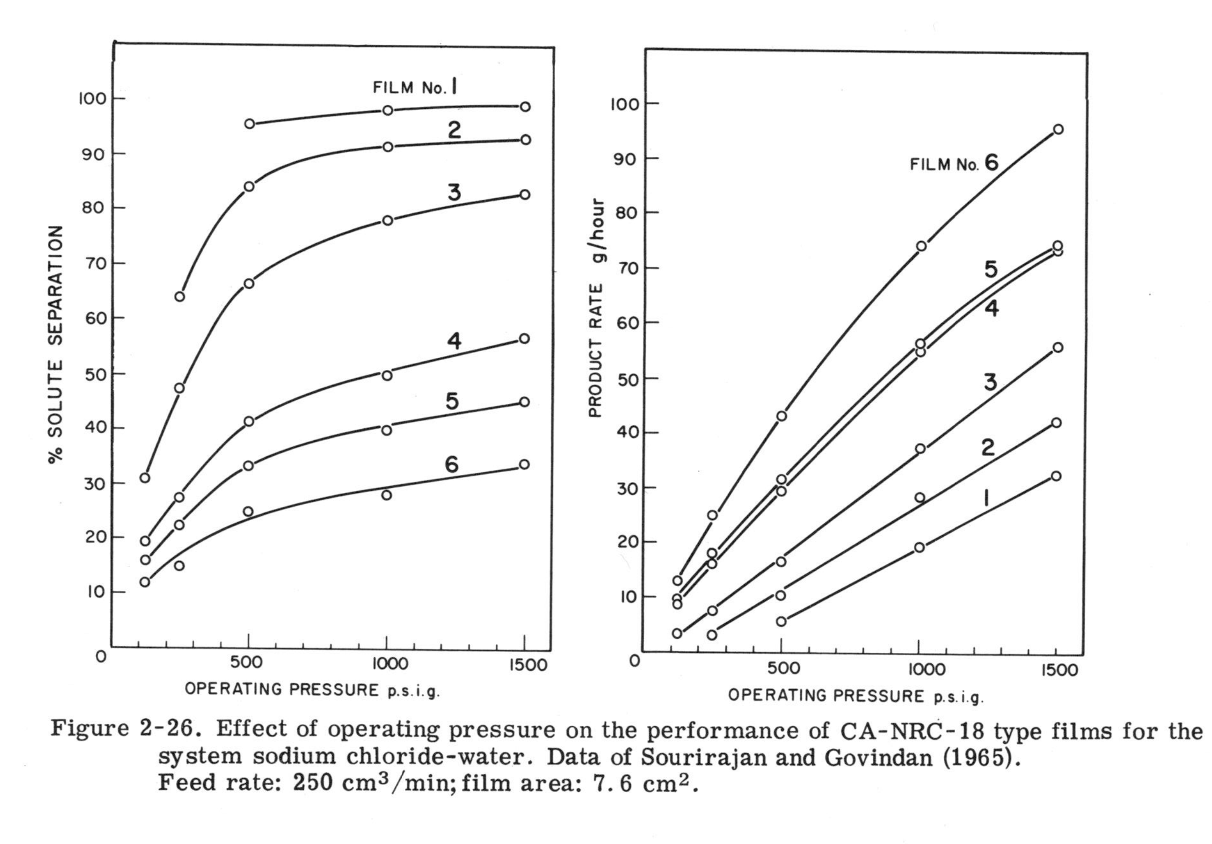

Figure 2-26. Effect of operating pressure on the performance of CA-NRC-18 type films for the system sodium chloride-water. Data of Sourirajan and Govindan (1965). Feed rate: 250 cm^3/min; film area: 7.6 cm^2.

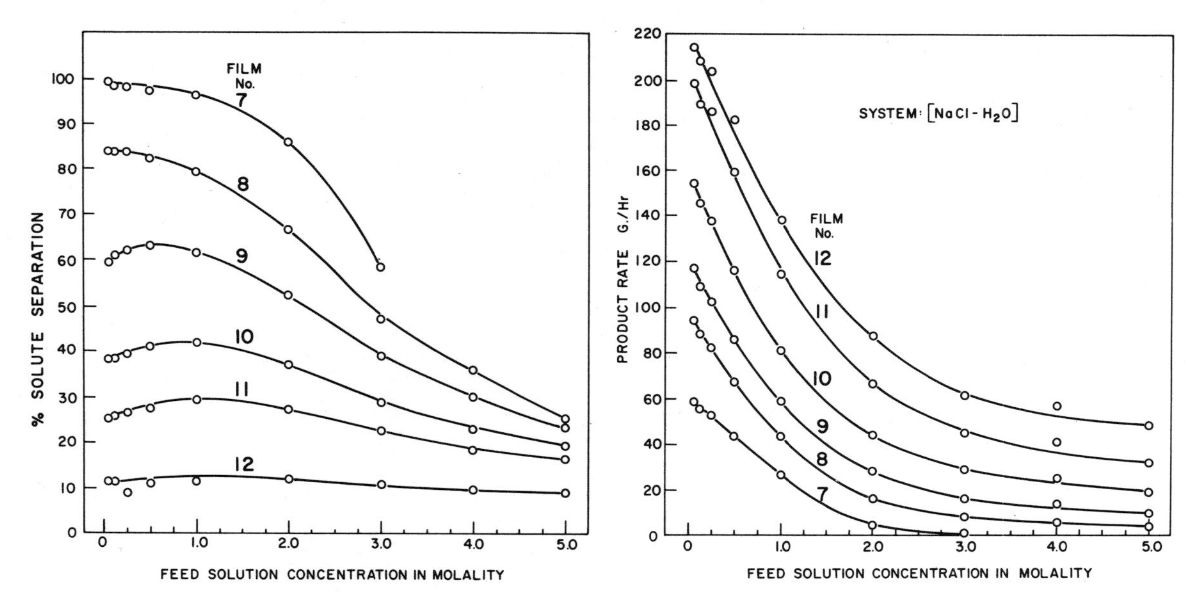

Figure 2-27. Effect of feed concentration on the performance of CA-NRC-18 type films for the system sodium chloride-water. Data of Sourirajan and Govindan (1965).

Feed rate: 250 cm^3/min; operating pressure: 1500 psig; film area: 7.6 cm^2.

Table 2-54. Effect of overall film thickness on the performance of CA-NRC-18 type membranes

Data of Sourirajan and Govindan (1965)

Film Thickness (in.)		[PWP]	Product rate	Solute
Before expt.	After expt.	g/hr	g/hr	separation (%)
0.0023	0.0020	76.8	47.4	98
0.0048	0.0030	71.1	42.8	99
0.0074	0.0038	67.6	42.4	96
0.0109	0.0049	72.0	44.8	92

PWP: pure water permeability

System: [NaCl-H_2O]
Feed molality: 0.5M
Feed rate: 300 cm^3/min
Operating pressure: 1500 psig
Film shrinkage temperature: 90°
Film area: 7.6 cm^2

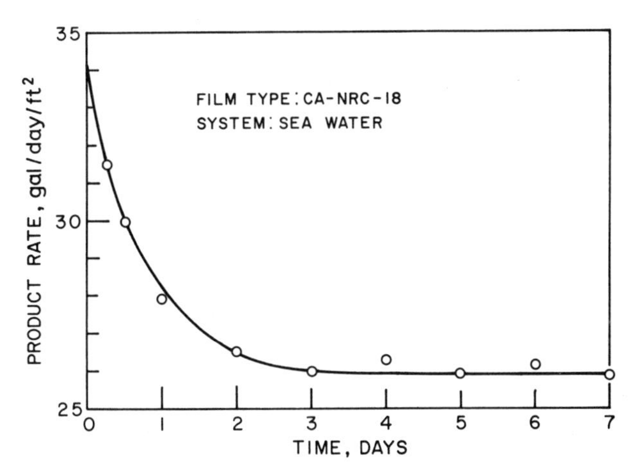

Figure 2-28. Performance of CA-NRC-18 type films for sea water conversion.

Feed rate: 400 cm^3/min; operating pressure: 1500 psig; product concentration: <300 ppm equivalent NaCl.

cellulose acetate membranes. In addition, as will be shown in the next chapter, generally Batch 18 films perform predictably in long-time continuous test runs in reverse osmosis. Consequently extensive work has been done with this type of membranes.

12. Structure of Porous Cellulose Acetate Membranes

Kopeček and Sourirajan (1969a) considered the problem of improving the performance of porous cellulose acetate membranes (i.e. increasing their product

Table 2-55. Performance of CA-NRC-18 type films for saline water conversion

Data of Sourirajan and Govindan (1965)

Film pretreatment							
Shrinkage temp. (°)	Pressure (psig)	Solute	Feed rate (cm^3/min)	Operating pressure (psig)	Solute concn. in feed (wt.%)	Solute concn. in product (ppm)	Product rate (gal/day/ft^2)
87	700	NaCl	300	600	0.25	400	52.4
88	700	NaCl	300	600	0.50	300	34.0
87	700	Na_2SO_4	300	600	0.25	150	53.3
87	700	Na_2SO_4	300	600	0.50	300	50.7
90	1700	NaCl	250	1500	1.44	<100	38.5
90	1700	NaCl	250	1500	2.84	280	31.3
90	1700	NaCl	250	1500	4.20	420	24.1
90	1700	NaCl	250	1500	5.53	550	18.2

rates without decreasing solute separation) for saline water conversion and similar applications.

The preferential sorption-capillary flow mechanism offers several approaches to the problem: (i) the effective thickness of the membrane (especially the thickness of the microporous surface layer controlling solute separation) may be reduced; (ii) the absolute number of pores of appropriate size on the microporous surface layer may be increased; (iii) the rigidity of the overall porous structure of the membrane may be altered so that membrane compaction is less during continuous operation under pressure; and/or (iv) the number, pore size and pore size distribution on the microporous surface layer of a given membrane may be altered such that increased permeability results without decreasing solute separation. The above approaches are not mutually exclusive, but any proposed technique for membrane preparation and/or membrane treatment may emphasise one or more of the above approaches.

The approaches (i) and (ii) are primarily concerned with the detailed steps involved in the process of making asymmetric Loeb-Sourirajan type membranes. With reference to Batch 18 type films, it is probable that the proportion (area of the pores)/(total area) on the surface layer side of the membrane is still very small; with further advances in our knowledge on the mechanism of pore formation in the process of casting membranes, it is reasonable to expect that cellulose acetate membranes developed in the future may give a 10-fold increase in product rate for a given level of solute separation under otherwise identical experimental conditions. The attempts of Merten et al. (1967) to reduce membrane compaction by the inclusion of inorganic or organic filler materials in the membrane casting solutions form part of approach (iii). The work of Kopeček and Sourirajan (1969a) summarised below is part of approach (iv).

Microporous Nature of Surface Layer in Reverse Osmosis Membrane. Some workers consider that the surface layer (the so called 'active' layer) in the reverse osmosis membrane must be nonporous in order to give high levels of solute separation (Lonsdale et al., 1965; Banks and Sharples, 1966c). On the other hand, the preferential sorption-capillary flow mechanism calls for a porous surface structure, and postulates a critical pore diameter on the membrane surface for maximum solute separation for any given membrane material-solution system. With reference to the Loeb-Sourirajan type porous cellulose acetate membranes, the concepts that the surface layer is microporous and heterogeneous, and that the transport of both the preferentially sorbed water and the bulk solution is through the capillary pores on the membrane surface at all levels of solute separation, are fundamentals arising from the mechanism. The practical consequence of the microporous nature of the surface layer in the reverse osmosis membrane constitutes the basis of the work of Kopeček and Sourirajan (1969a). This basis is also supported by the theoretical work of Glueckauf (1965) and Meares (1966), and the following three important experimental observations with respect to the Loeb-Sourirajan type porous cellulose acetate membranes.

(a) Shrinkage temperature v. solute separation and product rate. Using a 0.5M [NaCl-H_2O] feed solution, at an operating pressure of 1500 psig, a membrane without any prior thermal treatment gives very high product rate (>300 gal/day/ft^2) and very little solute separation. These data indicate that the surface of a typical untreated membrane is quite porous.

Solute separation can be progressively increased by heating the membrane under water for short periods of time, as stated earlier. The effect of temperature treatment then is to shrink the size of the pores on the membrane surface. When the shrinkage temperature is increased, the size of the pores on the membrane surface becomes smaller; consequently solute separation increases, and product rate decreases. This is illustrated in Figure 2-29 which gives the results obtained with several films shrunk at different temperatures.

On progressively increasing the shrinkage temperature, solute separation can be increased to levels >98 or 99 per cent, and still good product rates can be obtained. There is nothing to indicate however that at this stage the pores on the membrane surface have been closed, and the membrane surface has become

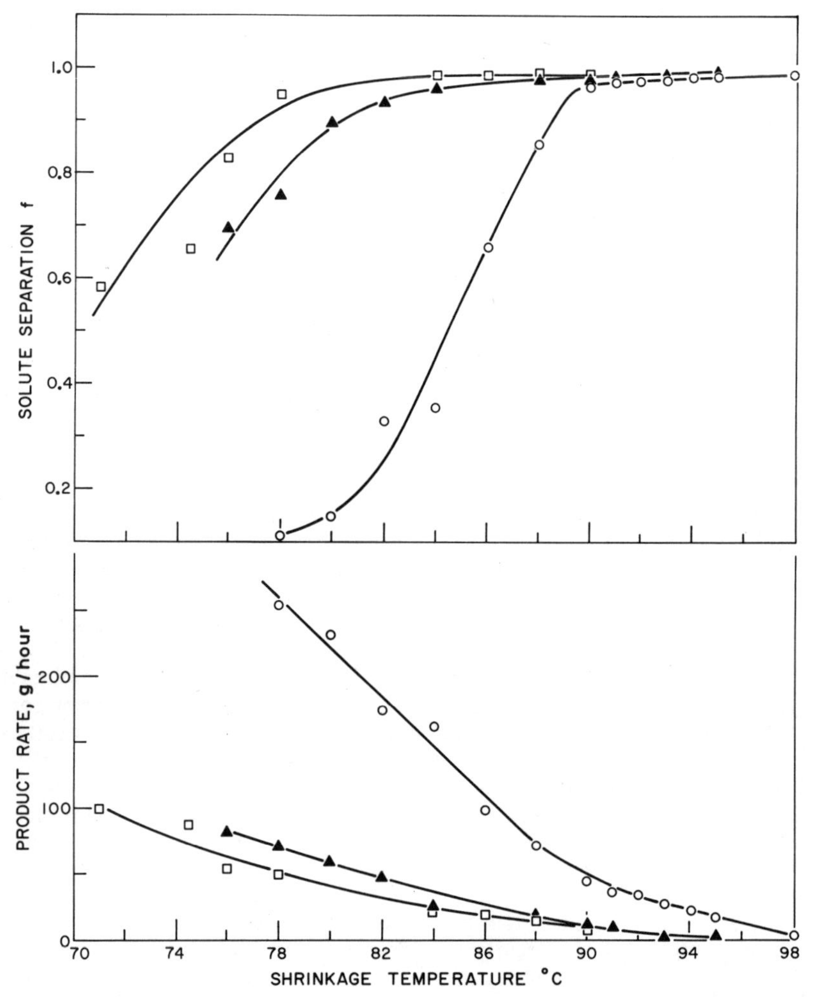

Figure 2-29. Effect of shrinkage temperature on the performance of different porous cellulose acetate membranes. Data of Kopeček and Sourirajan (1969a).

Film type ○ CA-NRC-18
▲ CA-NRC-47
□ CA-NRC-25

system: sodium chloride-water; feed concentration: 0.5M; feed rate: 250 cm^3/min; operating pressure: 1500 psig; film area: 7.6 cm^2.

essentially nonporous; on the contrary, the following data indicate that the membrane surface still remains microporous.

On increasing the shrinkage temperature still further after reaching the stage of 98 to ~100 per cent solute separation, the product rate does not remain constant (as one would expect if the membrane surface is nonporous), but it is reduced still further, indicating that the sizes of the pores on the membrane surface are getting still further reduced. This thermal pretreatment procedure can be carried far enough to close practically all the pores on the membrane surface and reduce the product rate to negligible levels. This is illustrated in Figure 2-30 with respect to the CA-NRC-18 type films.

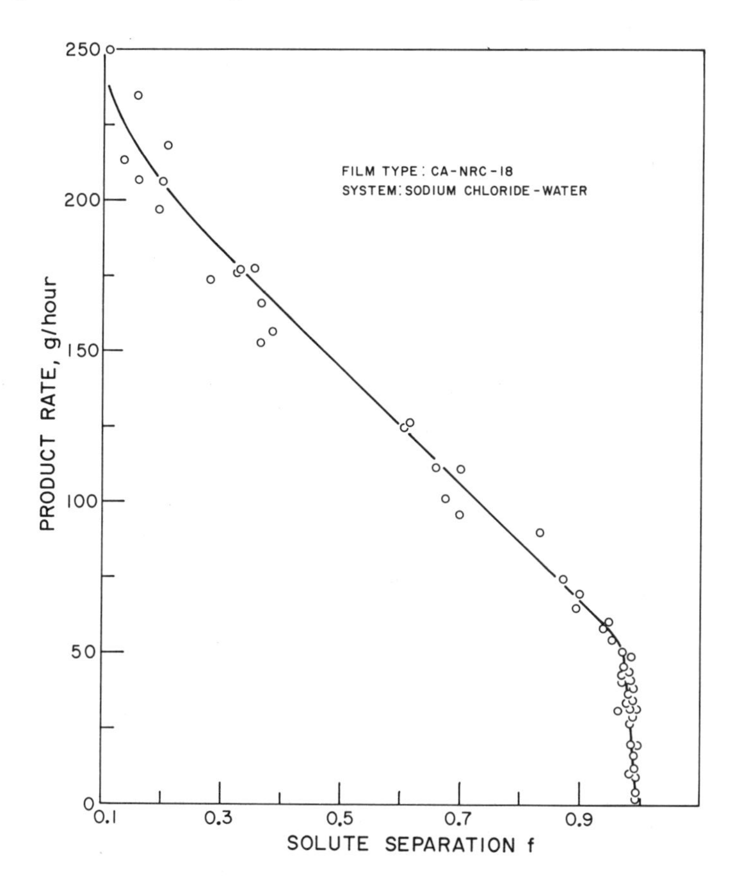

Figure 2-30. General performance of Batch 18 type membranes. Data of Kopeček and Sourirajan (1969a).

Feed concentration: 0.5M; feed rate: 250 cm^3/min; operating pressure: 1500 psig; film area: 7.6 cm^2.

The above shrinkage temperature-product rate profile obtained under conditions of high solute separation (near 100 per cent) cannot be attributed to any steep increase in effective film thickness brought about by high shrinkage-temperature, for at least two reasons. First, the shrinkage-temperature-product rate profile is different for films cast under different conditions; whatever be the profile, the behaviour of the films is similar. Secondly, any steep increase in thickness of the surface layer at any temperature should also result in an abrupt change in the mechanical properties of the film at that temperature; Keilin et al. (1965) found no abrupt changes in the mechanical properties of the porous cellulose acetate membranes in a wide range of temperature.

These observations indicate that even at the near-100 per cent level of solute separation, the Loeb-Sourirajan type porous cellulose acetate membranes have microporous surface structures.

(b) Effect of temperature on pure water permeability constant, A. The pure water permeability data (expressed as A in g. mole H_2O/cm^2 sec. atm) for a set of four different CA-NRC-18 type films were obtained in the operating temperature range 6 to 36°. The films were shrunk at different temperatures and hence capable of giving different levels of solute separation. With a 0.5M [NaCl-H_2O] feed solution at 1500 psig, the films 23, 24, 25, and 26 gave solute separations of 28.2, 49.0, 68.9, and 97.1 per cent respectively; the pure water permeability constant A increased with increase in operating temperature, and A multiplied by the viscosity of water at the operating temperature remained constant for all the films tested, as illustrated in Figure 3-61 in Chapter 3. These results indicate that fluid flow through each of the above films is essentially viscous, and the mechanism of fluid transport is the same for all the films tested, and hence all the above films have microporous surface structures, and they differ essentially in the magnitude of the average size of their surface pores.

(c) Pure water permeability constant, A, v. solute transport parameter, ($D_{AM}/K\delta$). The values of A (which expresses pure water permeability) and ($D_{AM}/K\delta$) for NaCl (which is analogous to a mass transfer coefficient for solute transport) for a number of CA-NRC-18 type films were determined by the Kimura-Sourirajan analysis discussed in Chapter 3. An increase in the value of A corresponds to an increase in the average pore size on the membrane surface. The films chosen covered a wide range of solute separation; for example, with respect to a 0.5M feed solution, films G1 and 12 gave 98.9 and 18.7 per cent solute separation respectively; the other films gave intermediate levels of solute separation. A log-log plot of A v. ($D_{AM}/K\delta$) is shown in Figure 2-31. The absence of any abrupt discontinuity in the A v. ($D_{AM}/K\delta$) correlation shows that the mechanism of solute and solvent transport in film G1 is no different from that in film 12, which again means that all the above films had microporous surface structures, and they differed only in the magnitude of the average size of their surface pores.

Effect of Temperature and Pressure Pretreatment on Porous Structure. The initial porous structure of the film before the temperature and pressure treatment, called here 'the primary gel structure', is a three-dimensional network with extensive physical crosslinks and macropores containing both swelling-water and nonswelling capillary-water in both the surface layer and the interior spongy mass of the film.

During the temperature (shrinkage) treatment, the polymer segments get closer together, thereby increasing physical crosslinks and decreasing the capillary spaces. Consequently both the swelling-water and the capillary-water in the entire film material are reduced, giving rise to a more dense surface as well as interior porous structure, called here 'the secondary gel structure'.

The pressure treatment compacts the entire porous structure bringing the polymer segments still closer together, further increasing the physical cross-

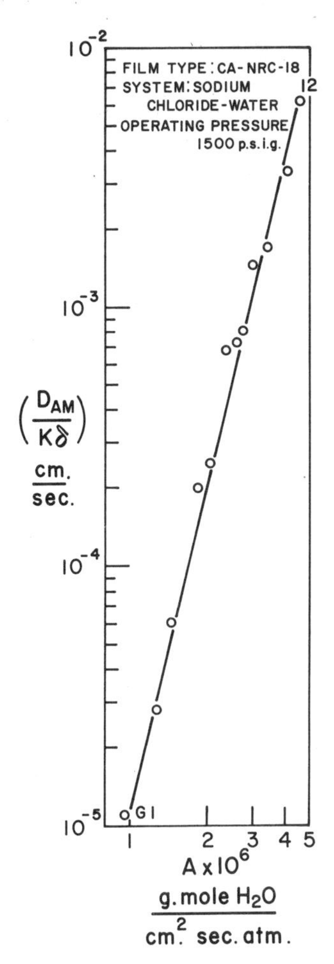

Figure 2-31. Pure water permeability constant v. solute transport parameter for sodium chloride. Data of Kopeček and Sourirajan (1969a).

links and decreasing the capillary spaces, resulting in a still more dense surface as well as interior structure, called here 'the tertiary gel structure'.

Performance of Membranes During Continuous Operation Under Pressure. Under the reverse osmosis experimental conditions, the product rate drops during continuous operation under pressure. Most of this drop occurs in the first 24 hours. Such film behaviour is illustrated in Table 2-56 which gives the data for 19 films covering a wide range of solute separation for the system sodium chloride-water at operating pressures of 600 and 1500 psig. The results show that in all the cases tested, the product rate after 24 or 72 hours of continuous operation is about 80 to 90 percent of the initial product rate while operating at 600 psig, and 70 to 80 per cent while operating 1500 psig, and the level of solute separation remains essentially the same or

Table 2-56. Performance of porous cellulose acetate membranes during continuous operation under pressure

Data of Kopeček and Sourirajan (1969a)

Film no.	Start $[PR]_1$ (g/hr)	Start $f \times 10^2$	After 24 hours $[PR]_2$ (g/hr)	After 24 hours $f \times 10^2$	$\frac{[PR]_2}{[PR]_1} \times 10^2$	After 72 hours $[PR]_3$ (g/hr)	After 72 hours $f \times 10^2$	$\frac{[PR]_3}{[PR]_1} \times 10^2$
Feed concentration: 0.5 wt.% NaCl; Operating pressure: 600 psig								
123JK	16.2	98.8	14.7	99.1	91.7	—	—	—
122JK	21.4	98.5	19.8	99.1	92.5	—	—	—
124JK	32.1	97.9	29.2	97.9	90.9	—	—	—
126JK	33.4	96.7	28.4	97.2	85.1	—	—	—
125JK	45.7	93.5	39.4	94.8	86.3	—	—	—
127JK	57.5	84.1	48.7	87.5	84.7	—	—	—
1JK	64.4	77.3	55.0	78.5	85.4	52.0	78.2	80.7
2JK	85.1	67.4	72.8	68.6	85.6	70.1	68.5	82.3
3JK	150.0	38.4	126.5	42.1	84.4	123.5	41.9	82.3
4JK	210.7	21.6	176.0	24.3	83.4	168.7	24.2	80.1
5JK	242.2	13.8	198.2	17.0	81.8	190.3	16.9	78.7
6JK	250.1	13.2	207.1	15.8	82.4	198.1	15.7	79.2
Feed concentration: 0.5M NaCl; Operating pressure: 1500 psig								
140JK	48.7	96.7	38.7	96.3	79.5	—	—	—
7JK	73.1	87.0	56.1	86.5	76.8	52.1	86.1	71.3
8JK	98.3	67.6	76.3	68.4	77.6	69.2	60.0	70.4
9JK	161.6	36.2	122.3	37.9	75.7	114.4	26.9	70.8
10JK	170.3	34.1	129.9	34.0	76.3	120.9	23.5	71.0
11JK	229.3	15.7	175.0	17.5	76.3	161.7	9.1	70.5
12JK	250.5	11.4	191.1	12.6	76.3	178.4	6.0	71.2

Film type: CA-NRC-18
System: [NaCl-H_2O]
Feed rate: 250 cm^3/min
Film area: 7.6 cm^2
[PR]: Product rate

slightly better at least during the first 24 hours. This result is considered to be due to normal membrane compaction under pressure during which the porous structure of the dense surface layer remains essentially intact, but the spongy bulk mass of the film material underneath the surface layer becomes progressively more dense and hence offers more resistance to fluid flow.

After 72 hours of continuous operation at 1500 psig, the changes in solute separation obtained for the last five films shown in Table 2-56 are more than those predictable (by the method discussed in Chapter 3) from compaction results (decrease in product rate) alone. Consequently, the data indicate that the surface pore structures of the above five films must have changed after the first 24 hours of continuous service. While such changes in surface pore structure need not, and do not, always occur during continuous operation under pressure, they may occur occasionally with some films (especially the less dense ones shrunk at lower temperatures) as shown in Table 2-56. On the basis of extensive experimental results, some of which are reported below, it can be concluded that most of the Batch 18 type films shrunk at temperatures higher than 85° exhibit stable surface structure in long-time experiments.

A Method For Improving Membrane Performance. During the temperature treatment, those pores on the surface layer of the membrane which were initially too big become smaller—which is desirable—and those which were initially small enough become still smaller, which is not desirable from the point of view of overall membrane performance for a given application. If the latter pores could be opened up again, after the temperature treatment, without widening the former ones too much, the product rate through the film could be increased without decreasing solute separation under a given set of operating conditions.

One possible technique for improving membrane performance based on the above approach is to force a fluid under pressure through the macropores on the reverse side of the film to enlarge the pores on the membrane surface advantageously.

Reverse osmosis experiments with the reverse side of the film in contact with the feed solution have been carried out first by Loeb and Sourirajan (1960a, 1961), and then by other workers (Banks and Sharples, 1966a; Keilin, 1963). The results are well known; under experimental conditions, the product rate increases but solute separation decreases very much. For example, in a short-time run, with a NaCl (0.5 wt.%)-H_2O feed solution at 600 psig operating pressure, a particular Batch 18-type film gave 90.2 per cent solute separation and 47.4 g/hr of product rate when the surface layer side of the film was in contact with the feed solution; when the experiment was repeated with the reverse side of the film in contact with the feed solution, the product rate increased to 92.7 g/hr and solute separation decreased to 3.5 per cent. Similar results were obtained at the operating pressure of 1500 psig using 0.5M NaCl-H_2O feed solution. For example, a particular membrane which gave 98.4 per cent solute separation and 26.3 g/hr product rate in a normal experiment gave 0.6 per cent solute separation and 330.5 g/hr product rate when the experiment was carried out with the reverse side of the film facing feed solution. These results are typical; they simply show that during the reverse osmosis operation with the reverse side of the film in contact with the feed solution at 600 psig or more, the pores on the membrane surface layer open too much.

Therefore, even though the idea of trying to enlarge the surface pores by forcing a fluid under pressure from the reverse side of the membrane seems workable, it is obvious that the level of pressure applied for the purpose should not be too great. Consequently any such workable back pressure treatment in many cases may have to be a membrane pretreatment technique independent of the reverse osmosis experimental operation, especially when the latter is at high pressures. Further, if any desirable changes in the structure of the surface layer can be brought about by a back pressure treatment, it is also desirable to make such changes permanent in the overall structure of the film material. The above considerations led Kopeček and Sourirajan (1969a) to the development of a general back pressure pretreatment method for improving the performance of Loeb-Sourirajan type porous cellulose acetate membranes for saline water conversion and other applications.

The method consists in pumping pure water past the reverse side of the film under just sufficient and necessary pressure for a sufficiently prolonged period of time; after such pretreatment, the membrane is used in the reverse osmosis experiment in the usual manner.

13. Performance of Back Pressure Treated Membranes

Reverse Osmosis Experiments at 600 psig Using 0.5 Wt.% NaCl in Aqueous Feed Solutions

Experimental procedure. All membranes discussed in this part of the investigation were initially shrunk at different temperatures, and subsequently pressure-treated in the usual manner at 700 psig for 1 hour. A set of at least two short-run reverse osmosis experiments was then carried out, after which they were subjected to a back pressure treatment. Unless otherwise stated, the back pressure treatment consisted of pumping pure water at the laboratory temperature past the reverse side of the film at 400 psig for 85 hours. The back pressure-treated membranes were then used in the reverse osmosis experiments. All reverse osmosis experiments were carried out in the normal manner, i.e. with the surface layer side of the membrane facing the feed solution.

Improvement in the performance of back pressure-treated membranes. In Figure 2-32, Line I gives the product rate v. solute separation data for a set of five films, shrunk at different temperatures (covering a solute separation range of 84 to 98 per cent), without back pressure treatment, and Line II gives the corresponding data for the same films with back pressure treatment.

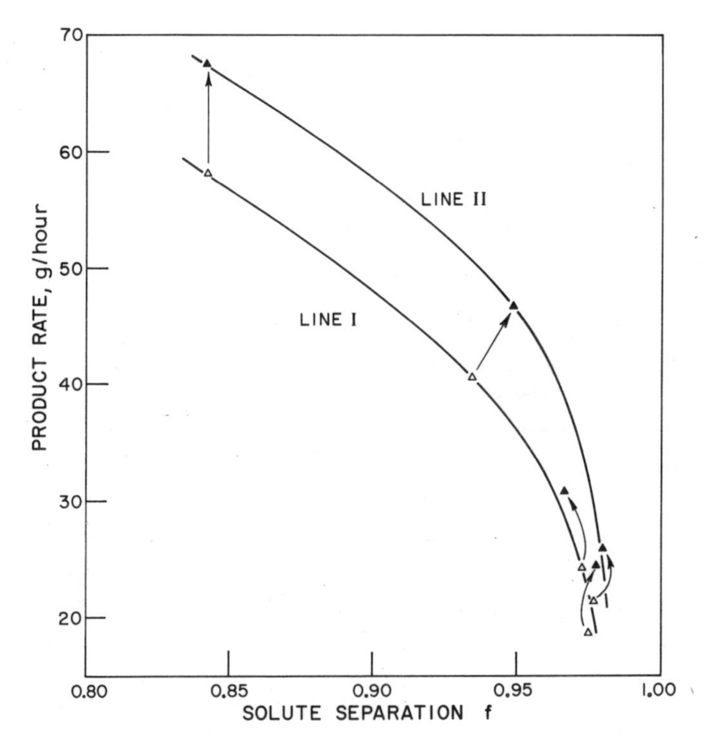

Figure 2-32. Performance of back pressure treated membranes. Data of Kopeček and Sourirajan (1969a). Film type: CA-NRC-18; System: sodium chloride-water; feed concentration: 0.5 wt.%; feed rate: 250 cm^3/min operating pressure: 600 psig; film area: 7.6 cm^2.

Line II is considerably above Line I, and the arrows indicate the change in performance of the individual films tested. The results show that as a result of back pressure treatment, the product rate increased by 15 to 30 per cent, and solute separation either remained the same or became better in all cases, except one in which solute separation decreased by 0.6 per cent. These data illustrate the essential validity of the approach used in this work and the practical success of the back pressure treatment technique for improving membrane performance.

The data given in Table 2-57 illustrate both the improvement in the performance of the back pressure treated films, and the compaction effect on the films during continuous operation under 600 psig for 72 hours in reverse osmosis experiments. The results show that the structure of the surface layer in the back pressure-treated membranes remains stable under these experimental conditions; further, the reduction in product rate obtained due to membrane compaction during extended operation is no more than that obtained with films which were not subjected to back pressure treatment. This means that the 15 to 30 per cent increase in product rate obtained by back pressure treatment is a net gain in the productivity of the CA-NRC-18 type films for saline water conversion and similar applications.

Figure 2-33 illustrates the effect of the duration of back pressure treatment on the change in the performance of the back pressure treated membranes. Three different sets of films were used in these experiments. Each set consisted of six films giving solute separations in the range 85 to 99 per cent. Each set of these films was subjected to a back pressure treatment at 400 psig for 50, 85 or 218 hours, and the changes in product rate and solute separation were then determined in the regular reverse osmosis experiments at 600 psig.

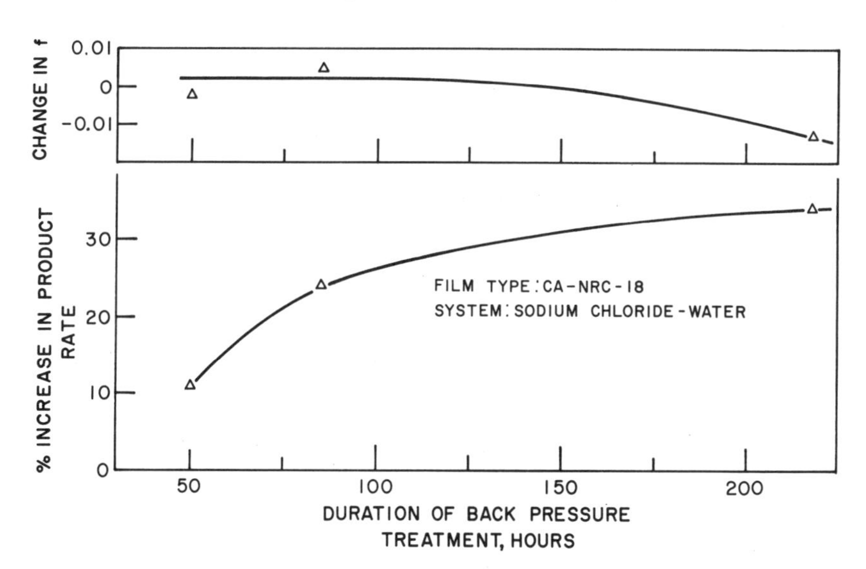

Figure 2-33. Effect of duration of back pressure treatment on membrane performance. Data of Kopeček and Sourirajan (1969a).

Feed concentration: 0.5 wt.%; feed rate: 250 cm^3/min; operating presure: 600 psig; back pressure: 400 psig.

In all cases, the product rate increased significantly, and solute separation changed less than 1 per cent. Figure 2-23 gives the average change obtained in the six films used in each set. The data showed that with a duration of 145 hours of back pressure treatment at 400 psig, the average increase in product rate was 30 per cent with no change in solute separation in the reverse osmosis experiments at 600 psig.

With 85 hours of back pressure treatment at 400 psig, the average increase in product rate was 23 per cent with a slight increase in solute separation. To be on the safer side, the duration of 85 hours was chosen for back pressure treatment in most of the experiments reported here.

Table 2-57. Performance of porous cellulose acetate membranes before and after back pressure treatment

Data of Kopeček and Sourirajan (1969a)

Film no.	Before treatment		After back pressure treatment at 400 psig for 85 hours								
	Start		Start		$\left[\frac{[PR]_2}{[PR]_1} - 1\right] \times 10^2$*	After 24 hours		$\frac{[PR]_3}{[PR]_2} \times 10^2$	After 72 hours		$\frac{[PR]_4}{[PR]_2} \times 10^2$
	$[PR]_1$ (g/hr)	$f \times 10^2$	$[PR]_2$ (g/hr)	$f \times 10^2$		$[PR]_3$ (g/hr)	$f \times 10^2$		$[PR]_4$ (g/hr)	$f \times 10^2$	
33JK	18.7	97.5	24.5	97.8	31.0	21.1	98.0	86.1	20.4	98.1	83.2
37JK	21.4	97.7	25.9	98.0	21.0	22.8	98.1	88.0	22.1	98.1	85.3
45JK	24.3	97.3	30.5	96.6	25.6	27.3	96.9	89.5	26.6	96.9	87.3
47JK	40.7	93.5	46.9	94.9	15.3	41.5	95.1	88.5	41.3	94.7	88.2
49JK**	58.3	84.3	67.5	84.3	15.9	60.1	86.3	89.0	58.8	85.4	87.2

* Per cent increase in product rate by back pressure treatment

** With 0.25% NaCl in feed solution Film 49 after back pressure treatment gave 72.1 g/hr product rate and 84% solute separation.

Film type: CA-NRC-18
System: Sodium chloride-water
Feed concentration: 0.5 wt.% NaCl
Feed rate: 250 cm^3/min
Operating pressure: 600 psig
Film area: 7.6 cm^2
[PR]: Product rate

Figure 2-34 illustrates the effect of the pressure used for back pressure treatment. Here again the results shown are the average of six films used in each of the three different sets of experiments. The results show that the gain in product rate increased with increase in pressure used for back pressure treatment, but pressures higher than 400 psig tend to decrease solute separation in the reverse osmosis experiments. Consequently a back pressure treatment at 400 psig for 85 hours seems to give the maximum increase (~20-25 per cent) in product rate with no decrease in solute separation in the reverse osmosis experiments at 600 psig. All these results are with respect to CA-NRC-18 type films capable of giving more than 80 per cent solute separation with respect to the system 0.5 wt.% $NaCl-H_2O$ at 600 psig.

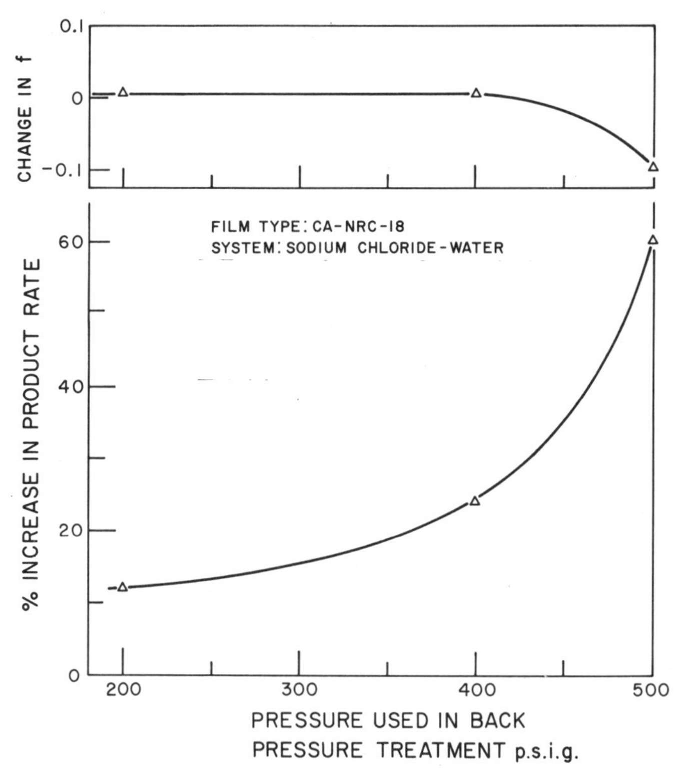

Figure 2-34. Effect of pressure used in back pressure treatment on membrane performance. Data of Kopeček and Sourirajan (1969a).

Feed concentration: 0.5 wt.%; feed rate: 250 cm^3/min; operating pressure: 600 psig; time of back pressure treatment: 85 hr.

A set of experiments were carried out using water at 40 ± 1° for back pressure treatment at 400 psig for 85 hours. The films subjected to this treatment also gave 19 to 30 per cent increase in product rate with no decrease in solute separation in the reverse osmosis experiments at 600 psig. Since similar results were obtained by using water at the laboratory temperature for back pressure treatment, there seems no particular advantage in using water at higher temperatures for such a purpose.

Figure 2-35 illustrates the effect of repeated back pressure treatment at 400 psig for 85 hours on films subjected to continuous reverse osmosis experi-

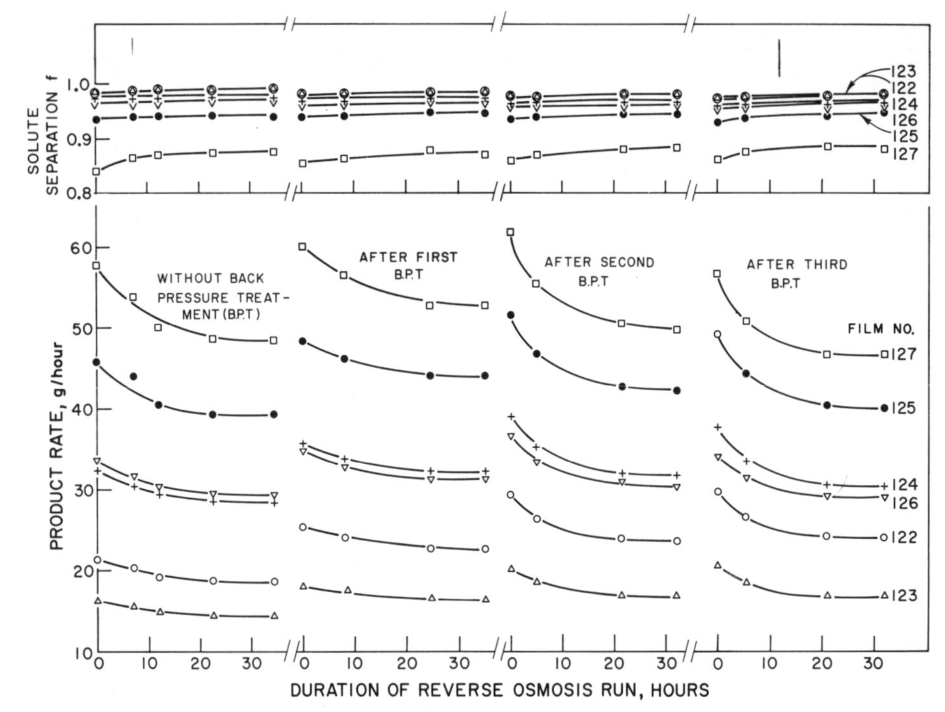

Figure 2-35. Effect of repeated back pressure treatment on membrane performance. Data of Kopeček and Sourirajan (1969a).

Film type: CA-NRC-18; system: sodium chloride-water; feed concentration: 0.5 wt.%; feed rate: 250 cm^3/min; operating pressure: 600 psig; film area: 7.6 cm^2.

ments at 600 psig for over 30 hours. In this set of experiments, a long-time reverse osmosis run was conducted initially with the film in the usual manner. The experiment was then stopped, and the film was subjected to the first back pressure treatment. A long-time reverse osmosis run was then conducted with the above film for another 30 hours. The experiment was stopped again, and the film was subjected to the second back pressure treatment which was followed by another long time reverse osmosis experiment. The experiment was stopped once again, and the film was subjected to the third back pressure treatment followed by another long time reverse osmosis experiment. The results of such tests carried out with six different films are plotted in Figure 2-35.

The results show that the back pressure treatment technique for improving membrane performance works even when applied repeatedly. After the first treatment, however, the gain in product rate is partly off-set by an increase in membrane compaction. Consequently, no more than two back pressure treatments seem advisable for obtaining the most benefit from the technique.

Effect of interruption and release of pressure during long-time reverse osmosis experiments. It is known (Sourirajan, 1967a) that interruption and release of pressure during long-time reverse osmosis experiments results in a temporary increase in product rate. Figure 2-36 gives the results of a set of reverse osmosis experiments where the long-time runs were interrupted for 2 to 3 days so that the results obtained could be compared with those given

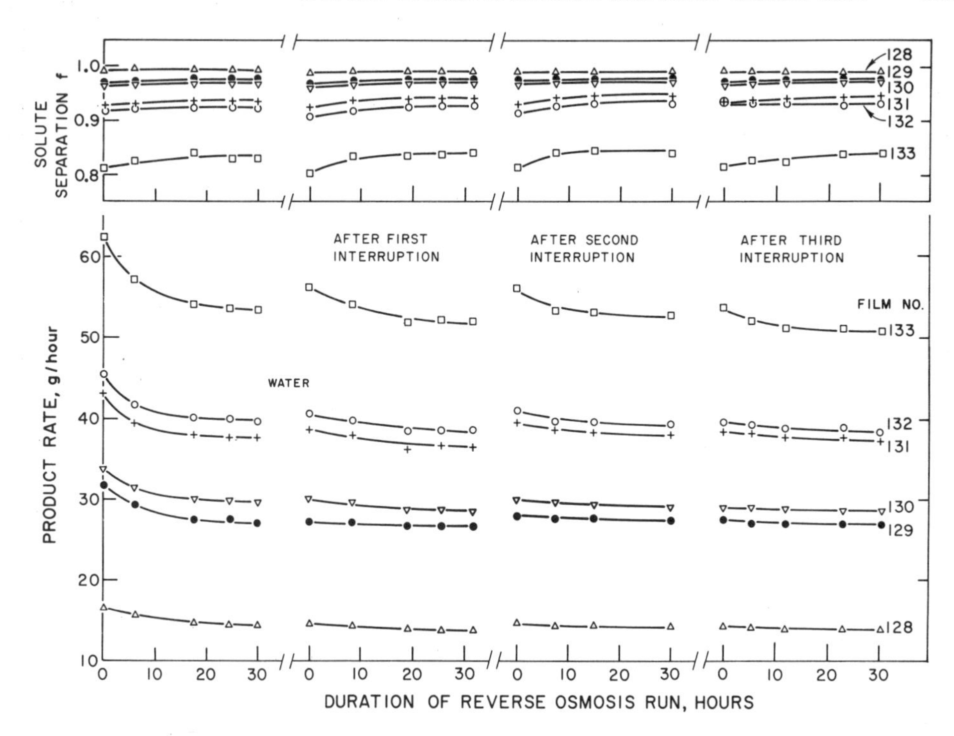

Figure 2-36. Effect of interruption during continuous reverse osmosis operation on membrane performance. Data of Kopeček and Sourirajan (1969a).

Film type: CA-NRC-18; system: sodium chloride-water; feed concentration: 0.5 wt.%; feed rate: 250 cm^3/min; operating pressure: 600 psig; film area: 7.6 cm^2.

in Figure 2-35. Figure 2-36 shows that the increase in product rate obtained by interruption and release of pressure in a long-time continuous reverse osmosis experiment is only about 3 to 5 per cent, and even that increase is temporary. On the other hand, the increase in product rate obtained by back pressure treatment is much higher (~20 to 25 per cent), and the latter is a net increase in the productivity of the film, as stated earlier. This is understandable on the basis that the mechanism of product rate increase is different in the two cases. In the former case, the increase is merely due to a temporary release from membrane compaction; in the latter case, the increase is due to a change in the porous structure of the surface layer.

In summary, the back pressure treatment at 400 psig for 85 hours offers a means of increasing the productivity of the Loeb-Sourirajan type porous cellulose acetate membranes by over 20 per cent without decreasing solute separation in reverse osmosis experiments operating at 600 psig. Hence this technique is of practical importance in the application of reverse osmosis for brackish water conversion.

Reverse Osmosis Experiments at 1500 psig Using 0.5M [$NaCl$-H_2O] Feed Solutions

Experimental procedure. All membranes discussed in this part of the investigation were initially shrunk at different temperatures. They were pressure-treated in the normal manner at 1700 psig for 1 hour. All reverse osmosis experiments were carried out with the surface layer side of the membrane facing the feed solution.

Performance of back pressure-treated membranes. Two sets of experiments were carried out to determine the effect of back pressure treatment on the performance of membranes in reverse osmosis experiments at 1500 psig.

In the first set of experiments, the back pressure treatment was given to films which were initially subjected to the usual temperature and pressure treatment. Several pressures in the range of 200 to 900 psig, and treatment periods in the range 65 to 150 hours were investigated for the back pressure treatment. In all cases it was found that the performance of the back pressure-treated film in the subsequent reverse osmosis experiment was either the same or worse (i.e. product rate increased and solute separation decreased) than its performance before back pressure treatment.

In the second set of experiments, back pressure treatment was given to films which were initially subjected to the usual temperature treatment but not the pressure treatment. The normal pressure treatment at 1700 psig with the surface layer of the film facing the feed water was given to the back pressure treated film. For the purpose of these experiments, three groups of films were used, each group consisting of five individual films. All the 15 films tested were cut from the same sheet of membrane subjected only to the temperature treatment. The five films used in each experiment were randomly chosen.

The usual reverse osmosis experiment was carried out with the first group of five films subjected to no back pressure treatment. The second group of films was given a back pressure treatment at 250 psig for 85 hours, after which the normal pressure treatment at 1700 psig was given. The third group of films was given a back pressure treatment at 400 psig for 85 hours, after which the normal pressure treatment at 1700 psig was given. A 24-hour continuous reverse osmosis test was given to the first and second group of films, and a short-run reverse osmosis test was given to the third group of films. The results of the tests are given in Table 2-58.

The results show that back pressure treatment of the preshrunk membranes before normal pressure treatment does increase product rate by an average amount of 10 to 15 per cent without any significant change in solute separation in reverse osmosis experiments at 1500 psig. The compaction effect for the back pressure-treated films is no different from that for films without any back pressure treatment. Therefore, the increase in product rate obtained by back pressure treatment is a net gain in productivity of the films under conditions of continuous operation. Back pressure treatment at 400 psig gives results better than those obtained with the same treatment at 250 psig; the optimum pressure for such treatment for films meant for reverse osmosis operation at 1500 psig has however not been investigated. The results given here are sufficient to show that back pressure treatment of preshrunk membranes improves their performance for reverse osmosis operation at 1500 psig.

14. Effect of Pressure on Pure Water Permeability Through Each Side of the Membrane During Cyclic Operations

The object of these experiments is to obtain some insight into the possible changes taking place in the overall structure of the film during back pressure treatment. The experiments consisted in determining the pure water permeability of the film as a function of increasing or decreasing pressure using

Table 2-58. Effect of back pressure treatment of preshrunk membranes for reverse osmosis service at 1500 psig

Data of Kopeček and Sourirajan (1969a)

Film group	Pressure used for BPT (psig)	Reverse osmosis experimental data: Initial values [PWP] (g/hr)	[PR] (g/hr)	$f \times 10^2$	Average values	Average values after 24 hours: $\frac{[PR]}{[PR]_{initial}}$	Change $f \times 10^2$
1	no BPT	75.0	50.0	97.7	[PWP] = 79.1		
		74.4	49.0	97.0			
		80.3	51.4	97.5	[PR] = 51.1	0.76	−0.6
		80.3	52.2	95.9	$f \times 10^2$ = 96.7		
		85.3	52.8	95.4			
2	250	87.6	57.3	97.6			
	250	89.1	57.3	97.8	[PWP] = 86.4		
	250	91.3	59.3	96.7	[PR] = 55.7	0.755	−0.3
	250	83.3	54.2	98.4			
	250	80.5	50.6	96.4	$f \times 10^2$ = 97.4		
3	400	90.1	58.5	96.8	[PWP] = 90.8		
	400	89.8	58.0	96.7			
	400	88.7	57.6	96.9	[PR] = 58.2	—	—
	400	93.7	60.1	96.5			
	400	91.8	56.9	96.2	$f \times 10^2$ = 96.6		

B.P.T.: Back pressure treatment

Film type: CA-NRC-18
System: Sodium chloride-water
Feed concentration: 0.5M
Feed rate: 250 cm^3/min
Operating pressure: 1500 psig
Film area: 7.6 cm^2
Duration of back pressure treatment: 85 hours
[PR]: Product rate
[PWP]: Pure water permeability

alternately the surface layer side of the film and the reverse side of the film to face the feed water under pressure. Two sets of films were used, one for operating pressures up to 600 psig, and the other for operating pressures up to 1500 psig. All the films were initially shrunk at different temperatures, and then pressure-treated in the normal manner at 700 and 1700 psig respectively During the experiments the pure water permeability data were taken at each operating pressure after the film had been under that pressure for at least 10 minutes. Operating pressures were changed from one to the next without stopping the run. The run was stopped only when the side of the film facing the feed water was changed. The experimental data obtained are plotted in Figures 2-37 and 2-38 for two typical films in each operating pressure range. The data on the top right-hand side of each plot are for the surface layer side operation, and those on the bottom left-hand side of each plot are for the reverse side operation. The numbers I, II, III, etc. represent the sequence of the cyclic operation, and the arrows indicate the direction of pressure change in each operation.

Similar experiments have been reported by Banks and Sharples (1966c). In the extent of work, object of the experiments, and the details of observation, the work of Kopeček and Sourirajan (1969a) is different from that of Banks and Sharples.

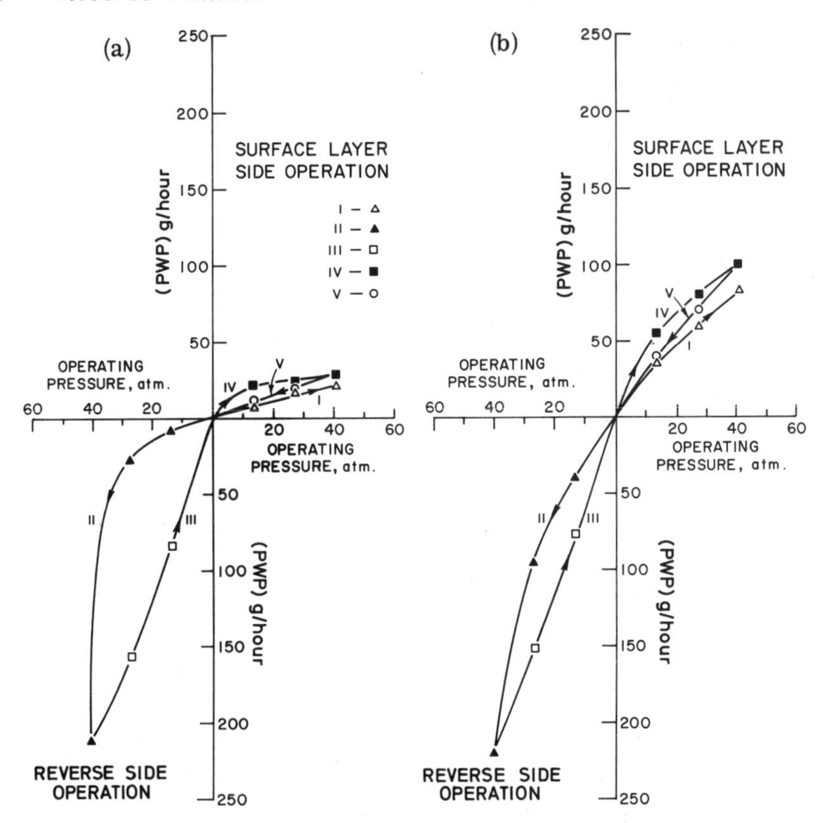

Figure 2-37. Effect of pressure on pure water permeability through each side of the membrane during cyclic operations up to 600 psig. Data of Kopeček and Sourirajan (1969a).

Film type: CA-NRC-18; system: water; film area: 7.6 cm^2.

Operating Pressures up to 600 psig. The films whose data are plotted in Figure 2-37(a) and 2-37(b) were preshrunk at 91° and 85° respectively, and both were pressure treated in the normal manner at 700 psig. The films gave solute separations of 98.4 per cent and 73.0 per cent respectively with 0.5 wt.% NaCl-H_2O feed solutions. Referring to Figure 2-37(a), the following observations can be made with respect to changes in the average pore size on the surface layer of the film during each of the five operations indicated in the cycle.

Operation I (surface layer side operation). Since pure water permeability is essentially proportional to operating pressure, no change in pore size is indicated.

Operation II (reverse side operation). Up to an operating pressure of about 14 atm, there seems no change in pore size; at higher operating pressures, pure water permeability increases disproportionately with pressure indicating widening and/or opening and redistribution of surface pores.

Operation III (reverse side operation). Since pure water permeability v. operating pressure is mostly a straight line in this decreasing pressure sequence, the pore size obtained at the end of Operation II seems to be maintained.

Operation IV (surface layer side operation). When the operating pressure is increased, the surface pores, which were widened and/or opened during Operation II, seem to shrink rapidly. However, at the highest operating pressure of 40.8 atm, the average pore size on the surface layer is significantly higher than that prevailing during Operation I.

Operation V (surface layer side operation). During this sequence of decreasing operating pressure, the pure water permeability v. pressure graph is a straight line indicating that the increase in average pore size observed at the end of Operation IV is still maintained.

Referring to Figure 2-37(b), the changes in the average pore size on the surface layer of the film during the above sequence of five operations is similar to those given above.

The data for both the above films indicate that the reverse side operation, with a film subjected to an initial pressure treatment at only 700 psig in the normal manner, ultimately results in a permanent increase in the average pore size on the surface layer in subsequent surface layer side experiments at 600 psig. Six films preshrunk at different temperatures were tested in the above cycle of operations, and the results obtained were similar in all cases.

Further, with respect to the first film (Figure 2-37(a)) whose initial average pore size was smaller, the pure water permeability changed from 21 to 212 g/hr at 40.8 atm pressure by changing the operation from surface layer side to reverse side of film; this represented a change of over 10-fold. But with respect to the second film (Figure 2-37(b)) whose initial average pore size was bigger than that of the first one, the corresponding change in pure water permeability was from 83 to 219 g/hr, representing a change of only about 2.6-fold. These data indicate that the reverse side operation widens and/or opens the smaller pores preferentially.

Operating Pressures up to 1500 psig. The films whose data are plotted in Figure 2-38(a) and 2-38(b) were preshrunk at 91° and 86° respectively, and both were pressure-treated in the normal manner at 1700 psig. The films gave solute separations of 98.9 per cent and 90.0 per cent respectively with 0.5M [$NaCl-H_2O$] feed solutions. Referring to Figure 2-38(a), the following observations can be made with respect to changes in the average pore size on the surface layer of the film during each of the nine operations indicated in the cycle.

Operation I (surface layer side operation). This includes a pressure range up to 40.8 atm; in this range, since pure water permeability is proportional to pressure, no change in pore size is indicated.

Operation II (reverse side operation). This also includes a pressure range up to 40.8 atm. Up to about 14 atm, the pure water permeability is proportional to pressure, indicating no change in pore size. At higher operating pressures, pure water permeability increases disproportionately with pressure indicating widening and/or opening, and redistribution of surface pores.

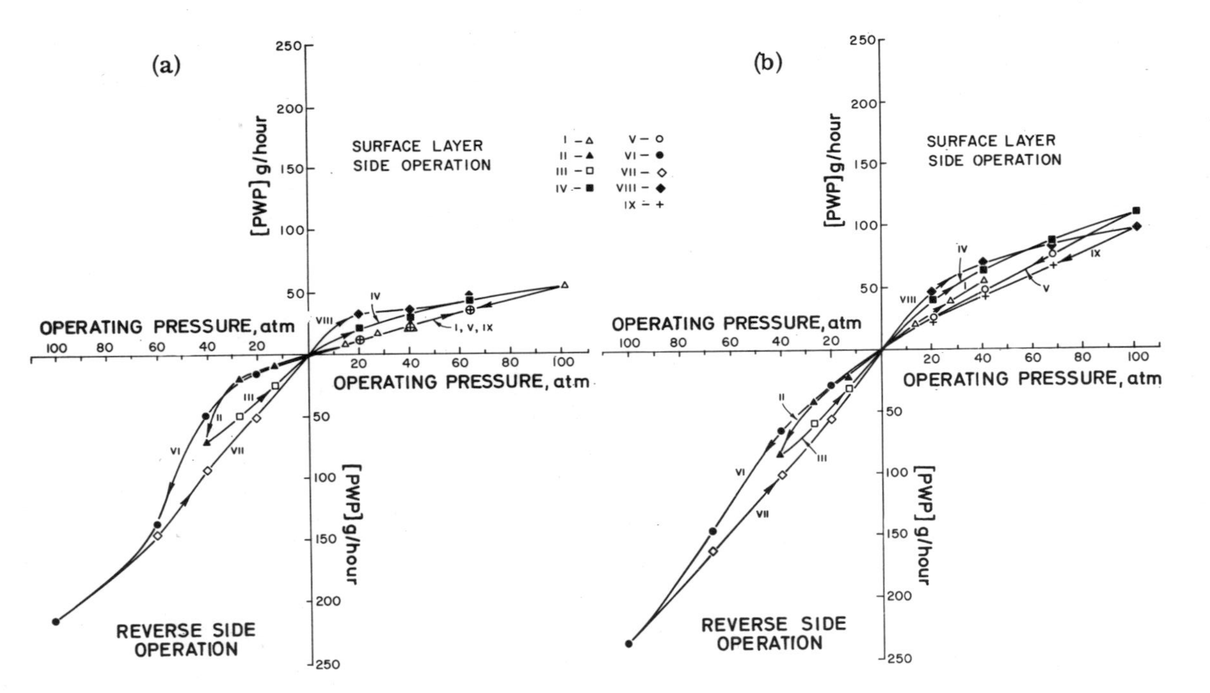

Figure 2-38. Effect of pressure on pure water permeability through each side of the membrane during cyclic operations up to 1500 psig. Data of Kopeček and Sourirajan (1969a).

Film type: CA-NRC-18; system: water; film area: 7.6 cm^2.

Operation III (reverse side operation). Since pure water permeability v. operating pressure is essentially a straight line in this decreasing pressure sequence, the pore size obtained at the end of Operation II seems to be maintained.

Operation IV (surface layer side operation). This includes a pressure range up to 102 atm. When the operating pressure is increased, the surface pores which were widened or opened during Operation II seem to shrink rapidly, reaching stable values at 102 atm.

Operation V (surface layer side operation). During this sequence of decreasing pressure, the pure water permeability v. pressure graph is a straight line which coincides with that obtained in Operation I. This indicates that the pore size obtained at the end of Operation IV is the same as that prevailing during Operation I.

Operation VI (reverse side operation). This also includes a pressure range up to 102 atm. Again there seems no change in pore size up to about 14 atm, above which pure water permeability increases steeply with pressure up to 60 atm, and less steeply thereafter. The data indicate that the surface pores are widened and/or opened during this operation.

Operation VII (reverse side operation). During this sequence of decreasing pressure in the pressure range 102 to 60 atm, the pore size partly tends to return to what it was at the same pressure during Operation VI; and at pressures less than 60 atm, there seems no change in pore size.

Operation VIII (surface layer side operation). This also includes a pressure range up to 102 atm. When the operating pressure is increased the surface pores which were widened and/or opened during Operation VI and remained so in Operation VII, seem to shrink rapidly. Above 60 atm the pure water permeability v. pressure curve coincides with that obtained in Operation IV. Thus at the end of Operation VIII, the pore size is the same as it was at the end of Operation IV at 102 atm.

Operation IX (surface layer side operation). During this sequence of decreasing pressure, the pure water permeability v. pressure curve is again a straight line coinciding with the results obtained in Operation V and I. This indicates that the average pore size at the end of Operation VIII is the same as that prevailing during Operation I.

Referring to Figure 2-38(b), the changes in the average pore size on the surface layer of the film during the above sequence of nine operations is essentially similar with one significant difference. The application of 102 atm of pressure during the surface layer side operation reduces the average pore size every time the experiment is done after a prior reverse side operation. Consequently, the reverse side operation with a film initially subjected to a normal pressure treatment at 1700 psig ultimately results in either no change or a decrease in the average pore size on the surface layer (or probably a slight increase in effective film thickness) in subsequent experiments at 1500 psig. Six films shrunk at different temperatures were tested in the same cycle of operations, and the results were similar in all cases; in no case was there an ultimate increase in the average pore size in the surface layer, as observed with films initially subjected to the normal pressure treatment at 700 psig, and subsequently operated at 600 psig.

Further, with respect to the first film (Figure 2-38(a)), the pure water permeability increased about 4-fold in reverse side operation compared to surface layer side operation at 1500 psig; the corresponding change with respect to the second film (Figure 2-38(b)) was only about 2-fold. These data again indicate that the reverse side operation widens and/or opens the smaller pores more than the larger ones.

15. Probable Structural Changes in the Film During Back Pressure Treatment

The Loeb-Sourirajan type porous cellulose acetate membrane has an asymmetric and amorphous structure with extensive physical cross links in the membrane structure. The initial temperature and pressure treatments create additional cross links which contribute to the rigidity of the surface layer.

During the reverse side operation under pressure, the pores on the surface layer are widened and/or opened; such change in pore size is more with respect to smaller pores than with respect to larger ones. This is clearly shown by the data in Figures 2-37 and 2-38, and this property is made use of in back pressure treatment.

Further, back pressure treatment brings about significant viscoelastic deformations in the macromolecular structure of the surface layer. The extent of such deformation depends of course on the secondary and tertiary gel structure existing in the membrane prior to back pressure treatment. Owing to such deformation, the polymer flows, and the macromolecules change their spatial distribution including a rearrangement of the physical cross links. The most probable distribution of end-to-end distances of polymer segments obtained finally is time-dependent. Due to the long duration involved in back pressure treatment, the creep recovery is reduced, and a degree of permanent deformation is set in the membrane structure.

Since all the membranes tested were made by the same method (and hence presumably had the same primary gel structure), the permanent deformation caused by back pressure treatment depends upon the secondary and tertiary gel structure.

The results obtained with back pressure-treated membranes, initially pressure-treated at 700 and 1700 psig respectively, show that the former ones were more susceptible to such deformation than the latter, which were probably too rigid to change by such treatment. Hence, in the latter case, back pressure treatment had to be done to change the secondary gel structure prior to normal pressure treatment.

The increase in product rate obtained (with no change in solute separation) by back pressure treatment of a normal membrane (initially temperature- and pressure-treated in the normal manner), subjected only to short-time reverse osmosis experiment, is proportionately about the same as the increase obtained by similar treatment of the same membrane which has been subjected to a continuous 30-hour reverse osmosis experiment. This means that the improvement in performance obtained with a normal membrane is about the same as that obtained with a compacted membrane, as a result of back pressure treatment. Further, the compaction effect on a normal membrane is essentially the same as that on a back pressure-treated membrane during continuous long-time experiments. These observations indicate that the deformations induced in the spongy macroporous mass of the film underneath the surface layer during back pressure treatment have no significant effect on the product rate and solute separation characteristics of the film. Consequently the deformation brought about in the surface layer by back pressure treatment seems to control the improvement obtained in membrane performance.

In the normal long-time experiments, the creep phenomenon seems responsible for closing pores and reducing product rate. By back pressure treatment, the same phenomenon is made use of to obtain a wider pore size distribution and improve membrane performance.

The compaction effect of the membrane increases with repeated back pressure treatment. This indicates some irreversible weakening of the membrane structure on such treatment similar to that observed by Haward (1942).

16. Significance of Results

The back pressure treatment technique offers a general method of improving the performance of asymmetric membranes in the reverse osmosis process. It can be used to change either the secondary or tertiary gel structure of the membrane. The duration, pressure, and sequence employed for back pressure treatment may be chosen to suit particular applications and the nature of the membrane structure prior to such treatment (Kopeček and Sourirajan, 1969b). The success of the back pressure treatment technique for improving the performance of the Loeb-Sourirajan type porous cellulose acetate membranes in reverse osmosis is yet another experimental evidence for the heterogeneous and microporous structure of their surface layer.

17. Summary

This chapter starts with details of the development of Loeb-Sourirajan type porous cellulose acetate membranes for saline water conversion and related reverse osmosis applications. The basic approach used was to: incorporate a suitable water-soluble additive in the film casting solution; leach out the additive from the membrane subsequently with water, thus creating the required porosity in the membrane structure; and adjust the film casting conditions such that a very thin microporous surface layer embedded on a macroporous spongy mass was created. The following are three different types of such membranes whose performance data are discussed in detail.

	Batch 25	*Batch 47*	*Batch 18*
Casting Solution Composition (wt.%)			
Cellulose acetate (acetyl content 39.8%)	*22.2*	*25*	*17*
Acetone	*66.7*	*45*	*68*
Magnesium perchlorate	*1.1*	—	*1.5*
Water	*10.0*	—	*13.5*
Formamide	—	*30*	—
Casting temperature	*−5° to −10°*	*Lab. temp. (~24°)*	*−10°*
Evaporation period	*4 min*	*<1 min*	*4 min*
Gelation period in ice-cold water	*1 to 2 hours in all cases*		
Nominal film thickness	*0.003 to 0.006 inch in all cases*		

The effects of variation of film material, solvent, additive, and film casting conditions have been extensively studied by several workers, and a wide variety of membranes having separation characteristics have been produced, but none is better in performance than the ones listed above.

Loeb and Manjikian conducted a successful 6-month field test with a laboratory-made porous cellulose acetate membrane (similar to Batch 25 type) for the conversion of Coalinga brackish waters. They also found that such membranes stored in water for 4½ months showed no deterioration in performance.

The porous structure of the preshrunk Batch 25 type membrane was examined under the electron microscope by Riley and others. They found that the membrane was asymmetric, consisting of a dense surface layer with a spongy porous mass underneath. The thickness of the surface layer was about 0.25μ compared to the total membrane thickness of 100μ. The surface layer showed no evidence of pores bigger than 100Å, whereas the spongy porous mass underneath the surface layer had pores of size 1000 to 4000Å.

Vos and others showed that the hydrolysis rate of the cellulose acetate material used in the Loeb-Sourirajan type membranes was minimum in the pH range 4 to 5.

The steps involved in the Loeb-Sourirajan technique for making porous cellulose acetate membranes were critically examined by Kesting and co-workers. They attributed the role of certain inorganic salts in increasing the flux of water through the membrane to the capacity of the component ions to swell the cellulosic substrate; according to them, coacervation and polymer desolvation during the gelation process gave rise to an open cell foam or pore structure in the resulting membrane. Meares estimated that 99 per cent of the transport of liquid water through the Loeb-Sourirajan type porous cellulose acetate membrane took place by viscous flow in tortuous channels 6 to 9Å in diameter.

Based on diffusion studies in cellulose acetate gels prepared from cellulose acetate-benzyl alcohol solution, Klemm and Friedman concluded that gel formation was the result of incomplete precipitation of the dispersed medium, and the resulting pore openings consisted of minute tubes of varying lengths, chaotically arranged but each intersecting others to form continuous open passages completely through the gel; they estimated that the average pore radius was 180Å at 2 per cent cellulose acetate concentration.

Using the Carnell-Cassidy technique, which consisted essentially of slowly drawing a clean glass plate out of a dilute solution of a polymer in a suitable solvent, Merten and co-workers made several 'thin' (500 to 3000Å) membranes of cellulose acetate, ethyl cellulose, ethyl cellulose-polyacrylic acid, cellulose butyrate and cellulose nitrate; Francis and Cadotte made several thin polysaccharide acetate membranes. The data on their performance in reverse osmosis experiments are given.

On drying in air at the laboratory temperature, the porous structure of the Loeb-Sourirajan type cellulose acetate membranes generally changes irreversibly. Merten and co-workers showed that the above membranes could be held 'dry' after treatment with a surface active agent such as a mixture of Tergitol 15S7 and glycerol in water.

Flowers and co-workers found that graphitic oxide membranes had desalination characteristics in reverse osmosis experiments. Using NaCl (0.5 wt.%)-H_2O feed solutions at 600 psig, water fluxes shown by good graphitic oxide membranes ranged from 0.7 to 5.8 gal/day/ft^2 with solute separations ranging from 65 to 83 per cent.

Several polysalt complex membranes have been prepared and tested in reverse osmosis experiments by Michaels and co-workers. Neutral polysalt mem-

branes gave solute separations of about 50 per cent at 1500 psig using aqueous sodium chloride (4 per cent NaCl) feed solutions with product rates of about 2 gal/day/ft^2 for a 1-mil thick film. Membranes showing water permeabilities of 80 to 100 gal/day/ft^2 at 100 psig with $>$90 per cent solute separation from dilute aqueous raffinose feed solutions have also been obtained.

Kraus and co-workers have found that salt-rejecting interfacial layers or membranes with frequently high product rates could be formed dynamically on porous bodies when certain additives are present in the pressurised feed solutions. Such membranes have been formed dynamically on substrates with pore diameters as high as 5 μ. After initial formation of the membranes, very low concentrations of additives, of the order of 1 ppm are frequently sufficient to maintain their properties. Examples of successful additives are: organic polyelectrolytes, colloidal dispersions of hydrous oxides, solutions of hydrolysable ions, ground-up low-cross linked ion-exchange beads, and certain natural products such as clays and humic acid. Membranes have also been formed from neutral (uncharged additives). Hydrous zirconium oxide membranes (prepared from boiled zirconium oxychloride solution) showed water fluxes of the order of 300 gal/day/ft^2 while giving significant solute separation (40 to 80 per cent) at 35 atm pressure using 0. 01 to 0. 05N sodium chloride feed solutions.

Three recent developments in the preparation and pretreatment of Loeb-Sourirajan type porous cellulose acetate membranes are then reviewed. These include the preparation and performance of Batch 47 type membranes, Loeb's gravity-drop technique for fabricating tubular membranes, and the emergence of Batch 18 type membranes for saline water conversion and similar reverse osmosis applications.

Experimental data supporting the microporous nature of the surface layer of the Loeb-Sourirajan type porous cellulose acetate membranes are given. A general method has been proposed by Kopeček and Sourirajan for improving the performance of the above membranes in reverse osmosis, by which product rates are increased without decreasing solute separation. The method consists in pumping pure water past the reverse side of the membrane under just enough pressure for a sufficiently prolonged period of time; after such pretreatment, the membrane is used in the reverse osmosis experiments in the normal manner with the surface layer side facing the feed solution. Back pressure treatment at 400 psig for 85 hours on preshrunk and normally pressure treated membranes increases the product rate by over 20 per cent, without decreasing solute separation in reverse osmosis experiments operating at 600 psig using 0. 5 wt. % NaCl-H_2O feed solutions; with a different sequence of back pressure treatment, similar results have been obtained in reverse osmosis experiments operating at 1500 psig also. The compaction effect of a normal membrane and that of a back pressure-treated membrane are the same during continuous reverse osmosis operation under 600 psig; the effects of back pressure treatment on a normal membrane and compacted membrane are also the same. The pure water permeability data obtained in cyclic experiments show that the smaller pores on the surface layer are opened more than the bigger ones during the reverse side operation. The probable structural changes taking place in the film during back pressure treatment are discussed.

CHAPTER 3

Transport through Reverse Osmosis Membranes

Scope of this chapter. The scope of this chapter is restricted to the discussion of the isothermal reverse osmosis separation process involving binary aqueous solution systems and Loeb-Sourirajan type porous cellulose acetate membranes having a preferential sorption for water from the aqueous solution under consideration. Unless otherwise stated, all data presented in this chapter are for 25°. As pointed out in Chapter 1, under isothermal conditions, the preferential transport of material through the membrane is always in the direction of lower chemical potential. Based on the transport of water from aqueous solutions due to the chemical potential gradient, a variety of membrane separation processes can be developed as indicated in Figure 3-1. The discussion in this chapter is restricted to the transport processes represented by lines a and b only in Figure 3-1, where line a represents pure water permeability through the membrane, and line b represents the transport of water through the membrane from the aqueous solution at above-atmospheric pressure to the less concentrated solution at atmospheric pressure on the other side of the membrane; lines c and d in Figure 3-1 are the limiting cases of b, representing 100 per cent and 0 per cent solute separation respectively. These restrictions also apply to the discussions in the subsequent chapters wherever the transport equations developed in this chapter are used.

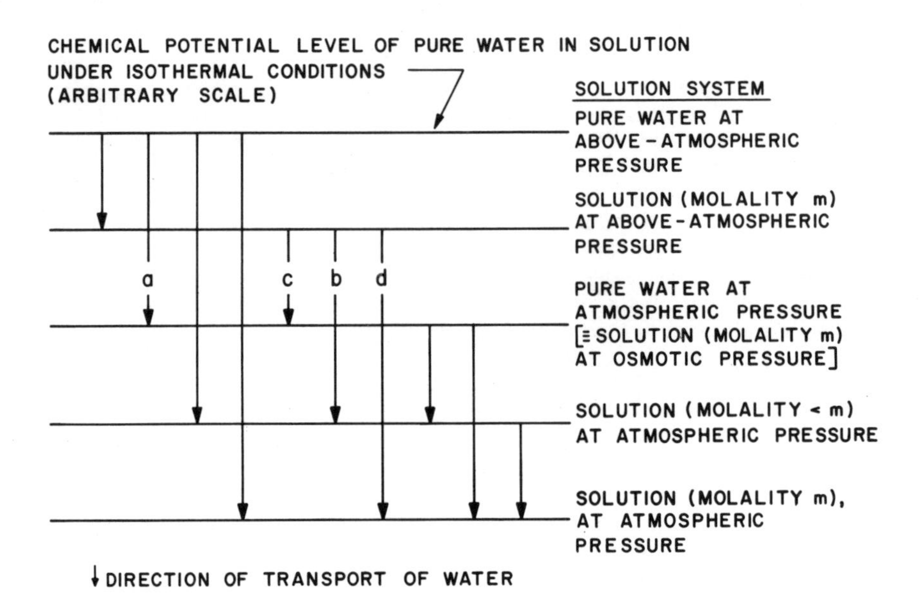

Figure 3-1. An arbitrary scale of chemical potential level of water in aqueous solutions.

The limitations on the scope of this chapter arise only out of the present state of knowledge on the subject, and not out of any inherent limitation on the scope of the reverse osmosis process itself. Even with these limitations, the scope of discussions in this and the subsequent chapters is extensive in view of the importance of the reverse osmosis treatment of aqueous solutions, and the potential applications of the Loeb-Sourirajan type porous cellulose acetate membranes.

This chapter is particularly concerned with the development of transport equations for reverse osmosis separation, and the correlations of reverse osmosis experimental data leading to practical methods of membrane specification, expressing membrane selectivity, and predicting membrane performance from a minimum of experimental data.

1. Transport Mechanisms

Several mechanisms of water and solute transport through the Loeb-Sourirajan type porous cellulose acetate membranes under reverse osmosis experimental conditions are discussed in the literature (Banks and Sharples, 1966c; Clark, 1963; Hoffer and Kedem, 1967; Jagur-Grodzinski and Kedem, 1966; Johnson et al., 1966; Lakshminarayanaiah, 1965; Lonsdale, 1966; Lonsdale et al., 1965; Meares, 1966; Merten, 1966a, 1966b; Michaels et al., 1965a; Sherwood et al., 1967; Spiegler and Kedem, 1966).

Merten and co-workers have developed transport equations based on a solution-diffusion model (Lonsdale et al., 1965); they make extensive use of the concept of 'perfect' and 'imperfect' membranes in their discussions. Perfect membranes are presumably those which have a completely nonporous surface structure. A definite 'theoretical' level of solute separation is calculated from solubility measurements for perfect membranes. For example, the theoretical salt separation is 99.82 per cent at 1500 psig for 1 per cent NaCl feed solutions at 25° for membranes made of cellulose acetate having an acetyl content of 39.8 per cent (Riley et al., 1967). Membranes having less than this theoretical level of solute separation are obviously imperfect according to them. Their transport equations then are apparently limited to their perfect membranes. According to Banks and Sharples (1966c) also, the transport mechanism in reverse osmosis is one of diffusive flow through the pore-free layer in the membrane. According to Michaels et al. (1965a), water transport is by molecular diffusion through polymer matrix, and ion transport is by three parallel flow mechanisms: (a) by sorption and activated diffusion within the polymer matrix governed solely by the ion-concentration difference across the membrane; (b) by pressure-biased activated diffusion of ions in near-molecular-sized pores in the membrane, governed both by the hydraulic gradient and the ion-concentration difference; and (c) by hydrodynamic flow of saline solution through larger pores. According to Sherwood et al. (1967), water and solute cross the membrane by the parallel processes of diffusion and pore flow. Transport equations based on non-equilibrium thermodynamics are given by several workers (Bennion, 1966; Bennion and Rhee, 1969; Johnson and Bennion, 1968; Johnson et al., 1966; Kedem and Katchalsky, 1958, 1961, 1963; Michaeli and Kedem, 1961; Podall, 1967, 1968a, 1968b; Spiegler, 1958; Spiegler and Kedem, 1966). Kedem and Katchalsky (1958, 1963) have pointed out that the solute and solvent fluxes through the membrane are governed by three coefficients representing solute-solvent, solute-membrane, and solvent-membrane interactions; in their approach coupling of solute and solvent flow is included as an independent parameter. Assuming constancy of three coefficients, namely the specific hydraulic permeability, the local solute permeability, and the reflection factor (which is said to be a quantitative index of solute separation), Spiegler and Kedem (1966) have given transport equations theoretically applicable for the entire 0 to 100 per cent range of solute separations, and their validity has been demonstrated experimentally by Jagur-Grodzinski and Kedem (1966).

This monograph presents a different approach to the subject consistent with the concept of the generality of this separation technique as formulated in Chapter 1. This approach has no place for the concept of perfect and imperfect membranes in the sense commonly used in the literature on reverse osmosis. That the membrane surface is porous and heterogeneous, and that the transport of both the preferentially sorbed water and the aqueous solution is through the capillary pores on the membrane surface, are concepts fundamental to this approach.

The basic characteristics of this separation process have been discussed in Chapter 1. With respect to a given membrane-solution system, provided the membrane material has a preferential sorption for water from the aqueous solution under consideration, some degree of solute separation (>0 per cent) will always occur when flow of the fluid through the membrane is established; further, since preferential sorption simply means a steep concentration gradient at the membrane-solution interface, the maximum solute separation attainable in practice with any film can only approach, but not quite reach, 100 per cent; thus solute separations of 0 per cent and 100 per cent are limiting values with respect to this separation process. Consequently the reverse osmosis process is best understood only in terms of the entire >0 per cent to <100 per cent range of solute separation. Further, it is the intermediate range of solute separation which offers the real testing ground for any hypothesis concerning the transport of material through porous membranes having separation characteristics. Therefore the approach presented in this chapter particularly involves the analysis of reverse osmosis data covering a wide range of solute separations.

The transport equations and the correlations of reverse osmosis experimental data presented in this chapter form the basis of reverse osmosis process design discussed in the next chapter. There is certainly no direct or indirect suggestion that the approach presented in this book is the only valid one. To the extent that they are consistent with the reverse osmosis experimental data, and to the extent that they predict membrane performance, all approaches are indeed equally valid.

2. Some Thermodynamic Considerations

Chemical Potential and Osmotic Pressure Relationships. It is the transport of the preferentially sorbed pure water through the membrane that gives rise to an effective solute separation in reverse osmosis. Consequently the chemical potential of pure water in aqueous solutions, and the osmotic pressures of the solutions are of interest. Hence some of the basic thermodynamic relations involving chemical potentials and osmotic pressure are summarised below.

The chemical potential (μ_i) of component i in a solution is defined in terms of the Gibbs free energy (G) by the relation

$$dG = -SdT + VdP + \sum_i \mu_i dN_i \tag{3-1}$$

where S is the entropy, T the absolute temperature, V the volume, P the pressure, and N_i the number of moles of component i. From Equation 3-1,

$$\mu_i = (\partial G/\partial N_i)_{T,P,N_j} \tag{3-2}$$

and

$$V = (\partial G/\partial P)_{T,N} \tag{3-3}$$

where N represents the entire set of N's, and N_j represents all N's except N_i. Differentiating Equation 3-3 with respect to N_i,

$$\left(\frac{\partial^2 G}{\partial N_i\ \partial P}\right)_{T,N_j} = \left(\frac{\partial \mu_i}{\partial P}\right)_{T,N} = \left(\frac{\partial V}{\partial N_i}\right)_{T,P,N_j} = \overline{V}_i \tag{3-4}$$

where $\overline{V}_i$ is the partial molal volume of component i. The activity (a_i) of component i is related to μ_i by the relation

$$\mu_i = \mu_i{}^0 + RT \ln a_i \tag{3-5}$$

where R is the gas constant, and $\mu_i{}^0$ is the standard chemical potential of i which at a given pressure, is dependent on temperature only and not on concentration.

In a water-solute binary system, let component i represent water (indicated by subscript w), and let $\mu_i{}^0 = \mu_w^* =$ chemical potential of pure water at the specified temperature.

The thermodynamic requirement for osmotic equilibrium is that the chemical potential of water in the solution phases be the same on both sides of the membrane. If there is just pure water at pressure P_1 on both sides of the membrane, the two phases will be in equilibrium (the chemical potential of water in both phases being μ_w^*), and there will be no net transfer of water through the membrane. If the pure water on one side of the membrane is replaced by an aqueous solution (both sides still being at pressure P_1), the chemical potential of water in the solution is less than that of pure water, and the former is given by

$$\mu_w = \mu_w^* + RT \ln a_w \tag{3-6}$$

where μ_w is the chemical potential of water in solution at pressure P_1. The equilibrium can be restored by increasing the pressure on the solution side to P_2 such that the chemical potential of water in the solution is raised to that of pure water, namely μ_w^*. The increase in chemical potential of water in the solution as the pressure is increased from P_1 to P_2 is obtained from Equation 3-4, as

$$\int_{P_1}^{P_2} (\partial \mu_w/\partial P)_{T,N}\, dP = \int_{P_1}^{P_2} \overline{V}_w dP \tag{3-7}$$

Since this increase, added to μ_w given by Equation 3-6, must restore the chemical potential of water in solution to that of pure water,

$$\mu_w + \int_{P_1}^{P_2} \overline{V}_w dP = \mu_w^* \tag{3-8}$$

Therefore

$$\int_{P_1}^{P_2} \overline{V}_w dP = \mu_w^* - \mu_w \tag{3-9}$$

$$= -RT \ln a_w \tag{3-10}$$

from Equation 3-6. If $\overline{V}_w$ is assumed constant,

$$\overline{V}_w(P_2 - P_1) = -RT \ln a_w \tag{3-11}$$

The pressure difference $(P_2 - P_1)$ is by definition the osmotic pressure of the solution, usually represented by Π. Thus

$$\overline{V}_w \Pi = -RT \ln a_w \tag{3-12}$$

or

$$\Pi = -\frac{RT}{\overline{V}_w} \ln a_w \tag{3-13}$$

For calculating the activity of water in solution, a_w, vapour pressure data are needed; often, a_w is calculated from the relation

$$a_w = p_w/p_w^* \tag{3-14}$$

where p_w is the vapor pressure of water in equilibrium with the solution, and p_w^* is that of pure water at the given temperature. If the solute is volatile, the partial vapor pressure of water must be used for p_w. The isopiestic method (Bousfield, 1918; Sinclair, 1933; Robinson and Sinclair, 1934; Scatchard et al., 1938) offers a convenient technique of obtaining accurate vapour pressure data as a function of solution molality. One can then obtain the activity of water as a function of solution molality using the Gibbs-Duhem equation (Lewis and Randal, 1966). Such activity data are available in the literature (Robinson and Stokes, 1959c; Scatchard et al., 1938) for a large number of aqueous solutions in terms of osmotic coefficients (ϕ) defined by the relation (Robinson and Stokes, 1959a)

$$\ln a_w = -\frac{\Sigma_i m M_B}{1000} \phi \tag{3-15}$$

where Σ_i is the total number of moles of ions given by one mole of the electrolyte solute ($\Sigma_i = 1$ for a non-electrolyte solute), m is the solution molality, and M_B is the molecular weight of water. Substituting Equation 3-15 in Equation 3-13,

$$\Pi = \frac{\Sigma_i RT M_B}{1000\overline{V}_w} m \phi \tag{3-16}$$

Appendix II gives the osmotic pressure data, along with other data, for several aqueous solution systems calculated from Equation 3-16 using the osmotic coefficient and density data available in the literature (Robinson and Stokes, 1959c).

Maximum Possible Solute Separations. The effective driving pressure (ΔP) for fluid flow through the capillary pores on the membrane surface may be given by the relation

$$\Delta P = P - \Delta\Pi \tag{3-17}$$

where P is the operating gauge pressure, and

$$\Delta\Pi = \Pi_F - \Pi_P \tag{3-18}$$

Π_F is the osmotic pressure of the concentrated boundary solution immediately next to the film on the high pressure side of the membrane, and Π_P is that of the membrane permeated product solution. For the limiting case when the mass transfer coefficient on the high pressure side of the membrane tends to infinity (or when the feed rate on the membrane surface tends to infinity), Π_F approaches the osmotic pressure of the feed solution.

When P is equal to or less than Π_F, Equation 3-17 gives also the maximum solute separation possible in this process, whatever the membrane used, under which condition

$$\Delta\Pi = P \text{ and } \Delta P = 0 \tag{3-19}$$

Using Equation 3-19, and letting Π_F = osmotic pressure of the feed solution, Figures 3-2 and 3-3 give the plots of the maximum solute separation possible (f_{max}) for the systems sodium chloride-water and glycerol-water respectively

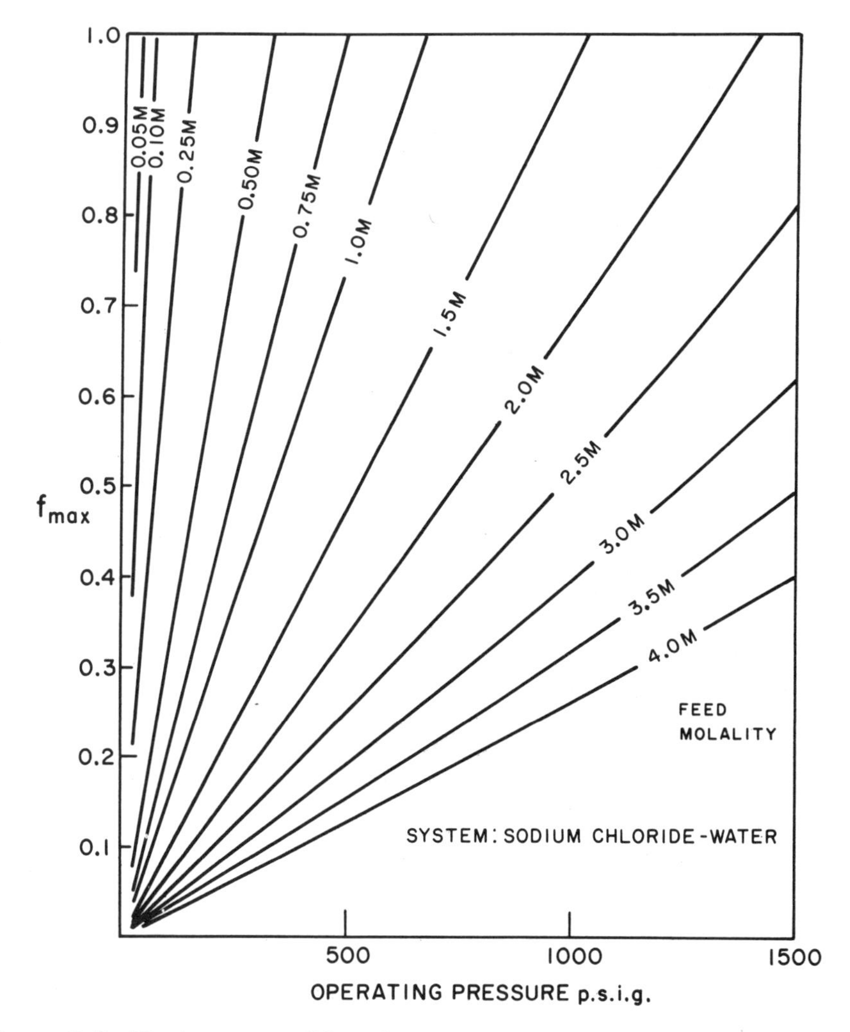

Figure 3-2. Maximum possible solute separations for the system sodium chloride-water in the reverse osmosis process at different operating pressures and feed concentrations.

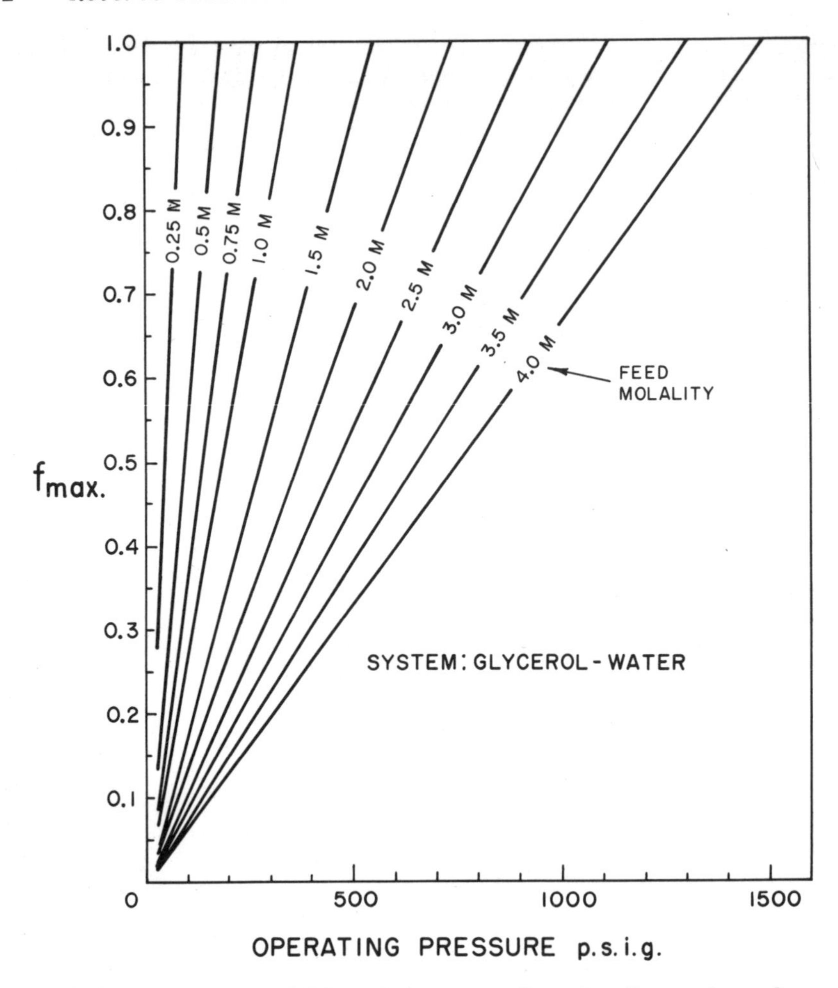

Figure 3-3. Maximum possible solute separations for the system glycerol-water in the reverse osmosis process at different operating pressures and feed concentrations. Data of Sourirajan and Kimura (1967).

at different operating pressures and feed concentrations. Figures 3-2 and 3-3 illustrate the thermodynamic significance of osmotic pressure as applied to this separation process.

When P is equal to or less than Π_F the values of f_{max} given in Figures 3-2 and 3-3 correspond to the condition that the product rate is zero; when P is greater than Π_F there is no thermodynamic limitation on the extent of solute separation and product rate obtainable in this process.

While Equation 3-17 is true whatever be the value of the osmotic pressure of the feed solution, it says nothing about the separation characteristics of any membrane. In other words, the actual performance of any particular membrane with respect to a given solution system depends, not only on the osmotic pressure of the feed solution, but on the physical and chemical nature of the membrane.

Minimum Work of Separation. The isothermal reversible work of separation (W) is the free energy change per mole (ΔG) given by the relation

$$-W = \Delta H - T\Delta S = \Delta G \tag{3-20}$$

where

ΔH = enthalpy of solution in final state — that in initial state, and

ΔS = entropy of solution in final state — that in initial state

both at the constant absolute temperature T at which the process takes place. Negative work (—W) refers to work done upon the system to cause the process to take place. The relation

$$-W = \Delta H - T_0\Delta S \tag{3-21}$$

where T_0 is the absolute temperature of the environment or the lowest temperature at which heat can be freely discharged, is applicable to any reversible process whether isothermal or not. When $T = T_0$, Equations 3-20 and 3-21 become identical; if T is far different from T_0, the difference between Equations 3-20 and 3-21 can be considerable. If the appropriate enthalpy and entropy data are available, the minimum work of separation can be calculated simply by using Equation 3-20 or 3-21.

With particular reference to saline water conversion to pure water, activity data for water (a_w) in different aqueous solutions are available in the literature (Robinson and Stokes, 1959c). Hence it is convenient to use the relation

$$-W = \Delta G = RT \ln a_w \tag{3-22}$$

for calculating W. From Equation 3-20,

$$-dW = \Delta G\, dn \tag{3-23}$$

where dn is the small increment of water removed from the system at any instance.

Therefore

$$-W = \int_{n_1}^{n_2} RT \ln a_w dn \tag{3-24}$$

where n_1 and n_2 are the initial and final number of moles of salt solution. Using Equations 3-22 and 3-24, Dodge and Eshaya (1960) and Johnson et al. (1966) calculated the values of W for various fractions of pure water recovery from sodium chloride-water and sea water solutions. Some of their data are summarised in Table 3-1. The practical energy requirements for the recovery of pure water from salt solutions in the reverse osmosis separation process are considerably higher than those given in Table 3-1; additional energy is needed to operate the system at high enough pressures, and to off-set the concentration polarisation and pressure drop effects.

3. Glueckauf Analysis

Glueckauf (1965) extended the preferential sorption-capillary flow model of Sourirajan, and attributed the desalination effect arising when aqueous electrolyte solutions are forced through a porous membrane to the repulsion which

Table 3-1. Minimum work of separation of pure water from aqueous solutions at 25°

Solution system	Feed concentration	% Recovery of pure water	Minimum work (kWhr/1000 gal)	Reference
NaCl-H_2O	3.5 Wt.% NaCl	0	2.98	
		50	4.15	Dodge and
		90	8.20	Eshaya (1960)
NaCl-H_2O	1.0 M	0	4.876	Johnson et
		25	5.650	al. (1966);
		50	6.899	Stoughton
		75	9.590	and Lietzke
		100	16.809	(1965)
NaCl-H_2O	0.5 M	0	2.404	
		25	2.770	
		50	3.351	
		75	4.532	
		100	10.080	
NaCl-H_2O	0.1 M	0	0.487	
		25	0.559	
		50	0.672	
		75	0.891	
		100	2.793	
NaCl-H_2O	0.05 M	0	0.245	
		25	0.283	
		50	0.340	
		75	0.449	
		100	1.566	
NaCl-H_2O	0.02 M	0	0.098	
		25	0.113	
		50	0.136	
		75	0.181	
		100	0.717	
Sea Water	—	0	2.680	
		25	3.095	
		50	3.736	
		75	5.133	
		100	11.699	

ions experience in the close vicinity of materials of low dielectric constant. The effect of dielectric constant difference in the membrane-solution system in reverse osmosis was considered earlier by Scatchard (1964), who concluded that the distribution of salt in the exterior phase very near the membrane surface did not affect solute separation. Glueckauf presented a different approach. His treatment is based on the following model. On entering a pore the potential of the ion steadily increases, and it may be expected to reach a maximum value at a distance (which is the same as the mean distance of the ionic cloud according to the Debye Huckel model); when this distance is exceeded, there is an appreciable chance that an ion of opposite charge would enter the pore, thus reducing considerably the potential of the first ion. One may then calculate the energy difference between the ion in solution and the ion in an aqueous hole, assumed to be in equilibrium with the bulk solution, and the probability of finding an ion at this energy level in the interior of the pore. Such calculations lead to a method of predicting the separation characteristics of a membrane for different electrolyte solution systems and feed concentrations.

Glueckauf's equation is:

$$\log_{10} \frac{\overline{m}}{m_x} = \frac{\epsilon^2 Z^2}{4.6\, DkT} \frac{(1 - \alpha)Q}{(R + \alpha b Q)} \tag{3-25}$$

where $\overline{m}$ = concentration of solution at the membrane surface

m_x = concentration of product solution

ϵ = electronic charge

Z = valency of the ion

D = dielectric constant of the medium (= 78.3 for water at 25°)

Q = $(D - D')/D$ where D' is the dielectric constant of the membrane material, taken here as 5 for cellulose acetate

k = Boltzmann constant

T = absolute temperature

R = membrane pore radius

α = $1 - (1 + K^2R^2)^{-1/2}$ where K is given by the relation below (Glasstone, 1951a)

K = $\sqrt{(4\pi\epsilon^2 \Sigma n_i Z_i^2/DkT)}$ where n_i = number of ions per cm^3

b = ionic radius

The above equation predicts that $\overline{m}/m_x$ decreases with increasing pore radius (R), decreases with increasing ionic radius (b), decreases with increasing solution concentration ($\overline{m}$), increases with increasing value of D/D', and increases with increasing ionic charge (Z).

Glueckauf found that solute separations calculated from Equation 3-25 were in good agreement with the experimental separation data of Sourirajan (1964b) for the S & S type porous cellulose acetate membranes. In view of the potential theoretical and practical importance of such an agreement, the general validity of Equation 3-25 was tested by Govindan and Sourirajan (1966) with the experimental separation data obtained with the Loeb-Sourirajan type (CA-NRC-18 type) porous cellulose acetate membranes.

Using Equation 3-25 and the experimental separation data for [0.5M NaCl-H_2O] feed solution, the effective pore radii were calculated for a set of films. Using the above pore radii, and the data on ionic radii available in the literature (Pauling, 1960; Robinson and Stokes, 1959b), the effects of the nature of the solute and of solute concentration on separation were investigated. In all cases it was assumed that the ion type giving higher separation would be the controlling one, in view of the fact that electrical neutrality must be maintained. The concentration of the solution at the membrane surface, $\overline{m}$, was assumed to be the same as that of the feed solution for the purpose of the calculations. This approximation was justified on the basis that at the feed rates used in the experiments there was little change in solute separation and product rate with change in feed flow rate (Sourirajan and Govindan, 1965).

The separation data predicted by Equation 3-25, along with the experimental results, for the systems [$NaCl-H_2O$], [$Na_2SO_4-H_2O$], [$MgCl_2-H_2O$], and [$MgSO_4-H_2O$] using four different films are given in Figure 3-4. While the equation predicts general trends at feed concentrations above approximately 1.0M, the actual agreement between the experimental and predicted data is not satisfactory. Figure 3-4 also shows the results calculated for Film 8 for the system [$NaCl-H_2O$] using the dielectric constant of the feed solution (Hasted et al., 1948) for D instead of that of water. The consequent shift in the theoretical curve does not make the agreement between the experimental and calculated values any better.

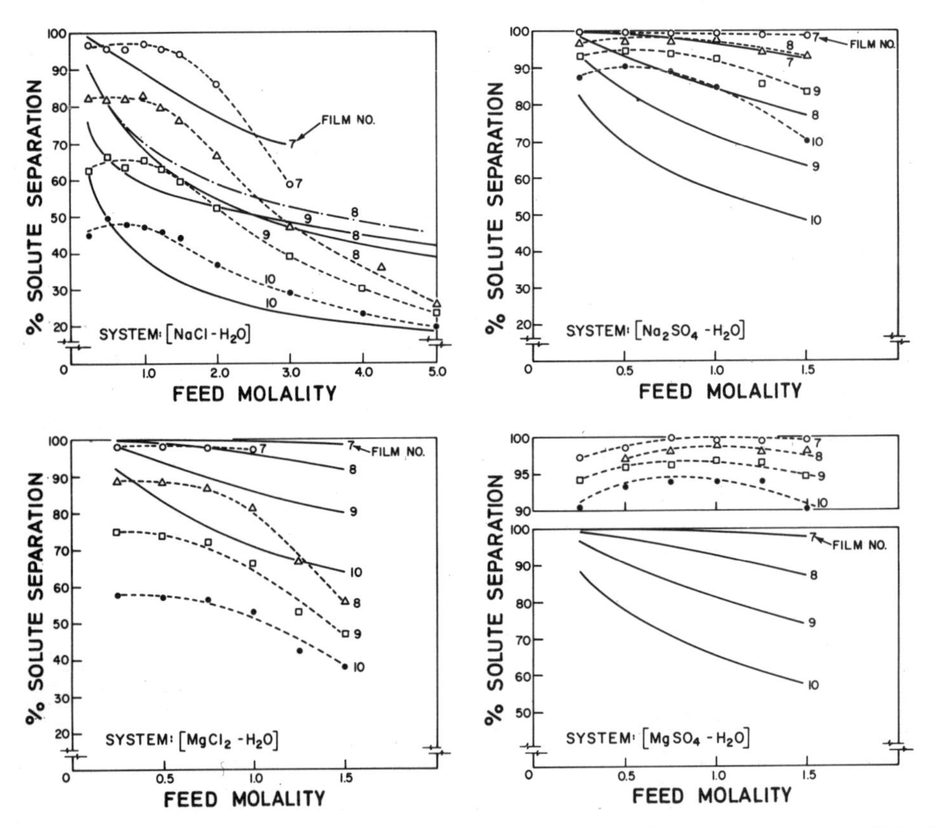

Figure 3-4. Comparison of experimental separation data with those predicted by Glueckauf's equation. Data of Govindan and Sourirajan (1966).

Film type: CA-NRC-18; feed rate: 250 cm^3/min; operating pressure: 1500 psig; —— predicted separation; ·—·—· predicted separation using D of solution; ------ Experimental.

Glueckauf's observations were based on calculated values of boundary concentrations of the feed. A set of calculations was made for one film (No. 9) using a procedure identical to that of Glueckauf. The calculated boundary layer concentrations, $\bar{m}$, and the experimental and calculated solute separations based on $\bar{m}$, along with those for the case $\bar{m}$ = feed concentration, for the systems [$NaCl-H_2O$] and [$Na_2SO_4-H_2O$] are given in Table 3-2. The results show that for the system [$NaCl-H_2O$], the calculated separation data based on $\bar{m}$ are

Table 3-2. Comparison of experimental separation data with those calculated from Equation 3-25 for Film 9 in Figure 3-4

Data of Govindan and Sourirajan (1966)

Feed conc. molality	Boundary conc. molality	Solute separation based on feed concn		Solute separation based on boundary concn	
		Exptl.	Calcd. from eqn. 3-25	Exptl.	Calcd. from eqn. 3-25
System [NaCl-H_2O]					
0.25	0.50	62.8	76.2	81.4	88.1
0.50[a]	0.84	66.4	66.4	80.0	80.0
0.75	1.13	63.5	60.9	75.8	74.1
1.0	1.37	65.3	59.0	74.7	70.0
1.5	1.87	59.3	55.3	67.3	64.2
2.0	2.31	52.4	53.6	58.8	59.9
3.0	3.25	39.1	49.1	43.8	53.0
4.0	4.26	30.0	44.7	34.2	48.1
5.0	5.22	23.5	42.0	26.7	44.5
System [Na_2SO_4-H_2O]					
0.25	0.70	93.2	83.6	97.6	94.1
0.50[a]	1.20	93.7	93.7	97.4	97.4
0.75	1.32	94.2	75.4	96.7	86.0
1.0	1.51	93.2	75.7	95.5	83.9
1.25	1.73	86.1	74.8	90.0	81.8
1.50	1.95	84.1	73.5	87.8	79.6

[a] Reference system

close to experimental values for feed concentrations up to 2.0M, but for higher feed concentrations, the agreement is poor; for the system [Na_2SO_4-H_2O], calculations based on $\bar{m}$ have not improved the agreement between the theoretical and experimental values.

The above results show that the Glueckauf analysis, though highly suggestive, is still insufficient to predict the separation characteristics of high-flow porous cellulose acetate membranes of the type employed in this study. There may be at least two reasons for this. The number of pores and pore size distribution on the membrane surface may be expected to affect solute separation; the product rates of the Loeb-Sourirajan type films are at least 20 times that of S & S type films giving the same level of solute separation under otherwise identical experimental conditions. Secondly, the dielectric constants of the membrane material, solvent, and solution may not be the only criteria governing solute separation. Nevertheless, the Glueckauf-analysis is one of real significance with respect to this separation process. It provides at least a partial answer to the fundamental question regarding the precise physicochemical criteria of preferential sorption at membrane-solution interfaces.

4. Some Empirical Correlations of Reverse Osmosis Experimental Data

Some empirical correlations of reverse osmosis experimental data have been established by Govindan and Sourirajan (1966). Since these correlations have been tested for a number of solution systems, and they cover a wide range of solute separations, they have a firm experimental basis. They may prove useful as rough approximations in preliminary design calculations in the absence of more precise information. Hence, they are summarised below.

Separation and Permeability Characteristics of Porous Cellulose Acetate Membranes for Some Related Solution Systems. The concept of related solution systems has been discussed in Chapter 1. Under experimental conditions of constant feed molality, operating pressure and temperature, and feed flow rate, if the separation data for two related systems and the product rate data for any one of them (reference system) are given, the product rate data for the other related system could be predicted from the relation

$$(PR)_2 = \left[(PR)_1 \frac{\mu_1}{\Delta P_1 \rho_1}\right]\left[\frac{\Delta P_2 \rho_2}{\mu_2}\right] \tag{3-26}$$

where

$$\Delta P = P - \Delta\Pi \tag{3-17}$$

Here P represents the operating gauge pressure and $\Delta\Pi$ the difference between the osmotic pressures of the feed and product solutions, μ and ρ refer to the viscosity and density of the product solutions, and subscripts 1 and 2 refer to the reference and related systems respectively. The validity of Equation 3-26 for the prediction of product rates for the system [$NaNO_3$-H_2O] using [$NaCl$-H_2O] as the reference system is illustrated in Figures 3-5 and 3-6. Equation 3-26 has also been found valid for the prediction of product rates for the systems [K_2SO_4-H_2O], [$CaCl_2$-H_2O], and [$CuSO_4$-H_2O] using [Na_2SO_4-H_2O], [$MgCl_2$-H_2O], and [$MgSO_4$-H_2O] as the respective reference systems at 1500 psig with 0.5M feed solutions.

Relative Separation and Permeability Data for Solution Systems involving Ions of Different Valencies. Experiments with the commercial S & S type porous cellulose acetate membranes showed that aqueous solutions containing ions of unequal valencies did not satisfy the criteria of related solution systems (Sourirajan, 1964b). The characteristics of the more porous CA-NRC-18 type cellulose acetate membranes, however, seem somewhat different. Figure 3-7 gives the relative separation and product rate data for the latter membranes for the systems [$NaNO_3$-H_2O], [$MgCl_2$-H_2O], [$AlCl_3$-H_2O], [Na_2SO_4-H_2O], and [$MgSO_4$-H_2O] under one set of operating conditions using [$NaCl$-H_2O] as the reference system. Two observations are significant. For any given membrane, the solute separation increases with increase in the valency of the ions; the effect is more pronounced with the variation of the valency of the anion. Secondly, even for solution systems involving ions of different valencies, the relative separation data fall on fairly unique lines characteristic of the membrane-solution system, similar to those obtained for solutions containing ions of the same valency; the relative product rate data, however, exhibit no such relationship.

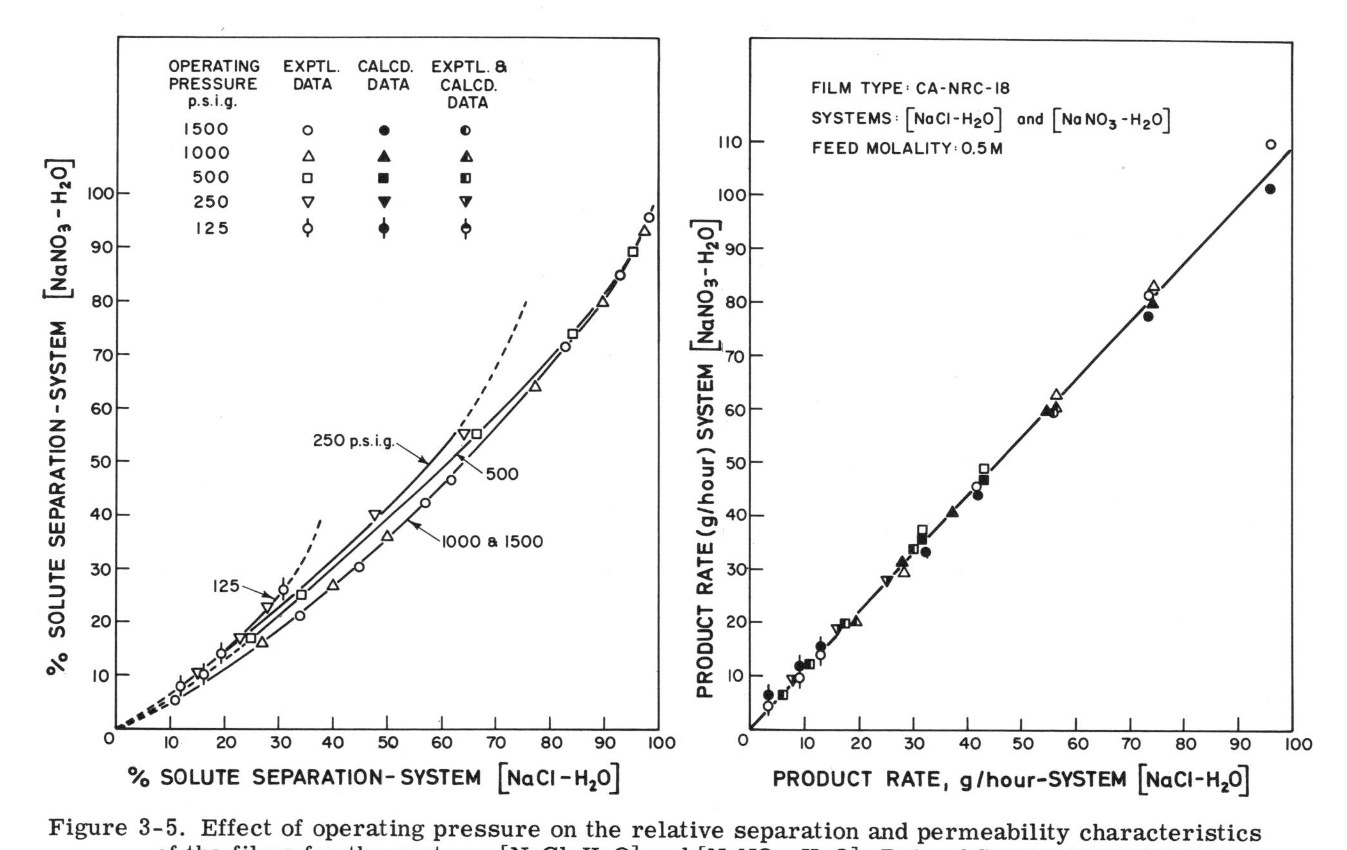

Figure 3-5. Effect of operating pressure on the relative separation and permeability characteristics of the films for the systems [NaCl-H_2O] and [$NaNO_3$-H_2O]. Data of Sourirajan and Govindan (1965).

Feed molality: 0.5M; feed rate: 250 cm^3/min; film area: 7.6 cm^2.

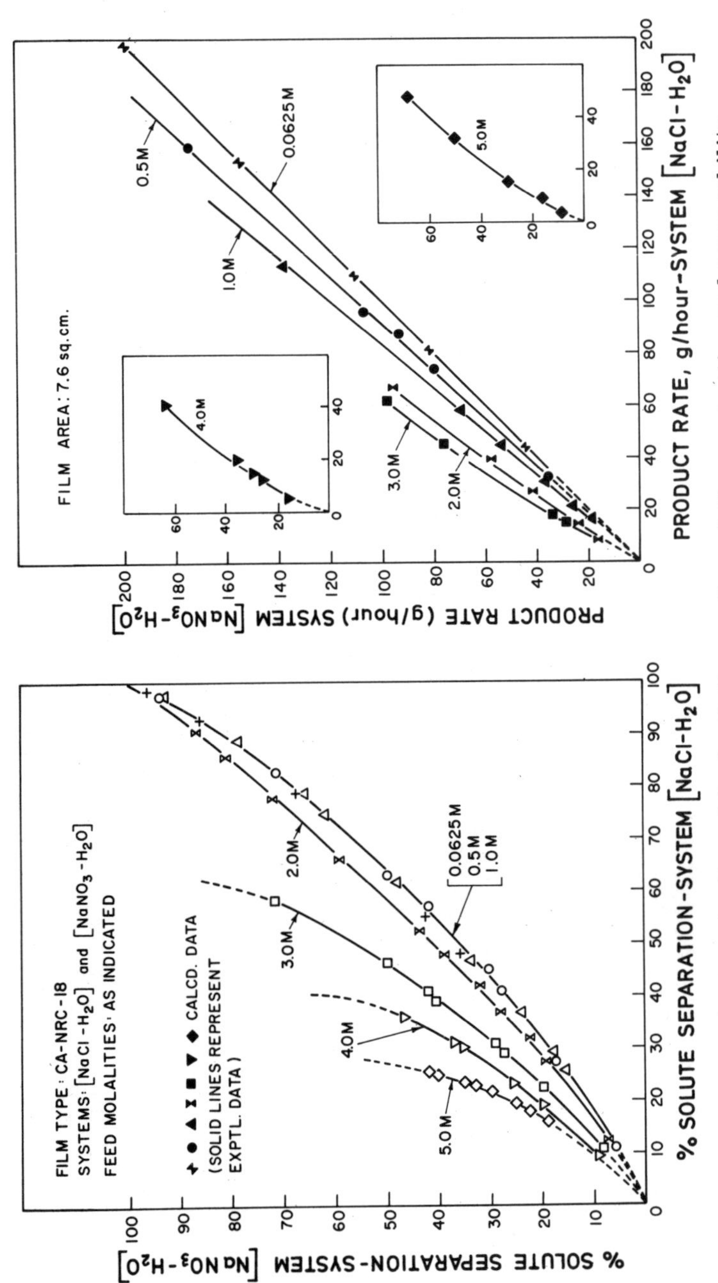

Figure 3-6. Effect of concentration of feed solution on the relative separation and permeability characteristics of the films for the systems [$NaCl$-H_2O] and [$NaNO_3$-H_2O]. Data of Sourirajan and Govindan (1965).

Feed rate: 250 cm^3/min; operating pressure: 1500 psig.

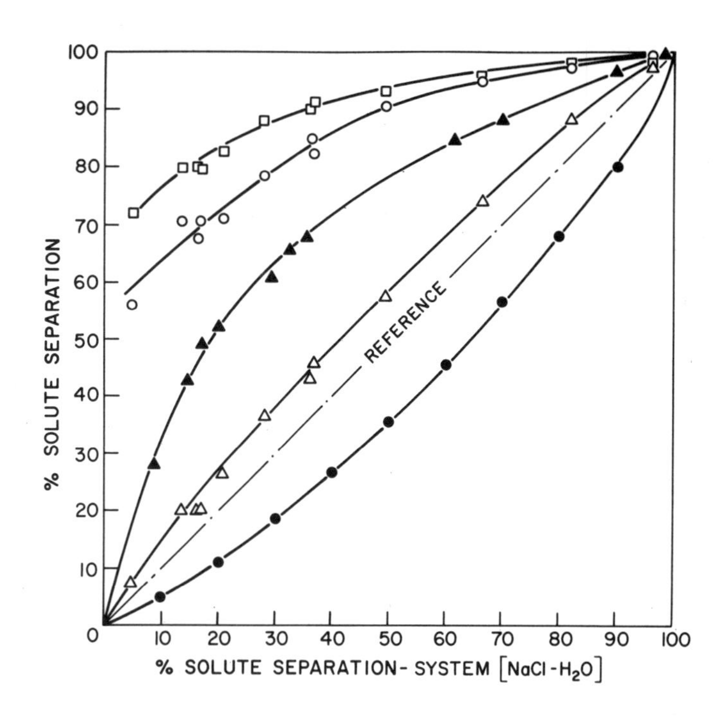

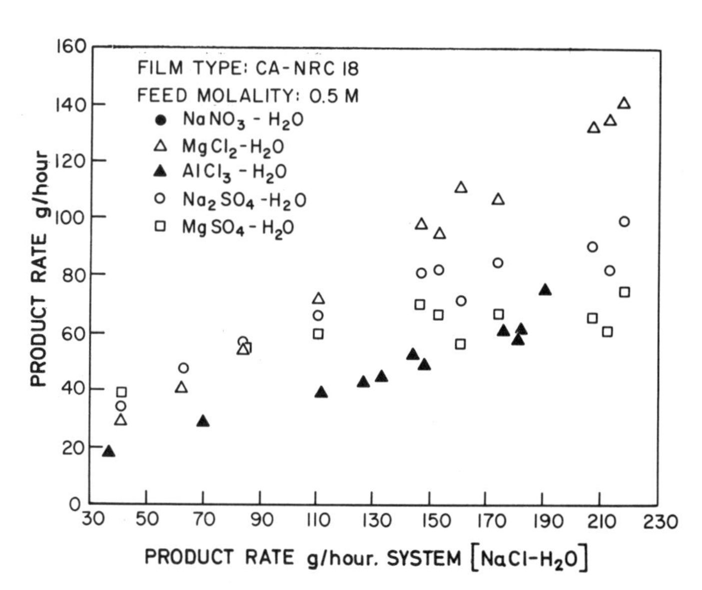

Figure 3-7. Relative separation and product rate data for solution systems involving ions of different valencies. Data of Govindan and Sourirajan (1966).

Feed rate: 250 cm^3/min; operating pressure: 1500 psig; film area: 7.6 cm^2.

Operating Pressure. The effect of operating pressure in the range 250 to 1500 psig on the separation characteristics of a set of films for the systems 0.5M [$NaCl$-H_2O], [Na_2SO_4-H_2O], and [$MgSO_4$-H_2O], 0.5M and 2.0M glycerol-water, and 0.5M sucrose-water is illustrated in Figures 3-8, 3-9 and 3-10 respectively; the corresponding product rate (PR) characteristics are plotted in Figures 3-11 to 3-13. Generally both solute separation and product rate increase with increase in operating pressure, and the following empirical equations fit the experimental data well:

$$f = aP/(bP + 1) \tag{3-27}$$

and

$$[PR] = [Ae^{-P/P_{max}} + B]\,\Delta P \frac{\rho}{\mu} \tag{3-28}$$

where f is fraction solute separation given by Equation 1-3, [PR] = product rate in grams per hour per given area, ρ and μ = density and viscosity respectively of product solution, P = operating pressure in psig, P_{max} = maximum pressure (psig) in the operating range, $\Delta P = P - \Delta\Pi$ (Equation 3-17), and a, b, A, and B are constants characteristic of the particular membrane and feed solution. When [PR] represents the pure water permeability [PWP], then $\Delta P = P$. The similarity of Equation 3-27 to the Langmuir adsorption isotherm is consistent with the preferential sorption-capillary flow mechanism.

Figure 3-14 shows the variation in the values of a and b of Equation 3-27 as a function of solute separation for the systems [$NaCl$-H_2O], [Na_2SO_4-H_2O], and [$MgSO_4$-H_2O]. The data indicate that for the systems tested, the values of a and b tend to become equal for solute separations above 80 per cent, in which case Equation 3-27 approximates to

$$f = aP/(aP + 1) \tag{3-29}$$

One may consider that the product rate [PR] is the sum of the preferentially sorbed pure water (PR_W), and the feed solution (PR_F). Figure 3-15 illustrates the variation of (PR_W) with ΔP for the systems [$NaCl$-H_2O], [Na_2SO_4-H_2O], and [$MgSO_4$-H_2O] for two different films. For the system 0.5M [$NaCl$-H_2O] the graph of PR_W v. ΔP is a straight line passing through the origin in the entire range of solute separations at operating pressures in the range 250 to 1500 psig. This correlation is of practical interest, since the separation and product rate data at any one pressure fix the relationship between (PR_W) and ΔP, which in conjunction with Equation 3-27 can be used to predict product rates for operating pressures in the range 250 to 1500 psig.

Feed Concentration. When the separation data are available, the product rate data can be correlated, at least in a limited feed concentration range, by the following empirical equation:

$$\frac{[PR]}{\Delta P}\frac{\mu}{\rho} = \frac{K_1}{K_2 + m^n} \tag{3-30}$$

where m is solute molality in feed, and K_1, K_2, and n are constants. When [PR] represents [PWP], Equation 3-30 becomes

$$\frac{[PWP]}{P}\cdot\frac{\mu_W}{\rho_W} = \frac{K_1}{K_2} \tag{3-31}$$

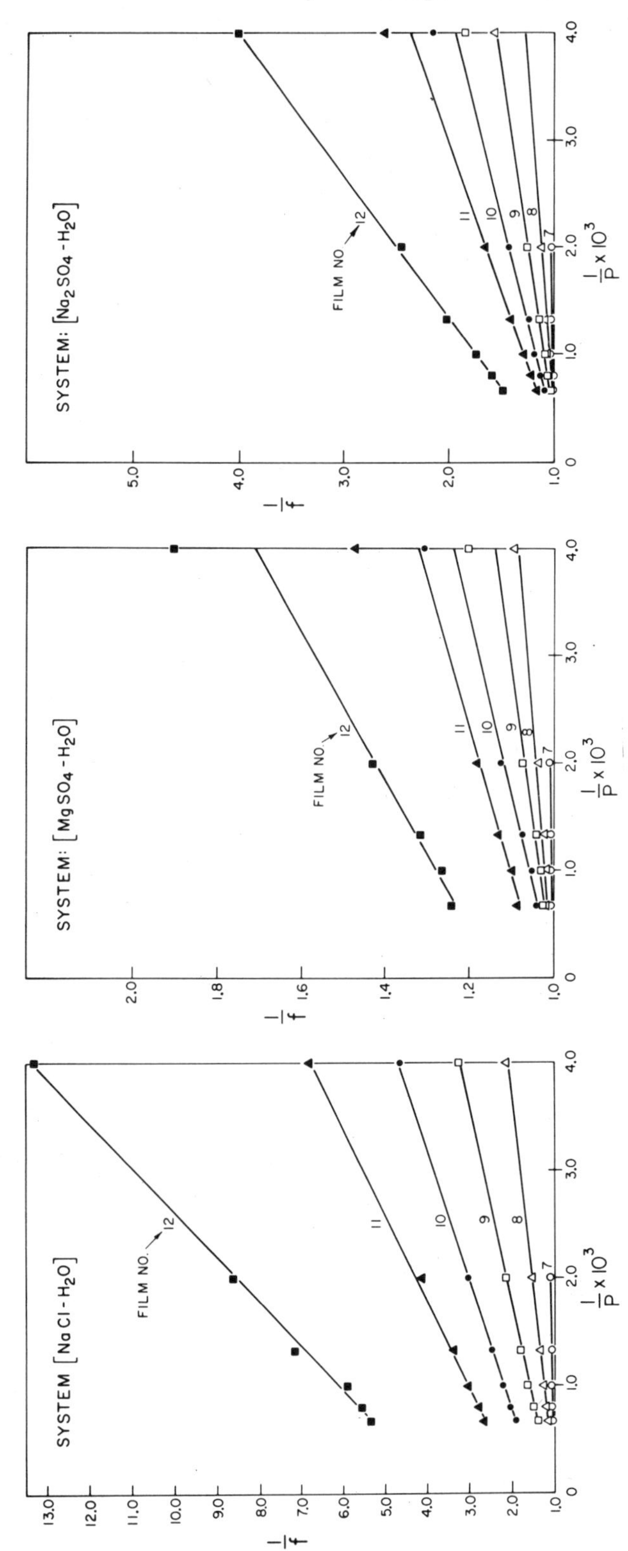

Figure 3-8. Effect of pressure on solute separation for systems [NaCl-H_2O], [Na_2SO_4-H_2O], and [$MgSO_4$-H_2O]. Data of Govindan and Sourirajan (1966).

Film type: CA-NRC-18; feed molality: 0.5M; feed rate: 250 cm^3/min.

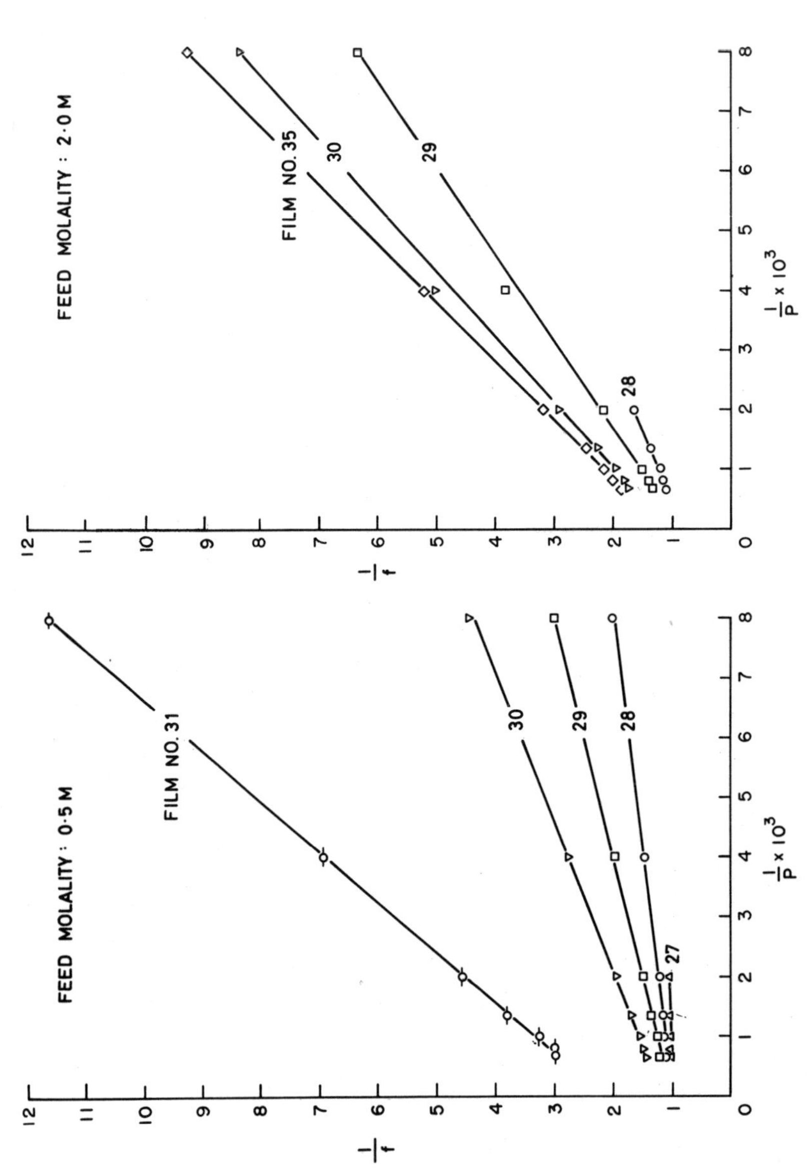

Figure 3-9. Effect of operating pressure on solute separation for the system glycerol-water. Data of Sourirajan and Kimura (1967).

Film type: CA-NRC-18; feed rate: 380 cm^3/min; operating pressure: P psig.

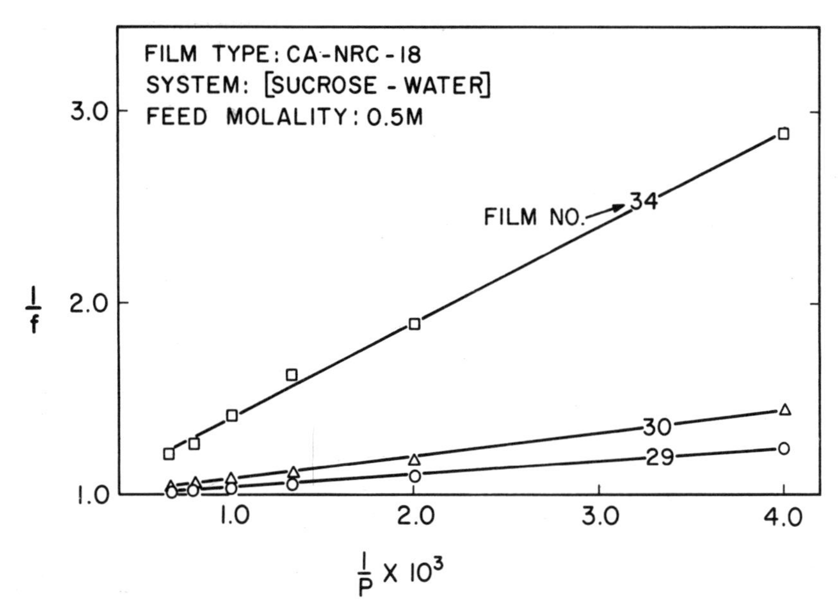

Figure 3-10. Effect of operating pressure on solute separation for the system sucrose-water. Data of Sourirajan (1967a).

Feed rate: 380 cm^3/min; operating pressure = P psig.

where μ_W and ρ_W represent respectively the viscosity and density of pure water. Combining Equations 3-30 and 3-31,

$$\left[\frac{\frac{[PWP]}{P} \cdot \frac{\mu_W}{\rho_W}}{\frac{[PR]}{\Delta P} \cdot \frac{\mu}{\rho}} - 1\right] = \frac{m^n}{K_2} \qquad (3\text{-}32)$$

Therefore the plot of

$$\left(\left[\frac{\frac{[PWP]}{P} \frac{\mu_W}{\rho_W}}{\frac{[PR]}{P} \frac{\mu}{\rho}}\right] - 1\right)$$

v. m on a log-log scale is a straight line. Such plots of the data for the systems [$NaCl$-H_2O], [Na_2SO_4-H_2O], [$MgCl_2$-H_2O], [$MgSO_4$-H_2O], [glycerol-water], and [sucrose-water] are illustrated in Figures 3-16, 3-17, and 3-18.

Operating Temperature. The experimental results on the effect of temperature in the range 10° to 35° on the pure water permeability through the films, and the separation and product rates for 0.5M and 2.0M [$NaCl$-H_2O] systems may be summarised as follows. For the system 0.5M [$NaCl$-H_2O], temperature has little effect on solute separation; but for the system 2.0M [$NaCl$-H_2O], solute separation at 35° is about 5 per cent less than at 10° for films giving more than 80 per cent solute separation. The [PWP] and [PR] values for both the systems increase with temperature. The [PWP] values,

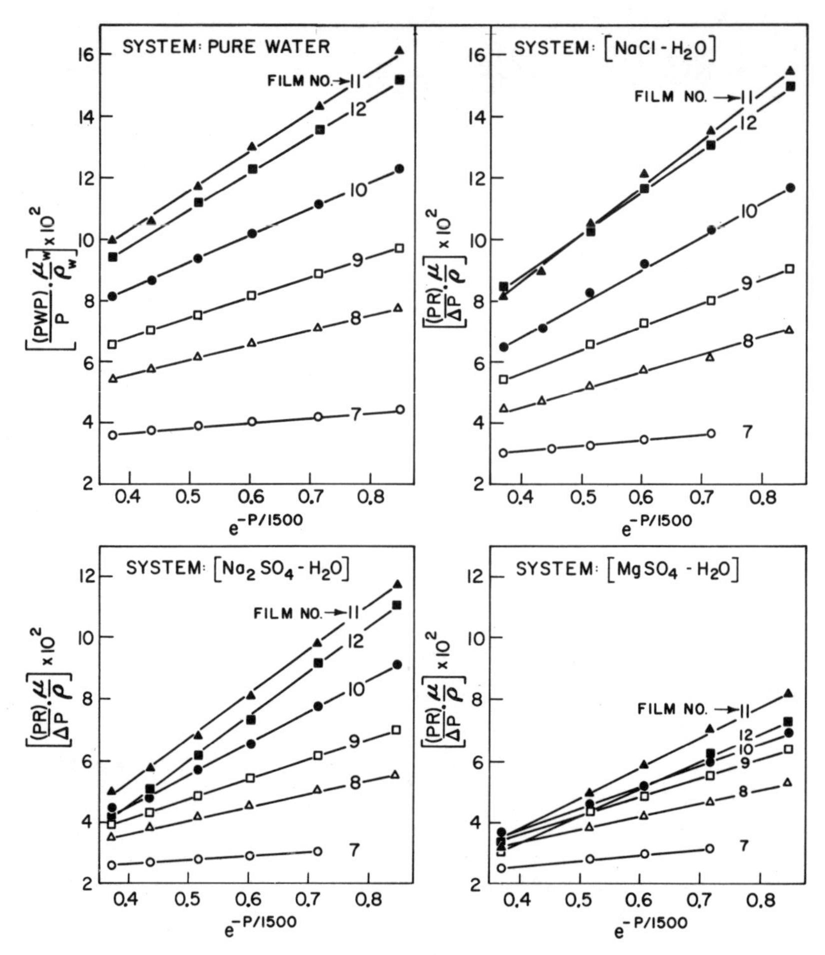

Figure 3-11. Effect of operating pressure on pure water permeability and product rates for the systems [NaCl-H_2O], [Na_2SO_4-H_2O], and [$MgSO_4$-H_2O]. Data of Govindan and Sourirajan (1966).

Film type: CA-NRC-18; feed molality: 0.5M; feed rate: 250 cm^3/min; film area: 7.6 cm^2.

and the [PR] values for the system 0.5M [NaCl-H_2O] when multiplied by the respective viscosities at the operating temperature, give essentially constant values for each film; but [PR] $\times$ μ for the system 2.0M[NaCl-H_2O] decreased with temperature. The latter results are illustrated in Figure 3-19.

5. Kimura-Sourirajan Analysis of Reverse Osmosis Experimental Data

This analysis (Kimura and Sourirajan, 1967) is based on a generalised capillary-diffusion model for the transport of solute through the membrane, and it is applicable for the entire possible range of solute separations in the reverse osmosis process. In this analysis, it is considered that when the size of the pores on the membrane surface is only a few times bigger than the size of the permeating molecules, and the interfacial forces are important

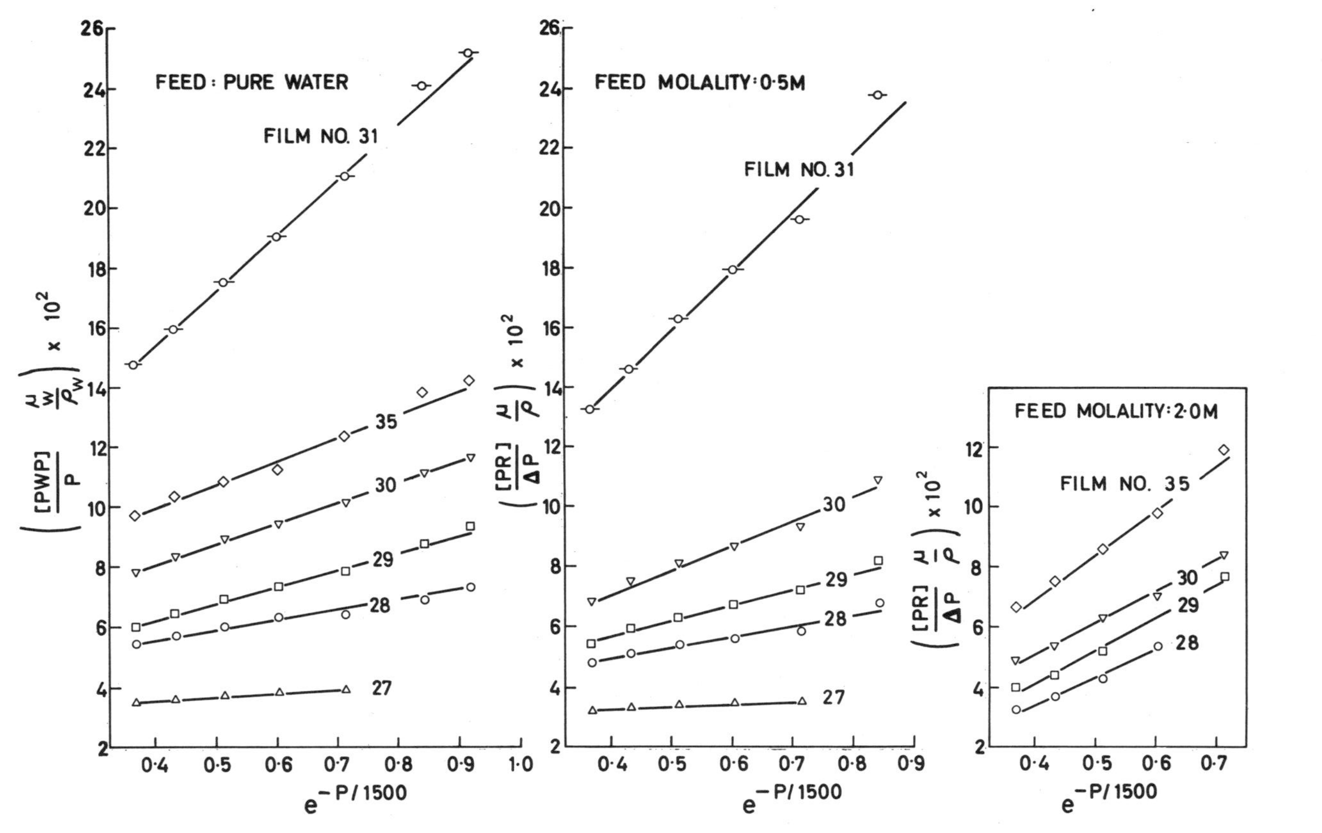

Figure 3-12. Effect of operating pressure on pure water permeability and product rate for the system glycerol-water. Data of Sourirajan and Kimura (1967).

Film type: CA-NRC-18; feed rate: 380 cm^3/min; operating pressure: P psig; film area: 7. 6 cm^2.

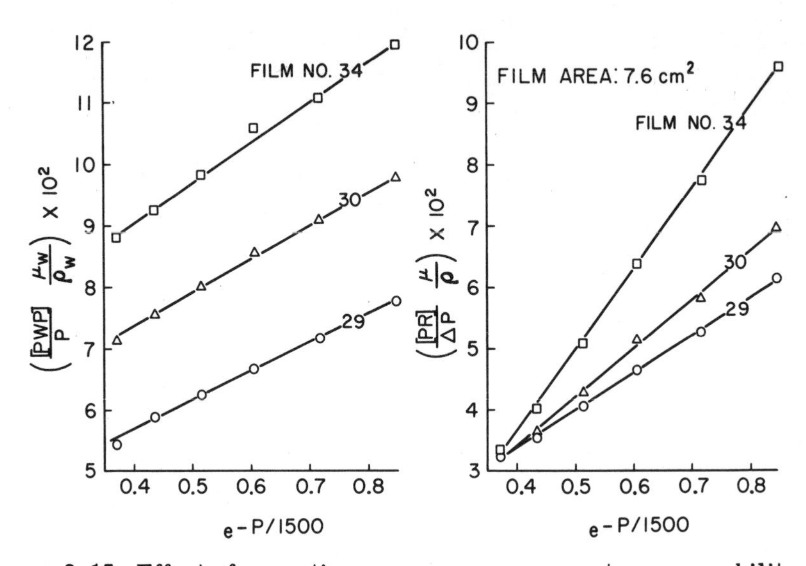

Figure 3-15. Effect of operating pressure on pure water permeability and product rate for the system sucrose-water. Data of Sourirajan (1967a).

Film type: CA-NRC-18; feed molality: 0·5M; feed rate: 380 cm^3/min; operating pressure = P psig.

enough to cause solute separation, the transport of solvent water through the porous membrane is proportional to the effective pressure, and that of the solute is due to pore diffusion and hence proportional to its concentration difference across the membrane.

This analysis has proved effective for the treatment of reverse osmosis data. It has led to a useful method of membrane specification, expressing membrane selectivity, and predicting membrane performance from a minimum of experimental data. It offers a rational basis for reverse osmosis process design as a general chemical engineering unit operation. The transport equations based on this analysis are derived and illustrated in detail in this and the next chapter.

Figure 3-20 represents a diagrammatic representation of a reverse osmosis process in continuous operation. The existence and the continuous withdrawal of the preferentially sorbed interfacial layer along with the bulk feed solution through the porous membrane give rise to a product solution which is less concentrated than the feed solution, and to a more concentrated boundary solution between the interfacial region on the membrane surface and the bulk feed solution. Consequently, there arises a concentration gradient between the boundary solution and the bulk feed solution. This concentration gradient, called concentration polarisation, has important effects on the performance of the reverse osmosis unit; this subject is discussed in detail in the next chapter. It is obvious that with respect to a given membrane-solution system, the magnitude of the interfacial region is a function of the concentration of the boundary solution, and the effective driving pressure for fluid flow through the membrane is the operating pressure, P, minus the difference between the osmotic pressures Π, of the concentrated boundary solution and of the membrane permeated product solution.

Let X, N, and c represent mole fraction, solute or solvent flux through the membrane (in g. mole/cm^2/sec), and molar density of solution (in g. mole/cm^3)

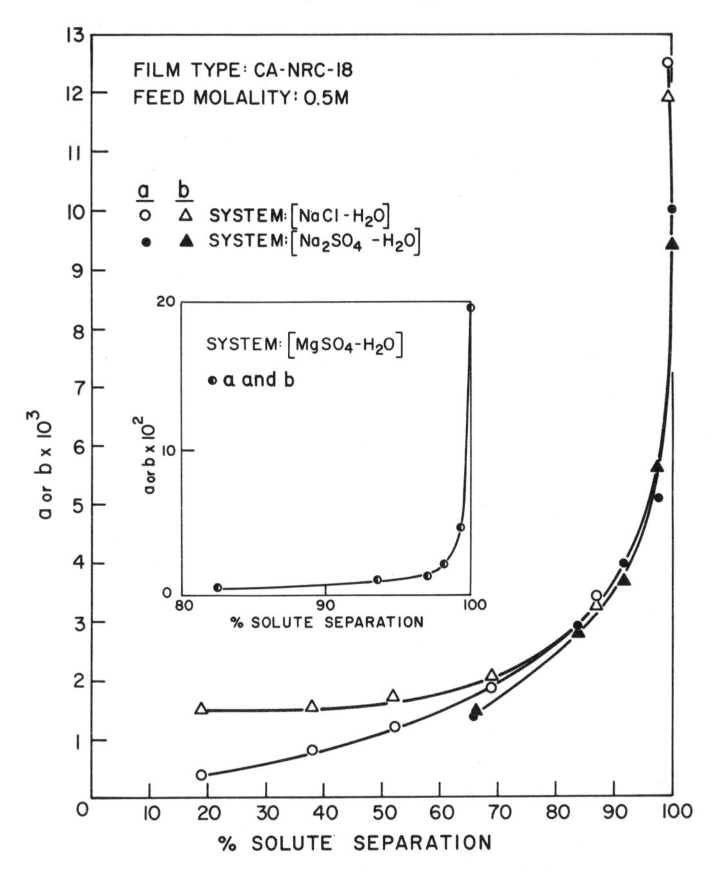

Figure 3-14. Variation of a and b in Equation 3-27 with solute separation. Data of Govindan and Sourirajan (1966).

Feed rate: 250 cm^3/min; operating pressure: 1500 psig.

respectively, and let subscripts A, B, M, 1, 2, and 3 represent solute, solvent-water, membrane phase, bulk feed solution, concentrated boundary solution, and the membrane permeated product solution respectively. Thus, X_{AM}, X_{A1}, X_{A2}, and X_{A3} are the mole fractions of solute in the membrane phase, bulk feed solution and the concentrated boundary solution on the high pressure side of the membrane, and the membrane permeated product solution on the atmospheric pressure side of the membrane respectively; the symbols for the molar densities c_M, c_1, c_2, and c_3 have similar meanings. Also N_A and N_B are the solute and solvent flux through the membrane respectively.

Pure Water Permeability Constant. For a given area of membrane surface (S cm^2), the pure water permeability, [PWP] grams/hour, is proportional to operating pressure (P atm); and the proportionality constant (given as g.mole H_2O/cm^2 sec.atm) is represented by the symbol A defined as follows:

$$A = \frac{[PWP]}{M_B \times S \times 3600 \times P} \tag{3-33}$$

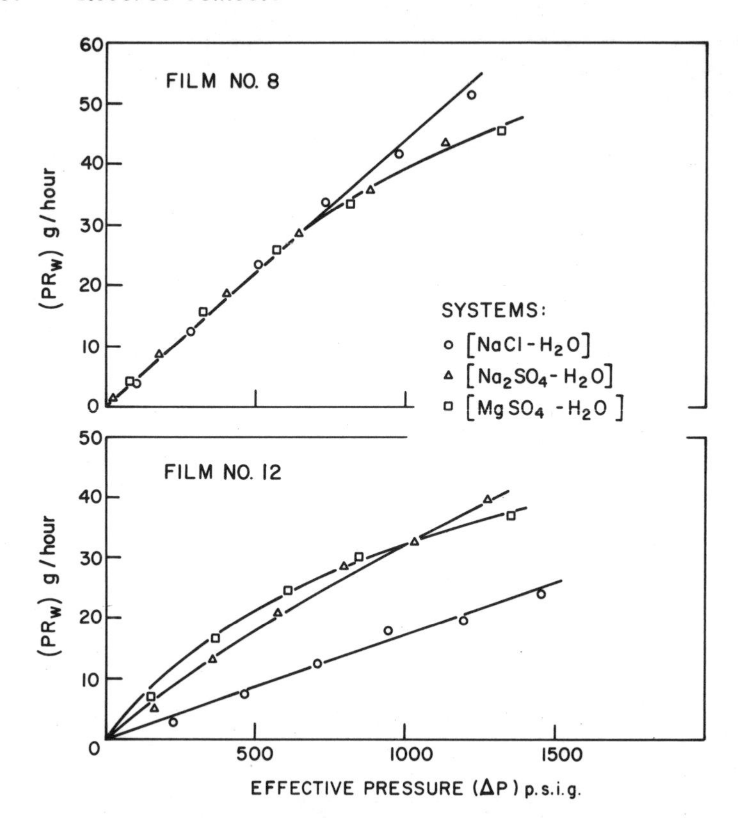

Figure 3-15. Variation of (PR_W) with effective pressure for the systems [NaCl-H_2O], [Na_2SO_4-H_2O], and [$MgSO_4$-H_2O]. Data of Govindan and Sourirajan (1966)

Film area: 7.6 cm^2; film type: CA-NRC-18; feed molality: 0·5M; feed rate: 250 cm^3/min.

where M_B is the molecular weight of water. The quantity A is a measure of the overall porosity of the film given in terms of the permeation rate of pure water for which the membrane material has a preferential sorption from the aqueous solution in the reverse osmosis process; further, A corresponds to conditions of zero concentration polarisation and it is independent of any solute under consideration.

Transport of Solvent Water Through the Porous Membrane. Solvent water transported through the membrane is the total water content in the product solution which includes the preferentially sorbed water. The above solvent-water flux (N_B) through the membrane is proportional to the effective pressure (ΔP) where the proportionality constant is A. Thus

$$N_B = A\,\Delta P = A[P - \{\Pi(X_{A2}) - \Pi(X_{A3})\}] \qquad (3\text{-}34)$$

$$= A[P - \Pi(X_{A2}) + \Pi(X_{A3})] \qquad (3\text{-}35)$$

where $\Pi(X_{A2})$ and $\Pi(X_{A3})$ represent the osmotic pressure Π corresponding to mole fractions of solute X_{A2} and X_{A3} respectively. Equation 3-35 is applicable for systems where the kinematic viscosity of the product solution is not too

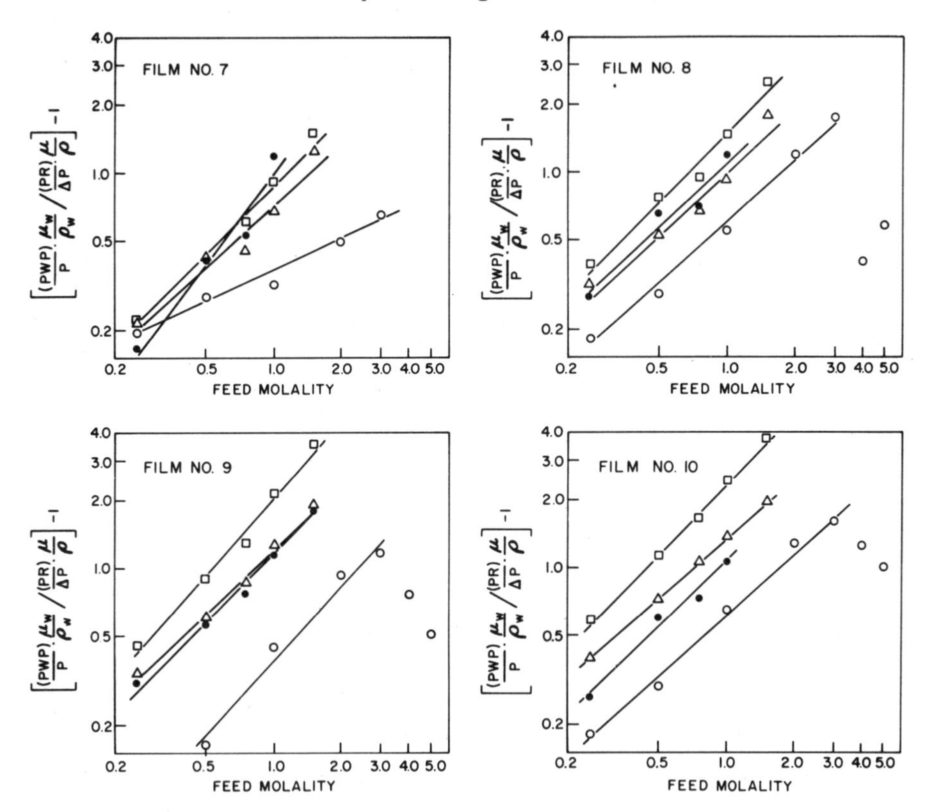

Figure 3-16. Effect of feed concentration on product rates for the systems [NaCl-H_2O], [Na_2SO_4-H_2O], [$MgCl_2$-H_2O], and [$MgSO_4$-H_2O]. Data of Govindan and Sourirajan (1966).

Film type: CA-NRC-18; feed rate: 250 cm^3/min; operating pressure: 1500 psig.

Systems: ○ [NaCl-H_2O]
△ [Na_2SO_4-H_2O]
● [$MgCl_2$-H_2O]
□ [$MgSO_4$-H_2O]

different from that of pure water. This condition is reasonably satisfied in most cases of practical interest.

The quantities N_B, A, P, and X_{A3} are obtained from the experimental data on [PWP], [PR], solute separation, operating pressure, and membrane area. Using the above data, one can calculate $\Pi(X_{A2})$, and hence X_{A2}, from Equation 3-35. Since the value of A must be known for obtaining X_{A2}, it is a fundamental quantity with respect to the membrane and to the reverse osmosis process.

At a given operating pressure, A is a constant, and it is independent of any solution under consideration; but N_B and X_{A3} are functions of feed concentration and feed flow rate (or degree of turbulence) on the membrane surface. Equation 3-35, by itself, does not suggest the existence or otherwise of any unique relationship (independent of feed concentration and feed flow rate) between X_{A2} and X_{A3}.

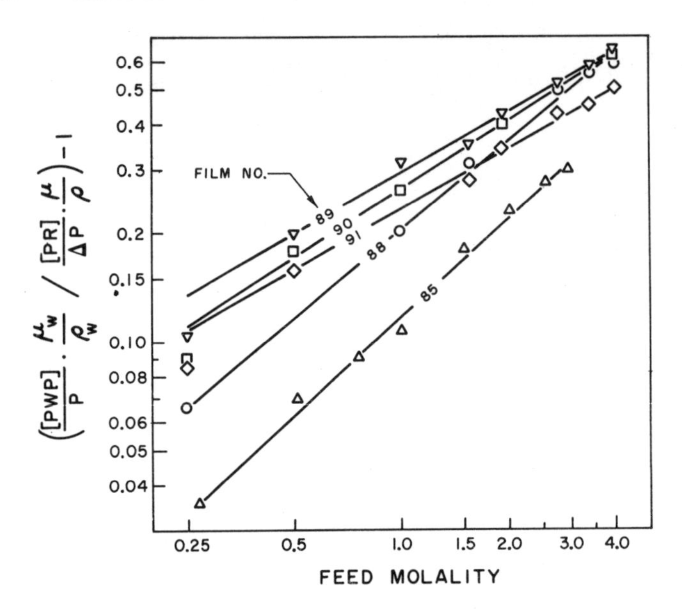

Figure 3-17. Effect of feed concentration on product rate for the system glycerol-water. Data of Sourirajan and Kimura (1967).

Film type: CA-NRC-18; feed rate: 380 cm^3/min; operating pressure: 1500 psig.

Transport of Solute Through the Membrane Phase. During the reverse osmosis process in continuous steady-state operation, a concentration difference exists on either side of the porous membrane; under such conditions, the transport of solute through the membrane phase is treated as being due to pore diffusion. Consequently, the solute flux through the membrane is proportional to the concentration difference on either side of the membrane phase. Thus

$$N_A = \frac{D_{AM}}{\delta}(c_{M2}X_{AM2} - c_{M3}X_{AM3}) \tag{3-36}$$

where X_{AM2} and X_{AM3} are mole fractions of solute in the membrane phase in equilibrium with X_{A2} and X_{A3} in the solution phases respectively, c_{M2} and c_{M3} are the molar densities corresponding to X_{AM2} and X_{AM3} in the membrane phase, D_{AM} is the diffusivity of the solute in the membrane phase, and δ is the effective thickness of the membrane.

None of the quantities on the right side of Equation 3-36 are known or easily measurable; and the dividing line in the membrane phase, between the regions corresponding to X_{AM2} and X_{AM3}, is only conceptual.

Equation 3-36 can be transformed into one containing measurable quantities and a group of unknown quantities, by assuming a simple linear relationship between X_A (concentration in the solution phase) and X_{AM} (concentration in the membrane phase). Thus let

$$cX_A = Kc_M X_{AM} \tag{3-37}$$

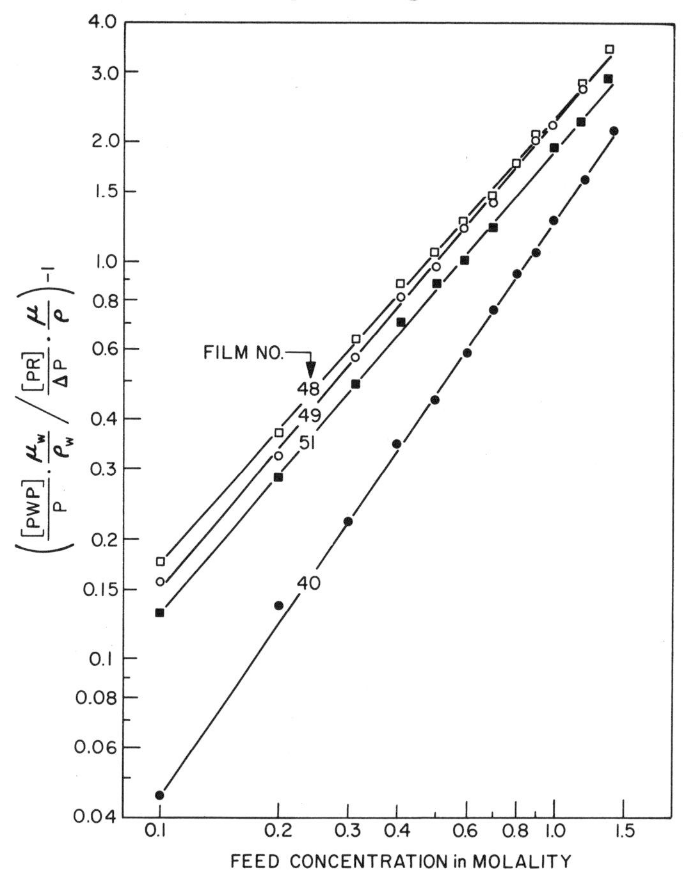

Figure 3-18. Effect of feed concentration on product rate for the system sucrose-water. Data of Sourirajan (1967a).

Film type: CA-NRC-18; feed rate: 380 cm^3/min; operating pressure: 1500 psig.

where K is a constant. Rewriting Equation 3-37 for equilibrium conditions on either side of the membrane

$$c_2 X_{A2} = K c_{M2} X_{AM2} \tag{3-38}$$

$$c_3 X_{A3} = K c_{M3} X_{AM3} \tag{3-39}$$

Equation 3-36 may now be written as

$$N_A = (D_{AM}/K\delta)\ (c_2 X_{A2} - c_3 X_{A3}) \tag{3-40}$$

Since N_A, c_2, c_3, and X_{A3} are measurable quantities and X_{A2} is obtainable from Equation 3-35, the quantity $(D_{AM}/K\delta)$, called the solute transport parameter, can be calculated from Equation 3-40. Since

$$N_A/(N_A + N_B) = X_{A3} \tag{3-41}$$

$$N_A = \frac{X_{A3}}{(1 - X_{A3})} N_B \tag{3-42}$$

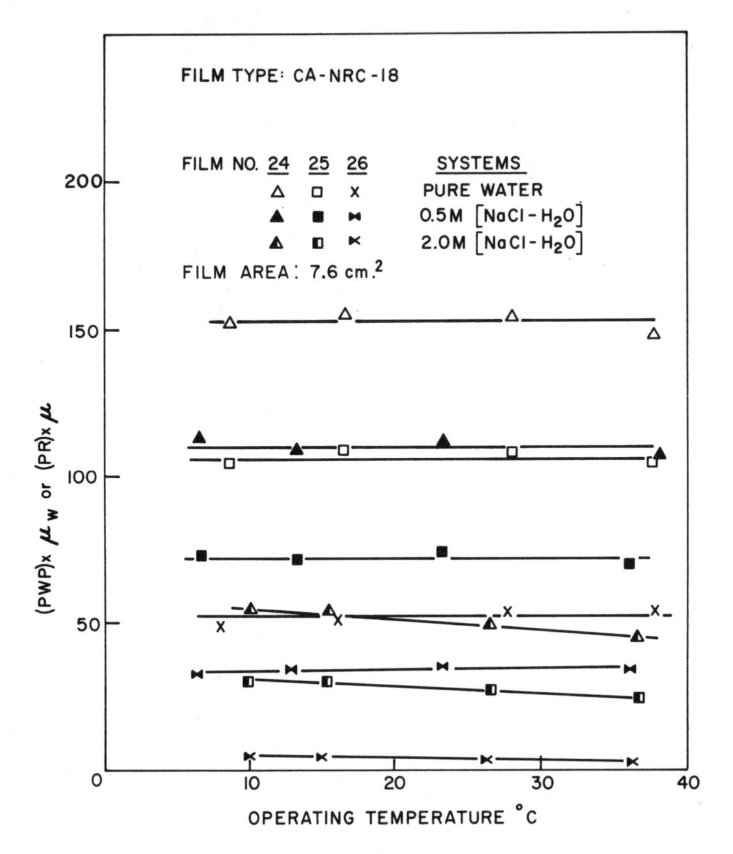

Figure 3-19. Effect of temperature on pure water permeability and product rate. Data of Govindan and Sourirajan (1966).

Feed rate: 250 cm^3/min; operating pressure: 1500 psig.

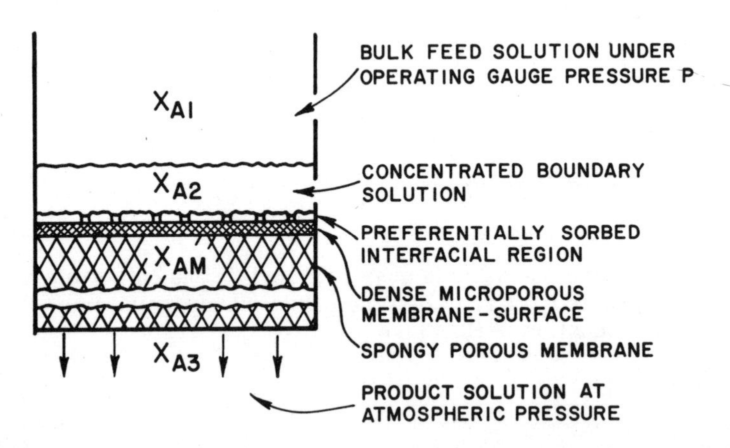

Figure 3-20. Reverse osmosis process under steady state conditions.

Substituting Equation 3-42 in Equation 3-40,

$$N_B = \left(\frac{D_{AM}}{K\delta}\right)\frac{(1 - X_{A3})}{X_{A3}}(c_2X_{A2} - c_3X_{A3}) \tag{3-43}$$

The parameter $(D_{AM}/K\delta)$ plays the role of a mass transfer coefficient with respect to solute transport through the membrane; hence it is treated as a single quantity for purposes of analysis. It must however be understood that $(D_{AM}/K\delta)$ is not a single factor, but is a combination of several inter-related factors none of which is, or need be, precisely known for chemical engineering calculations.

In the original derivation of Kimura and Sourirajan (1967) Equation 3-37 and hence Equations 3-38 and 3-39 were assumptions. Since X_{A2} and X_{A3} are bound to the membrane phase concentrations by the same relation 3-37, Equations 3-38 and 3-39 imply that with respect to any membrane, the values of X_{A2} and X_{A3} must be uniquely related, and this relationship must be independent of feed concentrations and feed flow rate past the surface of the membrane. That this is indeed true has been shown experimentally as illustrated with the data for the systems sodium chloride-water, glycerol-water, and urea-water in Figures 3-21, 3-22, and 3-23 respectively. Consequently, Equation 3-37 is no longer an assumption; it rests on the basis of extensive experimental results analysed in terms of Equations 3-33 and 3-35.

Mass Transfer on the High Pressure Side of the Membrane. Since the solute in the concentrated boundary solution also diffuses back to the less concentrated feed solution on the high pressure side of the membrane, a mass transfer coefficient, k, characteristic of the experimental conditions on the high pressure side of the reverse osmosis process can be calculated on the basis of the simple 'film theory'. The solute transfer from the concentrated boundary solution may be represented by the relation (Bird et al., 1960)

$$N_A = X_A(N_A + N_B) - D_{AB}c_1\, dX_A/dz \tag{3-44}$$

where D_{AB} is the diffusivity of solute in the aqueous feed solution, and z is the thickness of the concentrated boundary layer. Using Equation 3-41, Equation 3-44 can be written as

$$\frac{dX_A}{dz} - \frac{(N_A + N_B)}{c_1D_{AB}}X_A = -\frac{(N_A + N_B)}{c_1D_{AB}}X_{A3} \tag{3-45}$$

The boundary conditions for Equation 3-45 are:

when $z = o$, $X_A = X_{A1}$

and

when $z = l$, $X_A = X_{A2}$

Solving the simple differential Equation 3-45 with the above boundary conditions,

$$X_{A2} = X_{A3} + (X_{A1} - X_{A3})\exp\left[\frac{(N_A + N_B)}{c_1}\cdot\frac{l}{D_{AB}}\right] \tag{3-46}$$

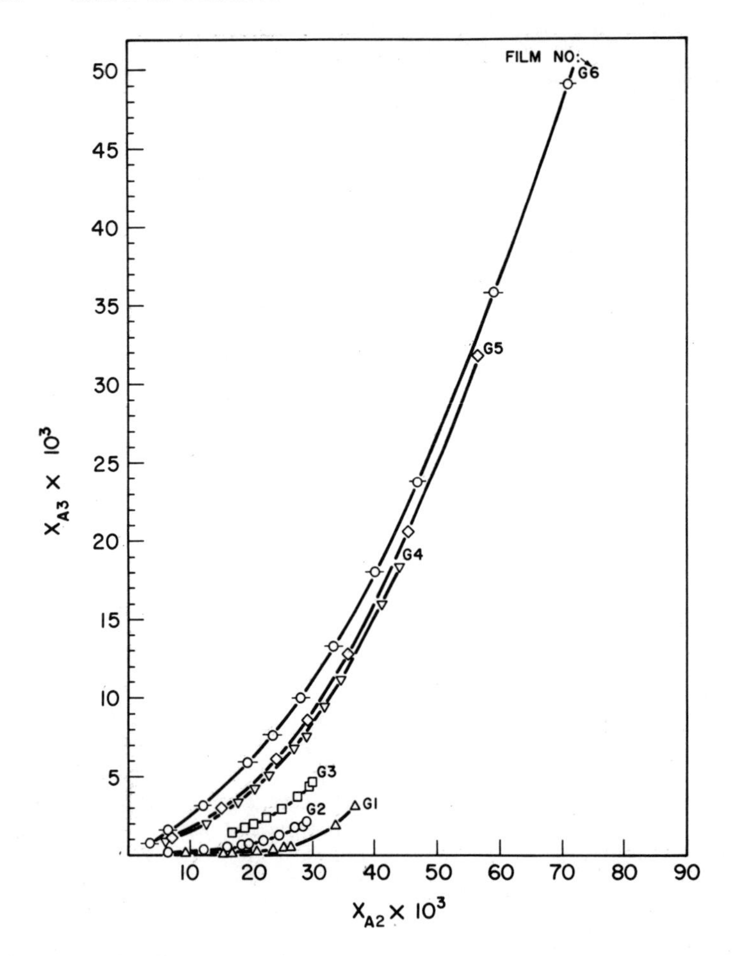

Figure 3-21. Concentration of solute in the boundary solution v. that in the product solution for the system sodium chloride-water.

Film type: CA-NRC-18; feed molality: 0.125-4.0M; feed rate: 25-400 cm^3/min; operating pressure: 1500 psig.

or

$$\ln\left(\frac{X_{A2} - X_{A3}}{X_{A1} - X_{A3}}\right) = \frac{(N_A + N_B)}{c_1} \cdot \frac{l}{D_{AB}} \tag{3-47}$$

Defining the mass transfer coefficient, k, on the high pressure side of the membrane in the conventional manner of the film theory (Sherwood, 1959; Sherwood and Pigford, 1952; Treybal, 1955),

$$k = D_{AB}/l \tag{3-48}$$

Equation 3-47 can be written as

$$\ln\left(\frac{X_{A2} - X_{A3}}{X_{A1} - X_{A3}}\right) = \frac{(N_A + N_B)}{kc_1} \tag{3-49}$$

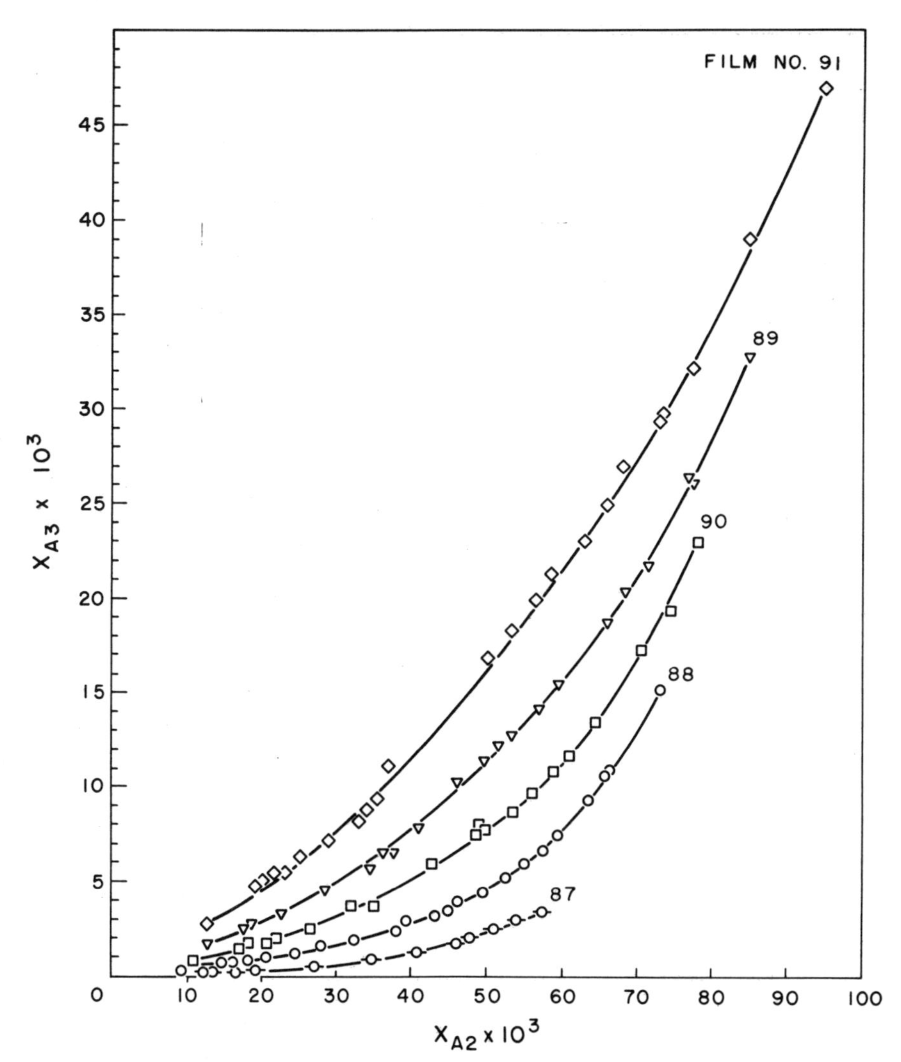

Figure 3-22. Concentration of solute in the boundary solution v. that in the product solution for the system glycerol-water. Data of Sourirajan and Kimura (1967).

Film type: CA-NRC-18; feed molality: 0.25-4.0M; feed rate: 25-725 cm^3/min; operating pressure: 1500 psig.

From Equation 3-42,

$$N_A + N_B = N_B/(1 - X_{A3}) \tag{3-50}$$

Substituting Equation 3-50 in Equation 3-49,

$$\ln\left(\frac{X_{A2} - X_{A3}}{X_{A1} - X_{A3}}\right) = \frac{N_B}{kc_1(1 - X_{A3})} \tag{3-51}$$

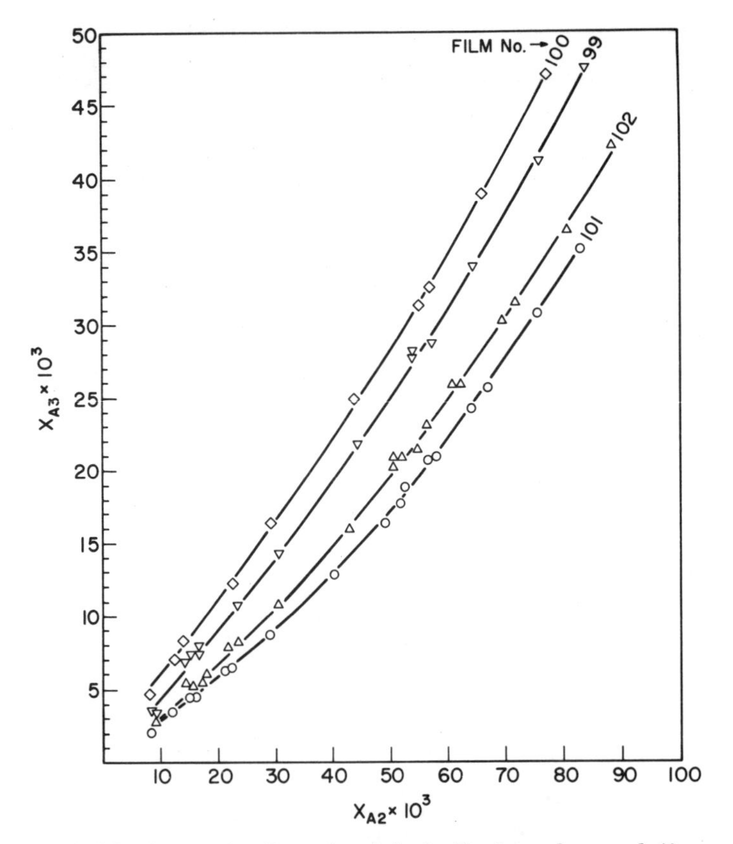

Figure 3-23. Concentration of solute in the boundary solution v. that in the product solution for the system urea-water. Data of Ohya and Sourirajan (1969a).

Film type: CA-NRC-18; feed molality: 0.25-4.0M:, feed rate: 45 to 420 cm^3/min; operating pressure: 1500 psig.

or

$$N_B = kc_1(1 - X_{A3}) \ln \left(\frac{X_{A2} - X_{A3}}{X_{A1} - X_{A3}} \right) \tag{3-52}$$

Basic Transport Equations. From the foregoing analysis, the relations

$$A = \frac{[PWP]}{M_B \times S \times 3600 \times P} \tag{3-33}$$

$$N_B = A[P - \Pi(X_{A2}) + \Pi(X_{A3})] \tag{3-35}$$

$$= \left(\frac{D_{AM}}{K\delta} \right) \frac{(1 - X_{A3})}{X_{A3}} (c_2 X_{A2} - c_3 X_{A3}) \tag{3-43}$$

$$= kc_1(1 - X_{A3}) \ln \left(\frac{X_{A2} - X_{A3}}{X_{A1} - X_{A3}} \right) \tag{3-52}$$

emerge as a set of basic equations describing the solute and solvent transport in a reverse osmosis process involving binary aqueous solutions, and membranes having a preferential sorption for water from such aqueous solutions.

In many aqueous solutions (such as sodium chloride-water) and in concentration ranges of practical interest, the molar densities c_1, c_2, c_3 may essentially be the same. In such cases, an average value can be used to represent the molar density (c) of the solution and it may be assumed that

$$c = c_1 = c_2 = c_3 \tag{3-53}$$

The set of basic transport Equations 3-35, 3-43, and 3-52 can then be written as

$$N_B = A[P - \Pi(X_{A2}) + \Pi(X_{A3})] \tag{3-35}$$

$$= c\left(\frac{D_{AM}}{K\delta}\right)\frac{(1 - X_{A3})}{X_{A3}}(X_{A2} - X_{A3}) \tag{3-54}$$

$$= kc(1 - X_{A3})\ \ln\left(\frac{X_{A2} - X_{A3}}{X_{A1} - X_{A3}}\right) \tag{3-55}$$

Solute Separation in Terms of Mole-fraction. Let W_A and W_B represent respectively the weight in grams of solute and solvent in an aqueous solution, and let M_A represent the molecular weight of the solute. Solute separation, f, can be expressed in terms of mole fractions (X_A) as follows

$$f = (m_1 - m_3)/m_1 = 1 - m_3/m_1 \tag{3-56}$$

$$m = \frac{W_A}{M_A} \cdot \frac{1000}{W_B} \tag{3-57}$$

$$X_A = \frac{W_A/M_A}{\dfrac{W_A}{M_A} + \dfrac{W_B}{M_B}} \tag{3-58}$$

$$1 - X_A = \frac{W_B/M_B}{\dfrac{W_A}{M_A} + \dfrac{W_B}{M_B}} \tag{3-59}$$

$$\frac{X_A}{1 - X_A} = \frac{W_A/M_A}{W_B/M_B} \tag{3-60}$$

Therefore

$$m = \frac{X_A}{1 - X_A} \cdot \frac{W_B}{M_B} \cdot \frac{1000}{W_B} = \frac{X_A}{1 - X_A} \cdot \frac{1000}{M_B} \tag{3-61}$$

Therefore

$$m_1 = \frac{X_{A1}}{1 - X_{A1}} \cdot \frac{1000}{M_B} \tag{3-62}$$

and

$$m_3 = \frac{X_{A3}}{1 - X_{A3}} \cdot \frac{1000}{M_B} \tag{3-63}$$

Therefore

$$f = 1 - \left(\frac{X_{A3}}{1 - X_{A3}}\right)\left(\frac{1 - X_{A1}}{X_{A1}}\right) \tag{3-64}$$

Product Rate. Let [PR] represent product rate in grams per hour per S cm^2 of film surface, and let Q' represent product rate in grams per second per cm^2 of film area; and let w_A and w_B be the grams of solute and solvent respectively in Q' grams of product solution. Then

$$Q' = w_A + w_B \ \frac{\text{grams}}{\text{sec. cm}^2} \tag{3-65}$$

$$[PR] = Q' \times S \times 3600 \ \frac{\text{grams}}{\text{hour}} \tag{3-66}$$

From Equation 3-56,

$$m_3 = m_1(1 - f) \tag{3-67}$$

Since

$$m_3 = \frac{w_A}{M_A} \cdot \frac{1000}{w_B} \tag{3-68}$$

$$\frac{w_B}{w_A} = \frac{1000}{m_1(1 - f)M_A} \tag{3-69}$$

$$Q' = w_A \ [1 + w_B/w_A] \tag{3-70}$$

$$= w_A \left[1 + \frac{1000}{m_1(1 - f)M_A}\right] \tag{3-71}$$

Therefore

$$w_A = Q' \Big/ \left(1 + \frac{1000}{m_1(1 - f)M_A}\right) \tag{3-72}$$

Since

$$w_B/M_B = N_B, \ w_B = N_B M_B \tag{3-73}$$

Also

$$w_B = Q' - w_A = Q'\,[1 - w_A/Q'] \tag{3-74}$$

$$= Q'\left[1 - 1\Big/\left(1 + \frac{1000}{m_1(1-f)M_A}\right)\right] \tag{3-75}$$

Therefore

$$Q' = \frac{N_B M_B}{\left[1 - 1\Big/\left(1 + \frac{1000}{m_1(1-f)M_A}\right)\right]} \tag{3-76}$$

Therefore

$$[PR] = \frac{N_B \cdot M_B \cdot S \cdot 3600}{\left[1 - 1\Big/\left(1 + \frac{1000}{m_1(1-f)M_A}\right)\right]} \tag{3-77}$$

Correlations of Reverse Osmosis Experimental Data. From any set of experimental [PWP], [PR], and f data at a given operating pressure, and from the physical properties of the feed solution, the values of A, $(D_{AM}/K\delta)$, and k can be calculated from Equations 3-33, 3-35, 3-43 or 3-54, and 3-52 or 3-55. The correlations of A, $(D_{AM}/K\delta)$, and k with operating pressure, feed concentration, feed flow rate, and the nature of solute are of practical interest from the point of view of specifying membranes and predicting membrane performance under different experimental conditions. Such correlations are discussed below for the Loeb-Sourirajan type porous cellulose acetate membranes and aqueous solution systems involving different solutes. All the experimental data discussed below were obtained from experiments carried out under flow conditions in the apparatus shown in Figures 1-16 and 1-17.

Effect of pressure on the pure water permeability constant. This is illustrated in Figure 3-24 by a set of experimental data for different films in the operating pressure range 125 to 1500 psig. All the above films were initially shrunk at different temperatures and subjected to a pure water pressure of 1700 psig for about 2 hours. The straight-line plots of log A v. P and log (A/A_0) v. P indicate that the variation of A as a function of operating pressure can be expressed by the relation

$$A = A_0 e^{-\alpha P} \tag{3-78}$$

where A_0 is the extrapolated value of A when P = 0, and α is a constant. Thus A_0 gives a measure of the initial porous structure of the film, and α expresses the compression effect on the pure water permeability characteristics of the film under pressure. The pure water feed flow rate has practically no effect on the [PWP] data. Consequently, the variation of A as a function of operating pressure for any particular film is completely specified by the two parameters A_0 and α, the determination of which needs the experimental [PWP] data at two different pressures. The values of A_0 and α may be expected to be different when the same sample of the film is subjected to different initial pressure treatments.

A set of log A v. P data is illustrated in Figure 3-25 in the operating pressure range 25 to 250 psig for a set of films initially shrunk at different temperatures and subjected to a pure water pressure of 300 psig for about 2 hours.

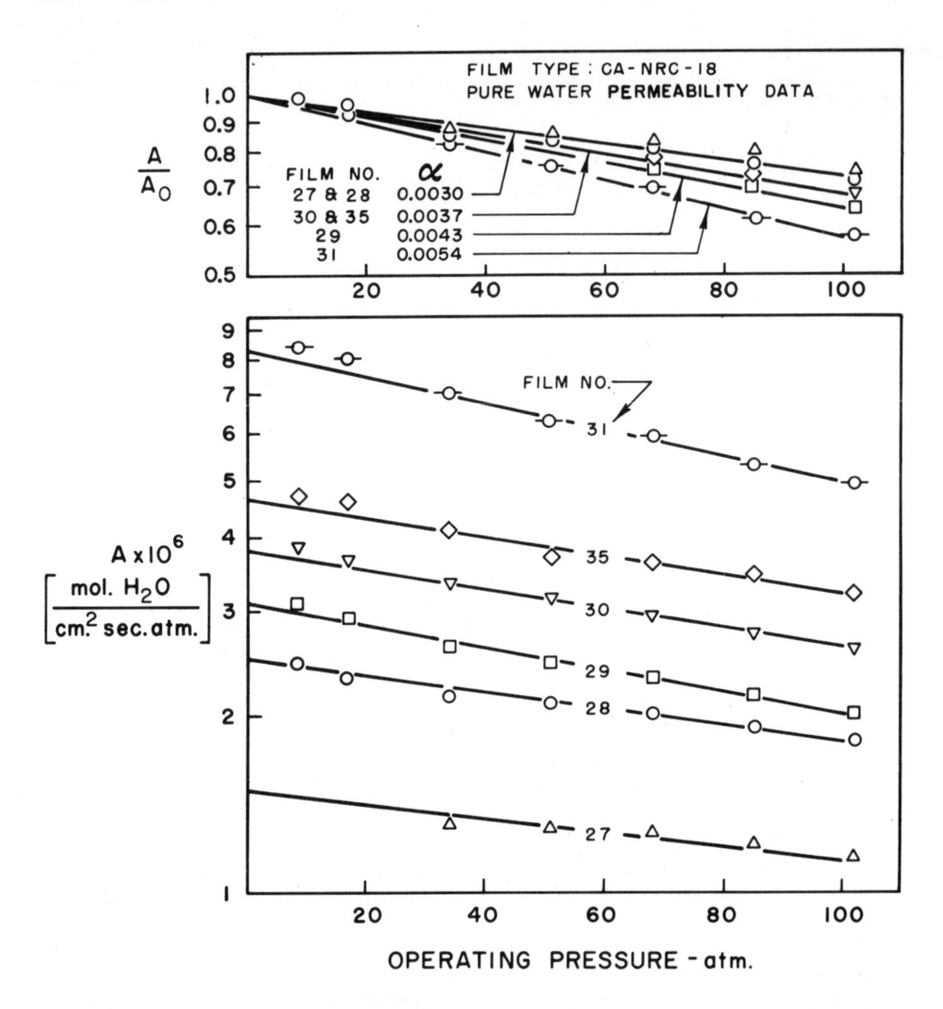

Figure 3-24. Effect of operating pressure on the pure water permeability contstant. Data of Sourirajan and Kimura (1967).

These data show that the relation expressed by Equation 3-78 is valid for low operating pressures also.

Correlations of solute transport parameter, $(D_{AM}/K\delta)$. The experimental data on the solute transport parameter for sodium chloride, glycerol, and urea are plotted in Figures 3-26 and 3-27, 3-28 and 3-29 respectively, for different films as a function of operating pressure in the range 125 to 1500 psig for 0.5M and 2.0M feed solutions. The data show that the graphs of log $(D_{AM}/K\delta)$ v. log P are straight lines, and that the data for both the feed concentrations fall on the same line for each film. Studies on the systems [$NaNO_3$-H_2O], [Na_2SO_4-H_2O], [$MgCl_2$-H_2O] and [$MgSO_4$-H_2O] also gave similar results. Consequently, for all the above systems, the variation of $(D_{AM}/K\delta)$ with P can be expressed by the relation

$$(D_{AM}/K\delta) \propto P^{-\beta} \tag{3-79}$$

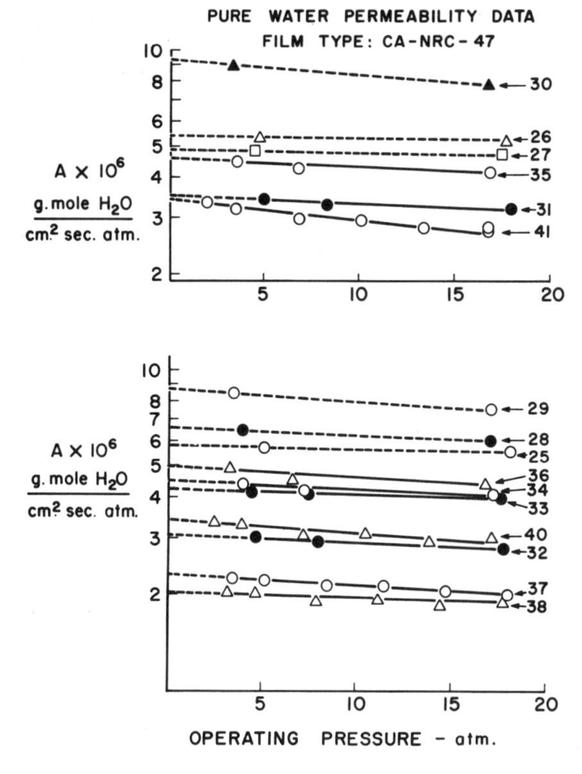

Figure 3-25. Effect of operating pressure on the pure water permeability constant (low pressure data).

or

$$\left(\frac{D_{AM}}{K\delta}\right) = \left(\frac{D_{AM}}{K\delta}\right)_r \left(\frac{P}{P_r}\right)^{-\beta} \tag{3-80}$$

where $(D_{AM}/K\delta)_r$ is the value of $(D_{AM}/K\delta)$ at the operating pressure P_r. Thus the parameter β, whose determination needs experimental $(D_{AM}/K\delta)$ values at two different pressures, describes the effect of operating pressure on the transport properties of the particular membrane for the given solute.

Since the value of β depends on the initial porous structure of the film, the correlation of A_0 v. β is particualrly interesting; this is illustrated in Figures 3-30, 3-31, and 3-32 for different solutes; the data show that β increases with increase in A_0 up to a point and then remains essentially constant. In the absence of precise experimental data for $(D_{AM}/K\delta)$, Figures 3-30, 3-31, and 3-32 can be used to evaluate $(D_{AM}/K\delta)$ at different pressures at least for the solution-membrane systems studied above.

The experimental data on the effect of feed concentration (in the range 0.125 to 4.0M), and feed flow rate (in the range 20 to 700 cm^3/min) on the values of $(D_{AM}/K\delta)$ for different solutes at a given operating pressure (1500 psig) using different samples of the Loeb-Sourirajan type porous cellulose acetate membranes covering a wide range of solute separations (9 to 99 per cent) are illustrated in Figures 3-33 to 3-43. The results show that, at a given operating

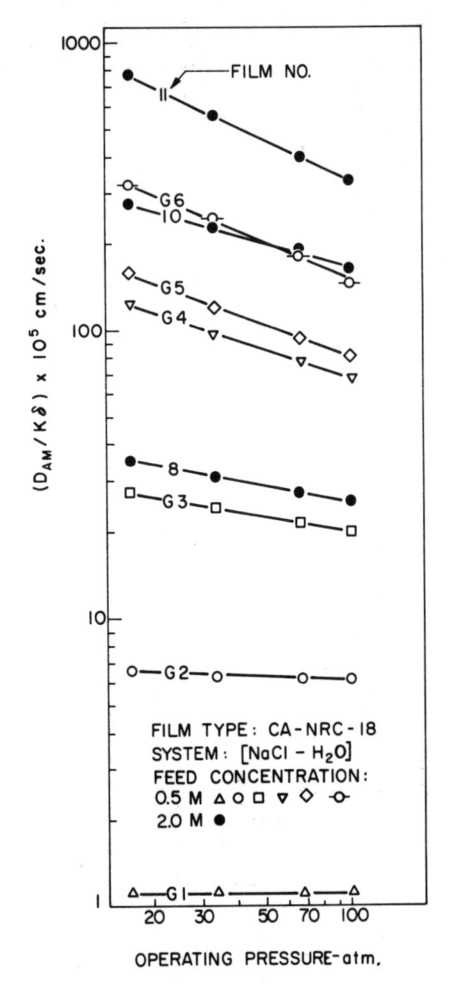

Figure 3-26. Effect of operating pressure on the solute transport parameter for the system sodium chloride–water.

Feed rate: 250 cm^3/min.

pressure, the values of $(D_{AM}/K\delta)$ for sodium chloride, sodium nitrate, sodium sulphate, magnesium sulphate, magnesium chloride, glycerol and urea depend only on the porous structure of the film and chemical nature of the solute, and they are independent of feed concentration and feed flow rate; similar results have been obtained for several other inorganic salts, such as lithium chloride, potassium chloride, ammonium chloride, lithium nitrate, potassium nitrate, potassium sulphate, calcium chloride, barium chloride and cupric sulphate. Figure 3-44 gives similar data for the system sodium chloride-water at low concentrations at an operating pressure of 200 psig.

The experimental fact that the values of $(D_{AM}/K\delta)$ for several solutes, such as sodium chloride, are independent of feed concentration and feed flow rate at any given operating pressure is one of great practical importance in the reverse osmosis separation process. This fact is made use of in establishing theoretical equations for reverse osmosis process design for saline water conversion and other applications (Chapter 4).

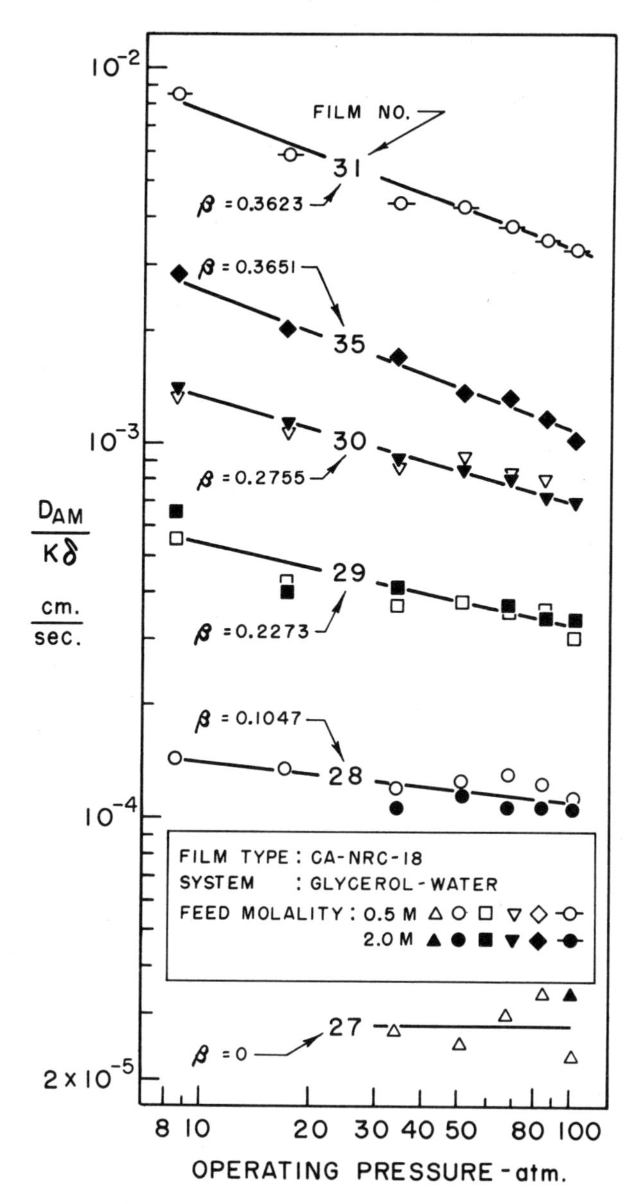

Figure 3-27. Effect of operating pressure on the solute transport parameter for the system glycerol-water. Feed rate: 380 cm^3/min. Data of Sourirajan and Kimura (1967).

Correlations of mass transfer coefficients, k, for systems such as sodium chloride-water, glycerol-water, and urea-water. The values of k are functions of feed rates and feed concentrations. Experimental results further show that operating pressures and boundary concentrations (X_{A2}) also affect the values of k; a 30 per cent variation in the value of k has been observed under experimental conditions of constant feed flow rate and feed concentration, with different films at different operating pressures. No definite trends, however, have been observed on the effect of operating pressure and boundary concentration on k.

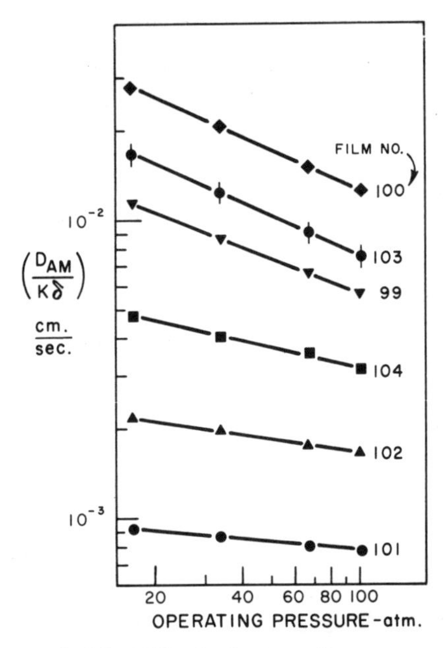

Figure 3-28. Effect of operating pressure on the solute transport parameter for the system urea-water. Data of Ohya and Sourirajan (1969a).

Film type: CA-NRC-18; feed molality: 2.0M; feed rate: 380 cm^3/min.

A 30 per cent variation in the experimental values of k does not appear to be too great from the point of view of establishing a practical criterion for calculating mass transfer coefficients. On the basis of the results discussed below, it seems reasonable to consider that k is primarily a function of feed flow rate and feed concentration only. Figure 3-45 illustrates the effect of these variables on the average values of k obtained with different films covering a wide range of solute separations for the system glycerol-water. Similar data have been obtained for the systems sodium chloride-water and urea-water. Such data can be expressed by the general relation

$$k \propto Q^n \tag{3-81}$$

where Q is feed rate in cm^3 per second, and n is a constant characteristic of the solution system and the apparatus used. For the systems sodium chloride-water, glycerol-water, and urea-water, the values of n obtained in the work discussed above are 0.78, 0.72, and 0.79 respectively.

Generalised correlation for mass transfer coefficients. The mass transfer coefficient, k, is usually defined as the ratio of a flux and a driving force; in order to find whether k so defined agrees with k (defined by Equation 3-48) obtained by the analysis of the reverse osmosis data, a set of diffusion current experiments was carried out by Sourirajan and Kimura (1967) as follows (Lin et al., 1951; Reiss and Hanratty, 1962).

The cathodic reduction of potassium ferricyanide was studied in a plexiglas cell identical in design to the stainless steel cell used in the reverse osmosis experiments. The electrolyte feed solution consisted of a mixture of potassium ferricyanide (0.0048 or 0.0039 gram mole per litre) and potassium ferrocyanide (0.0048 gram mole per litre) in 1N potassium nitrate solution. The

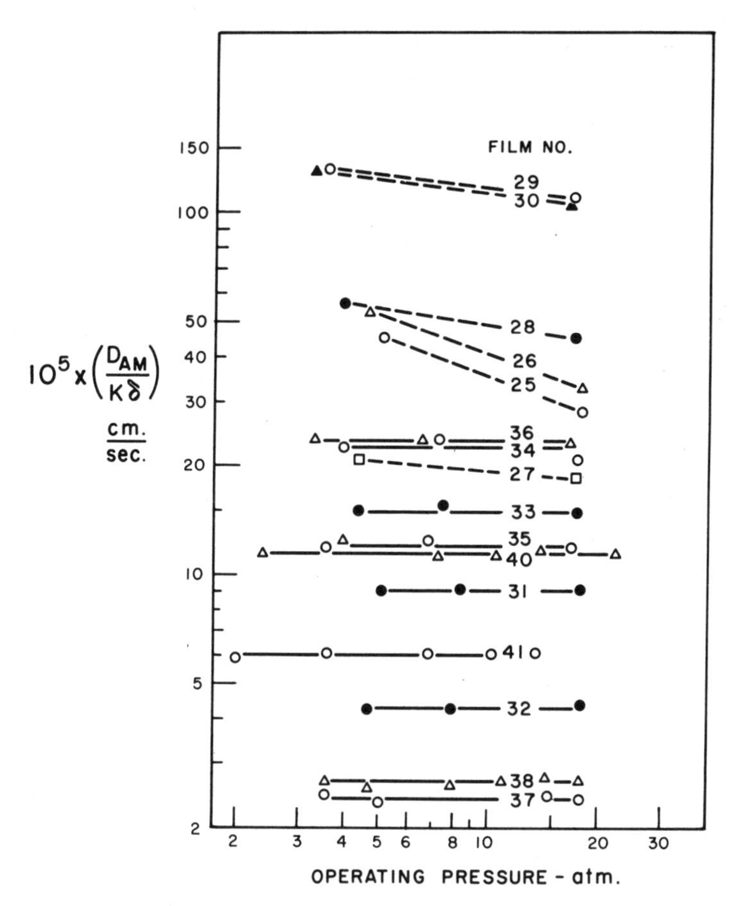

Figure 3-29. Effect of operating pressure on the solute transport parameter for the system sodium chloride-water (low pressure data).

Film type: CA-NRC-47; feed concentration: 2400 to 2700 ppm NaCl; feed rate: 300 cm^3/min.

feed solution was saturated with nitrogen before running, and kept in an atmosphere of nitrogen during the experiment. Platinum electrodes were used. The area of the cathode (A_c) was nearly that of the membrane area, and that of the anode was about twice that of A_c. The experiments were carried out at constant temperature using different feed flow rates, and one of the two ferricyanide concentrations in the feed in which the ferrocyanide concentration was kept constant. In each experiment, the currents in the circuit for different applied voltages were measured and the limiting current, i, corresponding to each feed flow rate was determined. The mass transfer coefficient, k, applicable for the transfer of ferricyanide ions to the surface of the cathode was calculated from the relation:

$$k = i/FA_c\bar{c}_i \tag{3-82}$$

The values of k were determined for two different concentrations of potassium ferricyanide ($\bar{c}_i$) at 5° temperature intervals in the range 10° to 45° for several feed flow rates in the range 30 to 750 cm^3 per minute to cover a wide range of Reynolds (N_{Re}), Schmidt (N_{Sc}) and Sherwood (N_{Sh}) numbers.

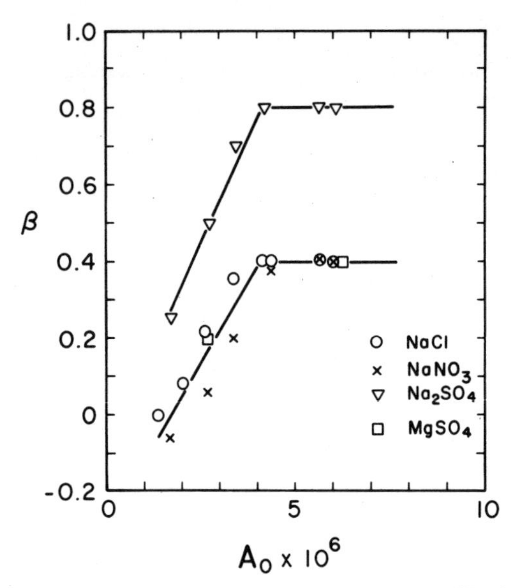

Figure 3-30. Variation of the pressure effect on the solute transport parameter with the initial porous structure of the membranes for aqueous solutions containing inorganic salts (Film type CA-NRC-18). Data of Kimura and Sourirajan (1967).

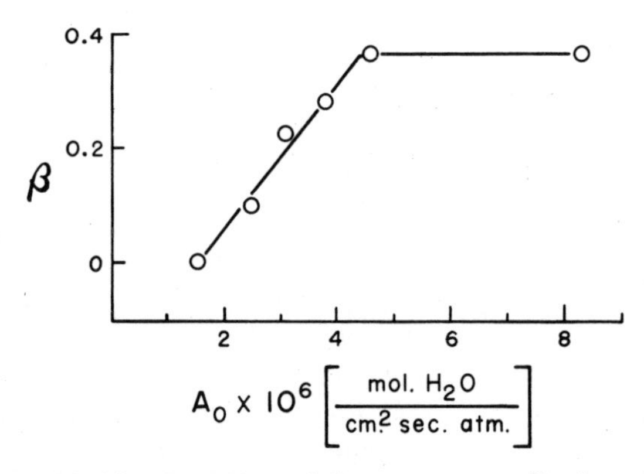

Figure 3-31. Variation of the pressure effect on the solute transport parameter with the initial porous structure of the membrane for the system glycerol-water. Film type: CA-NRC-18. Data of Sourirajan and Kimura (1967).

The Reynolds numbers were calculated arbitrarily, with reference to the flow rate across the central cross-section of the cell, using the following relationship:

$$N_{Re} = Q/h\nu \tag{3-83}$$

where Q, h and ν represent, respectively, the feed flow rate (cm^3/sec), depth of cell (=1.43 cm) and kinematic viscosity of solution (cm^2/sec). In view of the very low concentrations of the potassium ferri- and ferro-cyanides in the solution, the values of ν used were those of 1N potassium nitrate solution,

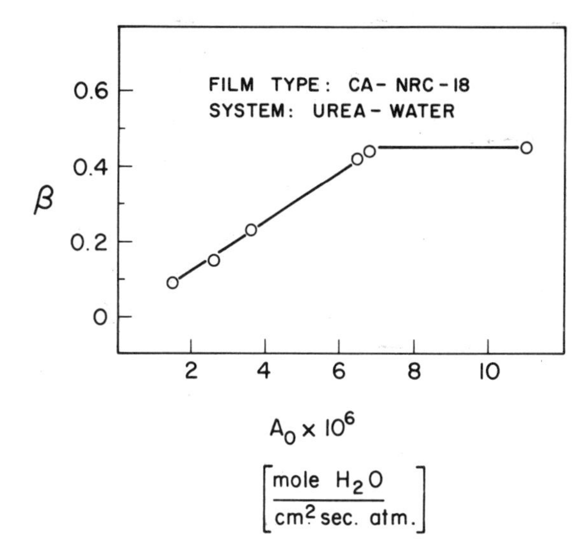

Figure 3-32. Variation of the pressure effect on the solute transport parameter with the initial porous structure of the membrane for the system urea-water. Data of Ohya and Sourirajan (1969a).

taken from the literature (International Critical Tables, 1928b; Chemical Handbook, 1958).

The diffusivity of the ferricyanide ion at infinite dilution at 25° is 0.89×10^{-5} cm^2/sec (Kolthoff and Lingane, 1952b). Assuming that diffusivity is inversely proportional to viscosity, the value of the diffusivity of the ferricyanide ion in 1N potassium nitrate solution (D_i) at 25° is 0.90×10^{-5} cm^2/sec. On the basis of the latter value, and considering that the diffusion current is directly proportional to $\sqrt{D_i}$ (Kolthoff and Lingane, 1952a), the values of D_i at other temperatures were calculated from the experimental measurements of the diffusion currents in a dropping mercury electrode. At the temperatures of 10°, 15°, 20°, 25°, 30°, 35°, 40°, and 45°, the values of D_i were found to be 0.60, 0.69, 0.79, 0.90, 1.04, 1.19, 1.39, and 1.51 ($\times 10^{-5}$ cm^2/sec), respectively. These values of D_i were used in the calculations of N_{Sc} and N_{Sh} defined as follows:

$$N_{Sc} = \nu/D_i \qquad (3\text{-}84)$$

$$N_{Sh} = kd/D_i \qquad (3\text{-}85)$$

where d = effective diameter of membrane surface.

The log-log plot of N_{Re} v. $N_{Sh}/N_{Sc}^{0.33}$ (Figure 3-46) correlates by a single line all the experimental data obtained in the diffusion current experiments; this line also coincides well with the similar plot (Fig 3-47) using the solute diffusivities given in the tables in Appendix II, and the average k values obtained in the reverse osmosis experiments using different films covering a wide range of solute separations for several aqueous solution systems. Hence the mass transfer correlation N_{Re} v. $N_{Sh}/N_{Sc}^{0.33}$ seems to have general validity for the reverse osmosis membrane separation process. The above data also show that the values of k obtained on the basis of the 'film theory' and pore diffusion model are applicable for the reverse osmosis process.

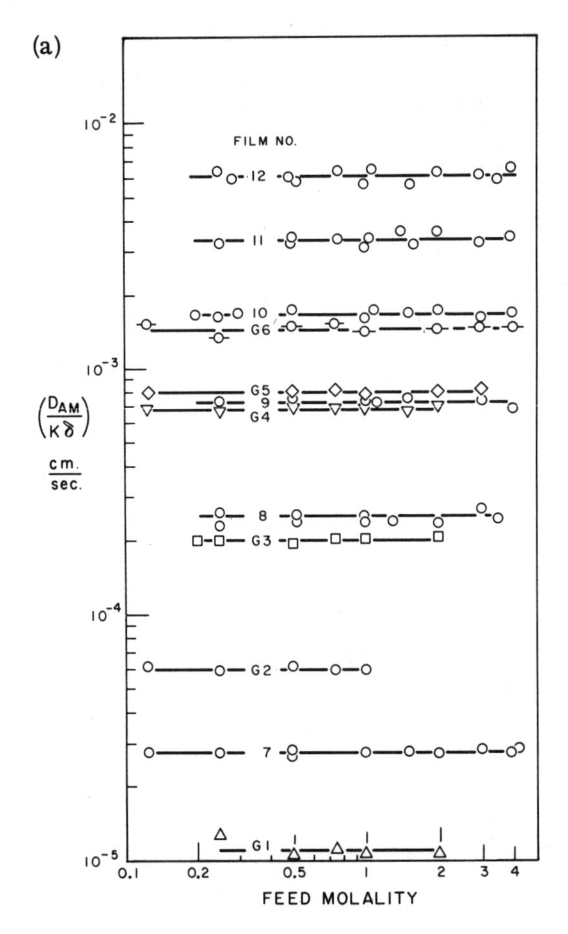

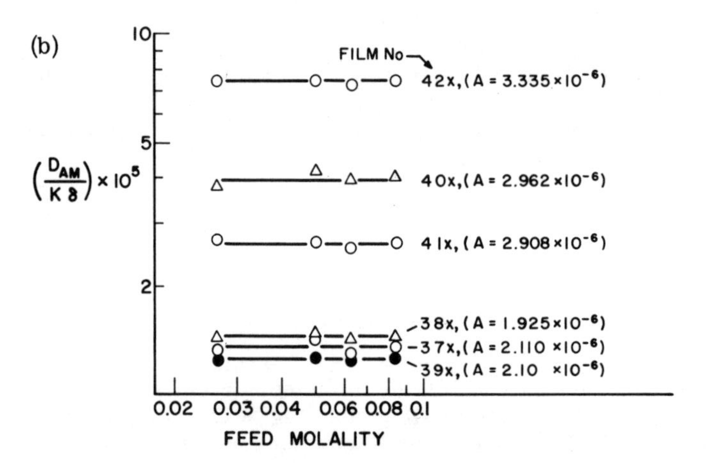

Figure 3-33. Effect of feed concentration on the solute transport parameter for the system sodium chloride-water: (a) 0.125M to 4.0M; film type: CA-NRC-18; feed rate: 260 cm^3/min; operating pressure: 1500 psig; (b) 0.025M to 0.08M range; film type: CA-NRC-47; feed rate: 300 cm^3/min; operating pressure: 200 psig.

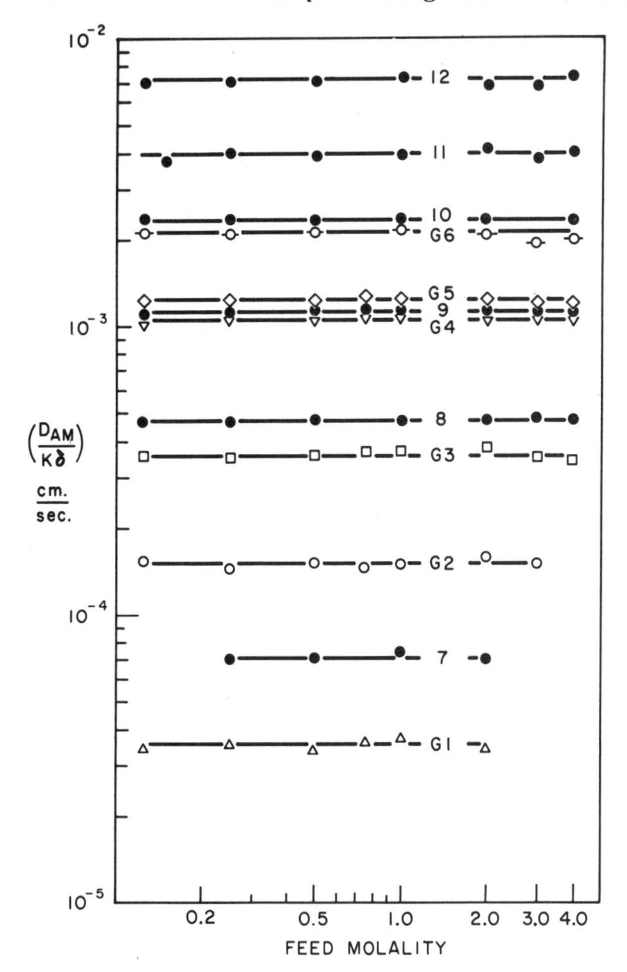

Figure 3-34. Effect of feed concentration on the solute transport parameter for the system sodium nitrate-water.

Film type: CA-NRC-18; feed rate: 260 cm^3/min; operating pressure: 1500 psig.

Figure 3-48 gives the above correlation for the system glycerol-water using the experimental values of k and the available diffusivity values for glycerol given in the literature (International Critical Tables, 1929). The figure shows that, in most cases, the value of k obtained experimentally is within 30 per cent of that obtained from the generalised correlation given by the diffusion current method. There are at least two reasons to consider that this agreement is close enough from the point of view of the usefulness of Figure 3-46 for predicting membrane performance. First, similar variations in the values of k are obtainable with films of different pore structures, even under the same experimental conditions of operating pressure, feed concentration and feed flow rate. Secondly, the experimental data presented in Tables 3-3 and 3-4 illustrate that the effects of a 30 per cent change in the value of k on membrane performance are not too great.

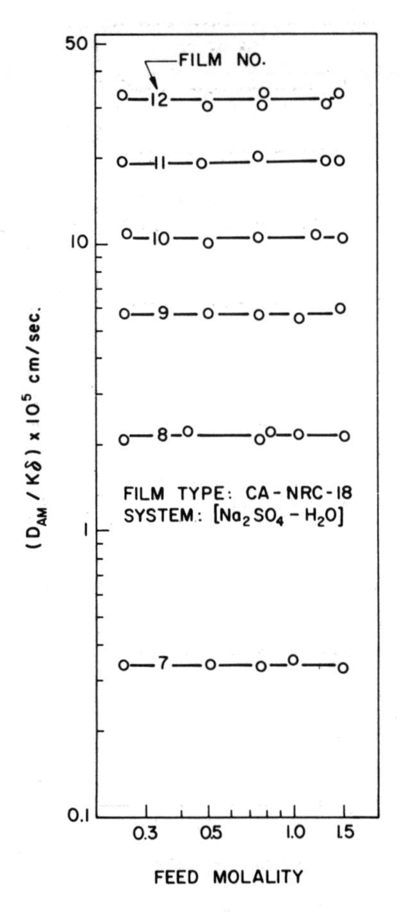

Figure 3-35. Effect of feed concentration on the solute transport parameter for the system sodium sulphate-water.

Feed rate: 250 cm^3/min; operating pressure: 1500 psig.

Referring to Table 3-3, Film 87 is typical of one capable of giving high levels of solute separations. For this film a 30 per cent change in k corresponds to <0.3 per cent change in f, and 5 per cent change in [PR]. Film 91 is typical of one capable of giving low levels of solute separations; for this film, a similar change in k corresponds to 3 per cent change in f and 4 per cent change in [PR]. Hence, for all films of the type employed in this work, the correspondence between the values of k for the system glycerol-water obtained from Figure 3-46 and the experimental values of f and [PR], is within 3 and 5 per cent, respectively; for films capable of giving high levels of solute separations, the correspondence in f values could be much closer. These observations are also valid for the data of other systems illustrated above. Consequently, the form of the generalised mass transfer correlation given in Figure 3-46 seems applicable for the reverse osmosis separation process.

6. Predictability of Membrane Performance

On the basis of the foregoing discussions, a single set of experimental pure water permeability, product rate and solute separation data at any operating

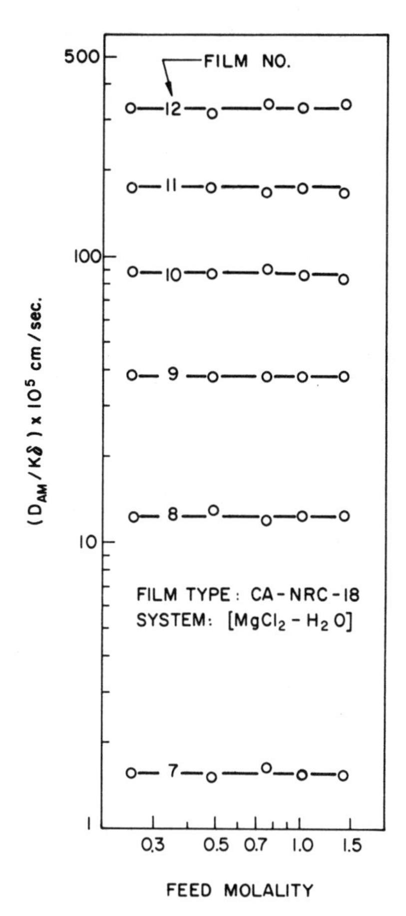

Figure 3-36. Effect of feed concentration on the solute transport parameter for the system magnesium chloride-water.

Feed rate: 250 cm^3/min; operating pressure: 1500 psig.

pressure enables the prediction of both solute separation and product rate at that pressure, for all feed concentrations and feed flow rates for which $(D_{AM}/K\delta)$ remains constant. The procedure for such prediction involves the following steps.

From the experimental data, the values of A and $(D_{AM}/K\delta)$ are calculated using Equations 3-33, 3-35, and 3-43; these values remain constant for all feed concentrations and feed flow rates.

The value of k is calculated for any pre-set conditions of the experiment from the generalised correlation shown in Figure 3-46 using the diffusivity data for the solute available from the literature; or k may also be obtained from predetermined experimental data of the type plotted in Figure 3-45.

Combining Equations 3-43 and 3-52,

$$\left(\frac{D_{AM}}{K\delta}\right) \frac{(c_2X_{A2} - c_3X_{A3})}{X_{A3}} = kc_1 \ln \left(\frac{X_{A2} - X_{A3}}{X_{A1} - X_{A3}}\right) \tag{3-86}$$

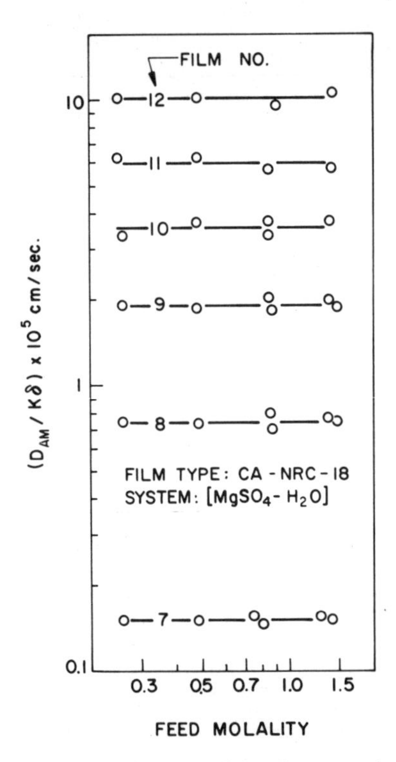

Figure 3-37. Effect of feed concentration on the solute transport parameter for the system magnesium sulphate-water.

Feed rate: 250 cm^3/min; operating pressure: 1500 psig.

Table 3-3. Mass transfer coefficient and membrane performance

Data of Sourirajan and Kimura (1967)

Film no.	Feed molality m_1	N_{Re}	f	[PR] (g/hr)	$k \times 10^4$ (cm/sec)
87	0.545	32	0.908	17.5	5.0
87	0.503	852	0.980	28.4	37.0
87	2.095	40	0.907	6.9	5.4
87	1.910	661	0.962	15.9	38.0
91	0.545	28	0.037	110.1	9.6
91	0.503	846	0.468	128.6	39.2
91	2.095	39	0.088	51.4	8.0
91	1.910	658	0.455	82.1	36.3

Film type : CA-NRC-18
System: Glycerol-water
Operating pressure: 1500 psig
Film area: 7.6 cm^2
[PR]: Product rate

Table 3-4. Comparison of experimental and predicted solute separation and product rate data for the system glycerol-water using Film 88

Data of Sourirajan and Kimura (1967)

				Predicted values			
		Experimental data		Based on av. k from Figure 3-46		Based on k from generalised correlation	
Feed molality m_1	Feed flow rate (cm^3/min)	$f \times 10^2$	[PR] (g/hr)	$f \times 10^2$	[PR] (g/hr)	$f \times 10^2$	[PR] (g/hr)
0.248	380	93.9	62.7	91.6	62.3	93.1	63.6
0.495	380	90.9	55.1	91.4	55.2	92.7	57.0
1.010	380	90.6	45.1	90.8	43.0	91.8	44.9
1.548	380	89.0	34.5	89.5	32.5	90.6	34.6
1.930	380	88.6	28.3	88.1	26.5	89.4	28.1
2.715	380	86.2	18.7	84.5	16.4	—	—
3.395	380	82.0	12.6	80.3	10.9	—	—
3.965	380	78.3	8.7	75.8	7.7	—	—
0.540	67	75.9	37.0	74.6	36.1	77.5	38.4
0.525	107	81.0	42.8	81.5	42.5	83.2	44.2
0.540	208	87.0	51.1	88.3	49.6	90.0	51.9
0.535	278	89.7	55.0	90.0	52.1	91.3	54.1
0.510	388	91.2	57.0	91.5	54.9	92.7	56.7
0.520	570	92.3	59.4	92.8	56.7	93.7	58.2
0.503	719	92.0	59.1	93.3	57.9	94.1	59.3
2.095	45	72.1	12.8	69.0	11.4	72.4	12.7
2.145	72	75.8	14.9	74.5	13.3	76.3	14.1
2.090	113	79.9	17.7	79.3	16.2	81.2	17.5
2.115	204	84.2	21.6	84.1	19.8	85.8	21.5
2.113	274	86.0	24.2	86.1	21.8	87.4	23.4
2.120	385	88.2	26.6	87.6	23.7	88.9	25.6
1.980	566	90.2	30.5	89.2	27.3	90.7	30.3
1.910	726	90.2	32.0	90.5	30.2	91.5	32.3

Operating pressure: 1500 psig

Film specifications at 1500 psig: $A = 1.4417 \times 10^{-6}$ $\frac{\text{g. mole } H_2O}{\text{cm}^2 \text{ sec. atm}}$; $\left(\frac{D_{AM}}{K\delta}\right) = 10.112 \times 10^{-5}$ cm/sec

Film area: 7.6 cm^2

[PR]: Product rate

or

$$\frac{(c_2/c_1)X_{A2} - (c_3/c_1)X_{A3}}{X_{A3}} = \lambda \ln\left(\frac{X_{A2} - X_{A3}}{X_{A1} - X_{A3}}\right) \tag{3-87}$$

where

$$\lambda = k/(D_{AM}/K\delta) \tag{3-88}$$

By trial and error, the combinations of X_{A2} and X_{A3} satisfying Equation 3-87 are first found: there are several.

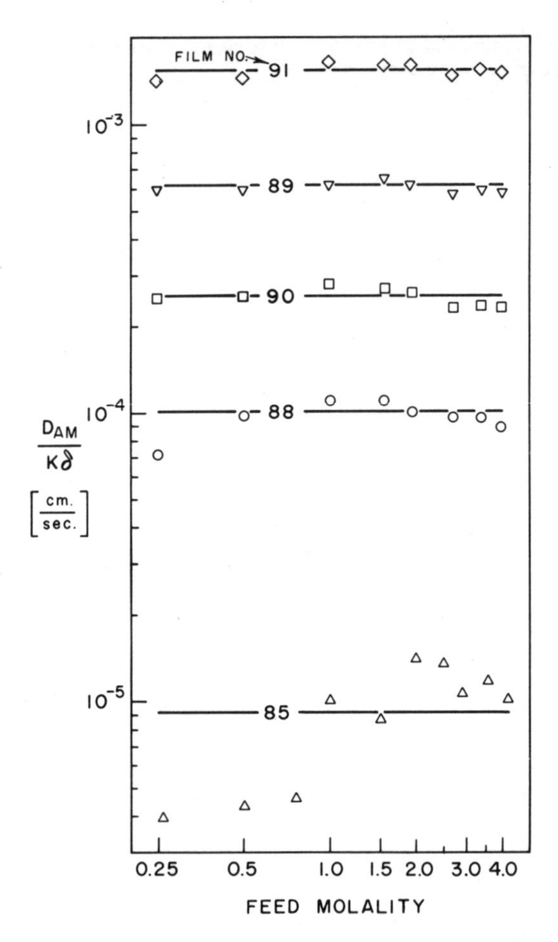

Figure 3-38. Effect of feed concentration on the solute transport parameter for the system glycerol-water. Data of Sourirajan and Kimura (1967).

Film type: CA-NRC-18; feed rate: 380 cm^3/min; operating pressure: 1500 psig.

Finally, that particular combination of X_{A2} and X_{A3} which satisfies the equality of Equations 3-35 and 3-43 is determined; this determines the value of N_B.

Using the values of X_{A3} and N_B determined above, and the known value of X_{A1} corresponding to the preset operating conditions, the values of solute separation f, and product rate [PR] are determined using Equations 3-64 and 3-77.

Table 3-4 compares the experimental solute separation and product rate data with those predicted by the procedure outlined above for a particular film chosen for illustration. The specified values of A and $(D_{AM}/K\delta)$ are the average of those obtained in several experiments. The results given in Table 3-4 show reasonably good agreement between the experimental data and the predicted ones, whether the latter are based on average k values taken from Figure 3-45 or obtained from the general mass transfer correlation (Figure 3-46) given by the diffusion current method. Similar agreement between the experimental and predicted solute separation and product rate data have been obtained with all the solution systems referred to in this chapter using the

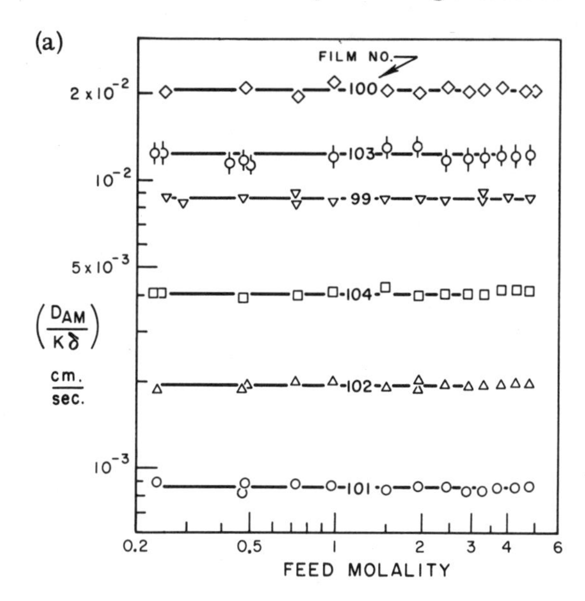

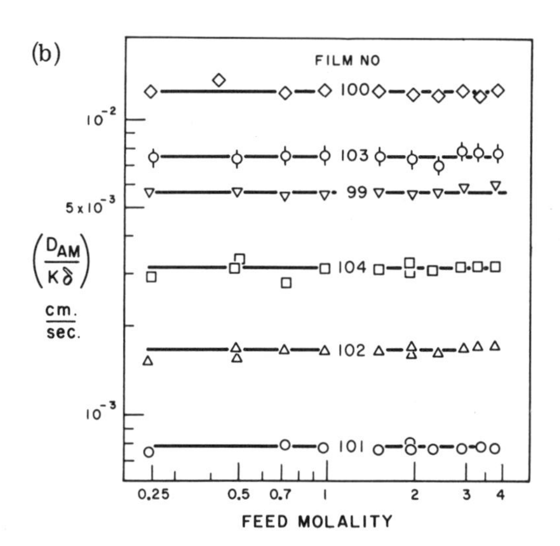

Figure 3-39. Effect of feed concentration on the solute transport parameter for the system urea-water at different operating pressures: (a) film type: CA-NRC-18; feed rate: 350-400 cm^3/min; operating pressure: 500 psig; (b) film type: CA-NRC-18; feed rate: 350-410 cm^3/min; operating pressure 1500 psig. Data of Ohya and Sourirajan (1969a).

Loeb-Sourirajan type porous cellulose acetate membranes; some of these data are illustrated in Figures 3-49 and 3-50.

7. Membrane Specifications

The Kimura-Sourirajan analysis leads to a useful method of membrane specification. With respect to a given solution system, the separation and product rate characteristics of a particular membrane are completely specified for the

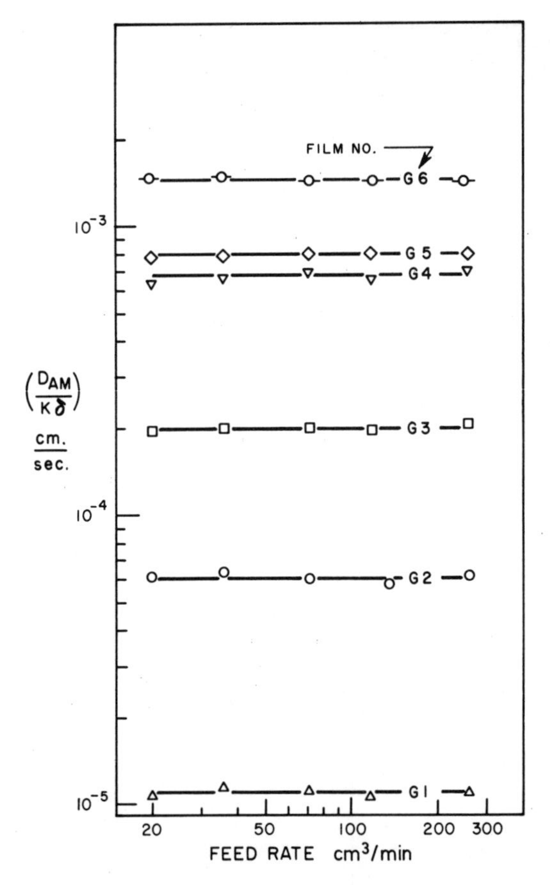

Figure 3-40. Effect of feed flow rate on the solute transport parameter for the system sodium chloride-water.

Film type: CA-NRC-18; feed molality: 0.5M; operating pressure: 1500 psig.

entire applicable range of operating pressures, feed concentrations and feed flow rates, by specifying the four parameters α, β, and A and $(D_{AM}/K\delta)$ at a particular operating pressure. The specification of the latter two parameters only is sufficient to specify membrane performance at that pressure for the entire applicable range of feed concentrations and feed flow rates. The determination of all the four parameters requires the experimental [PWP], [PR], and f data at two different operating pressures; the determination of the latter two parameters alone requires only a single set of experimental data at the specified operating pressure. Table 3-5 gives the specifications of some of the films whose performance is discussed in this work. These specifications are based on the data obtained from short run experiments, and represent the initial characteristics of the film. For information, the table also includes the temperatures at which the membranes were shrunk prior to use in the experiments; these temperatures have no precise significance from the point of view of membrane specification.

Table 3-5. Specifications of some CA-NRC-18 type porous cellulose acetate membranes
Data of Kimura and Sourirajan (1968b, 1968d)

Film no.	Film shrinkage temp. (°)	Solute	Operating pressure (psig)	$A \times 10^6$ $\left(\frac{\text{g. mole } H_2O}{\text{cm}^2 \text{sec. atm}}\right)$	$(D_{AM}/K\delta) \times 10^5$ (cm/sec)
1	90	NaCl	500	1. 142	0. 90
1	90	NaCl	1500	0. 97	0. 90
2	88	NaCl	500	1. 754	6. 00
2	88	NaCl	1500	1. 46	6. 00
3	86	NaCl	500	2. 329	23. 37
3	86	NaCl	1500	1. 87	20. 00
4	84	NaCl	500	3. 086	97. 43
5	82	NaCl	500	3. 246	117. 1
5	82	NaCl	1500	2. 37	75. 0
6	80	NaCl	1500	2. 93	140. 0
F1	87	NaCl	600	3. 773	18. 67
F2	88	NaCl	600	2. 670	4. 423
F3	90	NaCl	1500	1. 295	0. 891
F4	89	NaCl	600	2. 670	5. 825
29	86	NaCl	500	2. 365	52. 0
29	86	NaCl	1000	2. 066	42. 9
29	86	NaCl	1500	1. 838	40. 0
30	84	NaCl	500	3. 011	109. 0
30	84	NaCl	1000	2. 679	91. 1
30	84	NaCl	1500	2. 377	80. 7
34	80	NaCl	500	3. 700	720. 1
34	80	NaCl	1000	3. 311	483. 1
34	80	NaCl	1500	2. 841	350. 0
85	92. 5	Glycerol	1500	0. 6736	0. 918
87	92	Glycerol	1500	0. 6674	1. 896
88	88. 5	Glycerol	1500	1. 4417	10. 112
89	85	Glycerol	1500	2. 4491	61. 924
90	86. 5	Glycerol	1500	1. 8909	25. 311
91	82	Glycerol	1500	3. 1579	154. 434
94	80	NaCl	1500	1. 873	55. 13
105	86	NaCl	1500	1. 494	1. 832
106	80. 5	NaCl	1500	2. 962	41. 60
107	79. 5	NaCl	1500	3. 215	60. 58
108	78. 5	NaCl	1500	5. 356	476. 9
109	78. 5	NaCl	1500	4. 747	335. 8
111	86	NaCl	500	1. 574	2. 14
111	86	NaCl	1000	1. 475	2. 13
111	86	NaCl	1500	1. 393	2. 12
113	79. 5	NaCl	500	4. 216	148. 6
113	79. 5	NaCl	1000	3. 750	110. 0
113	79. 5	NaCl	1500	3. 385	94. 2
114	78	NaCl	500	3. 486	47. 7
114	78	NaCl	1000	3. 150	39. 5
114	78	NaCl	1500	2. 878	35. 4
115	78	NaCl	500	5. 685	524. 1
116	76. 5	NaCl	500	7. 045	1183. 0
116	76. 5	NaCl	1000	6. 100	770. 0
116	76. 5	NaCl	1500	5. 286	604. 5

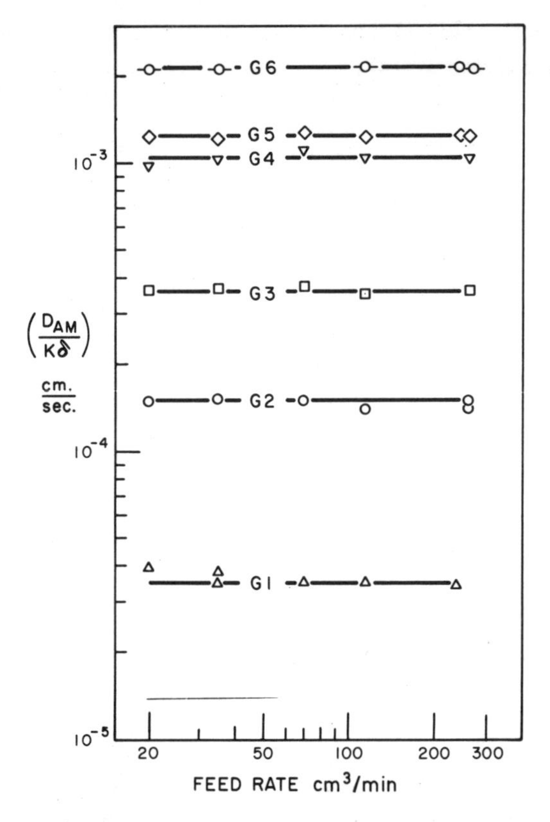

Figure 3-41. Effect of feed flow rate on the solute transport parameter for the system sodium nitrate-water.

Film type: CA-NRC-18; feed molality: 0. 5M; operating pressure: 1500 psig.

8. Extended Continuous Operation Under Pressure

Effect of Membrane Compaction. The performance of the Loeb-Sourirajan type porous cellulose acetate membranes during extended continuous operation under pressure in the reverse osmosis process is discussed in the literature (Loeb, 1966b). Lonsdale et al. (1964) conducted a 2-month laboratory test with recycled synthetic sea water. During this period the 'salt reduction factor' dropped from 66 to 12, and the 'membrane constant' decreased from about 8×10^{-6} to 4.5×10^{-6} g/cm^2 sec. atm. They also conducted a life test with 10 per cent sodium chloride solution, and observed a considerable decrease in the membrane's salt rejection capacity within about one week. Loeb and Manjikian (1964) conducted a 3-month laboratory test with sea water at an operating pressure of 1500 psig. The initial steady state flux was about 4×10^{-4} g/cm^2 sec, and the salt content of the product was about 500 ppm; after 3 months, the flux dropped to about 2×10^{-4} g/cm^2 sec, and the salt content of the product increased to over 4000 ppm. Loeb and Manjikian (1965) also conducted a 6-month field test of the membrane with brackish water at Coalinga, California. They found that the flux of desalinised water dropped from 24×10^{-4} to 8×10^{-4} g/cm^2 sec, and the salt content of the product increased from 150 to 300 ppm. Vos et al. (1966a) found that

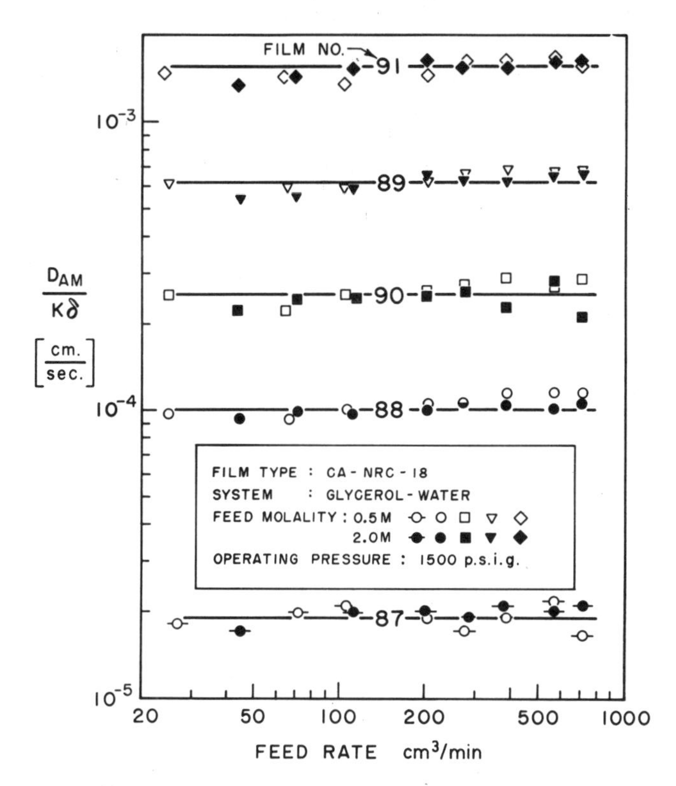

Figure 3-42. Effect of feed flow rate on the solute transport parameter for the system glycerol-water. Data of Sourirajan and Kimura (1967).

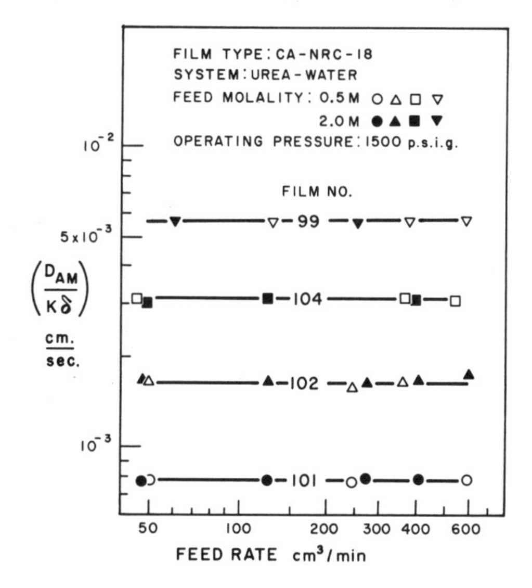

Figure 3-43. Effect of feed flow rate on the solute transport parameter for the system urea-water. Data of Ohya and Sourirajan (1969a).

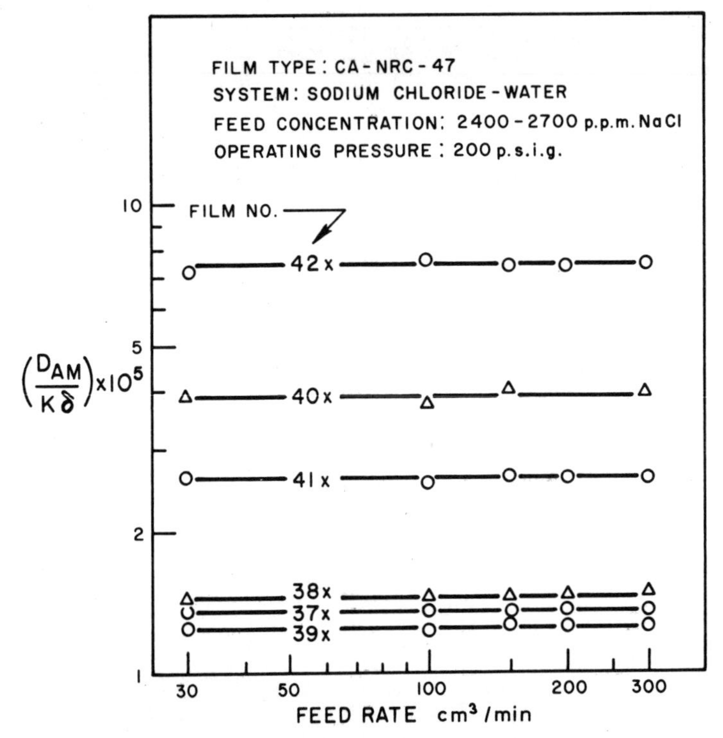

Figure 3-44. Effect of feed flow rate on the solute transport parameter for the system sodium chloride-water (low pressure data).

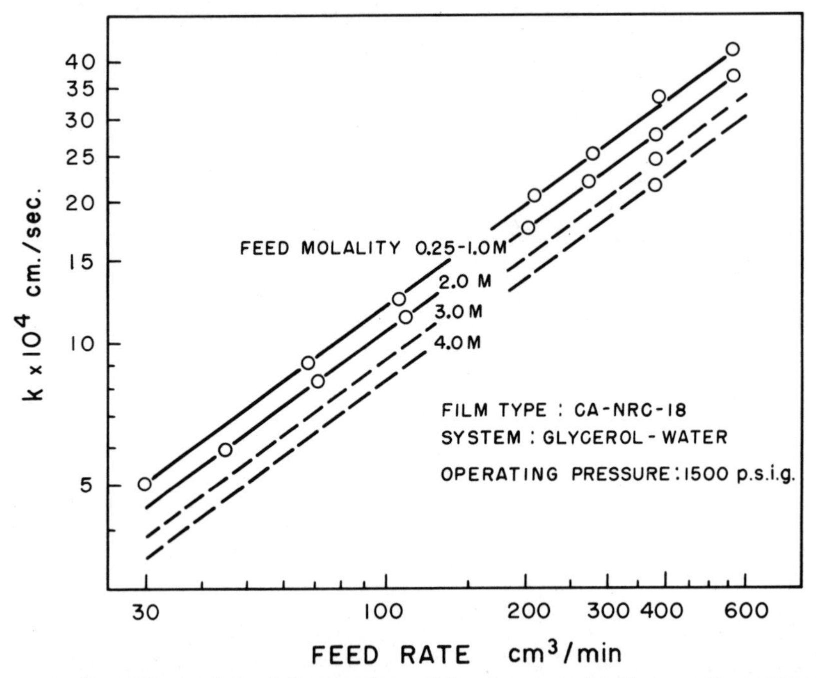

Figure 3-45. Effect of feed flow rate and feed concentration on the average mass transfer coefficients for the system glycerol-water. Data of Sourirajan and Kimura (1967).

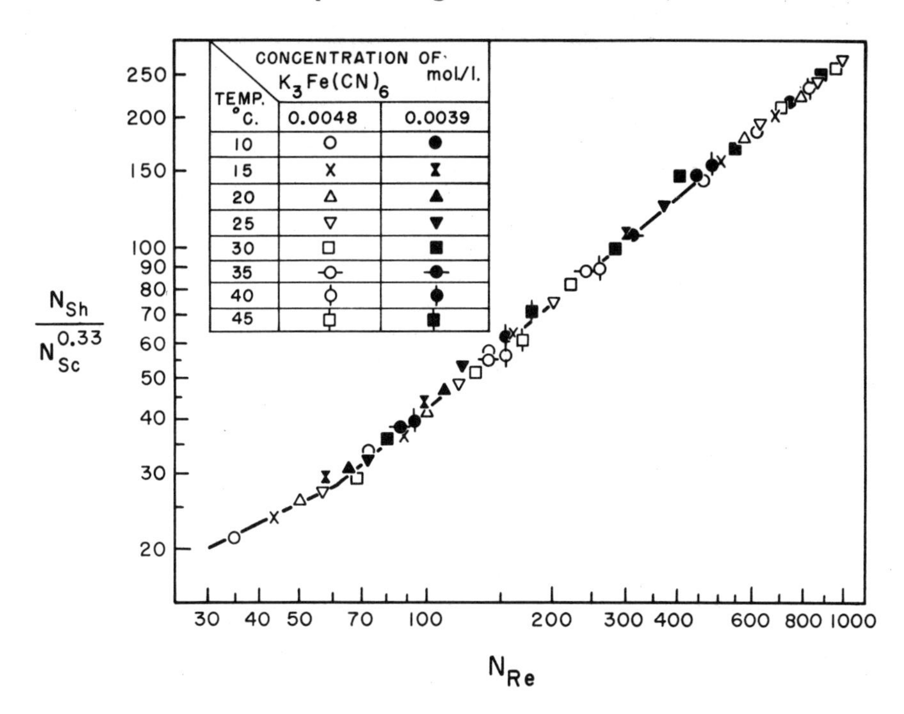

Figure 3-46. Generalised mass transfer coefficient correlation obtained by the diffusion current method. Data of Sourirajan and Kimura (1967).

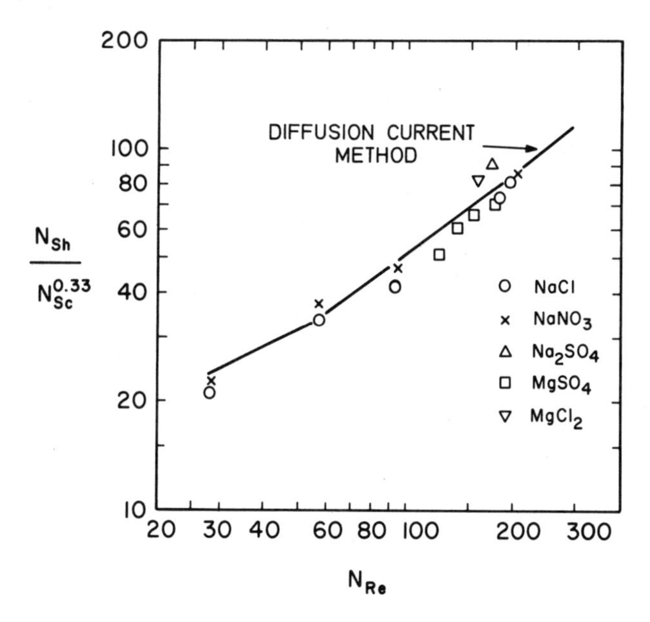

Figure 3-47. Mass transfer coefficient correlation obtained from experimental reverse osmosis data involving different inorganic solutes. Data of Kimura and Sourirajan (1967).

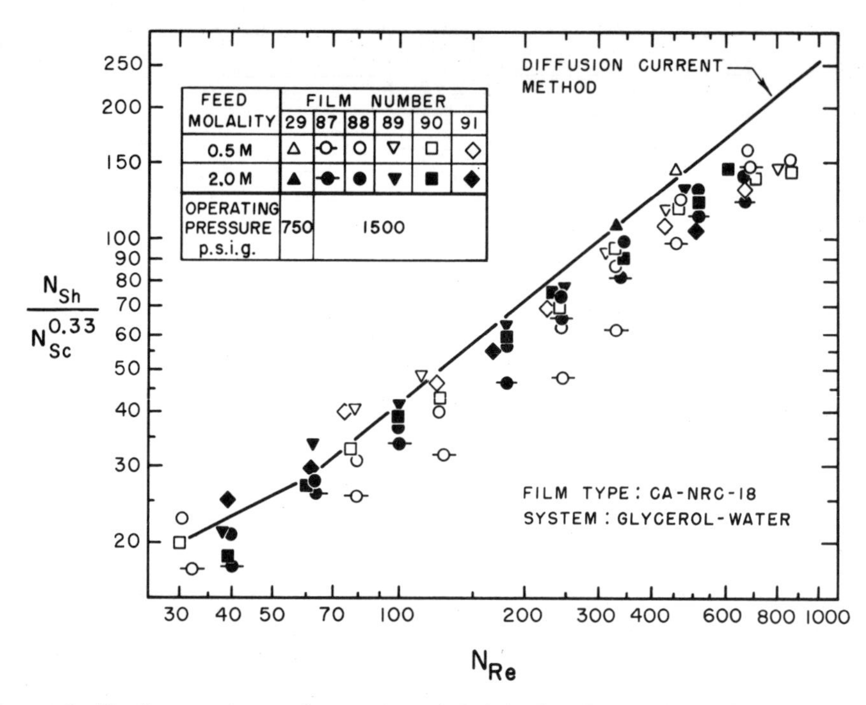

Figure 3-48. Comparison of experimental data for the system glycerol-water with the generalised mass transfer coefficient correlation. Data of Sourirajan and Kimura (1967).

membrane life was strongly dependent on feed water pH, and the above results might probably be attributed to partial hydrolysis of the membrane material.

From the point of view of process design, it would be of interest to be able to predict the variations in solute separations as the product rate slows down because of membrane compaction during continuous operation under pressure in the reverse osmosis process using the Loeb-Sourirajan type porous cellulose acetate membranes. The Kimura-Sourirajan analysis establishes a basis and offers an analytical means for prediction (Kimura and Sourirajan, 1968b) which is discussed below.

Relationships between A and ($D_{AM}/K\delta$) in short-time runs. On the basis of the preferential sorption-capillary flow mechanism (Sourirajan, 1963, 1967) it is clear that when the variations in A are due to changes in the average pore size on the membrane surface, the values of ($D_{AM}/K\delta$) should be extremely sensitive to changes in those of A. This is confirmed by the experimental observations. It has been shown (Kimura and Sourirajan, 1967; Sourirajan and Kimura, 1967) that, with respect to the type of films used in this work, ($D_{AM}/K\delta$) for sodium chloride is proportional to $A^{3.5}$, and ($D_{AM}/K\delta$) for glycerol is proportional to $A^{3.35}$. During conditions of extended continuous operation under pressure in the reverse osmosis process, the dense microporous surface structure of the film may be expected to remain intact, but the spongy porous structure of the interior of the film will become more densely compacted. The result is an increase in resistance to fluid flow and a decrease in A. It is of practical interest to determine the relationship between A and ($D_{AM}/K\delta$) when A decreases entirely due to membrane compaction during continuous operation under pressure. This was done experimentally from the

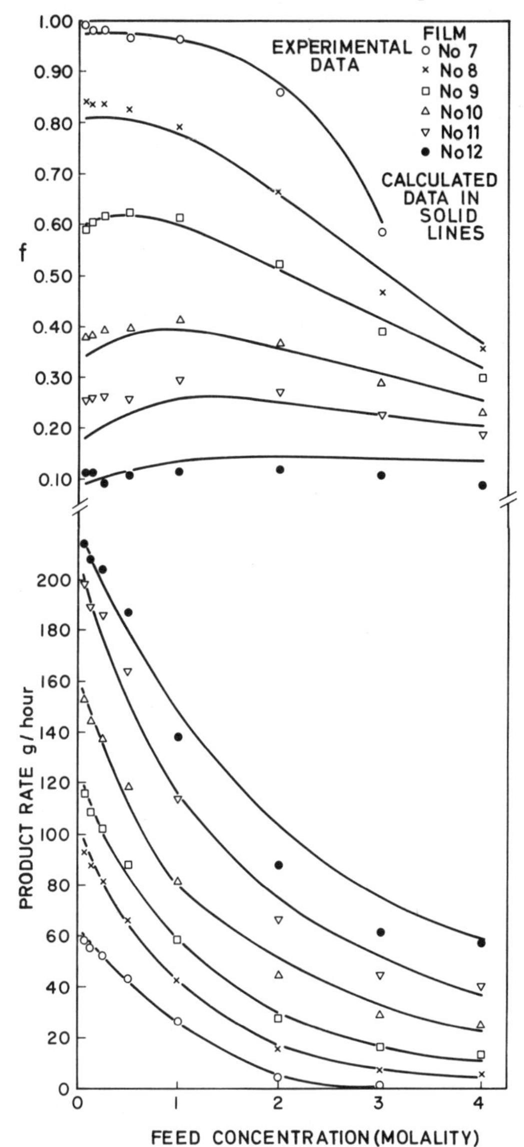

Figure 3-49. Comparison of experimental and calculated solute separation and product rate data for the system sodium chloride-water. Data of Kimura and Sourirajan (1967).

Film type: CA-NRC-18; operating pressure: 1500 psig; feed rate: 250 cm^3/min; film area: 7.6 cm^2.

extended continuous test run data for the systems glycerol-water and sodium chloride-water.

Experimental results with the system glycerol-water in long-time test runs. A 30-day long continuous test run was conducted with the system glycerol-water at an operating pressure of 1500 psig using five different films (87, 88, 89, 90, and 91) covering a wide range (51 to 98 per cent) of solute

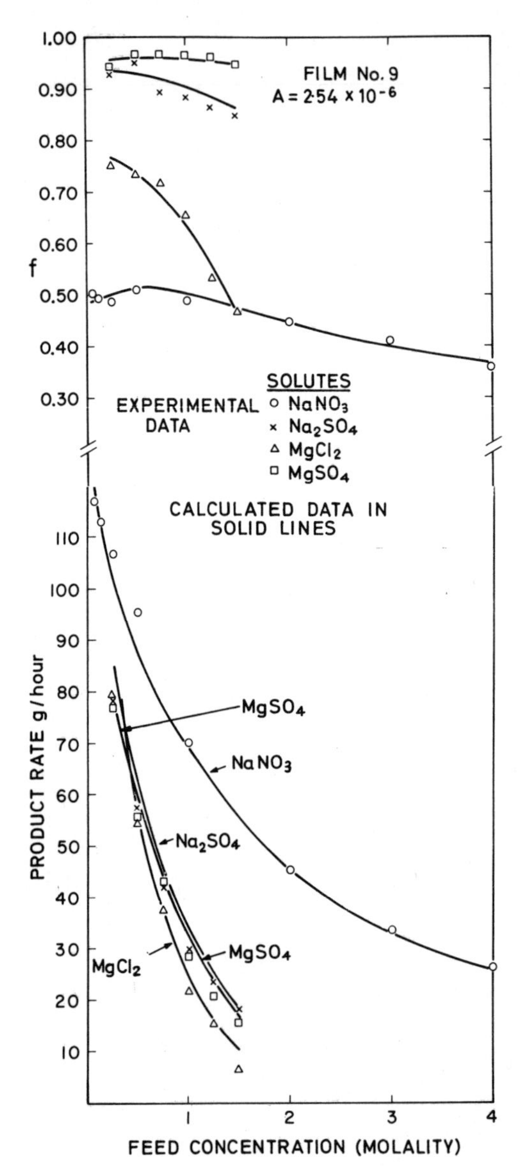

Figure 3-50. Comparison of experimental and calculated solute separation and product rate data for different solution systems. Data of Kimura and Sourirajan (1967).

Film type: CA-NRC-18; operating pressure: 1500 psig; feed rate: 250 cm^3/min; film area: 7.6 cm^2.

separations. The feed concentrations were maintained in the range 0.50 to 0.52M, and the feed flow rates in the range 580 to 640 cm^3/min during the run. The initial characteristics of the films used are specified in Table 3-5 in terms of A and $(D_{AM}/K\delta)$. During the continuous extended operation under pressure both product rate and solute separation changed with time. [PR] values decreased rapidly during the first 24 hours, and slowly thereafter. The values of f also decreased with decrease in [PR]. A set of typical data is given in Table 3-6. At different time intervals during the test run the experimental [PR] and f values were determined for each film; the corresponding feed concentrations and feed flow rates were also noted.

From the feed concentration, [PR] and f data, N_B, X_{A1} and X_{A3} can be determined; the mass transfer coefficient applicable for the conditions of the experiment can be obtained from the correlation of the experimental k data given in Figure 3-45. Therefore, by using Equations 3-35, 3-43, and 3-52, one can calculate the values of A and $(D_{AM}/K\delta)$ corresponding to every set of [PR] and f data obtained during the continuous test run. These calculations were performed with the experimental data obtained with the films 87, 88, 89, 90, and 91 during the 30 day-long test run, and the results are plotted in Figure 3-51. The results showed that even though the values of A decreased significantly during the test period (finally reaching about 57 per cent of the original value for Film 87 and 13 per cent of the original value for Film 91), $(D_{AM}/K\delta)$ remained essentially constant throughout for each one of the films tested.

Experimental results with the system sodium chloride-water in long-time test runs. A continuous test run extending for 72 hours was conducted with the system sodium chloride-water at an operating pressure of 1500 psig using six different films (94, 105, 106, 107, 108, and 109) covering a wide range (39.5 to 98.3 per cent) of solute separations. The feed flow rate was in the range 426 to 562 cm^3/min. The initial characteristics of the films used are specified in Table 3-5 in terms of A and $(D_{AM}/K\delta)$. As in the experiments with the system glycerol-water, both [PR] and f changed with time during the

Table 3-6. Variation of solute separation with product rate in reverse osmosis during continuous operation under pressure for the system glycerol-water

Data of Kimura and Sourirajan (1968b)

	Initial values		Values after several days of continuous operation	
Film no.	[PR] (g/hr)	f	[PR] (g/hr)	f
87	28.39	0.978	16.50	0.971
88	58.57	0.930	25.50	0.890
90	75.51	0.855	22.07	0.745
89	97.19	0.721	20.32	0.490
91	127.45	0.510	30.67	0.348

Film type: CA-NRC-18
Feed molality: 0.5M
Feed rate: 580 cm^3/min
Operating pressure: 1500 psig
Film area: 7.6 cm^2
[PR]: Product rate

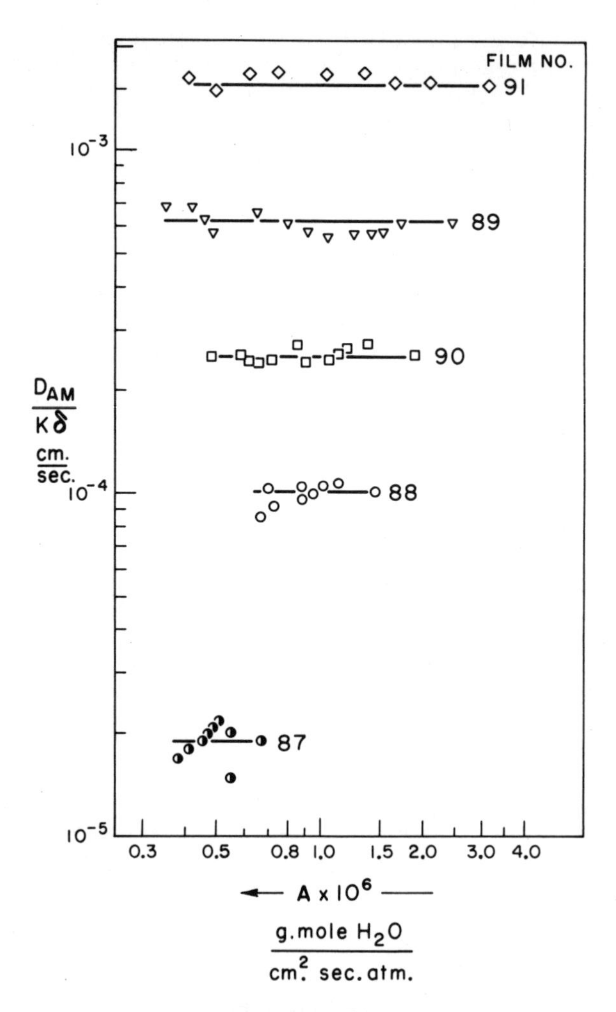

Figure 3-51. Variation of $(D_{AM}/K\delta)$ with decrease in A during 30 day-long continuous test runs with the system glycerol-water. Data of Kimura and Sourirajan (1968b).

Film type: CA-NRC-18; feed molality: 0.50 to 0.52; feed rate: 580 to 640 cm^3/min; operating pressure: 1500 psig.

progress of the run. At different intervals during the test run, the experimental [PR] and f values were determined for each film.

Using the generalised mass transfer coefficient correlation (Figure 3-46) for obtaining the k values, and Equations 3-35, 3-43, and 3-52, the values of A and $(D_{AM}/K\delta)$ were calculated for every set of [PR] and f data obtained during the continuous test run (Table 3-7).

The experimental data for the films 105 and 106 (which gave comparatively higher levels of solute separation) show that even a 40 per cent change in the value of $(D_{AM}/K\delta)$ corresponds to less than 1 per cent change in solute separation; similarly the data for films 109 and 108 (which gave comparatively lower

Table 3-7. Variation of $(D_{AM}/K\delta)$ with decrease in A in reverse osmosis during continuous operation under pressure with the system sodium chloride-water

Data of Kimura and Sourirajan (1968b)

Film no.	$A \times 10^6$ $\left(\frac{\text{g. mole } H_2O}{\text{cm}^2 \text{ sec. atm}}\right)$	$\left(\frac{D_{AM}}{K\delta}\right) \times 10^5$ (cm/sec)	Feed molality	Feed rate (cm^3/min)	[PR] (g/hr)	$f \times 10^2$	$\frac{[PR] \text{ final}}{[PR] \text{ initial}}$
105	1.494	2.000	1.0075	564	35.97	98.3	
	0.980	1.708	1.0070	556	24.89	98.0	
	0.885	1.895	1.0150	426	22.19	97.5	
	0.820	1.784	1.0150	426	20.73	97.5	
	0.803	1.774	1.0000	426	20.61	97.5	0.573
106	2.962	44.37	1.0075	563	72.98	82.1	
	1.914	40.46	1.0025	545	53.39	79.8	
	1.766	40.63	1.0075	546	50.09	78.9	
	1.754	40.93	1.0075	542	49.87	78.7	0.683
94	1.873	56.05	0.5030	487	72.67	78.1	
	1.485	54.21	0.4800	492	58.66	74.0	0.807
107	3.215	62.98	1.0075	562	82.22	77.8	
	1.921	59.39	1.0075	545	56.38	73.7	
	1.891	59.36	1.0075	541	55.69	73.5	0.677
109	4.747	347.5	1.0075	558	146.25	44.4	
	3.120	328.7	1.0025	542	109.11	42.4	
	3.080	321.9	1.0075	552	107.55	42.9	
	2.842	345.0	1.0075	537	102.80	40.4	0.703
108	5.536	488.6	1.0075	561	174.37	39.5	
	3.597	460.2	1.0025	543	130.83	37.9	
	3.513	457.7	1.0075	554	128.37	38.0	
	3.261	500.9	1.0075	544	124.56	34.5	0.714

Film type: CA-NRC-18
Operating pressure: 1500 psig
Film area: 7.6 cm^2
[PR]: Product rate

levels of solute separation) show that even an 8 per cent change in the value of $(D_{AM}/K\delta)$ corresponds to less than 1 per cent change in solute separation. Hence it is reasonable to conclude that within the limits of experimental error, $(D_{AM}/K\delta)$ values have remained constant for each film throughout the test period. Thus the data presented in Table 3-7 confirm that the observation that $(D_{AM}/K\delta)$ remains constant as A decreases because of membrane compaction during extended continuous service under pressure, is also valid for the system sodium chloride-water.

Probable general validity of the principle. For systems such as glycerol-water, and sodium chloride-water, the solute transport parameter $(D_{AM}/K\delta)$ is independent of X_{A2} which is the mole fraction of solute in the concentrated boundary solution on the high pressure side of the membrane (Kimura and Sourirajan, 1967; Sourirajan and Kimura, 1967); the foregoing results show that this relationship between $(D_{AM}/K\delta)$ and X_{A2} for the systems glycerol-water and sodium chloride-water remains unaffected during membrane compaction. Kimura and Sourirajan (1968d) have shown that for the system sucrose-water, $(D_{AM}/K\delta)$ is a particular function of X_{A2}, and again, this relationship is unaffected during membrane compaction. That the relationship between $(D_{AM}/K\delta)$ and X_{A2} remains unaffected as the pure water permeability constant A decreases because of membrane compaction is probably a basic principle of mass transport through reverse osmosis membranes. It is worthwhile to seek extensive experimental confirmation of this principle for the widest possible variety of membrane-solution systems. In any case, the experimental data presented here show that the principle is valid at least for the systems glycerol-water and sodium chloride-water using the Loeb-Sourirajan type porous cellulose acetate membranes, and hence it offers an analytical technique of predicting solute separation as a function of product rate in extended continuous operation of the reverse osmosis separation process involving the above systems. This technique has been applied for predicting the performance of a few typical films for the systems glycerol-water and sodium chloride-water, and the results are discussed below.

Prediction Technique. The A-factor, which is defined as the ratio of the A-value of the membrane at any time to its initial value specified in Table 3-5, may be considered as a relative measure of membrane compaction during continuous operation under pressure. Since $(D_{AM}/K\delta)$ remains constant during membrane compaction, the product rate and separation characteristics of any film are predictable for different A-factors at any feed concentration and feed flow rate. This is illustrated below for the systems glycerol-water and sodium chloride-water.

The membrane specifications given in Table 3-5 were used to represent the initial conditions, and $(D_{AM}/K\delta)$ was assumed constant during membrane compaction. The only additional data needed are the mass transfer coefficients applicable for the experimental conditions. For the system glycerol-water, the experimental k values were used (Figure 3-45); for the system sodium chloride-water, k values were obtained from the generalised mass transfer correlation obtained by the diffusion current method (Figure 3-46). By trial and error, the combination of X_{A2} and X_{A3} satisfying Equation 3-87 were first found; then that combination which would also satisfy Equations 3-35 and 3-43 was found for several assumed decremental values of A. Thus the values of N_B and X_{A3} corresponding to each assumed decremental value of A were obtained, from which f and [PR] values were calculated from the relations 3-64 and 3-77.

A set of such calculations for the system glycerol-water is illustrated in Figure 3-52 where A-factor denotes the assumed fraction of the initial A value specified for the particular film in Table 3-5. Figure 3-52 also contains the experimental solute separation data plotted against the [PR] data obtained during the progress of the test run. Figure 3-52 shows that the agreement between the experimental and calculated separation data corresponding to the different product rates obtained, is excellent for each one of the films tested.

Effect of membrane compaction on the performance of porous cellulose acetate membranes for the system sodium chloride-water. This is illustrated in Figure 3-53 for two different operating pressures (500 and 1500 psig) and seven different films (1, 2, 3, F3, 4, 5, and 6) involving a wide range (35.8 per cent to 99 per cent) of solute separations for the feed system 0.5M $NaCl$-H_2O, covering A-factors up to 0.15. The following observations are evident. Such calculations give directly solute separation as a function of product rate as the latter decreases due to membrane compaction; the [PR] v. f data are experimentally verifiable. Even though solute separation generally decreases with decrease in product rate due to membrane compaction, this is not always so. For example, at an operating pressure of 1500 psig, f remains essentially constant for Film 5 up to A-factor = 0.75; and for Film 6, f passes

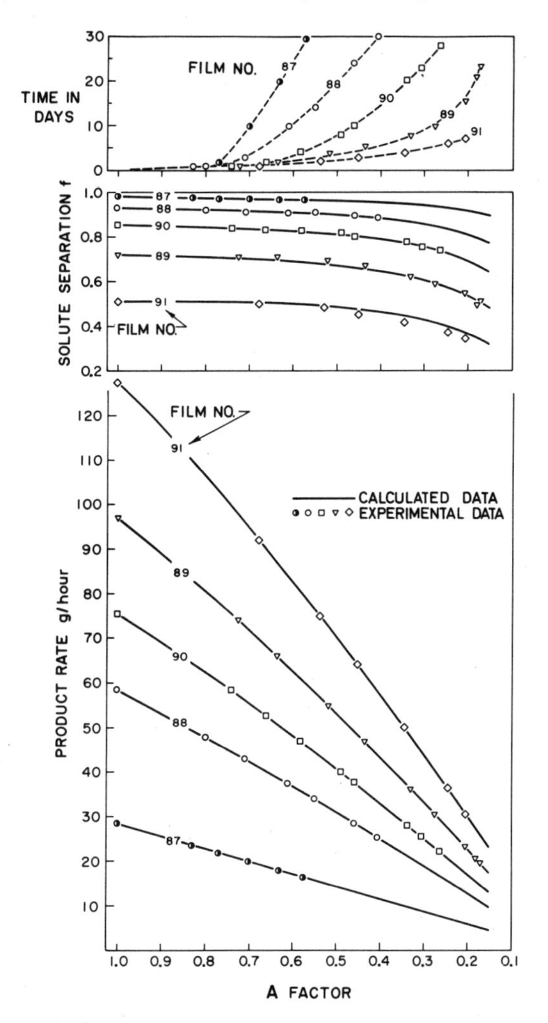

Figure 3-52. Comparison of experimental and calculated separation data corresponding to the product rates obtained during extended continuous test runs with the system glycerol-water. Data of Kimura and Sourirajan (1968b).

Film type: CA-NRC-18; feed molality: 0.5M, feed rate: 580 cm^3/min; operating pressure: 1500 psig; film area: 7.6 cm^2.

through a maximum at A-factor = 0.5. It would be interesting to obtain experimental confirmation of such film behaviour in extended continuous test runs. Further, for the 0.5M feed solution, the decrease in f due to membrane compaction appears to be generally steeper at lower operating pressure than at higher operating pressure, and for A-factors less than 0.5 for both the operating pressures investigated. These observations are of practical importance to reverse osmosis systems operating at different pressures. Films 1 and F3 are typical of the ones capable of giving high levels of solute separation at high operating pressures; f remains practically constant for these films at 1500 psig for A-factors up to 0.5.

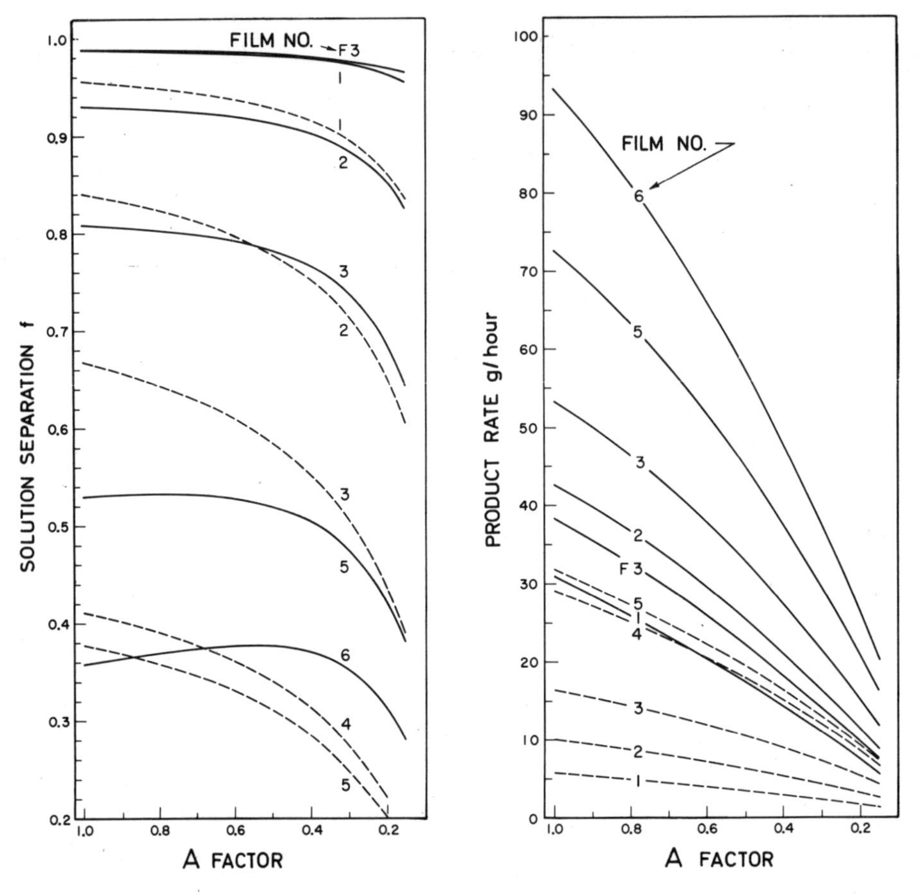

Figure 3-53. Effect of membrane compaction on the performance of porous cellulose acetate membranes for the system sodium chloride-water. Data of Kimura and Sourirajan (1968b).

Film type: CA-NRC-18; feed molality: 0.5M; feed rate: 250 cm^3/min; operating pressure: ——— 1500 psig — — — 500 psig; film area (right): 7.6 cm^2.

Effect of membrane compaction on the performance of porous cellulose acetate membranes suitable for brackish water conversion. Films F1, F2 and F4 are typical of those suitable for brackish water conversion. At an operating pressure of 600 psig, using a feed concentration of 2500 ppm NaCl, and a feed flow rate of 300 cm^3/min, Films F1 and F2 gave products containing 317 and 95 ppm of NaCl respectively; the corresponding product rates were 54.1 and 38.6 gal/day/ft^2 of film area. Under otherwise identical experimental conditions, using a feed solution containing 5000 ppm of NaCl, Film F4 gave a product containing 258 ppm of NaCl; the corresponding product rate was 35.8 gal/day/ft^2. The effect of membrane compaction on the separation and permeability characteristics of the films are of practical interest. Therefore, the variations of the product concentration and product rate for A-factors up to 0.15 were calculated, and the results are given in Figure 3.54. As the product rate slows down due to membrane compaction, the NaCl concentration in the product increases for all the above films. An important consequence of this analysis is that it is possible to calculate the allowable drop in product

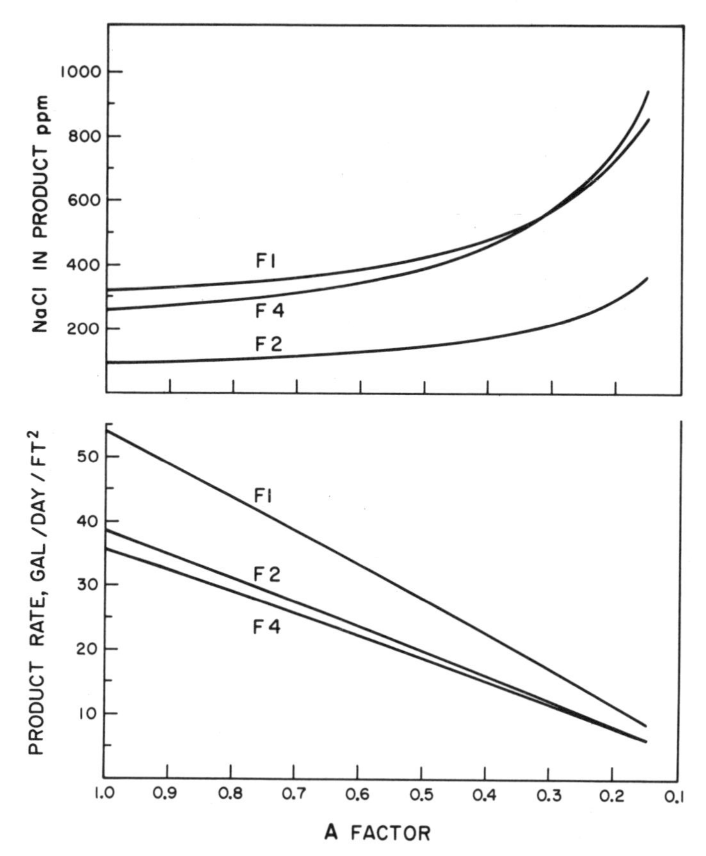

Figure 3-54. Effect of membrane compaction on the performance of porous cellulose acetate membranes suitable for brackish water conversion. Data of Kimura and Sourirajan (1968b).

Film type: CA-NRC-18; system: NaCl-H_2O; feed concentration: films F1 & F2: 2500 ppm NaCl; film F4: 5000 ppm NaCl; feed rate: 300 cm^3/min; operating pressure: 600 psig.

rate for a film to give a product of given quality. For example, in order to get a product water containing less than 500 ppm of NaCl under the conditions of the experiment given above, the product rates may be allowed to drop down to 21 and 13.5 gal/day/ft^2 respectively for the films F1 and F4, whereas Film F2 will give the required quality of the product water even when the product rate drops to 6 gal/day/ft^2.

Effect of Feed Flow Rate on the Effect of Membrane Compaction. Figure 3-55 shows the effect of feed flow rate and membrane compaction on the quality of the product and on the product rate obtainable with Film F1 at the operating pressure of 600 psig, using a feed solution containing 2500 ppm of NaCl. The results show that for getting water containing less than 500 ppm, it is desirable to operate at feed flow rates above 200 cm^3/min. The allowable level of membrane compaction to yield a product of the above quality increases with increase in feed flow rate; this is one possible way of increasing the service life of a given film to yield a product of given quality. The solute separation data at a feed flow rate of 100 cm^3/min show that product

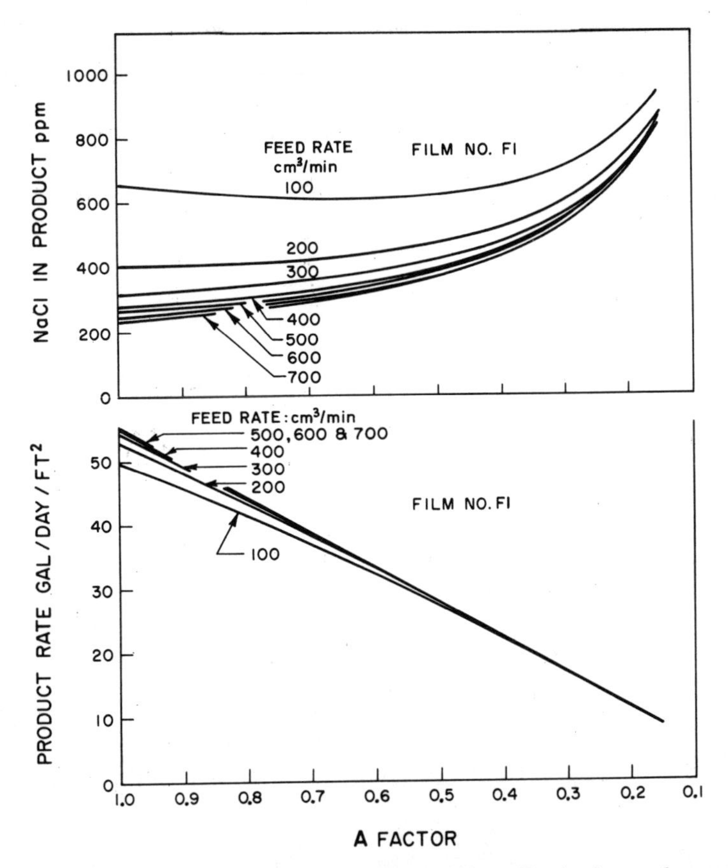

Figure 3-55. Effect of feed flow rate on the effect of membrane compaction. Data of Kimura and Sourirajan (1968b).

Film type: CA-NRC-18; system: NaCl-H_2O; feed concentration: 2500 ppm NaCl; operating pressure: 600 psig.

water containing less than 500 ppm cannot be obtained at that feed rate; but the data are interesting from another point of view. They give another example of a case where solute separation increases with decrease in product rate down to A-factor = 0.6. Obviously then, a particular combination of A, $(D_{AM}/K\delta)$ and feed flow rate can give rise to such film behaviour during compaction. The foregoing comments apply to the particular apparatus and experimental conditions used.

Relationship Between Mole Fraction of Solute in the Concentrated Boundary Solution and in the Product Solution. It has already been shown that, in the reverse osmosis separation process, there exists a unique relationship between the values of X_{A2} and X_{A3} where X_{A3} decreases as X_{A2} decreases. It is interesting to consider how membrane compaction affects this relationship. Membrane compaction increases the resistances to the flow both of the preferentially sorbed pure water and of the concentrated boundary solution through the membrane. Since concentration polarisation on the high pressure side of the membrane is the result of the withdrawal of the preferentially sorbed pure water whose rate decreases during membrane compaction, X_{A2} must decrease as the A-factor decreases. Since, in most cases, the decrease in A-factor results in an increase in X_{A3}, it is clear that membrane compaction should result in a shift in the characteristic X_{A2} v. X_{A3} curve. This is illustrated in Figure 3-56 in the case of Films 1 and 3 for the system NaCl-H_2O. The data for these curves were obtained by calculating the values of X_{A2} and X_{A3} for different feed flow rates in the range 100 to 700 cm^3/min at different A-factors. It may hence be concluded that the relationship between X_{A2} and X_{A3} is uniquely fixed by the overall porosity of the film and the chemical nature of the solute.

Significance of Results. The correlation that $(D_{AM}/K\delta)$ remains constant as A decreases due to membrane compaction during continuous service under pressure is an important one from several points of view. It establishes an experimental basis, and offers an analytical means, of predicting the variations in solute separation as the product rate slows down because of membrane

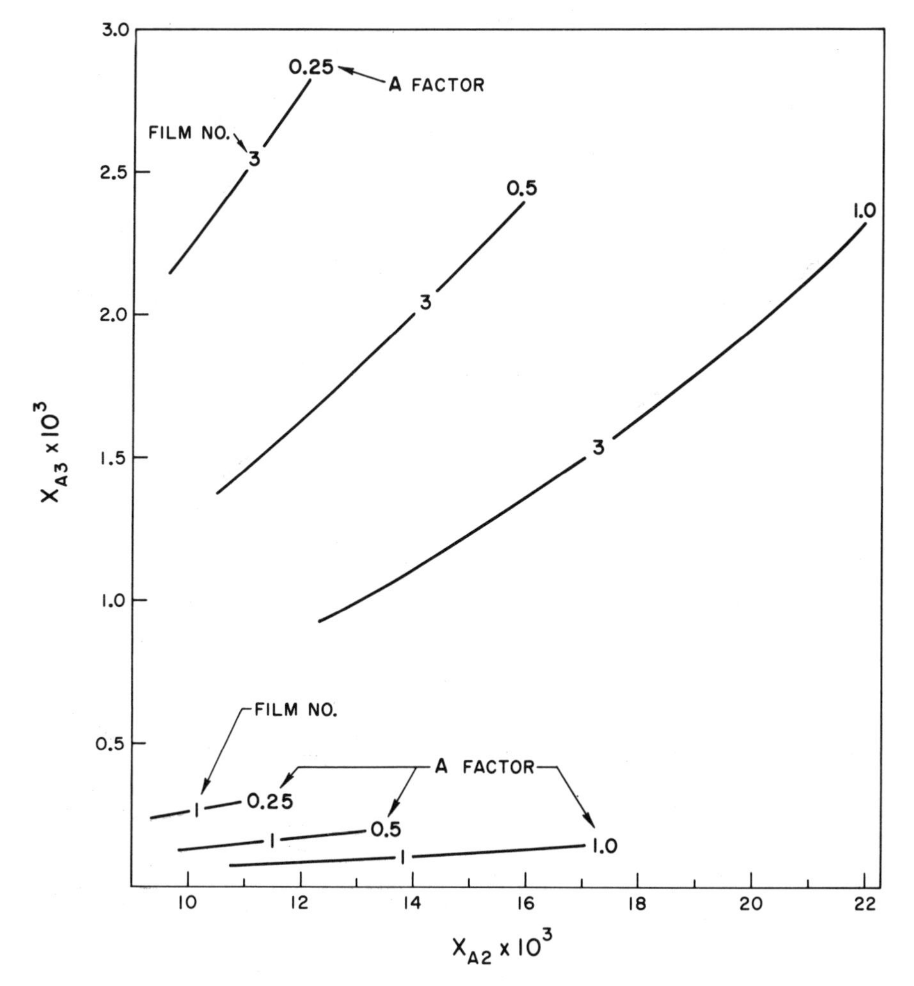

Figure 3-56. Effect of membrane compaction on the concentration of solute in the boundary solution v. that in the product for the system sodium chloride-water. Data of Kimura and Sourirajan (1968b).

Film type: CA-NRC-18; feed molality: 0.5M; feed rate: 100-700 cm^3/min; operating pressure: 1500 psig.

compaction during extended continuous operation under pressure in the reverse osmosis process using the Loeb-Sourirajan type porous cellulose acetate membranes. The experimental confirmation of the accuracy of prediction is a further verification of the method of analysis used, and it unfolds new approaches to process design and membrane development in reverse osmosis. For example, it is now clear that variations in solute separation with decrease in product rate must be expected, and can be calculated and allowed for, in practical reverse osmosis units. Such variations arise as a consequence of the effect of membrane compaction on the transport characteristics of reverse osmosis membranes, even in the absence of any deterioration of the membrane material due to chemical interactions such as hydrolysis. Therefore the solution to the problem of decrease in solute separation with decrease in product rate during extended service conditions lies essentially in reducing membrane compaction and altering the flow conditions suitably to give the desired result. The production of more rigid membranes is particularly important for low pressure service. Further, it is now possible to predict from the initial specifications of the film in terms of A and $(D_{AM}/K\delta)$, the allowable decrease in product rate in the continuous operating units to give a product of given quality, provided, of course, the applicable mass transfer correlation for the operating unit is available. However, such a prediction is possible only if the pore structure of the membrane surface remains intact, though the spongy interior structure of the film compacts under continuous pressure. Hence, it is necessary to produce films which do not change their surface pore structure during continuous service under pressure. The type of films used in this work seems to satisfy this criterion. If the surface pore structure changes, then $(D_{AM}/K\delta)$ will change with decrease in product rate under continuous service conditions. From the point of view of membrane development, it is desirable to investigate the film casting conditions under which $(D_{AM}/K\delta)$ will remain constant as A decreases to an acceptable fraction of its original value due to membrane compaction under continuous pressure, and to include such operating data in membrane specifications.

9. Specification, Selectivity, and Performance of Porous Cellulose Acetate Membranes

From the preceding discussions it is clear that a Loeb-Sourirajan type porous cellulose acetate membrane can be specified for reverse osmosis application with aqueous feed solutions, in terms of the pure water permeability constant A, and the solute transport parameter $(D_{AM}/K\delta)$ at a given operating pressure. For several inorganic and organic solutes, $(D_{AM}/K\delta)$ is independent of feed concentration and feed flow rate. This discussion is concerned only with such solutes. Membrane selectivity for such solutes can be expressed in terms of their relative values of $(D_{AM}/K\delta)$. Provided the mass transfer coefficient, k, applicable for the high pressure side of the membrane is available or can be calculated, the performance of a membrane, expressed in terms of solute separation and product rate, can be predicted from membrane specifications, as pointed out earlier. In this section, the data on membrane specification are related to those on membrane performance so that the significance of the specifying parameters can be recognised readily in practical terms of cause and effect (Agrawal and Sourirajan, 1969b).

Batch 18 Type Membranes. Figures 3-57 and 3-58 give three sets of A and $(D_{AM}/K\delta)$ data for the system [NaCl-H_2O] obtained from short-run experiments with CA-NRC-18 type porous cellulose acetate membranes (Agrawal and Sourirajan, 1969b). The data show that $(D_{AM}/K\delta)$ for NaCl is proportional to A^n where n is between 3.4 and 3.5. The three sets of films represented in

Figures 3-57 and 3-58 differ only in the pressure treatment given to them prior to the reverse osmosis experiments. Films used in reverse osmosis experiments at 1500 psig (Figure 3-57), and at 600 and 25 to 250 psig (Figure 3-58) were initially subjected to a pure water pressure of 1700, 700, and 300 psig respectively for 1 to 2 hours. From Figure 3-58, for a given value of $(D_{AM}/K\delta)$, the A value for a film subjected to 1700 psig initially is only about one-half of that of a film subjected initially to a pressure of 300 psig. This means that the initial pressure treatment results in a degree of permanent compaction of the porous structure of the film, and a consequent increase in resistance to fluid flow.

The solid lines in Figure 3-57 and 3-58 may be considered as the general specifications for the type of films used. Since every particular combination of A and $(D_{AM}/K\delta)$ for NaCl specifies a membrane at a particular operating pressure, their exact values are of practical importance in specifying reverse

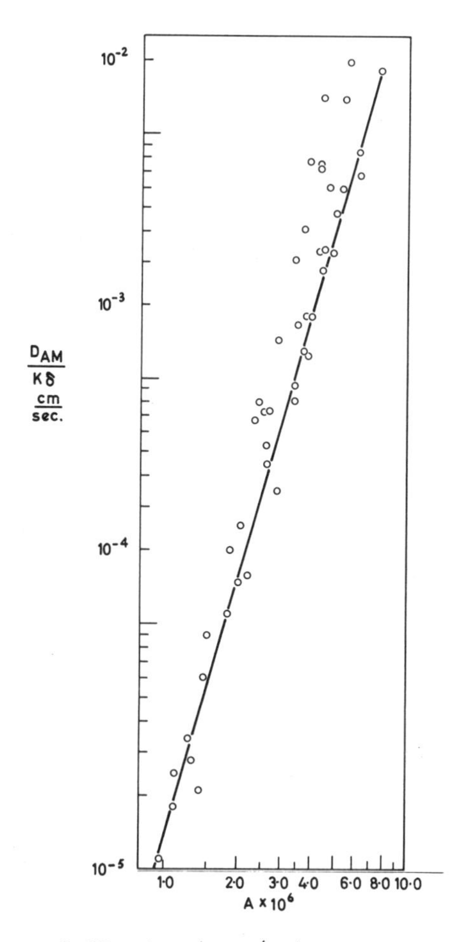

Figure 3-57. A v. $(D_{AM}/K\delta)$ for sodium chloride at 1500 psig for the CA-NRC-18 type porous cellulose acetate membranes. Data of Agrawal and Sourirajan (1969b).

System: sodium chloride-water; operating pressure: 1500 psig; film type: CA-NRC-18.

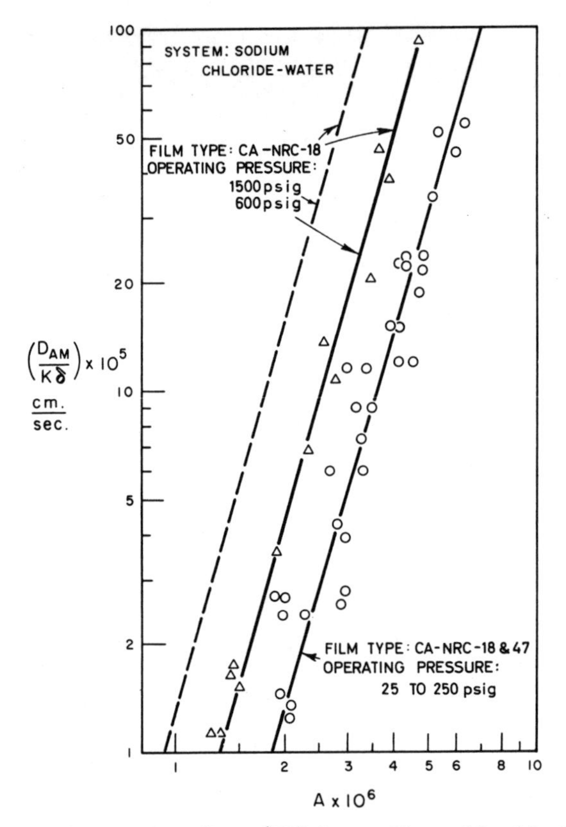

Figure 3-58. A v. $(D_{AM}/K\delta)$ for sodium chloride for the pressure range 25 to 600 psig for the CA-NRC-18 type porous cellulose acetate membranes. Data of Agrawal and Sourirajan (1969b).

osmosis systems and a wide variety of numerical calculations in reverse osmosis process design. Hence the numerical values of A and $(D_{AM}/K\delta)$ for NaCl obtained from the solid lines in Figures 3-57 and 3-58 are given in Table 3-8.

Correlation of experimental mass transfer coefficient data for the systems sodium chloride-water and urea-water. Figures 3-59 and 3-60 illustrate that the log-log plot of N_{Re} v. $N_{Sh}/N_{Sc}^{0.33}$, using the k values calculated from the results of reverse osmosis experiments in the operating pressure range 25 to 1500 psig (using different feed concentrations, feed flow rates, and different films involving a wide range of solute separations), agreed fairly well with the mass transfer coefficient correlation obtained by the diffusion current method (Figure 3-46) (Agrawal and Sourirajan, 1969b; Ohya and Sourirajan, 1969a). Agreement was observed with the data of all the other solution systems considered in this discussion.

Effect of Operating Temperature. For predicting membrane performance at any preset operating conditions, the values of A, $(D_{AM}/K\delta)$, and k are required. Hence the effect of temperature on the above parameters must be known in order to be able to predict membrane performance at different operating temperatures. To obtain the information, a set of reverse osmosis experiments were conducted in the temperature range 5 to 36° with the system

Table 3-8. General specifications of CA-NRC-18 type porous cellulose acetate membranes for reverse osmosis

Data of Agrawal and Sourirajan (1969b)

$(D_{AM}/K\delta) \times 10^5$ for NaCl at operating pressure (cm/sec)	$A \times 10^6$ (g. mole $H_2O/cm^2/sec/atm$) Operating pressure (psig)		
	1500[a]	600[b]	25-250[c]
1	0.93	1.35	1.85
2	1.12	1.61	2.27
3	1.26	1.81	2.55
4	1.37	1.95	2.77
5	1.45	2.08	2.95
6	1.53	2.20	3.12
7	1.60	2.28	3.27
8	1.66	2.38	3.39
9	1.72	2.45	3.50
10	1.77	2.53	3.61
20	2.15	3.06	4.44
30	2.42	3.43	5.00
40	2.62	3.71	5.40
50	2.80	3.97	5.75
60	2.94	4.17	6.10
70	3.08	4.35	6.40
80	3.20	4.50	6.60
90	3.30	4.69	6.82
100	3.40	4.81	7.00

[a] initial pressure treatment at 1700 psig
[b] initial pressure treatment at 700 psig
[c] initial pressure treatment at 300 psig

sodium chloride-water. Two different feed concentrations (0.5M and 2.0M) and four different films covering a wide range of solute separation (26 to 98 per cent were used at the operating pressure of 1500 psig. From the experimental [PWP], [PR], and f data at different operating temperatures, the values of A, $(D_{AM}/K\delta)$, and k were calculated. The results obtained are given in Figures 3-61, 3-62, and 3-63 (Agrawal and Sourirajan, 1969b). The data show that in the temperature (T) range 5 to 36°,

$$A \times \mu_W = \text{constant} \tag{3-89}$$

and for the system sodium chloride-water,

$$(D_{AM}/K\delta) \propto e^{0.005T} \tag{3-90}$$

and

$$k \propto e^{0.005T} \tag{3-91}$$

for all the films tested.

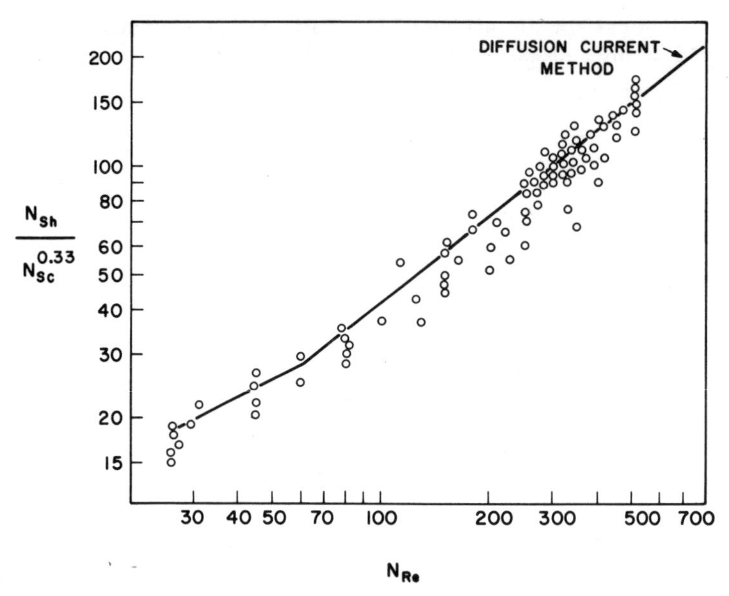

Figure 3-59. Comparison of experimental data for the system sodium chloride-water with the generalised mass transfer coefficient correlation. Data of Agrawal and Sourirajan (1969b).

Film type: CA-NRC-18; feed molality: 0. 0625M to 4. 0M; operating pressure 250 to 1500 psig.

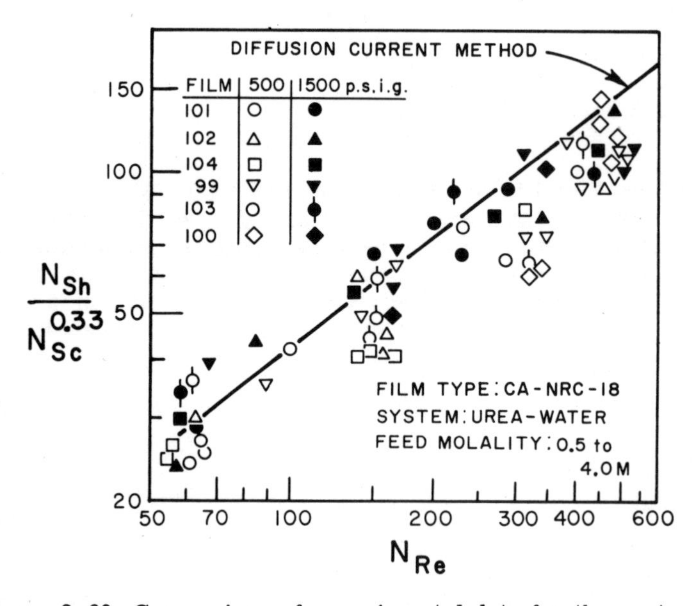

Figure 3-60. Comparison of experimental data for the system urea-water with the generalised mass transfer coefficient correlation. Data of Ohya and Sourirajan (1969a).

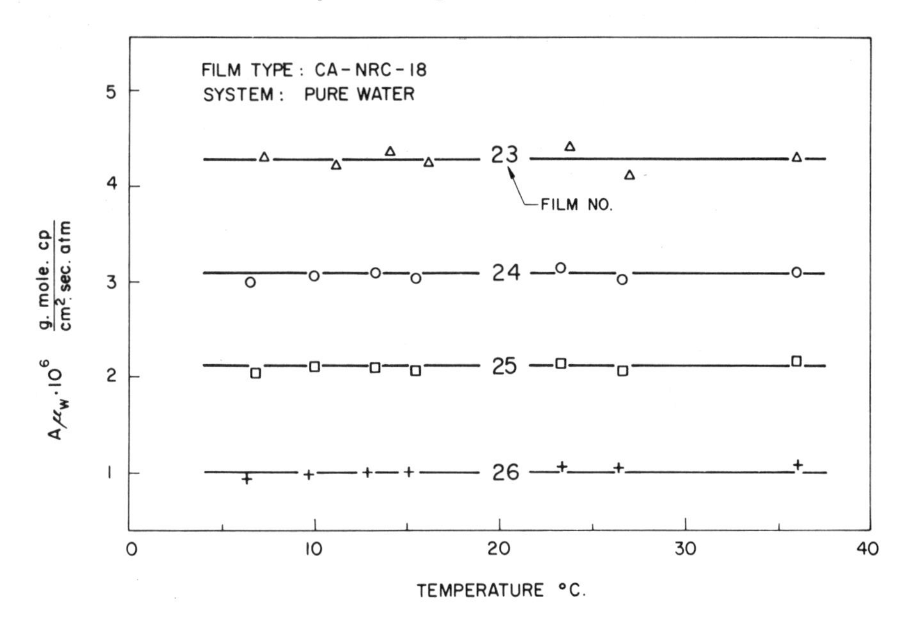

Figure 3-61. Effect of temperature on the pure water permeability constant. Data of Kopeček and Sourirajan (1969a).

Feed rate: 250 cm^3/min; operating pressure: 1500 psig.

A Relative Scale of Membrane Selectivity. Between the quantities A and $(D_{AM}/K\delta)$ for a solute, which specify a membrane at a given operating pressure, while the quantity A may change due to membrane compaction under conditions of continuous reverse osmosis operation, the quantity $(D_{AM}/K\delta)$ remains constant provided the surface pore-structure of the membrane remains the same. This has been illustrated and discussed before with particular reference to the systems sodium chloride-water and glycerol-water. The extensive experimental data obtained with all the other systems discussed here also point to the same conclusion. Hence the variation in the values of $(D_{AM}/K\delta)$ for different solutes obtained with a given membrane offers a method of expressing membrane selectivity for different solutes. Figure 3-64 gives a log-log plot of $(D_{AM}/K\delta)$ for NaCl v. $(D_{AM}/K\delta)$ for different solutes for several membranes of different surface porosities at the operating pressure of 1500 psig. All the data plotted are for 25°. In the range of data plotted, he above correlations are essentially straight lines with different slopes. Figure 3-64 represents a useful relative scale of membrane selectivity for different solutes. If A and $(D_{AM}/K\delta)$ for NaCl are given for a particular membrane at 1500 psig, the performance of that membrane for all the other systems at the same pressure can be predicted, using the data given in Figure 3-64, Table 3-8, and the generalised mass transfer coefficient correlation to obtain k. In view of the importance of the precise values of $(D_{AM}/K\delta)$ for reverse osmosis system specification and process design (Chapter 4), a set of numerical $(D_{AM}/K\delta)$ data is given in Table 3-9.

Figures 3-65 and 3-66 illustrate the effect of operating pressure on the relative values of $(D_{AM}/K\delta)$ for the systems [NaCl-H_2O], [$NaNO_3$-H_2O], and [urea-water]; such correlations for other systems have to be determined experimentally.

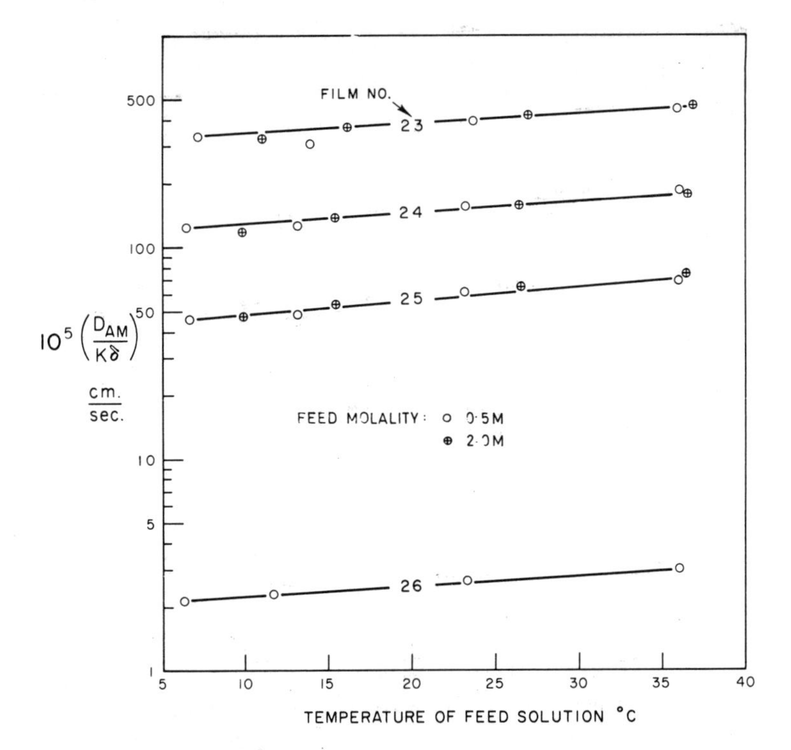

Figure 3-62. Effect of temperature on the solute transport parameter for the system sodium chloride-water. Data of Agrawal and Sourirajan (1969b).

Film type: CA-NRC-18; feed rate: 250 cm^3/min; operating pressure: 1500 psig.

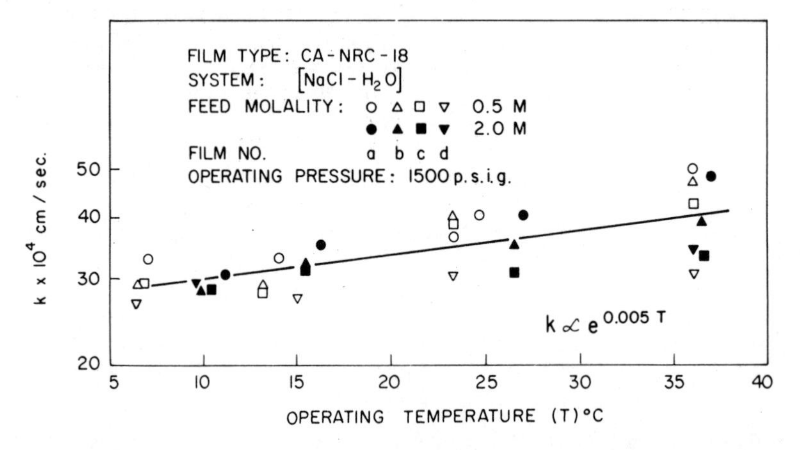

Figure 3-63. Effect of temperature on the mass transfer coefficient, k, for the system sodium chloride-water. Data of Agrawal and Sourirajan (1969b).

Table 3-9. A relative scale of membrane selectivity for different solutes

Data of Agrawal and Sourirajan (1969b)

Solute	$(D_{AM}/K\delta) \times 10^5$ (cm/sec)						
$MgSO_4$	0.07	0.24	0.41	1.42	2.42	4.1	5.6
Na_2SO_4	0.13	0.52	0.95	3.8	6.8	12.4	17.4
$BaCl_2$	0.28	1.66	3.55	20.5	42.1	95	148
$CaCl_2$	0.48	2.55	5.2	28	57.5	119	183
$MgCl_2$	0.56	2.76	5.5	27.1	54	107	160
LiCl	0.88	4.8	9.8	53	110	228	350
NaCl	1	5	10	50	100	200	300
KCl	1.27	6.7	13.6	72	146	300	455
NH_4Cl	1.66	8.6	17.5	90	184	372	560
$LiNO_3$	1.66	7.1	13.5	58	109	210	300
KNO_3	2.34	12	27.5	120	240	490	735
$NaNO_3$	3.0	12	21.3	84	151	275	390
Glycerol	3.5	12	20.5	71	120	220	287
Urea	42	95	136	310	445	635	780

Film type: CA-NRC-18
Operating pressure: 1500 psig

Prediction of Membrane Performance. Membrane performance, expressed in terms of solute separation (f), and product rate [PR] at preset operating conditions of pressure (P), feed concentration (X_{A1}), and feed flow rate (i.e. k), can be predicted from membrane specifications given in terms of A and $(D_{AM}/K\delta)$ by the technique described earlier using Equations 3-87, 3-35, 3-43, 3-64, and 3-77. When the values of A, $(D_{AM}/K\delta)$, and k at 25° are known, their values at other temperatures in the range 5 to 36° can be obtained from Equations 3-89, 3-90, and 3-91 respectively, and the effect of temperature on the performance of the film predicted; this procedure is valid at least for the system sodium chloride-water.

Mass transfer coefficients for prediction of membrane performance. The data on membrane performance illustrated below are based on the values of k at 25° obtained from the generalised mass transfer coefficient correlation given by the diffusion current method. From the above correlation, the values of k can be obtained as functions of feed flow rate (Q) and feed concentration as illustrated in Figure 3-67 for the system sodium chloride-water. Such k v. Q correlations can be expressed by the general relation $k \propto Q^n$ (Equation 3-81), where the value of n is between 0.78 and 0.80 for all the systems discussed below. The value of k is different for different solution systems even at the same values of feed concentration and feed flow rate. For example, for a 0.5M feed solution and a feed flow rate of 300 cm^3/min, the values of k for the systems involving the solutes $MgSO_4$, Na_2SO_4, $BaCl_2$, $CaCl_2$, $MgCl_2$, LiCl, NaCl, KCl, NH_4Cl, $LiNO_3$, KNO_3, $NaNO_3$, glycerol, and urea are 23, 33, 40.5, 43.5, 36,

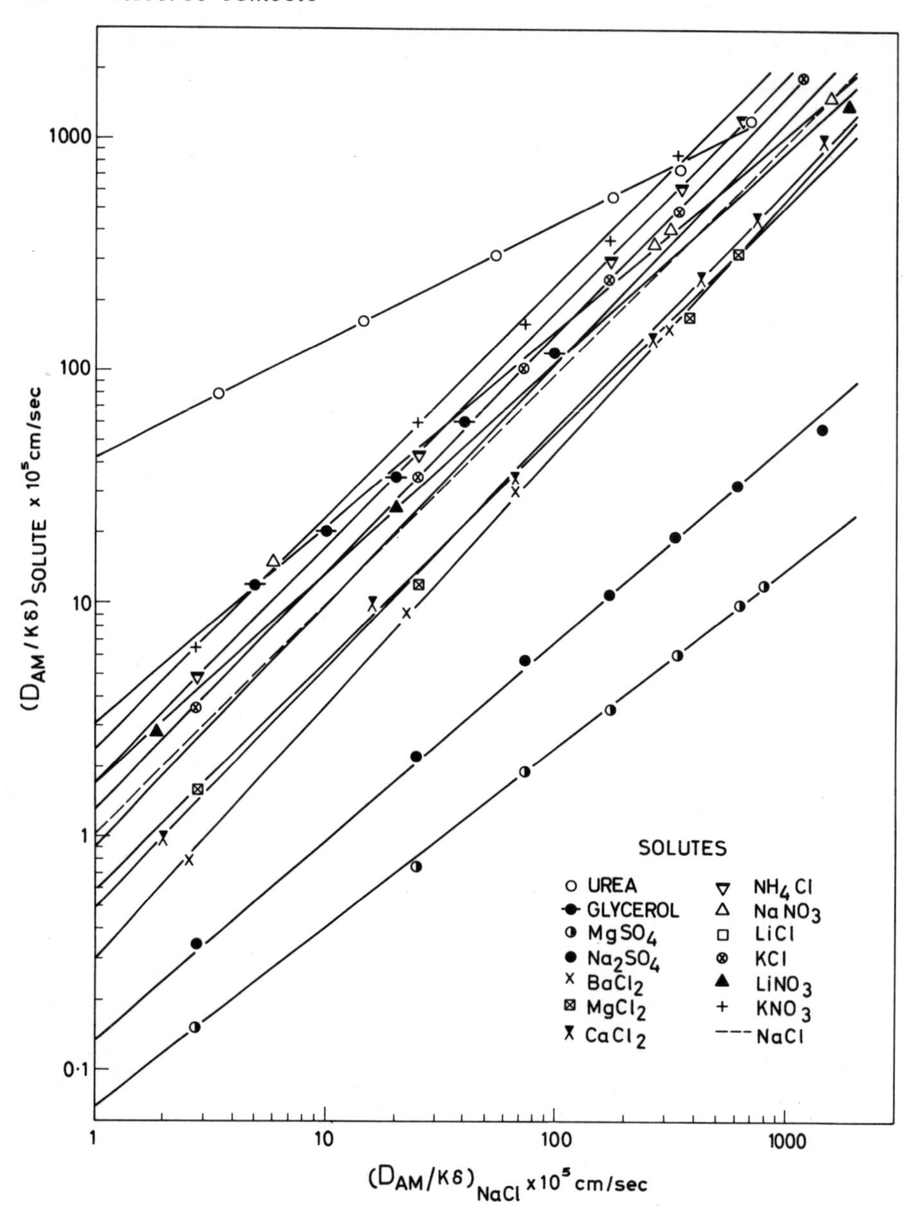

Figure 3-64. A relative scale of membrane selectivity for different solutes for the CA-NRC-18 type porous cellulose acetate membranes. Data of Agrawal and Sourirajan (1969b).

System: solute-water; operating pressure: 1500 psig.

43, 48.7, 57.1, 58, 44, 52, 42.5, 33.4, and 45 ($\times 10^{-4}$ cm/sec) respectively. These values are applicable only for the particular apparatus used.

Membrane performance data for the system sodium chloride-water. The effects of $(D_{AM}/K\delta)$ and feed molality, feed rate, operating temperature, and A-factor (compaction) on membrane performance for the system sodium chloride-water are illustrated in Figures 3-68, 3-69, 3-70, and 3-71 respectively.

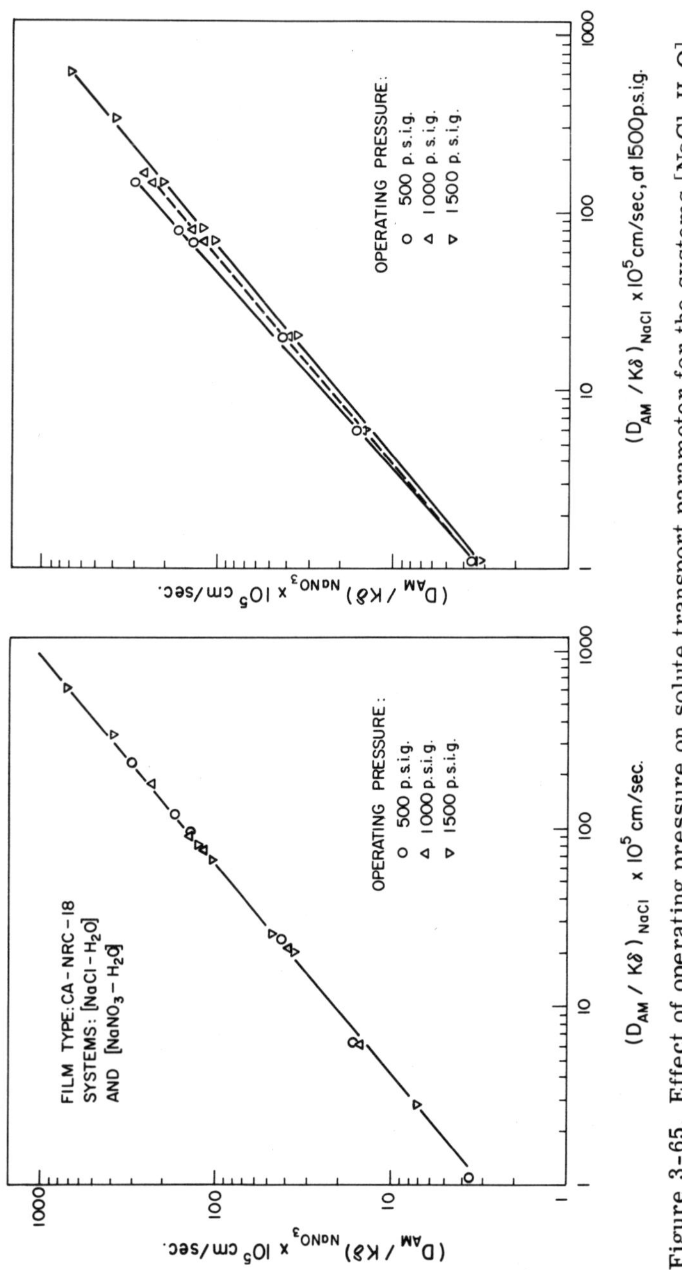

Figure 3-65. Effect of operating pressure on solute transport parameter for the systems [$NaCl$-H_2O] and [$NaNO_3$-H_2O]. Data of Agrawal and Sourirajan (1969b).

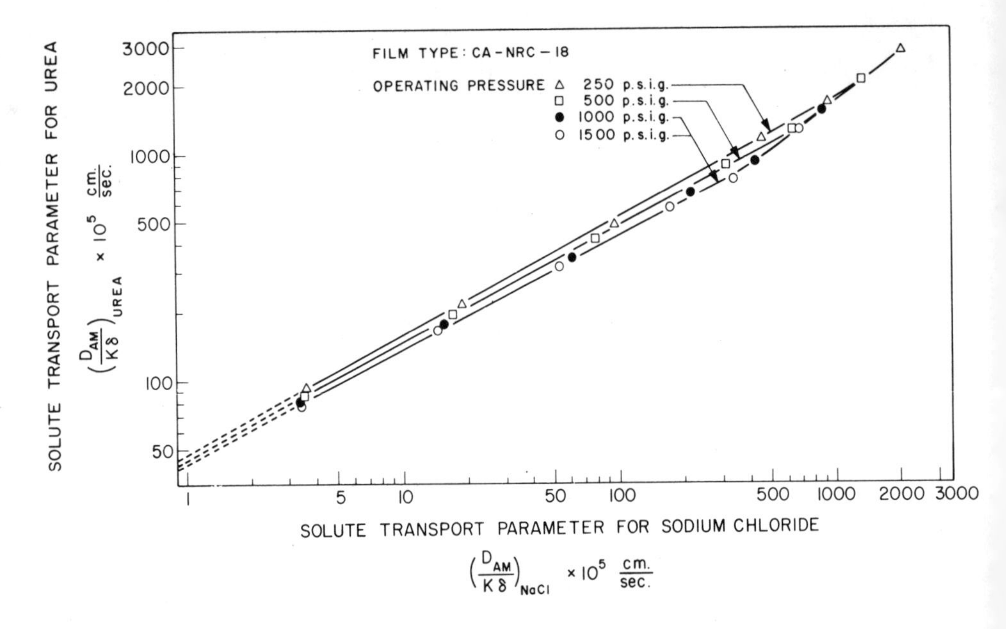

Figure 3-66. Relative selectivity of CA-NRC-18 type porous cellulose acetate membranes for sodium chloride and urea. Data of Ohya and Sourirajan (1969a).

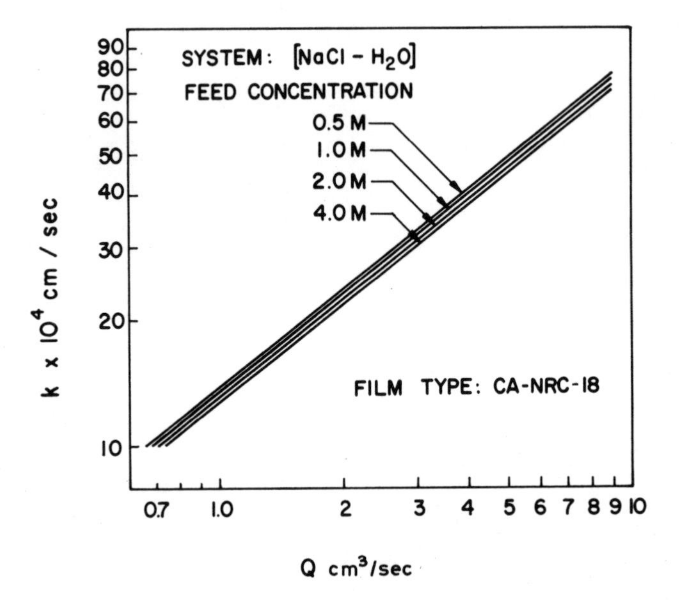

Figure 3-67. Effect of feed flow rate on mass transfer coefficient, k, for the system sodium chloride-water. Data of Agrawal and Sourirajan (1969)

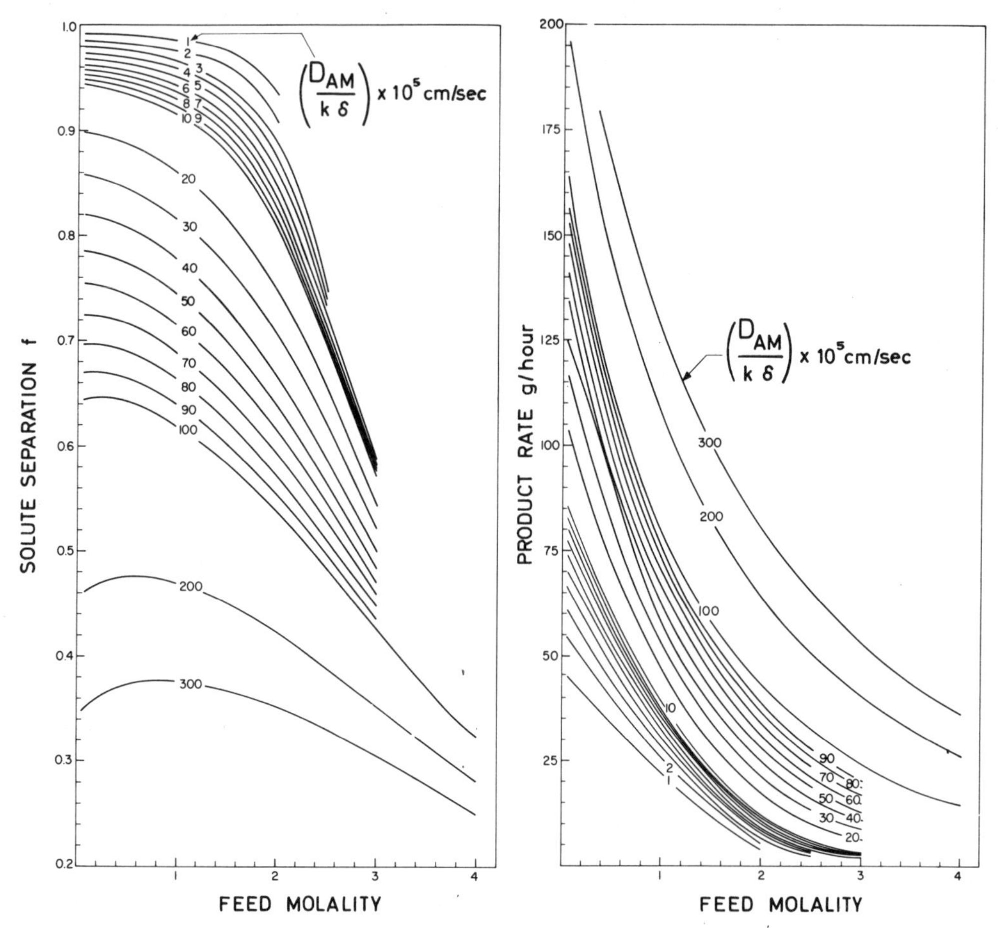

Figure 3-68. Performance data for the system sodium chloride-water using CA-NRC-18 type porous cellulose acetate membranes: effect of feed concentration. Data of Agrawal and Sourirajan (1969b).

Feed rate: 300 cm^3/min; operating pressure: 1500 psig; film area: 7.6 cm^2.

Figure 3-68 shows that both solute separation and product rate are functions of solute concentration in the feed solution; further solute separation as a function of feed concentration, passes through a slight maximum. Both these characteristics of reverse osmosis separation of sodium chloride in aqueous solution, are shown by the experimental results of Sourirajan and Govindan (1965). Figure 3-68 enables one to recognise readily the effect of the magnitude of $(D_{AM}/K\delta)$—(which is a function of the average pore size on the membrane surface)—on solute separation and product rate, and choose the appropriate values of $(D_{AM}/K\delta)$ for a given application for purposes of parametric and optimization studies.

Figure 3-69 illustrates the relative increase in solute separation and product rate obtainable by increase in feed flow rate. The effect of the latter on membrane performance is particularly pronounced with membranes whose $(D_{AM}/K\delta)$ values are higher or surface pores are larger. This is understandable because, for a given feed concentration, feed flow rate, and operating

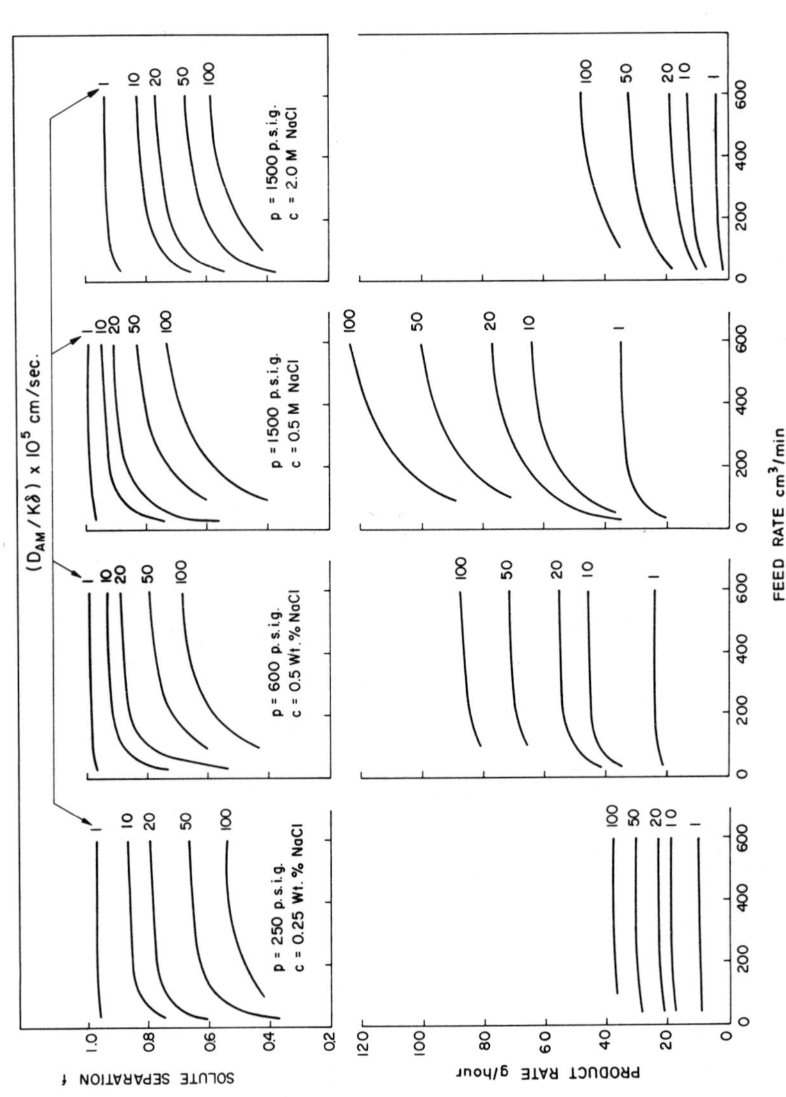

Figure 3-69. Performance data for the system sodium chloride-water using CA-NRC-18 type porous cellulose acetate membranes: effect of feed flow rate. Data of Agrawal and Sourirajan (1969b).

Film area = 7.6 cm^2; operating pressure: p; feed concentration: c.

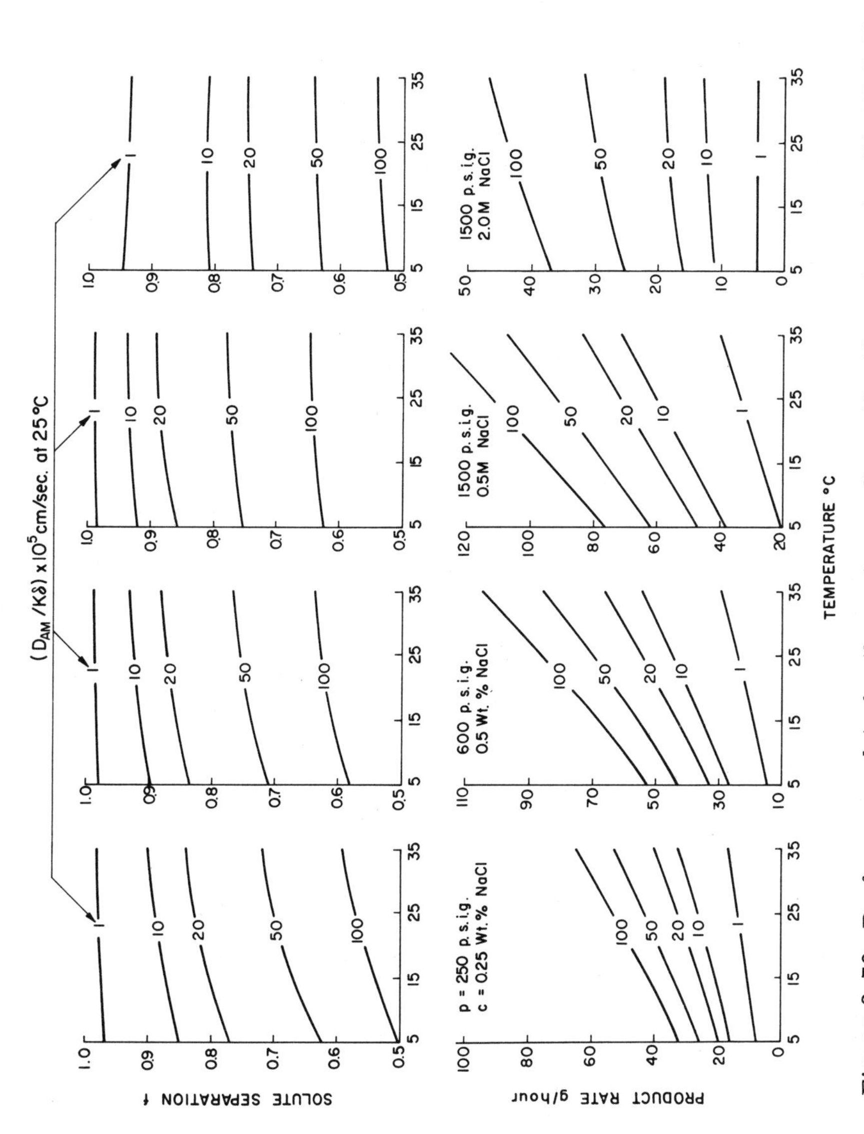

Figure 3-70. Performance data for the system sodium chloride-water using CA-NRC-18 type porous cellulose acetate membranes: effect of operating temperature. Data of Agrawal and Sourirajan (1969b).

Film area = 7.6 cm^2; operating pressure: p; feed concentration: c; feed rate: 300 cm^3/min.

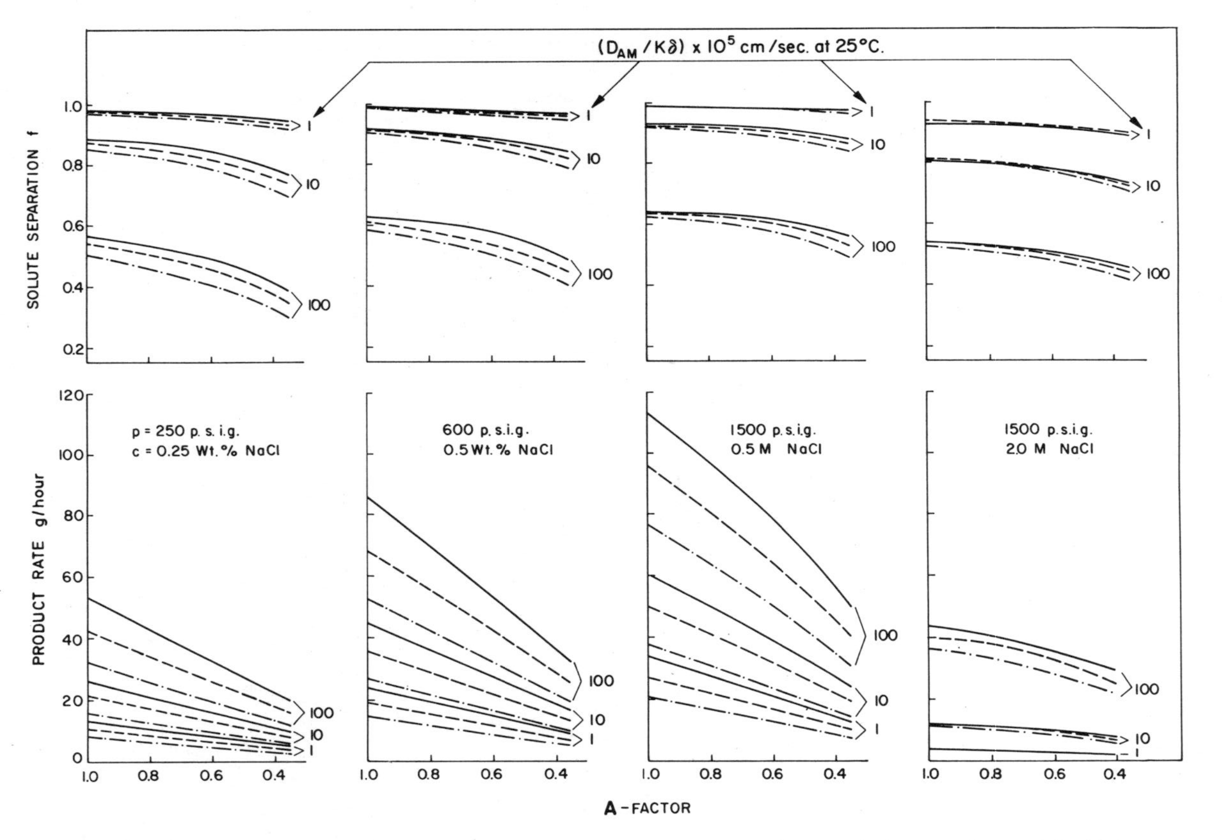

Figure 3-71. Performance data for the system sodium chloride-water using CA-NRC-18 type porous cellulose acetate membranes: effect of A-factor. Data of Agrawal and Sourirajan (1969b).

Operating pressure: p; feed concentration: c; film area: 7.6 cm^2; feed rate: 300 cm^3/min;

25°C 15°C 5°C.

pressure, concentration polarisation is higher with membranes whose ($D_{AM}/K\delta$) values are higher (Kimura and Sourirajan, 1968a); and a higher feed flow rate increases the value of k, and hence reduces concentration polarisation.

Figure 3-70 illustrates the effect of operating temperature on solute separation and product rate. For a film whose ($D_{AM}/K\delta$) = 1×10^{-5} cm/sec, increase in temperature has little effect on solute separation with feed concentrations less than 0.5M; when the feed concentration is 2.0M, solute separation slightly decreases with increase in temperature. The product rate increases with temperature, but the increase is relatively less with 2.0M feed solutions. The calculated data given in Figure 3-70 are consistent with the experimental results of Govindan and Sourirajan (1966) given earlier. For a film whose ($D_{AM}/K\delta$) is high (for example, 100×10^{-5} cm/sec) both solute separation and product rate increase with operating temperature, but here again the increase is relatively less with more concentrated feed solutions. The above results represent the combined effect of temperature on A, ($D_{AM}/K\delta$), k, osmotic pressure and molar density of the solution.

Figure 3-71 illustrates the effect of A-factor (compaction) on solute separation and product rate at different operating temperatures. The A-factor, which is defined as the ratio of the value of A for the membrane at any time to its initial value (Table 3-8), is a relative measure of membrane compaction as a result of continued reverse osmosis operation under pressure (Kimura and Sourirajan, 1968b). Product rate always decreases as a result of membrane compaction; solute separation also generally decreases. The latter change however is smaller with membranes whose ($D_{AM}/K\delta$) value is smaller. For example, with a 0.5M NaCl feed solution at 25° and an operating pressure of 1500 psig, the change in A-factor from 1.0 to 0.5 decreases solute separation from 99.0 to 98.3 per cent, and product rate from 33.6 to 17.6 grams per hour for a membrane (area = 7.6 cm^2) whose ($D_{AM}/K\delta$) = 1×10^{-5} cm/sec; for a membrane whose ($D_{AM}/K\delta$) = 100×10^{-5} cm/sec, the corresponding changes in solute separation and product rate are 64.3 to 60.0 per cent, and 112.9 to 66.7 grams per hour respectively. Compaction effect on solute separation is seen to be less at higher operating temperatures.

Membrane performance data for different solution systems-membrane selectivity and order of separation. The solute separation and product rate for five membranes of different porosities [($D_{AM}/K\delta$) for NaCl = 1, 10, 50, 100, and 200 ($\times 10^{-5}$ cm/sec)] are given in Table 3-10 for different solution systems for a particular set of operating conditions, namely, feed flow rate = 300 cm^3/min, and operating pressure = 1500 psig. The data illustrate the relative levels of solute separation and product rate obtainable for different solution systems for a given membrane, and the effect of pore size on membrane surface on the performance of the membrane for any given solution system and operating conditions. An increase in average pore size on the membrane surface steeply decreases solute separation, and, of course, increases product rate for all solution systems.

The order of solute separation in reverse osmosis with respect to a given membrane has been discussed from several points of view (Sourirajan, 1963, 1964; Erickson et al., 1966). The order of the lyotropic series is generally maintained in the order of solute separation especially with membranes having smaller pores on the membrane surface when used with low concentrations of feed solution. Reversal of such order is not uncommon with respect to membranes with larger surface pores and/or high feed concentrations.

Table 3-10. Membrane performance data for different solution systems

Data of Agrawal and Sourirajan (1969b)

Film	M1	M2	M3	M4	M5	M1	M2	M3	M4	M5
$\left(\frac{D_{AM}}{K\delta}\right) \times 10^5$ for NaCl (cm/sec)	1	10	50	100	200	1	10	50	100	200
Solute	Solute separation (%)					Product rate (g/hr)				
$MgSO_4$	99.9	99.5	98.3	97.2	95.2	36.3	58.6	74.3	79.8	84.7
Na_2SO_4	99.8	99.1	96.9	94.7	90.9	31.0	51.9	71.2	80.2	89.9
$BaCl_2$	99.7	97.0	87.1	77.7	61.8	27.7	46.1	65.9	78.4	97.7
$CaCl_2$	99.4	95.7	83.4	72.5	57.3	26.9	45.1	65.9	79.9	100.2
$MgCl_2$	99.2	94.8	81.6	70.3	55.5	25.2	41.2	59.0	71.2	88.7
LiCl	99.0	93.0	74.3	59.1	40.9	32.4	56.7	86.8	107.7	137.3
NaCl	98.9	93.4	77.4	64.3	47.6	33.6	59.9	91.9	112.9	141.8
KCl	98.8	92.0	73.1	58.8	41.6	34.4	62.6	98.8	122.9	155.7
NH_4Cl	98.4	90.1	69.0	53.7	36.9	34.5	63.2	101.0	126.8	160.4
$LiNO_3$	98.2	90.8	73.2	60.1	43.8	32.7	58.1	88.6	108.6	136.3
KNO_3	97.8	85.0	61.2	44.9	—	35.6	66.5	107.4	135.4	—
$NaNO_3$	97.0	87.1	67.1	53.9	38.8	34.5	62.9	99.0	122.2	152.7
Glycerol	96.5	85.8	64.3	50.3	—	38.3	68.7	104.4	126.4	—
Urea	72.5	53.4	35.3	—	—	41.3	78.4	124.8	—	—

Film type: CA-NRC-18
Feed concentration: 0.5M
Feed rate: 300 cm^3/min
Operating pressure: 1500 psig
Film area: 7.6 cm^2

A higher value of $(D_{AM}/K\delta)$ generally means a lower level of solute separation; but the order of solute separation with respect to any two solutes does not always correspond to the order of their values of $(D_{AM}/K\delta)$. This is illustrated by the data given in Tables 3-9, 3-10, and 3-11 and Figure 3-72. The data given in Table 3-10 and Figure 3-72 are based on the values of k corresponding to a constant feed flow rate of 300 cm^3/min for each system; the value of k so obtained is different for each system and feed concentration. The data presented in Table 3-11 are based on constant values of k for all the systems and feed concentrations considered. Referring to Table 3-9, reversal in the order of $(D_{AM}/K\delta)$ is shown for the solutes $CaCl_2$ and $MgCl_2$, LiCl and NaCl, KNO_3 and $NaNO_3$, $NaNO_3$ and glycerol, and NaCl and glycerol with increase in the average size of the surface pores; no such reversal of order is shown with respect to NaCl and $NaNO_3$ at least up to $(D_{AM}/K\delta)_{NaCl} = 300 \times 10^{-5}$ cm/sec. Referring to the performance data given in Table 3-10, for the 0.5M feed solution, no reversal in the order of solute separation is indicated except probably for the solutes LiCl and NaCl, and KNO_3 and $NaNO_3$. Figure 3-72 and Table 3-11 illustrate the effect of feed concentration on the order of solute separation. Figure 3-72 shows that the order of solute separation with respect to NaCl and $NaNO_3$

Table 3-11. Effect of feed concentration on membrane performance for different solution systems

Data of Agrawal and Sourirajan (1969b)

Film	M1					M2							M4						
$(D_{AM}/K\delta) \times 10^5$ for NaCl (cm/sec)	1					10							100						
Feed Molality	0.05	0.01	0.5	1	2	0.05	0.1	0.5	1	2	3	4	0.05	0.1	0.5	1	2	3	4
Solute	% Solute Separation																		
$MgSO_4$	99.9	99.9	99.9	99.9	99.8	99.8	99.8	99.7	99.7	99.3	—	—	98.7	98.7	98.7	98.5	96.8	—	—
Na_2SO_4	99.9	99.9	99.9	99.8	—	99.4	99.4	99.3	99.1	—	—	—	96.4	96.4	96.3	95.5		—	—
$BaCl_2$	99.8	99.8	99.7	99.3	—	97.9	97.9	97.3	95.0	—	—	—	81.2	81.5	80.2	74.0	—	—	—
$CaCl_2$	99.6	99.6	99.4	99.5	—	97.0	96.9	96.0	91.3	—	—	—	76.0	76.2	74.6	66.0	33.7	—	—
$MgCl_2$	99.5	99.5	99.3	97.8	—	96.8	96.7	95.7	89.5	—	—	—	77.1	77.3	75.4	65.4	29.3	—	—
LiCl	99.3	99.3	99.1	98.7	—	94.5	94.4	93.5	91.2	73.7	—	—	62.2	62.4	62.3	59.6	48.4	34.5	23.5
NaCl	99.2	99.2	99.0	98.6	93.4	94.4	94.3	93.5	91.7	81.0	57.3	—	64.4	64.6	64.6	62.5	54.2	43.1	32.7
KCl	99.0	98.9	98.7	98.3	94.6	92.5	92.4	91.5	89.5	80.8	63.7	47.9	55.4	55.6	56.0	54.8	49.6	42.8	36.0
NH_4Cl	98.6	98.6	98.4	97.8	93.7	90.6	90.5	89.4	87.2	78.4	63.1	48.9	49.6	49.8	50.5	49.7	45.7	40.4	35.1
$LiNO_3$	98.6	98.6	98.3	97.6	86.3	92.6	92.5	91.4	88.8	74.0	49.4	—	62.5	62.7	62.8	60.3	50.6	38.8	29.2
KNO_3	98.1	98.0	97.8	97.3	—	85.9	85.8	84.8	83.2	—	—	—	42.9	43.3	44.0	44.3	—	—	—
$NaNO_3$	97.6	97.5	97.1	96.3	92.4	88.7	88.6	87.4	85.4	78.5	68.3	58.0	54.5	54.7	55.2	54.4	51.0	46.5	42.2
Glycerol	97.2	97.2	97.0	96.6	95.5	89.2	89.2	88.8	88.1	85.6	81.3	74.8	60.1	60.3	61.6	62.0	60.9	58.3	54.8
Urea	74.25	74.1	73.2	72.1	69.5	55.5	55.4	55.0	54.4	53.1	51.6	49.9	—	—	—	—	—	—	—
Solute	Product Rate (g/hr)																		
$MgSO_4$	45.78	45.04	39.01	31.15	11.24	86.44	84.39	70.38	52.46	16.74	—	—	163.07	157.20	118.46	78.66	23.40	—	—
Na_2SO_4	44.98	43.20	32.41	21.69	—	84.30	80.19	56.86	36.52	—	—	—	156.45	145.67	94.29	57.99	—	—	—
$BaCl_2$	44.80	42.93	28.51	11.98	—	84.03	79.50	48.98	20.47	—	—	—	157.74	146.27	85.28	45.23	—	—	—
$CaCl_2$	44.78	42.79	27.44	8.85	—	83.98	79.29	46.94	16.23	—	—	—	157.97	147.18	83.87	40.94	8.86	—	—
$MgCl_2$	44.72	42.73	26.83	7.12	—	83.95	79.19	45.70	13.78	—	—	—	157.75	145.98	80.49	34.98	6.66	—	—
LiCl	45.31	43.83	32.88	20.06	—	85.33	82.17	58.45	35.25	7.55	—	—	162.79	155.67	111.22	75.09	35.85	17.77	9.70
NaCl	45.25	43.87	33.55	22.16	3.84	85.40	82.25	60.06	39.16	12.14	3.72	—	162.86	155.89	113.46	81.76	43.77	24.46	14.75
KCl	45.32	43.92	34.01	23.16	6.00	85.47	82.30	61.31	41.81	16.46	6.54	3.36	164.16	157.65	120.47	90.29	57.82	37.90	26.08
NH_4Cl	45.42	43.89	34.07	23.30	6.49	85.50	82.40	61.98	42.74	18.11	7.88	4.18	164.40	158.88	124.23	95.01	62.35	42.17	30.02
$LiNO_3$	45.41	43.88	33.20	20.79	2.72	85.69	82.16	59.31	37.23	10.51	3.57	—	162.91	155.65	111.65	78.25	39.42	21.42	13.11
KNO_3	45.33	44.09	35.39	27.10	—	85.74	83.06	66.14	52.14	—	—	—	165.41	161.02	134.76	121.16	—	—	—
$NaNO_3$	45.36	43.95	34.59	24.75	10.00	85.61	82.74	63.32	45.98	23.54	13.13	8.12	164.11	158.30	123.31	99.02	67.39	49.20	37.91
Glycerol	45.91	45.19	39.36	32.58	20.72	87.12	85.29	72.35	58.70	37.63	24.10	14.66	166.92	162.73	135.25	113.12	79.67	57.94	43.15
Urea	46.25	45.60	41.33	36.70	30.00	87.75	86.64	78.73	72.22	60.29	51.21	44.01	—	—	—	—	—	—	—

Film Type: CA-NRC-18
Film area: 7.6 cm^2
Operating pressure: 1500 psig
Mass transfer coefficient, $k = 50 \times 10^{-4}$ cm/sec

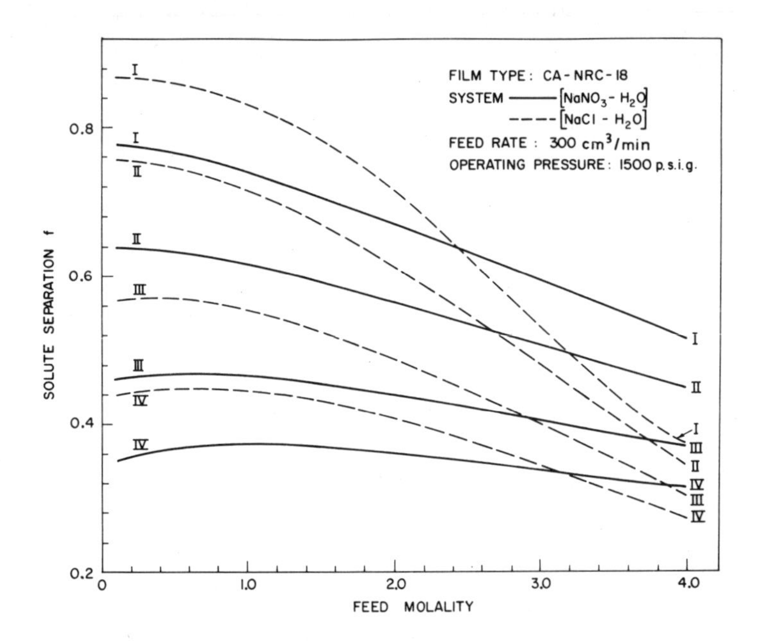

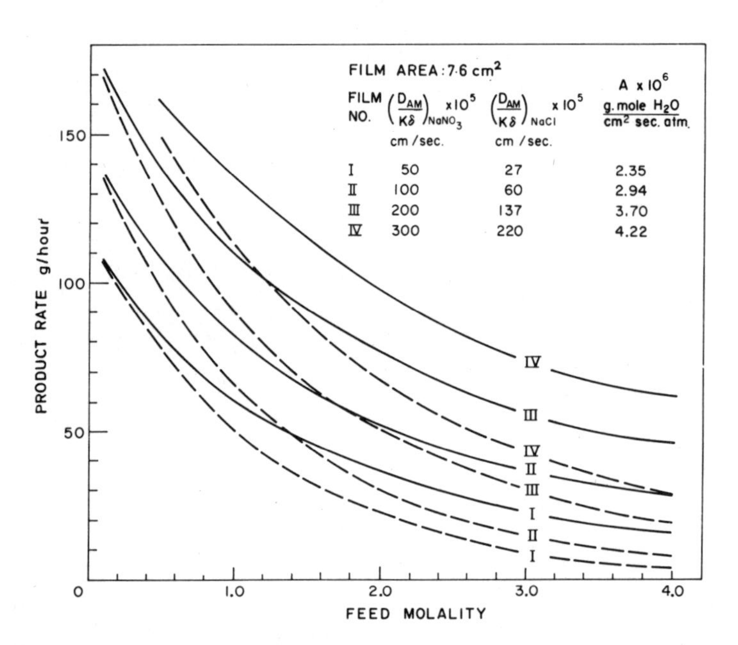

Figure 3-72. Performance data for the systems [NaCl-H_2O] and [$NaNO_3$-H_2O] using CA-NRC-18 type porous cellulose acetate membranes: effect of feed concentration. Data of Agrawal and Sourirajan (1969b).

is reversed by changing the feed concentration even though $(D_{AM}/K\delta)$ for NaCl is less that that for $NaNO_3$ for all the films investigated. Such results are obtained for other solution systems also even when the value of k is kept constant. This is illustrated in Table 3-11. For example, referring to the data presented in Table 3-11,

(a) $(D_{AM}/K\delta)_{CaCl_2} > (D_{AM}/K\delta)_{MgCl_2}$ for Film M4,
and $f_{CaCl_2} > f_{MgCl_2}$ for 2.0M feed solution;

(b) $(D_{AM}/K\delta)_{NaCl} > (D_{AM}/K\delta)_{LiCl}$ for Film M2,
and $f_{NaCl} > f_{LiCl}$ for 2.0M feed solution;

(c) $(D_{AM}/K\delta)_{KCl} > (D_{AM}/K\delta)_{NaCl}$ for Films M1, M2, and M4,
and $f_{KCl} > f_{NaCl}$ for 2.0M, 3.0M, and 4.0M feed solutions for Films M1, M2, and M4 respectively;

(d) $(D_{AM}/K\delta)_{NH_4Cl} > (D_{AM}/K\delta)_{KCl}$ for Film M2,
and $f_{NH_4Cl} > f_{KCl}$ for 4.0M feed solution;

(e) $(D_{AM}/K\delta)_{glycerol} > (D_{AM}/K\delta)_{NaNO_3}$ for Film M1,
and $f_{glycerol} > f_{NaNO_3}$ for 2.0M feed solution;

(f) $(D_{AM}/K\delta)_{glycerol} > (D_{AM}/K\delta)_{NaCl}$ for Films M1, M2, and M4,
and $f_{glycerol} > f_{NaCl}$ for 2.0 to 4.0M feed solutions.

On the basis of the above data, it seems reasonable to conclude that the order of $(D_{AM}/K\delta)$ and that of solute separation with respect to any two solutes must be considered separately. Since the values of $(D_{AM}/K\delta)$ for different solutes studied above are independent of feed concentration and feed flow rate, their relative values offer a firm basis for expressing membrane selectivity. The order of solute separation is a function not only of the relative values of $(D_{AM}/K\delta)$ for different solutes but also of the operating conditions of pressure, feed concentration, and the mass transfer coefficient, k.

10. Transport Characteristics of Membranes for the Separation of Sucrose in Aqueous Solutions

The analysis of the reverse osmosis separation data for the systems glycerol-water, urea-water, sodium chloride-water, and several others involving inorganic salts in aqueous solutions, using the Loeb-Sourirajan type porous cellulose acetate membranes, has been discussed in the preceding sections. All these solution systems are characterised by the fact that the solute transport parameter, $(D_{AM}/K\delta)$, with respect to the type of membrane used, is independent of the concentration of the boundary solution, (X_{A2}), at any given operating pressure for a wide range of feed molalities and feed flow rates. The system sucrose-water is an example of one for which $(D_{AM}/K\delta)$ depends on X_{A2} with respect to the same membranes. This is illustrated by the data of Kimura and Sourirajan (1968d).

Variation of Solute Transport Parameter. The values of $(D_{AM}/K\delta)$ and X_{A2} calculated from the experimental [PWP], [PR], and f data for the system sucrose-water using Equations 3-33, 3-35, and 3-43 are plotted in Figure 3-73 for the operating pressures of 500, 1000, and 1500 psig for four different films involving solute separations in the range 54 to 99 per cent. These data show that $(D_{AM}/K\delta)$ for sucrose decreases with increase in X_{A2}: the plot of log $(D_{AM}/K\delta)$ v. X_{A2} is a straight line, and this relationship is independent of the particular combination of the feed concentration and feed flow rate used; the slope of the line is a function of the operating pressure; the higher the operating pressure, the higher is the slope, and at a given operating pressure the slope is essentially the same for all the films tested. Consequently, the relationship between the solute transport parameter and the boundary concentration for the system sucrose-water can be expressed as

$$(D_{AM}/K\delta)_{\text{sucrose}} = (D_{AM}/K\delta)^*_{\text{sucrose}} \exp(-\epsilon X_{A2}) \qquad (3\text{-}92)$$

where $(D_{AM}/K\delta)^*_{\text{sucrose}}$ is the extrapolated value of $(D_{AM}/K\delta)$ sucrose at $X_{A2} = 0$, X_{A2} is the mole fraction of sucrose at the concentrated boundary layer calculated from Equations 3-33 and 3-35, and ϵ is a constant which is a function of the operating pressure.

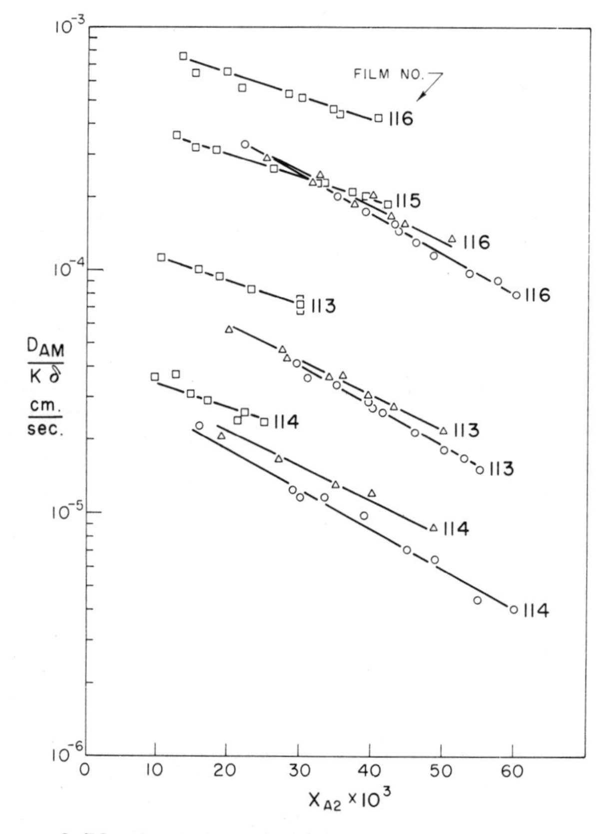

Figure 3-73. Variation of solute transport parameter for sucrose with its boundary concentration. Data of Kimura and Sourirajan (1968d).

Film type: CA-NRC-18; system: sucrose-water; feed rate: 120 to 560 cm^3/min; operating pressure: □ 500 psig, △ 1000 psig, ○ 1500 psig.

Figure 3-74 gives the plot of the average value of ϵ v. operating pressure for the data presented in Figure 3-73. The ϵ values given in Figure 3-74 are applicable for the type of membranes used for $X_{A2} \times 10^3$ values at least up to 30, 50, and 70 respectively at the operating pressure of 500, 1000, and 1500 psig.

A relative scale of membrane selectivity. Figure 3-75 relates the values of $(D_{AM}/K\delta)_{NaCl}$ with those of $(D_{AM}/K\delta)^*_{sucrose}$ as a function of operating pressure for several membranes of different surface porosities. As discussed

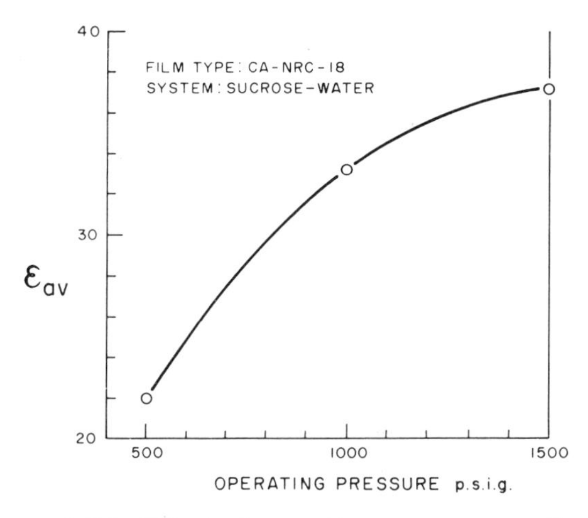

Figure 3-74. Effect of operating pressure on the slope of log $(D_{AM}/K\delta)_{sucrose}$ v. X_{A2} straight lines. Data of Kimura and Sourirajan (1968d).

earlier, such a relationship offers a method of expressing membrane selectivity on a relative scale. For the type of membranes used, Figure 3-75 shows that the log-log plot of $(D_{AM}/K\delta)_{NaCl}$ v. $(D_{AM}/K\delta)^*_{sucrose}$ is essentially a straight line whose slope depends on the operating pressure. Figures 3-74 and 3-75 are useful; they define Equation 3-92 for any particular membrane whose $(D_{AM}/K\delta)$ for NaCl is known.

Mass transfer coefficient correlation. The values of k obtained from the experimental [PWP], [PR], and f data using Equations 3-33, 3-35, and 3-52 are plotted in Figure 3-76 in the form N_{Re} v. $N_{Sh}/N_{Sc}^{0.33}$ on the log-log scale as before. The experimental data involved four different films, and a wide range of solute separations, feed flow rates and feed concentrations, and operating pressures of 500, 1000, and 1500 psig. Most of the data plotted in Figure 3-76 are considerably lower than those obtained by the diffusion current method (Figure 3-46). A mean straight line for the correlation N_{Re} v. $N_{Sh}/N_{Sc}^{0.33}$ represents the experimental data well. Based on the above straight line correlation, the values of k were calculated and plotted in Figure 3-77 as a function of the feed flow rate Q, and feed molality. Figure 3-77 shows that for a given feed concentration

$$k \propto Q^{0.865} \tag{3-93}$$

for the system sucrose-water under the experimental conditions used.

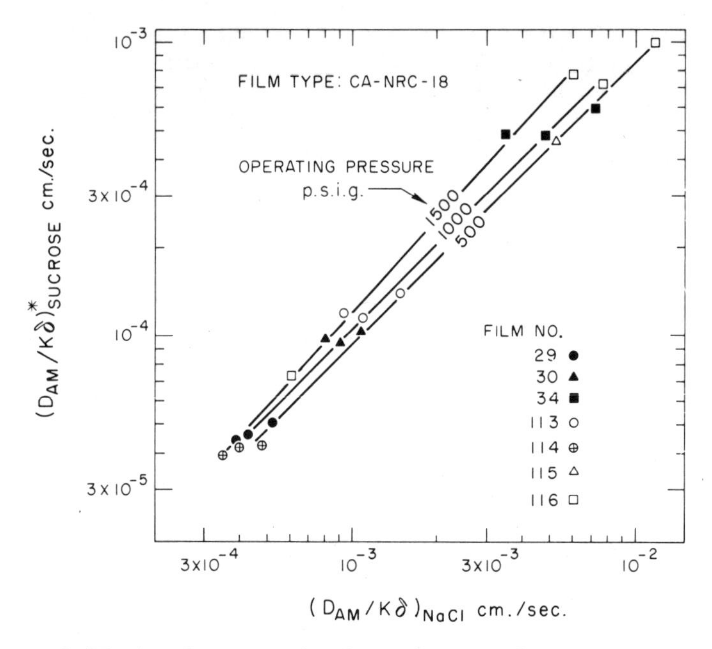

Figure 3-75. A relative scale of membrane selectivity for the systems sodium chloride-water and sucrose-water. Data of Kimura and Sourirajan (1968d).

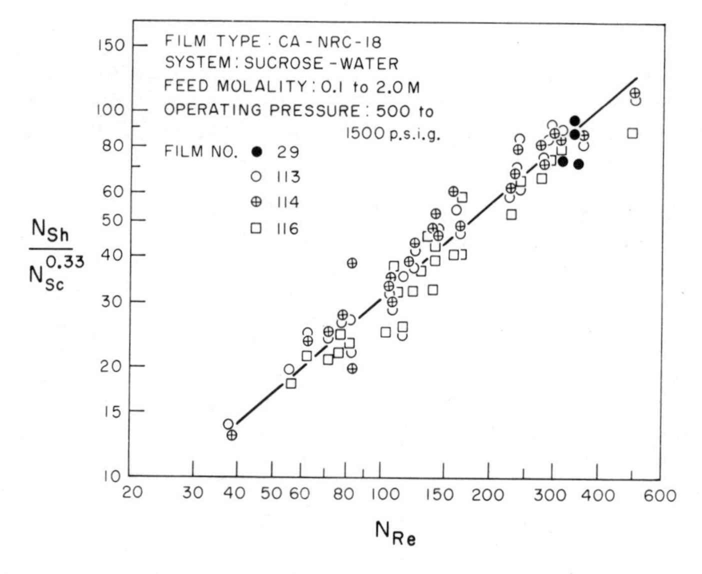

Figure 3-76. Experimental mass transfer coefficient data for the system sucrose-water. Data of Kimura and Sourirajan (1968d).

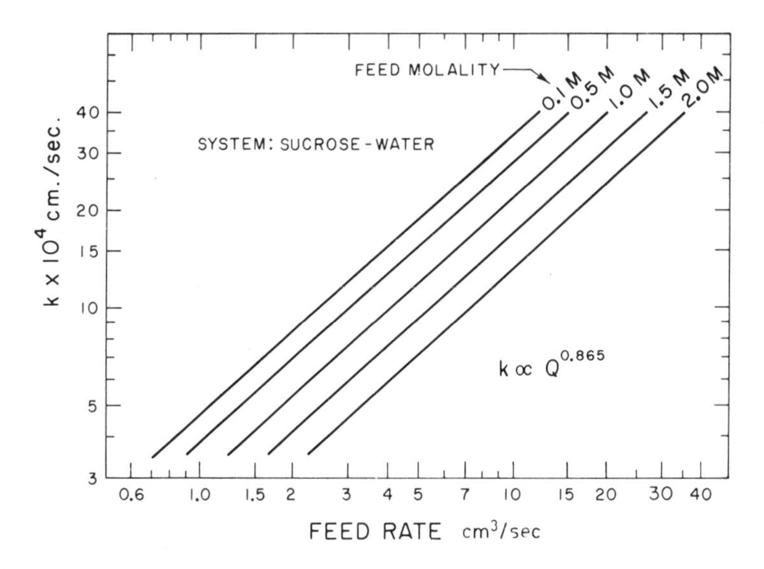

Figure 3-77. Variation of mass transfer coefficient with feed flow rate for the system sucrose-water. Data of Kimura and Sourirajan (1968d).

Predictability of Membrane Performance. If the applicable values of A, $(D_{AM}/K\delta)$, and k are given, the values of X_{A2}, X_{A3}, and N_B (and hence, f and [PR]) can be calculated as a function of operating pressure, feed molality, and feed rate using Equations 3-35, 3-43, and 3-52 by the procedure illustrated earlier.

The mass transfer coefficient k is a function of feed concentration, feed flow rate and the geometry of the apparatus used; for any given experimental conditions, the applicable mass transfer correlation has to be determined either theoretically (see Chapter 4) or experimentally. For purposes of illustration here, Figure 3-77 is used to obtain k for the system sucrose-water.

Since $(D_{AM}/K\delta)_{\text{sucrose}}$ is not a constant, Equation 3-92 should also be solved simultaneously to obtain X_{A3} and N_B. Equation 3-92 can be defined for this purpose from the experimental values of $(D_{AM}/K\delta)_{\text{sucrose}}$ at two different values of X_{A2} at any given operating pressure.

Consequently, the separation and permeability characteristics of a membrane for the system sucrose-water at a given operating pressure can be predicted as a function of feed molality and feed flow rate from two experimental values of [PR] and f (corresponding to two different values of X_{A2}) along with the [PWP] data, and the simultaneous solution of Equations 3-35, 3-43, 3-52, and 3-92. When the transport characteristics of a particular type of membrane for the system sucrose-water are given in the form of Figures 3-74 and 3-75, the performance of any particular membrane with the system sucrose-water can be predicted just from its specifications given in terms of A and $(D_{AM}/K\delta)_{NaCl}$ as given in Table 3-5. This is illustrated in Figures 3-78, 3-79, and 3-80 which give the effect of pressure, feed molality, and feed flow rate on solute separation and product rate for the system sucrose-water calculated for some arbitrarily chosen experimental conditions for several membranes

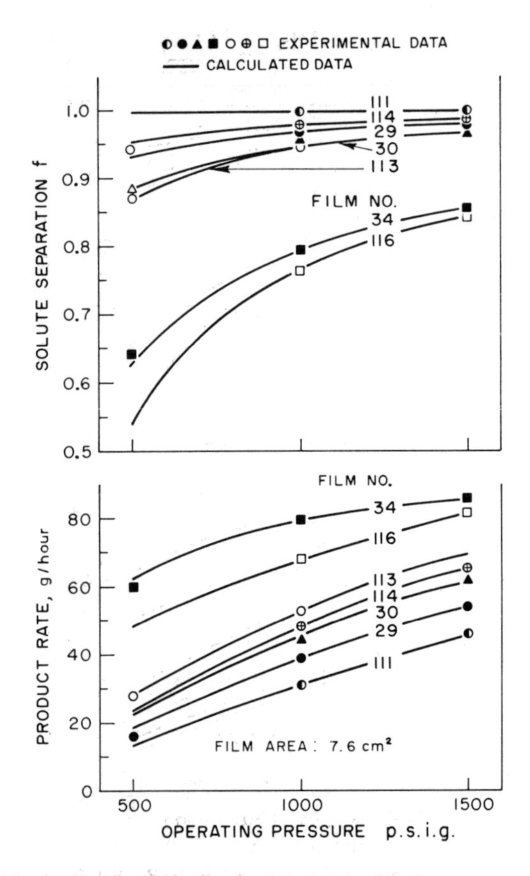

Figure 3-78. Effect of operating pressure on solute separation and product rate for the system sucrose-water. Data of Kimura and Sourirajan (1968d).

Film type: CA-NRC-18; feed molality: 0.5M; feed rate: 390 cm^3/min.

from their specifications given in Table 3-5; some of the available experimental data are also plotted in Figures 3-78, 3-79, and 3-80 for comparison. The excellent agreement between the experimental and the calculated results shows that the mass transfer coefficient correlation given in Figure 3-77 (based on the mean straight line correlation given in Figure 3-76) is good for practical purposes.

Membrane Compaction. In order to find the effect of A-factor (compaction effect) on the $(D_{AM}/K\delta)$ v. X_{A2} relationship given in Figure 3-81, continuous test runs were conducted with three different films at 1500 psig using a 0.5M sucrose-water feed solution for periods extending up to 7 days. From the experimental [PR] and f data obtained at different time intervals during the extended test runs, and the mass transfer coefficient correlations given in Figure 3-77, the corresponding values of A, $(D_{AM}/K\delta)$, and X_{A2} were calculated using Equations 3-35, 3-43, and 3-52; from the initial specifications

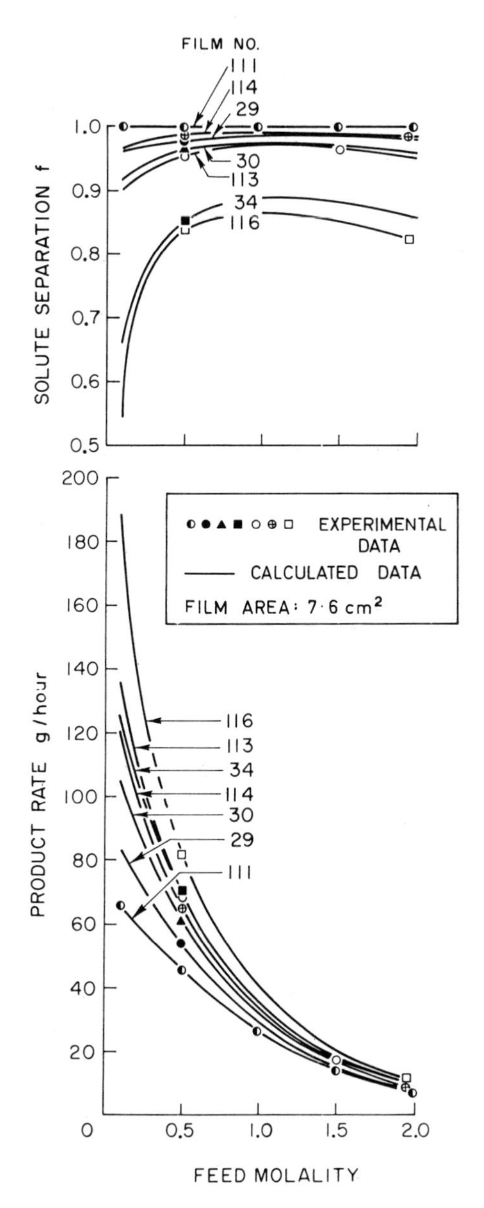

Figure 3-79. Effect of feed concentration on solute separation and product rate for the system sucrose-water. Data of Kimura and Sourirajan (1968d).

Film type: CA-NRC-18; feed rate: 390 cm^3/min; operating pressure: 1500 psig.

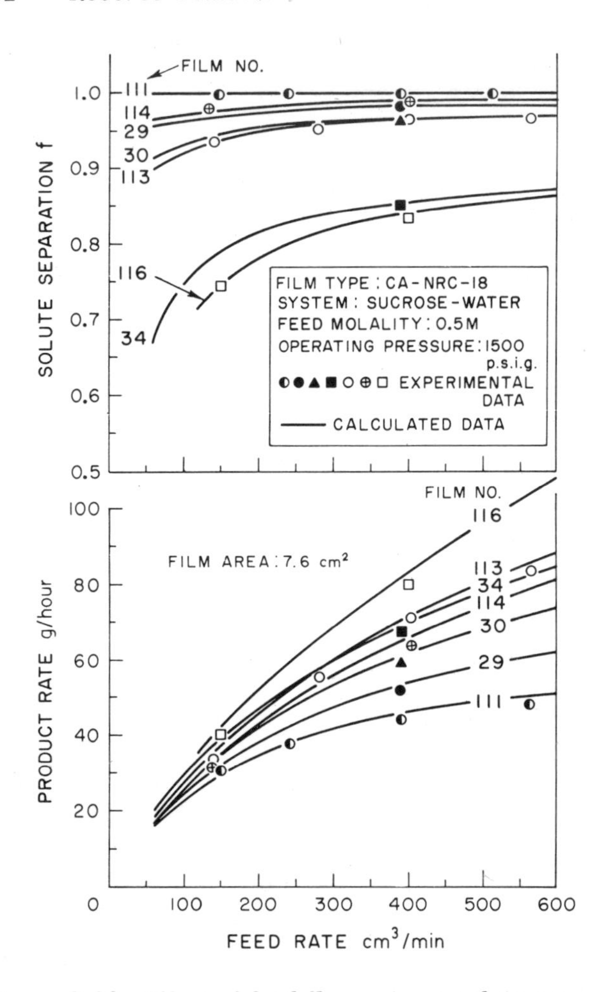

Figure 3-80. Effect of feed flow rate on solute separation and product rate for the system sucrose-water. Data of Kimura and Sourirajan (1968d).

of the films, the corresponding A-factors were also calculated. Some of the results are plotted in Figure 3-81. The solid lines are those given in Figure 3-73, and they correspond to A-factors in the range 1.0 to 0.97. Figure 3-81 shows that for a given value of X_{A2}, the $(D_{AM}/K\delta)$ values obtained at lower A-factors are the same as those obtained at A-factor $\approx$ 1.0. In other words, membrane compaction does not affect the $(D_{AM}/K\delta)$ v. X_{A2} relationship expressed by Equation 3-92 for the system sucrose-water.

Variations of solute separation. On the basis of the observation that Equation 3-92 is valid for all A factors, i.e. $(D_{AM}/K\delta)^*_{sucrose}$ and ϵ values are unaffecte by membrane compaction, one can calculate product rate and solute separation as a function of A-factor for the system sucrose-water from the initial specifications of the membrane given in Table 3-5 using Figures 3-74, 3-75 and

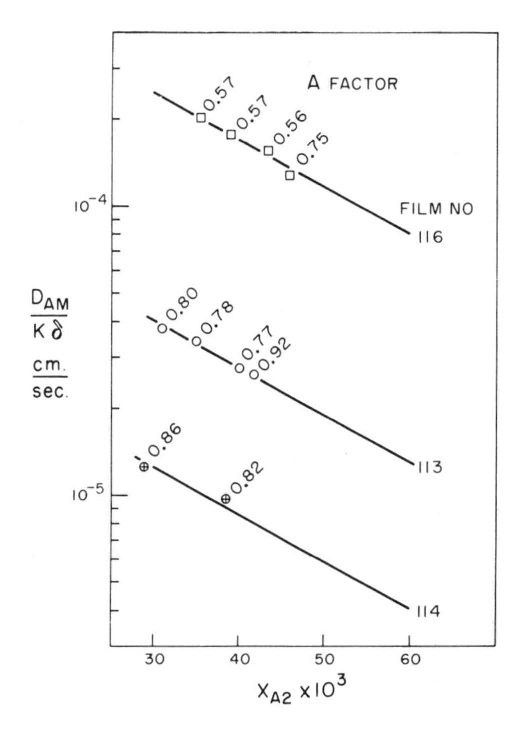

Figure 3-81. Effect of membrane compaction on the log $(D_{AM}/K\delta)_{sucrose}$ v. X_{A2} relationship. Data of Kimura and Sourirajan (1968d).

Film type: CA-NRC-18; system: sucrose-water; feed molality: 0.5M feed rate: 120-560 cm^3/min; operating pressure: 1500 psig.

3-77, and Equations 3-35, 3-43, 3-52 and 3-92. Such calculations are illustrated for three different membranes in Figure 3-82 which shows the variations in solute separation as the product rate decreases due to membrane compaction. Some of the experimental results obtained are also plotted in Figure 3-82. The excellent agreement between the experimental and the calculated results shows that the decrease in solute separation obtained with the decrease in product rate is a consequence of the effect of membrane compaction on the transport characteristics of the membrane, and is not due to membrane deterioration due to chemical interaction of any kind. These results again confirm the validity of the analysis. Further, it is clear that variations in solute separation with decrease in product rate due to membrane compaction must be expected, and can be calculated and allowed for in the design of practical reverse osmosis units for the system sucrose-water; this aspect is of particular interest in process design of the reverse osmosis concentration units for natural maple sap and other industrial sugar solutions.

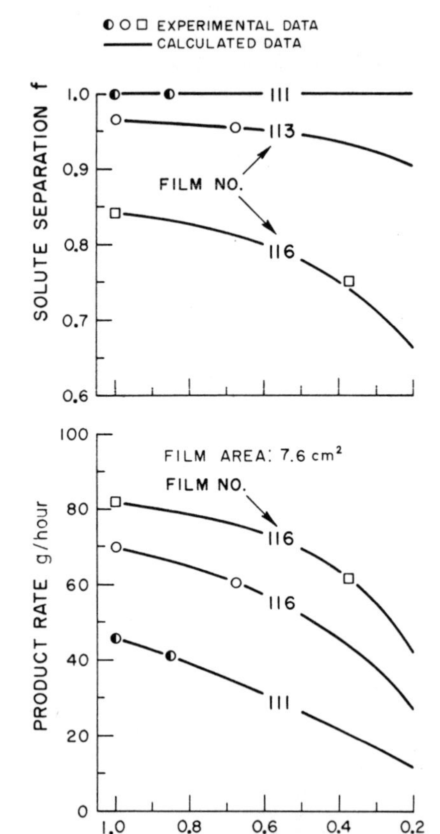

Figure 3-82. Effect of membrane compaction on solute separation for the system sucrose-water. Data of Kimura and Sourirajan (1968d).

Film type: CA-NRC-18; feed rate: 390 cm^3/min; operating pressure: 1500 psig.

Mole Fraction Relationships

The system sucrose-water. The data obtained with the system sucrose-water (Figure 3-83) also confirm the observation that there exists a unique relationship between X_{A2} and X_{A3} in reverse osmosis, and that relationship depends on the overall porous structure of the membrane and chemical nature of the solute but it is independent of feed concentration and feed flow rate.

Systems like sucrose-water. The fact that $(D_{AM}/K\delta)_{sucrose}$ is a unique function of X_{A2}, irrespective of the particular combination of feed concentration and feed flow rate used, makes the system sucrose-water different from those discussed in the preceding sections. It is possible that several similar system exist. The analysis and correlations given above are illustrative of the method applicable for the treatment of reverse osmosis separation data for systems like sucrose-water.

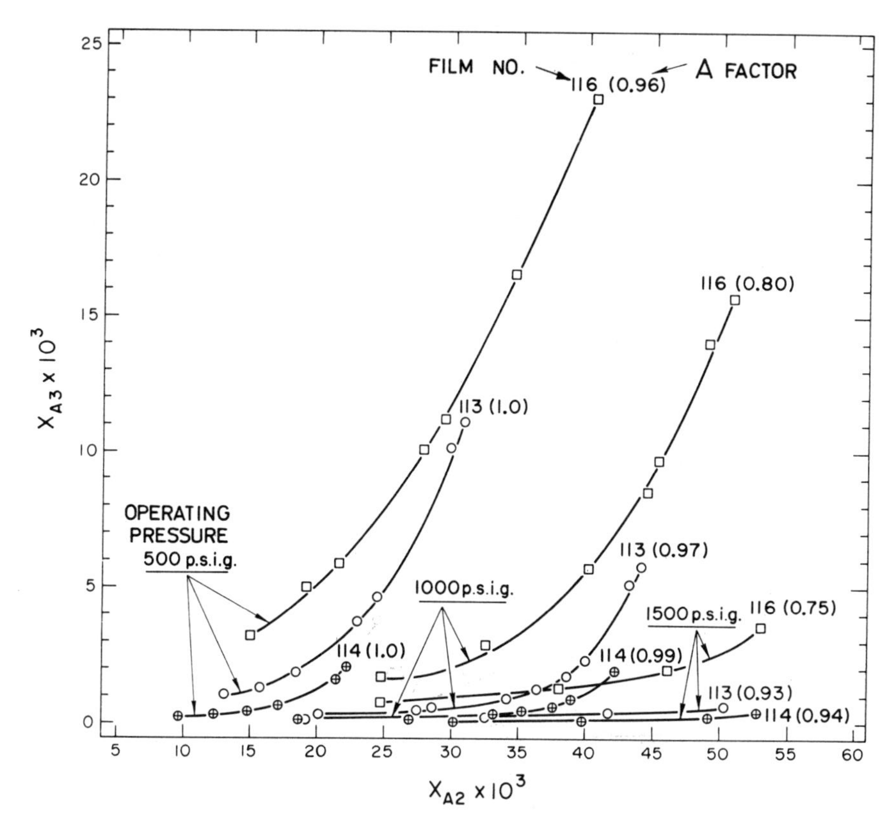

Figure 3-83. X_{A2} v. X_{A3} relationship for the system sucrose-water. Data of Kimura and Sourirajan (1968d).

Film type: CA-NRC-18; feed molality: 0.1 to 2.0M; feed rate: 120-560 cm^3/min.

11. Nomenclature for Chapter 3

a	=	constant
a_i	=	activity of component i in a system
a_w	=	activity of water in an aqueous solution
A	=	constant; or pure water permeability constant, $\frac{\text{g. mole } H_2O}{\text{cm}^2 \text{ sec. atm}}$
A_0	=	extrapolated value of the pure water permeability constant at P = 0
A_c	=	area of cathode surface, cm^2
b	=	constant
B	=	constant
c	=	molar density of solution, g. mole/cm^3

c_1, c_2, c_3	=	molar density of bulk solution and the concentrated boundary solution on the high pressure side of the membrane, and the membrane permeated product solution on the atmospheric pressure side of the membrane respectively, g. mole/cm^3
c_M	=	molar density of solution in the membrane phase, g. mole/cm^3
c_{M2}, c_{M3}	=	molar density of solution in the membrane phase in equilibrium with c_2 and c_3 respectively, g. mole/cm^3
$\bar{c}_i$	=	concentration of ion in the diffusion current experiment, g. mole/litre
d	=	diameter of cell, cm
D_i	=	diffusivity of ion, cm^2/sec
D_{AB}	=	diffusivity of solute, cm^2/sec
D_{AM}	=	diffusivity of solute in the membrane phase, cm^2/sec
f	=	solute separation defined by Equation 3-56
f_{max}	=	maximum possible solute separation
F	=	faraday
G	=	Gibbs free energy
h	=	depth of cell, cm
H	=	enthalpy of solution
i	=	limiting current
k	=	mass transfer coefficient, cm/sec
K	=	proportionality constant defined by Equation 3-37
K_1, K_2	=	constants
l	=	thickness of the concentrated boundary solution, cm
m	=	molality of solution
m_1, m_3	=	molality of bulk solution on the high pressure side of the membrane, and that of the product solution on the atmospheric pressure side of the membrane respectively
M_A	=	molecular weight of solute
M_B	=	molecular weight of water
n	=	constant; or number of moles of product water removed from the feed solution in reverse osmosis operation
N	=	number of moles of the entire set of components in a system
N_i	=	number of moles of component i in a system
N_j	=	number of moles of all the other components in a system except N_i
N_A	=	solute flux, $\frac{\text{g. mole}}{\text{cm}^2\ \text{sec}}$

N_B	=	solvent water flux, $\frac{\text{g. mole}}{\text{cm}^2\ \text{sec}}$
N_{Re}	=	Reynolds number defined by Equation 3-83
N_{Sc}	=	Schmidt number defined by Equation 3-84
N_{Sh}	=	Sherwood number defined by Equation 3-85
p_W	=	vapour pressure of water in equilibrium with the solution
p_W^*	=	vapour pressure of pure water
P	=	operating pressure, atm
P_{max}	=	maximum pressure used in the operating pressure range, atm
[PR]	=	product rate per given area of film surface, g/hour
[PWP]	=	pure water permeability per given area of film surface, g/hour
Q	=	feed rate, cm^3/sec
Q′	=	product rate, $\frac{\text{g. mole}}{\text{cm}^2\ \text{sec}}$
R	=	gas constant
S	=	entropy; or surface area, cm^2
T	=	absolute temperature, °K or operating temperature, °C
T_0	=	lowest absolute temperature of surroundings, °K
V	=	volume, cm^3
$\overline{V}_i$	=	partial molal volume of component i
$\overline{V}_W$	=	partial molal volume of water
w_A	=	amount of solute in Q′, grams
w_B	=	amount of solvent water in Q′, grams
W	=	isothermal reversible work of separation
W_A	=	amount of solute in a given solution, grams
W_B	=	amount of solvent water in a given solution, grams
X_A	=	mole fraction of solute
X_{A1}, X_{A2}, X_{A3}	=	mole fraction of solute in the bulk solution and in the concentrated boundary solution on the high pressure side of the membrane, and in the product solution on the atmospheric pressure side of the membrane respectively at any time
X_{AM}	=	mole fraction of solute in the membrane phase
X_{AM2}, X_{AM3}	=	Values of X_{AM} in equilibrium with X_{A2} and X_{A3} respectively
z	=	thickness of the concentrated boundary solution, cm

Greek Letters

α = constant

β = constant

δ = effective thickness of membrane, cm

ϵ = constant

μ = viscosity of solution, centipoise

μ_i = chemical potential of component i

μ_i^0 = standard chemical potential of component i

μ_W = viscosity of water, centipoise

μ_W^* = chemical potential of pure water

ν = kinematic viscosity of solution, cm^2/sec

$\Pi_F, \Pi_P, \Pi(X_A)$ = osmotic pressure of feed solution, product solution, and solution whose solute mole fraction is X_A respectively, atm

ρ = density of solution, g/cm^3

ρ_W = density of water, g/cm^3

Σ_i = total number of moles of ions given by one mole of the electrolyte

ϕ = osmotic coefficient

12. Summary

This chapter is concerned mainly with the development of the basic transport equations of Kimura and Sourirajan for the isothermal reverse osmosis separation process involving binary aqueous solutions, and Loeb-Sourirajan type porous cellulose acetate membranes. These equations, together with the correlations of the reverse osmosis experimental data, lead to practical methods of membrane specification, expressing membrane selectivity, and predicting membrane performance from a minimum of experimental data, and they constitute the basis for reverse osmosis process design discussed in the next chapter.

Several approaches to the subject of transport in reverse osmosis are currently in vogue. The approach presented in this chapter is consistent with the concept of the generality of the reverse osmosis separation technique as formulated in Chapter 1. That the membrane surface is porous and heterogeneous, and that the transport of both the preferentially sorbed water and the aqueous solution is through the capillary pores on the membrane surface, are concepts fundamental to this approach.

The chemical potential and osmotic pressure relationships which are of interest in reverse osmosis are summarised. The thermodynamic significance of osmotic pressure is illustrated by a set of data on the maximum possible solute separations in reverse osmosis for the systems sodium chloride-water and glycerol-water solutions. Useful numerical data on osmotic pressure, molar density, kinematic viscosity and solute diffusivity for several aqueous solution systems are given in Appendix II.

Glueckauf extended the preferential sorption capillary flow model of Sourirajan and attributed reverse osmosis desalination to the repulsion which ions experi-

enced in the close vicinity of membrane materials of low dielectric constant. Glueckauf's method of predicting the separation characteristics of a membrane for different electrolyte solution systems and feed concentrations is discussed in the light of experimental reverse osmosis data for several solution systems. The results show that the Glueckauf analysis, though highly suggestive, is still insufficient and needs improvement to predict the separation characteristics of the high permeability porous cellulose acetate membranes.

Some empirical correlations of reverse osmosis experimental data are given. The empirical equations of Govindan and Sourirajan predict solute separation and/or product rates in limited ranges of operating pressure, temperature, and feed concentrations.

An extensive discussion of the Kimura-Sourirajan analysis of reverse osmosis experimental data forms the most important part of this chapter. This analysis is based on a generalised capillary diffusion model for the transport of solute through the membrane, and it is applicable for the entire possible range of solute separations in the reverse osmosis process. In this analysis, the transport of solvent water through the porous membrane is proportional to the effective pressure, that of the solute is due to pore diffusion and hence proportional to its concentration difference across the membrane, and the mass transfer coefficient on the high pressure side of the membrane is obtained from a straight forward application of the 'film' theory.

The Kimura-Sourirajan analysis gives rise to a set of basic equations relating the pure water permeability constant, A, the transport of solvent water N_B *the solute transport parameter,* $(D_{AM}/K\delta)$*, and the mass transfer coefficient, k. In this analysis, A is a fundamental quantity; it is a measure of the overall porosity of the film; it corresponds to conditions of zero concentration polarisation; and it is independent of any solute under consideration. The parameter* $(D_{AM}/K\delta)$ *plays the role of a mass transfer coefficient with respect to solute transport through the membrane; it is a combination of several inter-related factors, none of which is or need be precisely known for chemical engineering calculations; hence it is treated as a single quantity for purposes of analysis. The mass transfer coefficient, k, on the high pressure side of the membrane is defined in the conventional manner of the 'film theory'; the values of k so defined and calculated from the experimental reverse osmosis data coincide satisfactorily with those calculated from a generalized mass transfer coefficient correlation obtained from the diffusion current method.*

The values of A, $(D_{AM}/K\delta)$*, and k for any system can be calculated from any single set of experimental pure water permeability, product rate, and solute separation data. Both A and* $(D_{AM}/K\delta)$ *depend on the porous structure of the membrane and hence they are different for different membranes. At a given operating pressure, the values of* $(D_{AM}/K\delta)$ *remain essentially constant for a wide range of feed concentrations, and feed flow rates for several solutes including* $MgSO_4$, Na_2SO_4, $BaCl_2$, $CaCl_2$, $MgCl_2$, *LiCl, NaCl, KCl,* NH_4Cl, $LiNO_3$, KNO_3, $NaNO_3$, *glycerol and urea. The mass transfer coefficient k for all the above systems is essentially a function of feed flow rate and feed concentration, and the values of k are well correlated by a generalised log-log plot of* N_{Re} *v.* $N_{Sh}/N_{Sc}^{0.33}$*. On the basis of the above Kimura-Sourirajan transport equations and correlations of experimental reverse osmosis data, a single set of experimental [PWP], [PR], and f data at any operating pressure specifies a film in terms of A and* $(D_{AM}/K\delta)$ *which makes possible the prediction of both solute separation and product rate at that pressure for all feed concentrations and feed flow rates for which* $(D_{AM}/K\delta)$ *remains constant.*

The experimental separation and product rate data obtained in the extended continuous test runs for the systems glycerol-water and sodium chloride-water show that the solute transport parameter, $(D_{AM}/K\delta)$, remains unchanged as the pure water permeability constant, A, decreases because of membrane compaction during continuous reverse osmosis operation under pressure. This observation establishes an experimental basis and offers an analytical means of predicting solute separation as the product rate decreases because of membrane compaction during continuous operation of the reverse osmosis process. This prediction technique is illustrated for the systems glycerol-water and sodium chloride-water, using a few typical films. The results show that variations in solute separation with decrease in product rate must be expected, and can be calculated and allowed for in practical reverse osmosis units. Such variations arise as a consequence of the effect of membrane compaction on the transport characteristics of reverse osmosis membranes even in the absence of any deterioration of the membrane material due to chemical interactions such as hydrolysis. Therefore the solution to the problem of decrease in solute separation with decrease in product rate during extended service conditions lies essentially in reducing membrane compaction and altering the flow conditions suitably to give the desired results.

The values of A and $(D_{AM}/K\delta)$ specify a membrane at a given operating pressure for all the solution systems mentioned. Provided the mass transfer coefficient, k, applicable for the high pressure side of the membrane, is available or can be calculated, the performance of a membrane, expressed in terms of solute separation and product rate, can be predicted from membrane specifications. The data on membrane specification are related to those on membrane performance so that the significance of the specifying parameters can be recognised readily in practical terms of cause and effect.

For a given membrane, the relative values of $(D_{AM}/K\delta)$ for different solutes offer a firm basis of expressing membrane selectivity. A relative scale of membrane selectivity for different solutes is given for one type of Loeb-Sourirajan type porous cellulose acetate membranes whose values of $(D_{AM}/K\delta)$ for NaCl range from 1 to 1000 $(\times 10^{-5})$ cm/sec. A higher value of $(D_{AM}/K\delta)$ generally means a lower level of solute separation; but the order of solute separation with respect to any two solutes does not always correspond to the order of their values of $(D_{AM}/K\delta)$. The order of solute separation is a function not only of the relative values of $(D_{AM}/K\delta)$ for different solutes but also of the operating conditions of pressure, feed concentration, and the mass transfer coefficient, k.

The reverse osmosis separation data for the system sucrose-water using a number of Loeb-Sourirajan type porous cellulose acetate membranes have been analysed. The results show that $(D_{AM}/K\delta)$ for sucrose decreases with increase in its boundary concentration X_{A2}. The plot of log $(D_{AM}/K\delta)$ v. X_{A2} is a straight line, and this relationship is independent of the particular combination of feed concentration and feed flow rate used. The slope of the above straight line is a function of the operating pressure; and at a given operating pressure, it is essentially the same for all the membranes tested. For every film, there also exists a unique relationship between the $(D_{AM}/K\delta)$ for NaCl and the extrapolated value of $(D_{AM}/K\delta)$ for sucrose at $X_{A2} = 0$, giving rise to a method of expressing membrane selectivity on a relative scale. The predictability of membrane performance for the reverse osmosis separation of sucrose in aqueous solution and the effect of membrane compaction on solute separation, from the initial specifications of the film given in terms of the pure water permeability constant A, and $(D_{AM}/K\delta)$ for NaCl, are illustrated and discussed.

CHAPTER 4

Concentration Polarisation Effects in Reverse Osmosis and Reverse Osmosis Process Design

As pointed out in the last chapter, the withdrawal of the preferentially sorbed interfacial fluid increases the concentration of the solution at the immediate vicinity of the membrane surface, and sets up a concentration gradient on the high pressure side of the membrane, causing the diffusion of the solute from the more to the less concentrated part of the solution. This phenomenon is called concentration polarisation. At a given operating pressure and in a given apparatus, concentration polarisation results in an increase in the effective osmotic pressure of the feed solution on the membrane surface, a consequent decrease in the effective pressure (ΔP) for fluid flow across the membrane, and progressive changes in the mass transfer coefficient, product rate, and solute separation along the length of the membrane in the direction of feed flow. The problem then is to predict the effect of concentration polarisation on solute separation and on product rate, under specified experimental conditions, with a given membrane-solution system, at any position or time, in a given reverse osmosis apparatus.

The practical importance of the above problem in reverse osmosis process design and development has been recognised, and several analytical studies reported with particular reference to saline water conversion (Brian, 1965a, 1965b, 1965c, 1966; Dresner, 1964; Fisher et al., 1965; Gill et al., 1965, 1966a, 1966b, 1966c; Johnson et al., 1966; Merten, 1963; Merten et al., 1964; Sherwood et al., 1964, 1965; Srinivasan et al., 1967; Tien and Gill, 1966). These studies assume that the membrane exhibits either complete solute separation, or incomplete solute separation at a constant level; such assumptions are, in general, invalid since solute separation can vary widely depending on feed concentration and feed flow rate even at the same operating pressure.

To be general and valid, the theory of concentration polarisation in reverse osmosis must be based on transport equations which are applicable for the entire range of solute separations and which can predict the effect of feed concentration and feed flow rate on both solute separation and product rate; these criteria are satisfied by the transport equations of Kimura and Sourirajan (1967) given in Chapter 3.

The distinguishing feature of the Kimura-Sourirajan analysis is the fact that for the Loeb-Sourirajan type porous cellulose acetate membranes, and for solution systems such as sodium chloride-water and glycerol-water, the pure water permeability constant A and the solute transport parameter $(D_{AM}/K\delta)$ completely specify a membrane with respect to this separation process; at a given operating pressure, both A and $(D_{AM}/K\delta)$ are independent of feed concentration and feed flow rate. For such membrane-solution systems, the effects of concentration polarisation under different operating conditions can be predicted from the specifications of the membrane given in terms of A and $(D_{AM}/K\delta)$, and a rational basis for reverse osmosis process design can be established (Kimura and Sourirajan, 1968a, 1968b, 1968c; Kimura et al., 1969). This is illustrated in this chapter with particular reference to the system sodium chloride-water, which is chosen because of its practical significance to saline water conversion.

1. Basic Transport Equations

For porous cellulose acetate membranes which have a preferential sorption for water from aqueous feed solutions, the Kimura-Sourirajan analysis gives rise to the following basic equations for the solvent and solute transport through the membrane at a given operating pressure.
Solvent water transport:

$$N_B = A[P - \Pi(X_{A2}) + \Pi(X_{A3})] \tag{4-1}$$

$$= kc_1(1 - X_{A3}) \ln\left(\frac{X_{A2} - X_{A3}}{X_{A1} - X_{A3}}\right) \tag{4-2}$$

Solute transport:

$$N_A = \left(\frac{D_{AM}}{K\delta}\right)(c_2X_{A2} - c_3X_{A3}) \tag{4-3}$$

In Equations 4-1, 4-2, and 4-3, A is the pure water permeability constant obtained from the pure water permeability data, and

$$X_{A3} = N_A/(N_A + N_B) \tag{4-4}$$

and subscripts 1, 2, and 3 refer, as before, to bulk solution, concentrated boundary solution, and membrane permeated product solution, respectively; the other symbols are the same as those given in Chapter 3 where the derivations of these equations are given.

For the purpose of present analysis, the following simplifying assumptions are made:

(i) the molar density of the solution is constant, i.e.

$$c_1 = c_2 = c_3 = c \tag{4-5}$$ *

(ii) the osmotic pressure of solution is proportional to the mole fraction of solute, i.e.

$$\Pi(X_A) = BX_A \tag{4-6}$$

where B is a constant; and

(iii) the solute flux across the membrane is small compared to solvent flux, i.e.

$$N_A \ll N_B \text{ or } X_{A3} \ll 1 \tag{4-7}$$

so that

$$X_{A3} = N_A/N_B \tag{4-8}$$

All the above assumptions are reasonably valid for the system sodium chloride-water at least up to a concentration of 1.0M (≡ 5.52 weight per cent

---

* In the rest of this chapter, the symbol c without a subscript denotes molar density of the solution, and c with a subscript denotes solute concentration.

NaCl). This is evident from the following data at 25° (Table A-17, Appendix II). The molar density of pure water is 5.535×10^{-2} mole/cm^3, and that of 1.0M solution is 5.530×10^{-2} mole/cm^3. The osmotic pressure of 0.1M ($X_A = 1.798 \times 10^{-3}$) solution is 67 lb/in^2, and that of 1.0M ($X_A = 17.693 \times 10^{-3}$) solution is 673 lb/in^2. For a 1.0M feed solution, even for 0 per cent solute separation, N_A is only about 1.8 per cent of N_B. For engineering calculations, the assumptions expressed by Equations 4-5 to 4-8 can be considered acceptable for the system NaCl-H_2O even up to a concentration of 2.0M (≡10.47 wt.% NaCl). The above assumptions do not restrict the scope of the following analysis, but they simplify the equations involved in illustrating the effect of process and design variables relating to sea water and brackish water conversion.

Using the above assumptions, Equations 4-1, 4-2, and 4-3 may be rewritten as follows:

$$N_B = A[P - \Pi(X_{A2}) + \Pi(X_{A3})] \qquad (4\text{-}1)$$

$$= kc(1 - X_{A3}) \ln\left(\frac{X_{A2} - X_{A3}}{X_{A1} - X_{A3}}\right) \qquad (4\text{-}9)$$

$$N_A = c(D_{AM}/K\delta)(X_{A2} - X_{A3}) \qquad (4\text{-}10)$$

When a particular membrane-solution system is specified in terms of A and $(D_{AM}/K\delta)$ at a given operating gauge pressure P, the experimental conditions can be specified in terms of three new dimensionless parameters, γ, θ, and λ defined as follows:

$$\gamma = BX^0_{A1}/P \qquad (4\text{-}11)$$

$$\theta = \frac{c}{AP}\left(\frac{D_{AM}}{K\delta}\right) \qquad (4\text{-}12)$$

and

$$\lambda = k/(D_{AM}/K\delta) \qquad (4\text{-}13)$$

where X^0_{A1} is the mole fraction of solute in the feed solution at membrane entrance, and k is the applicable mass transfer coefficient.

The above parameters can be built into the basic transport Equations 4-1, 4-9, and 4-10 as follows. Combining Equations 4-1 and 4-6,

$$N_B = A[P - B(X_{A2} - X_{A3})] \qquad (4\text{-}14)$$

$$= AP\left[1 - \frac{BX^0_{A1}}{P}\left(\frac{X_{A2}}{X^0_{A1}} - \frac{X_{A3}}{X^0_{A1}}\right)\right] \qquad (4\text{-}15)$$

$$= AP[1 - \gamma(C_2 - C_3)] \qquad (4\text{-}16)$$

where

$$C_i = X_{Ai}/X^0_{A1} \qquad (4\text{-}17)$$

which is the local salt concentration divided by the salt concentration in the brine feed at membrane entrance, i.e. $C_2 = X_{A2}/X^0_{A1}$ and $C_3 = X_{A3}/X^0_{A1}$. Combining Equations 4-8, 4-10, and 4-16,

$$X_{A3} = \frac{c\left(\frac{D_{AM}}{K\delta}\right)(X_{A2} - X_{A3})}{AP[1 - \gamma(C_2 - C_3)]} \tag{4-18}$$

or

$$C_3 = \frac{X_{A3}}{X^0_{A1}} = \frac{c\left(\frac{D_{AM}}{K\delta}\right)\left(\frac{X_{A2}}{X^0_{A1}} - \frac{X_{A3}}{X^0_{A1}}\right)}{AP[1 - \gamma(C_2 - C_3)]} \tag{4-19}$$

$$= \theta(C_2 - C_3)/[1 - \gamma(C_2 - C_3)] \tag{4-20}$$

Equation 4-20 can be written as

$$C_3 - \gamma C_2 C_3 + \gamma C_3^2 = \theta C_2 - \vartheta C_3 \tag{4-21}$$

or

$$\gamma C_3^2 + (1 + \theta - \gamma C_2)C_3 - \theta C_2 = 0 \tag{4-22}$$

Solving Equation 4-22 for C_3 and omitting the inapplicable negative root,

$$C_3 = \frac{\sqrt{[(1 + \theta - \gamma C_2)^2 + 4\gamma\theta C_2]} - (1 + \theta - \gamma C_2)}{2\gamma} \tag{4-23}$$

Since $X_{A3} \ll 1$, Equation 4-9 can be written as

$$N_B = kc \ln\left(\frac{X_{A2} - X_{A3}}{X_{A1} - X_{A3}}\right) \tag{4-24}$$

Combining Equations 4-8, 4-10, and 4-24,

$$X_{A3} = \frac{(D_{AM}/K\delta)}{k} \frac{(X_{A2} - X_{A3})}{\ln\left(\frac{X_{A2} - X_{A3}}{X_{A1} - X_{A3}}\right)} \tag{4-25}$$

or

$$C_3 = \frac{X_{A3}}{X^0_{A1}} = \frac{C_2 - C_3}{\lambda \ln\left(\frac{C_2 - C_3}{C_1 - C_3}\right)} \tag{4-26}$$

$$= \frac{C_2}{1 + \lambda \ln\left(\frac{C_2 - C_3}{C_1 - C_3}\right)} \tag{4-27}$$

Also

$$N_B = cv_W \tag{4-28}$$

where v_W is the fluid velocity in the y-direction, i.e., direction perpendicular to membrane surface. Combining Equations 4-16 and 4-28

$$v_W = \frac{N_B}{c} = \frac{AP}{c}\,[1 - \gamma(C_2 - C_3)] \tag{4-29}$$

Defining v_W^0 as the value of v_W for the hypothetical condition $C_2 = 1$, and $C_3 = 0$,

$$v_W^0 = \frac{AP}{c}\,(1 - \gamma) \tag{4-30}$$

the dimensionless velocity component in the y-direction is given by

$$V_W = \frac{v_W}{v_W^0} = \frac{[1 - \gamma(C_2 - C_3)]}{1 - \gamma} \tag{4-31}$$

Equations 4-23, 4-27, and 4-31 are applicable for both turbulent and laminar feed flow in any reverse osmosis cell, and for all experimental conditions defined by the parameters γ, θ, and λ, and hence they are general in scope for purposes of process design; they can be used to study the effects of concentration polarisation (C_2) on solute separation (which depends on X_{A3}) and product rate (which depends on V_W) under various experimental conditions of γ, θ, and λ, provided C_1 is known, and the applicable mass transfer coefficient values are used for k in defining λ. An illustration of the applicability of Equations 4-23, 4-27, and 4-31 for investigating the concentration polarisation effects in reverse osmosis is given below for the case of the flow of the brine solution in rectangular channels between flat parallel porous cellulose acetate membranes (Kimura and Sourirajan, 1968a).

2. Feed Flow Between Flat Parallel Membranes

In Figure 4-1, feed flow between flat parallel membranes is schematically represented. Let the length of the channel in the transverse direction be assumed very large compared to the membrane spacing (width). Let h be half-width of membrane spacing; consider unit length of channel in the transverse

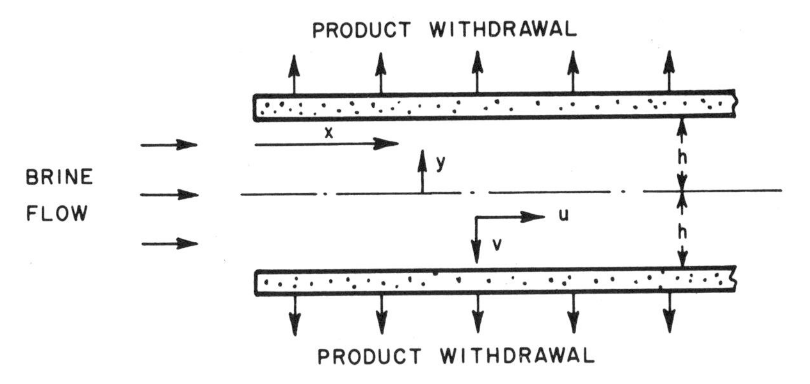

Figure 4-1. Brine flow through flat parallel membranes (Ohya and Sourirajan, 1969d).

direction. Product withdrawal is from both the upper and lower membranes, and brine flow is assumed symmetrical about the midplane of the channel width. The analysis is for brine containing a single salt with a constant diffusion coefficient, D. The volume change upon mixing salt solutions of different concentrations is assumed negligible.

Let $\bar{u}^0$ and $\bar{u}$ represent respectively the average velocity of brine solution at channel entrance, and at distance x along the longitudinal length of the channel respectively. Let the dimensionless longitudinal position X be defined as

$$X = \frac{v_w^0}{\bar{u}^0} \frac{x}{h} \tag{4-32}$$

Feed Flow as a Function of Longitudinal Position

Fraction product recovery. The volumetric fraction of feed stream that is recovered as the membrane permeated product in the entire longitudinal length 0 to x under consideration, $(= 1 - \bar{u}/\bar{u}^0)$, is obtained from the overall water-material balance which can be given as follows.

$$\bar{u}h = (\bar{u} + \frac{d\bar{u}}{dx} dx)h + v_w dx \tag{4-33}$$

Therefore

$$-d\bar{u}/dx = v_w/h \tag{4-34}$$

$$\int_{\bar{u}^0}^{\bar{u}} - du = \int_0^x \frac{v_w}{h} dx \tag{4-35}$$

$$\bar{u}^0 - \bar{u} = \int_0^x \frac{v_w}{h} dx \tag{4-36}$$

$$1 - \frac{\bar{u}}{\bar{u}^0} = \int_0^x \frac{v_w}{\bar{u}^0 h} dx \tag{4-37}$$

$$= \int_0^x \frac{v_w}{v_w^0} \cdot \frac{v_w^0}{\bar{u}^0 h} dx \tag{4-38}$$

$$= \int_0^X V_w dX' \tag{4-39}$$

Defining

$$\Delta = \int_0^X V_w dX' \tag{4-40}$$

and

$$\bar{u}/\bar{u}^0 = \bar{U} \tag{4-41}$$

the fraction product recovery for feed flow between flat parallel membranes is then given by

$$1 - \bar{u}/\bar{u}^0 = 1 - \bar{U} = \Delta. \tag{4-42}$$

Bulk concentration of solute. The dimensionless bulk concentration of solute, C_1, in the transverse direction, as a function of longitudinal position X is obtained from an over-all solute-material balance which can be given as follows.

$$(\bar{u}c_{A1})h = [\bar{u}c_{A1} + \frac{d}{dx}(\bar{u}c_{A1})dx]h + c_{A3}v_W dx \tag{4-43}$$

where c_{A1} and c_{A3} are the solute concentrations (g. mole/cm^3) in the bulk solution and product solution respectively; c_{A1} is strictly an average value in the transverse direction. From Equation 4-43,

$$-\frac{d}{dx}(\bar{u}c_{A1}) = \frac{c_{A3}v_W}{h} \tag{4-44}$$

$$\int_{\bar{u}^0}^{\bar{u}} - d(\bar{u}c_{A1}) = \int_0^x \frac{c_{A3}v_W}{h}\, dx \tag{4-45}$$

$$\bar{u}^0 c_{A1}^0 - \bar{u}c_{A1} = \int_0^x \frac{c_{A3}v_W}{h}\, dx \tag{4-46}$$

$$1 - \frac{\bar{u}}{\bar{u}^0} \cdot \frac{c_{A1}}{c_{A1}^0} = \int_0^x \frac{c_{A3}}{c_{A1}^0} \cdot \frac{v_W}{v_W^0} \cdot \frac{v_W^0}{\bar{u}^0 h}\, dx \tag{4-47}$$

When the molar density of solution is considered constant

$$c_{A1}/c_{A1}^0 = C_1 \tag{4-48}$$

and

$$c_{A3}/C_{A1}^0 = C_3 \tag{4-49}$$

Equation 4-47 can then be written in non-dimensional form as

$$1 - \bar{U}C_1 = \int_0^X C_3 V_W dX' \tag{4-50}$$

Therefore

$$C_1 = (1 - \int_0^X C_3 V_W dX')/\bar{U} \tag{4-51}$$

Combining Equations 4-42 and 4-51,

$$C_1 = (1 - \int_0^X C_3 V_W dX')/1 - \Delta) \tag{4-52}$$

Thus Equation 4-52 gives the dimensionless bulk solute concentration C_1 in the transverse direction, as a function of longitudinal position X for feed flow between flat parallel membranes.

Solute separation. The solute separation f is defined by the equation

$$f = \frac{m_1^0 - m_3}{m_1^0} = 1 - \left(\frac{1 - X_{A1}^0}{X_{A1}^0}\right)\left(\frac{X_{A3}}{1 - X_{A3}}\right) \tag{4-53}$$

where m_1^0 is the molality of the feed solution at channel entrance, and m_3 is the molality of the product solution at any longitudinal position X. Since

$$(1 - X_{A1}^0)/(1 - X_{A3}) \approx 1 \tag{4-54}$$

$$f = 1 - X_{A3}/X_{A1}^0 = 1 - C_3 \tag{4-55}$$

The average solute separation, $\bar{f}$, can be given as

$$\bar{f} = 1 - \bar{C}_3 \tag{4-56}$$

where $\bar{C}_3$ is the average concentration of the product solution given by the relation

$$\bar{C}_3 = \int_0^X C_3 V_W dX' / \int_0^X V_W dX' \tag{4-57}$$

$$= \frac{\int_0^X C_3 V_W dX'}{\Delta} \tag{4-58}$$

Combining Equations 4-52 and 4-58,

$$\bar{C}_3 = [1 - (1 - \Delta) C_1]/\Delta \tag{4-59}$$

In Equations 4-55, f and C_3 are local values; in Equation 4-56, $\bar{f}$ and $\bar{C}_3$ are average values for the entire length of the channel under consideration.

Some Illustrative Calculations. Equations 4-23, 4-27, 4-31, and 4-52 can be solved simultaneously to obtain C_1, C_2, C_3, and V_W as a function of X for any preset values of γ, θ, and λ defined by Equations 4-11, 4-12, and 4-13 respectively. From the values of C_1, C_3, and V_W thus obtained, Δ, $\bar{C}_3$, f and $\bar{f}$ can be calculated from Equations 4-40, 4-59, and 4-55 and 4-56 respectively. This procedure offers a means of studying concentration (C_2) and its effects on solute separation (f and $\bar{f}$) and product rate expressed as V_W along the longitudinal position in the channel for any preset operating conditions. One may also calculate N_B from V_W using Equations 4-28, 4-30, and 4-31.

Figures 4-2, 4-3, and 4-4 illustrate a set of such calculations for Film No. 1. Using 0.5 M ($NaCl$-H_2O] as the feed solution, and an operating pressure of 102 atm, and the membrane specifications given in Table 4-1, $\gamma = 0.22$, and $\theta = 0.005$ for the above film.

Figure 4-2 illustrates the variation of concentration polarisation as a function of X for various assumed constant values of λ which expresses the effect of the mass transfer coefficient k. The figure shows that C_2 is sensitive to λ, and for a given value of λ, the C_2 v. X curve is practically flat at low values of X.

Figure 4-3 shows the variation of solute separation (f and $\bar{f}$) as a function of X for various values of λ. Figure 4-4 gives the relative variations of C_2, f, and V_W as a function of λ for the condition $C_1 = 1$, i.e., for low values of X. The latter plots show that the values of f and V_W increase steeply with increase in λ at least up to a point. Calculations of the type represented in Figures 4-3 and 4-4 can help in deciding the optimum value of λ for the desired membrane performance under otherwise fixed experimental conditions.

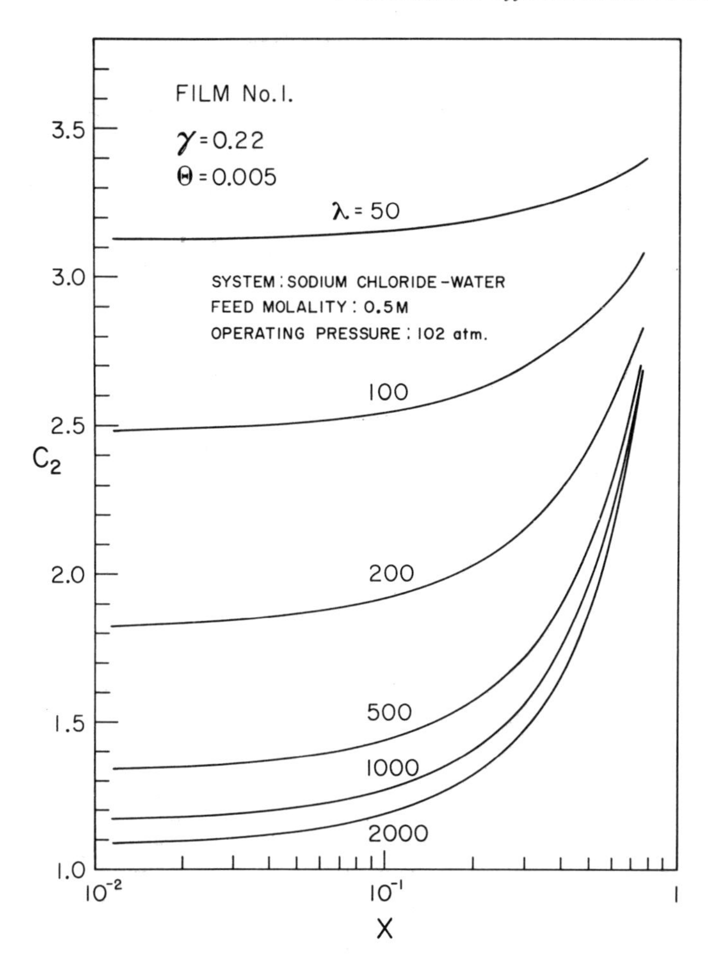

Figure 4-2. Effect of mass transfer coefficient on concentration polarisation v. X. Data of Kimura and Sourirajan (1968a).

Table 4-1. Membrane specifications

Film no.	Solute	Operating pressure (psig)	$A \times 10^6$ $\left(\frac{\text{g. mole } H_2O}{\text{cm}^2 \text{ sec atm}}\right)$	$\frac{D_{AM}}{K\delta} \times 10^5$ (cm/sec)
1	NaCl	1500	0.97	0.90
2	NaCl	1500	1.46	6.0
3	NaCl	1500	1.87	20.0
5	NaCl	1500	2.37	75.0
6	NaCl	1500	2.93	140.0

Film type: Loeb-Sourirajan type porous cellulose acetate membranes (CA-NRC-18)

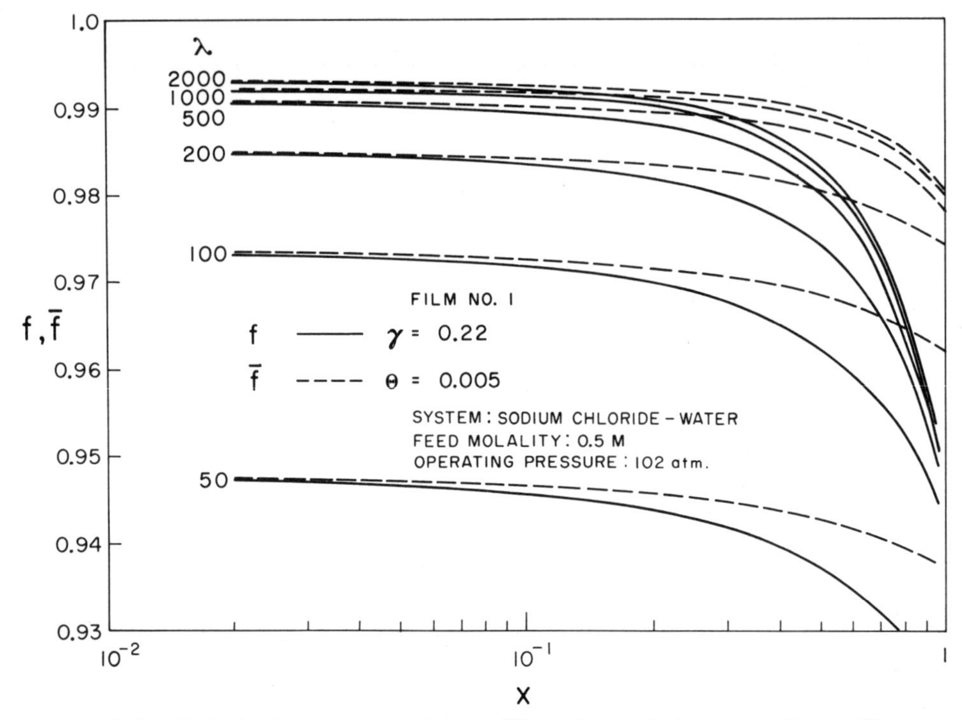

Figure 4-3. Effect of mass transfer coefficient on solute separation v. X. Data of Kimura and Sourirajan (1968a).

Since a large value of λ also implies a large feed flow rate, and hence power consumption, one may seek an optimum for λ from an economic standpoint. A detailed economic analysis is outside the scope of the present discussion; however, it is of interest to call attention to some factors governing such an analysis.

Product Rate v. Reynolds Number. Equations 4-23, 4-27, 4-31, and 4-52 apply to both turbulent and laminar feed flow between flat parallel membranes, provided the appropriate value of k is used to define λ.

For the case of turbulent flow between flat parallel membranes the following friction-factor-Reynolds-number relationships can be used (Knudsen and Katz, 1958d):

$$\text{Fanning friction factor, } f^* = \frac{2hg_c\Delta P}{\rho_1 \bar{u}^2 x} \tag{4-60}$$

$$\text{Reynolds number, } \quad N_{Re} = \frac{4h\bar{u}}{\nu} \tag{4-61}$$

where ΔP is the pressure drop, ρ_1 and ν are density and kinematic viscosity respectively of feed solution, and g_c is the conversion factor.

$$\text{Therefore pressure drop due to friction:} \quad \Delta P = \frac{f^* \rho_1 \bar{u}^2 x}{2hg_c} \tag{4-62}$$

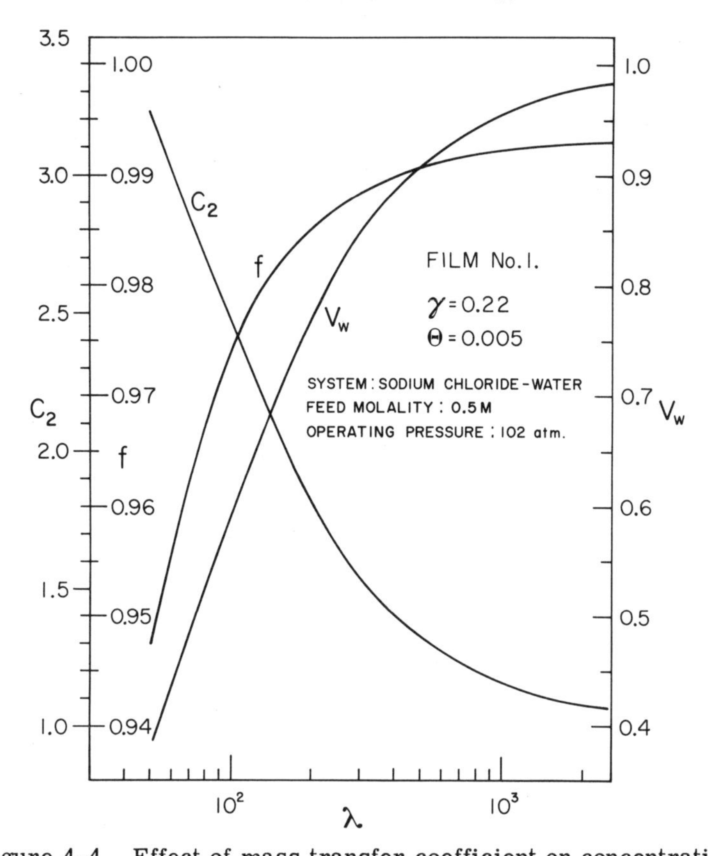

Figure 4-4. Effect of mass transfer coefficient on concentration polarisation, solute separation and product rate for low values of X. Data of Kimura and Sourirajan (1968a).

power consumption: $$\text{power} = \Delta P \cdot \bar{u} \cdot 2hW \tag{4-63}$$

$$= \frac{f^* \rho_1 \bar{u}^3 x W}{g_c} \tag{4-64}$$

where W is the transverse length of the membrane. The product rate from one side of the membrane, Q, is given by

$$Q = \int_0^X v_w dx \cdot W \tag{4-65}$$

$$= \left[\int_0^X \frac{v_w}{v_w^0} \frac{v_w^0}{\bar{u}^0 h} dx \right] \bar{u}^0 hW \tag{4-66}$$

$$= \int_0^X V_w dX' \cdot \bar{u}^0 hW \tag{4-67}$$

$$= \Delta \cdot \bar{u}^0 hW \tag{4-68}$$

$$= N_{Re} \frac{\nu}{4} W\Delta \qquad (4\text{-}69)$$

The total membrane throughput rate per unit power (F) is given by

$$F = \frac{2Q}{\text{Power}} = \frac{2g_c v_W^0}{f^* \rho_1 \bar{u}^3} \cdot \frac{\Delta}{X} \qquad (4\text{-}70)$$

$$= \frac{2g_c v_W^0}{f^* \rho_1} \cdot \frac{(4h)^3}{\nu^3 N_{Re}^3} \cdot \frac{\Delta}{X} \qquad (4\text{-}71)$$

In deriving Equations 4-69 and 4-71, $\bar{u}^0$ is assumed equal to $\bar{u}$ as an approximation. From an economic standpoint, one would seek a maximum for F. Assuming that the mass transfer coefficient in the brine solution between the membranes can be given by the relation

$$4hk/D = 0.023\,(N_{Re})^{0.8}(N_{Sc})^{0.33} \qquad (4\text{-}72)$$

which is the Colburn equation for a pipe, replacing diameter by an equivalent diameter for a rectangular channel, 4h, then

$$\lambda = \frac{0.023D}{4h(D_{AM}/K\delta)} (N_{Re})^{0.8}(N_{Sc})^{0.33} \qquad (4\text{-}73)$$

Figures 4-5 and 4-6 give the relationships between Δ/X and X, and Δ and X respectively for various values of λ. The plots show that at any value of X, a 40-fold increase in the value of λ results only in about a three-fold increase in the value of Δ/X. Consequently F is far more sensitive to changes in N_{Re} than to those in Δ/X. In order to obtain a high value of F, N_{Re} should be low according to Equation 4-71; but a low value of N_{Re} reduces the values of both Q (Equation 4-69) and Δ (Figure 4-6). Again one may find an optimum value for N_{Re} for maximum F. Also, when the brine feed rate at channel inlet remains constant (i.e., $\bar{u}^0 hW$ = constant), F will be high for high values of h, and consequently low values of $\bar{u}^0$ or W. These considerations indicate that the optimum product rate is obtainable probably at low Reynolds numbers.

Laminar Flow Between Flat Parallel Membranes. From the foregoing discussion, it is clear that the laminar flow case is of practical importance in reverse osmosis desalination; it is also of theoretical interest in view of the fact that the analysis (Kimura and Sourirajan, 1968a) does not involve the mass transfer coefficient, k, explicitly.

Under steady state conditions, the continuity equation for salt conservation can be written in the form

$$\frac{\partial}{\partial x}(uc_A) + \frac{\partial}{\partial y}\left(vc_A - D\frac{\partial c_A}{\partial y}\right) = 0 \qquad (4\text{-}74)$$

where u and v are fluid-velocity components in the x and y directions, respectively, and c_A is the molar concentration of the solute, and D its diffusion coefficient. This equation considers convection in the longitudinal direction

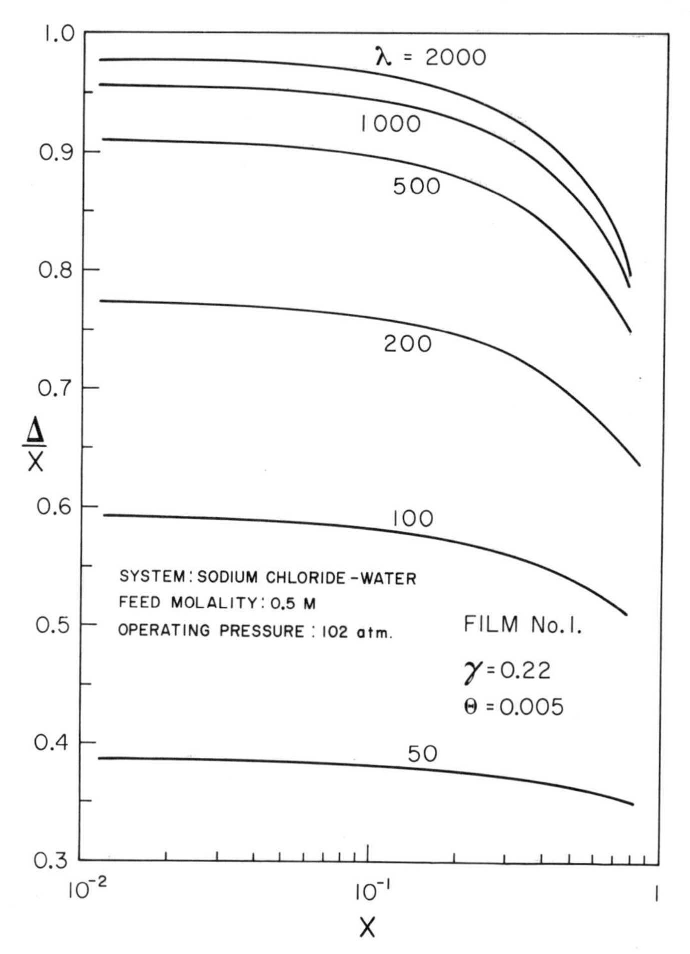

Figure 4-5. Effect of mass transfer coefficient on $\frac{\Delta}{X}$ v. X. Data of Kimura and Sourirajan (1968a).

and diffusion and convection in the transverse direction, and neglects longitudinal diffusion. The boundary conditions for Equation 4-74 are:

$$\text{at } x = 0, \quad c_A = c_A^0 \tag{4-75}$$

$$\text{at } y = 0, \quad \partial c_A/\partial y = 0 \tag{4-76}$$

$$\text{at } y = h, \quad \mathbf{D}(\partial \mathbf{c}_A/\partial y) = vc_A - N_A \tag{4-77}$$

Since

$$X_A = c_A/c \tag{4-78}$$

$$N_A = (D_{AM}/K\delta)(c_{A2} - c_{A3}) \tag{4-79}$$

Combining Equations 4-8, 4-28, 4-78 and 4-79,

$$X_{A3} = c_{A3}/c = N_A/N_B = (D_{AM}/K\delta)(c_{A2} - c_{A3})/cv_W \tag{4-80}$$

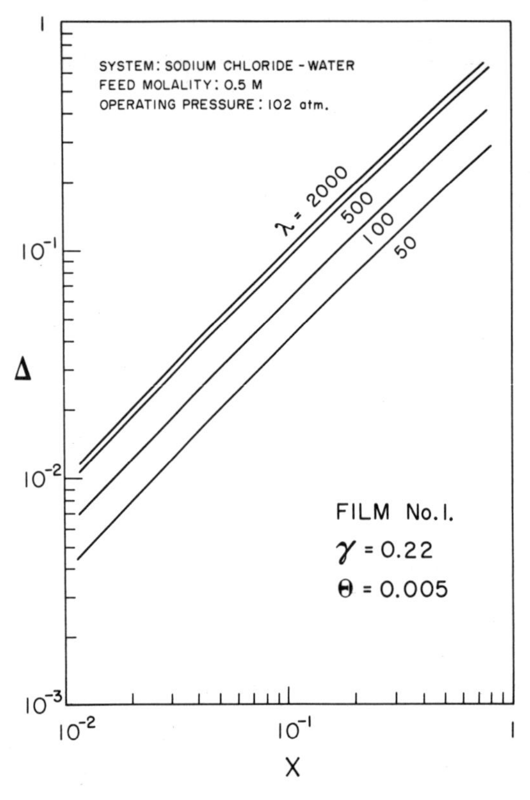

Figure 4-6. Effect of mass transfer coefficient on Δ v. X. Data of Kimura and Sourirajan (1968a).

$$c_{A3} v_W = (D_{AM}/K\delta)(c_{A2} - c_{A3}) \tag{4-81}$$

$$c_{A3}[v_W + (D_{AM}/K\delta)] = (D_{AM}/K\delta)c_{A2} \tag{4-82}$$

Therefore

$$c_{A3} = (D_{AM}/K\delta)c_{A2}/[v_W + (D_{AM}/K\delta)] \tag{4-83}$$

$$= (D_{AM}/K\delta)(c_{A2}/v_W)/[1 + (D_{AM}/K\delta)/v_W] \tag{4-84}$$

Combining Equations 4-79 and 4-84,

$$N_A = \left(\frac{D_{AM}}{K\delta}\right)\left[c_{A2} - \frac{\left(\frac{D_{AM}}{K\delta}\right)\frac{c_{A2}}{v_W}}{1 + \frac{(D_{AM}/K\delta)}{v_W}}\right] \tag{4-85}$$

$$= \left(\frac{D_{AM}}{K\delta}\right) c_{A2}\left[1 - \frac{(D_{AM}/K\delta)/v_W}{1 + \frac{(D_{AM}/K\delta)}{v_W}}\right] \tag{4-86}$$

$$= \left(\frac{D_{AM}}{K\delta}\right) c_{A2} \left[\frac{1}{1 + \frac{(D_{AM}/K\delta)}{v_W}}\right] \quad (4\text{-}87)$$

$$= v_W c_{A2} \left[\frac{(D_{AM}/K\delta)}{v_W + (D_{AM}/K\delta)}\right] \quad (4\text{-}88)$$

Using Equation 4-88, the boundary condition given by Equation 4-77 can be written as

$$D(\partial \mathbf{c}_A/\partial y) = v c_A [v/\{v + (D_{AM}/K\delta)\}] \quad (4\text{-}89)$$

at $y = h$.

Equation 4-74 can be written in nondimensional form as follows:

$$\frac{\partial}{\partial X}(UC) + \frac{\partial}{\partial Y}\left[VC - \alpha_0\left(\frac{\partial C}{\partial Y}\right)\right] = 0 \quad (4\text{-}90)$$

where

$$\alpha_0 = D/(v_W^0 h) \quad (4\text{-}91)$$

$$U = u/\bar{u}^0 \quad (4\text{-}41)$$

$$V = v/v_W^0 \text{ or } v_W/v_W^0 \quad (4\text{-}31)$$

$$X = \frac{v_W^0}{\bar{u}^0}\frac{x}{h} \quad (4\text{-}32)$$

$$Y = y/h \quad (4\text{-}92)$$

and

$$C = c_A/c_{A1}^0 \quad \text{or} \quad C_2 = c_{A2}/c_{A1}^0 \quad (4\text{-}93)$$

The boundary conditions for Equation 4-90 are

$$\text{at } X = 0, C = 1 \quad (4\text{-}94)$$

$$\text{at } Y = 0, \partial C/\partial Y = 0 \quad (4\text{-}95)$$

$$\text{at } Y = 1, \alpha_0 (\partial C/\partial Y) = V_W C_2 [V_W/(\kappa + V_W)] \quad (4\text{-}96)^*$$

---

* Using Equation 4-88, the boundary condition given by Equation 4-77 can be written as follows:

$$D\frac{\partial c_A}{\partial y} = v_W c_{A2}\left[\frac{v_W}{v_W + (D_{AM}/K\delta)}\right] = v_W c_{A2}\left[\frac{v_W/v_W^0}{\frac{v_W}{v_W^0} + \frac{(D_{AM}/K\delta)}{v_W^0}}\right]$$

Since $D\dfrac{\partial c_A}{\partial y} = \dfrac{D c_A^0}{h}\dfrac{\partial C}{\partial Y}$ $\quad v_W = V_W v_W^0$ $\quad c_{A2} = C_2 c_A^0$

and α_0 and κ are given by Equations 4-91 and 4-97 respectively, Equation 4-96 follows by substitution in the above equation.

where

$$\kappa = (D_{AM}/K\delta)/v_W^0 = \theta/(1-\gamma) \tag{4-97}$$

Equation 4-96 is obtained by combining Equations 4-77 and 4-88, and Equation 4-97 is obtained by combining Equations 4-12 and 4-30. Equations 4-90 to 4-96 are identical to those given by Brian (1965a) except the boundary condition given by Equation 4-96. In this boundary condition, Brian arbitrarily set solute separation f = constant; the corresponding term $[V_W/(\kappa + V_W)]$ in Equation 4-96 is not a constant.

In order to solve the differential Equation 4-90, the velocity field must first be specified. Following Brian (1965a) again, the velocity profiles are approximated in this work by the following equations:

$$U = \frac{3}{2}\,(1-\Delta)\,(1-Y^2) \tag{4-98}$$

$$V = V_W\,(Y/2)\,(3-Y^2) \tag{4-99}$$

Equations 4-98, and 4-99, given by Brian (1965a) are approximate forms of the solutions obtained by Berman (1953) for the velocity fields for the case in which the water flux through the membrane is uniform; the above equations assume that the water permeation Reynolds number ($= hv_W/\nu$) is small, the parabolic velocity profile in the longitudinal direction is not distorted, and it is already developed at x = 0. Gill et al. (1965) have pointed out that the assumption of constant transverse velocity across the membrane along the longitudinal direction may exaggerate the build-up of salt and, consequently, underestimate the productive capacity of the system. Srinivasan et al. (1967) have made a mathematical analysis of the simultaneous development of the velocity and concentration profiles in a reverse osmosis system consisting of two parallel flat membranes. Their analysis, based on boundary layer theory, assumes 100 per cent salt separation, but takes into account the non-linear effects created by the fact that water-flux varies in the longitudinal direction. However, the boundary concentration v. X relationship as obtained by the analysis of Srinivasan et al. (1967) is essentially the same as that obtained by Brian (1965a). According to Brian (1966), the effect of a permeation flux is simply to change the average longitudinal velocity with longitudinal position, but the general parabolic velocity profile is not distorted; hence it is reasonable to assume that Equations 4-98 and 4-99 would describe the longitudinal and transverse velocity profiles respectively, adequately even for the case in which the permeation velocity varies with longitudinal position.

Equations 4-23, 4-31, 4-52, and 4-90 apply for laminar flow between flat parallel membranes for experimental conditions specified by the dimensionless parameters γ, θ, and α_0 defined by Equation 4-91; the parameter λ which includes the mass transfer coefficient, k, is not required for solving the above equations.

Some illustrative calculations for laminar feed flow between flat parallel membranes. The system of equations represented by Equations 4-6, 4-23, 4-31, 4-52, 4-90, 4-91, 4-94 to 4-99, has been solved on a digital computer using the Fortran programme of Brian (1965b) appropriately modified to suit the given boundary conditions. Figures 4-7 to 4-10 illustrate some of the results (Kimura and Sourirajan, 1968a).

Figures 4-7 to 4-10 show the variation of C_2, V_W, f and $\bar{f}$, and Δ respectively as a function of X for Film 1. The different values of α_0 correspond to differ-

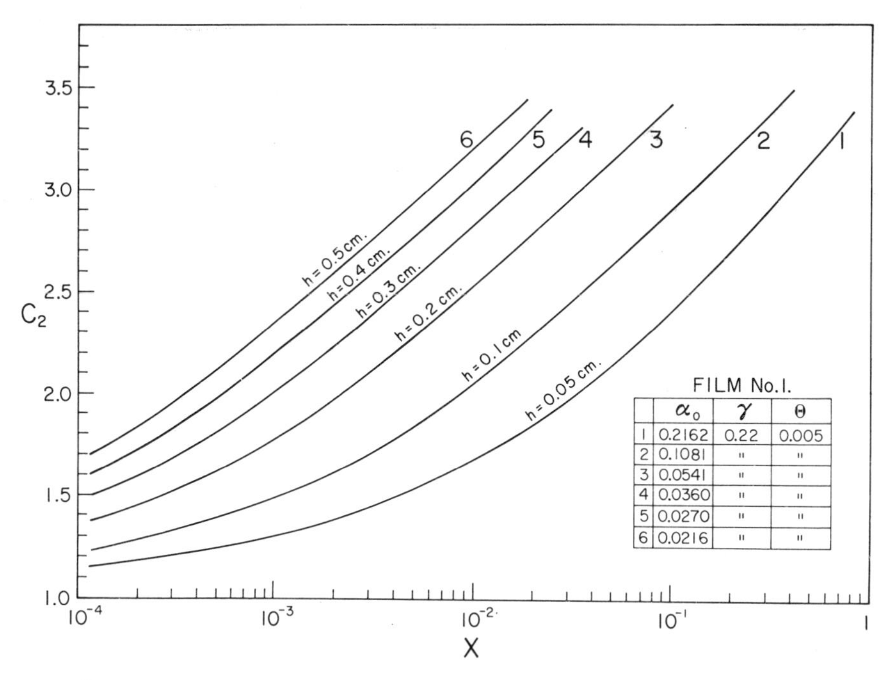

	α_0	γ	θ
1	0.2162	0.22	0.005
2	0.1081	"	"
3	0.0541	"	"
4	0.0360	"	"
5	0.0270	"	"
6	0.0216	"	"

Figure 4-7. Effect of variation of channel width on concentration polarisation v. X. Data of Kimura and Sourirajan (1968a).

System: sodium chloride-water; feed molality: 0.5M; operating pressure: 102 atm.

ent channel widths; thus $\alpha_0 = 0.2162, 0.1081, 0.0541, 0.0360, 0.0270$, and 0.0216 corresponds respectively to h = 0.05, 0.1, 0.2, 0.3, 0.4, and 0.5 cm for Film 1. The above figures show that the values of V_W, f and $\bar{f}$, and Δ decrease, and those of C_2 increase with increase in h for a given value of X.

The Fanning friction factor for isothermal laminar flow between infinite parallel planes is given by the relation (Knudsen and Katz, 1958a)

$$f^* = 24/N_{Re} \tag{4-100}$$

Substituting this value of f* in Equation 4-71, the total membrane throughput rate per unit power (F) is given by

$$F = \frac{16}{3} \frac{g_c v_W^0}{\mu_1 \nu^2} \left(\frac{h^3}{N_{Re}^2}\right)\left(\frac{\Delta}{X}\right) \tag{4-101}$$

Equation 4-101 shows that F increases with increase in h. Again one may find an optimum value for h to give the highest separation and membrane throughput rate for unit power consumption.

Films denoted by numbers 2, 3, 5, and 6 (Table 4-1) have comparatively larger surface pores than Film 1; consequently they have higher values of A and $(D_{AM}/K\delta)$ giving lower solute separation and higher product rate compared to Film 1 under any given set of experimental conditions. The concentration polarisation and its effects on the performance of the above films are illustrated

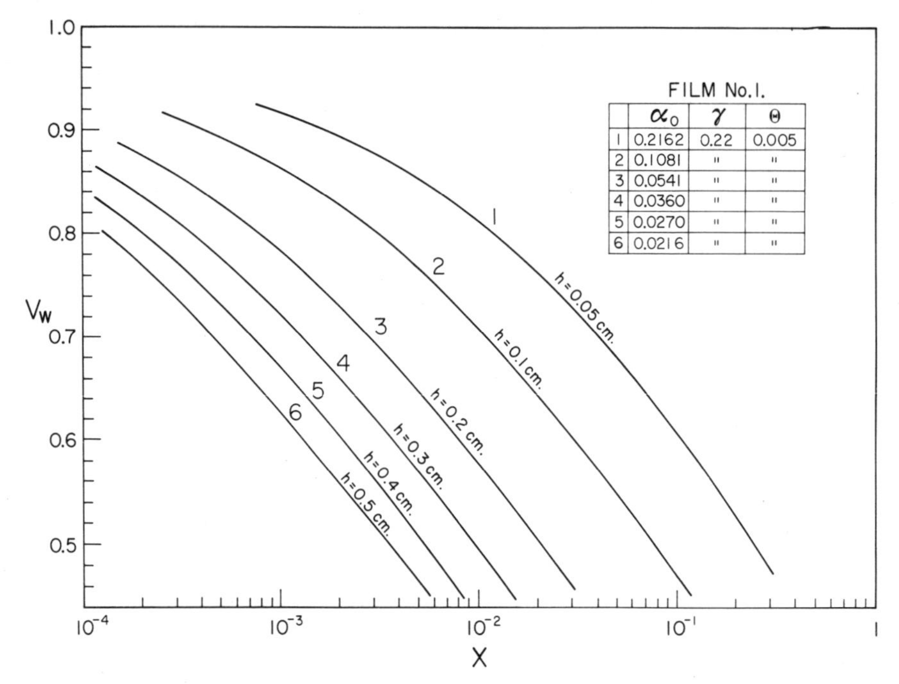

Figure 4-8. Effect of variation of channel width on V_W v. X. Data of Kimura and Sourirajan (1968a).

System: sodium chloride-water; feed molality: 0.5M; operating pressure: 102 atm.

in Figures 4-11 to 4-13. The corresponding data for Film 1 are also included in the figures for comparison. Since the values of v_W^0 (Equation 4-30) are different for different films, the variations of C_2, f and $\bar{f}$, and Δ are plotted against

$$X\left[\frac{(v_W^0)_r}{v_W^0}\right]$$

where $(v_W^0)_r$ is the value of v_W^0 for the reference Film 1. The plots are for the condition h = 0.1 cm. The figures show that the concentration build-up is larger for more porous films except at high values of X; further the fraction water withdrawn increases, and the decrease in solute separation is more steep for more porous films.

The above calculations illustrate the general applicability of the analysis for the Loeb-Sourirajan type porous cellulose acetate membranes capable of giving any level of solute separation.

An operating pressure of 102 atm was used in the above illustrations because the membrane specifications given in Table 4-1 are for this operating pressure. Similar calculations can be made for any other operating pressure provided the values of A and $(D_{AM}/K\delta)$ at that operating pressure are given. It was shown in Chapter 3 that the effect of operating pressure on A and $(D_{AM}/K\delta)$ for the system sodium chloride-water could be given by the relations

$$A \propto e^{-\alpha P} \qquad (4\text{-}102)$$

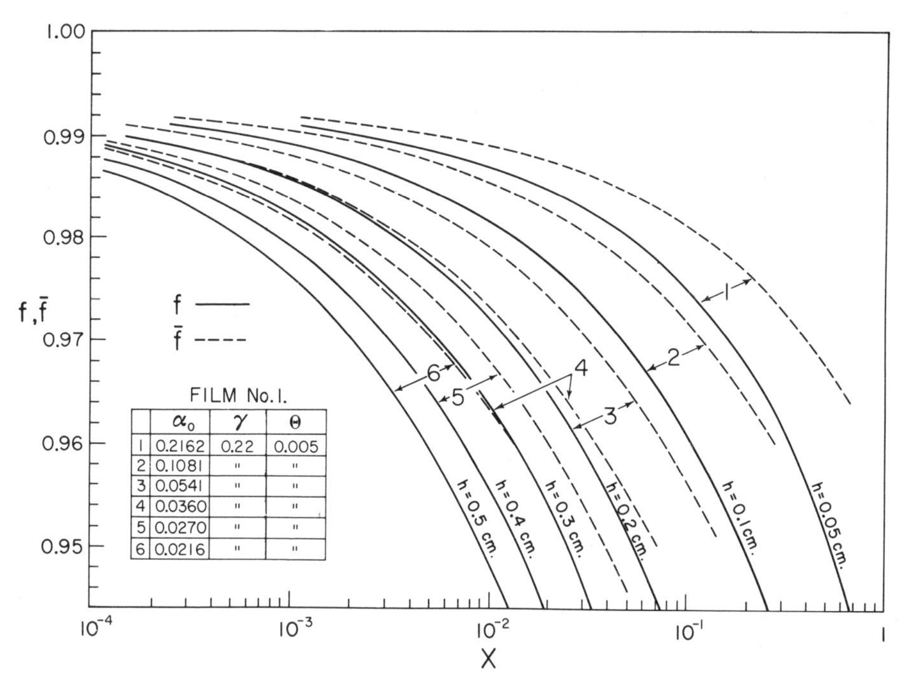

Figure 4-9. Effect of variation of channel width on solute separation v. X. Data of Kimura and Sourirajan (1968a).

System: sodium chloride-water; feed molality: 0.5M; operating pressure: 102 atm.

and

$$(D_{AM}/K\delta) \propto P^{-\beta} \tag{4-103}$$

where α and β are constants. The membrane specification, in terms of α, β, and A and $(D_{AM}/K\delta)$ at a particular operating pressure, will enable the calculation of the concentration polarisation effects in reverse osmosis for any operating pressure in the range for which the above relations are applicable.

3. Mass Transfer Coefficients For Use in Reverse Osmosis Process Design

From the foregoing illustrations, it is clear that the procedure for predicting the effect of concentration polarisation on membrane performance involves either the computer solution of a partial differential equation such as Equation 4-90 for laminar flow conditions, or the direct or indirect experimental determination of the mass transfer coefficient applicable for the particular reverse osmosis operating unit. From the point of view of reverse osmosis process design, it would be convenient if explicit expressions are available for calculating the applicable mass transfer coefficient, k, as a function of longitudinal position in the direction of feed flow; these values of k can then be used to define λ in Equation 4-27. Such expressions have been developed by Kimura and Sourirajan (1968c) for the flat and tubular membrane geometries for both

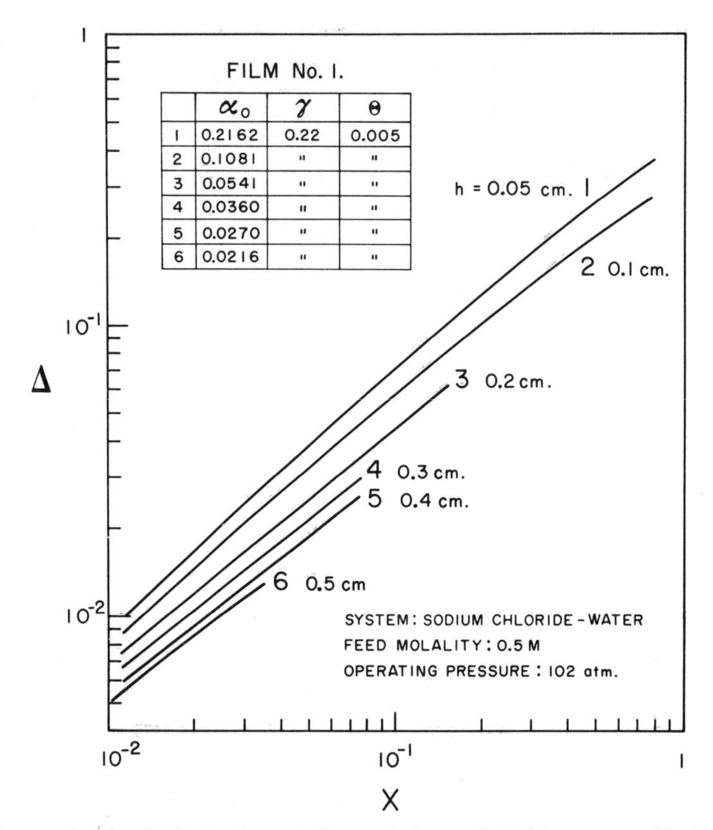

Figure 4-10. Effect of variation of channel width on Δ v. X. Data of Kimura and Sourirajan (1968a).

laminar and turbulent feed flow conditions; their results which involve correlations of the mass transfer coefficient, k, expressed in the form of Sherwood number, N_{Sh}, are discussed below.

Sherwood Numbers

Feed flow between flat parallel membranes. Let Sherwood number, N_{Sh}, for feed flow between flat parallel membranes be defined as

$$N_{Sh} = 4kh/D \tag{4-104}$$

From Equation 4-27,

$$\lambda = \frac{C_2 - C_3}{C_3 \ln\left(\dfrac{C_2 - C_3}{C_1 - C_3}\right)} \tag{4-105}$$

Using Equations 4-13, 4-91, 4-97, and 4-105, Equation 4-104 may be expressed as follows:

$$N_{Sh} = 4 \cdot \frac{k}{(D_{AM}/K\delta)} \cdot \frac{v_W^0 h}{D} \cdot \frac{(D_{AM}/K\delta)}{v_W^0} \tag{4-106}$$

$$= 4\lambda\kappa/\alpha_0 \tag{4-107}$$

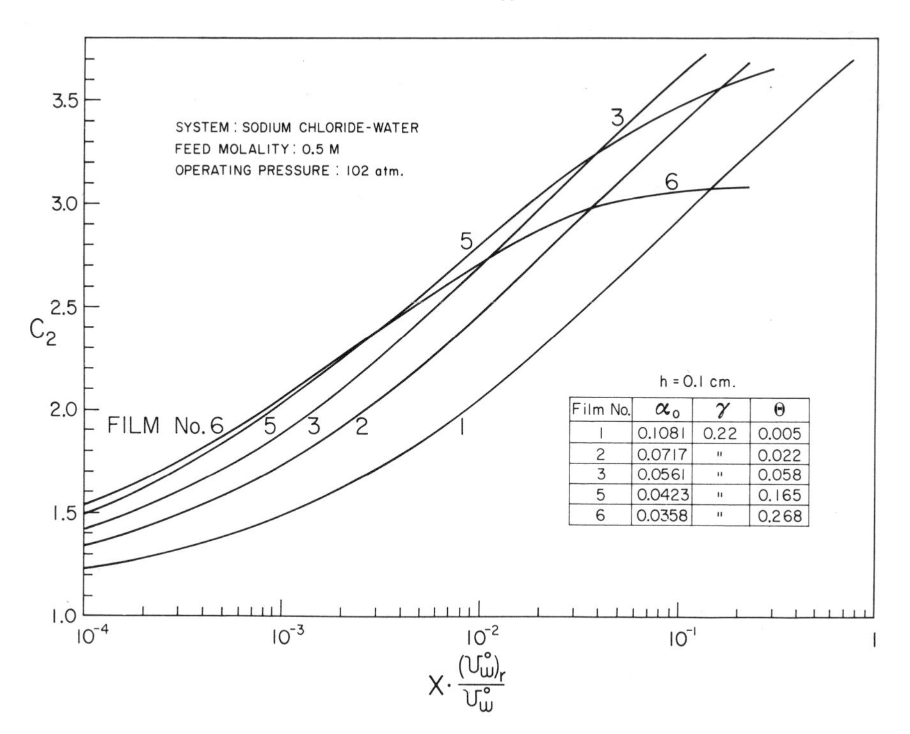

Film No.	α_0	γ	θ
1	0.1081	0.22	0.005
2	0.0717	"	0.022
3	0.0561	"	0.058
5	0.0423	"	0.165
6	0.0358	"	0.268

Figure 4-11. Concentration polarisation variations for membranes of different porosities. Data of Kimura and Sourirajan (1968a).

$$= \frac{4\kappa(C_2 - C_3)}{\alpha_0 C_3 \ln\left(\frac{C_2 - C_3}{C_1 - C_3}\right)} \tag{4-108}$$

where C_1 is given by Equation 4-52. Equation 4-108, along with Equation 4-52, is applicable for both laminar and turbulent flow between flat parallel membranes.

Feed flow in tubular membranes. Let Sherwood number, N_{Sh}, for feed flow in tubular membranes be defined as

$$N_{Sh} = 2kR/D \tag{4-109}$$

which can be expressed as

$$N_{Sh} = 2\lambda\kappa/\alpha_0 \tag{4-110}$$

$$= \frac{2\kappa(C_2 - C_3)}{\alpha_0 C_3 \ln\left(\frac{C_2 - C_3}{C_1 - C_3}\right)} \tag{4-111}$$

where C_1 is given by Equation 4-126 derived below for tubular membranes. Again, Equation 4-111, along with Equation 4-126, is applicable for both laminar and turbulent flow between flat parallel membranes.

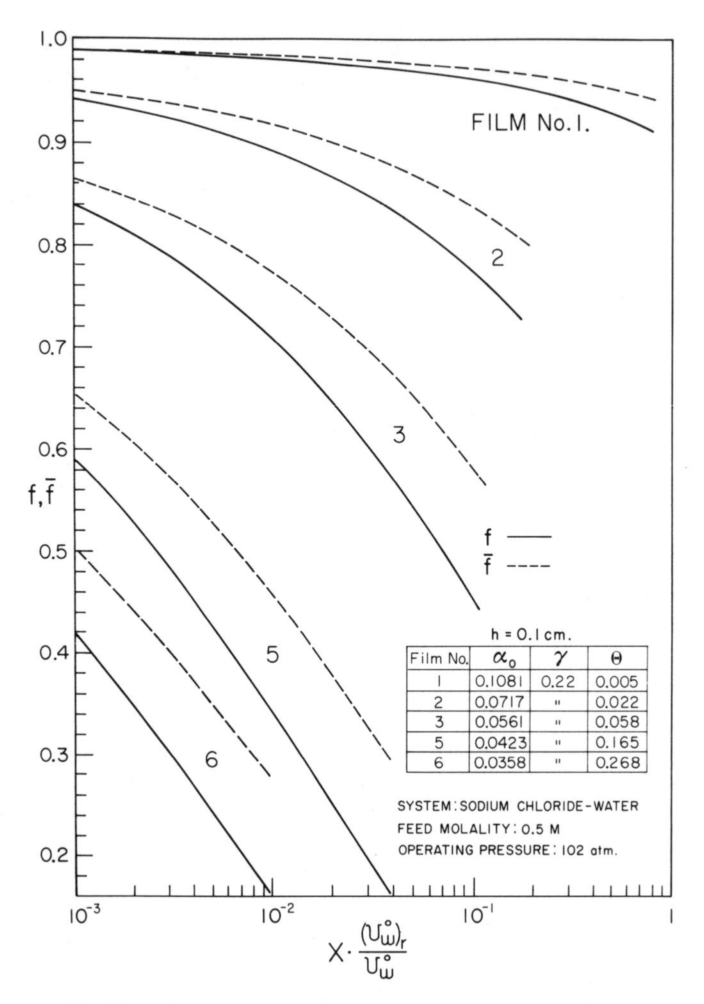

Figure 4-12. Variations in solute separation for membranes of different porosities. Data of Kimura and Sourirajan (1968a).

Expressions for volumetric fraction product recovery, and bulk solute concentration as functions of longitudinal position applicable for tubular membranes can be obtained by the method similar to that used for the flat membrane case.

Equations 4-23, 4-27, and 4-31 still apply. As before, let $\bar{u}^0$ and $\bar{u}$ represent respectively the average velocity of brine solution at membrane entrance, and at distance x along the longitudinal length of the tube in the direction of feed flow. Let the dimensionless longitudinal position be defined as

$$X = \frac{v_W^0}{\bar{u}^0} \cdot \frac{x}{R} \tag{4-112}$$

where v_W^0 is given by Equation 4-30, and R is the radius of the tubular membrane.

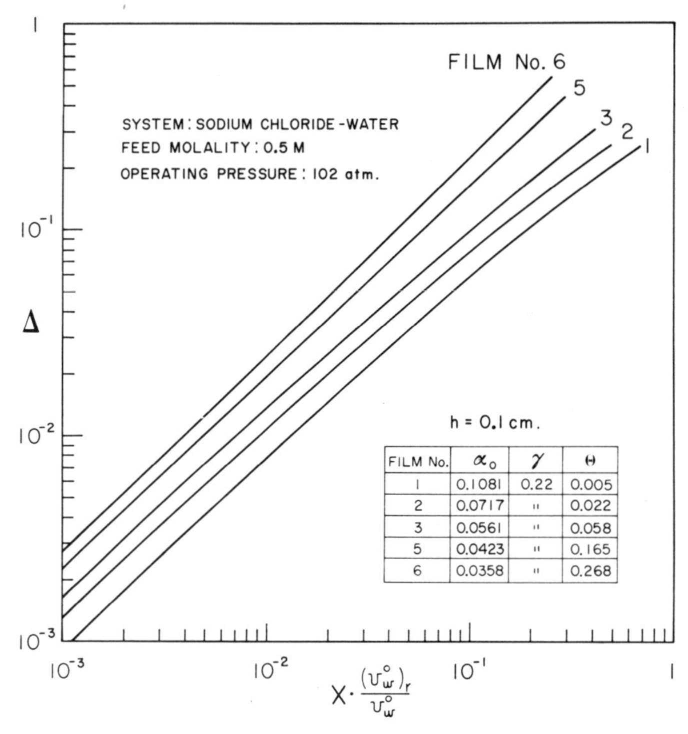

Figure 4-13. Variations in Δ for membranes of different porosities. Data of Kimura and Sourirajan (1968a).

From the water material balance,

$$\pi R^2 \bar{u} = \pi R^2 \left(\bar{u} + \frac{\partial \bar{u}}{\partial x} dx \right) + 2\pi R v_w dx \tag{4-113}$$

Therefore

$$\int_{\bar{u}^0}^{\bar{u}} - du = \int_0^x \frac{2}{R} v_w dx \tag{4-114}$$

or

$$\bar{u}^0 - \bar{u} = \int_0^x \frac{2}{R} v_w dx \tag{4-115}$$

Therefore

$$1 - \frac{\bar{u}}{\bar{u}^0} = \int_0^x 2 \cdot \frac{v_w}{v_w^0} \cdot \frac{v_w^0}{\bar{u}^0} \frac{dx}{R} \tag{4-116}$$

$$= 2 \int_0^X V_w dX' \tag{4-117}$$

$$= 2\Delta \tag{4-118}$$

since Δ is defined by Equation 4-40. Therefore, for the tubular membrane case, the volumetric fractional product recovery is given by Equation 4-118 which gives

$$\bar{u}/\bar{u}^0 = \bar{U} = 1 - 2\Delta \tag{4-119}$$

From solute material balance,

$$\pi R^2 \bar{u} c_{A1} = \pi R^2 [\bar{u} c_{A1} + \frac{\partial}{\partial x}(\bar{u} c_{A1}) dx] + 2\pi R v_W c_{A3} dx \tag{4-120}$$

Therefore

$$\int_{\bar{u}^0}^{\bar{u}} - d(\bar{u} c_{A1}) = \int_0^x \frac{2}{R} v_W c_{A3} dx \tag{4-121}$$

or

$$\bar{u}^0 c_{A1}^0 - \bar{u} c_{A1} = \int_0^x \frac{2}{R} v_W c_{A3} dx \tag{4-122}$$

Therefore

$$1 - \frac{\bar{u} c_{A1}}{\bar{u}^0 c_{A1}^0} = \int_0^x 2 \frac{c_{A3}}{c_{A1}^0} \cdot \frac{v_W}{v_W^0} \cdot \frac{v_W^0}{\bar{u}^0} \frac{dx}{R} \tag{4-123}$$

Therefore

$$1 - \bar{U} C_1 = 2 \int_0^X C_3 V_W dX' \tag{4-124}$$

Therefore

$$C_1 = \frac{1 - 2 \int_0^X C_3 V_W dX'}{\bar{U}} \tag{4-125}$$

$$= \frac{1 - 2 \int_0^X C_3 V_W dX'}{1 - 2\Delta} \tag{4-126}$$

Thus Equation 4-126 gives the dimensionless bulk solute concentration C_1 in the transverse direction, as a function of longitudinal position X for feed flow in tubular membranes.

Combining Equations 4-126 and 4-57, the average concentration of the product solution, $\bar{C}_3$, for the tubular membrane case is given by

$$\bar{C}_3 = [1 - (1 - 2\Delta) C_1]/2\Delta \tag{4-127}$$

The form of Equations 4-55 and 4-56 for solute separation applies for the tubular membrane case also.

The theoretical analysis of the laminar feed flow through tubular membranes can be done in a manner similar to that used for feed flow between flat parallel membranes. The partial differential equation describing steady state concentration field for the case of laminar flow in tubular membranes is given by (Fisher et al., 1965):

$$\frac{\partial}{\partial x}\left(uc_A - D\frac{\partial c_A}{\partial x}\right) + \frac{1}{r}\,\frac{\partial}{\partial r}\left[r\left(vc_A - D\,\frac{\partial c_A}{\partial r}\right)\right] = 0 \tag{4-128}$$

Equation 4-128 corresponds to Equation 4-90, given earlier, for the flat membrane case. Neglecting the second differential term, $D(\partial^2 c_A)/(\partial x^2)$, Equation 4-128 can be written in non-dimensional form as follows:

$$\frac{\partial}{\partial X}(UC) + \frac{VC}{Y} + \frac{\partial}{\partial Y}(VC)\,\frac{\alpha_0}{Y}\,\frac{\partial C}{\partial Y} - \alpha_0\,\frac{\partial^2 C}{\partial Y^2} = 0 \tag{4-129}$$

where

$$\alpha_0 = \frac{D}{v_W^0 R} \tag{4-130}$$

and

$$Y = r/R \tag{4-131}$$

where r is the transverse distance from the central axis of the tubular membrane. The boundary conditions for Equation 4-129 are the same as those given earlier for the flat membrane case, namely,

$$\text{at } X = 0,\ C = 1 \tag{4-94}$$

$$\text{at } Y = 0,\ \partial C/\partial Y = 0 \tag{4-95}$$

and

$$\text{at } Y = 1,\ \alpha_0(\partial C/\partial Y) = V_W C_2[V_W/(\kappa + V_W)] \tag{4-96}$$

Equation 4-129 can be solved if the velocity profiles are specified; for the purpose of this analysis, the velocity profiles are approximated by the following equations:

$$U = 2(1 - 2\Delta)(1 - Y^2) \tag{4-132}$$

and

$$V = V_W Y(2 - Y^2) \tag{4-133}$$

Equations 4-132 and 4-133 are approximate forms of those given by Yuan and Finkelstein (1956); the comments made earlier with respect to Equations 4-98 and 4-99 are also applicable to Equations 4-132 and 4-133.

Thus Equations 4-23, 4-31, 4-126, and 4-129 are applicable for laminar feed flow in tubular membranes for experimental conditions specified by the dimensionless parameters γ, θ, and α_0 defined by Equation 4-130.

4. Mass Transfer Coefficient Correlations

Laminar Flow

Laminar flow between flat parallel membranes. Kimura and Sourirajan (1968c) obtained computer solutions of the simultaneous Equations 4-23, 4-31,

4-52, and 4-90 with its boundary condition equations, for different experimental conditions defined by the parameters γ in the range 0 to 0.5, θ in the range 0 to 0.3, and α_0 (given by Equation 4-91) in the range 0.5 to 0.01, to give C_1, C_2, C_3, and V_W, and hence also N_{Sh} (given by Equation 4-108) as a function of X. All the results showed that the effect of variations in γ and θ on N_{Sh} was negligible in the calculated ranges. Consequently, the effect of water and solute permeation rate through the membrane on the mass transfer coefficient may be neglected in the range of γ and θ values studied.

The limiting case of both $\gamma = 0$, and $\theta = 0$ has been analysed by Dresner (1964), and Sherwood et al. (1965). The situation $\gamma = 0$ represents the limiting case of X^0_{A1} approaching zero, or the osmotic pressure of feed solution being negligible compared to the operating pressure; the situation $\theta = 0$ represents the limiting case of 100 per cent solute separation. From the point of view of this work, a consideration of the mass transfer coefficient at the above limiting case is of interest from the point of view of the generality of the analysis, and also for purposes of comparison of the results of the present analysis with those of the analysis of Sherwood et al. (1965).

When $\gamma = 0$, and $\theta \neq 0$, Equations 4-108 and 4-23 become respectively

$$N_{Sh} = \frac{4\theta(C_2 - C_3)}{\alpha_0 C_3 \ln\left(\dfrac{C_2 - C_3}{C_1 - C_3}\right)} \tag{4-134}$$

and

$$C_3 = C_2\theta/(1 + \theta) \tag{4-135}$$

Combining Equations 4-134 and 4-135,

$$\frac{N_{Sh}}{4} = \frac{(1 + \theta)(C_2 - C_3)}{\alpha_0 C_2 \ln\left(\dfrac{C_2 - C_3}{C_1 - C_3}\right)} \tag{4-136}$$

When $\gamma = 0$, and $\theta = 0$, C_3 becomes zero, and Equation 4-136 becomes

$$\frac{N_{Sh}}{4} = \frac{1}{\alpha_0 \ln (C_2/C_1)} \tag{4-137}$$

For the above case, according to Dresner (1964), Fisher et al. (1965), and Sherwood et al. (1965), the quantity (C_2/C_1) can be approximated as follows for small values of α_0. In the channel entrance (boundary layer) region

$$\frac{C_2}{C_1} = \begin{cases} 1 + 1.536\,(\xi_F)^{1/3} & \xi_F \lesssim 0.02 \\ 1 + \xi_F + 5[1 - \exp(-\sqrt{\xi_F/3})], & \xi_F \gtrsim 0.02 \end{cases} \tag{4-138}$$

where

$$\xi_F = \frac{X}{3\alpha_0^{\,2}} \tag{4-139}$$

and in the far downstream asymptotic (constant concentration profile) region

$$C_2/C_1 = 1 + \frac{1}{3\alpha_0^2} \tag{4-140}$$

Combining Equations 4-137, 4-138, and 4-140, in the channel entrance region

$$\frac{\alpha_0 N_{Sh}}{4} = \begin{cases} \dfrac{1}{\ln\left[1 + 1.536\,(\xi_F)^{1/3}\right]}, & \xi_F \lesssim 0.02 \\ \dfrac{1}{\ln\left[1 + \xi_F + 5\left\{1 - \exp\left(-\sqrt{\xi_F/3}\right)\right\}\right]}, & \xi_F \gtrsim 0.02 \end{cases} \tag{4-141}$$

and in the far downstream asymptotic region

$$\alpha_0 N_{Sh}/4 = 1/\ln\left[1 + \frac{1}{3\alpha_0^2}\right] \tag{4-142}$$

The values of N_{Sh} obtained by the simultaneous solution of Equations 4-23, 4-31, 4-52, and 4-90 for different experimental conditions are plotted in Figure 4-14 in the form suggested by Equations 4-141 and 4-142, i.e. $\alpha_0 N_{Sh}/4$ v. ξ_F. For the sake of convenience, only the results for the cases $\gamma = 0$ and $\theta = 0$, and $\gamma = 0.5$ and $\theta = 0.3$ are plotted in Figure 4-14 for different values of α_0; the rest of the results are in between the values plotted. For purposes of comparison, the results obtained from Equations 4-141 and 4-142 are also given in Figure 4-14 (line I and the asymptotic lines). The results show that line I and the solutions of the system of Equations 4-23, 4-31, 4-52, and 4-90 for the γ, θ, and α_0 values considered are reasonably close to each other. Further, the latter solutions can be well approximated by a straight line represented by the equation

$$\frac{\alpha_0 N_{Sh}}{4} = 0.65\,(\xi_F)^{-1/3} \tag{4-143}$$

The form of Equation 4-143 can also be obtained from Leveque solution, given by Equation 13-13 of Knudsen and Katz (1958f), which can be written as follows for the mass transfer case:

$$\frac{kx}{D} = \frac{x}{0.893}\left(\frac{b_F}{9Dx}\right)^{1/3} \tag{4-144}$$

where b_F is the velocity gradient at the wall. Multiplying both the sides of Equation 4-144 by (h/x),

$$\frac{kh}{D} = \frac{h}{0.893}\left(\frac{b_F}{9Dx}\right)^{1/3} \tag{4-145}$$

which can be written as

$$\frac{N_{Sh}}{4} = \frac{h}{0.893}\left(\frac{b_F}{9Dx}\right)^{1/3} \tag{4-146}$$

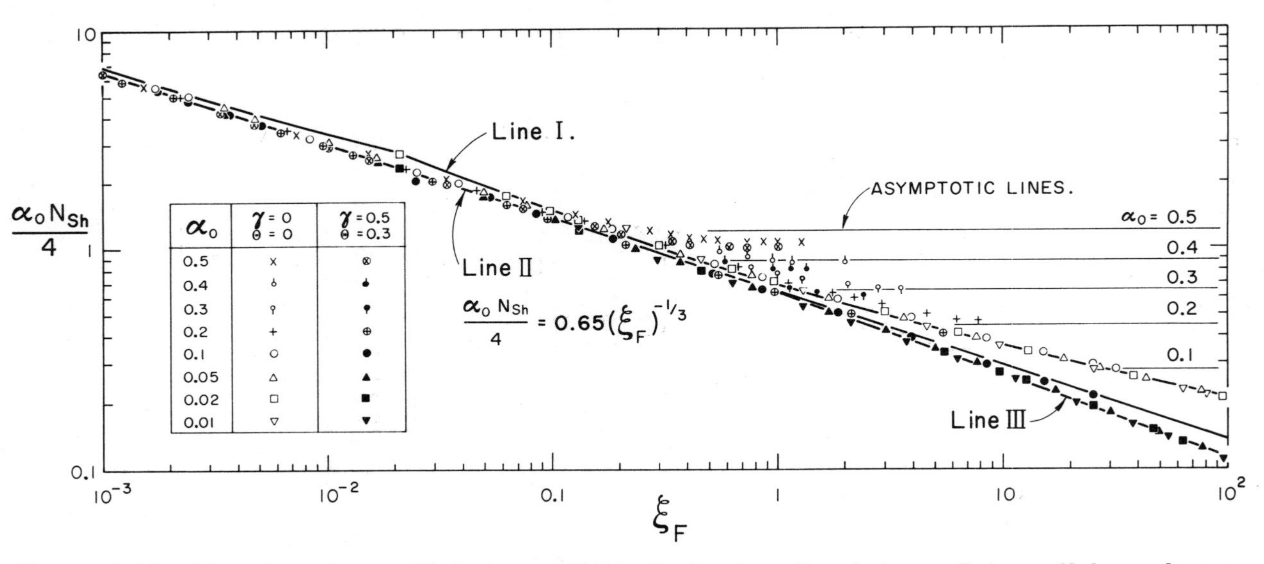

Figure 4-14. Mass transfer coefficient correlation for laminar flow between flat parallel membranes. Data of Kimura and Sourirajan (1968c).

The velocity gradient, b_F, may be expressed as

$$b_F = -(\partial u/\partial y)_{y=h} \quad (4\text{-}147)$$

Since

$$y = Yh \quad (4\text{-}92)$$

and

$$u = \bar{u}^0 U \quad (4\text{-}41)$$

and, from Equation 4-98,

$$u = \bar{u}^0 U = \frac{3}{2}\bar{u}\,(1 - Y^2) \quad (4\text{-}148)$$

$$b_F = -\frac{1}{h}\left(\frac{\partial u}{\partial Y}\right)_{Y=1} \quad (4\text{-}149)$$

$$= -\frac{1}{h}\left[\frac{\partial}{\partial Y}\left\{\frac{3}{2}\bar{u}\,(1 - Y^2)\right\}\right]_{Y=1} \quad (4\text{-}150)$$

$$= 3\bar{u}/h \quad (4\text{-}151)$$

Assuming $\bar{u}$ may be approximated by $\bar{u}^0$, Equation 4-151 can be written as

$$b_F = 3\bar{u}^0/h \quad (4\text{-}152)$$

Combining Equations 4-146, 4-152, and 4-91,

$$\alpha_0 N_{Sh}/4 = 0.54(\xi_F)^{-1/3} \quad (4\text{-}153)$$

Comparing Equations 4-143 and 4-153, even though the constant terms are slightly different, it is evident that Equation 4-143 has the same basis as the Leveque solution.

Considering Figure 4-14 again, the correlation represented by Equation 4-143 seems excellent up to $\xi_F = 1$. As ξ_F becomes larger than 1, discrepancy between line I and the line corresponding to the case $\gamma = 0.5$ and $\theta = 0.3$ (line III) becomes large; and the solutions of the partial differential equation scatter around line II depending on the values of γ and θ. There seems no simple way to express $\alpha_0 N_{Sh}$ as a function of γ and θ precisely. Consequently, in the following trial and error calculations involving Figure 4-14, the correlation represented by line II was considered sufficiently representative for the purpose.

Regarding the asymptotic region where the concentration profile no longer changes, $(\alpha_0 N_{Sh}/4)$ reaches a constant value depending on α_0; Equation 4-142 represents the asymptotic values reasonably well for the range of α_0 values considered.

Laminar flow in tubular membranes. Fisher et al. (1965) solved the partial differential equation 4-129 for the case when both γ and θ equal zero, and found that the results could be expressed by equations analogous to Equations 4-138 and 4-140 where ξ_F is replaced by ξ_T given by the equation

$$\xi_T = \frac{X}{4\alpha_0^2} \tag{4-154}$$

where X and α_0 are given by the Equations 4-112 and 4-130 respectively. This result is based on the consideration that the velocity gradient, b_T, at the wall of the tubular membrane is given by the equation

$$b_T = 4\bar{u}/R \qquad \text{(4-155)*}$$

For the tubular membrane case, the Leveque solution can be written as

$$\frac{kR}{D} = \frac{R}{0.893}\left(\frac{b_T}{9Dx}\right)^{1/3} \tag{4-156}$$

Equation 4-156 is analogous to Equation 4-145 given for the flat membrane case. When b_T is approximated by the equation

$$b_T = 4\bar{u}^0/R \tag{4-157}$$

and combined with Equations 4-156, 4-109, 4-112, and 4-130, the Leveque solution for flow in tubular membrane can be obtained as

$$\alpha_0 N_{Sh}/2 = 0.54\,(\xi_T)^{-1/3} \tag{4-158}$$

which corresponds to Equation 4-153 given for the flat membrane case.

---

* Equation 4-155 is obtained as follows:

$$b_T = -(\partial u/\partial r)_{r=R}$$

$$= -\frac{1}{R}\left(\frac{\partial u}{\partial Y}\right)_{Y=1} \qquad \text{since } r = RY \tag{4-131}$$

$$u = \bar{u}^0 \cdot U \tag{4-41}$$

$$= \bar{u}^0\, 2(1 - 2\Delta)(1 - Y^2) \qquad \text{(from Equation 4-132)}$$

$$= \bar{u}^0 \cdot 2 \cdot \frac{\bar{u}}{\bar{u}^0}(1 - Y^2) \qquad \text{(from Equation 4-119)}$$

$$= 2\,\bar{u}\,(1 - Y^2)$$

Therefore

$$(\partial u/\partial Y)_{Y=1} = -\,4\bar{u}$$

or

$$b_T = -\frac{1}{R}\left(\frac{\partial u}{\partial Y}\right)_{Y=1} = \frac{4\bar{u}}{R} \tag{4-155}$$

The finite difference solutions of the partial differential equation 4-129 were correlated with the parameter $(\alpha_0 N_{Sh}/2)$ v. ξ_T for the tubular membrane case; it was found that the correlation was identical with that given in Figure 4-14 for the flat membrane case. This is illustrated in Figure 4-15. Hence the correlation (corresponding to Equation 4-143) for the tubular membrane case is

$$\alpha_0 N_{Sh}/2 = 0.65\,(\xi_T)^{-1/3} \tag{4-159}$$

Based on Equation 4-154, the asymptotic values for the tubular membrane case may be given by the equation

$$\alpha_0 N_{Sh}/2 = \frac{1}{\ln\left[1 + \dfrac{1}{4\alpha_0^{\,2}}\right]} \tag{4-160}$$

Comparison of results for laminar feed flow using local and average mass transfer coefficients. Kimura and Sourirajan (1968c) compared the calculated results on membrane performance using local and average mass transfer coefficients. For this purpose the γ and θ values corresponding to two particular membranes designated earlier (Chapter 3) as Films 1 and F1 used with aqueous sodium chloride feed solutions were studied. Film 1 ($\gamma = 0.22$ and $\theta = 0.005$) was used in conjunction with a 0.5M feed solution at an operating pressure of 1500 psig; Film F1 ($\gamma = 0.05$ and $\theta = 0.07$) was used in conjunction with a feed solution containing 2500 ppm of NaCl at an operating pressure of 600 psig. The α_0 values considered for Film 1 were 0.5, 0.2, 0.1, 0.05, and 0.02 which correspond approximately to h or R values of 0.02, 0.05, 0.1, 0.2, and 0.5 cm, respectively. The α_0 values considered for Film F1 were 0.3, 0.2, 0.1, 0.06, and 0.01 which correspond approximately to h or R values of 0.02, 0.03, 0.06, 0.1, and 0.6 cm, respectively.

Values of $\bar{f}$, Δ (or 2Δ for the tubular membrane case), and C_2 were obtained solving the appropriate partial differential equations and compared with those obtained by the trial and error method using Equations 4-23, 4-27, 4-31, 4-32, 4-42, 4-52, and 4-59 for the flat membrane case, and Equations 4-23, 4-27, 4-31, 4-112, 4-118, 4-126, and 4-127 for the tubular membrane case. The local mass transfer coefficients used in the latter trial and error solutions were obtained from the equations given already (Equations 4-142, 4-143, 4-159, and 4-160) which can be expressed in a general form as follows:

$$\frac{\alpha_0 N_{Sh}}{Z_2} = \begin{cases} 0.65\,(\xi_i)^{-1/3}, & \xi_i < \xi_{AS} \\ \dfrac{1}{\ln\left[1 + \dfrac{1}{Z_1\alpha_0^{\,2}}\right]}, & \xi_i \geqslant \xi_{AS} \end{cases} \tag{4-161}$$

where $i \equiv F$, $Z_1 \equiv 3$, and $Z_2 \equiv 4$ for the flat membrane case, and $i \equiv T$, $Z_1 \equiv 4$, and $Z_2 \equiv 2$ for the tubular membrane case, and

$$\xi_{AS} = \left\{0.65\,\ln\left(1 + \frac{1}{Z_1\alpha_0^{\,2}}\right)\right\}^3 \tag{4-162}$$

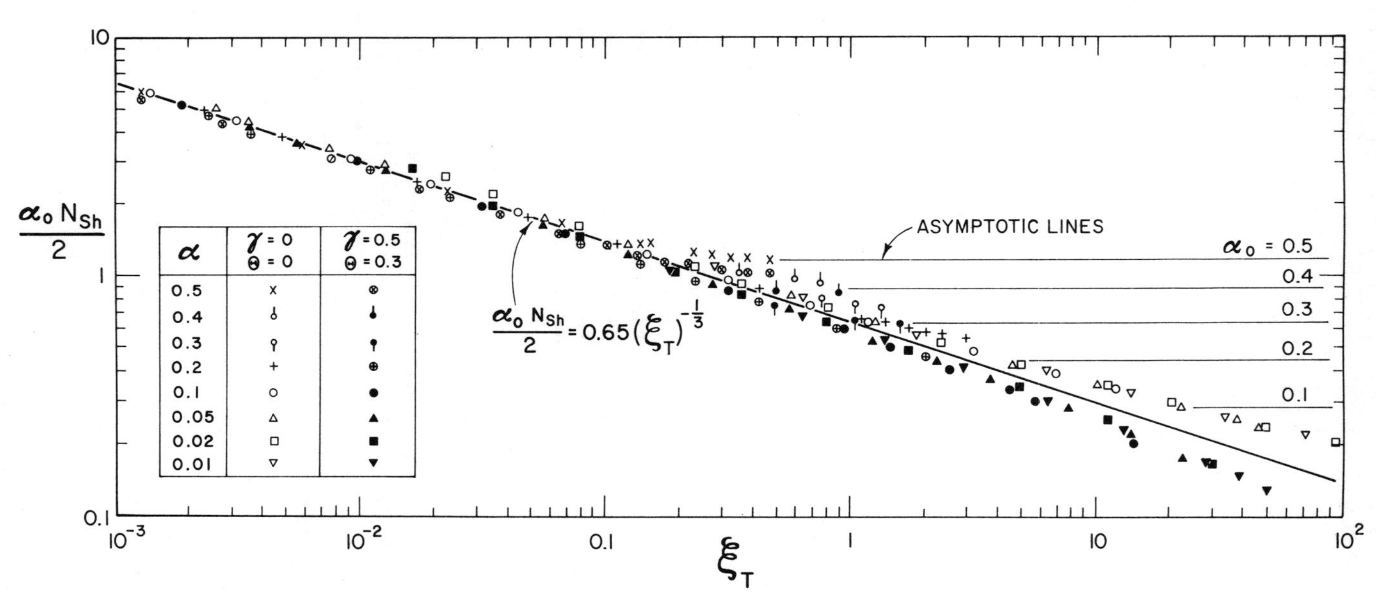

Figure 4-15. Mass transfer coefficient correlation for laminar flow in tubular membranes. Data of Kimura and Sourirajan.

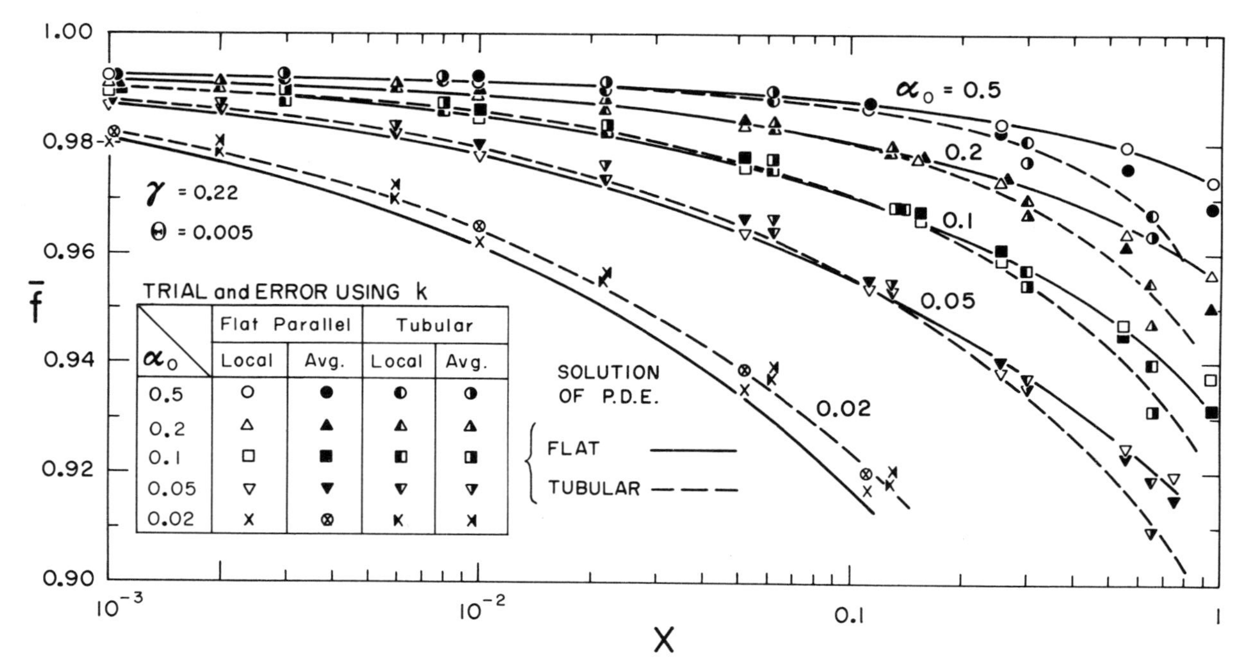

Figure 4-16. Comparison of the use of average and local mass transfer coefficients for calculating solute separation v. X. Data of Kimura and Sourirajan (1968c).

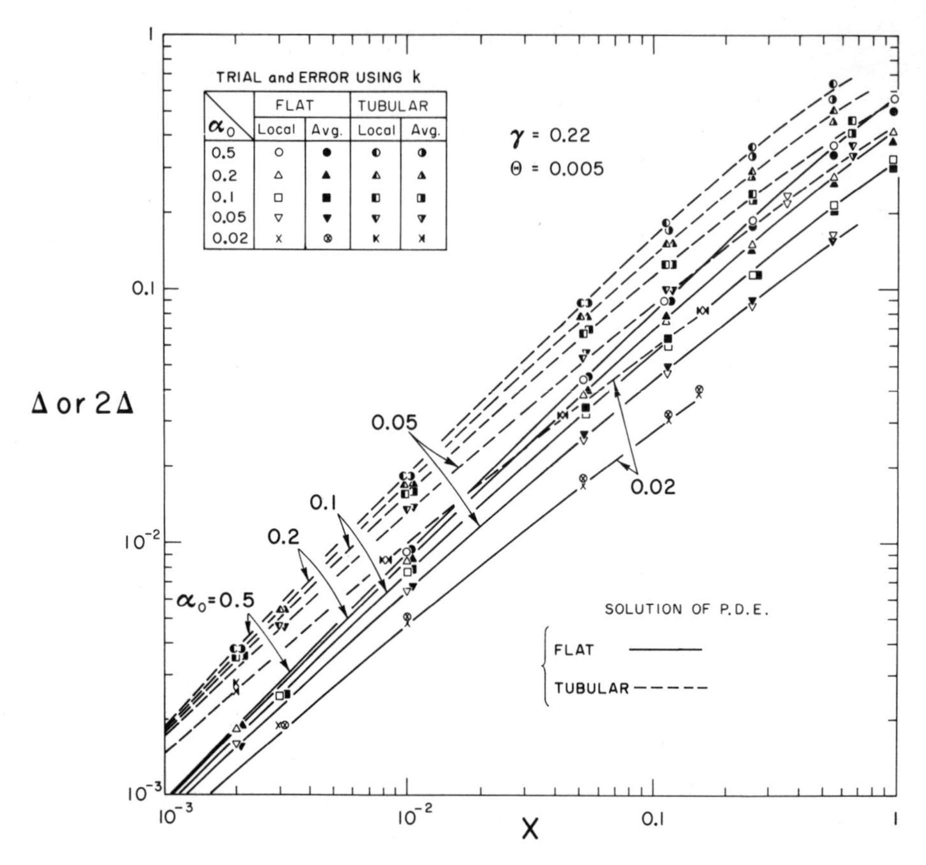

Figure 4-17. Comparison of the use of average and local mass transfer coefficients for calculating product recovery v. X. Data of Kimura and Sourirajan (1968c).

which represents the ξ value in the asymptotic region. The results obtained in the above calculations are plotted in Figures 4-16 to 4-21.

In the trial and error calculation, increments in X were first set equal to those in the finite difference solution of the partial differential equation where the increment started from 10^{-6} and increased logarithmically up to 0.1. This increment was later increased 10^3 times, and still the results remained essentially the same. All the results shown in Figures 4-16 to 4-21 are based on the latter increment.

It is particularly desirable to know whether one can use average mass transfer coefficients (instead of local values given by Equation 4-161) for predicting Δ (or 2Δ) and $\overline{C}_3$ which are the two important quantities needed to calculate product rate and average solute separation, $\bar{f}$, given by Equation 4-56. The use of the average mass transfer coefficient is particularly convenient for process design since it involves only one trial and error procedure to calculate Δ and $\overline{C}_3$. The average mass transfer coefficient can be obtained by taking average values of $(\alpha_0 N_{Sh}/4)$ for the flat membrane case and $(\alpha_0 N_{Sh}/2)$ for the tubular membrane case over the entire ξ_F and ξ_T respectively. For example, for the

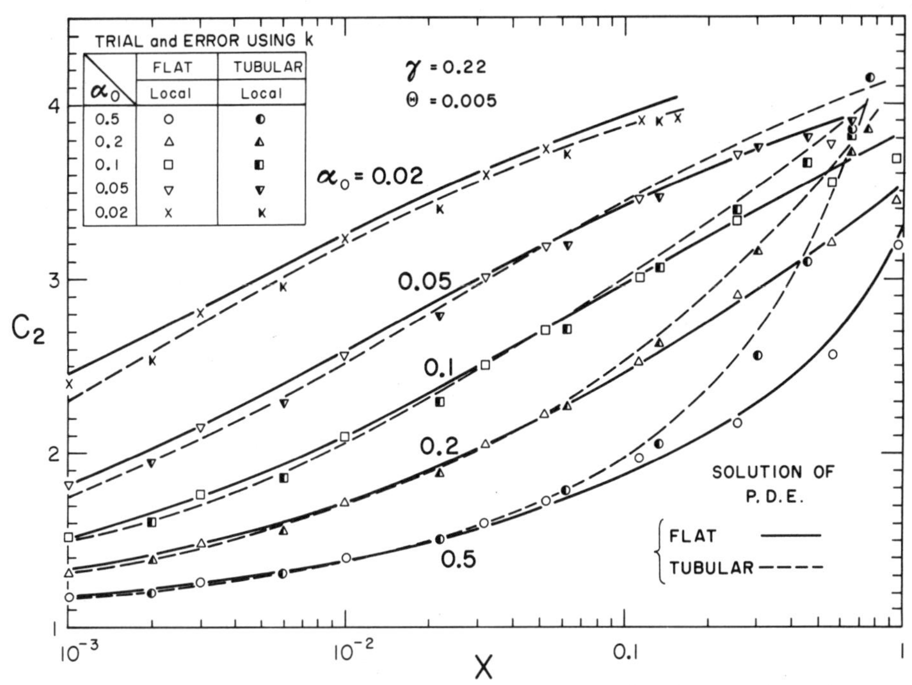

Figure 4-18. Comparison of the use of average and local mass transfer coefficients for calculating concentration polarisation v. X. Data of Kimura and Sourirajan (1968c).

flat membrane case, in the channel entrance region, the average value of $(\alpha_0 N_{Sh}/4)$ is given by

$$\frac{\alpha_0 \overline{N}_{Sh}}{4} = \frac{1}{\xi_F} \int_0^{\xi_F} \left(\alpha_0 N_{Sh}/4\right) d\xi' \tag{4-163}$$

$$= \frac{1}{\xi_F} \int_0^{\xi_F} 0.65\ (\xi_F)^{-1/3}\ d\xi' \tag{4-164}$$

$$= (0.65) \cdot \left(\frac{3}{2}\right) \cdot \xi_F^{-1/3} \tag{4-165}$$

$$= 0.975\ \xi_F^{-1/3} \tag{4-166}$$

Similarly, in the asymptotic region,

$$\frac{\alpha_0 \overline{N}_{Sh}}{4} = \left[\int_0^{\xi_{AS}} 0.65\ (\xi_F)^{-1/3}\ d\xi' + \int_{\xi_{AS}}^{\xi_F} \frac{1}{\ln\left(1 + \frac{1}{3\alpha_0^2}\right)}\ d\xi' \right] \frac{1}{\xi_F} \tag{4-167}$$

$$= \left\{ 0.975\ (\xi_{AS})^{2/3} + \left[(\xi_F - \xi_{AS}) / \ln\left(1 + \frac{1}{3\alpha_0^2}\right) \right] \right\} \frac{1}{\xi_F} \tag{4-168}$$

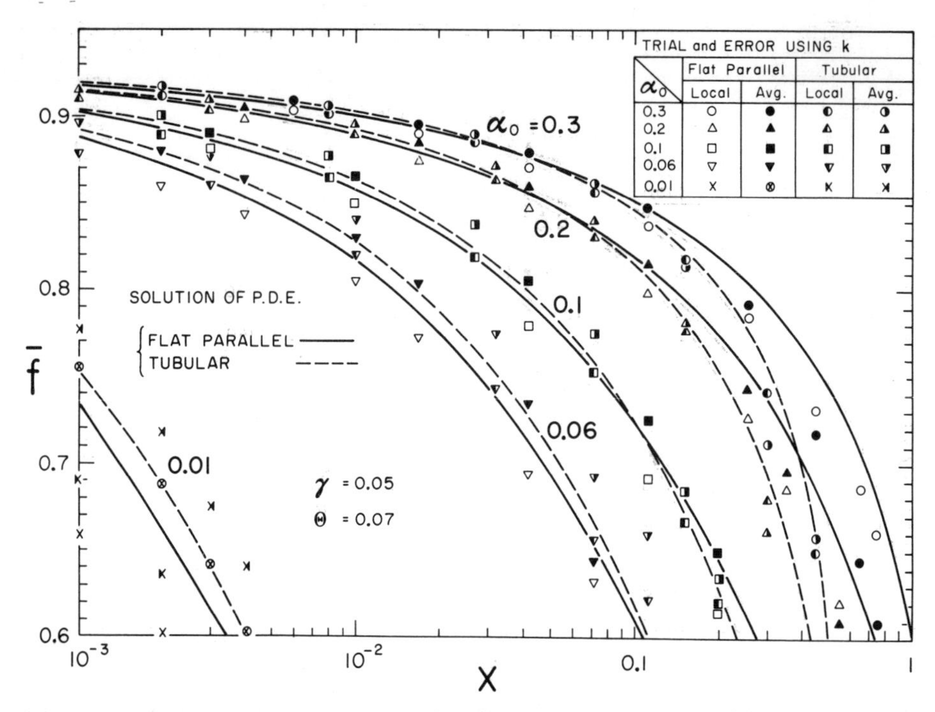

Figure 4-19. Comparison of the use of average and local mass transfer coefficients for calculating solute separation v. X. Data of Kimura and Sourirajan (1968c).

The corresponding expressions for the average mass transfer coefficient for the tubular membrane case are similar. Hence the general expressions for the average mass transfer coefficients can be given as

$$\frac{\alpha_0 \overline{N}_{Sh}}{Z_2} = \begin{cases} 0.975\,(\xi_i)^{-1/3}, & \xi_i < \xi_{AS} \\ \left\{0.975\,(\xi_{AS})^{2/3} + \left[(\xi_i - \xi_{AS})/\ln\left(1 + \dfrac{1}{Z_1 \alpha_0^2}\right)\right]\right\}\dfrac{1}{\xi_i}, & \xi_i \geqslant \xi_{AS} \end{cases} \quad (4\text{-}169)$$

In Equations 4-169, the symbols i, Z_1, Z_2, and ξ_{AS} have the same meaning as that given in Equations 4-161 and 4-162. Using Equation 4-169, an average value of λ can be obtained for a given value of X, which can then be used for the trial and error solution to obtain $\overline{C}_1$, $\overline{C}_2$, $\overline{C}_3$, and $\overline{V}_W$, in which case

$$\Delta = \overline{V}_W X \quad (4\text{-}170)$$

Figure 4-16 to 4-21 give a comparison of the results obtained from the solutions of the partial differential equations with those obtained by the 'trial and error' method using local and average mass transfer coefficients both for the flat and tubular membrane cases. Figures 4-16, 4-17, and 4-18 give $\bar{f}$, Δ (or 2Δ) and C_2 v. X for Film 1, and Figures 4-19, 4-20, and 4-21 give similar results for Film F1. In Figures 4-18 and 4-21, C_2 v. X are given using only the local mass transfer coefficients.

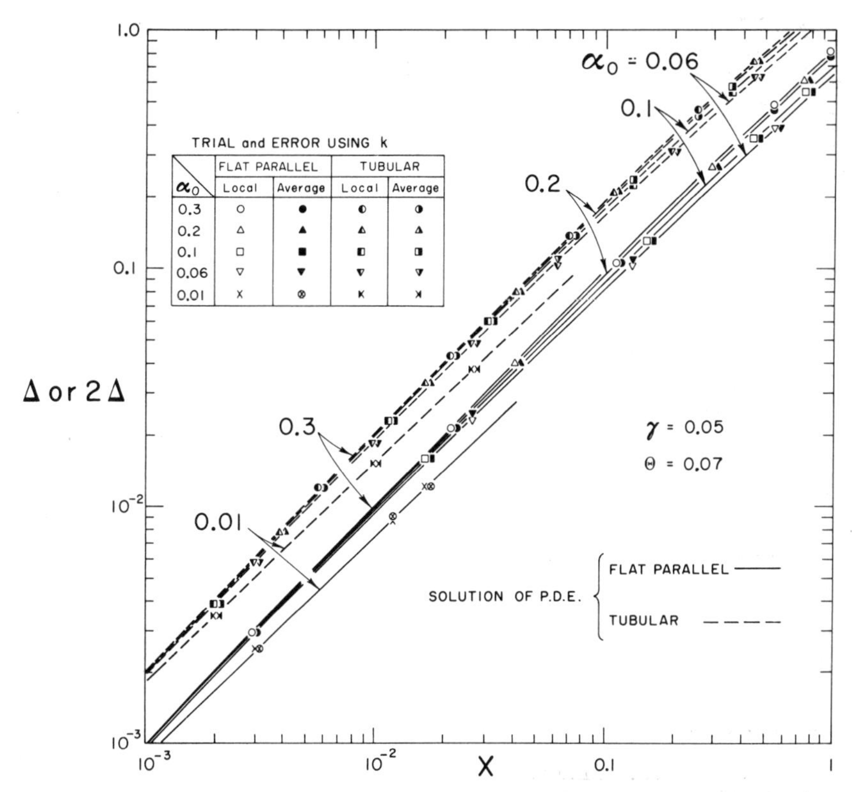

Figure 4-20. Comparison of the use of average and local mass transfer coefficients for calculating product recovery v. X. Data of Kimura and Sourirajan (1968c).

There is good agreement among the results obtained by the solution of the appropriate partial differential equations, and by the trial and error method using the local or average mass transfer coefficients for all values of ξ_F or ξ_T up to 1; the results also show that average mass transfer coefficients calculated from Equation 4-169 can be used with good accuracy for design purposes in the trial and error calculations. When α_0 is as small as 0.01, the deviations among results (Figure 4-19) become significant, but this situation corresponds to the case $\xi > 1$ when Equations 4-161 and 4-169 are not very accurate. Figure 4-21 shows that when α_0 is small C_2 becomes very large, and hence it is desirable to operate at lower values of X which also means lower values of ξ where the agreement is much better. In general, when $\xi > 1$, $\gamma < 0.1$, and $\theta < 0.05$, it seems prudent to use Equation 4-141 instead of Equation 4-143 since, under such conditions, the mass transfer correlation approaches more nearly to line 1 in Figure 4-14.

Turbulent Flow

Turbulent flow in tubular membranes. Even though a fluid may enter the tubular (or the flat parallel) membrane in fully developed turbulent flow, concentration polarisation develops from the entrance edge of the membrane. Consequently the entrance effect in the turbulent flow case is as important as in the laminar flow case.

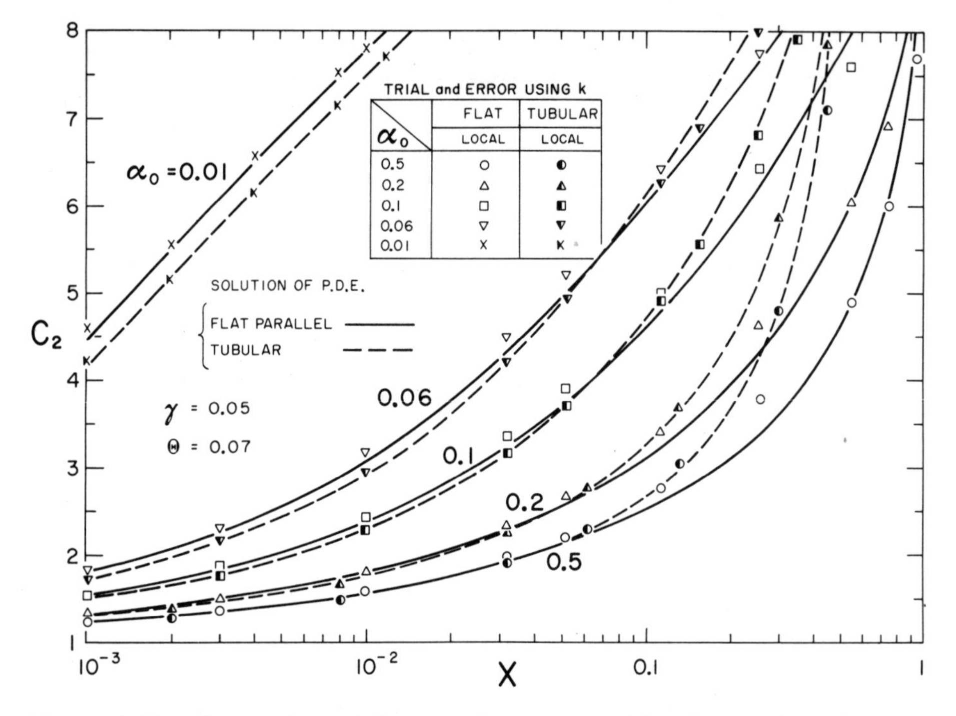

Figure 4-21. Comparison of the use of average and local mass transfer coefficients for calculating concentration polarisation v. X. Data of Kimura and Sourirajan (1968c).

Shaw et al. (1963) presented data from diffusion current experiments on mass transfer coefficients in pipe line flow where the Reynolds number (N_{Re}) was varied from 1000 to 75, 000, and the Schmidt number (N_{Sc}) was constant at about 2400. They used 10 mass transfer sections with lengths of from 0.0177 to 4.31 times diameter. Their results showed that the average Stanton number ($\overline{N}_{St}$) in the pipe entry region can be expressed by the following equation derived by the method of Linton and Sherwood (1950).

$$\overline{N}_{St} = 0.276\,(N_{Re})^{-0.42}\,(N_{Sc})^{-2/3}\,(x/d)^{-1/3} \tag{4-171}$$

Equation 4-171 can be transformed to the one similar to Equation 4-169 as follows:

$$\overline{N}_{St} = \bar{k}/\bar{u}^0 \tag{4-172}$$

$$\overline{N}_{Sh} = \bar{k}d/D = \bar{k}/\bar{u}^0 \cdot \bar{u}^0 d/\nu \cdot \nu/D = \overline{N}_{St} \cdot N_{Re} \cdot N_{Sc} \tag{4-173}$$

$$= 0.276\,(N_{Re})^{0.58}\,(N_{Sc})^{1/3}\,(x/d)^{-1/3} \tag{4-174}$$

$$= 0.276\,(N_{Re})^{0.58}\left(\frac{\nu}{D} \cdot \frac{d}{x}\right)^{1/3} \tag{4-175}$$

$$\alpha_0 \overline{N}_{Sh} = 0.276\,(N_{Re})^{0.58}\left(\alpha_0{}^2 \cdot \frac{\bar{u}^0 R}{v_W^0 x} \cdot \frac{\nu}{\bar{u}^0 d} \cdot \frac{d^2}{R^2}\right)^{1/3} \tag{4-176}$$

$$= 0.276\,(N_{Re})^{0.58}\left(\alpha_0^{2}\cdot\frac{1}{X}\cdot\frac{1}{N_{Re}}\cdot 4\right)^{1/3} \quad (4\text{-}177)$$

$$= 0.276\,(N_{Re})^{0.25}\,(\xi_T)^{-1/3} \quad (4\text{-}178)$$

Therefore

$$\alpha_0\overline{N}_{Sh}/2 = 0.138\,(N_{Re})^{0.25}\,(\xi_T)^{-1/3} \quad (4\text{-}179)$$

Also, since

$$\alpha_0\overline{N}_{Sh}/2 = \frac{1}{\xi_T}\int_0^{\xi_T}\frac{\alpha_0 N_{Sh}}{2}\,d\xi'_T \quad (4\text{-}180)$$

$$\alpha_0 N_{Sh}/2 = 0.092\,(N_{Re})^{0.25}\,(\xi_T)^{-1/3} \quad (4\text{-}181)$$

Thus Equations 4-179 and 4-181 can be used to calculate average and local mass transfer coefficients respectively for turbulent flow in the tube entry region.

Turbulent flow between flat parallel membranes. The Leveque solution (Equation 4-146) with the appropriate expression for the velocity gradient at the wall, b_F, offers a means of analysing this case. An expression for b_F can be obtained as follows.

Equation 13-3 of Knudsen and Katz (1958e) can be written for the mass transfer case as follows:

$$b_F y(\partial c/\partial x) = D(\partial^2 c/\partial y^2) \quad (4\text{-}182)$$

where $\partial c/\partial x$ and $\partial^2 c/\partial y^2$ are the appropriate concentration gradients. The following equation of Linton and Sherwood (1950) also provides a suitable description of the mass transfer process for short transfer sections and large Schmidt numbers (Shaw et al., 1963):

$$\frac{(u^*)^2}{\nu}\,y\,\frac{\partial c}{\partial x} = D\,\frac{\partial^2 c}{\partial y^2} \quad (4\text{-}183)$$

where u^* is the friction velocity. Comparing Equations 4-182 and 4-183,

$$b_F = (u^*)^2/\nu \quad (4\text{-}184)$$

Using the following relations for friction velocity (u^*), friction factor (f^*), and Reynolds number (N_{Re}), (Knudsen and Katz, 1958b, 1958c, 1958d),

$$u^* = \bar{u}^0\sqrt{\frac{f^*}{2}} \quad (4\text{-}185)$$

$$f^* = 0.079\,(N_{Re})^{-1/4} \quad (4\text{-}186)$$

$$(N_{Re}) = 4h\bar{u}^0/\nu \quad (4\text{-}61)$$

the velocity gradient, b_F, can be given as

$$b_F = (\bar{u}^0)^2\,(0.079)\,(N_{Re})^{-1/4}/2\nu \quad (4\text{-}187)$$

Equation 4-146 then becomes

$$\frac{N_{Sh}}{4} = \frac{h}{0.893}\left[\frac{(\bar{u}^0)^2\,(0.079)\,(N_{Re})^{-1/4}}{2\nu \cdot 9Dx}\right]^{1/3} \tag{4-188}$$

Therefore

$$\frac{\alpha_0 N_{Sh}}{4} = \frac{(0.079)^{1/3}}{(0.893)\,(2 \times 9 \times 12)^{1/3}}\,(\xi_F)^{-1/3}\,(N_{Re})^{1/4} \tag{4-189}$$

$$= 0.08\,(N_{Re})^{1/4}\,(\xi_F)^{-1/3} \tag{4-190}$$

The average value of $(\alpha_0 N_{Sh}/4)$ can be expressed as

$$\frac{\alpha_0 \overline{N}_{Sh}}{4} = \frac{3}{2}\left(\frac{\alpha_0 N_{Sh}}{4}\right) \tag{4-191}$$

$$= 0.12\,(N_{Re})^{1/4}\,(\xi_F)^{-1/3} \tag{4-192}$$

Thus at the entrance region, the local and average mass transfer coefficients for turbulent flow between flat parallel membranes can be obtained from Equations 4-190 and 4-192 respectively.

The average Stanton number, analogous to Equation 4-171, for the flat membrane case can also be obtained from Equation 4-192 as follows:

$$\overline{N}_{Sh} = \frac{0.48}{\alpha_0}\,(N_{Re})^{1/4}\,(\xi_F)^{-1/3} \tag{4-193}$$

$$= 0.48\,(N_{Re})^{1/4}\left(\frac{3}{\alpha_0 X}\right)^{1/3} \tag{4-194}$$

$$= (0.48)\,(3)^{1/3}\,(N_{Re})^{1/4}\left(\frac{v_W^0 h}{D} \cdot \frac{\bar{u}^0 h}{v_W^0 x}\right)^{1/3} \tag{4-195}$$

$$= (0.48)\,(3)^{1/3}\,(N_{Re})^{1/4}\left(\frac{1}{4} \cdot \frac{4\bar{u}^0 h}{\nu} \cdot \frac{\nu}{D} \cdot \frac{h}{x}\right)^{1/3} \tag{4-196}$$

$$= (0.48)\left(\frac{3}{4}\right)^{1/3}\,(N_{Re})^{1/4}\,(N_{Re} \cdot N_{Sc} \cdot h/x)^{1/3} \tag{4-197}$$

$$= 0.44\,(N_{Re})^{7/12}\,(N_{Sc})^{1/3}\,(h/x)^{1/3} \tag{4-198}$$

Combining Equations 4-173 and 4-198,

$$\overline{N}_{ST} = \frac{\overline{N}_{Sh}}{N_{Re} \cdot N_{Sc}} \tag{4-199}$$

$$= 0.44\,(N_{Re})^{-5/12}\,(N_{Sc})^{-2/3}\,(x/h)^{-1/3} \tag{4-200}$$

5. Stage-wise Reverse Osmosis Process Design

The transport equations and the mass transfer coefficient correlations presented above offer a rational basis for reverse osmosis process design. The reverse osmosis process can be operated either as a single-stage process, or as a multistage process where the membrane permeated product from one stage constitutes the feed for another stage. The general effect of concentration polarisation is to bring about a continuous change in product rate and solute separation in the direction of feed flow from the start to the end of the reverse osmosis operating unit. Consequently a multistage operation may be necessary for at least some applications; further, it may also be necessary to build a number of inner stages in series within each primary stage to break the boundary concentration profile by effecting complete mixing of the feed fluid between every two adjacent inner stages. Thus whatever be the number of primary stages, and inner stages within each primary stage, which may be needed for a particular application, stage-wise analysis offers a general approach to reverse osmosis process design. The basic equations governing stage-wise reverse osmosis process design have been developed and their application to saline water conversion has been illustrated by Kimura, Sourirajan, and Ohya (1969) whose treatment of the subject is given below.

Some definitions. For the purpose of this analysis, the basic transport equations are written in terms of molar concentrations of the solute (c_A) instead of mole fractions (X_A) using the relations

$$X_A = c_A/c \tag{4-78}$$

and

$$\Pi(X_A) = BX_A = Bc_A/c = B'c_A \tag{4-201}$$

and two new parameters v_W^*, and γ^* are defined as follows:

$$v_W^* = AP/c \tag{4-202}$$

and

$$\gamma^* = \frac{B}{cP} \tag{4-203}$$

Combining Equations 4-12 and 4-202,

$$\theta = (D_{AM}/K\delta)/v_W^* \tag{4-204}$$

v_W^* is essentially an alternate expression for the pure water permeability characteristic of the film.

Basic Transport Equations. Using Equations 4-7, 4-8, and 4-28, Equations 4-1, 4-9, and 4-10 can be written as

$$v_W = \frac{N_B}{c} = \frac{AP}{c}\left[1 - \frac{Bc_{A2}}{cP} + \frac{Bc_{A3}}{cP}\right] \tag{4-205}$$

$$= k \ln\left[\frac{c_{A2} - c_{A3}}{c_{A1} - c_{A3}}\right] \tag{4-206}$$

$$= (D_{AM}/K\delta)\,[(c_{A2} - c_{A3})/c_{A3}] \tag{4-207}$$

Using Equations 4-202 and 4-203, Equation 4-205 can be written as

$$v_W = v_W^*[1 - \gamma^* c_{A2} + \gamma^* c_{A3}] \tag{4-208}$$

Using Equation 4-204, and equating the right side of Equations 4-207 and 4-208,

$$\theta\,[(c_{A2} - c_{A3})/c_{A3}] = 1 - \gamma^* c_{A2} + \gamma^* c_{A3} \tag{4-209}$$

or

$$c_{A2} = [(1 + \theta)c_{A3} + \gamma^*(c_{A3})^2]/(\gamma^* c_{A3} + \theta) \tag{4-210}$$

Equation 4-210 shows that for any given solution system, c_{A2} and c_{A3} (hence X_{A2} and X_{A3}) are uniquely related for specified values of A and $(D_{AM}/K\delta)$ at a given operating pressure. This has been illustrated experimentally for a number of membrane-solution systems in Chapter 3. From Equation 4-210,

$$(c_{A2} - c_{A3})/c_{A3} = 1/(\gamma^* c_{A3} + \theta) \tag{4-211}$$

Equating the right side of Equations 4-206 and 4-207,

$$(D_{AM}/K\delta)\,[(c_{A2} - c_{A3})/c_{A3}] = k \ln[(c_{A2} - c_{A3})/(c_{A1} - c_{A3})] \tag{4-212}$$

or

$$(c_{A2} - c_{A3})/c_{A3} = \lambda \ln[(c_{A2} - c_{A3})]/(c_{A1} - c_{A3})] \tag{4-213}$$

where

$$\lambda = k/(D_{AM}/K\delta) \tag{4-13}$$

Combining Equations 4-211 and 4-213

$$\frac{1}{(\gamma^* c_{A3} + \theta)} = \lambda \ln\left[\frac{c_{A3}}{(c_{A1} - c_{A3})(\gamma^* c_{A3} + \theta)}\right] \tag{4-214}$$

or

$$\exp\left[\frac{1}{\lambda(\gamma^* c_{A3} + \theta)}\right] = \frac{c_{A3}}{(c_{A1} - c_{A3})(\gamma^* c_{A3} + \theta)} \tag{4-215}$$

or

$$c_{A1} - c_{A3} = \frac{c_{A3}}{(\gamma^* c_{A3} + \theta)} \exp\left[-\frac{1}{\lambda(\gamma^* c_{A3} + \theta)}\right] \tag{4-216}$$

or

$$c_{A1} = c_{A3}\left[1 + \frac{1}{(\gamma^* c_{A3} + \theta)} \exp\left\{-\frac{1}{\lambda(\gamma^* c_{A3} + \theta)}\right\}\right] \tag{4-217}$$

or

$$c_{A1} = c_{A3} q \tag{4-218}$$

where

$$q = 1 + \frac{1}{(\gamma^* c_{A3} + \theta)} \exp\left[-\frac{1}{\lambda(\gamma^* c_{A3} + \theta)}\right] \quad (4\text{-}219)$$

Equation 4-218 is an important one; it expresses directly the relationship between c_{A1} and c_{A3} at any point on the membrane surface as a function of γ^*, θ, and λ without involving the term c_{A2}.

As pointed out earlier, average mass transfer coefficients can be used with reasonable accuracy for a wide range of operating conditions for predicting the product rate and solute separation in a reverse osmosis operating unit. When the average mass transfer coefficient is used for obtaining λ in Equation 4-219, and c_{A1} corresponds to the outlet concentration of brine on the high pressure side of the membrane, the inlet solute concentration, c^0_{A1}, can be obtained by considering the mass balance for the solute and solvent water for the reverse osmosis unit under consideration as follows. For feed flow between two flat parallel membranes, the solute material balance is given by the equation

$$c^0_{A1} = c_{A1}(1 - \Delta) + c_{A3}\Delta \quad (4\text{-}220)$$

where Δ is fraction product recovery given by Equation 4-42; c_{A3} is overall product concentration corresponding to Δ. The solute material balance for the flat membrane case is also given by the relation

$$2hW\bar{u}^0 c^0_{A1} = 2hW\bar{u}c_{A1} + 2Wxv_W c_{A3} \quad (4\text{-}221)$$

where 2h is the spacing (width) between the flat membranes, W is the length of the membrane in the transverse direction, x is the longitudinal distance along the length of the membrane, and $\bar{u}^0$ and $\bar{u}$ represent the average fluid velocity in the direction of feed flow at the inlet and outlet respectively on the high pressure side of the membrane*. From Equation 4-221,

$$c^0_{A1} = \frac{\bar{u}}{\bar{u}^0} c_{A1} + \frac{v_W x}{\bar{u}^0 h} c_{A3} \quad (4\text{-}222)$$

$$= (1 - \Delta)\, c_{A1} + \frac{v_W}{\bar{u}^0} \cdot \frac{x}{h} c_{A3} \quad (4\text{-}223)$$

Comparing Equations 4-220 and 4-223,

$$\Delta = \frac{v_W}{\bar{u}^0} \frac{x}{h} \quad (4\text{-}224)$$

Let the dimensionless longitudinal position X be defined as

$$X = \frac{v^*_W}{\bar{u}^0} \frac{x}{h} \quad (4\text{-}225)$$

---

* In Equation 4-221, v_W and c_{A3} represent average values for the membrane section under consideration.

Equation 4-224 can be written as

$$\Delta = \frac{v_W^*}{\bar{u}^0} \cdot \frac{x}{h} \cdot \frac{v_W}{v_W^*} \tag{4-226}$$

From Equations 4-208, and 4-211,

$$v_W/v_W^* = 1 - \gamma^* (c_{A2} - c_{A3}) \tag{4-227}$$

$$= 1 - \left[\gamma^* c_{A3}/(\gamma^* c_{A3} + \theta)\right] \tag{4-228}$$

$$= \theta/(\gamma^* c_{A3} + \theta) \tag{4-229}$$

Therefore, Equation 4-226 becomes

$$\Delta = \theta X/(\gamma^* c_{A3} + \theta) \tag{4-230}$$

Combining Equations 4-218 and 4-220,

$$c_{A1}^0 = c_{A3} q(1 - \Delta) + c_{A3}\Delta \tag{4-231}$$

$$= c_{A3}[q(1 - \Delta) + \Delta] \tag{4-232}$$

or

$$c_{A1}^0 = c_{A3} p \tag{4-233}$$

where

$$p = q(1 - \Delta) + \Delta \tag{4-234}$$

$$= 1 + \frac{1 - \Delta}{(\gamma^* c_{A3} + \theta)} \exp\left[-\frac{1}{\lambda(\gamma^* c_{A3} + \theta)}\right] \tag{4-235}$$

Equations 4-218 and 4-233 are important design equations for stage-wise reverse osmosis process design. They express the relationships connecting the concentrations of the feed solution entering (c_{A1}^0) and leaving (c_{A1}) the unit stage on the high pressure side of the membrane, and the average concentration of the membrane-permeated product solution (c_{A3}) leaving the atmospheric pressure side of the membrane, as functions of A, ($D_{AM}/K\delta$), operating pressure, molar density of the solution, average mass transfer coefficient, and the volumetric fraction of product recovery.

Unit Stage. The basic unit stage of the reverse osmosis process is shown in Figure 4-22. Here the feed solution enters the stage i at solute concentration c_i^0 and feed rate L_i^0 at one end, and leaves at solute concentration c_i'' and rate L_i'' at the other end on the high pressure side of the membrane; and the product leaves the stage on the atmospheric pressure side of the membrane at solute concentration c_i' and product rate L_i'. These symbols are adopted here in view of their common usage in stage-wise design. The relationships between the new symbols and the ones used in the basic equations given above are as follows:

c_i^0 is identical with c_{A1}^0

c_i'' is identical with c_{A1}

c_i' is identical with c_{A3}

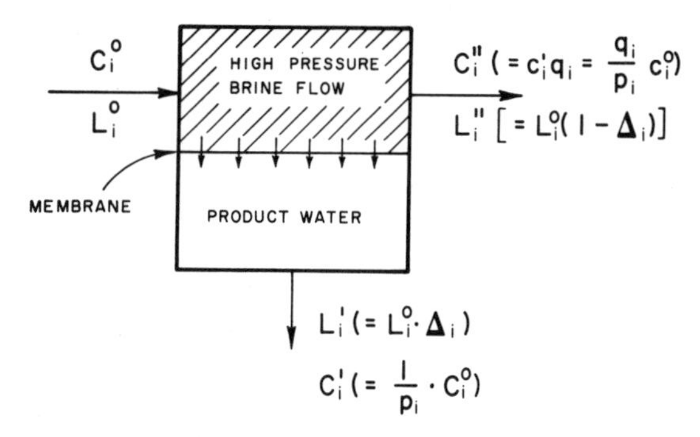

Figure 4-22. Basic unit stage for the reverse osmosis process (Kimura, Sourirajan, and Ohya, 1969).

where the subscript i indicates the ith unit stage. Using Equations 4-218 and 4-233 the following equations can be written for the unit stage.

Product concentration: $$c_i' = \frac{1}{p_i} c_i^0 \qquad (4\text{-}236)$$

Product rate: $$L_i' = L_i^0 \Delta_i \qquad (4\text{-}237)$$

Concentration of feed leaving stage: $$c_i'' = c_i' q_i = \frac{q_i}{p_i} c_i^0 \qquad (4\text{-}238)$$

Rate of feed leaving stage: $$L_i'' = L_i^0(1 - \Delta_i) \qquad (4\text{-}239)$$

General Cascade Equations. Figure 4-23 represents a conceptual multistage reverse osmosis separation system. The stages are numbered from 1 at the top to N at the bottom. The feed enters somewhere in an intermediate stage. The upper stages are called 'purification stages' where the object is to obtain desalinised water and the lower stages are called 'concentration stages' where the object is to obtain concentrated brine solution. The desalinised product water leaves stage 1 and the concentrated brine leaves stage N. The product from stage (i + 1) and the concentrated brine from stage (i — 1) constitute the feed to stage i; product water leaves stage 1 at concentration c_p $(= c_1')$ at the rate **P**, and concentrated brine leaves stage N at concentration c_W $(= c_N'')$ at the rate **W**. L_i' and L_i'' represent respectively the flow rates of the product water and the concentrated brine from any stage i. Following the formalism of the fractional distillation process, the general cascade equations can be given as follows.
Material balance relations:

$$\mathbf{P} c_p + L_i'' c_i'' = L_{i+1}' c_{i+1}' \qquad (4\text{-}240)$$

$$\mathbf{P} + L_i'' = L_{i+1}' \qquad (4\text{-}241)$$

$$\mathbf{W} c_W + L_j' c_j' = L_{j-1}'' c_{j-1}'' \qquad (4\text{-}242)$$

$$\mathbf{W} + L_j' = L_{j-1}'' \qquad (4\text{-}243)$$

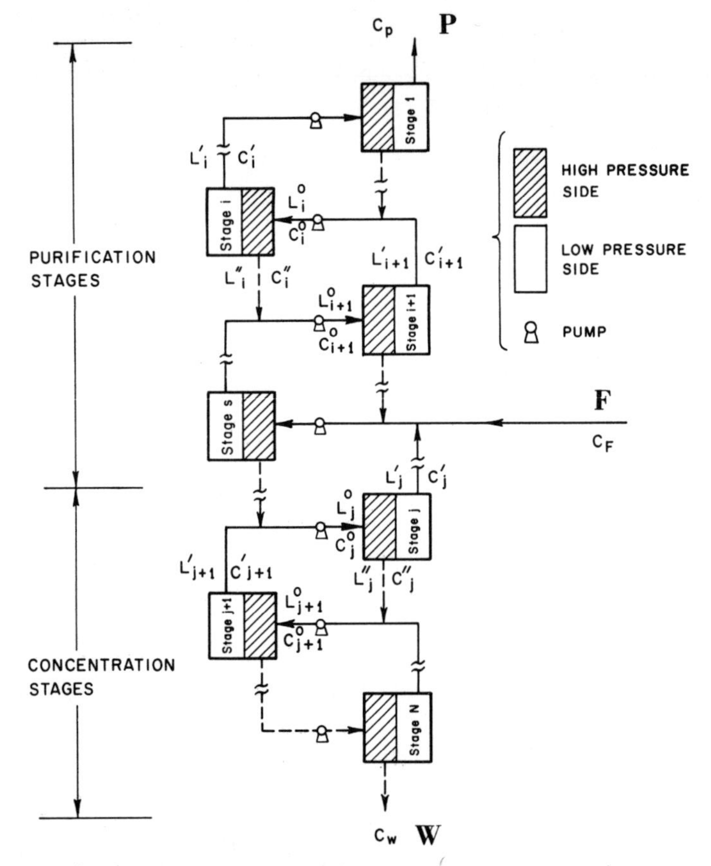

Figure 4-23. Conceptual multi-stage reverse osmosis separation system (Kimura, Sourirajan, and Ohya, 1968).

The number of stages required to separate feed into product and concentrated brine of specified composition is a minimum at total reflux for which the condition is (Benedict and Pigford, 1957a)

$$c_i'' = c_{i+1}' \tag{4-244}$$

Using Equation 4-238, the condition for total reflux is given by

$$c_{i+1}' = c_i' q_i \tag{4-245}$$

Hence

$$c_2' = q_1 c_1' = q_1 c_p \tag{4-246}$$

Similarly

$$c_3' = q_2 c_2' \tag{4-247}$$

$$c_4' = q_3 c_3' \tag{4-248}$$

$$\vdots \quad \vdots$$

$$c_N' = q_{N-1} c_{N-1}' \tag{4-249}$$

$$c_W = q_N c'_N \tag{4-250}$$

Therefore

$$c_W = q_1 \cdot q_2 \cdot q_3 \cdot \ldots \cdot q_N c_p \tag{4-251}$$

Defining

$$\bar{q}_N = {}^N\sqrt{(q_1 \cdot q_2 \cdot q_3 \cdot \ldots \cdot q_N)} \tag{4-252}$$

$$c_W = (\bar{q}_N)^N c_p \tag{4-253}$$

Therefore

$$\frac{c_W}{c_p} = (\bar{q}_N)^N \tag{4-254}$$

$$N_{min} = \frac{\ln(c_W/c_p)}{\ln(\bar{q}_N)} \tag{4-255}$$

where N_{min} is the minimum number of stages required, which corresponds to conditions of total reflux.

At total reflux, the difference in composition between corresponding streams on adjacent stages is a maximum. As the reflux ratio is decreased the difference in composition decreases and reaches zero at minimum reflux. Thus the condition for minimum reflux is (Benedict and Pigford, 1957b)

$$c'_{i+1} = c'_i \tag{4-256}$$

Combining Equations 4-240 and 4-241,

$$c_p + \frac{L''_i}{\mathbf{P}} c''_i = \left(\frac{\mathbf{P} + L''_i}{\mathbf{P}}\right) c'_{i+1} \tag{4-257}$$

Using Equation 4-256, Equation 4-257 becomes

$$c_p + \frac{L''_i}{\mathbf{P}} c''_i = \frac{L''_i}{\mathbf{P}} c'_i + c'_i \tag{4-258}$$

Therefore

$$\frac{L''_i}{\mathbf{P}} (c''_i - c'_i) = c'_i - c_p \tag{4-259}$$

Therefore

$$(L''_i/\mathbf{P})_{min} = (c'_i - c_p)/(c''_i - c'_i) \tag{4-260}$$

$$= \frac{1 - (c_p/c'_i)}{(c''_i/c'_i) - 1} \tag{4-261}$$

$$= \frac{1 - (c_p/c''_i) q_i}{q_i - 1} \tag{4-262}$$

since

$$c_i' = c_i''/q_i \qquad (4\text{-}238)$$

In Equation 4-262, $(L_i''/\mathbf{P})_{min}$ is the minimum value of $(L_i''/\mathbf{P})$ which occurs under conditions of minimum reflux.

In the so-called 'ideal cascade' design (which leads to minimum total inter-stage flow, and which is approximated by all isotope separation plants designed for minimum cost), the product solution and the concentrated brine fed to each stage have the same composition (Benedict and Pigford, 1957c), i.e.

$$c_{i+1}' = c_{i-1}'' = c_i^0 \qquad (4\text{-}263)$$

or

$$c_{i+2}' = c_i'' = c_{i+1}^0 \qquad (4\text{-}264)$$

Combining Equations 4-236 and 4-238, and 4-263 and 4-264

$$c_i'' = c_i^0 q_i/p_i \qquad (4\text{-}265)$$

$$= c_{i+1}^0 \qquad (4\text{-}266)$$

$$= c_{i+1}' p_{i+1} \qquad (4\text{-}267)$$

$$= c_i^0 p_{i+1} \qquad (4\text{-}268)$$

Equating the right hand side of Equations 4-265 and 4-268

$$p_{i+1} = q_i/p_i \qquad (4\text{-}269)$$

Equation 4-269 is an important design equation based on the 'ideal cascade' concept, and it can be used for multistage reverse osmosis process design.

One may also calculate the number of stages in the ideal cascade design as follows.
Rewriting Equation 4-236,

$$c_i^0 = c_i' p_i \qquad (4\text{-}236)$$

Therefore

$$c_1^0 = c_p p_1 \qquad (4\text{-}270)$$

$$c_2^0 = c_2' p_2 \qquad (4\text{-}271)$$

$$\vdots \quad \vdots$$

$$c_F = c_s' p_s \qquad (4\text{-}272)$$

$$= p_1 \cdot p_2 \cdot p_3 \cdots p_s c_p \qquad (4\text{-}273)$$

where s is the stage where the feed of concentration c_F enters the system. Defining

$$\bar{p}_s = \sqrt[N_s]{(p_1 \cdot p_2 \cdot p_3 \cdots\cdot p_s)} \qquad (4\text{-}274)$$

$$c_F = (\bar{p}_S)^{N_S} c_p \tag{4-275}$$

or

$$\frac{c_F}{c_p} = (\bar{p}_S)^{N_S} \tag{4-276}$$

or

$$N_S = \ln (c_F/c_p)/\ln (\bar{p}_S) \tag{4-277}$$

where N_S is the total number of stages in the purification section. Similarly the total number of stages N for both the purification and concentration sections for ideal cascade design can be expressed as

$$N = \ln (c_W/c_p)/\ln (\bar{p}_N) \tag{4-278}$$

where

$$\bar{p}_N = \sqrt[N]{(p_1 \cdot p_2 \cdot p_3 \cdot \ldots \cdot p_N)} \tag{4-279}$$

The ideal cascade design is popular in the design of isotope separation plants. Generally far better separation is obtained in each stage of the reverse osmosis process, and hence the number of stages that may be needed in a reverse osmosis plant is far smaller than that in an isotope separation plant. Consequently one may consider that Equation 4-277 or 4-278 to give the value of $\bar{p}_S$ and $\bar{p}_N$ respectively for a fixed number of stages N_S or N; from the value of $\bar{p}_S$ or $\bar{p}_N$ thus obtained, individual p_i may be calculated using Equation 4-269.

Unit or Single Stage Process Design. A unit stage i may itself be considered as a combination of several inner stages (i, 1), (i, 2), (i, 3).... (i, m), ... (i, M) connected in series as illustrated in Figure 4-24. Complete mixing of the concentrated brine between adjacent inner stages is assumed; consequently, the concentration profile of the boundary solution is broken between inner stages. The detailed design of the unit stage, which includes the inner stages, is obviously of decisive importance in the overall perform-

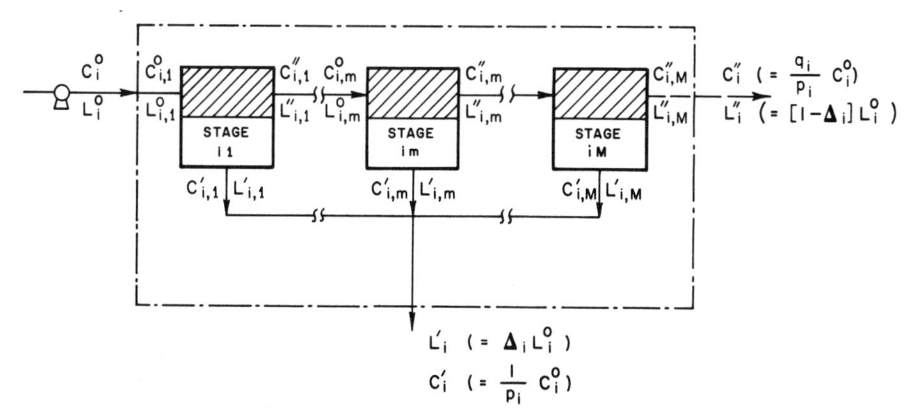

Figure 4-24. Inner stages in a unit stage (Kimura, Sourirajan, and Ohya, 1969).

ance of a reverse osmosis desalination plant; hence some equations governing the design of a unit or single stage are derived below.

Let the unit stage be considered as a combination of several inner stages (i, 1), (i, 2), ... (i, m), ... (i, M) identical in design and connected in series with some mixing device in between adjacent inner stages. Let the design considered here be for feed flow between flat parallel membranes.

As a result of product withdrawal, the velocity of the feed brine progressively decreases from the beginning to the end in the entire unit stage. Let $\bar{u}^0$ and $\bar{u}$ be the average velocities of the brine solution entering and leaving the unit stage i respectively; and, let $\Delta_i, \Delta_{i,1}, \Delta_{i,2}, \ldots \Delta_{i,m}, \ldots \Delta_{i,M}$ be the fraction product withdrawal in the entire unit stage i and in the inner stages (i, 1), (i, 2), ... (i, m), ... (i, M) respectively. Then

$$\bar{u} = \bar{u}^0(1 - \Delta_{i,1})(1 - \Delta_{i,2}) \ldots (1 - \Delta_{i,m}) \ldots (1 - \Delta_{i,M}) \tag{4-280}$$

or

$$\bar{u}/\bar{u}^0 = (1 - \Delta_i) = (1 - \Delta_{i,1})(1 - \Delta_{i,2}) \ldots (1 - \Delta_{i,m}) \ldots (1 - \Delta_{i,M}) \tag{4-281}$$

Similarly, applying the basic unit stage Equation 4-238 to the inner stages,

$$c_i'' = c_i^0 \left(\frac{q_{i,1}}{p_{i,1}}\right)\left(\frac{q_{i,2}}{p_{i,2}}\right) \cdots \left(\frac{q_{i,m}}{p_{i,m}}\right) \cdots \left(\frac{q_{i,M}}{p_{i,M}}\right) \tag{4-282}$$

or

$$\frac{c_i''}{c_i^0} = \frac{q_i}{p_i} = \left(\frac{q_{i,1}}{p_{i,1}}\right)\left(\frac{q_{i,2}}{p_{i,2}}\right) \cdots \left(\frac{q_{i,m}}{p_{i,m}}\right) \cdots \left(\frac{q_{i,M}}{p_{i,M}}\right) \tag{4-283}$$

Applying Equation 4-225 to the inner-stages,

$$X_{i,m-1} = \frac{v_w^*}{\bar{u}_{i,m-1}^0} \cdot \frac{x}{h} \tag{4-284}$$

and

$$X_{i,m} = \frac{v_w^*}{\bar{u}_{i,m-1}^0[1 - \Delta_{i,m-1}]} \cdot \frac{x}{h} \tag{4-285}$$

$$= \frac{X_{i,m-1}}{1 - \Delta_{i,m-1}} \tag{4-286}$$

If however the inner stages do not have the same dimensions and if they are so defined as to give the same feed velocity at each inlet, then

$$X_{i,1} = X_{i,m} = X_{i,M} \tag{4-287}$$

Equations 4-225, 4-230, 4-234, 4-235, 4-236, 4-238, 4-281, 4-283, and 4-286 offer a means of parametric study of the unit or single stage reverse osmosis process. Several such studies are possible. The procedural technique is illustrated below for a simple case.

From the membrane specifications given in terms of A and $(D_{AM}/K\delta)$ at a particular operating pressure, and the properties of the feed solution, the parameters v_W^*, γ^*, and θ can be calculated using Equations 4-202, 4-203, and 4-204 respectively. When the average mass transfer coefficient applicable for the system is also known (i.e. when λ is known), one may then calculate product concentration (c_i') and the concentration of brine leaving the unit stage (c_i'') on the high pressure side of the membrane for a given feed concentration (c_i^0) as a function of product recovery (Δ_i), and the total number of inner stages (M_i) for various values of membrane-spacing (h), assuming that the inner stages have the same dimensions. This is done as follows.

For the inner stage (i, 1), Equations 4-225, 4-230, 4-234, 4-235, and 4-236 may be written as

$$X_{i,1} = \frac{v_W^*}{\bar{u}^0} \cdot \frac{x}{h} \tag{4-288}$$

$$\Delta_{i,1} = \frac{\theta X_{i,1}}{\gamma^* c_{i,1}' + \theta} \tag{4-289}$$

$$q_{i,1} = \frac{p_{i,1} - \Delta_{i,1}}{1 - \Delta_{i,1}} \tag{4-290}$$

$$p_{i,1} = 1 + \frac{\left(1 - \dfrac{\theta X_{i,1}}{\gamma^* c_{i,1}' + \theta}\right)}{(\gamma^* c_{i,1}' + \theta)} \exp\left[-\frac{1}{\lambda(\gamma^* c_{i,1}' + \theta)}\right] \tag{4-291}$$

and

$$c_{i,1}^0 = c_{i,1}' \left[1 + \frac{\left(1 - \dfrac{\theta X_{i,1}}{\gamma^* c_{i,1}' + \theta}\right)}{(\gamma^* c_{i,1}' + \theta)} \exp\left\{-\frac{1}{\lambda(\gamma^* c_{i,1}' + \theta)}\right\}\right] \tag{4-292}$$

Let $c_{i,1}^0$ be given (as, for example, 3.5 wt.% NaCl). Let Δ_i and M_i be preset. Then the calculation steps are as follows:

(i) First assume a value for $X_{i,1}$.

(ii) Calculate $c_{i,1}'$ from Equation 4-292 by trial and error.

(iii) Calculate $\Delta_{i,1}$ from Equation 4-289.

(iv) Calculate $p_{i,1}$ from Equation 4-291.

(v) Calculate $q_{i,1}$ from Equation 4-290.

(vi) Next, for the inner stage (i, 2), using Equations 4-238 and 4-286, calculate $c_{i,2}^0$ and $X_{i,2}$ from the relations

$$c_{i,2}^0 = c_{i,1}'' = c_{i,1}^0 \frac{q_{i,1}}{p_{i,1}} \tag{4-293}$$

and

$$X_{i,2} = \frac{X_{i,1}}{1 - \Delta_{i,1}} \tag{4-294}$$

Knowing $c^0_{i,2}$ and $X_{i,2}$ calculate $c'_{i,2}$, $\Delta_{i,2}$, $p_{i,2}$, and $q_{i,2}$ following the procedure exactly similar to that used for determining the corresponding values for the inner stage (i, 1).

(vii) Continue similar calculations up to the stage (i, M), and obtain the values of $\Delta_{i,1}, \Delta_{i,2}, \ldots \Delta_{i,M}$, and $p_{i,1}, p_{i,2}, \ldots p_{i,M}$, and $q_{i,1}$, $q_{i,2}, \ldots q_{i,M}$.

(viii) Then calculate the overall Δ_i for the entire unit from Equation 4-281; if this value of Δ_i is not the same as the preset value, repeat the calculations until they are the same.

(ix) Then calculate (q_i/p_i) from Equation 4-283.

(x) The concentration of the effluent brine solution leaving the unit stage is then calculated from the relation

$$c''_i = c^0_i(q_i/p_i) \tag{4-238}$$

(xi) From Equation 4-234,

$$(q_i/p_i)\,(1 - \Delta_i) + \Delta_i/p_i = 1 \tag{4-295}$$

Calculate p_i for the unit stage from Equation 4-295.

(xii) Finally, the overall product concentration c'_i leaving the atmospheric pressure side of the unit stage is calculated from the relation

$$c'_i = c^0_i/p_i \tag{4-236}$$

The above calculation technique involves the use of the appropriate values for λ which can be obtained as follows. For the flat membrane case, from Equation 4-13,

$$\lambda = \frac{k}{(D_{AM}/K\delta)} = \frac{4kh}{D} \cdot \frac{D}{4hv^*_W} \cdot \frac{v^*_W}{(D_{AM}/K\delta)} \tag{4-296}$$

$$= \frac{\alpha_0 N_{Sh}}{4\theta} \tag{4-297}$$

where

$$N_{Sh} = \frac{4hk}{D} \tag{4-104}$$

and

$$\alpha_0 = \frac{D}{v^*_W h} \tag{4-298}$$

Comparing Equations 4-30 and 4-202, $v^0_W = v^*_W$ when $\gamma = 0$. Referring to the mass transfer correlation given in Figure 4-14, the line corresponding to $\gamma = 0$ (line I) and that corresponding to $\gamma = 0.5$ (line II) are nearly the same. It hence seems reasonable to assume that the mass transfer correlation given by line II in Figure 4-14 is valid even when X and α_0 are defined by Equations 4-225 and 4-298 instead of by Equations 4-32 and 4-91 respectively.

Consequently the average mass transfer coefficient for use in stage-wise design calculations can be obtained from the following relations given earlier:

$$\frac{\alpha_0 \overline{N}_{Sh}}{4} = 0.975\ (\xi_F)^{-1/3} \tag{4-166}$$

where

$$\xi_F = \frac{X}{3\alpha_0^2} \tag{4-139}$$

and X and α_0 are given by Equations 4-225 and 4-298. Combining Equations 4-297, 4-166, and 4-139,

$$\lambda = 0.975\ \frac{(3\alpha_0^2)^{1/3}}{X^{1/3}\theta} \tag{4-299}$$

Using Equation 4-299 for obtaining the appropriate value of λ, some results of the above calculation technique are illustrated in Figures 4-25 to 4-30 for the single stage desalination of 3.5 per cent aqueous sodium chloride solution at an operating pressure of 1500 psig using a typical Loeb-Sourirajan type porous cellulose acetate membrane whose specifications at 1500 psig are given as follows:

$$A = 0.97 \times 10^{-6}\ \text{g.mole}\ H_2O/\text{cm}^2\ \text{sec atm}$$

$$(D_{AM}/K\delta)_{NaCl} = 0.9 \times 10^{-5}\ \text{cm/sec}$$

During continuous operation of the reverse osmosis unit under pressure, membrane compaction may occur as a result of which A decreases as a function of time but $(D_{AM}/K\delta)$ remains constant provided the surface pore structure of the film does not change; this is generally the case with the Loeb-Sourirajan type porous cellulose acetate membranes (refer to Chapter 3). Thus A-factor (=value of A at any time/initial value of A) is a measure of membrane compaction; an A-factor of 0.5 may be expected to occur in about six months of continuous service under an operating pressure of 1500 psig.

Figures 4-25 and 4-26 illustrate the effects of A-factor, spacing between membranes (h), the total number of inner stages (M_i) and the total fraction product recovery (Δ_i), on the concentration of product water (c_i'), and on that of the effluent concentrated brine solution (c_i''). The data illustrate the importance of the number of inner stages required, and some of the conditions under which potable water can be obtained from the 3.5 per cent [NaCl-H_2O] feed solution.

Figures 4-27 and 4-28 give the values of $X_{i,1}$ corresponding to the data presented in Figures 4-25 and 4-26 respectively.

The total membrane area (S) per unit product rate (**P**) may be calculated as follows. Assuming equal area per each inner stage,

$$S/\mathbf{P} = 2WxM_i/\mathbf{P} \tag{4-300}$$

$$= \frac{2WxM_i}{W.\,2h.\,\bar{u}^0\Delta_i} \tag{4-301}$$

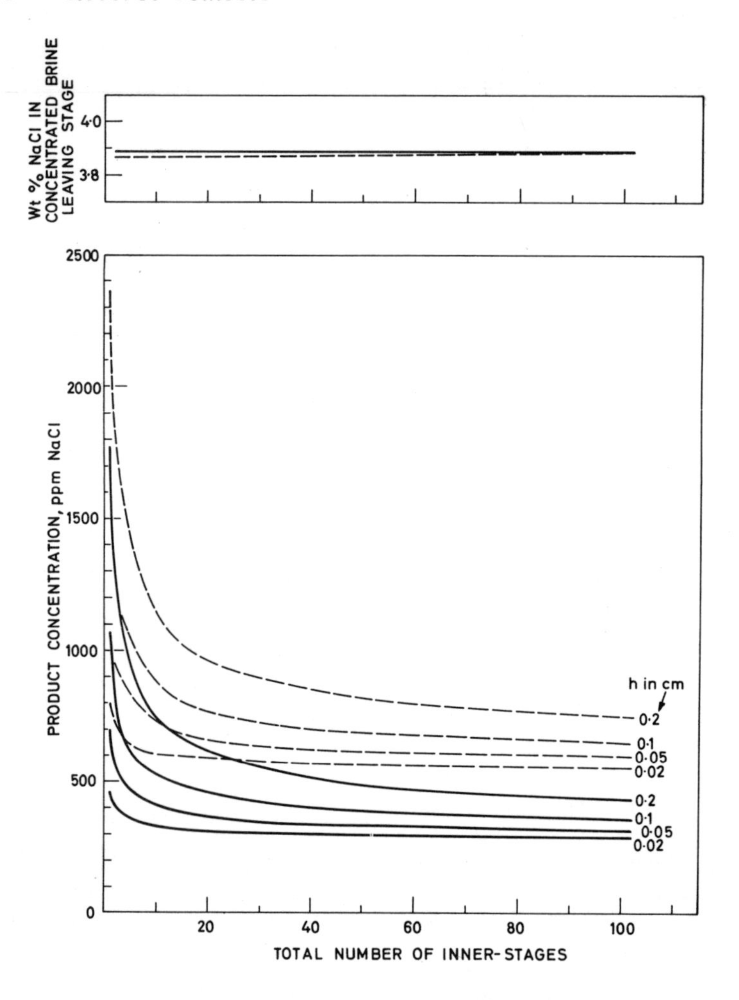

Figure 4-25. Effects of A-factor, spacing between membranes, and the total number of inner stages on the concentration of product water and that of the effluent brine solution in a single-stage desalination process. Data of Kimura, Sourirajan, and Ohya (1969).

System: sodium chloride-water; operating pressure: 1500 psig; feed flow condition: laminar flow between flat parallel membranes; film type: CA-NRC-18.

Film specifications at 1500 psig:

$$A = 0.97 \times 10^{-6} \frac{\text{g. mole } H_2O}{\text{cm}^2 \text{ sec atm}}$$

$$\left(\frac{D_{AM}}{K\delta}\right)_{NaCl} = 0.9 \times 10^{-5} \text{ cm/sec}$$

Data for $\Delta = 0.1$ (Fraction Total Product Recovery)
A Factor = 1.0 ———
A Factor = 0.5 – – – –

Feed concentration = 3.5 wt.% NaCl.

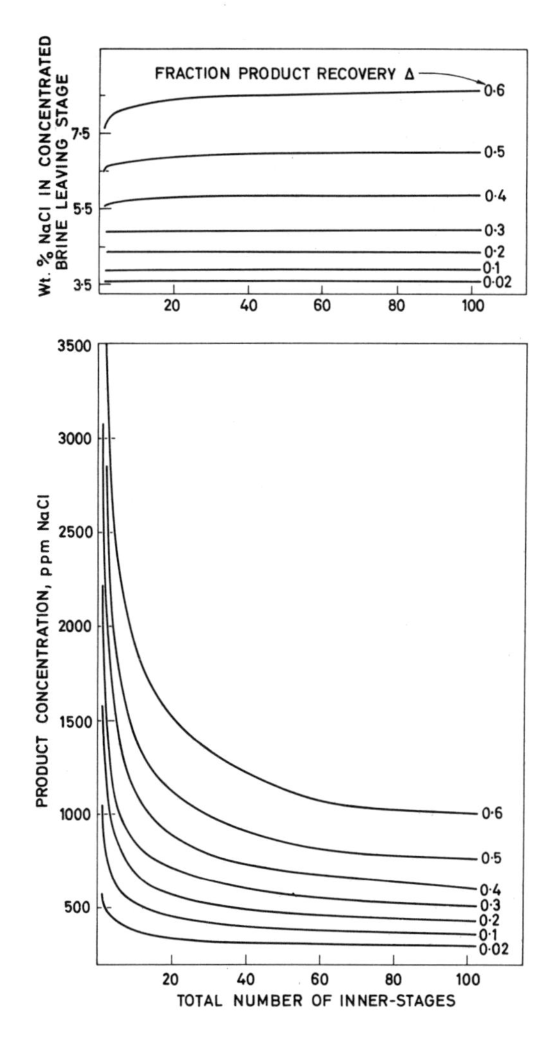

Figure 4-26. Effects of fraction product recovery, and the total number of inner stages on the concentration of product water and that of the effluent brine solution in a single-stage desalination process. Data of Kimura, Sourirajan, and Ohya (1969).

System: sodium chloride-water; operating pressure: 1500 psig; feed flow condition: laminar flow between flat parallel membranes; film type: CA-NRC-18.

Film specifications at 1500 psig:

$$A = 0.97 \times 10^{-6} \frac{\text{g. mole } H_2O}{\text{cm}^2 \text{sec atm}}$$

$$\left(\frac{D_{AM}}{K\delta}\right)_{NaCl} = 0.9 \times 10^{-5} \text{ cm/sec}$$

Data for A factor = 1   h = 0.1 cm.

Feed concentration = 3.5 wt.% NaCl

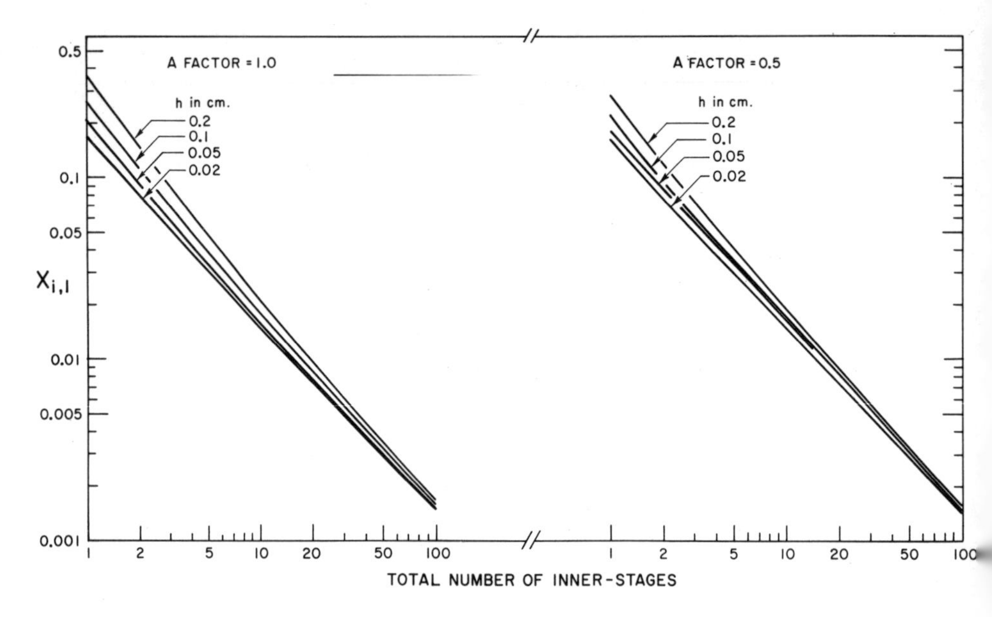

Figure 4-27. Values of $X_{i,1}$ corresponding to data presented in Figure 4-25. Data of Kimura, Sourirajan, and Ohya.

$$= \frac{M_i}{\Delta_i}\left(\frac{v_w^*}{\bar{u}^0} \cdot \frac{x}{h}\right)\frac{1}{v_w^*} \tag{4-302}$$

$$= \frac{X_{i,1}M_i}{\Delta_i v_w^*} \tag{4-303}$$

where M_i is the total number of inner stages in the unit stage i.

Figure 4-29 illustrates the variations in the values of S/**P** as functions of h, M_i, and Δ_i; the data again illustrate the importance of the number of inner stages in reducing the total membrane area requirement for a given product rate.

The power consumption per unit product rate due to the frictional pressure drop in the entire unit stage in the laminar flow range can be calculated as follows. From Equation 4-84 of Knudsen and Katz (1958a), the pressure drop due to friction in any inner stage (i, m) is given by the relation

$$\text{pressure drop} = \frac{3\mu\bar{u}_{i,m}^0 x}{g_c h^2} \tag{4-304}$$

$$\text{Flow rate through the inner stage} = 2hW\bar{u}_{i,m}^0 \tag{4-305}$$

Therefore power consumption due to pressure drop per inner stage

$$= \text{flow rate} \times \text{pressure drop} = \frac{6\mu Wx(\bar{u}_{i,m}^0)^2}{g_c h} \tag{4-306}$$

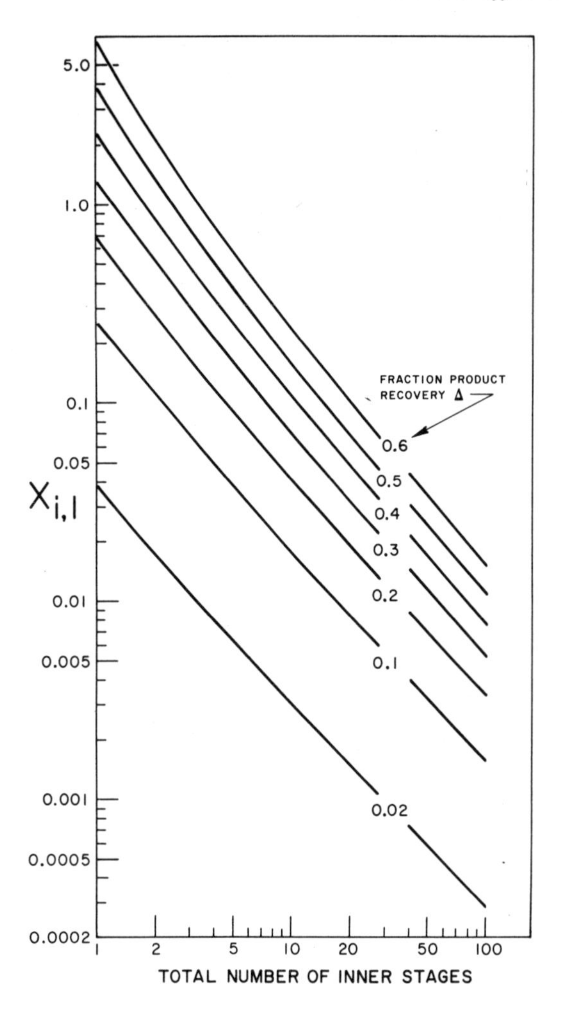

Figure 4-28. Values of $X_{i,1}$ corresponding to data presented in Figure 4-26. Data of Kimura, Sourirajan, and Ohya.

In the inner stages,

$$\bar{u}^0_{i,m} = \bar{u}^0_{i,m-1}\left[1 - \Delta_{i,m-1}\right] \tag{4-307}$$

Therefore the total power consumption due to the frictional pressure drop in the entire unit stage containing M_i inner stages is given by

$$[\text{Power consumption}]_{\text{total}} = \frac{6\mu Wx}{g_c h}\left[(\bar{u}^0_{i,1})^2 + (\bar{u}^0_{i,2})^2 + \ldots + (\bar{u}^0_{i,m})^2\right] \tag{4-308}$$

$$= \frac{6\mu Wx}{g_c h}(\bar{u}^0_{i,1})^2 \cdot E \tag{4-309}$$

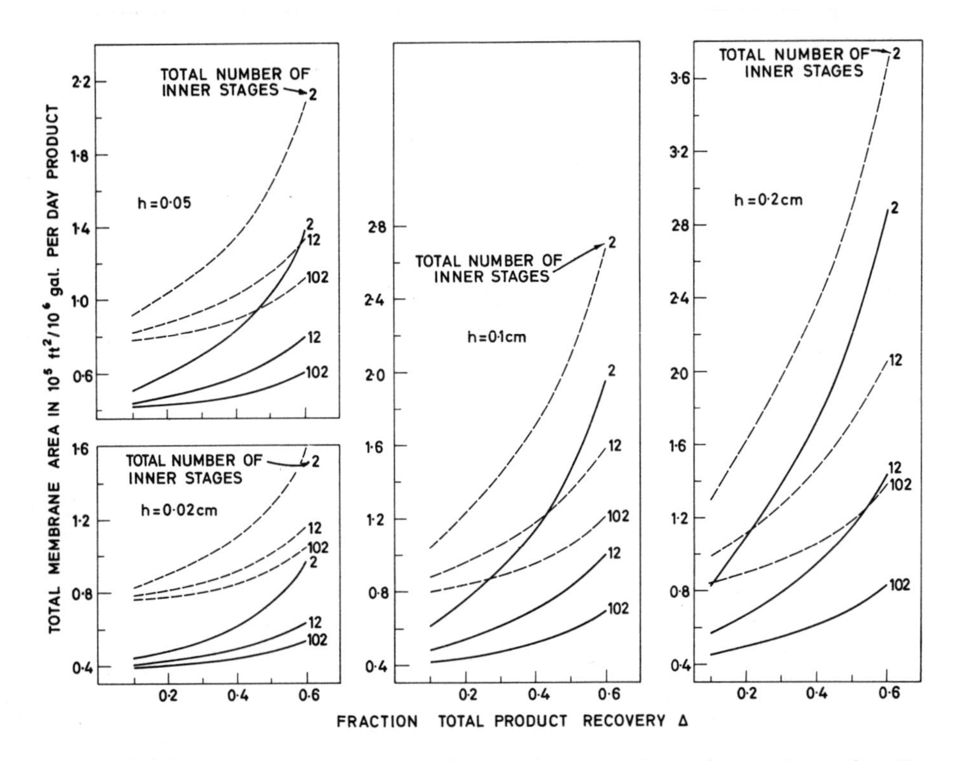

Figure 4-29. Membrane area requirements per unit product rate as functions of fraction product recovery, membrane spacing and A-factor in a single-stage desalination process. Data of Kimura, Sourirajan, and Ohya (1969).

System: sodium chloride-water; operating pressure: 1500 psig; feed flow condition; laminar flow between flat parallel membranes; film type: CA-NRC-18.

Film specifications at 1500 psig:

$$A = 0.97 \times 10^{-6} \frac{\text{g. mole } H_2O}{\text{cm}^2 \text{sec atm}}$$

$$\left(\frac{D_{AM}}{K\delta}\right)_{NaCl} = 0.9 \times 10^{-5} \text{ cm/sec}$$

Data for A factor = 1.0 ———

A factor = 0.5 – – – –

Feed concentration = 3.5 wt.% NaCl

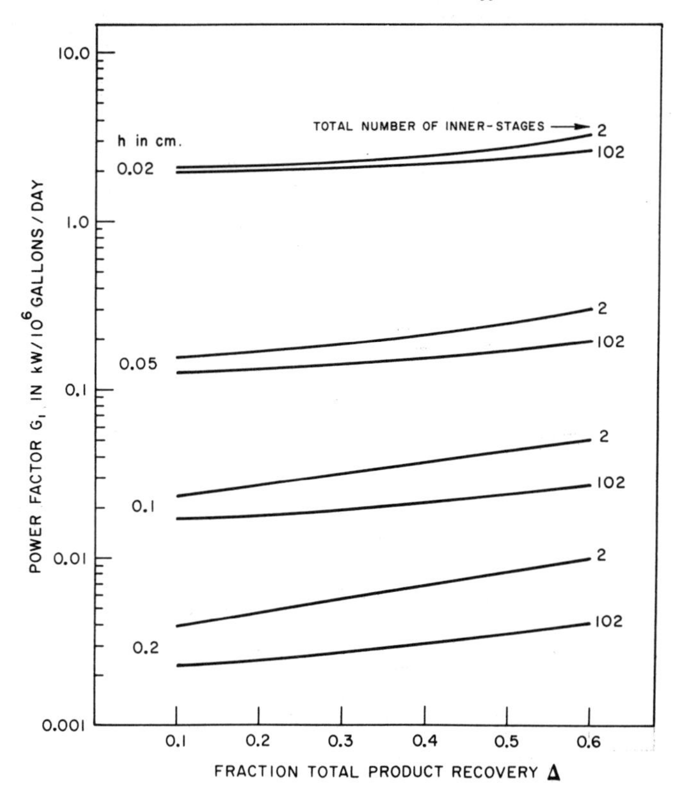

Figure 4-30. Effect of fraction product recovery and total number of inner-stages on power factor in a single-stage desalination process. Data of Kimura, Sourirajan, and Ohya (1969).

System: sodium chloride-water; operating pressure: 1500 psig; feed flow condition: laminar flow between flat parallel membranes; film type: CA-NRC-18.

Film specifications at 1500 psig:

$$A = 0.97 \times 10^{-6} \frac{\text{g. mole } H_2O}{\text{cm}^2 \text{sec atm}}$$

$$\left(\frac{D_{AM}}{K\delta}\right)_{NaCl} = 0.9 \times 10^{-5} \text{ cm/sec}$$

Data for A factor = 1.0; feed concentration = 3.5 wt.% NaCl

where

$$E = 1 + (1 - \Delta_{i,1})^2 + (1 - \Delta_{i,2})^2 + \ldots + (1 - \Delta_{i,m-1})^2 \quad (4\text{-}310)$$

The total product rate for the unit stage is given by

$$\mathbf{P} = 2hW \cdot (\bar{u}^0_{i,1}) \cdot \Delta_i \quad (4\text{-}311)$$

Therefore

$$\frac{[\text{Power consumption}]_{\text{total}}}{\mathbf{P}} = \frac{3\mu x \bar{u}^0_{i,1}}{g_c h^2 \Delta_i} \cdot E \quad (4\text{-}312)$$

From Equation 4-225

$$x = \bar{u}^0_{i,1} h X_{i,1} / v^*_W \quad (4\text{-}313)$$

From Equation 4-298,

$$h = \frac{D}{v^*_W \alpha_0} \quad (4\text{-}314)$$

From Equation 4-311,

$$\bar{u}^0_{i,1} = \frac{\mathbf{P}}{2hW\Delta_i} \quad (4\text{-}315)$$

Substituting for x, h, and $u^0_{i,1}$ in Equation 4-312,

$$\frac{[\text{Power consumption}]_{\text{total}}}{\mathbf{P}} = \frac{3\mu}{4g_c D^3} \left(\frac{\mathbf{P}}{W}\right)^2 \frac{(v^*_W)^2 (\alpha_0)^3 X_{i,1}}{(\Delta_i)^3} \cdot E \quad (4\text{-}316)$$

The Reynolds number at membrane entrance for the flat membrane case may be given as (Knudsen and Katz, 1958d)

$$N_{Re} = 4h(\bar{u}^0_{i,1})/\nu \quad (4\text{-}317)$$

Substituting for $\bar{u}^0_{i,1}$ from Equation 4-317,

$$N_{Re} = \frac{2}{\Delta_i \nu} \left(\frac{\mathbf{P}}{W}\right) \quad (4\text{-}318)$$

or

$$\mathbf{P}/W = \Delta_i \nu N_{Re}/2 \quad (4\text{-}319)$$

Substituting for (**P**/W), Equation 4-316 becomes

$$\frac{[\text{Power consumption}]_{\text{total}}}{\mathbf{P}}$$

$$= (N_{Re})^2 \cdot \frac{3}{16} \frac{\rho}{g_c} (N_{Sc})^3 \frac{(v^*_W)^2 (\alpha_0)^3 X_{i,1}}{\Delta_i} \cdot E \quad (4\text{-}320)$$

$$= (N_{Re})^2 \cdot G_i \quad (4\text{-}321)$$

where

$$G_i = \frac{3}{16} \frac{\rho}{g_c} (N_{Sc})^3 \frac{(v^*_W)^2 (\alpha_0)^3 X_{i,1}}{\Delta_i} \cdot E \quad (4\text{-}322)$$

The symbol G_i, expressed by Equation 4-322, may be called the 'power factor' for the unit stage i. Equation 4-321 shows that the total power consumption due to frictional pressure drop in the unit stage is proportional to $(N_{Re})^2$ in laminar feed flow. If N_{Re} is small, total power consumption becomes small, and from Equation 4-318, W becomes large for given values of **P** and Δ_i. Also from Equation 4-312, power consumption becomes less when h becomes large when all the other terms remain constant. Equations 4-312 and 4-320 show that there are several interdependent variables which govern total power consumption. Figure 4-30 illustrates a set of calculations expressing power factor (G_i) as a function of total product recovery (Δ_i) for different values of h and M_i. The data again show the importance of the total number of inner stages in reducing power consumption.

The results given in Figures 4-25 to 4-30 are merely intended to illustrate the calculation procedure for single-stage reverse osmosis process design; they do not necessarily indicate the optimum results obtainable for saline water conversion.

Two-Stage Process Design. A schematic diagram of a two-stage reverse osmosis process for saline water conversion is shown in Figure 4-31. Both stages are assumed to be operated at 1500 psig. Subscripts 1 and 2 replace the letter i in the unit stage design and refer to Stage 1 (final product stage), and Stage 2 (feed stage) respectively. As before, c^0 and c'' refer to the concentration of the feed solution entering the stage and that of the concentrated brine leaving the stage on the high pressure side of the membrane, and c' refers to the concentration of the product leaving the stage on the low pressure side of the membrane; L^0, L'', and L' refer to the corresponding flow rates.

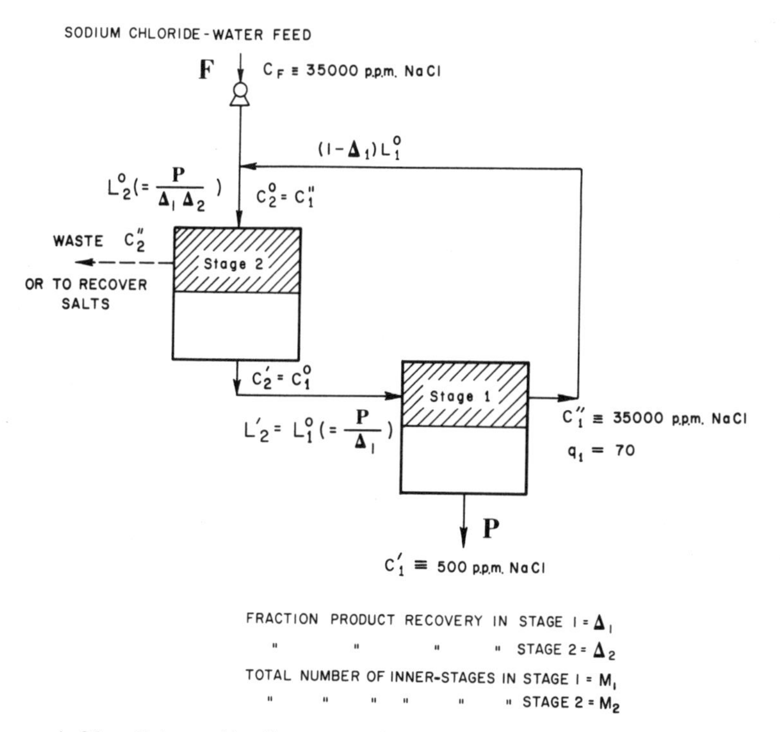

Figure 4-31. Schematic diagram of a two-stage desalination process (Kimura, Sourirajan, and Ohya, 1969).

For purposes of illustration, the desalination of 3.5 per cent aqueous sodium chloride solution is considered. The concentration of product leaving Stage 1 is fixed as 500 ppm of salt, and that of the concentrated brine leaving Stage 1 is fixed as 35,000 ppm of salt, which is the same as the concentration of the feed solution to Stage 2. The specifications of the membrane used in Stage 1 are the same as those given above for the single-stage process. A more porous membrane is used for the feed stage; the specifications of the membrane used in the feed stage at 1500 psig are as follows:

$$A = 2.1 \times 10^{-6} \frac{\text{g . mole } H_2O}{\text{cm}^2 \text{ sec atm}}$$

$$(D_{AM}/K\delta) = 38.6 \times 10^{-5} \text{ cm/sec}$$

The membranes chosen for the two stages are arbitrary. For purposes of calculation an A-factor = 0.5 is assumed.

Stage 2 (feed stage) is designed first for preset values of Δ_2 and the total number of inner stages (M_2) for the feed solution, whose concentration is fixed as 35,000 ppm. Following the procedure identical to that given above for unit stage design, the concentrations of the product leaving Stage 2 (c_2') and p_2 are calculated. Figure 4-32 illustrates the variations of c_2' and $X_{2,1}$ as functions of Δ_2 and M_2.

For Stage 1 (final product stage),

$$q_1 = c_1''/c_1' = 70 \tag{4-323}$$

Using Equation 4-234, and the ideal cascade criterion given by Equation 4-269, the value of p for Stage 2 is given by

$$p_2 = q_1/p_1 = \frac{q_1}{q_1(1-\Delta_1)+\Delta_1} \tag{4-324}$$

$$= \frac{70}{70(1-\Delta_1)+\Delta_1} \tag{4-325}$$

Using the value of p_2 obtained in the feed stage design, and Equation 4-325, Δ_1 is then calculated.

Stage 1 is then designed on the basis of the above values of Δ_1 and the concentration of the feed solution to Stage 1 (c_1^0) which is the same as the concentration of the product solution from Stage 2 (c_2') given in Figure 4-32.

From Equations 4-234 and 4-323,

$$q_1 = (p_1 - \Delta_1)/(1-\Delta_1) = 70 \tag{4-326}$$

Since Δ_1 is known, p_1 is calculated from Equation 4-326. Therefore, for Stage 1, p_1, q_1, Δ_1, product concentration (c_1') and effluent brine concentration (c_1'') are known. The problem then is simply to find $X_{1,1}$ for preset values of M_1; this is done by trial and error procedure discussed for unit stage design.

The total recovery of the final product water (Δ_{Total} = **P/F**) may be obtained as follows:

$$L_1^0 = L_2' = L_2^0\Delta_2 \tag{4-327}$$

$$\mathbf{P} = L_1^0\Delta_1 = L_2^0\Delta_1\Delta_2 \tag{4-328}$$

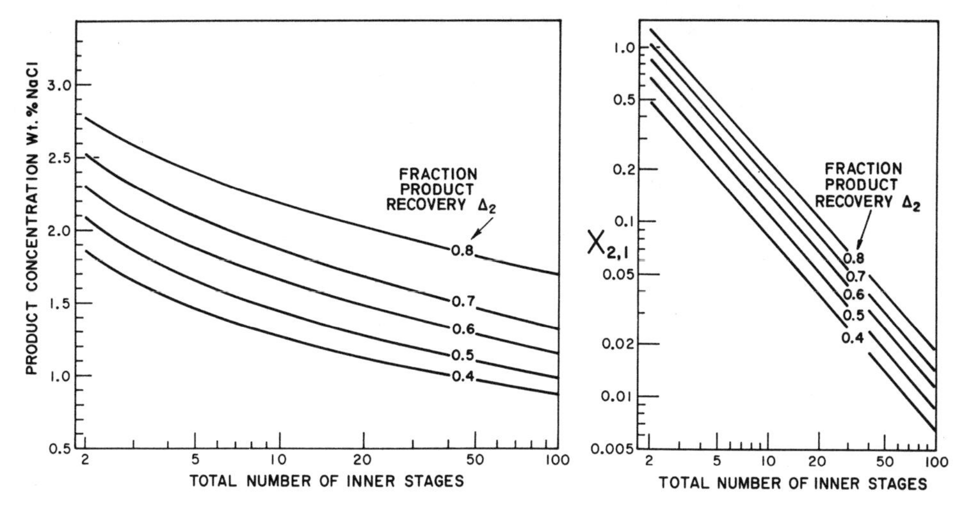

Figure 4-32. Effect of fraction product recovery and total number of inner stages on product concentration and $X_{2,1}$ in the feed-stage in a two-stage desalination process. Data of Kimura, Sourirajan, and Ohya.

System: sodium chloride-water; operating pressure: 1500 psig; feed flow condition: laminar flow between flat parallel membranes; h = 0.1 cm.

Feed concentration: 3.5% NaCl; film type: CA-NRC-18.

Film specifications at 1500 psig:

$$A = 2.1 \times 10^{-6} \frac{\text{g. mole } H_2O}{\text{cm}^2 \text{ sec atm}}$$

$$\left(\frac{D_{AM}}{K\delta}\right)_{NaCl} = 38.6 \times 10^{-5} \text{ cm/sec}$$

Data for A factor = 0.5.

Therefore

$$L_2^0 = \frac{P}{\Delta_1 \Delta_2} \tag{4-329}$$

$$= F + (1 - \Delta_1) L_1^0 \tag{4-330}$$

or

$$\frac{P}{\Delta_1 \Delta_2} = F + (1 - \Delta_1)\frac{P}{\Delta_1} \tag{4-331}$$

or

$$\Delta_{total} = \frac{P}{F} = \frac{\Delta_1 \Delta_2}{1 - \Delta_2(1 - \Delta_1)} \tag{4-332}$$

One may also calculate total recovery as a function of recovery in feed stage (Δ_2) and the concentration of the product leaving the feed stage ($c_2' = c_1^0$) for fixed values of the initial feed and final product concentrations. For the case represented in Figure 4-31, Δ_{total} may be obtained as follows. Expressing concentrations, c_i^0, c_F, and c_W as weight fractions w_i^0, w_F and w_W respectively, the solute material balance for Stage 1 and Stage 2 may be written respectively as

$$w_1^0 = 5 \times 10^{-4}\Delta_1 + 0.035(1 - \Delta_1) \tag{4-333}$$

$$w_F = 0.035 = w_1^0\Delta_2 + (1 - \Delta_2)w_W \tag{4-334}$$

Therefore

$$\Delta_1 = (70 - 2000w_1^0)/69 \tag{4-335}$$

$$\Delta_2 = [1 - (0.035/w_W)]/(1 - w_1^0/w_W) \tag{4-336}$$

and, using Equation 4-332,

$$\Delta_{total} = \frac{(70 - 2000w_1^0)\Delta_2}{69 - \Delta_2(2000w_1^0 - 1)} \tag{4-337}$$

Figure 4-33 expresses the relation 4-337 for various values of w_1^0 and Δ_2.

From material balance relations one may also calculate the weight per cent NaCl in the concentrated brine leaving Stage 2 (w_W) for various values of **P/F**; this is illustrated in Figure 4-34.

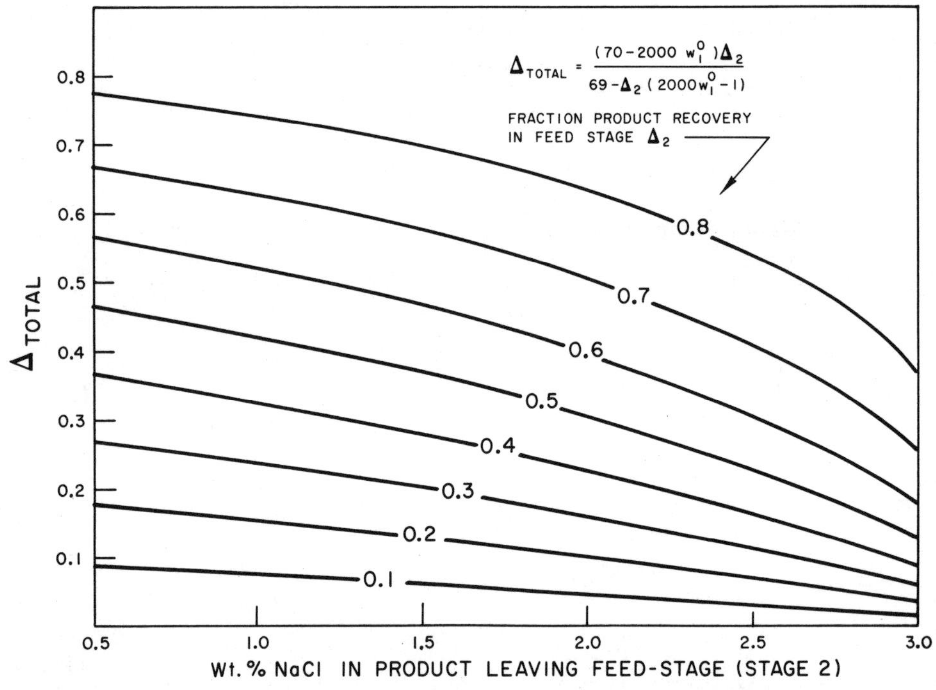

Figure 4-33. Effect of feed stage product concentration and fraction product recovery on total product recovery in a two-stage desalination process. Data of Kimura, Sourirajan, and Ohya.

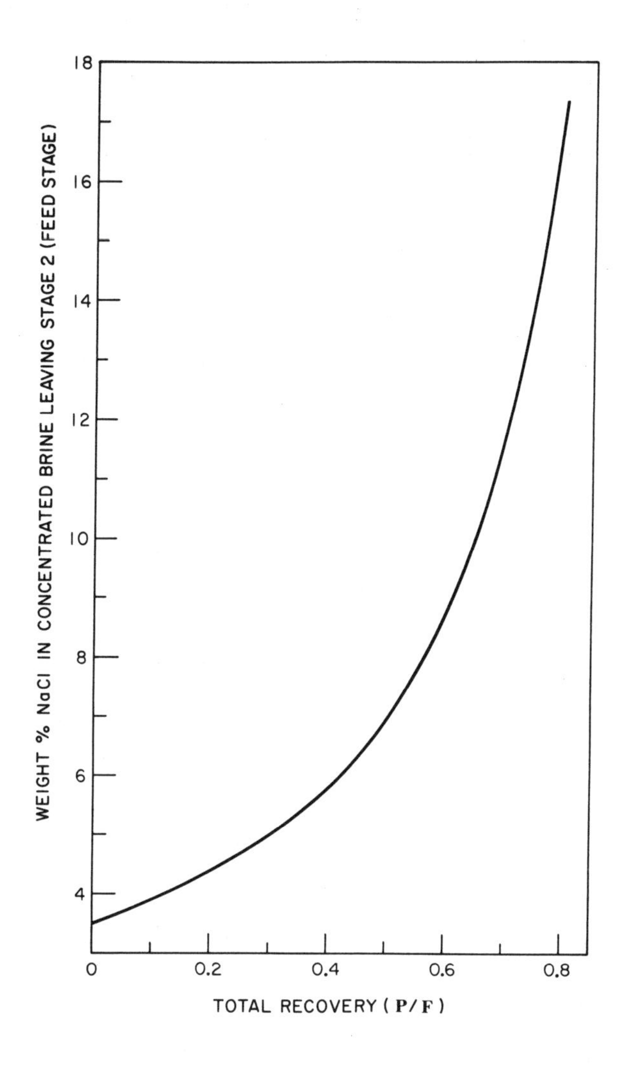

Figure 4. 34. Total fraction product recovery v. salt concentration in brine leaving feed stage in a two-stage desalination process. Data of Kimura, Sourirajan, and Ohya (1969).

System: sodium chloride-water; operating pressure: 1500 psig; feed flow condition: laminar flow between flat parallel membranes; film type: CA-NRC-18.

Feed concentration = 3. 5 wt.% NaCl;

Product concentration: 500 ppm NaCl.

The membrane area and total power consumption due to frictional pressure drop per unit product rate for Stages 1 and 2 may be calculated as before; using the form of Equations 4-303 and 4-320, for Stage 1

$$\left(\frac{S}{P}\right)_1 = \frac{X_{1,1}M_1}{\Delta_1 v_W^*} \tag{4-338}$$

and

$$\left[\frac{(\text{Power consumption})_{\text{total}}}{P}\right]_1 = (N_{Re})_1^2 G_1 \tag{4-339}$$

where

$$(N_{Re})_1 = \frac{2}{\Delta_1 \nu}\left(\frac{P}{W}\right)_1 \tag{4-340}$$

and

$$G_1 = \frac{3}{16}\frac{\rho}{g_c}(N_{Sc})^3 \frac{(v_W^*)^2(\alpha_0)^3 X_{1,1}}{\Delta_1} E \tag{4-341}$$

and M_1 represents the total number of inner stages in Stage 1. For Stage 2, the corresponding equations are similar but Δ_1 is replaced by $\Delta_1\Delta_2$ in view of Equation 4-329. Thus

$$\left(\frac{S}{P}\right)_2 = \frac{X_{2,1}M_2}{\Delta_1\Delta_2 v_W^*} \tag{4-342}$$

and

$$\left[\frac{(\text{Power consumption})_{\text{total}}}{P}\right]_2 = (N_{Re})_2^2 G_2 \tag{4-343}$$

where

$$(N_{Re})_2 = \frac{2}{\Delta_1\Delta_2 \nu}\left(\frac{P}{W}\right)_2 \tag{4-344}$$

and

$$G_2 = \frac{3}{16}\frac{\rho}{g_c}(N_{Sc})^3 \frac{(v_W^*)^2(\alpha_0)^3 X_{2,1}}{\Delta_1\Delta_2} \cdot E \tag{4-345}$$

and M_2 represents the total number of inner stages in Stage 2.

Figure 4-35 illustrates the total membrane area requirements for Stage 2 (feed stage) per unit final product rate in Stage 1, as a function of total product recovery (**P/F**) for different preset values of Δ_2, h, and M_2. Figure 4-36 gives total membrane area for Stage 1 per unit product rate in Stage 1 as a function of product recovery $\mathbf{\Delta_1}$ for different preset values of h and M_1. These data again illustrate the importance of the number of inner stages in each unit stage.

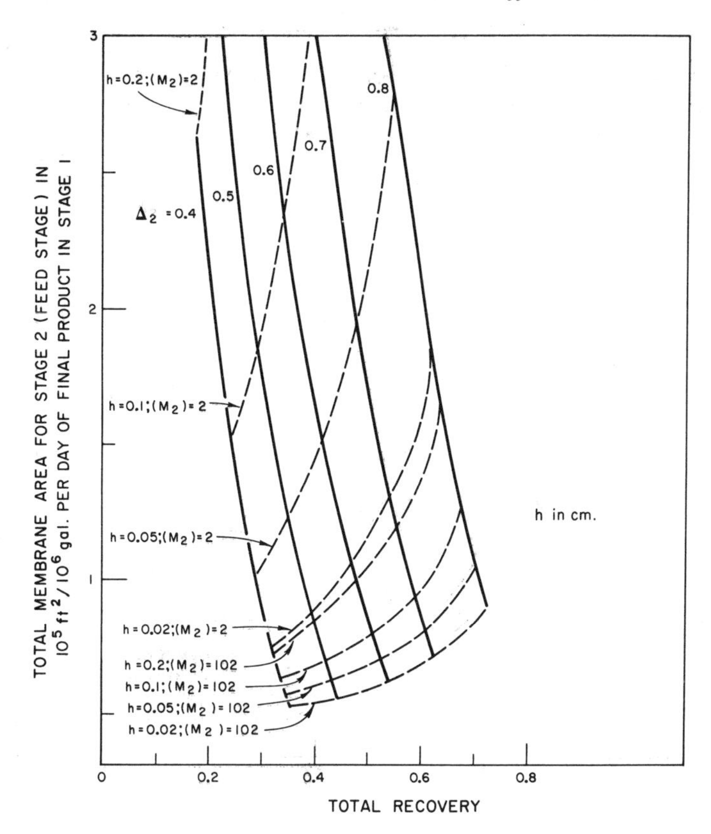

Figure 4-35. Membrane area requirements in feed-stage per unit final product rate in a two-stage desalination process. Data of Kimura, Sourirajan, and Ohya (1969).

System: sodium chloride-water; operating pressure: 1500 psig; feed flow condition: laminar flow between flat parallel membranes; film type: CA-NRC-18.

Film specifications at 1500 psig:

$$A = 2.1 \times 10^{-6} \frac{\text{g. mole } H_2O}{\text{cm}^2 \text{ sec atm}}$$

$$\left(\frac{D_{AM}}{K\delta}\right)_{NaCl} = 38.6 \times 10^{-5} \text{ cm/sec}$$

Data for A factor = 0.5; feed concentration to stage 2 = 3.5 wt.% NaCl; total number of inner stages in Stage 2 = (M_2).

Detailed optimisation of the various design variables for the reverse osmosis desalination process is beyond the scope of this discussion. The results of some preliminary studies on the multistage reverse osmosis process for saline water conversion have been reported by Bray and Menzel (1966), and Johnson et al. (1967). The foregoing discussion offers a different approach to the subject; it establishes a rational basis for stage-wise reverse osmosis

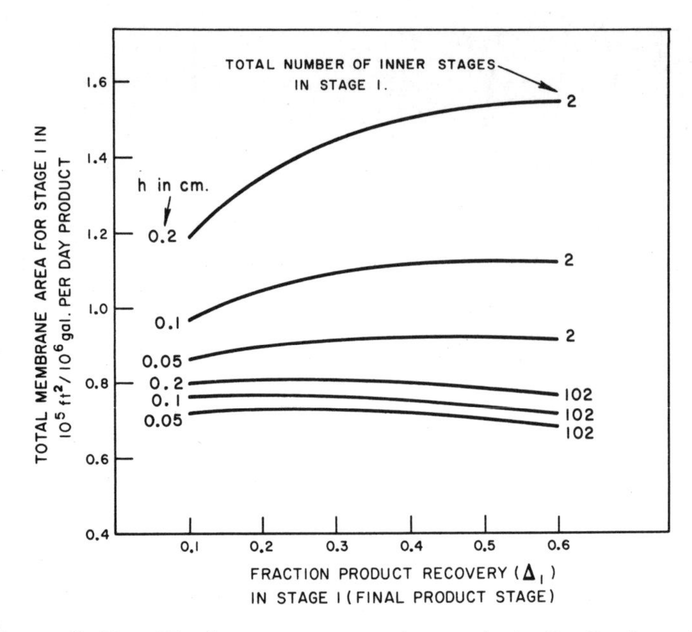

Figure 4-36. Membrane area requirements in the final product stage per unit product rate in a two-stage desalination process. Data of Kimura, Sourirajan, and Ohya (1969).

System: sodium chloride-water; operating pressure: 1500 psig; feed flow condition: laminar flow between flat parallel membranes; film type: CA-NRC-18.

Film specifications at 1500 psig:

$$A = 0.97 \times 10^{-6} \frac{\text{g. mole } H_2O}{\text{cm}^2 \text{ sec atm}}$$

$$\left(\frac{D_{AM}}{K\delta}\right)_{NaCl} = 0.9 \times 10^{-5} \text{ cm/sec}$$

Data for A factor = 0.5; product concentration for stage 1 = 500 ppm NaCl; NaCl in concentrated brine leaving stage 1 = 3.5 wt.%.

process design in all its applications based on the fundamental equations governing transport through reverse osmosis membranes.

6. Some General Equations for Reverse Osmosis Process Design

On the basis of the various equations governing transport through reverse osmosis membranes presented in the preceding sections, Ohya and Sourirajan (1969d) have derived analytical expressions, in nondimensional forms, for instantaneous values of solute separation, product rate, and the other related process variables, for solution-membrane-operating systems characterised by the nondimensional parameters γ, θ, and λ defined Equations 4-11, 4-204, and 4-13 respectively. Their analysis offers a generalised approach to reverse osmosis process design.

Their analysis applies to membranes for which $(D_{AM}/K\delta)$ is independent of feed concentration and feed flow rate, and to solution systems for which

Equations 4-5 to 4-8 are valid. These restrictions do not limit the scope of the analysis; they only simplify the analytical expressions involved.

The analysis of Ohya and Sourirajan (1969d) is given below. The case of batch-wise reverse osmosis operation is considered first, and the scope of the analysis is then extended to the flow case.

System specification. The solution-membrane-operating system is specified by the fundamental nondimensional design parameters γ, θ, and λ defined earlier as follows:

$$\gamma = \frac{BX^0_{A1}}{P} = \frac{B}{cP}c^0_{A1} = \frac{\text{osmotic pressure of feed solution}}{\text{operating pressure}} \tag{4-11}$$

$$\theta = (D_{AM}/K\delta)/v^*_W = \frac{\text{solute transport parameter}}{\text{pure water permeation velocity}} \tag{4-204}$$

$$\lambda = k/(D_{AM}/K\delta) = \frac{\text{mass transfer coefficient on high-pressure side of membrane}}{\text{solute transport parameter}} \tag{4-13}$$

Definitions. The dimensionless solute concentrations given by the general relation $C = c_A/c^0_A$ are expressed specifically as follows:

$$C_1 = c_{A1}/c^0_{A1} \tag{4-48}$$

$$C^0_1 = c^0_{A1}/c^0_{A1} = 1 \tag{4-48a}$$

$$C_2 = c_{A2}/c^0_{A1} \tag{4-93}$$

$$C^0_2 = c^0_{A2}/c^0_{A1} \tag{4-93a}$$

$$C_3 = c_{A3}/c^0_{A1} \tag{4-49}$$

$$C^0_3 = c^0_{A3}/c^0_{A1} \tag{4-49a}$$

$$¢ = C_3/C^0_3 = c_{A3}/c^0_{A3} \tag{4-346}$$

where, as before, the subscripts 1 and 2 refer to the bulk solution and the concentrated boundary solution respectively on the high pressure side of the membrane, the subscript 3 refers to the membrane permeated product solution, and the superscript zero refers to the initial condition in the batch process or to the condition at membrane entrance in the flow process. Combining Equations 4-11 and 4-203,

$$\gamma = \gamma^* c^0_{A1} \tag{4-347}$$

and

$$\gamma^* c_{A3} = \gamma C_3 \tag{4-348}$$

Basic equations. The following basic equations have been derived in the preceding sections:

$$v_W = N_B/c = v^*_W[1 - \gamma(C_2 - C_3)] \tag{4-29}$$

$$C_1 = C_3 q \tag{4-349}$$

where

$$q = 1 + \frac{1}{(\gamma C_3 + \theta)} \exp\left[-\frac{1}{\lambda(\gamma C_3 + \theta)} \right] \tag{4-350}$$

$$C_2 = [1 + 1/(\gamma C_3 + \theta)]\, C_3 \tag{4-351}$$

$$C_3 = \frac{\sqrt{[(1 + \theta - \gamma C_2)^2 + 4\gamma\theta C_2]} - (1 + \theta - \gamma C_2)}{2\gamma} \tag{4-23}$$

Equations 4-349, 4-350, and 4-351 are nondimensional forms of Equations 4-218, 4-219, and 4-210 respectively.

Equations 4-29, 4-349, 4-350, 4-351, and 4-23 are applicable at any point on the membrane surface at any time in the reverse osmosis separation system.

7. Analysis of Batch-wise Reverse Osmosis Process

The object of this analysis is to obtain analytical expressions for the change of volume of solution on the high-pressure side of the membrane (V_1/V_1^0), the concentration of the bulk solution (C_1) and that of the concentrated boundary solution (C_2) on the high pressure side of the membrane, the change in the permeating velocity of water through the membrane $[v_w/(v_w)^0]$, solute separation, and the other related quantities, at any instant; each of these expressions, in terms of dimensionless quantities, is required as a function of the concentration of the product solution (C_3) on the atmospheric pressure side of the membrane, or of time from the start of the operation for solution-membrane-operating systems specified by the parameters γ, θ, and λ.

General case. At any time t, let V_1 and C_1 be the volume and concentration respectively of the bulk solution on the high pressure side of the membrane, and let V_3 and C_3 be the corresponding values for the membrane-permeated product solution on the atmospheric pressure side of the membrane. At time $t = 0$,

$$C_1 = C_1^0 = 1 \tag{4-352}$$

$$C_3 = C_3^0 \tag{4-353}$$

$$V_1 = V_1^0 \tag{4-354}$$

At any time t,

$$-dV_1 = dV_3 \tag{4-355}$$

$$-d(V_1C_1) = C_3 dV_3 = -C_3 dV_1 \tag{4-356}$$

Therefore

$$dV_1/V_1 = -dC_1/(C_1 - C_3) \tag{4-357}$$

Integrating Equation 4-357,

$$\ln V_1/V_1^0 = -\int_1^{C_1} dC_1/(C_1 - C_3) = -Z \tag{4-358}$$

where

$$Z = \int_1^{C_1} dC_1/(C_1 - C_3) \tag{4-359}$$

From Equation 4-349,

$$C_1 - C_3 = C_3(q - 1) \tag{4-360}$$

Differentiating Equations 4-349 and 4-350, and using Equation 4-360, Z may be expressed as

$$Z = \int_{C_3^0}^{C_3} \left[\frac{\gamma}{\lambda(\gamma C_3 + \theta)^2} - \frac{\gamma}{(\gamma C_3 + \theta)} + \left\{ \frac{(\gamma C_3 + \theta)}{C_3} \exp \frac{1}{\lambda(\gamma C_3 + \theta)} \right\} + \frac{1}{C_3} \right] dC_3 \tag{4-361}$$

On integration Equation 4-361 becomes

$$Z = -\frac{1}{\lambda}\left[\frac{1}{(\gamma C_3 + \theta)} - \frac{1}{(\gamma C_3^0 + \theta)}\right] - \left(1 + \theta e^{\frac{1}{\lambda\theta}} - \frac{1}{\lambda} - \theta\right) \ln \frac{(\gamma C_3 + \theta)}{(\gamma C_3^0 + \theta)}$$

$$+ \left(1 + \theta e^{\frac{1}{\lambda\theta}}\right) \ln \frac{C_3}{C_3^0} + \gamma(C_3 - C_3^0)$$

$$+ \frac{1}{\lambda} \sum_{m=1}^{\infty} \frac{\theta^m}{m} \left\{ \left(\frac{1}{\gamma C_3 + \theta}\right)^m - \left(\frac{1}{\gamma C_3^0 + \theta}\right)^m \right\} \sum_{n=1}^{\infty} \frac{1}{(m + n + 1)!} \frac{1}{(\lambda\theta)^{m+n}} \tag{4-362}$$

Even though Equation 4-362 is a general solution for Z, it is inconvenient for practical use. Simpler expressions for Z will be derived later in this discussion for some cases of practical interest.

The permeating velocity of solvent water (v_w) through the membrane can be expressed as

$$v_w = -\frac{1}{S} \frac{dV_1}{dt} \tag{4-363}$$

where S is the area of the membrane surface.

Combining Equation 4-29 and 4-363,

$$-\frac{1}{S} \frac{dV_1}{dt} = v_w^* \left[1 - \gamma(C_2 - C_3)\right] \tag{4-364}$$

From Equations 4-358 and 4-359,

$$V_1 = V_1^0 \exp(-Z) \tag{4-365}$$

Therefore

$$dV_1 = - V_1^0 \exp(-Z) dZ \tag{4-366}$$

Combining Equations 4-364 and 4-366

$$\frac{Sv_W^*}{V_1^0}\,dt = \frac{\exp(-Z)dZ}{1-\gamma(C_2 - C_3)} \tag{4-367}$$

Let a dimensionless parameter τ, representing time, be defined as

$$\tau = Sv_W^* t/V_1^0 \tag{4-368}$$

From Equation 4-367

$$\tau = \int_{C_3^0}^{C_3} \frac{\exp(-Z)dZ/dC_3}{1-\gamma(C_2 - C_3)}\,dC_3 \tag{4-369}$$

From Equation 4-351,

$$C_2 - C_3 = C_3/(\gamma C_3 + \theta) \tag{4-370}$$

and

$$1 - \gamma(C_2 - C_3) = \theta/(\gamma C_3 + \theta) \tag{4-371}$$

Substituting for dZ/dC_3 from Equation 4-361,

$$\tau = \int_{C_3^0}^{C_3} \left[\frac{\gamma}{\lambda(\gamma C_3 + \theta)} - \gamma + \frac{(\gamma C_3 + \theta)}{C_3}\left\{\left[(\gamma C_3 + \theta)\exp\frac{1}{\lambda(\gamma C_3 + \theta)}\right] + 1\right\}\right]\frac{\exp(-Z)}{\theta}\,dC_3 \tag{4-372}$$

It does not seem possible to obtain an analytical solution to Equation 4-372, but τ may be evaluated by numerical integration as functions of γ, θ, and λ for any assumed value of C_3. For this integration, C_3^0 is obtained from the simultaneous solutions of the following equations obtained from Equations 4-351 and 4-23 which, at $t = 0$, become

$$C_2^0 = [1 + 1/(\gamma C_3^0 + \theta)]\,C_3^0 \tag{4-373}$$

and

$$C_3^0 = \frac{\sqrt{[(1+\theta-\gamma C_2^0)^2 + 4\gamma\theta C_2^0]} - (1+\theta-\gamma C_2^0)}{2\gamma} \tag{4-374}$$

The instantaneous solute separation f may be defined as

$$f = 1 - C_3/C_1 \tag{4-375}$$

Using Equation 4-349 again,

$$f = 1 - 1/q \tag{4-376}$$

or

$$f = \frac{\exp\left[-\frac{1}{\lambda(\gamma C_3 + \theta)}\right]}{(\gamma C_3 + \theta) + \exp\left[-\frac{1}{\lambda(\gamma C_3 + \theta)}\right]} \tag{4-377}$$

Also, from Equations 4-29 and 4-370,

$$(v_w/v_w^*) = \theta/(\gamma C_3 + \theta) \tag{4-378}$$

$$[v_w/(v_w)^o] = \frac{1 - \gamma(C_2 - C_3)}{1 - \gamma(C_2^o - C_3^o)} \tag{4-379}$$

Thus for the general case of the batch-wise process, the dimensionless quantities (V_1/V_1^o), Z, τ, C_1, C_2, C_2^o, C_3^o, q, f, (v_w/v_w^*), and $[v_w/(v_w)^o]$ are expressed by Equations 4-358, 4-362, 4-372, 4-349, 4-351, 4-373, 4-374, 4-350, 4-377, 4-378, and 4-379 respectively, and their magnitudes can be calculated for any assumed value of C_3 with reference to any particular system specified by the parameters γ, θ, and λ. The correlations of (V_1/V_1^o), C_1, C_3, $\bar{C}_3$, q, and $[v_w/(v_w)^o]$ as functions of τ are particularly convenient for reverse osmosis process design. A set of such correlations is illustrated in Figures 4-37 to 4-42. An extensive compilation of such correlations may prove very useful for design purposes.

8. Some Special Cases of the Batch-wise Process

Case 1. $\gamma = 0$, and $\lambda \to \infty$. This limiting case is of interest both for purposes of comparison with actual cases, and as an approximation to those practical cases where the osmotic pressure and concentration polarisation effects are negligible. Using Equations 4-29, 4-349, 4-350, 4-358, 4-367, and 4-376 for the special case, the following relationships are obtained.

$$(v_w/v_w^*) = 1 \tag{4-380}$$

$$[v_w/(v_w)^o] = 1 \tag{4-381}$$

$$q = (1 + \theta)/\theta \tag{4-382}$$

$$f = 1/(1 + \theta) \tag{4-383}$$

$$C_1 = C_3(1 + \theta)/\theta \tag{4-384}$$

$$C_2 = C_1 \tag{4-385}$$

$$C_2^o = C_1^o = 1 \tag{4-386}$$

$$C_3 = C_1\theta/(1 + \theta) \tag{4-387}$$

$$C_3^o = C_1^o\theta/(1 + \theta) = \theta/(1 + \theta) \tag{4-388}$$

$$C_1 - C_3 = C_1/(1 + \theta) \tag{4-389}$$

$$\ln \frac{V_1}{V_1^o} = -\int_1^{C_1} (1 + \theta)\frac{dC_1}{C_1} \tag{4-390}$$

$$= -(1 + \theta) \ln C_1 \tag{4-391}$$

Therefore

$$Z = (1 + \theta) \ln C_1 \tag{4-392}$$

Figure 4-37.
τ v. (V_1^0/V_1) for $\theta = 0.01$, $\gamma = 0.1$ to 0.5, and $\lambda = 200$ to ∞. Data of Ohya and Sourirajan (1969d).

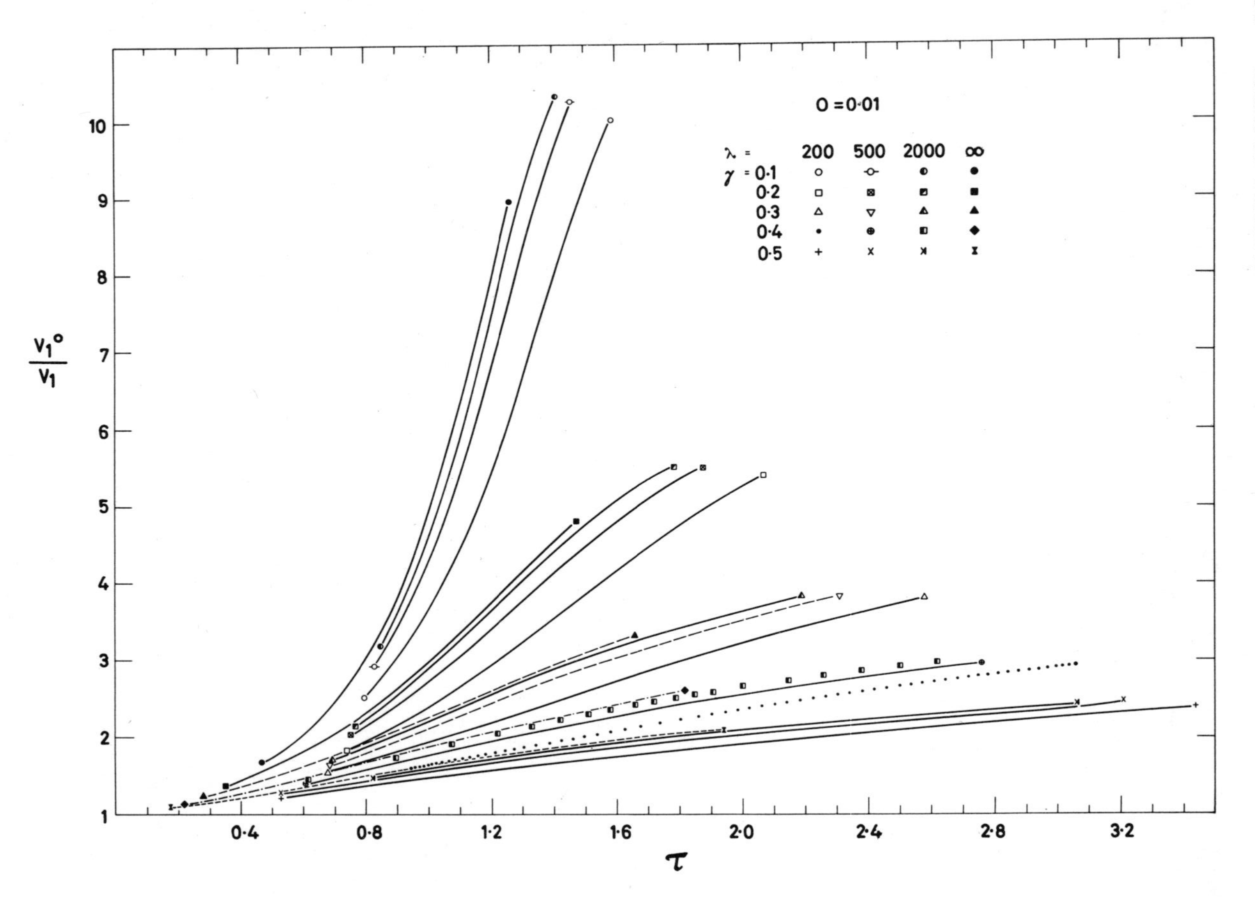

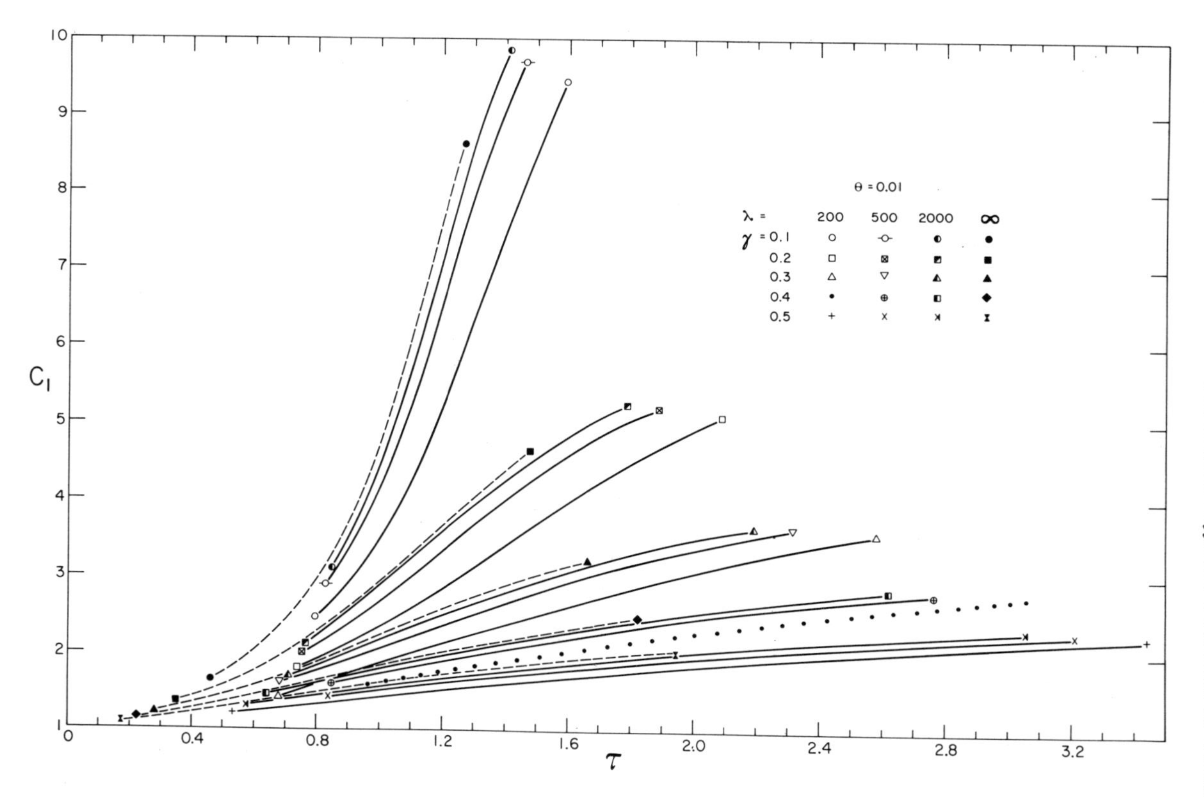

Figure 4-38.
τ v. C_1 for
$\theta = 0.01$, $\gamma = 0.1$ to 0.5
and $\lambda = 200$ to ∞. Data
of Ohya and Sourirajan
(1969d).

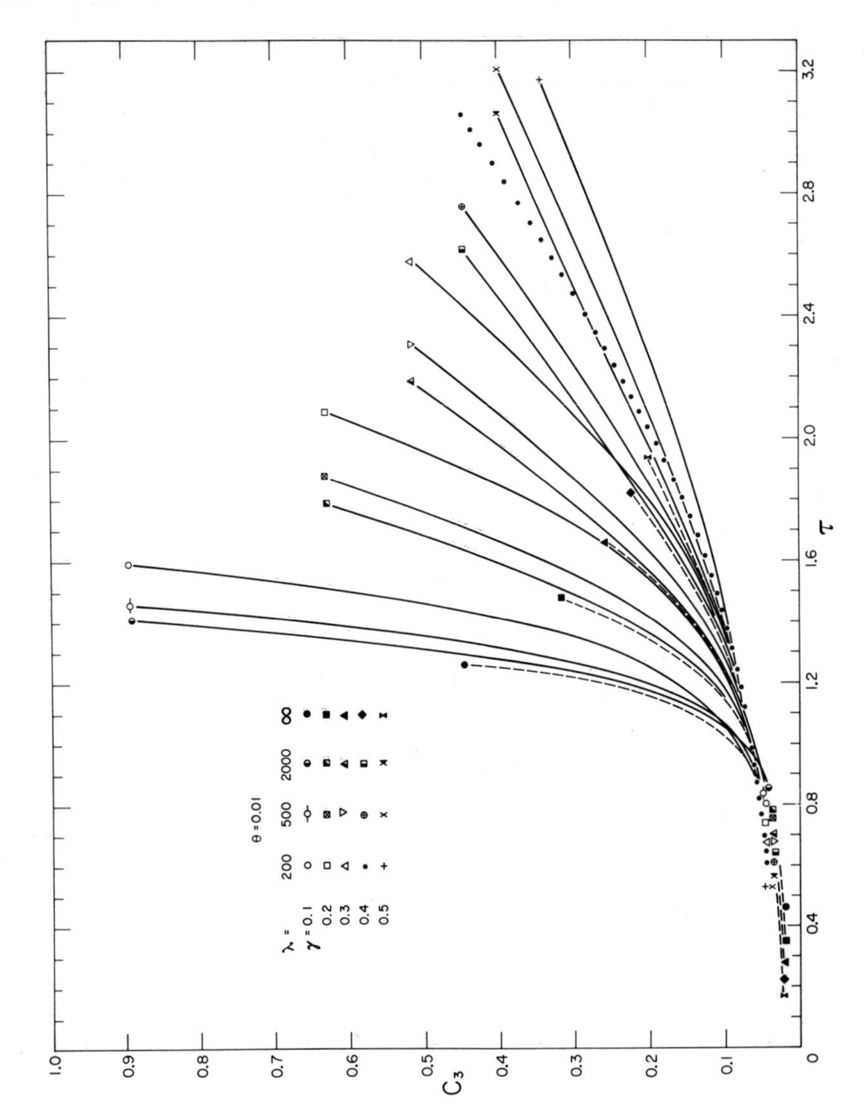

Figure 4-39.
τ v. C_3 for
$\theta = 0.01, \gamma = 0{\cdot}1$ to
0.5 and $\lambda = 200$ to ∞.
Data of Ohya and
Sourirajan (1969d).

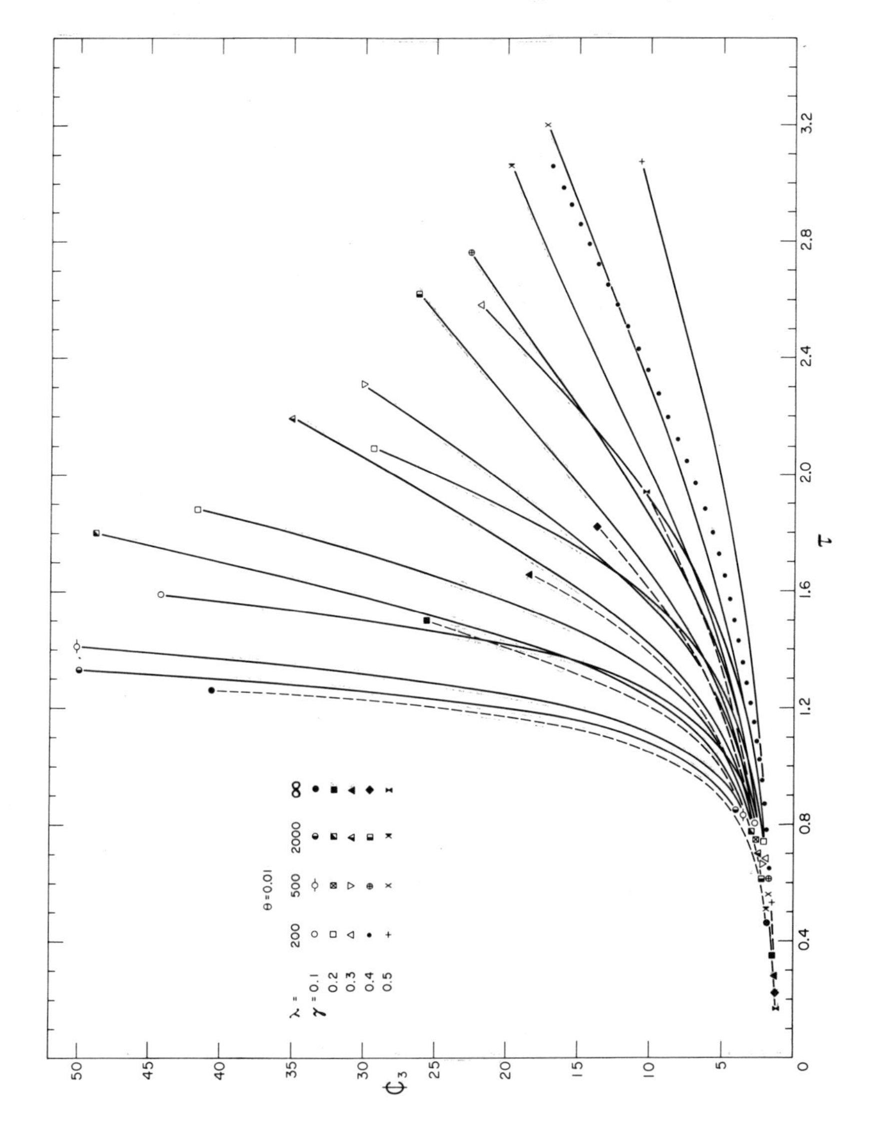

Figure 4-40.
τ v. Ȼ_3 for
$\theta = 0.01, \gamma = 0.1$ to 0.5 and $\lambda = 200$ to ∞. Data of Ohya and Sourirajan (1969d).

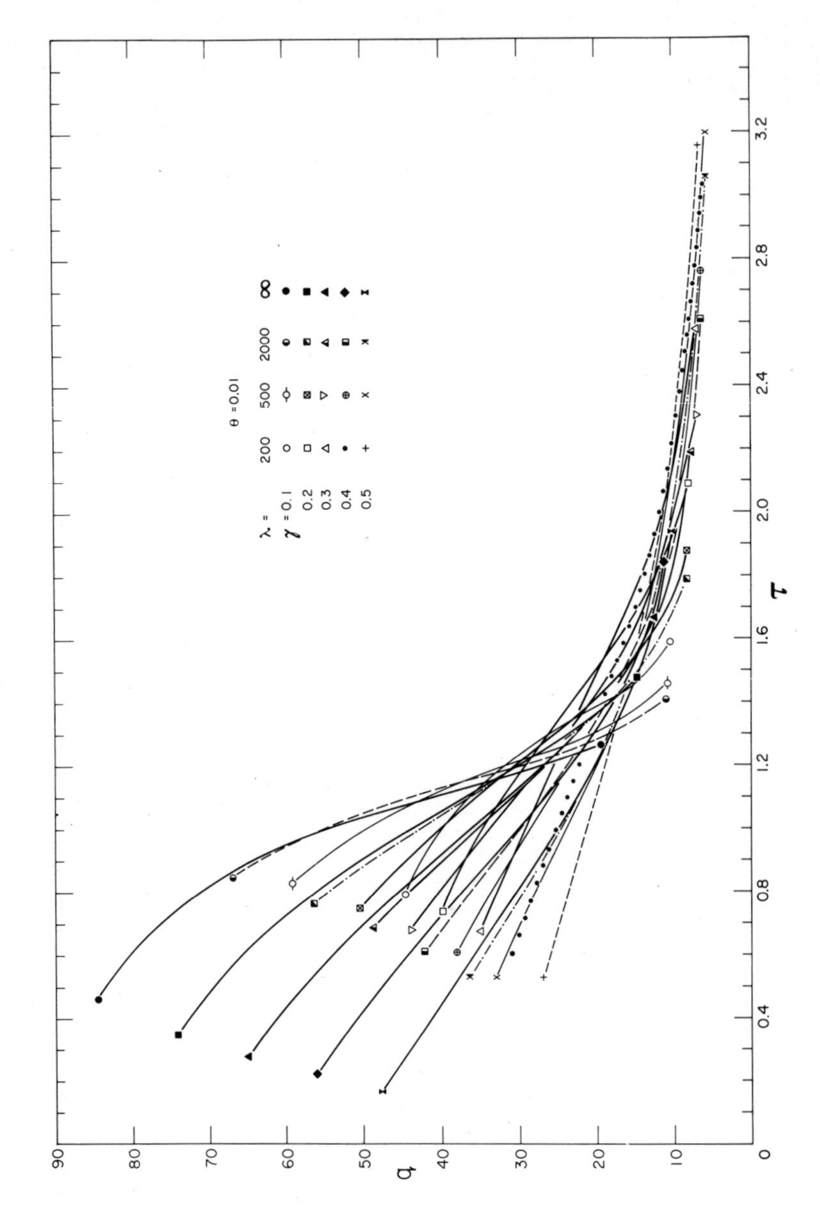

Figure 4-41. τ v. q for $\theta = 0.01$, $\gamma = 0.1$ to 0.5 and $\lambda = 200$ to ∞. Data of Ohya and Sourirajan (1969d).

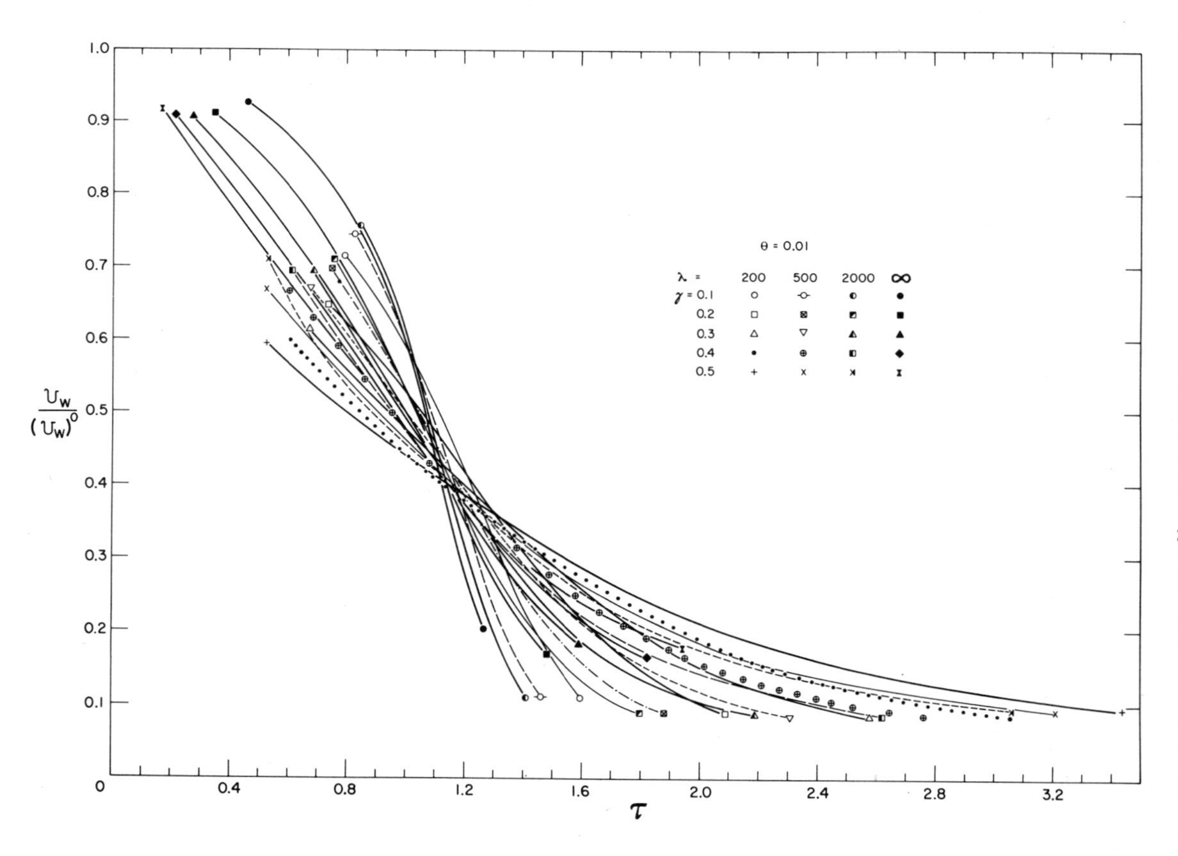

Figure 4-42.
τ v. $[v_W/(v_W)^0]$ for $\theta = 0.01, \gamma = 0.1$ to 0.5, $\lambda = 200$ to ∞. Data of Ohya and Sourirajan (1969d).

$$dZ = (1 + \theta) \frac{dC_1}{C_1} \tag{4-393}$$

$$\tau = \int \exp(-Z)dZ \tag{4-394}$$

$$= \int_1^{C_1} (1 + \theta)\, C_1^{-(1+\theta)} \frac{dC_1}{C_1} \tag{4-395}$$

$$= 1 - C_1^{-(1+\theta)} \tag{4-396}$$

Thus for the special case $\gamma = 0$, and $\lambda \to \infty$, the quantities (V_1/V_1^0), Z, τ, C_1, C_2^0, C_3^0, q, f, (v_W/v_W^*) and $[v_W/(v_W)^0]$ are expressed by Equations 4-391, 4-392, 4-396, 4-384, 4-385, 4-386, 4-388, 4-382, 4-383, 4-380, and 4-381 respectively, and their magnitudes can be calculated for any assumed value of C_3 with reference to any particular system specified by the parameter θ. Further, from Equation 4-396,

$$C_1 = \left(\frac{1}{1-\tau}\right)^{1/(1+\theta)} \tag{4-397}$$

Using Equations 4-387 and 4-388,

$$C_3 = C_3^0 \left(\frac{1}{1-\tau}\right)^{1/(1+\theta)} \tag{4-398}$$

Therefore

$$¢_3 = \frac{C_3}{C_3^0} = \left[\frac{1}{1-\tau}\right]^{1/(1+\theta)} \tag{4-399}$$

The form of Equation 4-399 is particularly interesting for practical design purposes, and it will be used in the next section.

Case 2. $\lambda \to \infty$. This limiting case is of interest both for purposes of comparison with actual cases, and also as an approximation to practical cases involving extremely high mass transfer coefficients, or very efficient stirring in the vicinity of the membrane surface on the high pressure side. Using Equations 4-29, 4-349, 4-350, 4-351, 4-23, and 4-361 for the above special case, the following equations are obtained.

$$q = 1 + 1/(\gamma C_3 + \theta) \tag{4-400}$$

$$f = 1 - 1/q = 1/(\gamma C_3 + \theta + 1) \tag{4-401}$$

$$C_2 = C_1 \tag{4-385}$$

$$C_2^0 = C_1^0 = 1 \tag{4-386}$$

$$C_1 = [1 + 1/(\gamma C_3 + \theta)]\, C_3 \tag{4-402}$$

$$C_3 = \frac{\sqrt{[(1 + \theta - \gamma C_1)^2 + 4\gamma\theta C_1]} - (1 + \theta - \gamma C_1)}{2\gamma} \tag{4-403}$$

$$C_3^0 = \frac{\sqrt{[(1 + \theta - \gamma)^2 + 4\gamma\theta]} - (1 + \theta - \gamma)}{2\gamma} \tag{4-404}$$

$$(v_w/v_w^*) = 1 - \gamma(C_1 - C_3) \tag{4-405}$$

$$[v_w/(v_w)^0] = \frac{1 - \gamma(C_1 - C_3)}{1 - \gamma(1 - C_3^0)} \tag{4-406}$$

$$Z = \gamma(C_3 - C_3^0) + (1 + \theta)\ln\frac{C_3}{C_3^0} - \ln\left[\frac{C_3 + \frac{\theta}{\gamma}}{C_3^0 + \frac{\theta}{\gamma}}\right] \tag{4-407}$$

Therefore

$$\ln\frac{V_1}{V_1^0} = -\left[\gamma(C_3 - C_3^0) + (1 + \theta)\ln\frac{C_3}{C_3^0} - \ln\frac{\left(C_3 + \frac{\theta}{\gamma}\right)}{\left(C_3^0 + \frac{\theta}{\gamma}\right)}\right] \tag{4-408}$$

From Equations 4-367 and 4-371

$$\frac{Sv_w^* dt}{V_1^0} = \frac{\exp(-Z)dZ}{1 - \gamma(C_1 - C_3)} \tag{4-409}$$

and

$$1 - \gamma(C_1 - C_3) = \theta/(\gamma C_3 + \theta) \tag{4-410}$$

Differentiating Equation 4-407, and substituting in Equation 4-409,

$$\tau = \int_{C_3^0}^{C_3}\left[\frac{(\gamma C_3 + \theta)^2}{C_3\theta} + \frac{1}{C_3}\right]\exp(-Z)dC_3 \tag{4-411}$$

where

$$\exp(-Z) = \left[\frac{C_3}{C_3^0}\right]^{-(1+\theta)}\left[\frac{C_3 + \frac{\theta}{\gamma}}{C_3^0 + \frac{\theta}{\gamma}}\right]\exp[-\gamma(C_3 - C_3^0)] \tag{4-412}$$

Thus for the special case $\lambda \to \infty$, the quantities (V_1/V_1^0), Z, τ, C_1, C_2, C_2^0, C_3^0, q, f, (v_w/v_w^*), and $[v_w/(v_w)^0]$ are expressed by Equations 4-408, 4-407, 4-411, 4-402, 4-385, 4-386, 4-404, 4-400, 4-401, 4-405, and 4-406 respectively, and their magnitudes can be calculated for any assumed value of C_3 with reference to any particular system specified by the parameters γ and θ.

Further, following the form of Equation 4-399, one may express

$$\mathcal{C}_3 = \frac{C_3}{C_3^0} = \left[\frac{1}{1 - F(\tau)}\right]^{\frac{1}{1+\theta}} \tag{4-413}$$

and establish a correlation between τ and $F(\tau)$ for different values of γ and θ for the case $\lambda \to \infty$. A set of such correlations is illustrated in Figures 4-43, 4-44, and 4-45. Such correlations are extremely useful for a quick evaluation of $\mathcal{C}_3$ for any given value of τ.

Case 3. λ is large and finite, and $\lambda\theta > 1$. In most cases of practical interest, the mass transfer coefficient is sufficiently large so that λ becomes large and $\lambda\theta$ is greater than 1. When λ is large,

$$\exp\left[-\frac{1}{\lambda(\gamma C_3 + \theta)}\right] \approx 1 - \frac{1}{\lambda(\gamma C_3 + \theta)} \tag{4-414}$$

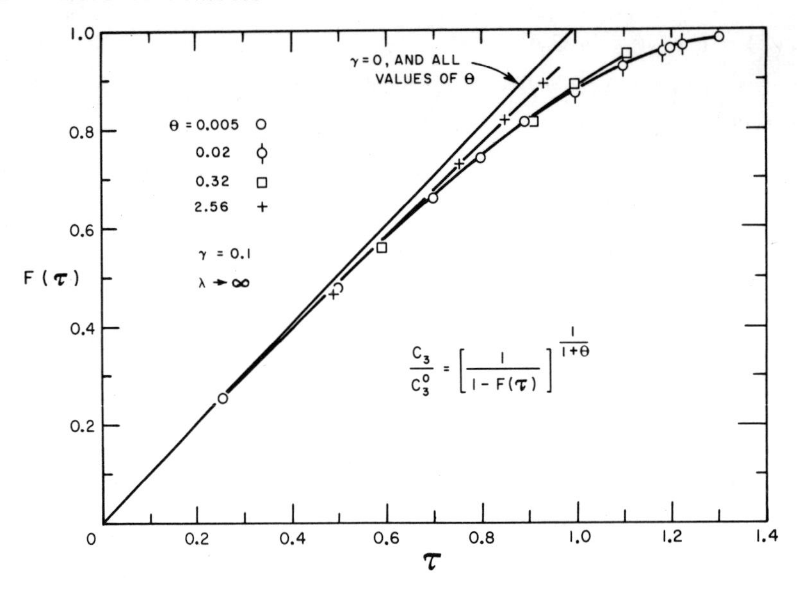

Figure 4-43 τ v. $F(\tau)$ for $\gamma = 0.1$, $\theta = 0.005$ to 2.56 and $\lambda = \infty$. Data of Ohya and Sourirajan (1969d).

Therefore

$$q = 1 + \frac{1}{(\gamma C_3 + \theta)}\left[1 - \frac{1}{\lambda(\gamma C_3 + \theta)}\right] \tag{4-415}$$

$$= 1 + \frac{1}{(\gamma C_3 + \theta)} - \frac{1}{\lambda(\gamma C_3 + \theta)^2} \tag{4-416}$$

or

$$q - 1 = \frac{\lambda(\gamma C_3 + \theta) - 1}{\lambda(\gamma C_3 + \theta)^2} \tag{4-417}$$

Therefore

$$f = 1 - 1/q = (q - 1)/q \tag{4-418}$$

$$= \left[\frac{\lambda(\gamma C_3 + \theta) - 1}{\lambda(\gamma C_3 + \theta)^2}\right] \Big/ \left[1 + \frac{1}{(\gamma C_3 + \theta)} - \frac{1}{\lambda(\gamma C_3 + \theta)^2}\right] \tag{4-419}$$

$$C_1 = C_3 q \tag{4-349}$$

$$= \left[1 + \frac{1}{(\gamma C_3 + \theta)} - \frac{1}{\lambda(\gamma C_3 + \theta)^2}\right] C_3 \tag{4-420}$$

Differentiating Equation 4-420, and evaluating C_3 (q − 1),

$$dC_1 = \left[1 + \frac{\theta}{(\gamma C_3 + \theta)^2} - \frac{\theta - \gamma C_3}{\lambda(\gamma C_3 + \theta)^3}\right] dC_3 \tag{4-421}$$

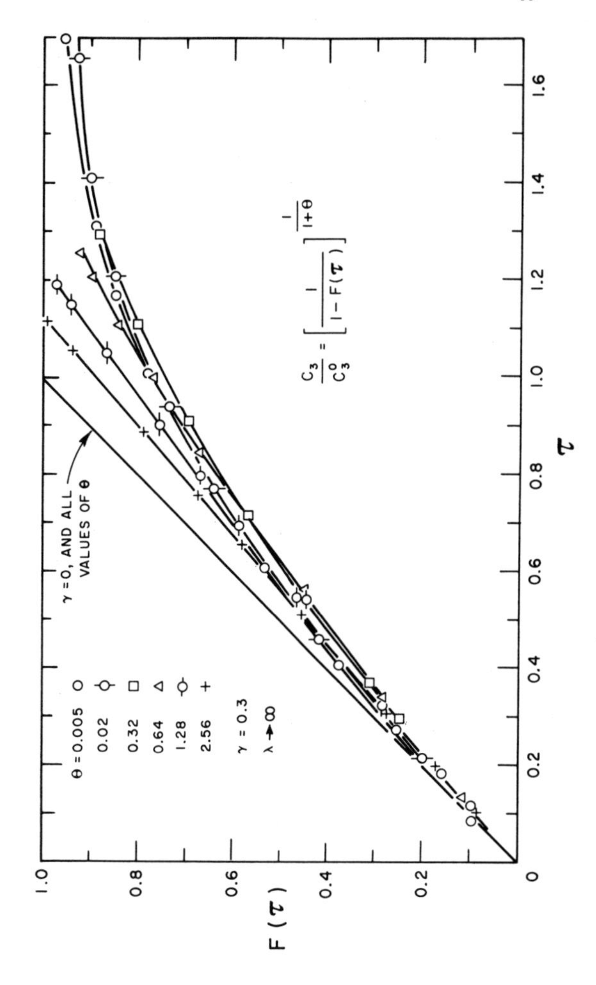

Figure 4-44.
τ v. $F(\tau)$ for $\gamma = 0.3$, $\theta = 0.005$ to 2.56 and $\lambda = \infty$. Data of Ohya and Sourirajan (1969d).

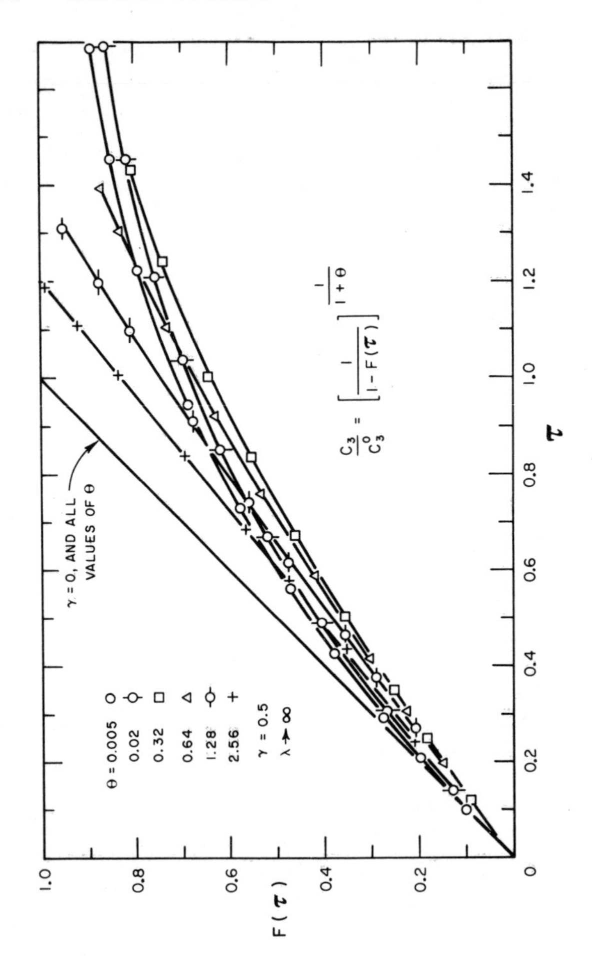

Figure 4-45. τ v. $F(\tau)$ for $\gamma = 0.5, \theta = 0.005$ to 2.56 and $\lambda = \infty$. Data of Ohya and Sourirajan (1969d).

$$\frac{1}{C_3(q-1)} = \frac{\lambda(\gamma C_3 + \theta)^2}{C_3[\lambda(\gamma C_3 + \theta) - 1]} \cdot \tag{4-422}$$

$$Z = \int_{C_3^0}^{C_3} \frac{dC_1}{C_3(q-1)} \tag{4-359}$$

$$= \int_{C_3^0}^{C_3} \left[\frac{\lambda(\gamma C_3 + \theta)^2}{C_3[\lambda(\gamma C_3 + \theta) - 1]} + \frac{\lambda\theta + 1}{C_3[\lambda(\gamma C_3 + \theta) - 1]} - \frac{2\theta}{C_3(\gamma C_3 + \theta)[\lambda(\gamma C_3 + \theta) - 1]} \right] dC_3 \tag{4-423}$$

$$= \gamma(C_3 - C_3^0) + \left[1 + \theta + \frac{\theta}{(\lambda\theta - 1)}\right] \ln \frac{C_3}{C_3^0} - 2 \ln \left[\frac{C_3 + \frac{\theta}{\gamma}}{C_3^0 + \frac{\theta}{\gamma}}\right] + \left[1 - \frac{1}{\lambda(\lambda\theta - 1)}\right] \ln \left(\frac{C_3 + \frac{\theta}{\gamma} - \frac{\theta}{\gamma\lambda}}{C_3^0 + \frac{\theta}{\gamma} - \frac{1}{\gamma\lambda}}\right) \tag{4-424}$$

Therefore

$$\ln \frac{V_1}{V_1^0} = -\gamma(C_3 - C_3^0) - \left[1 + \theta + \frac{\theta}{(\lambda\theta - 1)}\right] \ln \frac{C_3}{C_3^0} + 2 \ln \left[\frac{C_3 + \frac{\theta}{\gamma}}{C_3^0 + \frac{\theta}{\gamma}}\right] - \left[1 - \frac{1}{\lambda(\lambda\theta - 1)}\right] \ln \left(\frac{C_3 + \frac{\theta}{\gamma} - \frac{1}{\gamma\lambda}}{C_3^0 + \frac{\theta}{\gamma} - \frac{1}{\gamma\lambda}}\right) \tag{4-425}$$

Differentiating Equation 4-424,

$$\frac{dZ}{dC_3} = \gamma - \frac{2}{C_3 + \frac{\theta}{\gamma}} + \left[1 + \theta + \frac{\theta}{(\lambda\theta - 1)}\right] \frac{1}{C_3} + \left[1 - \frac{1}{\lambda(\lambda\theta - 1)}\right] \frac{1}{\left(C_3 + \frac{\theta}{\gamma} - \frac{1}{\gamma\lambda}\right)} \tag{4-426}$$

Let

$$dZ/dC_3 = F_1(C_3) \tag{4-427}$$

$$\tau = \int_{C_3^0}^{C_3} \frac{\exp(-Z)\, dZ/dC_3}{1 - \gamma(C_2 - C_3)}\, dC_3 \tag{4-369}$$

Using Equations 4-370 and 4-427,

$$\tau = \int_{C_3^0}^{C_3} \frac{(\gamma C_3 + \theta)}{\theta} F_1(C_3) \exp(-Z) dC_3 \tag{4-428}$$

Thus for the case λ is large and $\lambda\theta > 1$, the quantities (V_1/V_1^0), Z, τ, C_1, q, and f are expressed by Equations 4-425, 4-424, 4-428, 4-420, 4-416, and 4-419 respectively; and C_2, C_2^0, C_3^0, (v_w/v_w^*), and $[v_w/(v_w)^0]$ are expressed by Equations 4-351, 4-373, 4-374, 4-378 and 4-379 respectively as in the general case.

Further following the form of Equations 4-399 and 4-413, one may express

$$\text{¢}_3 = \frac{C_3}{C_3^0} = \left[\frac{1}{1 - F(\tau)}\right]^{\frac{1}{1 + \theta \exp(1/\lambda\theta)}} \qquad (4\text{-}429)$$

and establish a correlation between τ and $F(\tau)$ for different values of γ, θ, and λ. A set of such correlations is illustrated in Figures 4-46, 4-47 and 4-48 which show that the values of $F(\tau)$ are essentially the same for a given value of $\lambda\theta$. The latter observation suggests that the quantity $\lambda\theta$ is a particularly useful design parameter. Equation 4-429, by itself, is not, and does not need to be, limited to $\lambda\theta > 1$.

Case 4. $\gamma = 0$. This case is an approximation to practical cases where the osmotic pressure of the solution is negligible. From Equations 4-350, 4-361, and 4-369, q, Z, and ¢_3 for the above case may be expressed as

$$q = 1 + \frac{1}{\theta} \exp(-1/\lambda\theta) \qquad (4\text{-}430)$$

$$Z = [1 + \theta \exp(1/\lambda\theta)] \ln \text{¢}_3 \qquad (4\text{-}431)$$

and

$$\text{¢}_3 = \left[\frac{1}{1 - \tau}\right]^{\frac{1}{1 + \theta \exp(1/\lambda\theta)}} \qquad (4\text{-}432)$$

9. Equations for the Flow Process

Consider the reverse osmosis process for feed flow through a channel. The geometry of the channel is immaterial for this discussion. Let 1/h be defined as membrane area per unit volume of channel space. Let $\bar{u}^0$ and $\bar{u}$ be the average velocity of feed at channel entrance, and at a longitudinal distance x from channel entrance respectively. From the water and solute material balance over a differential length dx at distance x from channel entrance, the following equations are obtained.

$$\bar{u} = \bar{u} + \frac{d\bar{u}}{dx} dx + \left(\frac{v_w}{h}\right) dx \qquad (4\text{-}433)$$

Therefore

$$-\frac{d\bar{u}}{dx} = \frac{v_w}{h} \qquad (4\text{-}434)$$

$$\bar{u}C_1 = \bar{u}C_1 + \frac{d}{dx}(\bar{u}C_1)dx + C_3\left(\frac{v_w}{h}\right) dx \qquad (4\text{-}435)$$

Therefore

$$-\frac{d}{dx}(\bar{u}C_1) = \frac{C_3 v_w}{h} \qquad (4\text{-}436)$$

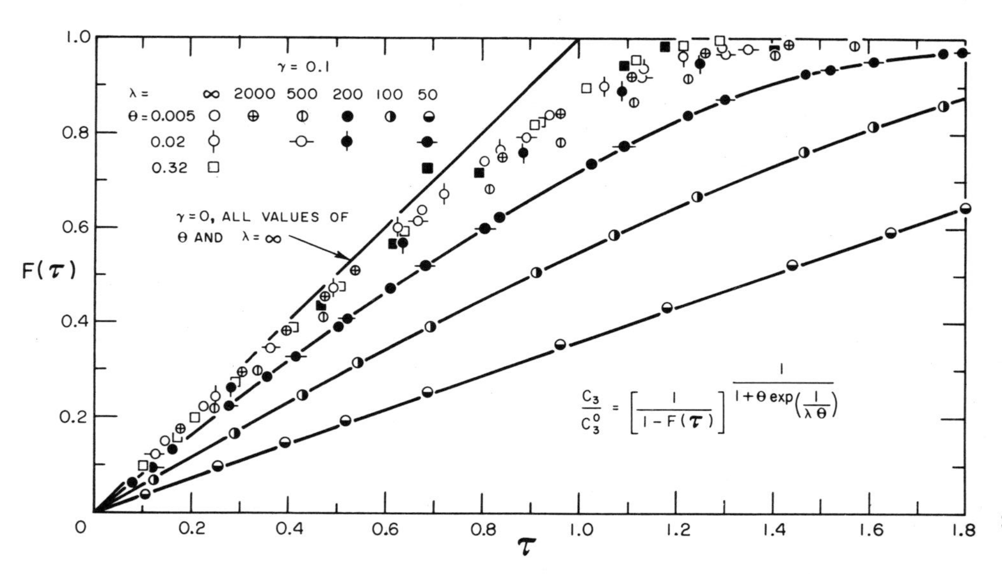

Figure 4-46.
τ v. $F(\tau)$ for
$\gamma = 0.1$, $\theta = 0.005$ to
0.32, and $\lambda = 50$ to ∞.
Data of Ohya and Sourirajan (1969d).

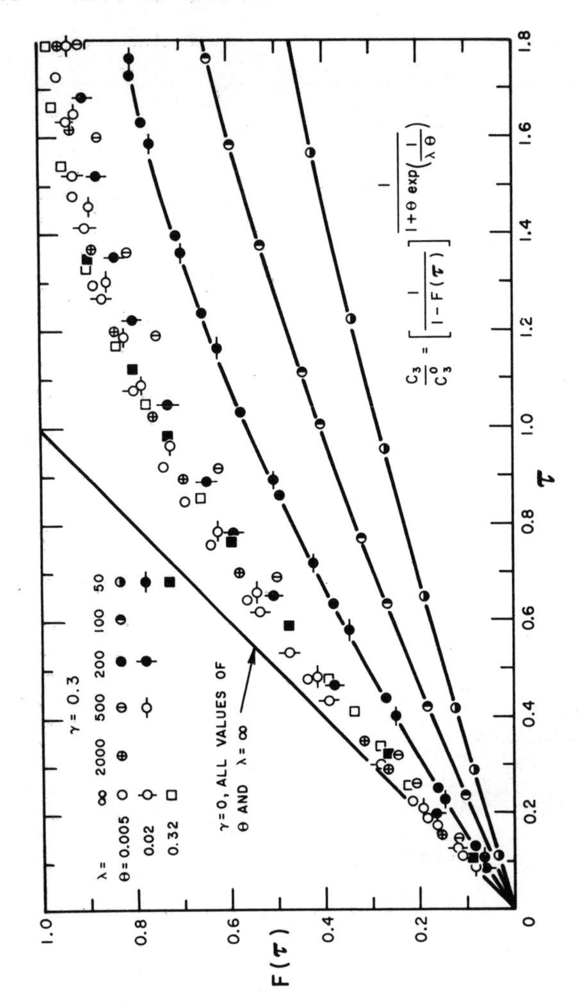

Figure 4-47.
τ v. $F(\tau)$ for $\gamma = 0.3$, $\theta = 0.005$ to 0.32, and $\lambda = 50$ to ∞. Data of Ohya and Sourirajan (1969a).

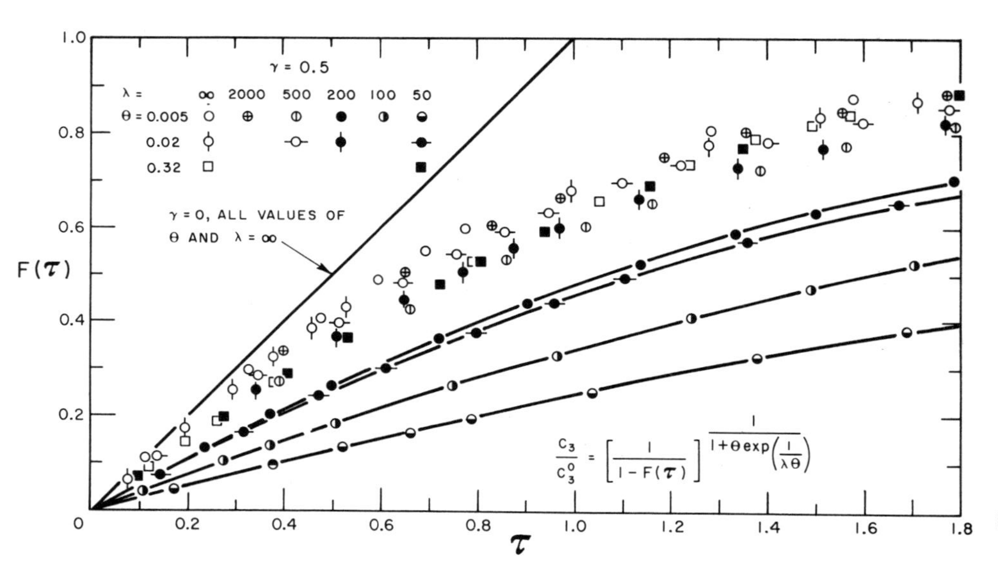

Figure 4-48.
τ v. $\mathbf{F}(\tau)$ for
$\gamma = 0.5$, $\theta = 0.005$ to
0.32, and $\lambda = 50$ to ∞.
Data of Ohya and Sourirajan (1969a).

$$= - C_3 \frac{d\bar{u}}{dx} \tag{4-437}$$

$$\bar{u} \frac{dC_1}{dx} + C_1 \frac{d\bar{u}}{dx} = C_3 \frac{d\bar{u}}{dx} \tag{4-438}$$

$$\bar{u}\,(dC_1/dx) = -\,(C_1 - C_3)\,d\bar{u}/dx \tag{4-439}$$

$$d\bar{u}/\bar{u} = -\,dC_1/(C_1 - C_3) \tag{4-440}$$

The right-hand side of Equation 4-440 is identical to that of Equation 4-357 given for the batch-wise operation. Comparing the dimensionless parameters

$$X = \frac{v_w^*}{\bar{u}^0} \frac{x}{h} \tag{4-225}$$

and

$$\tau = Sv_w^* t/V_1^0 \tag{4-368}$$

it is evident that at time $t = 0$, $1/h$ in Equation 4-225 is the same as membrane area per unit initial volume of feed solution, S/V_1^0, in Equation 4-368, and distance x from channel entrance divided by velocity at channel entrance, $x/\bar{u}^0$, in Equation 4-225 is the same as time elapsed from the start of the operation, t, in Equation 4-368. Consequently when S/V_1^0 and t in the batch-wise operation are replaced by $1/h$ and $(x/\bar{u}^0)$ respectively for the flow process, τ for the batch process corresponds identically to X for the flow process; and, on the basis of Equation 4-440, all the correlations developed above for the batch process when expressed as functions of τ become applicable for the flow process when expressed as functions of X.

10. Basic Correlations For Reverse Osmosis Process Design

The quantities C_1 (solute concentration in the concentrated effluent from the reverse osmosis unit), C_2 (concentration polarisation), C_3 and $\overline{C}_3$ (local and average concentration of the product solution) and τ (or X) defined by Equations 4-368 and 4-225 as functions of Δ (fraction product recovery) are the basic correlations of practical interest with reference to the design and performance of any reverse osmosis unit. The quantities $\bar{f}$ (average solute concentration) and $\bar{v}_w$ (average velocity of product water) are also of practical interest. The equations establishing τ v. V_1/V_1^0, C_1, C_2, and C_3 correlations have already been given. From the τ v. V_1/V_1^0 and τ v. C_1 correlations, the quantities Δ, $\overline{C}_3$, $\bar{f}$, and $\bar{v}_w$ can be calculated as follows.

$$\Delta = 1 - V_1/V_1^0 = 1 - \bar{u}/\bar{u}^0 \tag{4-441}$$

From Equation 4-220,

$$1 = C_1(1 - \Delta) + \overline{C}_3\Delta \tag{4-442}$$

$$\bar{f} = 1 - \overline{C}_3 \tag{4-56}$$

$$\bar{v}_w = -\frac{1}{S} \frac{\int_{V_1^0}^{V_1} \left(dV_1/dt\right) dt}{\int_o^t dt} \tag{4-443}$$

$$= \frac{V_1^0}{St} \Delta \tag{4-444}$$

or,

$$\frac{\bar{v}_W}{v_W^*} = \frac{V_1^0}{Sv_W^* t} \Delta = \frac{\Delta}{\tau} \tag{4-445}$$

When V_1/V_1^0 and C_1 are known as functions of τ (or X), then $\Delta, \overline{C}_3$, f, and v_W/v_W^* can be calculated from Equations 4-441, 4-442, 4-56, and 4-445 respectively. Since $\bar{f}$ and $\bar{v}_W$ can be calculated from $\overline{C}_3$, Δ v. $C_1, C_2, C_3, \overline{C}_3$, and τ (or X) for given values of γ, θ, and λ are the basic correlations for reverse osmosis process design. A set of such correlations is illustrated in Appendix III.

11. The System Urea-Water

Another illustration of stage-wise reverse osmosis process design. The possibility of separating urea from aqueous solutions in a stage-wise reverse osmosis operation is of practical interest from the point of view of water-renovation in space ships. The use of the general equations of process design derived in the preceding sections is illustrated below for the system urea-water as shown by Ohya and Sourirajan (1969a).

Some General Equations for Stage-wise Reverse Osmosis Process Design. For the purpose of this analysis, let Figure 4-49 represent a unit stage. Considering an initial unit quantity of solution, the solute material balance for the unit stage may be written as

$$c_{A1}^0 = \Delta \bar{c}_{A3} + (1 - \Delta) c_{A1} \tag{4-220}$$

or

$$1 = \Delta \overline{C}_3 + (1 - \Delta) C_1 \tag{4-442}$$

where

$$\overline{C}_3 = \bar{c}_{A3} / c_{A1}^0 \tag{4-49}$$

and

$$C_1 = c_{A1} / c_{A1}^0 \tag{4-48}$$

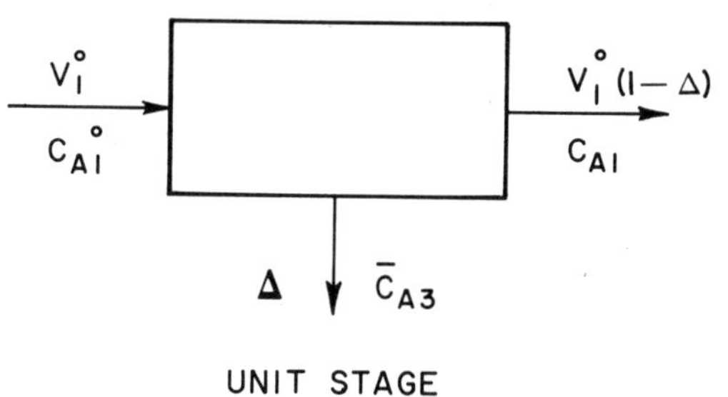

Figure 4-49. Unit stage in a multi-stage reverse osmosis unit.

[$\bar{c}_{A3}$ and $\bar{C}_3$ are simply the overall or average values of c_{A3} and C_3 respectively corresponding to fraction product recovery Δ]. For a given reverse osmosis system, of the four quantities c^0_{A1}, Δ, c_{A3}, and c_{A1}, in Equation 4-220, only any two of them are independent variables; or, referring to Equation 4-442, of the three quantities Δ, $\bar{C}_3$, and C_1, only any one of them is an independent variable. This should be clear from the preceding discussions.

A multistage reverse osmosis unit can also be represented by Figure 4-50. It consists of a feed stage, and a number of unit stages in the concentration and purification sections. The subscript f outside the brackets refers to the feed stage; the superscripts 1, 2, . . . j, . . . w outside the brackets refer to the consecutive unit stages in the concentration section; and the subscripts, 1, 2, . . . i, . . . p outside the brackets refer to the consecutive unit stages in the purification section. c^0_{A1}, c_{A1}, and $\bar{c}_{A3}$ refer to the concentrations indicated in Figure 4-49 for each unit stage. Let **P** and $[\bar{c}_{A3}]_p$ be the quantity of product (expressed in moles or cm³), and the average product concentration (expressed in mole fraction or g. mole/cm³) respectively in the final stage p in the purification section; and let **W** and $[c_{A1}]^W$ be the quantity (in moles or cm³) and concentration (in mole fraction or g. mole/cm³) respectively of the concentrated

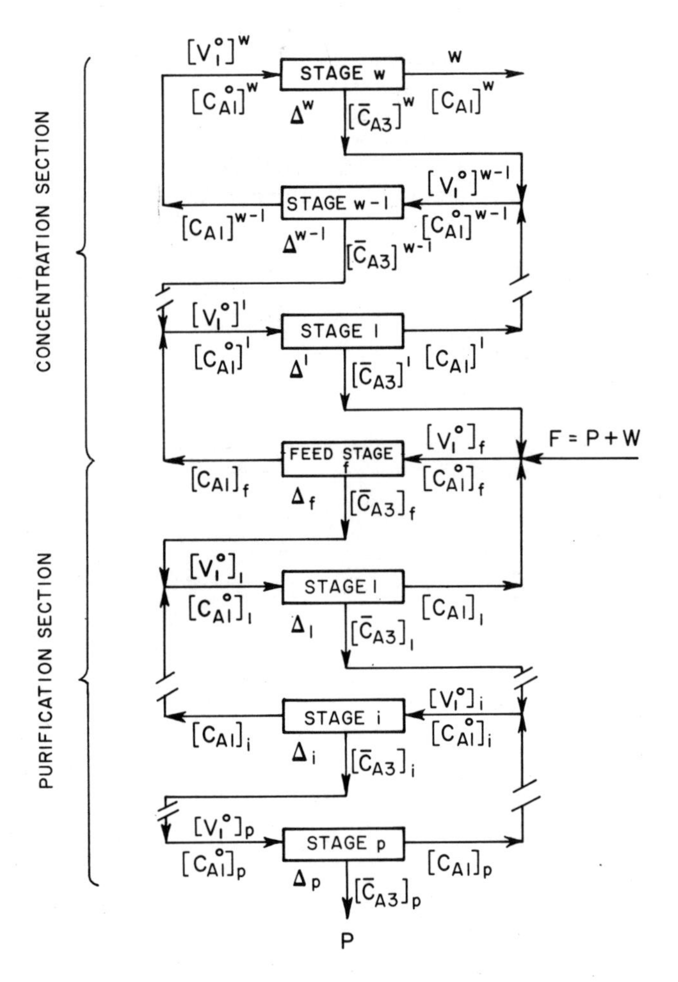

Figure 4-50. A multi-stage reverse osmosis unit.

solution on the high pressure side of the membrane, leaving the final stage w in the concentration section. Let **F** (= **P** + **W**) be the quantity (in moles or cm³) of the make-up feed solution. The solute material balance equations for the multistage unit may be written as follows

$$\left.\begin{aligned}
[c^0_{A1}]^w &= \Delta^w [\bar{c}_{A3}]^w + (1 - \Delta^w) [c_{A1}]^w \\
&- \; - \; - \; - \; - \; - \; - \\
[c^0_{A1}]^j &= \Delta^j [\bar{c}_{A3}]^j + (1 - \Delta^j) [c_{A1}]^j \\
&- \; - \; - \; - \; - \; - \; - \\
[c^0_{A1}]_f &= \Delta_f [\bar{c}_{A3}]_f + (1 - \Delta_f) [c_{A1}]_f \\
&- \; - \; - \; - \; - \; - \; - \\
[c^0_{A1}]_i &= \Delta_i [\bar{c}_{A3}]_i + (1 - \Delta_i) [c_{A1}]_i \\
&- \; - \; - \; - \; - \; - \; - \\
[c^0_{A1}]_p &= \Delta_p [\bar{c}_{A3}]_p + (1 - \Delta_p) [c_{A1}]_p
\end{aligned}\right\} \qquad (4\text{-}446)$$

The total (solvent and solute) material balance for the multistage unit may be written as follows

$$\left.\begin{aligned}
\mathbf{W} &= (1 - \Delta^w) [V^0_1]^w \\
[V^0_1]^w &= (1 - \Delta^{w-1}) [V^0_1]^{w-1} \\
[V^0_1]^{w-1} &= \Delta^w [V^0_1]^w + (1 - \Delta^{w-2}) [V^0_1]^{w-2} \\
&- \; - \; - \; - \; - \; - \; - \\
[V^0_1]_f &= \Delta^1 [V^0_1]^1 + (1 - \Delta_1) [V^0_1]_1 + \mathbf{F} \\
&- \; - \; - \; - \; - \; - \; - \\
[V^0_1]_i &= \Delta_{i-1} [V^0_1]_{i-1} + (1 - \Delta_{i+1}) [V^0_1]_{i+1} \\
&- \; - \; - \; - \; - \; - \; - \\
[V^0_1]_p &= \Delta_{p-1} [V^0_1]_{p-1} \\
\mathbf{P} &= \Delta_p [V^0_1]_p
\end{aligned}\right\} \qquad (4\text{-}447)$$

Assuming that the solution streams entering each unit stage have the same composition for economic operation, the following additional equations can be written

$$\left.\begin{aligned}
[c^0_{A1}]^w &= [c_{A1}]^{w-1} \\
[c^0_{A1}]^{w-1} &= [\bar{c}_{A3}]^w = [c_{A1}]^{w-2} \\
&- \; - \; - \; - \; - \\
[c^0_{A1}]^j &= [\bar{c}_{A3}]^{j+1} = [c_{A1}]^{j-1} \\
&- \; - \; - \; - \; - \\
[c^0_{A1}]_f &= [\bar{c}_{A3}]^1 = [c_{A1}]_1 \\
&- \; - \; - \; - \\
[c^0_{A1}]_i &= [\bar{c}_{A3}]_{i-1} = [c_{A1}]_{i+1} \\
&- \; - \; - \; - \; - \\
[c^0_{A1}]_p &= [\bar{c}_{A3}]_{p-1}
\end{aligned}\right\} \qquad (4\text{-}448)$$

Equation 4-446 has 2(w + p + 1) degrees of freedom; when the constraints of Equation 4-448 are applied to Equation 4-446, the latter has only two degrees of freedom.

Referring to Figure 4-50 again,

$$[\overline{C}_3]^w = \frac{[\bar{c}_{A3}]^w}{[c^0_{A1}]^w} = \frac{[\bar{c}_{A3}]^w}{[c_{A1}]^{w-1}} \tag{4-449}$$

and

$$[C_1]^{w-1} = \frac{[c_{A1}]^{w-1}}{[c^0_{A1}]^{w-1}} = \frac{[c_{A1}]^{w-1}}{[\bar{c}_{A3}]^w} \tag{4-450}$$

Therefore

$$[\overline{C}_3]^w \cdot [C_1]^{w-1} = 1$$

— — —

Similarly

$$[\overline{C}_3]^{w-1} \cdot [C_1]^{w-2} = 1$$

— — — —

$$[\overline{C}_3]^j \cdot [C_1]^{j-1} = 1$$

— — — —

$$[\overline{C}_3]^1 \cdot [C_1]_f = 1$$

— — — —

$$[\overline{C}_3]_f \cdot [C_1]_1 = 1$$

— — — —

$$[\overline{C}_3]_i \cdot [C_1]_{i+1} = 1$$

— — — —

and $[\overline{C}_3]_{p-1} \cdot [C_1]_p = 1$ (4-451)

Further, since

$$[c_{A1}]^w = [C_1]^w \, [c^0_{A1}]^w$$

$$[c_{A1}]^{w-1} = [C_1]^{w-1} \, [c^0_{A1}]^{w-1}$$

and

$$[c^0_{A1}]^w = [c_{A1}]^{w-1} \tag{4-452}$$

$$[c_{A1}]^w = [C_1]^w \, [C_1]^{w-1} \, [c^0_{A1}]^{w-1} \tag{4-453}$$

or

$$[c_{A1}]^w = [C_1]^w \, [c_{A1}]^{w-1} \tag{4-454}$$

Equations similar to Equation 4-454 can be written for each unit stage in the multistage unit. It then follows that

$$[c_{A1}]^w/[c_{A1}]_p = [C_1]^w \cdot [C_1]^{w-1} \ldots\ldots [C_1]_f \cdot [C_1]_1 \ldots\ldots [C_1]_{p-1} \tag{4-455}$$

and

$$[\bar{c}_{A3}]_p = [\bar{C}_3]_p\,[c^0_{A1}]_p \tag{4-456}$$

$$= [\bar{C}_3]_p\,[\bar{c}_{A3}]_{p-1} \tag{4-457}$$

$$= [\bar{C}_3]_p\,[\bar{C}_3]_{p-1}\,[c^0_{A1}]_{p-1} \tag{4-458}$$

$$= [\bar{C}_3]_p\,[\bar{C}_3]_{p-1}\,[\bar{c}_{A3}]_{p-2} \tag{4-459}$$

$$= [\bar{C}_3]_p\,[\bar{C}_3]_{p-1}\,[\bar{C}_3]_{p-2}\cdots[\bar{C}_3]_f\,[\bar{C}_3]^1\cdots[\bar{C}_3]^w\,[c^0_{A1}]^w \tag{4-460}$$

$$= [\bar{C}_3]_p\,[\bar{C}_3]_{p-1}\,[\bar{C}_3]_{p-2}\cdots[\bar{C}_3]_f\,[\bar{C}_3]^1\cdots[\bar{C}_3]^w\,[c_{A1}]^w/[C_1]^w \tag{4-461}$$

The application of one or more of Equations 4-220, 4-442, 4-446 to 4-461, combined with the other general equations given in the preceding sections, offers a convenient method for stage-wise reverse osmosis process design for a specified reverse osmosis system.

Illustrative Calculations. A hypothetical problem of water renovation from aqueous urea solution is considered here for the purpose of some illustrative calculations (Ohya and Sourirajan, 1969a). The make-up feed solution is assumed to contain 277.77 moles, the mole fraction of urea being 7×10^{-3}; these data correspond to 2.35 wt.% urea, and about 5070 cm^3 of solution at 25°. The object is to recover water containing less than 200 ppm of urea from the above solution. The mole fraction of urea in the final concentrated solution is assumed to be 6×10^{-2} (= 17.55 wt.%) and the mole fraction of urea in the final product water is 6×10^{-5} (≡ 200 ppm). As pointed out already, only one of the above two concentrations can be fixed independently in the reverse osmosis process; for the sake of convenience, the former is fixed in these calculations. Further, the molar density of the solution is assumed constant, concentrations are expressed in mole fraction, and the total quantities of solution entering and leaving the system are expressed in moles.

The number of stages needed to accomplish the above separation by the reverse osmosis process can be found by a set of step-wise calculations using a trial and error procedure. For this purpose, the reverse osmosis operating system must first be specified in terms of γ, θ, and λ.

In the concentration range of interest, the plot of osmotic pressure v. mole fraction of solute for the system urea-water is approximately a straight line, and the osmotic pressure of a 2.35 wt.% urea solution is about 140 lb/in^2 (Appendix II, A-23). Hence, at an operating pressure of 1500 psig, the value of γ for the feed stage is $\frac{140}{1500} = 0.093$. The average molar density (c) of the solution may be taken as 5.53×10^{-2} mole/cm^3. From the data on membrane specifications given in Chapter III (Film 1: $A = 1.142 \times 10^{-6}$, $(D_{AM}/K\delta)_{urea} = 41 \times 10^{-5}$; Film 101: $A = 1.258 \times 10^{-6}$, $(D_{AM}/K\delta)_{urea} = 78 \times 10^{-5}$), the values of θ for the films 1 and 101 are 0.19 and 0.34 respectively at the operating pressure of 1500 psig.

For purposes of illustration, the performance of two films whose θ values are 0.32 and 0.20 are considered with γ for the feed stage assumed equal to 0.1. These values are close to those given above for films 101 and 1 respectively. If a high mass transfer coefficient can be assumed on the high pressure side of the membrane, λ may be approximated as ∞. The performance data for the reverse osmosis systems specified as $\gamma = 0$, 0.1, and 0.3, $\theta = 0.32$ and 0.20

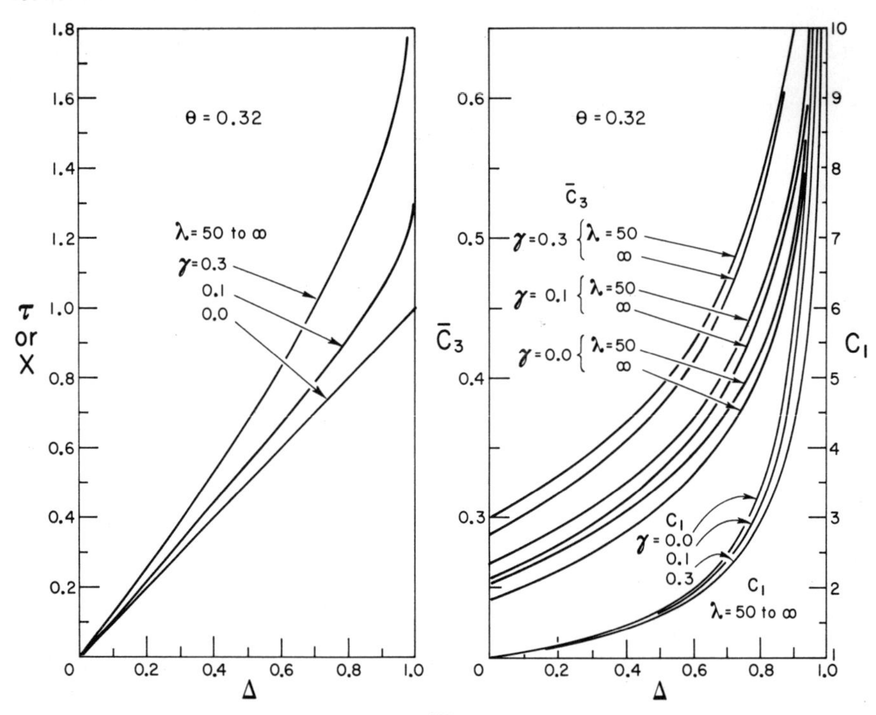

Figure 4-51. Δ v. τ (or X), C_1, and $\overline{C}_3$ for $\gamma = 0, 0.1$, and 0.3, $\theta = 0.32$, and $\lambda = 50$, and ∞. Data of Ohya and Sourirajan (1969a).

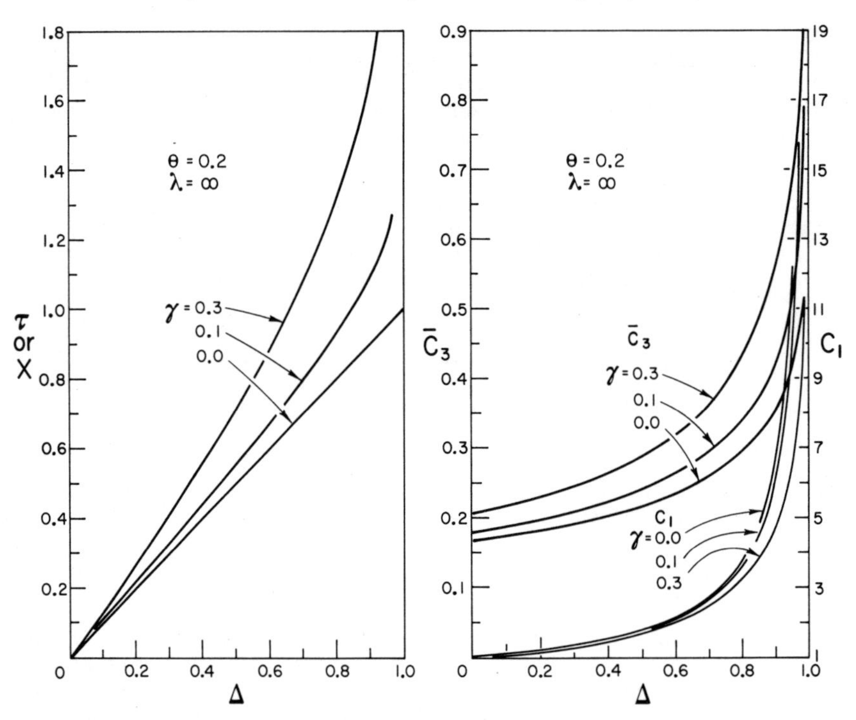

Figure 4-52. Δ v. τ (or X), C_1, and $\overline{C}_3$ for $\gamma = 0, 0.1$, and 0.3, $\theta = 0.20$, and $\lambda = \infty$. Data of Ohya and Sourirajan (1969a).

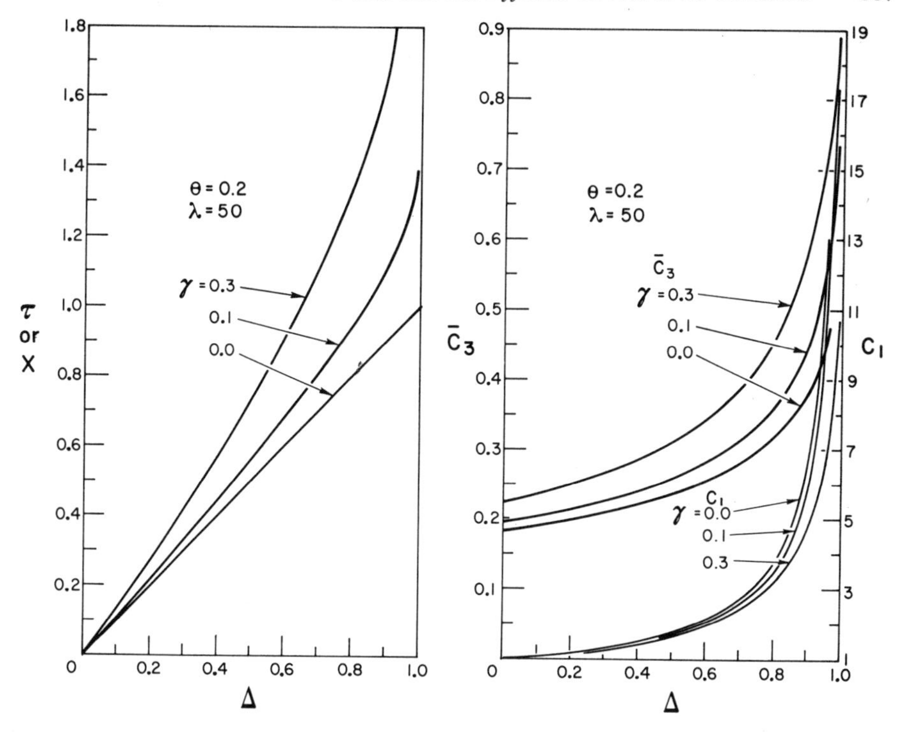

Figure 4-53. Δ v. τ (or X), C_1, and $\overline{C}_3$ for $\gamma = 0, 0.1$, and 0.3, $\theta = 0.20$, and $\lambda = 50$. Data of Ohya and Sourirajan (1969d).

and $\lambda = 50$ and ∞ are given in Figures 4-51, 4-52, and 4-53 as Δ v. C_1, $\overline{C}_3$, and τ (or X). These data are used in the calculations involved.

A detailed set of calculations is given below for the reverse osmosis system specified as $\gamma = 0.1$ for the feed stage, $\theta = 0.32$, and $\lambda = \infty$.

The first step is to make a very rough estimate of the number of stages required. This may be done as follows. For the case $\gamma = 0, \lambda = \infty$,

$$C_3/C_1 = \theta/(1 + \theta) \tag{4-387}$$

when $\theta = 0.32$ and $C_3/C_1 = 0.24$ for each stage, or $(0.24)^n$ for n stages. Approximating C_1 to initial feed concentration, the C_3/C_1 ratio needed for the purification section is 1/100, and approximating C_3 to the initial concentration, the C_3/C_1 ratio needed for the concentration section is about 1/10, indicating four stages for the purification section and less than two stages for the concentration section.

Therefore, for the first trial, a feed stage and a number of purification stages are assumed. Figure 4-54(a) represents such a multistage system. The performance of each unit stage in the system is then determined step by step as follows. Figure 4-51 is used to obtain Δ, $\overline{C}_3$, and τ values corresponding to any given value of C_1.

Feed stage

$$[C_1]_f = \frac{[c_{A1}]_f}{[c^0_{A1}]_f} = \frac{6 \times 10^{-2}}{7 \times 10^{-3}} = 8.6$$

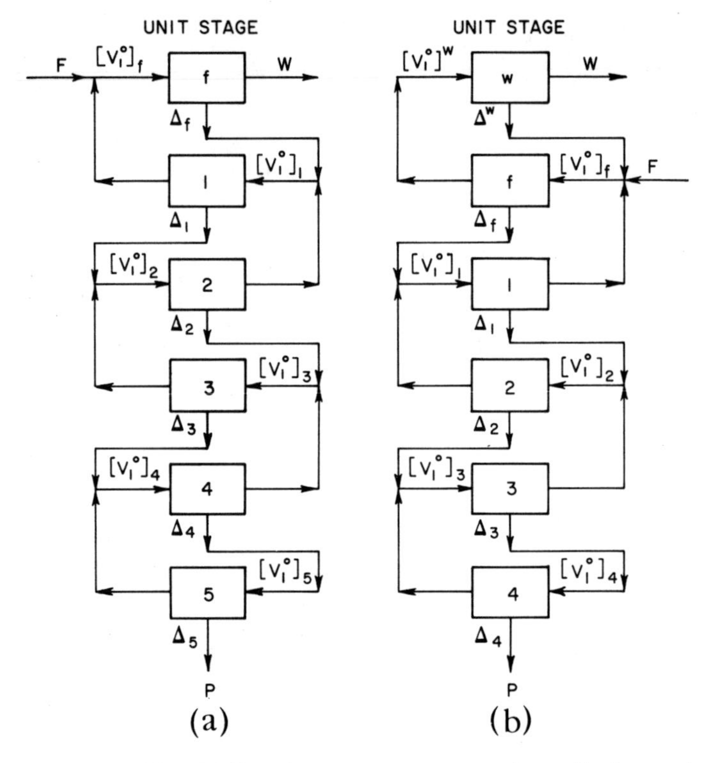

Figure 4-54. Multi-stage reverse osmosis units for water renovation from aqueous area solutions.

$\gamma = 0.1$

$\Delta_f = 0.957$

$$[\overline{C}_3]_f = \frac{[\bar{c}_{A3}]_f}{[c^0_{A1}]_f} = 0.655$$

$$[\bar{c}_{A3}]_f = [\overline{C}_3]_f \cdot [c^0_{A1}]_f$$

$$= 0.655\, [c^0_{A1}]_f = 4.585 \times 10^{-3}$$

$\tau = 1.165$

Stage 1

$$[C_1]_1 \cdot [C_3]_f = 1$$

Therefore

$$[C_1]_1 = 1/[\overline{C}_3]_f = 1/0.655 = 1.527$$

$\gamma = 0.0655$

$\Delta_1 = 0.43$

$[\overline{C}_3]_1 = 0.311$

$[\bar{c}_{A3}]_1 = [\overline{C}_3]_1 \cdot [\overline{C}_3]_f \cdot [c^0_{A1}]_f$

$= 0.2037\ [c^0_{A1}]_f = 1.426 \times 10^{-3}$

$\tau = 0.455$

Stage 2

$[C_1]_2 \cdot [\overline{C}_3]_1 = 1$

Therefore

$[C_1]_2 = 1/[\overline{C}_3]_1 = 1/0.311 = 3.215$

$\gamma = 0.0204$

$\Delta_2 = 0.79$

$[\overline{C}_3]_2 = 0.4075$

$[\bar{c}_{A3}]_2 = [\overline{C}_3]_2 \cdot [\overline{C}_3]_1 \cdot [\overline{C}_3]_f \cdot [c^0_{A1}]_f$

$= 0.083\ [c^0_{A1}]_f = 0.5811 \times 10^{-3}$

$\tau = 0.82$

Stage 3

$[C_1]_3 \cdot [\overline{C}_3]_2 = 1$

$[C_1]_3 = 1/[\overline{C}_3]_2 = 1/0.4075 = 2.454$

$\gamma = 0.0083$

$\Delta_3 = 0.695$

$[\overline{C}_3]_3 = 0.365$

$[\bar{c}_{A3}]_3 = [\overline{C}_3]_3 \cdot [\overline{C}_3]_2 \cdot [\overline{C}_3]_1 \cdot [\overline{C}_3]_f \cdot [c^0_{A1}]_f$

$= 0.0303\ [c^0_{A1}]_f = 2.121 \times 10^{-4}$

$\tau = 0.70$

Stage 4

$[\bar{C}_1]_4 \cdot [\overline{C}_3]_3 = 1$

$[C_1]_4 = 1/[\overline{C}_3]_3 = 1/0.365 = 2.740$

$\gamma = 0.0030$

$\Delta_4 = 0.738$

$[\overline{C}_3]_4 = 0.375$

$[\bar{c}_{A3}]_4 = [\overline{C}_3]_4 \cdot [\overline{C}_3]_3 \cdot [\overline{C}_3]_2 \cdot [\overline{C}_3]_1 \cdot [\overline{C}_3]_f \cdot [c^0_{A1}]_f$

$= 0.0114\ [c^0_{A1}]_f = 7.95 \times 10^{-5}$

$\tau = 0.74$

Stage 5

$[C_1]_5 \cdot [\overline{C}_3]_4 = 1$

$$[C_1]_5 = \frac{1}{[\overline{C}_3]_4} = \frac{1}{0.375} = 2.67$$

$\gamma = 0.0011$

$\Delta_5 = 0.726$

$[\overline{C}_3]_5 = 0.360$

$[\bar{c}_{A3}]_5 = [\overline{C}_3]_5 \cdot [\overline{C}_3]_4 \cdot [\overline{C}_3]_3 \cdot [\overline{C}_3]_2 \cdot [\overline{C}_3]_1 \cdot [\overline{C}_3]_f \cdot [c^0_{A1}]_f$

$= 0.00408\ [c^0_{A1}]_f = 2.856 \times 10^{-5}$

$\equiv 95$ ppm

$\tau = 0.726$

Since the concentration of urea in the product leaving stage 5 is only 95 ppm, no further stages are necessary. From a material balance the following data can be obtained:

F = 277.77 moles, mole fraction of urea = 7×10^{-3}

W = 32.27 moles, mole fraction of urea = 6×10^{-2}

P = 245.50 moles, mole fraction of urea = 2.856×10^{-5}

To find the V^0_1 value for each unit stage, Equation 4-447 can be rewritten as follows for the particular multistage unit under consideration.

$$[V^0_1]_f = \frac{\mathbf{W}}{(1 - \Delta_f)} \quad \text{(a)}$$

$$[V^0_1]_f = (1 - \Delta_1)[V^0_1]_1 + \mathbf{F} \quad \text{(b)}$$

$$[V^0_1]_1 = \Delta_f[V^0_1]_f + (1 - \Delta_2)[V^0_1]_2 \quad \text{(c)}$$

$$[V^0_1]_2 = \Delta_1[V^0_1]_1 + (1 - \Delta_3)[V^0_1]_3 \quad \text{(d)}$$

$$[V^0_1]_3 = \Delta_2[V^0_1]_2 + (1 - \Delta_4)[V^0_1]_4 \quad \text{(e)}$$

$$[V^0_1]_4 = \Delta_3[V^0_1]_3 + (1 - \Delta_5)[V^0_1]_5 \quad \text{(f)}$$

$$[V^0_1]_5 = \Delta_4[V^0_1]_4 \quad \text{(g)}$$

$$\mathbf{P} = \Delta_5[V^0_1]_5 \quad \text{(h)}$$

Any six of the above eight equations can be used to calculate the six values of V_1^0. Since the Δ values were read from the graphical correlations, some differences should be expected in the V_1^0 values depending on the choice of the six equations used to determine them. In this illustration, Equations (h), (g), (f), (e), (d), and (b) above are used to calculate $[V_1^0]_5$, $[V_1^0]_4$, $[V_1^0]_3$, $[V_1^0]_2$, $[V_1^0]_1$, and $[V_1^0]_f$ respectively, and Equations (a) and (c) are used to check the values of Δ_f and $[V_1^0]_f$ respectively. The value of V_1^0 obtained for each stage following the above procedure is given in Table 4-2. The values of Δ_f and $[V_1^0]_f$ obtained from Equations (a) and (c) above are 0.9568 and 746.05 which check well with those given in Table 4-2, Example 1.

Table 4-2. Some results of numerical calculations for stage-wise reverse osmosis process design for water renovation from aqueous urea solutions

Data of Ohya and Sourirajan (1969a)

Example	Stage	System specification			C_1	$\bar{c}_{A3} \times 10^5$ mole fraction	Δ	V_1^0 (moles)	τ	S (cm²)
		γ	θ	λ						
1	f	0.1	0.32	∞	8.6	458.50	0.957	746.2	1.165	1882
	1	0.066	0.32	∞	1.527	142.60	0.430	821.9	0.455	810
	2	0.020	0.32	∞	3.215	58.11	0.790	513.8	0.820	912
	3	0.008	0.32	∞	2.454	21.21	0.695	526.0	0.700	797
	4	0.003	0.32	∞	2.740	7.95	0.738	458.2	0.740	734
	5	0.001	0.32	∞	2.670	2.86	0.726	338.2	0.726	532
						(= 95 ppm)				
2	f	0.1	0.20	∞	8.6	346.50	0.935	492.0	1.170	1370
	1	0.05	0.20	∞	2.020	84.89	0.585	516.1	0.615	756
	2	0.012	0.20	∞	4.082	28.01	0.855	389.4	0.870	807
	3	0.004	0.20	∞	3.030	8.05	0.775	388.6	0.775	717
	4	0.001	0.20	∞	3.478	2.42	0.815	301.2	0.815	584
						(=81 ppm)				
3	f	0.1	0.20	50	8.6	364.00	0.940	528.3	1.178	1482
	1	0.052	0.20	50	1.923	92.82	0.550	556.6	0.585	775
	2	0.013	0.20	50	3.922	32.50	0.850	404.4	0.867	835
	3	0.005	0.20	50	2.857	9.71	0.754	399.5	0.755	718
	4	0.001	0·20	50	3.344	3.15	0.815	301.2	0.815	584
						(= 105 ppm)				
4	w	0.17	0.32	∞	5.108	700.00	0.910	355.6	1.240	955
	f	0.10	0.32	∞	1.684	236.25	0.510	725.8	0.560	880
	1	0.034	0.32	∞	2.963	95.68	0.765	530.1	0.820	941
	2	0.013	0.32	∞	2.469	34.92	0.695	524.6	0.705	801
	3	0.005	0.32	∞	2.734	13.18	0.740	457.7	0.745	738
	4	0.002	0.32	∞	2.649	4.90	0.725	338.7	0.727	533
						(= 163 ppm)				

Basis: One hour operation
Make up feed: 277.77 moles
Mole fraction of urea in feed: 7×10^{-3}
Mole fraction of urea in concentrate: 6×10^{-2}
Operating pressure: 1500 psig

The membrane area required in each stage can be calculated from the τ values (defined by Equation 4-368). For this purpose the values of t and v_W^* must be fixed. Assuming an operational time t = 1 hour, and $v_W^* = 2.32 \times 10^{-3}$ cm/sec (which corresponds to data of Film 101), and an average molar density of solution = 5.53×10^{-2} mole/cm^3, the membrane area required in each stage has been calculated, and the data are given in Table 4-2, Example 1.

The results of similar calculations for three other examples are also given in Table 4-2.

In Examples 2 and 3, the performance of a film whose θ value is 0.20, and v_W^* value is 2.11×10^{-3} cm/sec, is given for conditions $\gamma = 0.1$ for feed stage, and $\lambda = \infty$ and $\lambda = 50$. In both the cases, a total of 5 stages are needed, and the membrane area required in the latter case is only about 4 per cent more than that required for the limiting case $\lambda = \infty$.

Example 4 is a recalculation of Example 1 where the multistage unit is made up of a concentration stage w, in addition to the feed stage f, and a number of purification stages [Figure 4-54(b)]. For this case, the calculation procedure for the concentration stage is as follows.

$$[c_{A1}]^W/[\bar{c}_{A3}]^W = [c_{A1}]^W/[c_{A1}]_f \cdot [c_{A1}]_f/[\bar{c}_{A3}]^W = [C_1]^W \cdot 1/[\bar{C}_3]^W$$

$$= \frac{6 \times 10^{-2}}{7 \times 10^{-3}} = 8.6$$

A trial and error procedure is used to find Δ^W initially assuming an appropriate value for γ for the concentration stage, which is checked subsequently by the results obtained. Let $\gamma = 0.17$ as a first assumption.

From Figure 4-51, by trial and error, $\Delta^W = 0.91$ corresponding to $(C_1/\bar{C}_3) =$ 8.6. Using the relation

$$1 = \Delta^W[\bar{C}_3]^W + (1 - \Delta^W)[C_1]^W$$

$$[\bar{C}_3]^W = 0.594$$

and

$$[C_1]^W = 8.6[\bar{C}_3]^W = 5.108$$

Now check for the γ value:

$$[c_{A1}^0]^W = \frac{[\bar{c}_{A3}]^W}{[\bar{C}_3]^W} = \frac{1}{[\bar{C}_3]^W} \cdot [c_{A1}^0]_f$$

$$= 1.68\ [c_{A1}^0]_f$$

Therefore, the γ value for the concentration stage is 1.68 times that for the feed stage. Since γ for the feed stage is 0.1, γ for the concentration stage = 0.168 which is close to the assumed value of 0.17. For $\gamma = 0.17$, and $\Delta^W = 0.91$, $\tau = 1.24$ from Figure 4-51.

The calculations for the feed and the successive purification stages are then similar to those illustrated for Example 1.

The results of Example 4 show that even though a total of six stages is still needed, the total membrane area requirement is considerably less in the new configuration compared to that given in Example 1.

The examples illustrate an application of the general equations of Ohya and Sourirajan (1969a, d) for stage-wise reverse osmosis process design.

12. Unit Operation in Chemical Engineering

The dimensionless quantities γ, θ, and λ are the basic design parameters for the reverse osmosis membrane separation process. When any solution-membrane-operating system is specified by these parameters, the general equations presented above offer a means of predicting membrane performance for different operating conditions, and establishing the optimum conditions of operation for a given performance. Thus the reverse osmosis membrane separation process can be treated as a general unit operation in chemical engineering.

13. Nomenclature for Chapter 4

A = pure water permeability constant, $\frac{\text{g. mole } H_2O}{\text{cm}^2 \text{ sec atm}}$

b_F, b_T = fluid velocity gradient at the wall for the flat membrane case, and the tubular membrane case respectively, sec^{-1}

B = proportionality constant defined by Equation 4-6, atm

B' = proportionality constant defined by Equation 4-201, $\frac{\text{cm}^3 \text{ atm}}{\text{g. mole}}$

c = molar density of solution, g. mole/cm^3

c_1, c_2, c_3 = molar density of feed solution, concentrated boundary solution, and the product solution respectively, g. mole/cm^3

c_1^0, c_2^0, c_i^0 = solute concentration in feed solution entering Stage 1, Stage 2, and Stage i respectively, g. mole/cm^3 or mole fraction

c_1', c_2', c_i' = solute concentration in product solution leaving Stage 1, Stage 2, and Stage i respectively, on the atmospheric pressure side of the membrane, g. mole/cm^3 or mole fraction

c_1'', c_2'', c_i'' = solute concentration in the concentrated solution leaving Stage 1, Stage 2, and Stage i respectively, on the high pressure side of the membrane, g. mole/cm^3 or mole fraction

$c_{i,1}^0$, $c_{i,2}^0$, $c_{i,m}^0$ = solute concentration in feed entering inner stages 1, 2, and m respectively in Unit-Stage i, g. mole/cm^3 or mole fraction

$c_{i,1}'$, $c_{i,2}'$, $c_{i,m}'$ = solute concentration in product leaving inner stages 1, 2, and m respectively on the atmospheric pressure side of the membrane in Unit-Stage i, g. mole/cm^3 or mole fraction

$c_{i,1}''$, $c_{i,2}''$, $c_{i,m}''$ = solute concentration in the concentrated solution leaving inner stages 1, 2, and m respectively on the high pressure side of the membrane in Unit-Stage i, g. mole/cm^3 or mole fraction

c_A = solute concentration, g. mole/cm^3 or mole fraction

c_{A1}, c_{A2}, c_{A3}	= solute concentration in the bulk solution and the concentrated boundary solution on the high pressure side of the membrane, and in the product solution on the atmospheric pressure side of the membrane respectively, at any point on the membrane surface at any time, g. mole/cm^3 or mole fraction
$c^0_{A1}, c^0_{A2}, c^0_{A3}$	= values of c_{A1}, c_{A2}, c_{A3} respectively at time t = 0 in batch-wise operation, or at membrane entrance in the flow process, g. mole/cm^3 or mole fraction
$\bar{c}_{A3}$	= average solute concentration in product solution corresponding to a given Δ, g. mole/cm^3 or mole fraction
C	= $(c_A/c^0_{A1}) = (X_A/X^0_{A1})$
C_1, C_2, C_3, C_i	= (c_{A1}/c^0_{A1}), (c_{A2}/c^0_{A1}), (c_{A3}/c^0_{A1}), and (c_{Ai}/c^0_{A1}) respectively
$C^0_1, C^0_2, C^0_3, C^0_i$	= values of C_1, C_2, C_3, and C_i respectively at time t = 0 in batch-wise operation, or at membrane entrance in the flow process
$\bar{C}_3$	= $\bar{c}_{A3}/c^0_{A1}$
$₵_3$	= $(C_3/C^0_3) = (c_{A3}/c^0_{A3})$
d	= diameter of membrane, cm
D	= diffusivity of solute in water, cm^2/sec
$(D_{AM}/K\delta)$	= solute transport parameter, cm/sec
E	= quantity defined by Equation 4-310
f	= solute separation defined by Equation 4-53, 4-55, or 4-375 as indicated
$\bar{f}$	= average solute separation defined by Equation 4-56.
f*	= Fanning friction factor
F	= total product rate per unit power
F	= flow rate of feed solution entering the reverse osmosis system, cm^3/sec or moles per unit time
g_c	= conversion factor
G_1, G_2, G_i	= quantities defined by Equation 4-341, 4-345, and 4-322 respectively
h	= half-width of channel spacing, cm
(1/h)	= membrane area per unit volume of fluid space, cm^{-1}
$k, \bar{k}$	= local and average mass transfer coefficients respectively, cm/sec
L^0_1, L^0_2, L^0_i	= feed flow rate at membrane entrance in Stage 1, Stage 2 and Stage i respectively, cm^3/sec or moles per unit time
L'_1, L'_2, L'_i	= product rate leaving Stage 1, Stage 2, and Stage i respectively on the atmospheric pressure side of the membrane, cm^3/sec or moles per unit time
L''_1, L''_2, L''_i	= flow rate of concentrated solution leaving Stage 1, Stage 2, and Stage i respectively on the high pressure side of the membrane, cm^3/sec or moles per unit time

m_1^0, m_3 = molalities of the initial feed solution and the final product solution respectively

M_1, M_2, M_i = total number of inner stages in Stage 1, Stage 2, and Stage i respectively

N = total number of stages

N_A = solute flux through membrane, $\frac{g.\,mole}{cm^2 sec}$

N_B = solvent water flux through membrane, $\frac{g.\,mole}{cm^2 sec}$

N_{min} = minimum number of stages

N_S = number of stages in the purification section

N_{Pe} = Peclet number, $N_{Re} \cdot N_{Sc}$

N_{Re} = Reynolds number, $\frac{4h\bar{u}^0}{\nu}$ or $\frac{d\bar{u}^0}{\nu}$

N_{Sc} = Schmidt number, $\frac{\nu}{D}$

N_{Sh} = Sherwood number, $\frac{4kh}{D}$ or $\frac{kd}{D}$

N_{St} = Stanton number, $\frac{k}{\bar{u}^0}$

p = quantity given by Equation 4-233; or number of purification stages if so stated

p_1, p_2, p_i, p_N, p_s = values of p for Stage 1, Stage 2, Stage i, Stage N, and Stage s respectively

$\bar{p}_s, \bar{p}_N$ = average values of p defined by Equations 4-274 and 4-279 respectively

$p_{i,1}, p_{i,2}, p_{i,m}$ = values of p for the inner stages 1, 2, and m respectively in the Unit-Stage i

P = operating pressure, atm

ΔP = pressure drop, atm

P = final product rate, cm^3/sec or moles per unit time

q = quantity defined by Equation 4-219

q_1, q_2, q_i, q_N = values of q for Stage 1, Stage 2, Stage i, and Stage N respectively

$\bar{q}_N$ = average value of q defined by Equation 4-252

$q_{i,1}, q_{i,2}, q_{i,m}$ = values of q for the inner stages 1, 2, and m respectively in the Unit-Stage i

Q = total product rate, cm^3/sec

r = transverse distance from the centre of the tubular membrane, cm

R = radius of tubular membrane, cm

S = membrane surface area, cm^2

t = time, sec

$u, \bar{u}$ = local and average fluid velocities respectively at any given position, cm/sec

$\bar{u}^0$ = average fluid velocity at membrane entrance, cm/sec

u^* = friction velocity defined by Equation 4-185, cm/sec

$\bar{u}^0_{i,1}, \bar{u}^0_{i,2}, \bar{u}^0_{i,m}$ = average fluid velocity at membrane entrance in the inner stages 1, 2, and m respectively in the Unit-Stage i, cm/sec

$U, \bar{U}$ = $(u/\bar{u}^0)$ and $(\bar{u}/\bar{u}^0)$ respectively

v_W = permeating velocity of solvent water through the membrane, cm/sec

v_W^0 = quantity defined by Equation 4-30, cm/sec

$(v_W)^0$ = value of v_W at time = 0 in batch-wise operation or at membrane entrance in the flow process, cm/sec

v_W^* = quantity defined by Equation 4-202 cm/sec

V_W = (v_W/v_W^0)

V_1, V_3 = quantity of solution on the high pressure side of the membrane, and that of the membrane permeated product on the atmospheric pressure side of the membrane, respectively, at any time in the batch-wise operation, cm^3 or moles

V_1^0 = value of V_1 at time = 0 in the batch-wise operation, cm^3 or moles

w = total number of concentration stages exluding feed stage.

w_1^0 = weight fraction of solute in the solution entering Stage 1 (final product stage) in the two-stage process

w_F = weight fraction of solute in the feed solution entering Stage 2 (feed stage) in the two-stage process

w_W = weight fraction of solute in the concentrated solution leaving Stage 2 (feed stage) in the two-stage process

W = length of membrane in the transverse direction, cm

W = flow rate of the final concentrated effluent from the reverse osmosis system, cm^3/sec or moles per unit time

x = longitudinal distance along the length of the membrane from channel entrance, cm

X = quantity defined by Equation 4-32, or Equation 4-225 as indicated

X_A = mole fraction of solute

X_{A1}, X_{A2}, X_{A3} = mole fraction of solute in the bulk solution and the concentrated boundary solution on the high pressure side of the membrane, and in the product solution on the atmospheric pressure side of the membrane respectively, at any point on the membrane surface at any time

$X_{A1}^0, X_{A2}^0, X_{A3}^0$ = values of X_{A1}, X_{A2}, and X_{A3} respectively at $t = 0$ in batch-wise operation, or at membrane entrance in the flow process

$X_{i,1}, X_{i,2}, X_{i,m}$ = value of X for the inner stages 1, 2, and m respectively, in the Unit-Stage i

y = length in the transverse direction for channel mid-plane, cm

Y = (y/h) or (r/R)

Z = quantity defined by the Equation 4-359

Z_1 = 3 or 4 as indicated

Z_2 = 4 or 2 as indicated

Greek letters

α = constant

α_0 = quantity defined by Equation 4-91 or 4-298

β = constant

γ = quantity defined by Equation 4-11

γ^* = quantity defined by Equation 4-203

Δ = fraction product recovery defined by Equation 4-441 or quantity defined by Equation 4-40 as indicated

Δ_i = fraction product recovery for the Unit-Stage i

$\Delta_{i,1}, \Delta_{i,2}, \Delta_{i,m}$ = fraction product recovery for the inner stages 1, 2, and m respectively in the Unit-Stage i

θ = quantity defined by Equation 4-12

κ = quantity defined by Equation 4-97

λ = quantity defined by Equation 4-13

μ = viscosity of solution, g/cm/sec

ν = kinematic viscosity, cm^2/sec

ξ_F = $X/3\alpha_0^2$

ξ_T = $X/4\alpha_0^2$

ξ_{AS} = quantity defined by Equation 4-162

$\Pi(X_A)$ = osmotic pressure corresponding to X_A, atm

ρ = density of solution, g/cm^3

τ = quantity defined by Equation 4-368

Subscripts f, 1, 2, ... i, ... p indicate feed stage, and purification stages 1, 2, ... i, ... p respectively

Superscripts 1, 2, ... j, ... w indicate concentration stages 1, 2, ... j, ... w respectively

14. Summary

Four major topics are discussed in this chapter, and the results are discussed with particular reference to saline water conversion or water renovation from aqueous urea solutions.

Analytical expressions for predicting concentration polarisation effects in reverse osmosis have been developed for solution-membrane-operating systems specified by the dimensionless parameters γ (ratio of osmostic pressure of feed solution to the operating pressure), θ (ratio of solute transport parameter to the pure water permeation velocity for the membrane), and λ (ratio of the mass transfer coefficient to the solute transport parameter).

The partial differential equations describing the concentration profile on the brine side of the membrane have been solved for laminar flow case for different values of α_0 (a dimensionless diffusivity parameter), γ, and θ, and the results are transformed such that the mass transfer coefficient is expressed in the form of Sherwood number (N_{Sh}) as a function of longitudinal position (X). The correlation ($\alpha_0 N_{Sh}/4$) v. $\xi_F(= X/3\alpha_0^2)$ for the flat membrane case which is identical with that of ($\alpha_0 N_{Sh}/2$) v. $\xi_T(= X/4\alpha_0^2)$ for the tubular membrane case, is in good agreement with that of Sherwood et al. given for the limiting case $\gamma = 0$, and $\theta = 0$. Analysis of the characteristics of two typical films shows that the variations of average solute separation, fractional water withdrawal, and boundary concentration with X obtained by the solution of the trial and error solution of the basic transport equations using local or average mass transfer coefficients obtained from the above correlations. Using the experimental results of Shaw et al. and the Leveque solution, the turbulent flow case is analysed. On the basis of these analyses, explicit expressions for calculating local and average mass transfer coefficients are given for both laminar and turbulent flow between flat parallel membranes and in tubular membranes.

The basic equations governing the stage-wise design of the reverse osmosis process have been developed. Expressions connecting the concentrations of the feed solution entering and leaving the unit stage on the high pressure side of the membrane, and the concentration of the membrane-permeated product solution leaving the atmospheric pressure side of the membrane are derived. Following the formalism of the multistage distillation process, the cascade theory is applied to the multistage reverse osmosis process, and the expressions for the minimum number of stages, and the minimum reflux ratio are derived. The ideal cascade theory is then used to establish a practical criterion for multistage reverse osmosis process design. The application of the design equations is illustrated by a set of calculations for the desalination of aqueous sodium chloride solutions. The calculations show the possibility of obtaining potable water from sea water both by a single-stage and by a two-stage reverse osmosis process using the Loeb-Sourirajan type porous cellulose acetate membranes; the calculations also illustrate the importance of including several inner stages in each unit-stage design, in order to reduce product concentration, membrane area per unit product rate and the power consumption due to pressure drop in each stage.

Finally, a generalised approach to reverse osmosis process design is presented. Analytical expressions are derived in terms of dimensionless quantities, for the change of volume of solution, concentration of the bulk solution and that of the concentrated boundary solution on the high pressure side of the membrane, the change in the permeating velocity of solvent water through the membrane, solute separation, and other related quantitites, at any instant, as a function of concentration of the product solution on the atmospheric pres-

sure side of the membrane, or a function of time from the start of the operation, for reverse osmosis systems specified by γ, θ, *and* λ. *The equations presented are applicable to membranes for which* $(D_{AM}/K\delta)$ *is independent of solute concentration and feed flow rate, and for aqueous feed solutions whose molar density can be assumed constant, and whose osmotic pressure is proportional to mole fraction. The equations are developed first for the case of batch-wise operation and then extended to the flow case. An application of this generalised approach is illustrated by a set of detailed numerical calculations for stage-wise reverse osmosis process design with particular reference to the problem of water renovation from aqeous urea solutions.*

The entire approach presented in this chapter offers a means of treating the reverse osmosis membrane separation process as a unit operation in chemical engineering.

CHAPTER 5

Reverse Osmosis Process as a General Concentration Technique

The development of the reverse osmosis separation process as a practical concentration technique has many potential scientific and industrial applications. As an analytical tool, this technique can be used generally for the concentration of dilute solutions so as to facilitate the isolation, identification, and analysis of the solutes involved (Andelman and Suess, 1966). Blatt et al. (1965) used this technique for the concentration of protein solutions with no evidence of denaturation, using Diaplex (polyion complex) membranes (Michaels, 1965); it is hence particularly promising for the concentration of biological fluids and biochemical solutions where the usual methods such as precipitation, lyophilisation, evaporation, and freeze drying are likely to affect the stability, property, and chemical nature of the solution. The technique may be expected to find extensive application in food processing and fermentation industries for the concentration of heat-sensitive fruit juices, coffee, tea, milk, etc. (Lowe et al., 1968; Marshall et al., 1968; Merson and Morgan, 1968; Morgan et al., 1965), for the clarification, sterilisation, and concentration of liquid foods and beverages, and for the recovery of food products and industrial chemicals. The most extensive application is still likely to be in the general field of concentration of industrial solutions, waste recovery, and pollution control.

Even at the present state of its development, the Loeb-Sourirajan type porous cellulose acetate membranes appear promising for the economic concentration of industrial aqueous solutions in general, and sugar solutions and natural maple sap in particular (Sourirajan, 1967a). For this reason this type of application is discussed below in some detail.

1. Parameters of Process Design

Membranes Capable of Giving Essentially Complete Solute Separation from Aqueous Solutions

Volume change and solute recovery during concentration. In a batchwise reverse osmosis concentration process, it is of practical interest to know the increase in solute molality, m, obtainable in the concentrate starting with a given feed solution as its volume, V, is being reduced by the amount of liquid permeated through the membrane. Let the process be carried out at constant operating pressure, and let solute separation, f, be assumed constant in the entire concentration range involved in the process. At any time, for the differential flow of water and solute transfer across the membrane,

$$(1-f) = \frac{W_W}{dW_W} \cdot \frac{dW_S}{W_S} \tag{5-1}$$

or

$$(1-f)\,\frac{dW_W}{W_W} = \frac{dW_S}{W_S} \tag{5-2}$$

where W_W and W_S are the weight in grams of water and solute respectively on the high pressure side of the membrane. Let suffixes i and f represent, respectively, the initial and final states of the given feed solution on the high pressure side of the membrane. Integrating Equation 5-2,

$$(1 - f) \log \frac{(W_w)_f}{(W_w)_i} = \log \frac{(W_s)_f}{(W_s)_i} \tag{5-3}$$

Since

$$m = \frac{W_s}{M_s} \cdot \frac{1000}{W_w} \tag{5-4}$$

$$(m)_i = \frac{(W_s)_i}{M_s} \cdot \frac{1000}{(W_w)_i} \tag{5-5}$$

and

$$(m)_f = \frac{(W_s)_f}{M_s} \cdot \frac{1000}{(W_w)_f} \tag{5-6}$$

where M_s is the molecular weight in grams of solute,

$$\log \frac{(W_s)_f}{(W_s)_i} = \log \frac{(m)_f}{(m)_i} + \log \frac{(W_w)_f}{(W_w)_i} \tag{5-7}$$

Combining Equations 5-3 and 5-7,

$$\log \frac{(m)_f}{(m)_i} = f \log \frac{(W_w)_i}{(W_w)_f} \tag{5-8}$$

Also since

$$(V)_i = [(W_w)_i + (W_s)_i]/(\rho_1)_i \tag{5-9}$$

and

$$(V)_f = [(W_w)_f + (W_s)_f]/(\rho_1)_f \tag{5-10}$$

where $(\rho_1)_i$ and $(\rho_1)_f$ denote the initial and final density of the feed solution being concentrated,

$$\frac{(W_w)_i}{(W_w)_f} = \frac{(V)_i}{(V)_f} \frac{(\rho_1)_i \left[1 + \dfrac{(m)_f M_s}{1000}\right]}{(\rho_1)_f \left[1 + \dfrac{(m)_i M_s}{1000}\right]} \tag{5-11}$$

Combining Equations 5-8 and 5-11,

$$\log \frac{(V)_i}{(V)_f} = \frac{1}{f} \log \frac{(m)_f}{(m)_i} - \log \left[\frac{(\rho_1)_i \left\{1 + \dfrac{(m)_f M_s}{1000}\right\}}{(\rho_1)_f \left\{1 + \dfrac{(m)_i M_s}{1000}\right\}}\right] \tag{5-12}$$

Let $\overline{F} = (W_s)_f/(W_s)_i$ be the fraction solute recovery in the concentrate during the process. From Equations 5-7 and 5-8,

$$\log \overline{F} = \left(1 - \frac{1}{f}\right) \log \frac{(m)_f}{(m)_i} \tag{5-13}$$

For the concentration of aqueous sugar solutions by the reverse osmosis technique, it is logical to consider the case where the membrane is capable of giving essentially complete solute separation. Figure 5-1 illustrates the application of Equations 5-11, 5-12, and 5-13 for the concentration of 0.1M aqueous sucrose solution using membranes capable of giving 99 per cent or more of solute separation. Thus in the reverse osmosis concentration process using a membrane having a constant f value in the concentration range involved, the values of $(V)_i/(V)_f$, $(W_w)_f/(W_w)_i$, and $(W_s)_f/(W_s)_i$ are uniquely fixed for each value of $(m)_f/(m)_i$.

Processing capacity of a membrane for solute concentration. The processing capacity of a membrane, $[= (V)_i/\overline{A}t]$, is defined as the volume of charge (feed solution) that 1 square foot of the film surface can handle per day in a batch concentration process to increase the solute molality from $(m)_i$ to $(m)_f$ at a

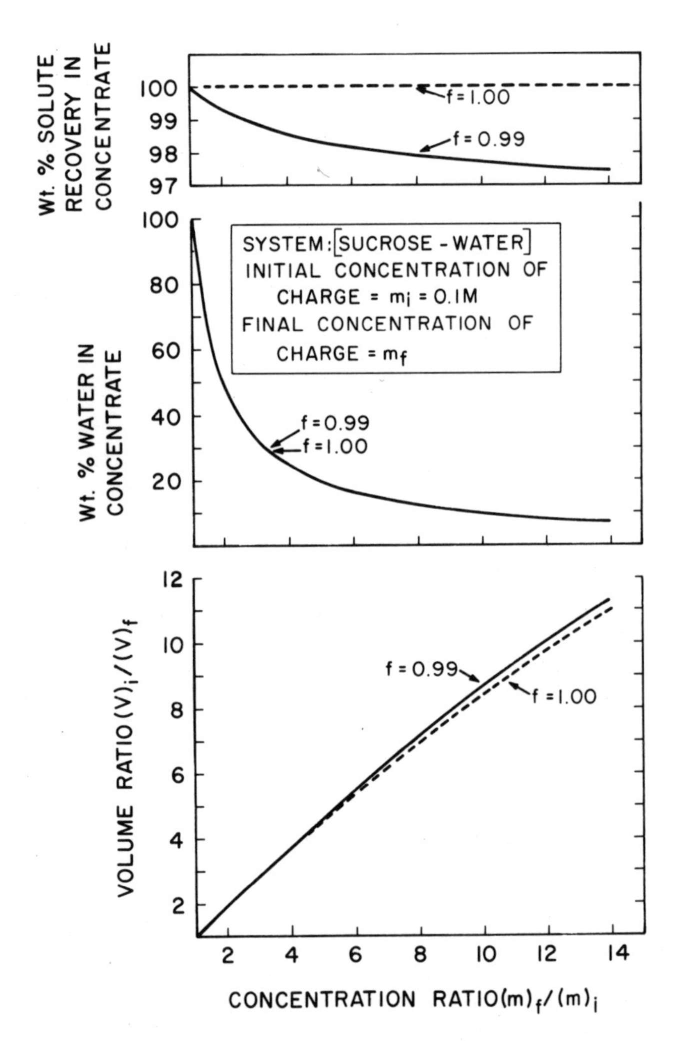

Figure 5-1. Volume change, water removal, and solute recovery during concentration of 0.1M aqueous sucrose solution. Data of Sourirajan (1967a).

given operating pressure. The calculation of this processing capacity for the case $f = 1$ is illustrated as follows.

Let W and Z represent, respectively, the total weight of solution and weight fraction of solute on the high pressure side of the membrane, and let $\bar{A}$ and q represent, respectively, the total area of membrane surface and the product rate in weight units per unit area per unit time (t). Then,

$$WZ = \text{constant} = \bar{K} = (W)_i(Z)_i \qquad (5\text{-}14)$$

$$q = -\frac{1}{\bar{A}}\frac{dW}{dt} \qquad (5\text{-}15)$$

$$\frac{d}{dt}(WZ) = 0 \qquad (5\text{-}16)$$

Differentiating Equation 5-16, and substituting from Equations 5-14 and 5-15,

$$\int_{(Z)_i}^{(Z)_f} \frac{1}{q}\frac{dZ}{Z^2} = \frac{\bar{A}}{(W)_i\,(Z)_i}\int_0^t dt \qquad (5\text{-}17)$$

or

$$\int_{(Z)_f}^{(Z)_i} \frac{1}{q}\, d\left(\frac{1}{Z}\right) = \frac{\bar{A}t}{(W)_i(Z)_i} = \frac{\bar{A}t}{\bar{K}} \qquad (5\text{-}18)$$

Setting

$$\int_{(Z)_f}^{(Z)_i} \frac{1}{q}\, d\left(\frac{1}{Z}\right) = S \qquad (5\text{-}19)$$

$$(W)_i/\bar{A}t = 1/(Z)_i S \qquad (5\text{-}20)$$

since

$$(W)_i = (V)_i(\rho_1)_i \qquad (5\text{-}21)$$

$$(V)_i/\bar{A}t = 1/[(\rho_1)_i\,(Z)_i S] \qquad (5\text{-}22)$$

Sometimes it may be of interest to calculate the average product rate, q_{av}, obtained during the concentration process, as follows:

$$q_{av} = \frac{1}{\bar{A}t}[(W)_i - (W)_f] \qquad (5\text{-}23)$$

$$= \frac{\bar{K}}{\bar{A}t}\left[\frac{1}{(Z)_i} - \frac{1}{(Z)_f}\right] \qquad (5\text{-}24)$$

$$= \frac{1}{S}\left[\frac{1}{(Z)_i} - \frac{1}{(Z)_f}\right] \qquad (5\text{-}25)$$

since, from Equations 5-18 and 5-19,

$$\bar{K}/\bar{A}t = 1/S \qquad (5\text{-}26)$$

Also

$$(W)_i - (W)_f = (V)_i(\rho_1)_i - (V)_f(\rho_1)_f \tag{5-27}$$

$$= (V)_i \left[\frac{\alpha_V(\rho_1)_i - (\rho_1)_f}{\alpha_V} \right] \tag{5-28}$$

where

$$\alpha_V = (V)_i/(V)_f \tag{5-29}$$

Therefore

$$\frac{(V)_i}{\bar{A}t} = q_{av} \left[\frac{\alpha_V}{\alpha_V(\rho_1)_i - (\rho_1)_f} \right] \tag{5-30}$$

Performance of Membranes for Sucrose Concentration. The performance of three typical films (40, 38, and 36) which can be used for the concentration of aqueous sucrose solutions is illustrated in Figures 5-2, 5-3, and 5-4. These experiments were conducted in the apparatus shown in Figures 1-16 and 1-17. The three films were made in the same manner, but shrunk at different temperatures—namely, 88°, 89°, and 90°C, respectively—each for 10 minutes. Figure 5-2 gives the separation and product rate characteristics of the films for aqueous sucrose solutions at 1500 psig operating pressure for feed concentrations in the range 0.05 to 1.4M. All the three films gave essentially the same solute separations, which were >99 per cent for all feed solutions in the range 0.1 to 1.4M. The product rate characteristics of the films were particularly interesting. Even though, for a given feed solution, the product rate was generally more for the film shrunk at the lower temperature, the difference in product rates obtained with any two films diminished rapidly with increase in feed concentration.

For the purpose of calculating the processing capacities of membranes 40, 38, and 36 for a given concentration job, the product rate v. feed molality data given in Figures 5-2 were first replotted in the form $1/q$ v. $1/Z$, and the values of the parameter S, defined by Equation 5-19, evaluated graphically. Then the capacities of the membranes in gallons of 0.1M sucrose-water feed solution that the films can handle per square foot per day to accomplish a given increase in solute molality were calculated using Equation 5-22 and the results are given in Figure 5-3.

The processing capacities of the films given in Figure 5-3 are based on product rates obtained in short-run experiments. Under conditions of continuous operation for longer periods of time, the films yield lower product rates due to membrane compaction. The latter yields are the ones of practical interest from the point of view of process design. Hence a typical one-month-long continuous test run was conducted to assess the performance of the films under such conditions. During this period, the operating pressure was 1500 psig, and the feed concentration was maintained constant at 0.5M. The separation and product rate characteristics of the films were periodically determined (Figure 5-4). The solute separation remained essentially constant at >99 per cent throughout the entire test period, but the product rate dropped under conditions of continuous operation. At the end of 24 hours of operation, the product rates were 75 to 78 per cent of those obtained at the start of the run; at the end of 16 days of operation, the product rates were 62 to 76 per cent of those obtained at the start of the run, and there was a negligible drop in product rates thereafter.

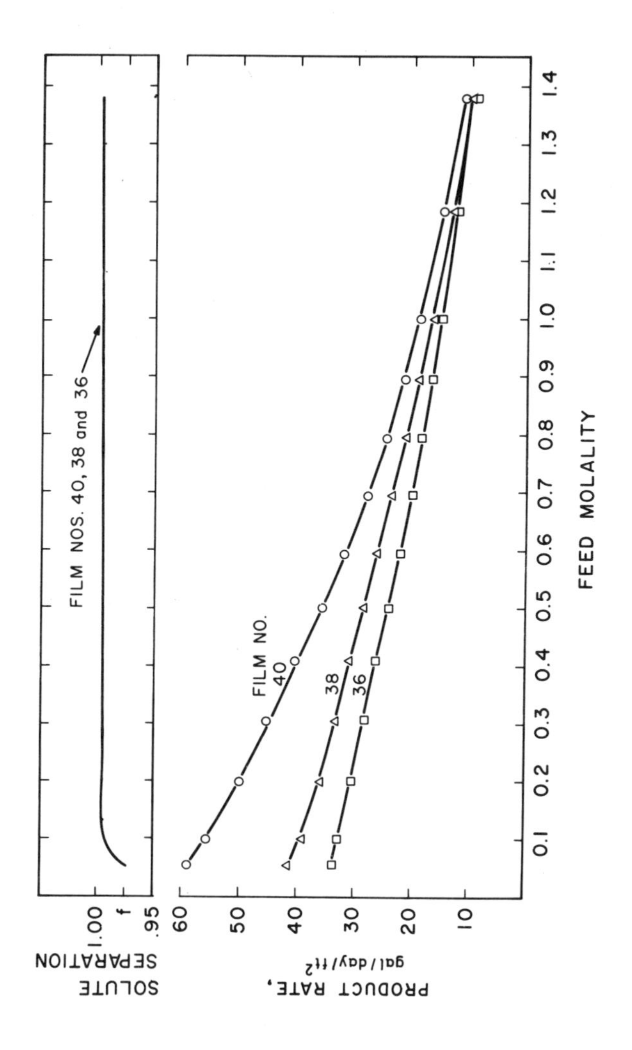

Figure 5-2. Effect of feed concentration on product rates for the system sucrose-water using different films giving greater than 99% solute separations. Data of Sourirajan (1967a).

Film type: CA-NRC-18; feed rate: 380 cm^3/min; operating pressure: 1500 psig.

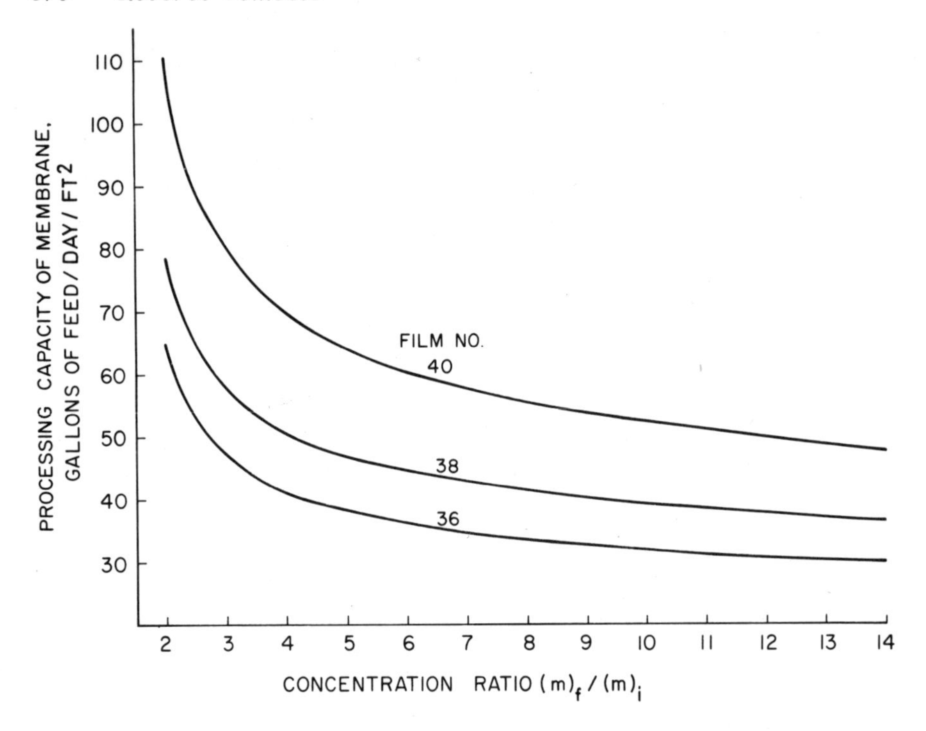

Figure 5-3. Processing capacities of membranes for the concentration of 0.1M aqueous sucrose solution. Data of Sourirajan (1967a).

Film type: CA-NRC-18; system: [sucrose-water]; initial concentration of charge $m_i = 0.1M$; final concentration of charge: m_f; operating pressure: 1500 psig; feed rate: 380 cm^3/min.

After 30 days of continuous operation, the run was stopped, and the pressure released for 24 hours. Then a short run was conducted, the results of which are also shown in Figure 5-4. In the latter test, the solute separation still remained the same, but the product rate showed a significant increase for each film. These results indicate that even after 30 days of continuous operation, the films did not become permanently set, and, on release of pressure, they regained at least part of their original porous structure.

On the basis of the results plotted in Figure 5-4, it seems reasonable to suggest that the actually realisable processing capacities of films 40, 38, and 36, under conditions of 24-hour batch operation carried out for an extended period of time, may be expected to be 60 to 70 per cent of those given in Figure 5-3. Thus starting with 0.1M sucrose-water feed solutions (comparable to natural maple sap), for obtaining a concentration ratio $(m)_f/(m)_i = 14$, i.e. increasing the sucrose concentration from 3.3 to 32.4 wt. %, the porous cellulose acetate membranes of this type may be expected to have practical processing capacities of 20 to 30 gallons per day per square foot of film area at an operating pressure of 1500 psig. The above processing capacity for the membrane may justify further engineering and economic studies of this separation process as a practical concentration technique for industrial sugar solutions.

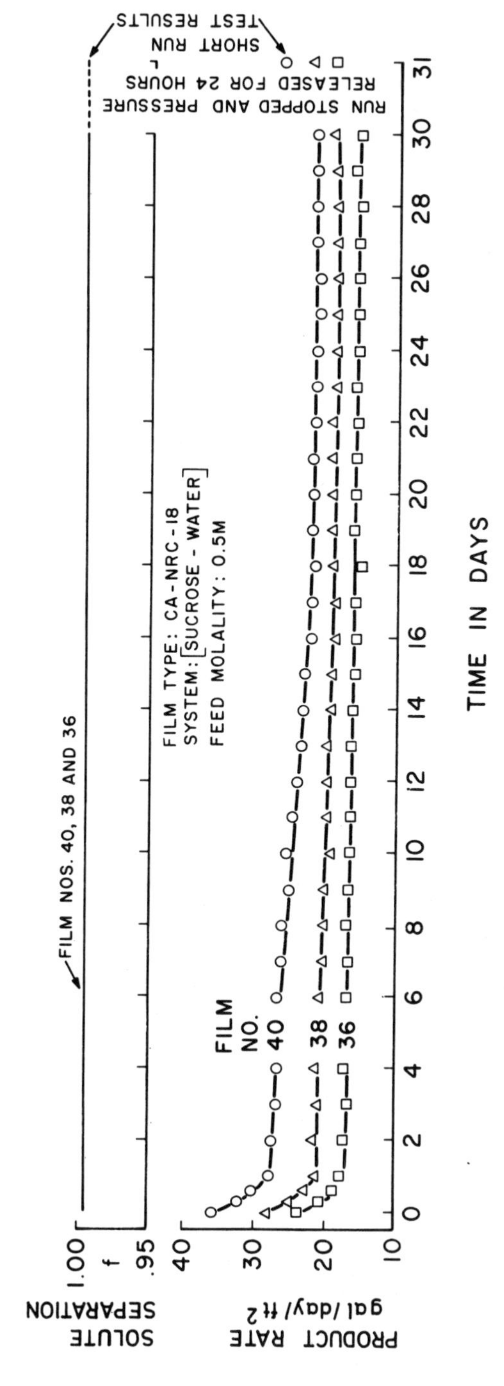

Figure 5-4. Membrane performance with the system sucrose-water in a month-long continuous test run. Data of Sourirajan (1967a).

Feed rate: 380 cm^3/min; operating pressure: 1500 psig.

Concentration of Natural Maple Sap. A few experiments were carried out by Sourirajan (1967a) on the batchwise concentration of natural maple sap in the static cell shown in Figure 1-24. A cellulose acetate film shrunk at 90° was used in these experiments. In the particular apparatus employed, the effective area of the film was 9.6 cm^2. The natural maple sap used was clear and colourless, and had a solute concentration of 3.76 per cent by weight (0.114M) equivalent sucrose.

The cell was initially charged with 250 cm^3, $(V)_i$, of sap, and then pressurised with nitrogen gas which was maintained at 1000 psig during the experiment. The solution in the cell was kept well stirred during the run. In each experiment, the duration of the run, the amount of product, final volume $(V)_f$, of the concentrate, the molality of the product and that of the concentrate (in terms of the equivalent aqueous sucrose solutions) were determined; and from Equation 5-30, the processing capacity of the membrane was estimated in gallons of natural maple sap it can handle per day per square foot of film area to accomplish different levels of solute molality, $(m)_f/(m)_i$, in the concentrate at the operating pressure of 1000 psig. The density data for aqueous sucrose solutions were used. The results are given in Figure 5-5.

The duration of each run varied from 4 to 12 hours in these experiments. This duration is more than that of the short-run experiments described earlier. Between the runs, the film was kept in contact with pure water at atmospheric pressure for 12 hours or more. The product rate decreased slowly from run to run. The order in which these experiments were carried out is given by the number in brackets in Figure 5-5. The plot of the processing capacity data for the membrane reflects the effect of duration of run on the permeability characteristics of the film.

The refractive index of the product was identical to that of pure water. On evaporation to dryness at 110°, the total solid content in the product (maple sap water) was found to be only 38 ppm. Thus the solute recovery was essentially 100 per cent with the particular film used. Further, the experimental results showed that a membrane processing capacity of about 10 gallons per day per square foot of film area was attainable at an operating pressure of 1000 psig for concentrating natural maple sap from 0.114M to 1.545M—i.e. 3.76 to 34.6 per cent by weight—equivalent sucrose solution.

A sample of the above 34.6 per cent maple sap concentrate was further concentrated on a hot plate to about 65 per cent equivalent sucrose solution; a light-coloured yellowish maple syrup was obtained with its characteristic flavour and taste indicating the possible practical applicability of the reverse osmosis concentration technique in the maple syrup industry.

Further experiments with natural maple sap. The concentration of natural maple sap was further investigated making use of different samples of CA-NRC-18 type membranes. It was found that solute separation from natural maple sap was always higher than that obtained from the equivalent aqueous sucrose solution. For example, a membrane capable of giving 93 per cent solute separation from a 0.1M aqueous sucrose solution gave $>$99 per cent solute separation from an equivalent natural maple sap. Processing capacities of the membranes for maple sap concentration were calculated from Equation 5-30 using the actual density data for concentrated maple sap given in Figure 5-6; the results obtained were essentially the same as those given in Figure 5-5.

Some preliminary experiments on the concentration of natural maple sap using a larger apparatus containing Loeb-Sourirajan type porous cellulose acetate

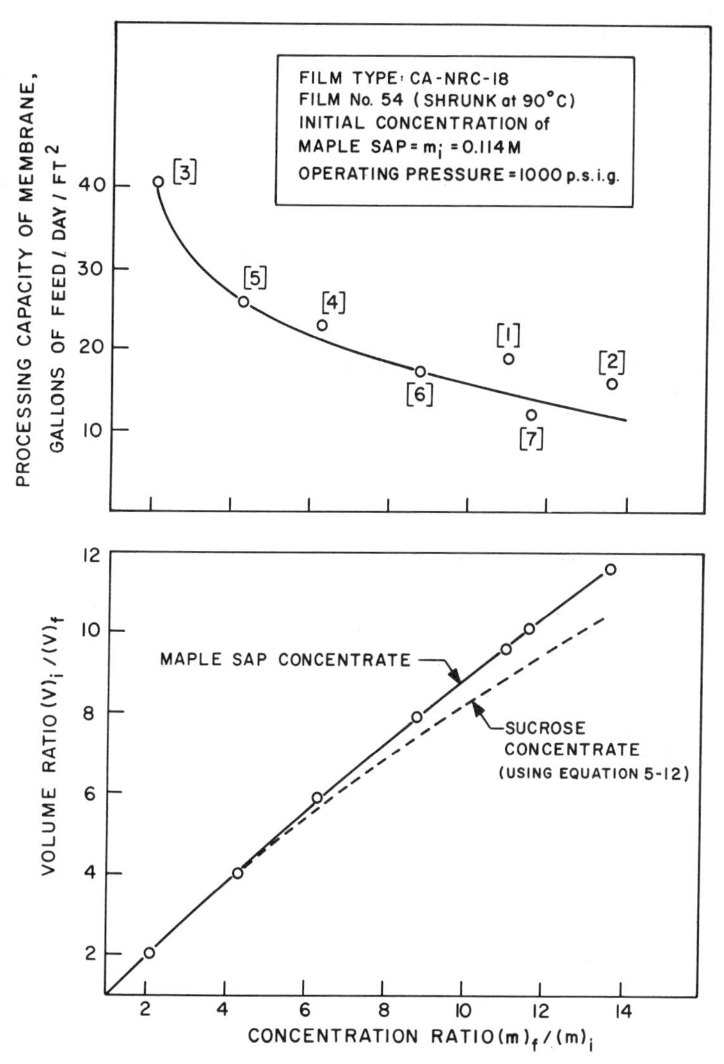

Figure 5-5. Concentration of natural maple sap. Data of Sourirajan (1967a).

membranes in modular configuration, have been reported by Willits et al. (1967). The apparatus (supplied by General Dynamics Corp.) contained 4 modules, each containing 3.7 square feet of film area. The operating pressure employed was in the range 200 to 600 psig. The results involved concentrations up to about 8 per cent starting with 2.5 to 3.0 per cent sap. The total solids in product water was in the range 0.3 to 0.09 mg/100 ml. The pH of the sap increased from 6.8 to 7.1 during the concentration process.

Concentration of Wood Sugar Solution. Some preliminary experiments were carried out on the reverse osmosis concentration of wood sugar solution (a by-product in the production of hardboard) which usually contains pentosans and hexosans in a wide range of molecular weights. These experiments were similar to those described above for the concentration of natural maple sap. The wood sugar solution, supplied by Masonite Corporation, was used with preshrunk CA-NRC-18 type porous cellulose acetate membranes. The concentration technique was successful. The initial concentration of the wood sugar solution was 3.95 weight per cent total solids; a ten-fold increase in concentration was attainable by the reverse osmosis technique. The membrane permeated product was slightly acidic to litmus. Using Equation 5-30, and the density of the concentrated wood sugar solutions given in Figure 5-6, it was found that the processing capacity of the membranes was of the order of 5 to 7 gallons per day per square foot of film area for the concentration of the Masonite wood sugar solution from 3.95 to 35 wt.% at an operating pressure of 1000 psig with a solute recovery of over 99 per cent. This preliminary result seems interesting enough to undertake more detailed investigations on the subject.

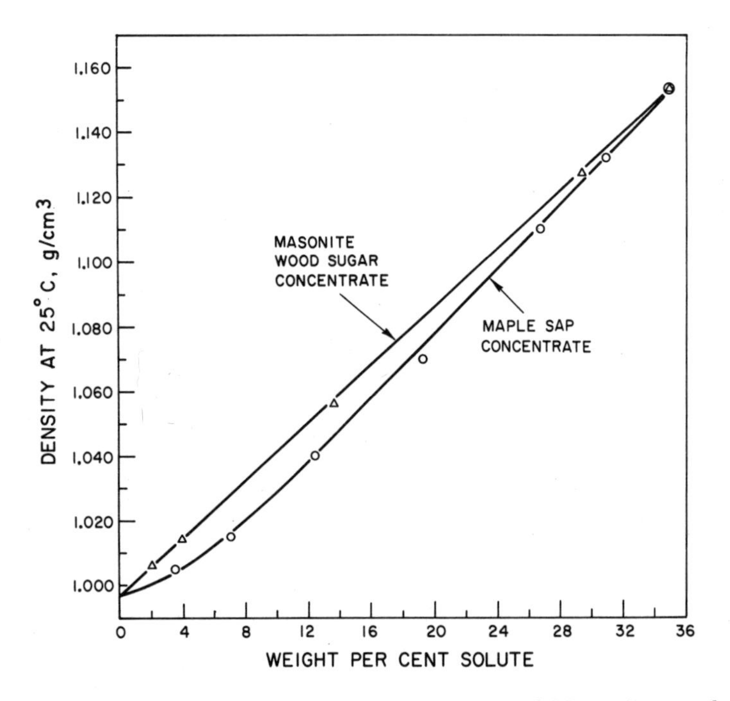

Figure 5-6. Densities of natural maple sap and Masonite wood sugar solution concentrates.

Membranes Capable of Giving Different Levels of Solute Separation from Aqueous Solutions. The general case where solute separation, f, is not equal to 1, and changes as a function of feed concentration, is considered below (Sourirajan and Kimura, 1967).

Volume change and solute recovery during concentration. Let the process be carried out at constant operating pressure, and let solute separation be some function of solution molality represented by the equation

$$f = f(m) \qquad (5\text{-}31)$$

At any time, for the differential flow of water and solute across the membrane,

$$[1 - f(m)]dW_W/W_W = dW_S/W_S \qquad (5\text{-}32)$$

where W_w and W_s are the weight in grams of water and solute respectively on the high pressure side of the membrane. Since

$$m = \frac{W_s}{M_s}\frac{1000}{W_w} \tag{5-4}$$

$$dm = \frac{1000}{M_s} d\left(\frac{W_s}{W_w}\right) \tag{5-33}$$

$$= m\left(\frac{dW_s}{W_s} - \frac{dW_w}{W_w}\right) \tag{5-34}$$

Therefore

$$dm/m = dW_s/W_s - dW_w/W_w \tag{5-35}$$

or

$$dW_s/W_s = dm/m + dW_w/W_w \tag{5-36}$$

Combining Equations 5-32 and 5-36

$$-\frac{dW_w}{W_w} = \frac{dm}{m\,f(m)} \tag{5-37}$$

which, on integration, gives

$$\ln\frac{(W_w)_i}{(W_w)_f} = \int_{(m)_i}^{(m)_f} \frac{dm}{m\,f(m)} \tag{5-38}$$

where, as before, suffixes i and f represent, respectively, the initial and final states of the given feed solution on the high pressure side of the membrane. Integrating Equation 5-36, the fraction solute recovery, $\overline{F} = (W_s)_f/(W_s)_i$, is given by

$$\ln\overline{F} = \ln\frac{(m)_f}{(m)_i} - \int_{(m)_i}^{(m)_f} \frac{dm}{m\,f(m)} \tag{5-39}$$

Combining Equations 5-11 and 5-38, the volume ratio $(V)_i/(V)_f$ is given by

$$\ln\frac{(V)_i}{(V)_f} = \int_{(m)_i}^{(m)_f} \frac{dm}{m\,f(m)} - \ln\left[\frac{(\rho_1)_i\left(1 + \dfrac{(m)_f M_s}{1000}\right)}{(\rho_1)_f\left(1 + \dfrac{(m)_i M_s}{1000}\right)}\right] \tag{5-40}$$

Thus the variations of volume change $(V)_i/(V)_f$, water removal $(W_w)_f/(W_w)_i$, and solute recovery $(W_s)_f/(W_s)_i$, with the concentration ratio $(m)_f/(m)_i$ are expressed by Equations 5-40, 5-38, and 5-39 respectively. The right side of

Equation 5-38 can be evaluated either analytically or graphically from the experimental f v. m data. Two simple cases are particularly useful. When

$$f(m) = f = \text{constant} \tag{5-41}$$

$$\int_{(m)_i}^{(m)_f} \frac{dm}{m f(m)} = \frac{1}{f} \ln \frac{(m)_f}{(m)_i} \tag{5-42}$$

and Equations 5-40, 5-38, and 5-39 reduce to Equations 5-12, 5-8, and 5-13 respectively. When

$$f(m) = B_1 + B_2 m \tag{5-43}$$

where B_1 and B_2 are constants,

$$\int_{(m)_i}^{(m)_f} \frac{dm}{m\, f(m)} = \frac{1}{B_1} \ln \left[\frac{(m)_f}{(m)_i} \frac{\left((m)_i + \frac{B_1}{B_2}\right)}{\left((m)_f + \frac{B_1}{B_2}\right)} \right] \tag{5-44}$$

The former case was illustrated in Figure 5-1 with the experimental data for the system sucrose-water (Sourirajan, 1967a). The latter case is illustrated in Figure 5-8 for the system glycerol-water (Sourirajan and Kimura, 1967) with the experimental data given in Figure 5-7 which shows that the f v. m curve is essentially a straight line in the concentration range 0.25 to 3.0M under the specified experimental conditions.

Processing capacity of a membrane for solute concentration. The calculation of the processing capacity is illustrated below for the general case for which the experimental solute separation and product rate data are available as a function of feed molality (Sourirajan and Kimura, 1967).

Let W and Z represent, respectively, the total weight of solution and the weight fraction of solute on the high pressure side of the membrane; and let q and a represent, respectively, the product rate in weight units per unit area per unit time (t), and the weight fraction of solute in the membrane permeated product. Then

$$q = -\frac{1}{A} \frac{dW}{dt} \tag{5-15}$$

$$qa = -\frac{1}{A} \frac{d}{dt} (WZ) \tag{5-45}$$

Substituting Equation 5-15 in Equation 5-45, and differentiating the latter,

$$-dW/W = dZ/(Z - a) \tag{5-46}$$

which, on integration, gives

$$\ln \frac{(W)_i}{W} = \int_{(Z)_i}^{Z} \frac{dZ'}{Z' - a} \tag{5-47}$$

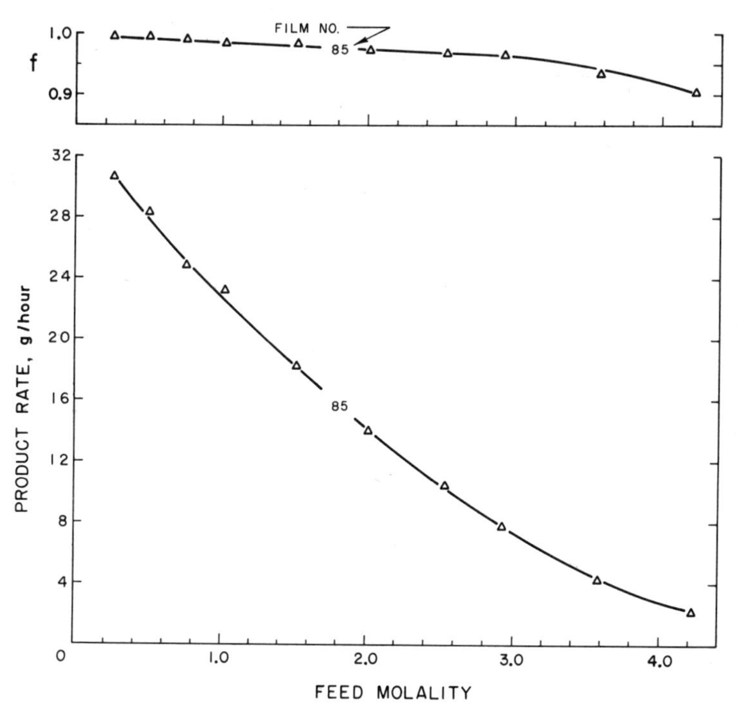

Figure 5-7. Experimental separation and product rate data for the system glycerol-water with Film 85. Data of Sourirajan and Kimura (1967).

Film type: CA-NRC-18; feed rate: 380 cm^3/min; operating pressure: 1500 psig; film area: 7.6 cm^2.

or

$$W = (W)_i \exp\left(-\int_{(Z)_i}^{Z} \frac{dZ'}{Z' - a}\right) \tag{5-48}$$

Equation 5-15 can then be written as

$$q = -\frac{(W)_i}{\bar{A}} \frac{d}{dt}\left[\exp\left(-\int_{(Z)_i}^{Z} \frac{dZ'}{Z' - a}\right)\right] \tag{5-49}$$

or

$$\frac{\bar{A}dt}{(W)_i} = -\frac{1}{q} d\left[\exp\left(-\int_{(Z)_i}^{Z} \frac{dZ'}{Z' - a}\right)\right] \tag{5-50}$$

Integrating Equation 5-50,

$$\frac{\bar{A}t}{(W)_i} = \int_{(Z)_i}^{(Z)_f} \frac{\exp\left(-\int_{(Z)_i}^{Z} \frac{dZ'}{Z' - a}\right)}{q(Z - a)} dZ \tag{5-51}$$

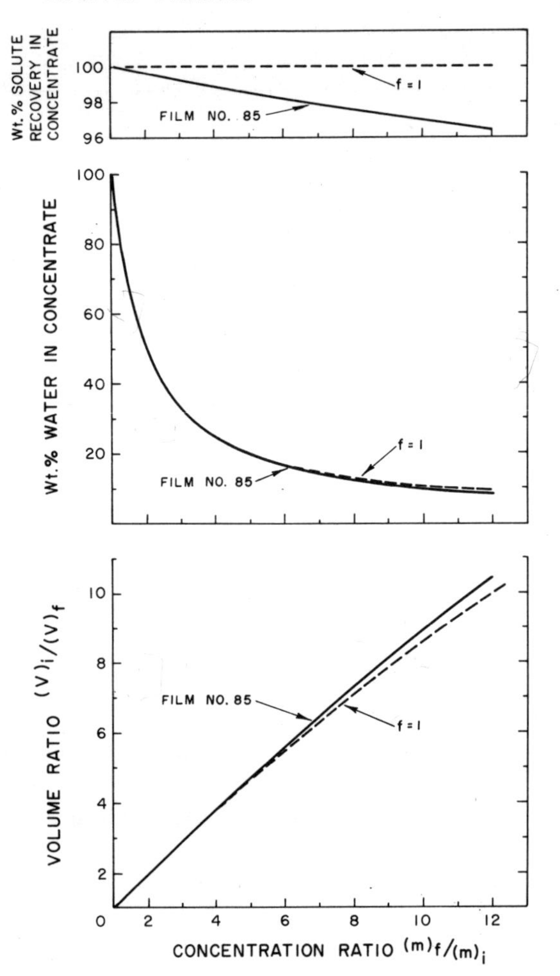

Figure 5-8. Volume change, water removal, and solute recovery during concentration of 0.25M aqueous glycerol solution. Data of Sourirajan and Kimura (1967).

Film type: CA-NRC-18; system: glycerol-water; initial concentration of charge = $(m)_i$ = 0.25M; final concentration of charge = $(m)_f$; feed rate: 380 cm^3/min; operating pressure: 1500 psig.

Since

$$(W)_i = (V)_i(\rho_1)_i \tag{5-21}$$

$$\frac{(V)_i}{\bar{A}t} = \frac{1}{(\rho_1)_i \displaystyle\int_{(Z)_i}^{(Z)_f} \frac{\exp\left(-\displaystyle\int_{(Z)_i}^{Z} \frac{dZ'}{Z' - a}\right)}{q(Z - a)}\, dZ} \tag{5-52}$$

For the special case a = 0, Equation 5-52 becomes identical with Equation 5-22.

Figure 5-9 illustrates the applicability of Equation 5-52 for the concentration of aqueous glycerol solutions using the experimental data given in Figure 5-7.

For the purpose of the processing capacity calculations, the experimental solute separation and product rate data were first expressed in terms of q, z, and a; they were then plotted in the forms of Z v. $1/(Z - a)$, and

$$Z \text{ v.} \exp\left(-\int_{(Z)_i}^{Z} \frac{dZ'}{Z' - a}\right) \Big/ q(Z - a),$$

and the integrations were carried out graphically.

The processing capacities of the film given in Figure 5-9 are based on product rates obtained in short-run experiments. A 22-day-long continuous test run was conducted with a typical film (No. 87) to assess its performance under such conditions, for the system glycerol-water. During this period, the operating pressure was 1500 psig, and the feed concentration was maintained at 0.5M. The separation and product rate characteristics of the film were

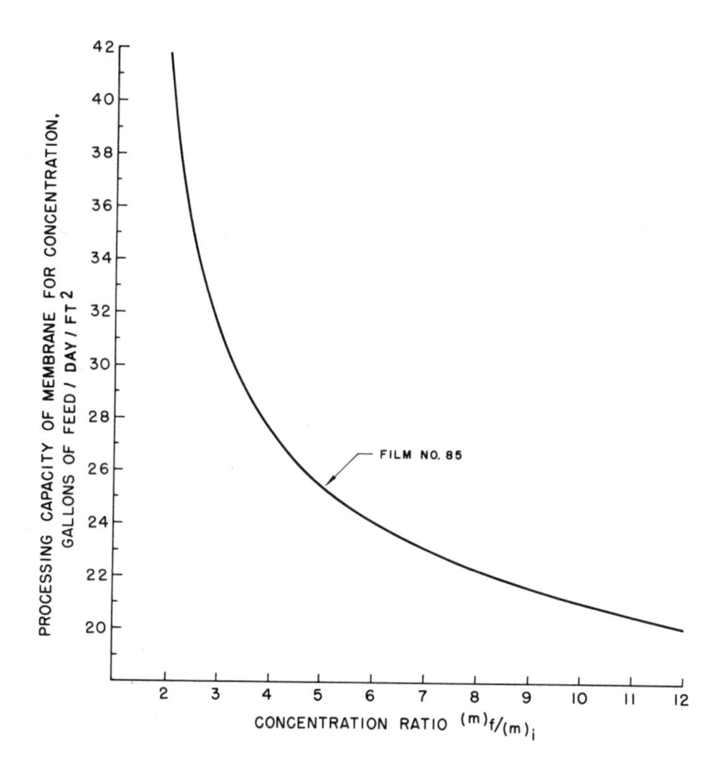

Figure 5-9. Processing capacity of a membrane for concentration of 0.25M aqueous glycerol solution. Data of Sourirajan and Kimura (1967).

Film type: CA-NRC-18; system: glycerol-water; initial concentration of charge = $(m)_i$ = 0.25M; final concentration of charge = $(m)_f$; feed rate: 380 cm^3/min; operating pressure: 1500 psig.

periodically determined (Figure 5-10). The solute separation remained essentially constant (~98 per cent) throughout the entire test period, and the product rates dropped with time. At the end of 1, 16 and 22 days of continuous operation, the product rates were, respectively, 81.5, 65.2, and 63.8 per cent of those obtained at the start of the run.

On the basis of the results given in Figure 5-10, the actually realisable processing capacity of Film 85, under conditions of 24-hour batch operation carried out for an extended period of time, may be expected to be about 60 per cent of those given in Figure 5-9. Thus, starting with a 0.25M glycerol-water feed solution, for obtaining a concentration ratio $(m)_f/(m)_i = 12$ (i.e. increasing the glycerol concentration from 2.25 to 21.65 wt. %), Film 85 may

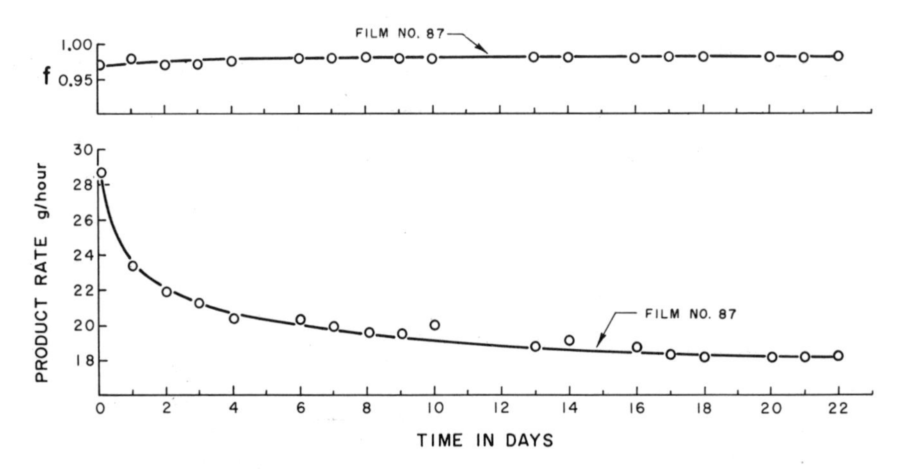

Figure 5-10. Performance of a typical membrane in a 22-day-long continuous test run with the system glycerol-water. Data of Sourirajan and Kimura (1967).

Film type: CA-NRC-18; feed molality: 0.5M; feed rate: 380 cm^3/min; operating pressure: 1500 psig; film area: 7.6 cm^2.

be expected to have a practical processing capacity of 12 gallons per day per square foot of film area, and yield a solute recovery of 96 per cent in the concentrate at an operating pressure of 1500 psig. These data are illustrative of the procedure involved in calculating some parameters of process design in the batchwise reverse osmosis concentration process.

2. Effect of Membrane Compaction on the Variations of Solute Separation and Product Rate

The calculations illustrated for the systems sucrose-water and glycerol-water do not take into account the effect of membrane compaction on solute separation. From the membrane specifications given in terms of the pure water permeability constant A, and the appropriate solute transport parameter $(D_{AM}/K\delta)$ at the operating pressure, the effect of membrane compaction (A factor) on solute separation and product rate can be predicted, by the method described in Chapter 3, on the basis that the relationship between $(D_{AM}/K\delta)$ and X_{A2} remains unchanged during compaction. Figures 5-11 and 5-12 illustrate the results of such prediction calculations on the effect of A-

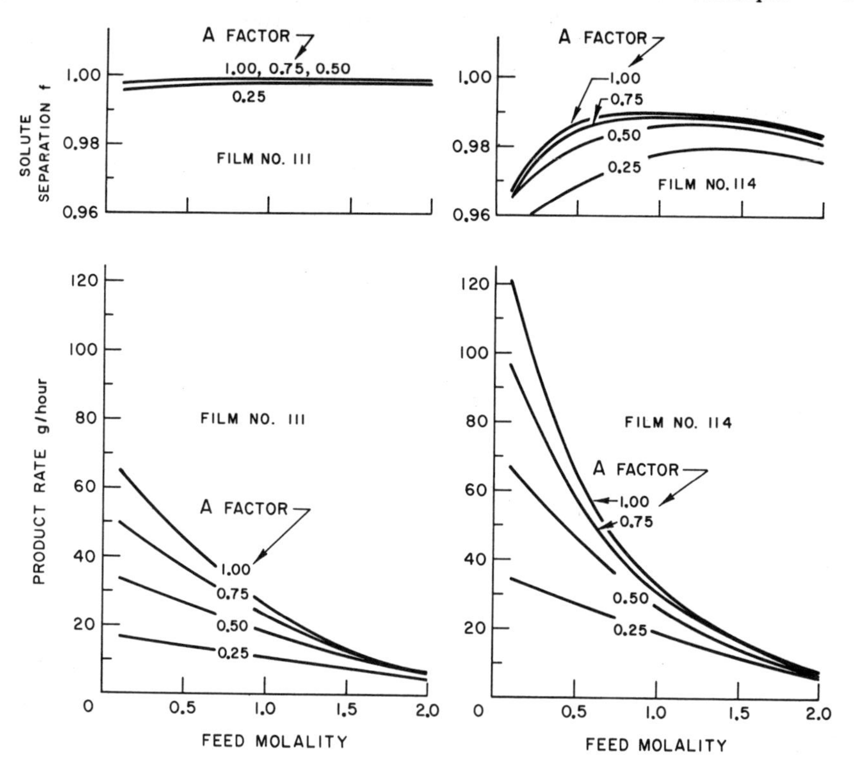

Figure 5-11. Effect of membrane compaction on solute separation and product rate for the system sucrose-water. Data of Kimura and Sourirajan (1968d).

Film type: CA-NRC-18; feed rate: 390 cm^3/min; operating pressure: 1500 psig; area of film: 7.6 cm^2.

factor on solute separation and product rate as a function of feed concentration for the systems sucrose-water and glycerol-water respectively, using three typical films (Kimura and Sourirajan, 1968b, 1968d); these data were obtained using the mass transfer coefficient data given in Figures 3-77 and 3-45. The results show that even with A-factors as low as 0.5, decrease in solute separation, f, is less than 1 per cent with the particular films tested, for both the systems sucrose-water and glycerol-water for feed concentrations in the range 0.1 to 2.0M; the effect of A-factor on f becomes more significant for higher feed concentrations for the system glycerol-water. Such prediction calculations can be made for any film whose specifications are given, and for which the applicable mass transfer coefficient data are available; the solute recovery and the processing capacity calculations can then be repeated with the actual solute separations and the corresponding product rates obtainable at various levels of membrane compaction. These results may then be taken into account in the overall design of the reverse osmosis concentration system.

3. Reverse Osmosis Concentration Technique

Recovery of Industrial Wastes and Water Pollution Control. The majority of industrial and municipal wastes and polluted waters resulting

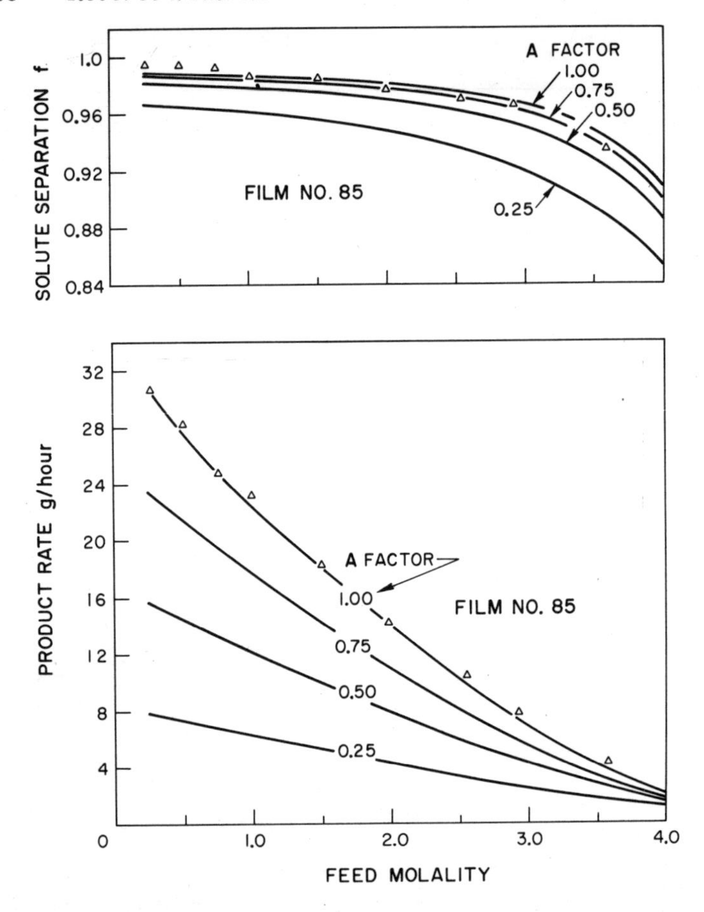

Figure 5-12. Effect of membrane compaction on solute separation and product rate for the system glycerol-water. Data of Kimura and Sourirajan (1968b).

Film type: CA-NRC-18; feed rate: 380 cm^3/min; operating pressure: 1500 psig. Δ experimental data (1.00 A); —calculated; data film area: 7.6 cm^2.

therefrom contain valuable or objectionable inorganic and/or organic solutes. They are amenable to reverse osmosis separation and concentration treatment by which over 90 per cent of the water can be decontaminated and reused; the concentrated waste water can then be chemically treated or evaporated either for the recovery of valuable chemicals or for the disposal of the residue after rendering it harmless.

Pulp and paper-mill waste streams (particularly spent bleach liquors and diluted pulp digester streams) can be treated advantageously by the reverse osmosis process (Ironside and Sourirajan, 1967). The quality of the membrane-permeated product water obtained in the process may often be satisfactory for reuse by the mill, thereby greatly reducing fresh water consumption and stream-pollution loads; further, the concentrated retentates from the process can be economically reduced by evaporation and burned, or useful chemicals (lignins, sugars, etc.) recovered.

Effluents from food processing plants (particularly dairies, meat packing houses, canneries) can also be treated profitably by the reverse osmosis process. While, again, the major objective of such treatment is pollution control, the concentrates obtained from such treatment are likely to be rich in nutritive components (proteins, sugars, etc.) which can be profitably recovered for use as fertilizer and animal food.

Numerous similar potential applications for the reverse osmosis concentration process can be cited; in many such applications, the Loeb-Sourirajan type porous cellulose acetate membranes can be profitably employed.

4. Concentration By Induced Reverse Osmosis

The reverse osmosis concentration of economically valuable aqueous solutions such as maple sap or fruit juices using the Loeb-Sourirajan type porous cellulose acetate membranes can be greatly facilitated by keeping the low pressure side of the membrane in contact with a suitable external solution such as aqueous sucrose solution. The difference between the chemical potential of pure water in the respective solutions on the high pressure and atmospheric pressure sides of the membrane is increased (see Figure 3-1) even at a given operating pressure. The result is an increased effective pressure, and hence product rate, and a consequent increase in the capacity of the membrane for a given level of solute concentration at any particular operating pressure. This technique is obviously a combination of 'osmosis' and 'reverse osmosis', and it is called here Induced Reverse Osmosis to distinguish it from Reverse Osmosis which does not involve an external solution. Even though no experimental data have appeared in the literature using the above technique, it holds promise at least for some applications in the reverse osmosis concentration process.

5. Nomenclature for Chapter 5

a = weight fraction of solute in product

A = pure water permeability constant, gram mole H_2O/cm^2 sec atm

$\overline{A}$ = area of membrane surface, ft^2

B_1 = constant

B_2 = constant

$\left(\frac{D_{AM}}{K\delta}\right)$ = parameter for the transport of solute through the membrane phase, cm/sec

f = solute separation $= \frac{m_1 - m_3}{m_1}$

$\overline{F}$ = fraction solute recovery in the batchwise concentration process

$\overline{K}$ = constant

m = solute molality

m_1 = solute molality in feed solution

m_3 = solute molality in product solution

M = concentration of solution in molality unit

M_S = molecular weight of solute, grams per mole

[PR] = product rate, g/hr/given film area

[PWP] = pure water permeability, g/hr/given film area

q = product rate, grams per day per square foot of film area

S = quantity defined by Equation 5-19

t = time, days

V = volume of solution on the high pressure side of membrane during the batchwise concentration process, cm^3

W = weight of solution on the high pressure side of membrane during the batchwise concentration process, grams

W_S = weight of solute in the above solution, grams

W_W = weight of water in the above solution, grams

X_{A1} = mole fraction of solute in the feed solution

X_{A2} = mole fraction of solute in the concentrated boundary solution

X_{A3} = mole fraction of solute in the product

Z = weight fraction of solute in the solution on the high pressure side of the membrane during the batchwise concentration process

α_V = density of solution on the high pressure side of membrane in the batchwise concentration process, g/cm^3

Suffixes

i = initial state of feed solution

f = final state of feed solution

Superscript

$'$ = dummy variable in definite integral

6. Summary

The reverse osmosis membrane separation process may be expected to develop as a practical concentration technique of great significance in science and industry. Some parameters of process design including volume change and solute recovery during concentration, and the processing capacity of a membrane for a given level of solute concentration, applicable for the reverse osmosis concentration of solutes in aqueous solution, have been developed, and illustrated with the experimental data for the systems sucrose-water and glycerol-water. Available information on the reverse osmosis concentration of natural maple sap and wood sugar solution is also summarised. The advantage of the induced reverse osmosis technique for the concentration of valuable aqueous solutions is pointed out.

CHAPTER 6

Reverse Osmosis Separation of Mixed Solutes in Aqueous Solution

The transport equations and correlations of reverse osmosis experimental data using the Loeb-Sourirajan type porous cellulose acetate membranes discussed in Chapters 3, 4, and 5 are for aqueous solutions containing one solute only. The extension of the Kimura-Sourirajan analysis to mixed solute systems is complicated by the general non-availability of the applicable osmotic pressure data, and the possibility of ionic interactions in such systems. Most natural waters, and industrial aqueous solutions contain more than one solute. The application of reverse osmosis for the separation and fractionation of such mixed solutes is of great practical interest. Hence the development of suitable methods for predicting membrane performance for such mixed solute systems from the specifications of the membrane, given in terms of the pure water permeability constant and solute transport parameter for the single solutes, is an area of fundamental importance in reverse osmosis transport. However, very few studies have been reported on the subject. Some of the available data on the reverse osmosis separation of mixed inorganic solutes in aqueous solution are presented in this chapter.

Some early results. Some of the earliest investigations with mixed solute systems were made by Sourirajan (1963c, 1964b) using Schleicher and Schuell (S & S) porous cellulose acetate membranes. Table 6-1 gives some of the results. In these studies the following systems were briefly tested with respect to the degree of separation of individual solutes: $NaCl$-$NaNO_3$-H_2O, Na_2SO_4-$NaCl$-H_2O, Na_2SO_4-$NaNO_3$-H_2O, and $BaCl_2$-$NaCl$-H_2O at a total molality of ~0.5M. It was found that the relative moles of the solutes in the product solutions were different from those in the feed solutions, depending on the nature of the mixed solute system, total molality of the feed, and the porous structure of the membrane. With respect to each salt at a given feed molality, the extent of separation was generally less in a mixed solute system than in the single solute system. The following results are illustrative. With a particular film, at the same operating pressure and feed flow rate, solute separations with respect to $NaNO_3$ in the feed systems [$NaNO_3$ (0.1M)-H_2O], [$NaCl$ (0.4M)-$NaNO_3$ (0.1M)-H_2O], and [Na_2SO_4 (0.4M)-$NaNO_3$ (0.1M)-H_2O] were 54.0, 51.8, and 3.0 per cent respectively; also, solute separations with respect to $NaCl$ in the feed systems [$NaCl$ (0.1M)-H_2O], [$NaCl$ (0.1M)-$NaNO_3$ (0.4M)-H_2O], [$BaCl_2$ (0.4M)-$NaCl$ (0.1M)-H_2O], and [Na_2SO_4 (0.4M)-$NaCl$ (0.1M)-H_2O] were 69.0, 59.2, 36.6, and 32.3 per cent respectively. Table 6-2 gives some data of Erickson (1966) on the reverse osmosis separation of solutes in several mixed solute systems using the Loeb-Sourirajan type porous cellulose acetate membranes. From the point of view of predictability of membrane performance for aqueous feed solutions containing mixed solutes, all the above data are only qualitative in scope.

1. Inorganic Salts in Aqueous Solution Containing Two Solutes With a Common Ion

Agrawal and Sourirajan (1969c) have studied the reverse osmosis separation of solutes using feed solution systems sodium nitrate-sodium chloride-water, sodium chloride-magnesium chloride-water, and sodium chloride-barium

Table 6-1. Some data on reverse osmosis separation of mixed solutes in aqueous solution using porous cellulose acetate membranes

Data of Sourirajan (1963c)

System [A1-A2-H_2O]	Feed solution molality (m_1)			Product solution		
	$(m_1)_{A1}$	$(m_1)_{A2}$	$\frac{(m_1)_{A1}}{(m_1)_{A2}}$	Sepn. of A1(%)	Sepn. of A2(%)	$\frac{(m_3)_{A1}}{(m_3)_{A2}}$
NaCl-$NaNO_3$-H_2O	0.501	—	—	63.3	—	—
	0.409	0.107	3.822	63.6	51.8	2.888
	0.307	0.199	1.543	61.2	50.5	1.209
	0.208	0.298	0.698	60.7	49.5	0.543
	0.107	0.398	0.269	59.2	48.2	0.212
	—	0.499	—	—	48.5	—
Na_2SO_4-NaCl-H_2O	0.502	—	—	95.4	—	—
	0.401	0.103	3.893	92.7	32.3	0.420
	0.301	0.202	1.490	95.5	43.6	0.118
	0.202	0.301	0.671	95.9	50.8	0.056
	0.103	0.397	0.259	98.4	58.4	0.010
	—	0.500	—	—	63.5	—
Na_2SO_4-$NaNO_3$-H_2O	0.501	—	—	93.7	—	—
	0.401	0.098	4.092	94.7	3.0	0.222
	0.301	0.200	1.503	95.3	23.0	0.093
	0.201	0.297	0.675	96.6	37.7	0.037
	0.101	0.400	0.253	96.1	46.3	0.019
	—	0.498	—	—	48.6	—
$BaCl_2$-NaCl-H_2O	0.499	—	—	88.5	—	—
	0.399	0.099	4.034	90.9	36.6	0.576
	0.300	0.200	1.500	90.5	47.6	0.270
	0.196	0.317	0.618	92.4	53.6	0.101
	0.096	0.416	0.230	94.6	61.3	0.032
	—	0.501	—	—	63.5	—

Film type: S & S ultrafine, superdense (preshrunk)
Operating pressure: 1500 psig
Feed rate: 15 cm^3/min

chloride-water, and Loeb-Sourirajan type porous cellulose acetate membranes. Their results lead to a simple method of predicting membrane performance for aqueous feed solutions containing two inorganic salts with a common ion, from the specifications of the membrane given in terms of the pure water permeability constant, and the respective solute transport parameters along with the mass transfer coefficient correlation applicable for single solute systems.

Experimental Details. Agrawal and Sourirajan (1969c) used Batch 18-type porous cellulose acetate membranes shrunk at different temperatures to give different levels of solute separation at a given set of operating conditions. The experiments were of the short run type, each lasting for about 2 hours, they were carried out at the laboratory temperature, and the reported product rates are those corrected to 25°, using the relative viscosity and density data for pure water. In each experiment, the pure water permeability [PWP], and product rate [PR] in grams/hour/7.6 cm^2 of effective film area

Table 6-2. Some data on reverse osmosis separation of mixed solutes in aqueous solution using porous cellulose acetate membranes

Data of Erickson (1966)

Film no.	Solution system	Solute	Solute concn. (ppm) in feed	Solute concn. (ppm) in product	Product rate (gal/day/ft^2)
II	$NaCl-H_2O$	NaCl	26, 000	10, 250	30. 5
	$MgCl_2-H_2O$	$MgCl_2$	4980	955	39. 5
	$Na_2SO_4-H_2O$	Na_2SO_4	3920	356	50. 0
	$CaCl_2-H_2O$	$CaCl_2$	1100	300	50. 5
	$NaHCO_3-H_2O$	$NaHCO_3$	192	38	59. 0
II	$NaCl-MgCl_2-NaHCO_3-H_2O$	Na^+	10, 300	3500	14. 5
		Mg^{++}	1270	50	
		Cl^-	18, 250	6020	
		HCO_3^-	142	58	
XII	$NaCl-Na_2SO_4-CaCl_2-H_2O$	Na^+	11, 270	4850	38. 4
		Ca^{++}	400	100	
		Cl^-	14, 700	7600	
		SO_4^{--}	2650	360	
XII	$NaCl-NaHCO_3-CaCl_2-H_2O$	Na^+	10, 300	5700	40. 7
		Ca^{++}	400	127	
		Cl^-	15, 452	8150	
		HCO_3^-	142	50	
XII	$CaCl_2-Na_2SO_4-H_2O$	Na^+	1267	248	49. 4
		Ca^{++}	400	43	
		Cl^-	702	321	
		SO_4^{--}	2650	165	
S	Sea Water	Na^+	10, 560	1070	13. 1
		K^+	380	50	
		Mg^{++}	1270	50	
		Ca^{++}	400	17	
		Cl^-	18, 980	1900	
		HCO_3^-	142	12	
		Br^-	65	11	
		SO_4^{--}	2649	10	
		TDS*	34, 446	3120	
S	Rosewell water	Na^+	4830	280	25. 0
		Mg^{++}	300	13	
		Ca^{++}	550	12	
		Cl^-	8064	502	
		HCO_3^-	120	15	
		SO_4^{--}	1528	30	
		TDS*	15, 392	852	

* TDS: total dissolved solids

Film type: Batch 47
Film shrinkage temperature: 75°
Operating pressure: 600 psig

at preset operating conditions were determined, along with solute separations (f) defined as follows with respect to each solute:

$$f = \frac{\text{molality of feed } (m_1)_A - \text{molality of product } (m_3)_A}{\text{molality of feed } (m_1)_A}$$

For single solute systems, the concentrations of the solute in the feed and product solutions were determined by refractive index measurements using a precision Bausch and Lomb refractometer, or by specific resistance measurements. For the system $NaNO_3$-$NaCl$-H_2O, total salt content was determined by evaporation, and NaCl was determined by conductometric titration with standard silver nitrate. For the system $NaCl$-$MgCl_2$-H_2O, magnesium was determined by EDTA titration, and total chloride was determined by the Mohr method. For the system $NaCl$-$BaCl_2$-H_2O, total chloride and barium were determined by conductometric titrations using standard silver nitrate and potassium chromate respectively. For the systems $NaCl$-KCl-$MgCl_2$-H_2O and $NaCl$-KCl-$MgCl_2$-$BaCl_2$-H_2O atomic absorption methods of analysis were used (Slavin, 1968). In each case, the analytical method was established by means of standard solutions of mixed solutes. The results of reverse osmosis experiments were as follows.

Product rate v. Fractional solute molality. At a given set of operating conditions of pressure, total feed molality, and feed flow rate, the product rate obtained with a mixed solute system was intermediate between those obtained for the respective single solute systems at the same total feed molality. Representing the mixed solute system as A1-A2-H_2O (where A1 and A2 are the solutes under consideration), the graph of the product rate at total molality v. fractional molality of solute (molality of A2/total molality) was a straight line. For example, the product rate obtained with the system [$NaNO_3$ (0.5M)-NaCl (0.5M)-H_2O] was the mean of the product rates obtained with the systems [$NaNO_3$ (1.0M)-H_2O] and [NaCl (1.0M)-H_2O] under otherwise identical experimental conditions. This type of correlation is illustrated in Figure 6-1 with a typical set of experimental data for the systems $NaNO_3$-$NaCl$-H_2O, $NaCl$-$MgCl_2$-H_2O, and $NaCl$-$BaCl_2$-H_2O for total feed molalities of 1M and/or 2M.

Separation data for mixed solute systems. This correlation between product rate and fractional solute molality, together with the separation data at the given total molality for single solute systems, offers a means of calculating solute separations for mixed solute systems. From the solute separation and product rate data obtained for the single solute systems [A1-H_2O] and [A2-H_2O] at the given total molality, N_{A1} and N_{B1}, and N_{A2} and N_{B2} can be calculated. Defining

$$x_1 = \frac{(m_1)_{A1}}{(m_1)_{A1} + (m_1)_{A2}} \tag{6-1}$$

and

$$x_2 = \frac{(m_1)_{A2}}{(m_1)_{A1} + (m_1)_{A2}} \tag{6-2}$$

the correlation given in Figure 6-1 may be expressed as

$$[PR]_{Mix} = x_1[PR]_{A1} + x_2[PR]_{A2} \tag{6-3}$$

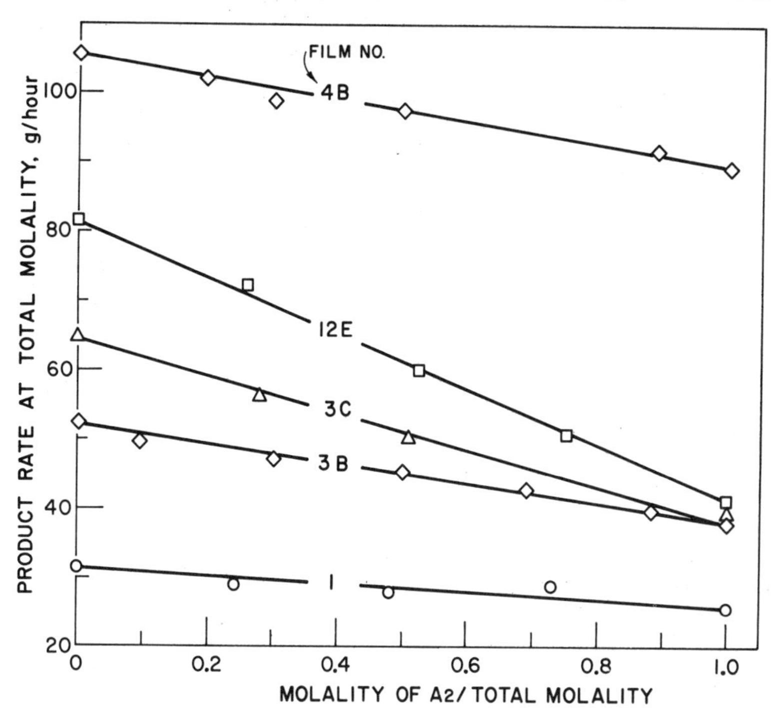

Figure 6-1. Product rate v. fractional solute molality for feed solutions containing two inorganic salts with a common ion. Data of Agrawal and Sourirajan (1969c).

Film type: CA-NRC-18; system: (A1-A2-H_2O); [total feed molality]:
$NaNO_3$-NaCl-H_2O ○ 1M ◇ 2M
NaCl-$BaCl_2$-H_2O △ 1M
NaCl-$MgCl_2$-H_2O □ 1M
Feed rate: 360 cm^3/min; operating pressure: 1500 psig; film area: 7.6 cm^2.

where $[PR]_{Mix}$, $[PR]_{A1}$, and $[PR]_{A2}$ represent respectively the product rate for the mixed solute system, and the product rates for the single solute systems [A1-H_2O] and [A2-H_2O] at the same feed molality. Since product rate is proportional to effective pressure, Equation 6-3 simply means that, in a system containing mixed solutes with a common ion, the difference between the osmotic pressure of the boundary solution on the high pressure side of the membrane and that of the product solution on the atmospheric pressure side of the membrane is a linear function of the corresponding values for the single solute systems at constant total feed concentration.

Since

$$[PR]_{A1} = k_{A1}N_{A1} + k_B N_{B1} \tag{6-4}$$

and

$$[PR]_{A2} = k_{A2}N_{A2} + k_B N_{B2} \tag{6-5}$$

therefore

$$[PR]_{Mix} = x_1 k_{A1} N_{A1} + x_2 k_{A2} N_{A2} + k_B (x_1 N_{B1} + x_2 N_{B2}) \tag{6-6}$$

The molalities of the respective solutes in the product solution from the mixed solute system are then given by

$$(m_3)_{A1} = 1000 \cdot \frac{x_1 k_{A1} N_{A1}}{M_{A1} k_B (x_1 N_{B1} + x_2 N_{B2})} \tag{6-7}$$

$$= 55.5 \cdot \frac{x_1 N_{A1}}{x_1 N_{B1} + x_2 N_{B2}} \tag{6-8}$$

$$(m_3)_{A2} = 1000 \cdot \frac{x_2 k_{A2} N_{A2}}{M_{A2} k_B (x_1 N_{B1} + x_2 N_{B2})} \tag{6-9}$$

$$= 55.5 \cdot \frac{x_2 N_{A2}}{x_1 N_{B1} + x_2 N_{B2}} \tag{6-10}$$

using the relations

$$\frac{k_{A1}}{k_B} = \frac{M_{A1}}{M_B} \tag{6-11}$$

$$\frac{k_{A2}}{k_B} = \frac{M_{A2}}{M_B} \tag{6-12}$$

and

$$\frac{1000}{M_B} = 55.5 \tag{6-13}$$

The solute separations in the mixed solute systems can then be calculated from the relations:

$$\text{Separation with respect to A1} = \frac{(m_1)_{A1} - (m_3)_{A1}}{(m_1)_{A1}} \tag{6-14}$$

and

$$\text{Separation with respect to A2} = \frac{(m_1)_{A2} - (m_3)_{A2}}{(m_1)_{A2}} \tag{6-15}$$

Experimental solute separation and product rate data in single solute and mixed solute systems. Table 6-3 gives the experimental solute separation and product rate data for several films (covering a wide range of solute separations) for the single solute systems $NaNO_3$-H_2O, NaCl-H_2O, $MgCl_2$-H_2O, and $BaCl_2$-H_2O at total molalities of ~1M or 2M and at two different feed rates. The operating pressure was kept constant in all cases. Table 6-4 gives similar data for the mixed solute system $NaNO_3$-NaCl-H_2O; this table also gives the product rate data calculated from Equation 6-3 and the respective solute

Table 6-3. Reverse osmosis data for some single solute systems using porous cellulose acetate membranes

Data of Agrawal and Sourirajan (1969c)

Film no.	Feed solution system	Feed molality	Solute separation (%)	Product rate (g/hr)
Feed rate: 360 cm^3/min				
1	Sodium chloride-water	1.052	97.5	25.8
1	Sodium nitrate-water	0.995	95.8	31.4
1C	Sodium chloride-water	1.052	96.9	27.0
1C	Barium chloride-water	0.976	98.5	15.5
2	Sodium chloride-water	1.052	86.6	49.5
2	Sodium nitrate-water	0.995	80.4	58.7
2A	Sodium chloride-water	1.025	86.2	49.3
2A	Sodium nitrate-water	0.970	78.6	54.6
2B	Sodium chloride-water	2.030	69.7	20.3
2B	Sodium nitrate-water	1.945	70.1	33.7
2C	Sodium chloride-water	1.052	86.0	48.2
2C	Barium chloride-water	0.976	90.5	27.1
3	Sodium chloride-water	1.052	73.4	67.3
3	Sodium nitrate-water	0.995	66.3	79.4
3B	Sodium chloride-water	2.030	52.9	38.0
3B	Sodium nitrate-water	1.945	52.2	52.3
3C	Sodium chloride-water	1.052	71.9	64.6
3C	Barium chloride-water	0.976	78.1	37.6
4B	Sodium chloride-water	2.030	27.3	89.5
4B	Sodium nitrate-water	1.945	26.0	105.4
4C	Sodium chloride-water	1.052	42.5	118.1
4C	Barium chloride-water	0.976	51.6	75.9
12	Sodium chloride-water	2.030	42.8	58.0
12	Sodium nitrate-water	2.010	42.9	73.0
12A	Sodium chloride-water	2.020	47.0	50.9
12A	Sodium nitrate-water	1.995	44.7	68.5
12E	Sodium chloride-water	1.010	55.3	81.9
12E	Magnesium chloride-water	1.010	56.3	41.7
13	Sodium chloride-water	2.030	52.7	43.6
13	Sodium nitrate-water	2.010	53.7	56.9
13A	Sodium chloride-water	2.020	56.2	38.4
13A	Sodium nitrate-water	1.995	55.5	52.5
13E	Sodium chloride-water	1.010	64.5	67.8
13E	Magnesium chloride-water	1.010	64.4	33.8
16	Sodium chloride-water	2.030	70.1	21.9
16	Sodium nitrate-water	2.010	66.4	37.3
16A	Sodium chloride-water	2.020	69.9	21.8
16A	Sodium nitrate-water	1.995	66.7	36.6
16B	Sodium chloride-water	1.013	84.4	53.1
16B	Sodium nitrate-water	0.994	75.8	61.0
16E	Sodium chloride-water	1.010	79.5	52.6
16E	Magnesium chloride-water	1.010	82.7	20.6
Feed rate: 180 cm^3/min				
12C	Sodium chloride-water	1.040	50.5	73.8
12C	Sodium nitrate-water	1.020	41.2	85.2
13C	Sodium chloride-water	1.040	62.7	60.4
13C	Sodium nitrate-water	1.020	54.4	69.0
16C	Sodium chloride-water	1.040	79.6	44.6
16C	Sodium nitrate-water	1.020	67.9	54.1

Film type: CA-NRC-18
Operating pressure: 1500 psig
Film area: 7.6 cm^2

Table 6-4. Reverse osmosis data for mixed solute system sodium nitrate-sodium chloride-water

Data of Agrawal and Sourirajan (1969c)

Film no.	Total molality $(m_1)_{A1} + (m_1)_{A2}$	$\frac{(m_1)_A}{(m_1)_{A1} + (m_1)_{A2}}$		Solute separation (%)				Product rate (g/hr)		Solute sepn. (%) in single solute system molality $(m_1)_A$	
				$NaNO_3$		NaCl					
		$NaNO_3$	NaCl	Exptl.	Calcd.	Exptl.	Calcd.	Exptl.	Calcd.	$NaNO_3$	NaCl
Feed rate = 360 cm^3/min											
1	1.084	0.76	0.24	94.3	96.3	99.6	98.0	28.9	30.0	96.9	98.7
1	1.046	0.52	0.48	94.2	96.0	98.2	97.8	27.9	28.7	96.9	98.3
1	1.013	0.27	0.73	93.6	95.7	98.0	97.6	29.0	27.3	96.1	98.0
2	1.046	0.52	0.48	77.7	80.4	88.5	87.9	52.0	54.3	83.8	89.4
3	1.046	0.52	0.48	64.9	65.6	75.7	75.8	69.8	73.5	69.4	77.3
2A	0.953	0.10	0.9	74.7	75.1	87.3	85.4	53.1	53.7	86.7	87.2
16B	1.015	0.75	0.25	74.8	75.6	87.3	86.0	57.5	58.5	76.6	89.7
16B	1.012	0.50	0.50	73.4	74.8	86.4	85.4	57.1	56.6	80.4	85.9
16B	1.018	0.26	0.74	71.3	74.0	85.3	85.1	53.1	54.6	83.1	85.8
2B	2.025	0.905	0.095	68.8	70.0	83.7	81.5	31.7	32.5	72.1	90.4
2B	2.007	0.700	0.300	65.8	66.8	81.5	79.6	29.3	29.7	75.6	89.2
2B	2.028	0.500	0.500	62.5	63.7	79.6	77.6	27.5	27.0	78.6	86.2
2B	2.039	0.312	0.688	59.4	57.5	77.3	76.5	25.2	24.5	82.5	82.4
2B	2.016	0.116	0.884	—	54.1	74.2	71.8	23.0	21.8	85.0	76.3
3B	2.025	0.905	0.095	51.2	52.7	65.0	65.5	49.6	51.0	54.0	76.0
3B	2.007	0.700	0.300	48.0	49.2	63.2	63.0	47.0	48.0	55.8	71.6
3B	2.028	0.500	0.500	45.3	46.4	60.2	60.7	45.4	45.2	58.6	67.3
3B	2.039	0.312	0.688	—	—	58.9	60.1	42.8	42.5	61.5	62.7

3B	2.016	0.116	0.884	—	37.8	55.6	54.7	40.2	39.7	69.0	57.7
4B	2.025	0.905	0.095	25.8	27.8	—	—	101.7	103.9	27.1	40.0
4B	2.007	0.700	0.300	23.2	24.5	33.5	35.2	98.7	100.6	28.3	38.2
4B	2.028	0.500	0.500	21.4	22.8	32.2	33.4	97.7	97.4	27.5	35.4
4B	2.016	0.116	0.884	—	16.5	28.1	28.3	92.0	91.3	35.0	30.3
16	2.019	0.750	0.250	61.4	62.8	78.6	80.0	32.9	33.5	70.2	85.9
16	2.021	0.500	0.500	58.2	58.4	76.0	77.5	29.0	29.6	73.0	83.5
16	2.066	0.260	0.740	54.2	53.0	—	75.0	25.9	25.9	79.1	79.4
13	2.019	0.750	0.250	49.1	51.2	61.7	60.7	50.9	53.6	56.2	69.4
13	2.021	0.500	0.500	45.8	48.6	58.2	58.4	48.6	50.2	57.0	66.0
13	2.066	0.260	0.740	49.1	51.2	55.3	56.8	46.3	47.0	56.2	61.5
12	2.019	0.750	0.250	38.5	40.3	51.8	50.9	66.4	69.3	45.4	59.2
12	2.021	0.500	0.500	35.0	37.8	48.2	48.5	63.8	65.5	45.0	54.9
12	2.066	0.260	0.740	32.6	34.9	45.8	47.1	60.3	61.9	49.6	51.1
16A	2.046	0.500	0.500	—	58.5	76.4	77.2	29.4	29.2	74.2	83.4
13A	2.046	0.500	0.500	48.3	49.0	61.5	62.5	45.4	45.4	60.2	68.0
12A	2.046	0.500	0.500	37.6	37.4	51.7	53.4	59.9	59.7	47.6	57.3
Feed rate = 180 cm^3/min											
16C	1.057	0.497	0.503	65.2	66.3	82.1	81.6	48.6	49.3	71.7	82.8
13C	1.057	0.497	0.503	51.9	53.4	65.6	65.4	64.0	64.6	57.5	66.5
12C	1.057	0.497	0.503	39.2	39.7	54.3	54.2	78.7	79.4	43.4	54.4

Film type: CA-NRC-18
System (A1-A2-H_2O): $NaNO_3$-NaCl-H_2O
Solute molality in feed: $(m_1)_A$
Operating pressure: 1500 psig
Film area: 7.6 cm^2

separation data calculated from Equations 6-8, 6-10, 6-14 and 6-15. Table 6-5 gives similar experimental and calculated solute separation and product rate data for the mixed solute systems NaCl-$MgCl_2$-H_2O, and NaCl-$BaCl_2$-H_2O. The data given in Tables 6-4 and 6-5 show that there is excellent agreement between the experimental and calculated values.

The last two columns in Tables 6-4 and 6-5 give the experimental separation data in single solute systems at the same feed molality as that in the mixed solute system with respect to the particular solute under consideration. A comparison of these data with solute separations obtained in the mixed solute system shows that solute separations are generally less in mixed solute systems. This, however, need not be the case under all conditions. Solute separations in the mixed solute systems depend on the nature of the solutes, total molality, and the porous structure of the membrane, in addition to operating pressure and feed flow rate.

Method For Predicting Membrane Performance For Some Mixed Solute Systems. The foregoing analysis, and the agreement between the experimental and calculated separation and product rate data illustrated in Tables 6-4 and 6-5, offer a simple means of predicting membrane performance for low concentrations of aqueous feed solution systems involving two inorganic solutes with a common ion. For such systems, the prediction technique is as follows.

From the specifications of the membrane given in terms of the pure water permeability constant, A, and solute transport parameters, $(D_{AM}/K\delta)$, and the applicable mass transfer coefficient correlation, calculate solute separation and product rate for the single solute systems at the given total molality following the Kimura-Sourirajan analysis.

Calculate N_{A1}, N_{B1}, N_{A2}, and N_{B2} from the above solute separation and product rate data for the single solute systems.

Calculate the product rate for the mixed solute system using Equation 6-3.

Finally, calculate solute separations with respect to each solute using Equations 6-8, 6-10, 6-14, and 6-15.

The accuracy of the above prediction technique depends on the accuracy of the membrane specifications and the applicable mass transfer coefficient correlation. That the Kimura-Sourirajan analysis predicts membrane performance well for single solute systems has been illustrated extensively in Chapter 3. Figure 6-2 is another illustration. Figure 6-2 gives the experimental and calculated solute separation and product rate data for the single solute systems $NaNO_3$-H_2O and NaCl-H_2O as functions of feed concentrations. The calculated values are based on membrane specifications for the film 16D given in Table 6-6 (which also gives the specifications of all the films used in this work) and the generalised mass transfer coefficient correlation given in Figure 3-46 in Chapter 3. Figure 6-2 illustrates not only the excellent predictability of membrane performance for single solute systems by using the Kimura-Sourirajan analysis, but also the change in the order of solute separations with respect to sodium nitrate and sodium chloride at different operating conditions as discussed in Chapter 3.

Some further studies. Sourirajan and Agrawal (1969) recently reported the results given in Table 6-7 on the mixed solute systems KCl-NaCl-H_2O and $NaNO_3$-NaCl-H_2O at total feed molalities of ~2M and 3M respectively.

Table 6-5. Reverse osmosis data for some mixed solute systems

Data of Agrawal and Sourirajan (1969c)

Film no.	Total molality $(m_1)_{A1} + (m_1)_{A2}$	$\frac{(m_1)_A}{(m_1)_{A1} + (m_1)_{A2}}$		Solute separation (%)				Product rate (g/hr)		Solute sepn. (%) in single solute system, molality $(m_1)_A$	
				A1		A2					
		A1	A2	Exptl.	Calcd.	Exptl.	Calcd.	Exptl.	Calcd.	A1	A2
System (A1-A2-H_2O) = NaCl-$MgCl_2$-H_2O											
16E	1.005	0.247	0.753	60.2	61.3	86.2	87.2	27.9	28.4	83.9	88.8
16E	1.003	0.496	0.504	68.5	70.0	89.6	90.1	36.1	36.2	82.7	91.7
16E	1.026	0.742	0.258	74.0	75.2	91.0	91.9	45.1	43.4	81.3	93.4
13E	1.005	0.247	0.753	41.2	41.5	68.8	71.2	41.2	41.7	69.4	71.1
13E	1.003	0.496	0.504	53.7	51.6	74.0	76.2	49.9	50.4	68.0	75.8
13E	1.026	0.742	0.258	59.6	58.6	76.7	79.8	59.8	58.2	66.3	79.0
System (A1-A2-H_2O) = NaCl-$BaCl_2$-H_2O											
1C	1.052	0.255	0.745	—	—	99.0	98.8	16.3	18.4	98.9	99.0
1C	1.019	0.490	0.510	96.1	96.2	99.1	98.9	19.2	21.2	98.5	99.4
1C	1.059	0.720	0.280	97.0	96.4	99.2	99.0	23.0	23.7	98.2	99.5
2C	1.052	0.255	0.745	—	—	90.9	92.5	29.7	32.3	91.8	92.8
2C	1.019	0.490	0.510	84.6	82.5	93.8	93.4	37.3	37.6	89.5	94.7
2C	1.052	0.720	0.280	82.0	83.7	93.1	93.9	42.1	42.1	88.5	95.4
3C	1.019	0.490	0.510	—	—	84.9	85.4	50.1	51.1	77.1	85.8
3C	1.052	0.720	0.280	69.4	67.4	85.4	86.7	56.2	57.0	75.0	88.5
4C	1.019	0.490	0.510	—	—	62.4	63.6	94.4	97.0	46.7	58.0
4C	1.052	0.720	0.280	34.3	35.2	—	—	104.0	106.2	45.2	62.2

Film type: CA-NRC-18
System type: A1-A2-H_2O
Solute molality in feed: $(m_1)_A$
Feed rate: 360 cm^3/min
Operating pressure: 1500 psig
Film area: 7.6 cm^2

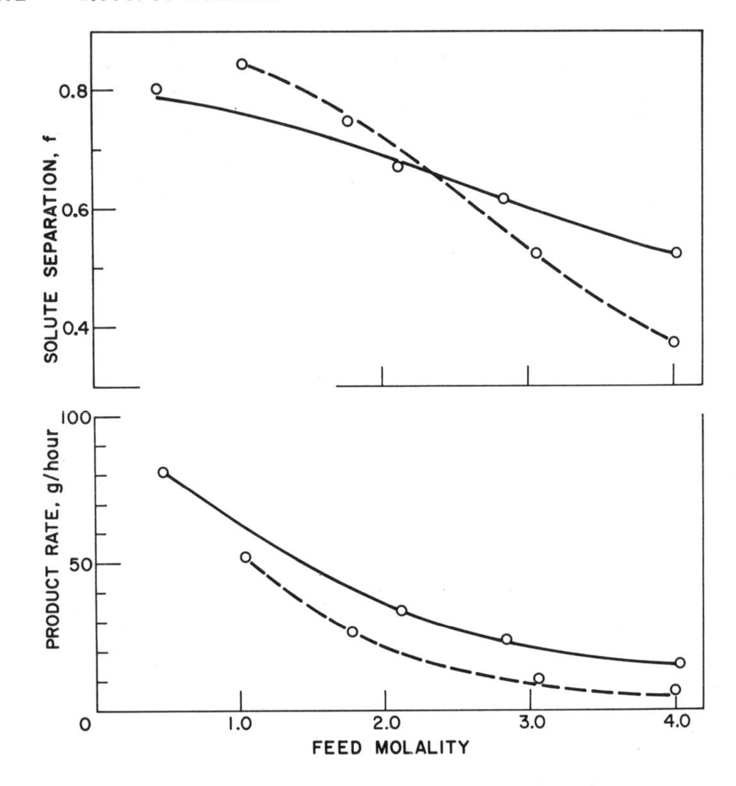

Figure 6-2. Comparison of experimental and calculated solute separation and product rate data for systems [$NaNO_3$–H_2O] and [NaCl–H_2O] as a function of feed concentration. Data of Agrawal and Sourirajan (1969c).

Film type: CA-NRC-18; film no.: 16D; system: —— $NaNO_3$-H_2O; – – – NaCl-H_2O; feed rate: 360 cm^3/min; operating pressure: 1500 psig; ○ Experimental; —— – – – } Calculated; film area: 7.6 cm^2.

The data for the system KCl-NaCl-H_2O are of interest from the point of view of the practical separations of KCl and NaCl found in admixture in natural rock formations. The data given in Table 6-7 confirm the procedure outlined above for the prediction of membrane performance for such mixed solute systems.

The data for the system $NaNO_3$-NaCl-H_2O given in Table 6-7 are of importance in view of the high total concentration of the feed solutions involved. They indicate the need for a slightly modified procedure for the prediction of membrane performance for such mixed solute systems as outlined below.

Prediction procedure for membrane performance for mixed solute systems containing a common ion, at high feed concentrations. The experimental data for the system $NaNO_3$-NaCl-H_2O at the total feed molality of ~3M (Table 6-7) showed that, under such conditions, Equation 6-3 was valid only when x_1 and

Table 6-6. Membrane specifications

Data of Agrawal and Sourirajan (1969c)

Film no.	$A \times 10^6$ $\left(\frac{\text{g. mole } H_2O}{\text{cm}^2\text{sec atm}}\right)$	$(D_{AM}/K\delta) \times 10^5$ (cm/sec)			
		NaCl	$NaNO_3$	$MgCl_2$	$BaCl_2$
1	1.202	2.02	4.20	—	—
1C	1.186	1.70	—	—	0.70
2	2.147	21.86	37.43	—	—
2A	1.920	24.00	44.70	—	—
2B	1.920	22.70	48.50	—	—
2C	2.056	21.30	—	—	8.70
3	2.708	62.90	92.90	—	—
3B	2.290	89.50	136.2	—	—
3C	2.550	63.50	—	—	28.00
4B	3.330	441.0	570.0	—	—
4C	3.850	336.0	—	—	145.0
12	2.712	173.0	232.8	—	—
12A	2.614	149.2	201.4	—	—
12C	2.613	144.8	201.0	—	—
12E	2.600	162.0	—	92.10	—
13	2.456	97.30	131.8	—	—
13A	2.354	83.90	113.1	—	—
13C	2.354	81.50	113.0	—	—
13E	2.270	99.10	—	57.50	—
16	2.200	28.90	51.4	—	—
16A	2.160	27.80	50.3	—	—
16B	2.163	27.10	47.6	—	—
16C	2.159	27.80	50.3	—	—
16D	2.183	27.10	47.8	—	—
16E	2.080	36.72	—	14.40	—

Film type: CA-NRC-18
Operating pressure: 1500 psig

x_2 (Equations 6-1 and 6-2) were defined in terms of molarity, instead of molality, as illustrated in Figure 6-3. Since molarity and molality are not too different up to about 2M for the systems $NaNO_3$-H_2O and NaCl-H_2O, it is probably more general to express the product rate correlation in terms of molarities, as illustrated in Figure 6-3. Under such conditions, the prediction procedure is slightly different and it is illustrated below with a numerical example.

First calculate the effective total molality for the single solute systems corresponding to the total molarity of the mixed solute system. For example, referring to Table 6-7, Run no. 556, Film no. 25, for the system $NaNO_3$-NaCl-H_2O [A1-A2-H_2O],

$$(m_1)_{A1} = 0.794 \text{ molal} \equiv 0.78 \text{ molar}$$
$$(m_1)_{A2} = 2.236 \text{ molal} \equiv 2.14 \text{ molar}$$

Table 6-7. Reverse osmosis data for some mixed solute systems

Data of Sourirajan and Agrawal (1969)

Run no.	Film no.	Total feed molality	$\frac{(m_1)_A}{(m_1)_{A1} + (m_1)_{A2}}$ A1	A2	Solute separation A1 Exptl.	A1 Calcd.	A2 Exptl.	A2 Calcd.	Product rate (g/hr) Exptl.	Calcd.
System: $KCl-NaCl-H_2O$										
570	22	2.0	0.8	0.2	64.7	63.6	72.3	71.0	37.2	37.6
570	26	2.0	0.8	0.2	43.8	43.2	52.4	54.0	44.0	44.4
571	22	2.0	0.6	0.4	62.0	61.8	68.9	69.6	35.5	35.8
571	26	2.0	0.6	0.4	42.8	40.8	50.8	52.1	41.8	42.5
572	22	2.0	0.4	0.6	59.8	59.9	65.2	68.1	32.8	34.0
572	26	2.0	0.4	0.6	37.0	38.1	45.2	49.9	39.8	40.6
573	22	2.0	0.2	0.8	59.1	57.7	63.7	66.3	32.2	32.2
System: $NaNO_3-NaCl-H_2O$										
552	22	3.03	0.505	0.495	42.3	39.6	57.2	57.5	24.1	23.9
552	25	3.03	0.505	0.495	16.8	13.9	22.3	19.5	63.1	61.6
556	25	3.03	0.262	0.738	13.3	13.4	18.3	16.3	60.3	61.0
563	22	3.08	0.746	0.254	48.2	47.0	60.9	61.5	26.8	26.2
563	25	3.08	0.746	0.254	16.4	14.8	21.5	18.0	63.2	62.8
563	26	3.08	0.746	0.254	34.5	32.8	45.8	49.4	33.3	32.6

Film type: Batch 18
System type: $A1-A2-H_2O$
Solute molality in feed: $(m_1)_A$
Feed rate: 360 cm^3/min
Operating pressure: 1500 psig
Film area: 7.6 cm^2

Total molality = 3.03; total molarity = 2.92
2.92 molar $NaNO_3$ ≡ 3.28 molal $NaNO_3$
2.92 molar NaCl ≡ 3.10 molal NaCl

Therefore the effective total molalities for the single solute systems [$NaNO_3-H_2O$] and [$NaCl-H_2O$] are 3.28 and 3.10 respectively.

Next, calculate solute separations, and product rates $[PR]_{A1}$ and $[PR]_{A2}$, for the single solute systems at their corresponding total effective molalities calculated above from data on membrane specifications and the applicable mass transfer coefficient correlation; or such data can be obtained experimentally.

Then calculate $N_{A1}, N_{B1}, N_{A2}, N_{B2}$ from the above solute separation and product rate data obtained for the single solute systems.

Find x_1 and x_2 from Equations 6-1 and 6-2 using molarity instead of molality; for the above examples,

$$x_1 = \frac{0.78}{0.78 + 2.14} = 0.267$$

$$x_2 = \frac{2.14}{0.78 + 2.14} = 0.733$$

Finally, calculate as before the product rate for the mixed solute system, $[PR]_{Mix}$ from Equation 6-3, and solute separations with respect to each solute using Equations 6-8, 6-10, 6-14, and 6-15.

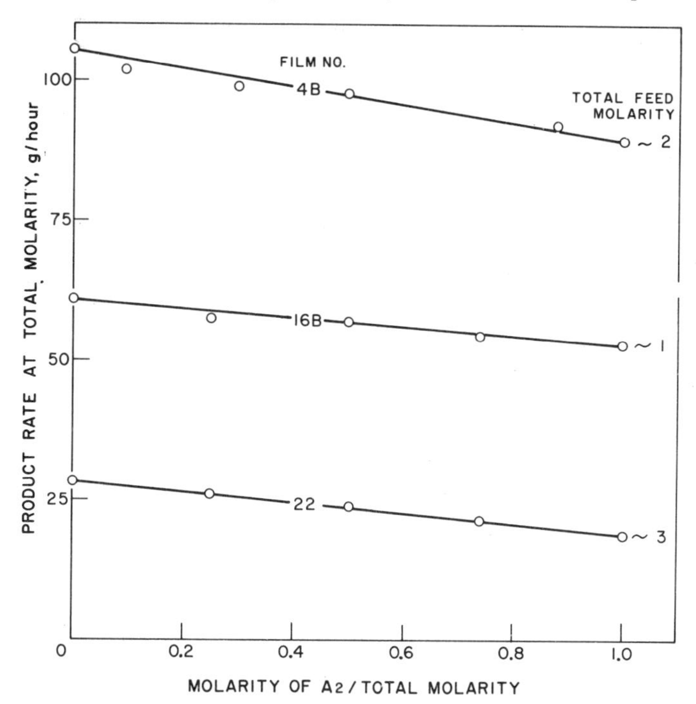

Figure 6-3. Product rate v. fractional solute molarity correlation for the system $NaNO_3$-$NaCl$-H_2O. Data of Sourirajan and Agrawal (1969).

Film type: CA-NRC-18; system: $NaNO_3$-$NaCl$-H_2O [A1-A2-H_2O]; feed rate: 360 cm^3/min; operating pressure: 1500 psig; film area: 7.6 cm^2.

A set of such results of calculations for the system $NaNO_3$-$NaCl$-H_2O at the total feed molality of ~3M is illustrated in Table 6-7 covering a wide range of solute separations which show good agreement between experimental and calculated data.

Systems containing more than two solutes with a common ion. The foregoing analysis can be extended to systems containing any number of solutes with a common ion. Equations 6-1, 6-2, and 6-3 can be written in general forms as

$$x_i = \frac{(m_1)_{Ai}}{\Sigma (m_1)_{Ai}} \tag{6-16}$$

and

$$[PR]_{Mix} = \Sigma \, x_i \, [PR]_{Ai} \tag{6-17}$$

where x_i, $(m_1)_{Ai}$, and $[PR]_{Ai}$ represent respectively the fractional molality of any solute Ai, molality of the feed solution with respect to Ai, and the product rate for the single solute system [Ai-H_2O] at the total molality of the mixed solute system. Using the data on single solute systems and Equations 6-16 and 6-17, the performance of a membrane for low concentrations of mixed solute systems containing any number of solutes with a common ion can be predicted

Table 6-8. Reverse osmosis data for systems containing more than two solutes with a common ion

Data of Agrawal and Sourirajan (1969c)

Film no.	Total feed molality $\Sigma(m_1)_{Ai}$	$\frac{(m_1)_{Ai}}{\Sigma(m_1)_{Ai}}$				Solute separation (%)								Product rate (g/hr)	
						A1		A2		A3		A4			
		A1	A2	A3	A4	Exptl.	Calcd.	Exptl.	Calcd.	Exptl.	Calcd.	Exptl.	Calcd.	Exptl.	Calcd.
System: NaCl-KCl-$MgCl_2$-H_2O (A1-A2-A3-H_2O)															
31	1.002	0.264	0.269	0.467	—	58.5	59.2	52.2	49.2	79.0	79.4	—	—	59.4	59.8
33	1.002	0.264	0.269	0.467	—	79.7	81.1	73.0	74.9	89.7	90.2	—	—	30.9	30.7
31	0.977	0.504	0.249	0.247	—	62.9	64.1	54.3	55.5	78.4	81.7	—	—	69.4	69.9
33	0.977	0.504	0.249	0.247	—	82.2	84.2	77.4	78.8	90.9	91.6	—	—	36.0	36.8
31	0.975	0.260	0.513	0.227	—	65.3	65.7	58.2	57.6	80.1	82.3	—	—	71.5	73.3
33	0.975	0.260	0.513	0.227	—	82.5	84.7	77.6	79.5	91.0	91.9	—	—	37.3	38.0
31	0.989	0.337	0.337	0.326	—	61.6	62.7	55.2	53.7	79.9	80.9	—	—	66.3	66.4
33	0.989	0.337	0.337	0.326	—	80.2	82.6	76.0	77.0	91.3	90.8	—	—	34.6	33.3
System: NaCl-KCl-$MgCl_2BaCl_2$-H_2O (A1-A2-A3-A4-H_2O)															
31	1.035	0.256	0.249	0.238	0.257	59.3	60.9	49.8	51.6	77.4	79.9	79.1	81.2	59.0	60.8
33	1.035	0.256	0.249	0.238	0.257	78.9	81.7	74.0	76.0	90.2	90.4	89.8	92.1	30.8	30.8

Film type: CA-NRC-18
Solute molality in feed = $(m_1)_{Ai}$
Feed rate: 360 cm^3/min
Operating pressure: 1500 psig
Film area: 7.6 cm^2

following the procedure similar to that outlined before for two solute systems. The data presented in Table 6-8 for the systems $NaCl$-KCl-$MgCl_2$-H_2O and $NaCl$-KCl-$MgCl_2$-$BaCl_2$-H_2O at total feed molalities of ~1.0M illustrate the accuracy of such predictions for three-solute and four-solute systems. In these experiments the data on the single solute systems were obtained experimentally (Agrawal and Sourirajan, 1969c).

The General Problem of Predictability. The techniques just described offer a general approach to the problem of predicting membrane performance for feed solutions containing inorganic solutes with a common ion. The same techniques may be applicable for aqueous feed solutions containing mixed nonionic organic solutes. Further experimental and analytical studies are needed for a complete answer to the general problem of predictability of membrane performance for mixed solute systems involving inorganic solutes with no common ions, and mixed inorganic and organic solutes.

2. Nomenclature for Chapter 6

A = pure water permeability constant, $\frac{\text{g. mole } H_2O}{\text{cm}^2\text{sec atm}}$

$(D_{AM}/K\delta)$ = solute transport parameter, cm/sec

f = solute separation

k_{A1} = $7.6 \times M_{A1} \times 3600, \frac{\text{cm}^2\text{sec.g}}{\text{g.mole}}$

k_{A2} = $7.6 \times M_{A2} \times 3600, \frac{\text{cm}^2\text{sec.g}}{\text{g.mole}}$

k_B = $7.6 \times M_B \times 3600, \frac{\text{cm}^2\text{sec.g}}{\text{g.mole}}$

$(m_1)_{A1}$ = molality of feed solution with respect to solute A1

$(m_1)_{A2}$ = molality of feed solution with respect to solute A2

$(m_1)_{Ai}$ = molality of feed soltuion with respect to solute Ai

$(m_3)_{A1}$ = molality of product solution with respect to solute A1

$(m_3)_{A2}$ = molality of product solution with respect to solute A2

M_{A1} = molecular weight of solute A1

M_{A2} = molecular weight of solute A2

M_B = molecular weight of water

N_{A1} = flux of solute A1 through membrane for feed system $[A1\text{-}H_2O], \frac{\text{g.mole}}{\text{cm}^2\text{sec}}$

N_{A2} = flux of solute A2 through membrane for feed system $[A2\text{-}H_2O], \frac{\text{g.mole}}{\text{cm}^2\text{sec}}$

N_{B1} = flux of solvent water through membrane for the feed system $[A1\text{-}H_2O], \frac{\text{g.mole}}{\text{cm}^2\text{sec}}$

N_{B2} = flux of solvent water through membrane for the feed system $[A2\text{-}H_2O], \frac{g.mole}{cm^2 sec}$

$[PR]_{A1}$ = product rate for feed system $[A1\text{-}H_2O]$, g/hr/7.6 cm^2 of film area

$[PR]_{A2}$ = product rate for feed system $[A2\text{-}H_2O]$, g/hr/7.6 cm^2 of film area

$[PR]_{Mix}$ = product rate for feed system $[A1\text{-}A2\text{-}H_2O]$, g/hr/7.6 cm^2 of film area, or quantity defined by Equation 6-17

[PWP] = pure water permeability, g/hour/7.6 cm^2 of film area

x_1, x_2, x_i, = fractional solute concentration ratios defined by Equations 6-1, 6-2 and 6-16 respectively.

3. Summary

Available data on the reverse osmosis separation of mixed solutes in aqueous solutions are summarised. A simple method is described for predicting membrane performance for feed solution systems involving two or more inorganic solutes with a common ion; this method needs only experimental data on single solute systems, or membrane specifications together with the applicable mass transfer coefficient correlation as discussed in Chapter 3 for single solute systems.

CHAPTER 7

Reverse Osmosis Separation of Mixtures of Organic Liquids

The reverse osmosis technique has potentially a very wide field of application in the separation, concentration, and fractionation problems encountered in the organic chemical industry, particularly in the petrocarbon and petrochemical industry. These applications mainly depend on the successful development of porous membranes made of different materials appropriate for the purpose. Cellulose acetate membranes cannot be expected to be appropriate for all reverse osmosis applications; generally, they are not suitable for organic feed mixtures, even though the Loeb-Sourirajan type porous cellulose acetate membranes appear good enough for at least some applications. In any case, current literature on the subject is neither extensive nor intensive. This situation will certainly change as the chemical industry becomes more aware of the general applicability of the reverse osmosis separation technique and its potentialities. This chapter again illustrates this point of view with particular reference to the separation of mixtures of organic liquids.

The consequences of the preferential sorption-capillary flow mechanism of reverse osmosis have been discussed in Chapter 1. With respect to a given solution containing components A and B, the membrane permeated product can be enriched either in A or in B depending upon the chemical nature of the membrane surface in contact with the solution. This is illustrated by the data of Sourirajan (1964c) given in Table 7-1. For the system xylene-ethyl alcohol, cellulose acetate membranes have preferential sorption for ethyl alcohol, and polyethylene membranes exhibit preferential sorption for xylene. For the system xylene-n-heptane, both cellulose acetate and polyethylene films exhibit a preferential sorption for xylene. For the system n-heptane-ethyl alcohol, the membrane-permeated product shows alcohol enrichment with the cellulose acetate membranes and n-heptane enrichment with the polyethylene membranes. The data also illustrate that azeotropic composition is no limitation to this separation process, that the chemical nature of the membrane surface determines the direction of separation for a given feed solution system, and that different levels of separation and product rates are obtainable for the same membrane material depending on the porous structure of the films used.

Kopeček and Sourirajan (1969c, 1969d) extended the above work and further illustrated the applicability of the reverse osmosis technique for the separation of mixtures of alcohols and/or hydrocarbons (including azeotropic and isomeric mixtures) using CA-NRC-18 type porous cellulose acetate membranes. The details of this work are given below.

1. Experimental Details

The experiments were carried out at the laboratory temperature (23°-25°) in the static cell described in Chapter 1 (Figures 1-23 and 1-24). Reagent grade organic compounds were used in the feed solution. The shrunk membranes were immersed in absolute alcohol several times to remove the water from their pores; they were then subjected to pure ethyl alcohol pressures of about 1200 psig for at least an hour prior to use in reverse osmosis experiments. The porous structure of each film was characterised by its initial pure ethyl alcohol permeability at 1000 psig, expressed as grams of ethyl alcohol per hour per 9.6 cm^2 of film area. About 250 cm^3 of feed solution

Table 7-1. Separation of some organic liquid mixtures using porous cellulose acetate and polyethylene membranes

Data of Sourirajan (1964c)

Expt. no. (a)	Film type	Components in mixture	Composition (mole %) Feed	Composition (mole %) Product	Product rate (g/hr)
11	CA-NRC-18	Xylene	51.0	39.0	2.03
		EtOH	49.0	61.0	
12	CA-NRC-18	Xylene	51.0	34.0	1.67
		EtOH	49.0	66.0	
13	CA-NRC-18	Xylene	51.0	28.0	0.43
		EtOH	49.0	72.0	
14	CA-NRC-18	Xylene	50.5	84.0	0.18
		n-Heptane	49.5	16.0	
15	CA-NRC-18	n-Heptane	33.0*	8.5	2.16
		EtOH	67.0	91.5	
16	CA-NRC-18	n-Heptane	33.0*	12.5	25.21
		EtOH	67.0	87.5	
17	Polyethylene	Xylene	51.5	97.5	0.17
		EtOH	48.5	2.5	
18	Polyethylene	Xylene	51.5	98.5	0.12
		EtOH	48.5	1.5	
19	Polyethylene	Xylene	51.0	98.0	0.18
		EtOH	49.0	2.0	
20	Polyethylene	Xylene	50.5	53.5	0.47
		n-Heptane	49.5	46.5	
21	Polyethylene	Xylene	50.5	54.5	0.47
		n-Heptane	49.5	45.5	
22	Polyethylene	n-Heptane	33.0*	61.5	0.18
		EtOH	67.0	38.5	

* Azeotropic mixture

(a) A different film was used for each experiment

Operating pressure: 1000 psig
Film area: 9.6 cm^2

were used in each run. The feed solution was kept well stirred during the experiment by means of a magnetic stirrer fitted in the cell about 1/4 inch above the membrane surface. The quantity of liquid removed by membrane permeation was small compared to the amount of feed solution in the pressure chamber. The compositions of the feed and product solutions were analysed by means of a precision Bausch and Lomb refractometer.

The porous structure of the membrane is affected, reversibly or irreversibly, by the nature of the organic liquids in contact with it. Consequently, before each experiment, a small volume of feed solution was permeated through the membrane under pressure, and the film was kept in contact with the feed solution under atmospheric pressure for 12 hours or more until constant values were obtained for solute separation and product rate in the reverse osmosis experiments. Thus the separation and product rate data reported here represent the steady state values obtained for the particular membrane-solution system under study.

Pure ethyl alcohol permeability data. Figure 7-1 gives the effect of shrinkage temperature on the pure ethyl alcohol permeability of the films used. Table 7-2 gives the pure ethyl alcohol permeability data obtained with the particular films employed in the work of Kopeček and Sourirajan (1969c, 1969d); these data are the initial values obtained with the membranes before use in reverse osmosis experiments, and they give a relative measure of their initial porous structure.

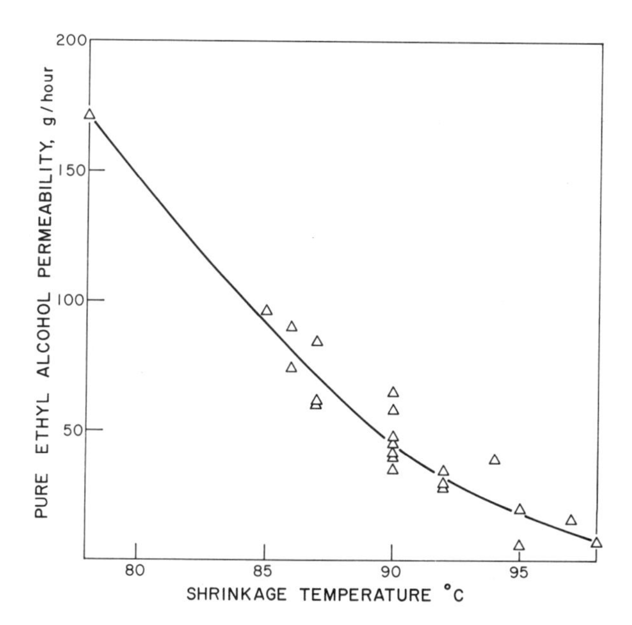

Figure 7-1. Effect of shrinkage temperature on pure ethyl alcohol permeability of films. Data of Kopeček and Sourirajan (1969d).

Film type: CA-NRC-18; operating pressure: 1000 psig; film area: 9.6 cm^2.

2. Some Binary Mixtures of Organic Liquids

Separation factor. The separation factor α_{AB} is defined by the relation

$$\alpha_{AB} = \frac{Y_A/Y_B}{X_A/X_B}$$

where Y_A and Y_B are the mole fractions of components A and B in the product, and X_A and X_B are the corresponding mole fractions in the feed solution respectively. The terms 'product' and 'product rate' refer to the membrane-permeated product solution.

Table 7-2. Pure ethyl alcohol permeability data at 25°

Data of Kopeček and Sourirajan (1969c, 1969d)

Film no.	P. Alc. P.* (g/hr)	Film no.	P. Alc. P.* (g/hr)	Film no.	P. Alc. P.* (g/hr)
2	6.3	31	38.6	203	20.3
3	157.0	32	44.5	204	39.4
5	171.1	33	41.3	205	15.9
6	43.6	36	45.7	206	7.3
7	97.4	41	22.8	207	37.4
19	83.9	44	28.0	208	63.8
20	61.6	121	89.6	209	42.2
21	61.2	122	73.4	210	59.3
25	29.7	123	123.6	291	29.3
26	35.7	201	65.0		
27	59.4	202	34.3		

* Initial values of pure ethyl alcohol permeability

Film type: CA-NRC-18
Operating pressure: 1000 psig
Film area: 9.6 cm^2

Separation of some binary mixtures of organic liquids. Tables 7-3, 7-4, and 7-5 illustrate the reverse osmosis separation of azeotropic mixtures, compounds in the same homologous series, and isomers respectively. The separation technique was successful in all the cases tested. No simple generalisations seem possible regarding the direction of separation. For example, in the case of alcohol-hydrocarbon feed mixtures, cellulose acetate material might be expected to have a preferential sorption for alcohol. While this is so in many systems, the opposite also seems true in systems such as n-PrOH-toluene, iso-PrOH-benzene, iso-PrOH-toluene, and iso-BuOH-toluene at least for the particular compositions given in Table 7-3. In the case of mixtures of compounds in the same homologous series, while the product is enriched in the lower member of the series in some systems such as benzene-toluene and n-heptane-n-decane, the opposite is observed in systems such as p-xylene-benzene and p-xylene-toluene (Table 7-4). Similar comments can be made with respect to the separation of isomers also. In the system n-PrOH-iso-PrOH, the product is enriched in iso-PrOH; and, in the system n-BuOH-iso-BuOH, the product is enriched in n-BuOH (Table 7-5). These observations only emphasise the complex nature of the physico-chemical criteria of preferential sorption in reverse osmosis.

Tables 7-3, 7-4, and 7-5 present data for only one composition of the feed mixture for each system, and the above comments regarding the direction of separation apply only to the particular compositions given in the tables.

Table 7-3. Reverse osmosis separation of azeotropic mixtures

Data of Kopeček and Sourirajan (1969d)

Film no.	System A-B	Mole % A in Feed	Mole % A in Product	α_{AB}	Product rate (g/hr)
32	MeOH-Benzene	61.4	71.5	1.58	0.7
121	EtOH-n-Heptane	67.0	84.0	2.59	68.9
122	EtOH-CCl_4	38.5	48.5	1.50	56.3
36	EtOH-Benzene	44.8	48.0	1.14	2.4
123	EtOH-Cyclohexane	44.5	74.0	3.55	41.5
33	EtOH-Toluene	81.0	84.0	1.23	27.6
206	n-PrOH-Benzene	20.9	26.1	1.34	2.7
31	n-PrOH-Benzene	20.9	24.5	1.23	16.2
206	n-PrOH-Toluene	60.0	58.6	0.94	3.5
206	iso-PrOH-Benzene	39.3	33.7	0.79	2.9
19	iso-PrOH-Benzene	39.3	37.5	0.93	34.9
206	iso-PrOH-Toluene	77.0	71.9	0.76	2.6
20	iso-PrOH-Toluene	77.0	74.5	0.87	30.0
21	n-BuOH-Cyclohexane	11.0	41.5	5.74	3.7
27	n-BuOH-Toluene	37.0	39.0	1.09	36.2
26	iso-BuOH-Benzene	10.0	13.0	1.35	28.8
25	iso-BuOH-Toluene	50.0	43.0	0.75	8.4
206	iso-BuOH-Toluene	50.0	42.5	0.74	1.4

Film type: CA-NRC-18
Operating pressure: 1000 psig
Film area: 9.6 cm^2

3. Effect of Organic Liquids on the Porous Structure of the Membrane

The porous structure of the cellulose acetate membrane changes reversibly or irreversibly by changes in the composition of the feed solution in contact with it. The following pure ethyl alcohol, and pure p-xylene permeability data obtained with film 201 are particularly illustrative. For this film, the initial pure alcohol permeability was 65.0 g/hr under conditions given in Table 7-2. It was then held in contact with p-xylene for 24 hours, after which its pure xylene permeability was 54.0 g/hr. When the film was kept in contact with pure p-xylene for another 24 hours, its xylene permeability dropped to 41.6 g/hr. The film was then held in contact with pure alcohol for 24 hours, after which its alcohol permeability was 59.5 g/hr. The film was then kept in contact with n-heptane-p-xylene mixtures for a period of 7 days, after which its xylene

Table 7-4. Reverse osmosis separation of some compounds in the same homologous series

Data of Kopeček and Sourirajan (1969d)

Film no.	System A-B [Mole ratio in feed = 50/50]	α_{AB}	Product rate (g/hr)
291	Benzene-Toluene	1.14	7.4
204	p-Xylene-Benzene	1.25	10.8
291	p-Xylene-Toluene	1.20	4.4
44	n-Heptane-n-Decane	1.17	1.1

Film type: CA-NRC-18
Operating pressure: 1000 psig
Film area: 9.6 cm^2

Table 7-5. Reverse osmosis separation of isomers

Data of Kopeček and Sourirajan (1969d)

Film no.	System A-B [Mole ratio in feed = 50/50]	α_{AB}	Product rate (g/hr)
206	n-PrOH-iso-PrOH	0.92	2.0
206	n-BuOH-iso-BuOH	1.30	0.8
41	n-BuOH-tert-BuOH	1.13	1.60
41	tert-BuOH-sec-BuOH	1.04	1.64
203	p-Xylene-o-Xylene	1.19	1.0
203	m-Xylene-o-Xylene	1.30	0.6

Film type: CA-NRC-18
Operating pressure: 1000 psig
Film area: 9.6 cm^2

permeability dropped to 3.8 g/hr. The membrane was then held in contact with pure ethyl alcohol for 24 hours after which its alcohol permeability was 29.2 g/hr. The membrane was again kept in contact with p-xylene for 4 weeks after which its xylene permeability was 10.4 g/hr. On continued contact with p-xylene for another six weeks, the xylene permeability of the membrane dropped to 2.9 g/hr. These data indicate different degrees of collapse of the porous structure of the membrane when held in contact with organic liquids for which the membrane material has different degrees of sorption. Similar results have been reported by Breitenbach and Forster (1952).

4. Equisorptic Compositions in Reverse Osmosis

On the basis of the preferential sorption-capillary flow mechanism, an appropriate chemical nature of the film surface in contact with the solution, together with the existence of pores of appropriate size on the area of the porous film at the interface, is an indispensable twin requirement for the success of this separation technique. There are two circumstances under which little or no separation will be obtained in the reverse osmosis process even when the chemical nature of the membrane surface is such that it has some degree of sorption for each of the components of a binary mixture. When the size of the pores on the membrane surface is too big compared to the thickness of the preferentially sorbed layer, separation will obviously be negligible under the reverse osmosis conditions. Further, preferential sorption is a function of the composition of the solution for any given membrane material. If, at a particular composition of the solution, the membrane material has equal sorption for either component in a binary mixture, then under that condition, preferential sorption is zero, and consequently there can be no separation in reverse osmosis whatever be the porous structure of the membrane surface. Kopeček and Sourirajan (1969c) showed the existence of such equisorptic compositions in reverse osmosis.

Figures 7-2, 7-3, and 7-4 give the experimental separation and product rate data obtained with membranes of different porosities using different compositions of feed solution systems isopropyl alcohol-benzene and isobutyl alcohol-toluene. The numbers by the side of the product rate data refer to the order in which the experiments were carried out.

The separation data for both the feed solution systems illustrate the existence of equisorptic compositions in reverse osmosis. On one side of the equisorptic composition, the membrane material has a preferential sorption for the alcohol, and on the other side, it has a preferential sorption for the hydrocarbon. With respect to the particular cellulose acetate membrane material and feed solutions studied, the equisorptic compositions (mole ratios) are 26/74 for isorpropyl alcohol-benzene, and 21/79 for isobutyl alcohol-toluene systems; for comparison, the azeotropic compositions of the mixtures are 39.3/60.7 and 50/50 respectively. The experimental separation data obtained with different films (Figures 7-2 and 7-3) further show that the equisorptic composition is the same whatever be the porous structure of the membrane surface; and, at least for the system isobutyl alcohol-toluene, it is also independent of the operating pressure up to 1000 psig (Figure 7-4).

The product rate data for the system isopropyl alcohol-benzene (Figure 7-2) illustrate the partial and irreversible collapse of the porous structure of the membrane by feed solutions containing more than about 25 mole per cent benzene. Such collapse of porous structure however did not affect separation data significantly, indicating thereby that the porous structure of the dense surface layer of the membrane remained essentially unaffected during the process. The product rate data for the system isobutyl alcohol-toluene (Figure 7-3) do not show any indication of irreversible changes in the porous structure of the membrane as a result of changes in the composition of the feed solution. In any case, it is evident that the changes in the porous structure of the membrane brought about by changes in the composition of the feed, do not affect the equisorptic compositions.

At a given operating pressure and temperature, the equisorptic compositions are obviously functions of the chemical nature of the membrane material and that of the feed solution. They are hence significant physical constants in reverse osmosis, analogous to azeotropic compositions in distillation. Their existence adds a new dimension to the science and engineering of reverse osmosis (Kopeček and Sourirajan, 1968c).

5. Alcohols and Alcohol-Hydrocarbon Systems

The reverse osmosis separation and product rate data for the binary mixtures of MeOH, EtOH, n-PrOH, and n—BuOH using two membranes shrunk at different

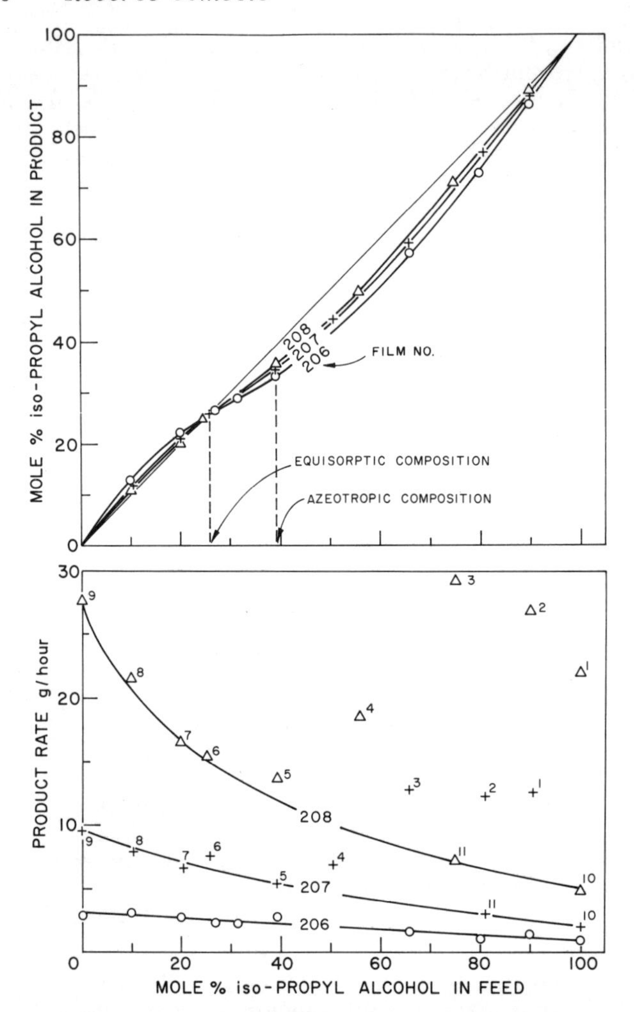

Figure 7-2. Separation and product rate data for the system isopropyl alcohol-benzene using porous cellulose acetate membranes. Data of Kopeček and Sourirajan (1969c).

Film type: CA-NRC-18; operating pressure: 1000 psig; film area: 9.6 cm^2.

temperatures are illustrated in Table 7-6 and Figure 7-5. The results show that mixtures of alcohols can be separated by the reverse osmosis technique even though separation factors obtainable in one stage using porous cellulose acetate membranes are low. With respect to feed mixtures of two consecutive alcohols, the following order of separation was observed (Table 7-6):

MeOH-EtOH < EtOH-n-PrOH < n-PrOH-n-BuOH

Further, the greater the difference between the carbon numbers of the alcohols in the feed mixture, the greater was the magnitude of separation under otherwise identical experimental conditions. The product rate data shown in Figure 7-5 are particularly interesting. Here x_1 and n_{C1} are the mole fraction and

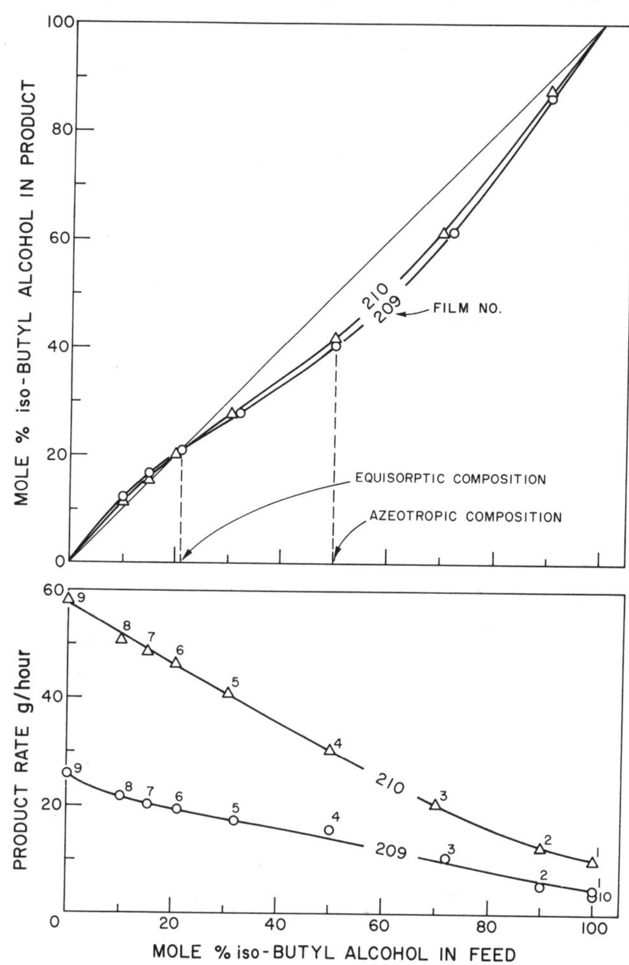

Figure 7-3. Separation and product rate data for the system isobutyl alcohol-toluene using porous cellulose acetate membranes. Data of Kopeček and Sourirajan (1969c).

Film type: CA-NRC-18; operating pressure: 1000 psig; film area: 9.6 cm^2.

carbon number respectively of component 1 in the feed mixture, and x_2 and n_{c2} are the corresponding numbers with respect to component 2. For example, pure EtOH, pure n-PrOH, and mole ratio 1/1 EtOH/n-PrOH mixture are represented by the numbers 2, 3, and 2.5 respectively in the x-axis in Figure 7-5. The results show that the product rate data for the pure components and their binary mixtures fall on a unique line for each membrane studied. This indicates that the changes taking place in the porous structure of the membrane as a result of polymer-solution interaction are completely reversible, and the sorption characteristic of the membrane for the alcohol solutions is a linear function of the mole fraction of their components in the feed mixture.

System ethyl alcohol-p-xylene. Cellulose acetate membranes have a preferential sorption for ethyl alcohol from ethyl alcohol-p-xylene mixtures.

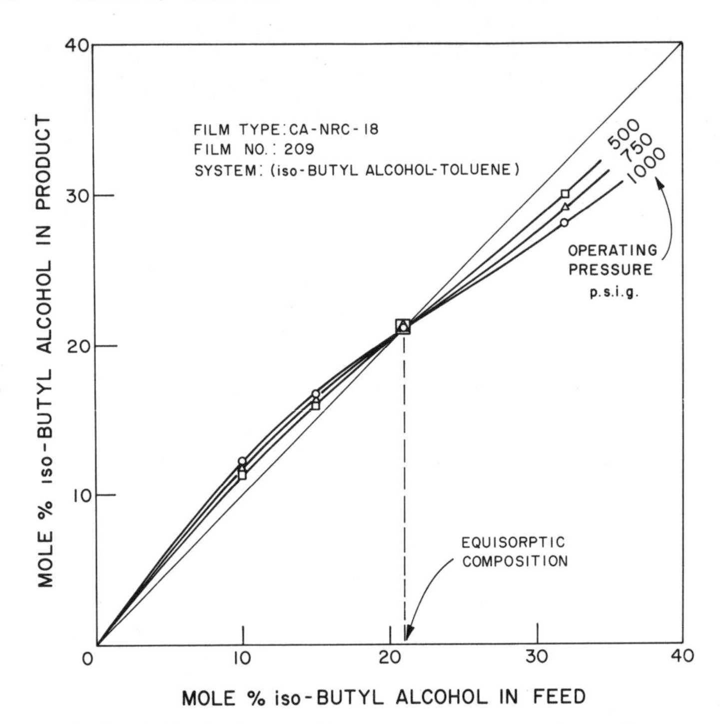

Figure 7-4. Effect of operating pressure on equisorptic composition for the system isobutyl alcohol-toluene. Data of Kopeček and Sourirajan (1969c).

Figure 7-6 gives the separation and product rate data for the above system for three membranes shrunk at different temperatures. A membrane shrunk at lower temperature has bigger (average) surface pores, gives higher product rate and lower separation at a given operating pressure. The curves showing separation data v. mole per cent ethyl alcohol in feed are not symmetrical, and the corresponding product rate data correlations are not straight lines. The feed mixtures containing lower mole fractions of alcohol are relatively more enriched in alcohol in the product. The above observations again indicate that the size of the pores on the membrane surface and/or the preferential sorption characteristics of the membrane material are affected by the composition of the feed solution. An increase in the mole fraction of that component in the feed mixture for which the membrane has a less tendency for sorption obviously tends to increase the number of physical cross links in the membrane structure, reduce the average pore size, and collapse the pore structure of the membrane to different extents, depending on the chemical nature of the components of the feed mixture with respect to that of the membrane material.

With feed solutions containing more than about 25 mole% ethyl alcohol, the separation and product rate data obtained with each film were reproducible whatever be the order in which the experiments were carried out. This indicates that, under those experimental conditions, the changes in the pore structure of the membrane brought about by changes in the composition of the feed solution were reversible. When the alcohol content in the feed solution was less than about 25 mole%, irreversible changes in the porous structure of the membrane took place on continued contact with the feed solution. Figure 7-7 shows the effect of operating pressure on separation and product rate.

Table 7-6. Reverse osmosis separation of alcohols

Data of Kopeček and Sourirajan (1969d)

System A-B [Mole ratio in feed = 50/50]	Separation factor α_{AB}	
	Film 204	Film 205
MeOH-EtOH	1.004*	1.033*
MeOH-n-PrOH	1.062	1.092
MeOH-n-BuOH	1.101	1.160
EtOH-n-PrOH	1.033	1.079
EtOH-n-BuOH	1.088	1.146
n-PrOH-n-BuOH	1.062	1.106

Film type: CA-NRC-18
Operating pressure: 1000 psig
* ±0.004

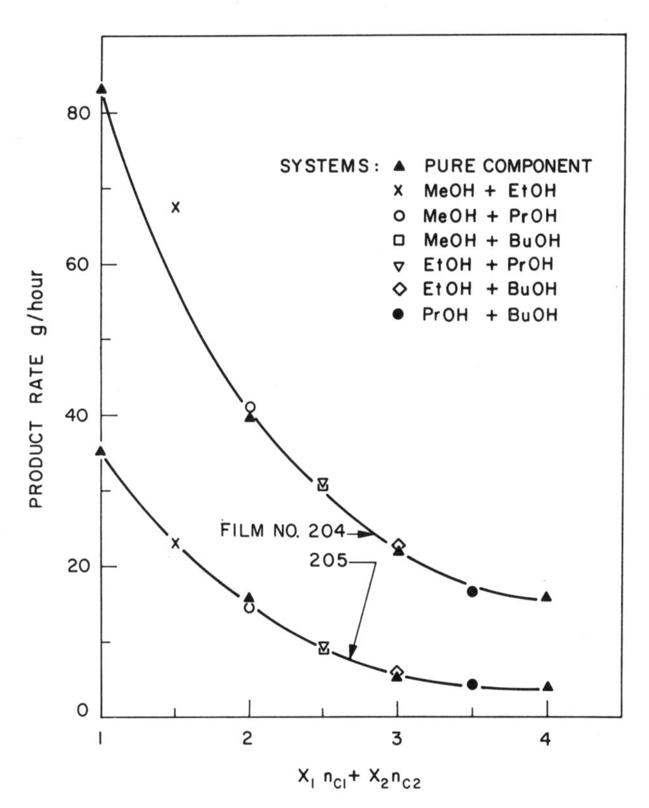

Figure 7-5. Product rate data for alcohol systems. Data of Kopeček and Sourirajan (1969d).

Film type: CA-NRC-18; operating pressure: 1000 psig; film area: 9.6 cm^2.

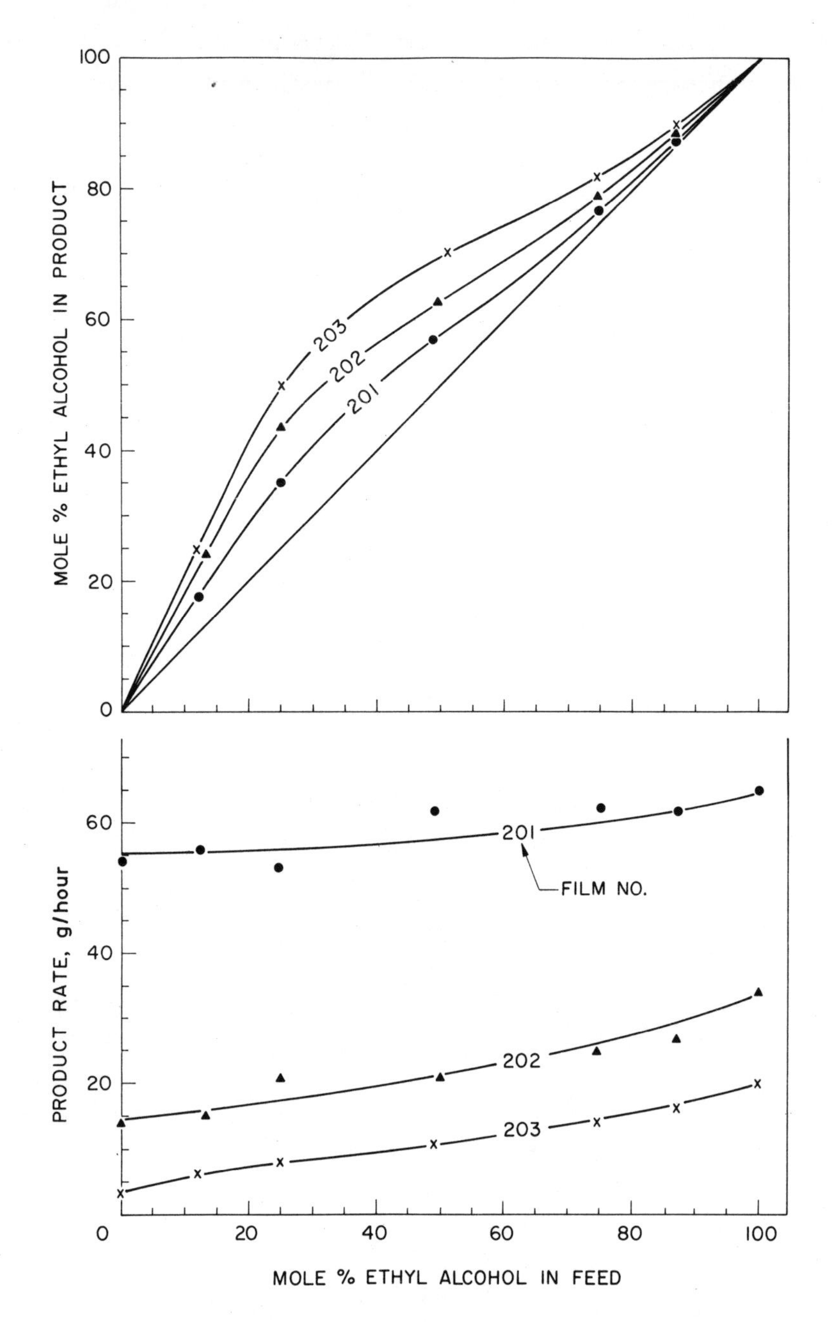

Figure 7-6. Performance data for the system ethyl alcohol-p-xylene—effect of porosity of membrane surface. Data of Kopeček and Sourirajan (1969d).

Film type: CA-NRC-18; operating pressure: 1000 psig; film area: 9.6 cm^2.

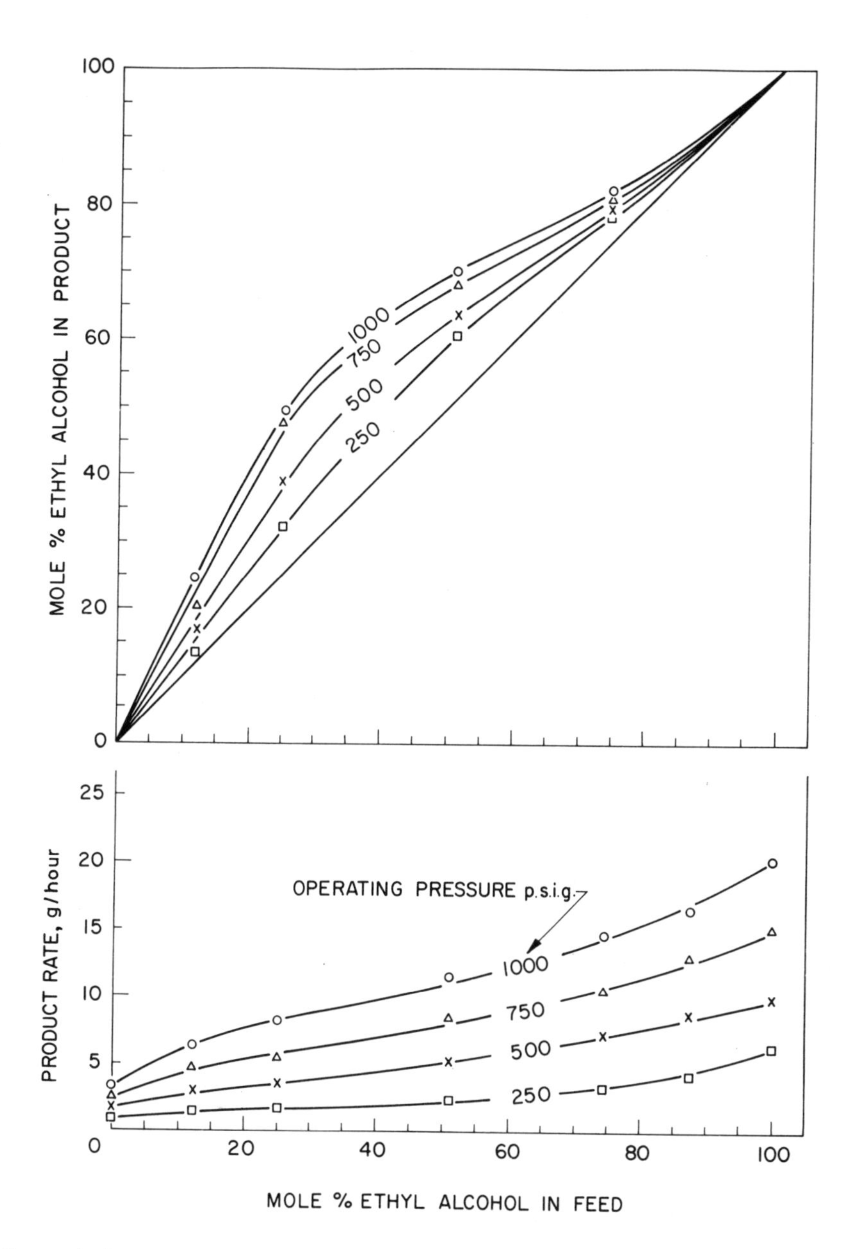

Figure 7-7. Performance data for the system ethyl alcohol-p-xylene—effect of operating pressure. Data of Kopeček and Sourirajan (1969d).
Film type: CA-NRC-18; film no: 203; film area: 9.6 cm^2.

As in the case of aqueous solution systems (Sourirajan and Govindan, 1965; Sourirajan and Kimura, 1967), increase of operating pressure increased both separation and product rate. This effect was relatively more for the membrane whose surface pores were smaller, as illustrated by the data in Figure 7-8.

System ethyl alcohol-n-heptane. Cellulose acetate membranes have a preferential sorption for ethyl alcohol from ethyl alcohol-n-heptane mixtures. Figure 7-9 gives the separation and product rate data for the above system for four membranes of different pore structures obtained by shrinkage at different temperatures. The curves showing the separation data v. mole% ethyl alcohol in feed, and those representing the corresponding product rate data, are similar to those described for the system ethyl alcohol-p-xylene. The performance of porous cellulose acetate membranes for the reverse osmosis separation for the system ethyl alcohol-n-heptane is far more favourable than that for the system ethyl alcohol-p-xylene. For example, even with a film such as 5 in Figure 7-9 (whose pure ethyl alcohol permeability at 1000 psig is 171.1g/hr/9.6 cm^2 of film area), the separation and product rate data obtained are such that they can be considered significant for industrial exploitation. With membranes having smaller surface pores, the separations are

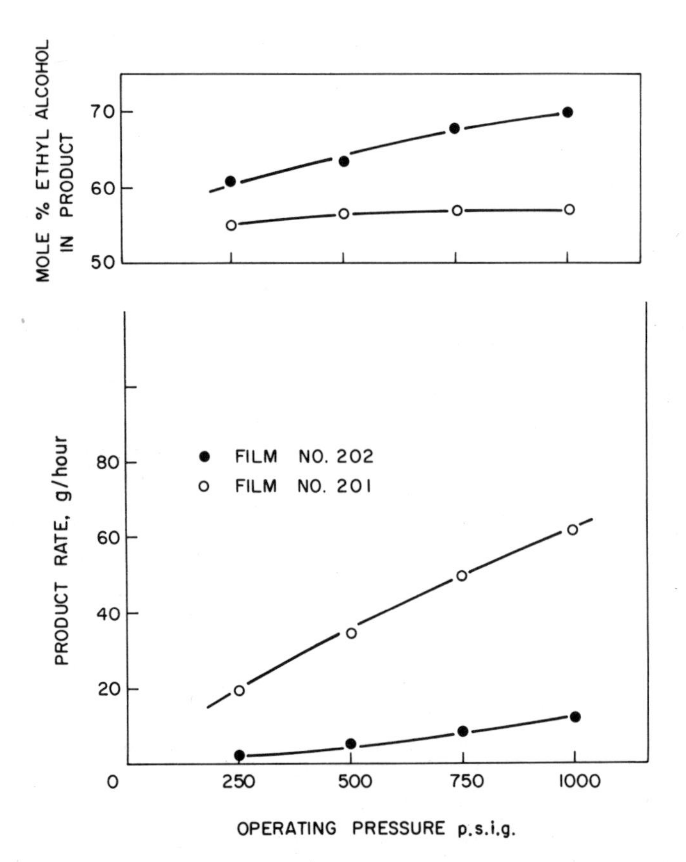

Figure 7-8. Performance data for the system ethyl alcohol-p-xylene—effect of operating pressure. Data of Kopeček and Sourirajan (1969d).

Film type: CA-NRC-18; system: (ethyl alcohol 50 mole %) (p-xylene 50 mole %); film area: 9.6 cm^2.

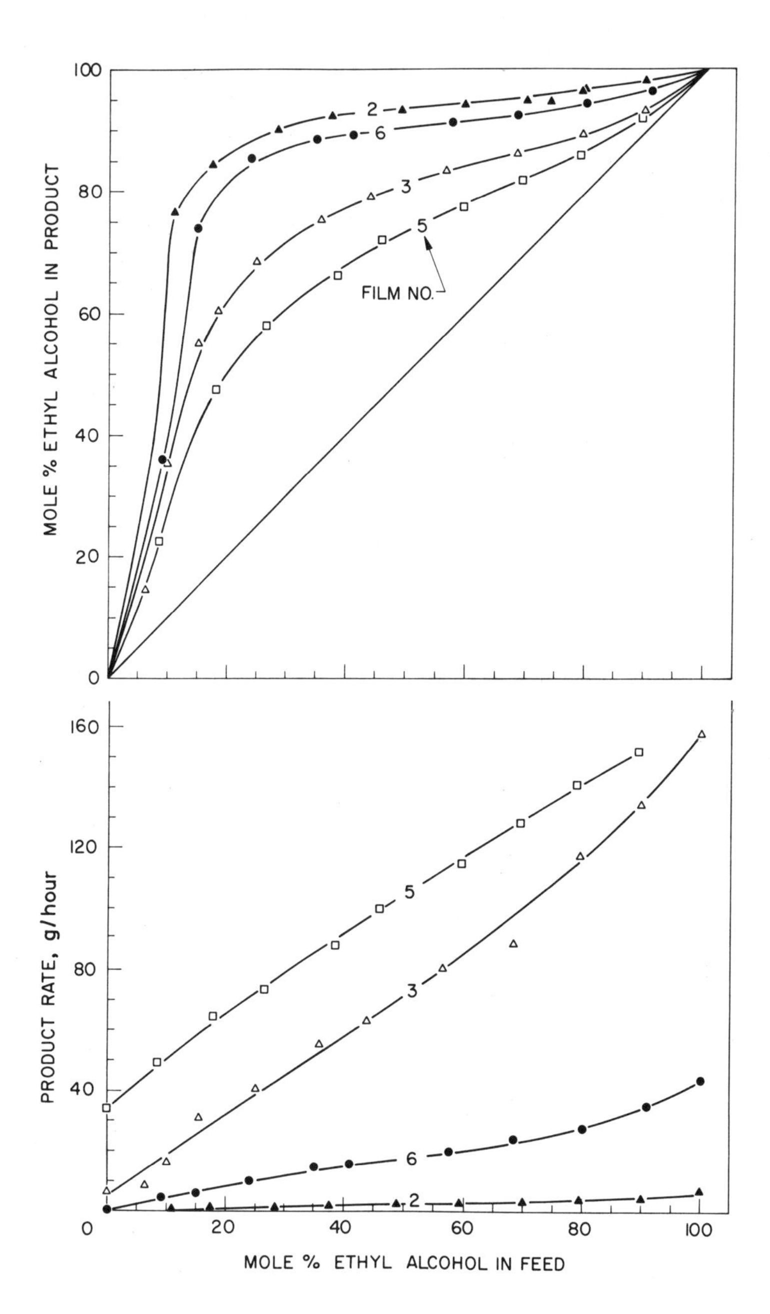

Figure 7-9. Performance data for the system ethyl alcohol-n-heptane—effect of porosity of membrane surface. Data of Kopeček and Sourirajan (1969d).

Film type: CA-NRC-18; operating pressure: 1000 psig; film area: 9.6 cm^2.

even more favourable; the product rates of course are lower. In this system also, with feed solutions containing more than about 25 mole% ethyl alcohol, the separation and product rate data obtained for different feed concentrations were reproducible, whatever the order in which the experiments were carried out. This again indicates that, under those conditions, the changes in the porous structure of the membrane brought about by the changes in the composition of the feed solution were reversible.

Figure 7-10 shows the effect of pressure used in the initial pressure treatment of the film on its subsequent performance in reverse osmosis. The results illustrate that the initial pressure treatment affects the overall porous structure of the membrane.

Figure 7-11 shows the effect of operating pressure on separation and product rate for the system ethyl alcohol-n-heptane. Again, both separation and product rate increased with operating pressure. Further, if the product is treated simply as a mixture of the feed solution and the preferentially sorbed alcohol, the plot of the latter v. operating pressure is essentially a straight line in the pressure range 250 to 1000 psig.

System p-xylene-n-heptane. This system is of interest in view of the low tendency of cellulose acetate material for the sorption of either component in the feed solution. Three membranes with different initial porous structures (as shown by their pure alcohol permeability values, Table 7-2) were tested with the above system. The product was enriched in p-xylene and the Y_{xylene} v. X_{xylene} graph was essentially symmetrical; the product rates were low (less than 3 g/hr per 9.6 cm^2 of film area) in all the cases. Separation increased with operating pressure (Figure 7-12). The levels of separation obtained with the three films tested were not too different from each other, even though their initial porous structures were very different (Figure 7-13). The above observations indicate the collapse of the porous structure of the film with the above feed system, and the tendency of the film to approach some kind of a limiting structure not necessarily indentical in all cases.

6. Criteria of Preferential Sorption

The effects of various intermolecular forces (such as dispersion, repulsion, and polar forces, and dipole interactions) on solubility of nonelectrolytes have been reviewed in the literature (Burrell, 1955; Hildebrand, 1949; Small, 1953). Since solution and sorption are related phenomena, and one may consider solution simply as the limit of sorption, the physico-chemical criteria of solubility based on intermolecular forces may be expected to lead to valid criteria of preferential sorption in reverse osmosis.

This is suggested by a consideration of the cohesive energy density concept which is based principally on dispersion forces, and the polar attraction of molecules based principally on the hydrogen bonding forces. The solubility parameter gives a quantitative measure of the cohesive energy density; no such quantitative measure is available for the hydrogen bonding forces which may, however, be arbitrarily represented by the iodine-bonding number of Small (1953). At least with respect to nonelectrolyte binary mixtures containing components whose solubilities are governed primarily either by the hydrogen bonding or dispersion forces, one may expect to be able to predict the direction of separation in reverse osmosis. Since cellulose acetate itself has a hydrogen-bonded structure, the membrane-permeated product may be expected to be enriched in the more hydrogen-bonded component of the feed mixture. The data on the separation of the mixtures of alcohols (Table 7-6) justify

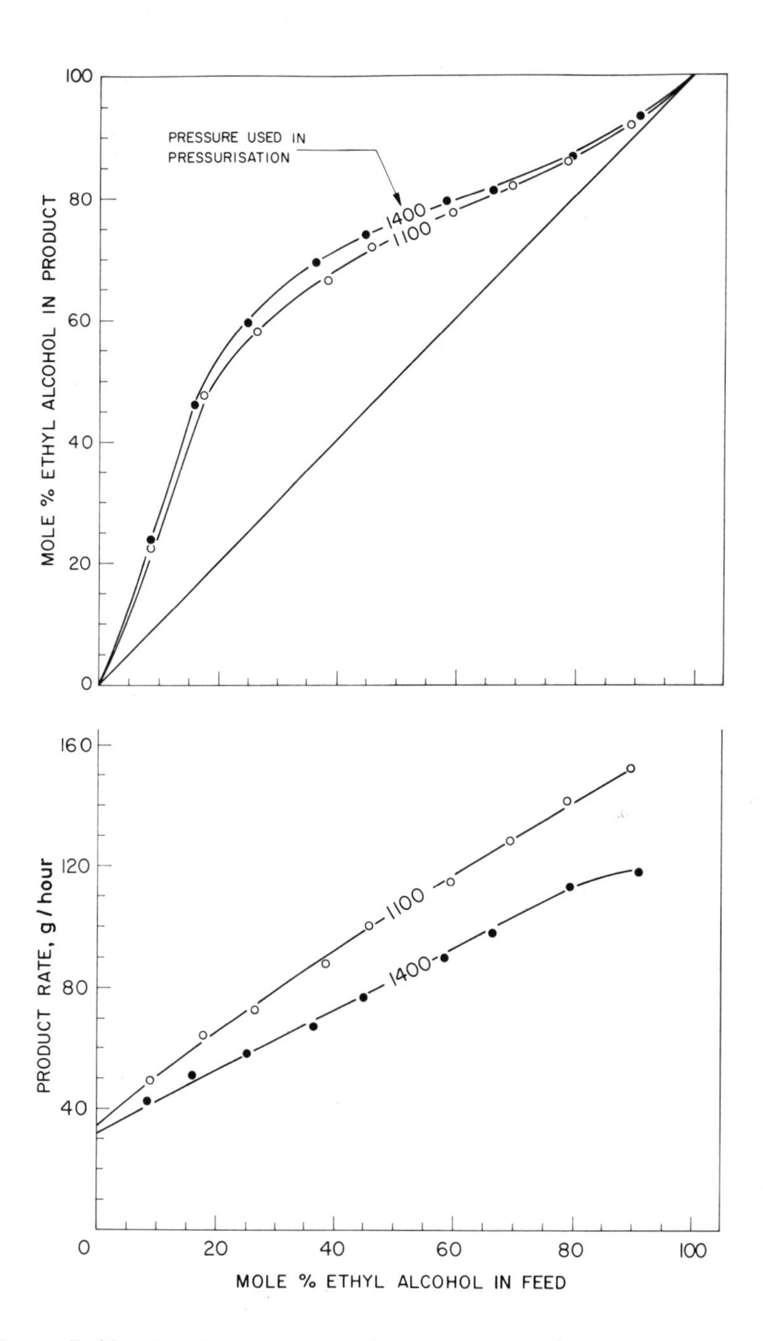

Figure 7-10. Performance data for the system ethyl alcohol-n-heptane—effect of pre-treatment pressure. Data of Kopeček and Sourirajan (1969d).

Film type: CA-NRC-18; film no.: 5; operating pressure: 1000 psig; film area: 9.6 cm^2.

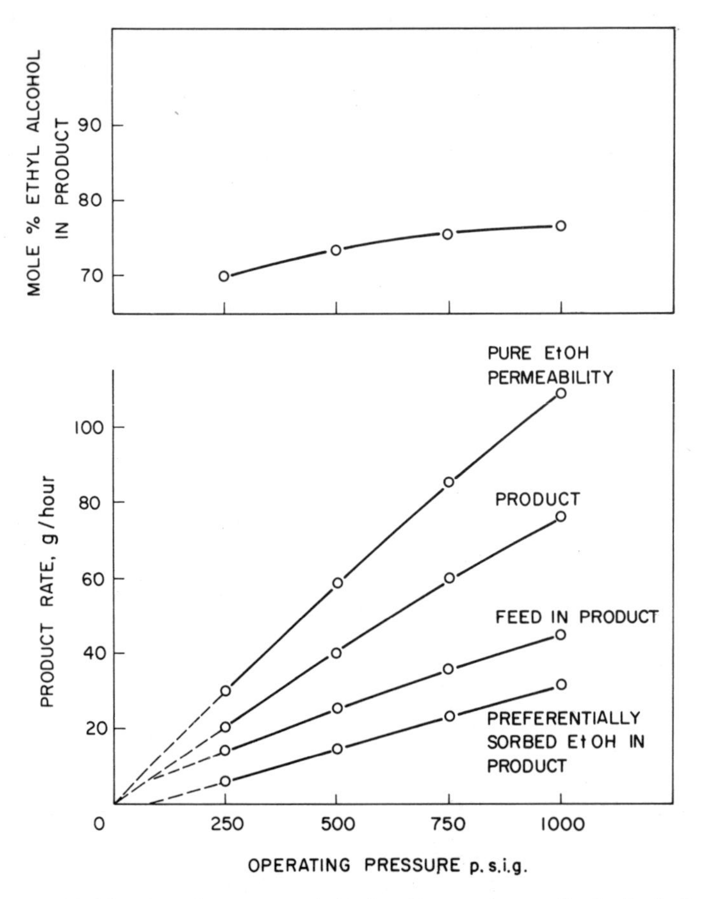

Figure 7-11. Performance data for the system ethyl alcohol-n-heptane—effect of operating pressure. Data of Kopeček and Sourirajan (1969d).

Film type: CA-NRC-18; film no. 5; system: (ethyl alcohol 50 mole %) (n-heptane 50 mole %); film area: 9.6 cm^2.

this suggestion; the product is always enriched in the lower member of the homologous series, which is also the more hydrogen-bonded one in the mixture (Small, 1953). When the feed mixture consists of components whose solubilities are governed primarily by dispersion forces, the solubility parameter of the component with respect to that of the membrane material may be expected to offer a valid criterion for preferential sorption. Since a decrease in the difference between the solubility parameters of any two substances indicates an increase in their mutual tendency for solubility, it may be expected that the membrane-permeated product in reverse osmosis will be enriched in that component for which $(\delta_m - \delta_c)$ is lower, where δ_m and δ_c are the values of the solubility parameters for the membrane material and the component in the feed mixture respectively.

The reverse osmosis data for the systems p-xylene-n-heptane and benzene-toluene given in Figures 7-12 and 7-13 and Table 7-4 justify this expectation. The solubility parameters for cellulose acetate, p-xylene, n-heptane, benzene, and toluene are 10.9, 8.8, 7.5, 9.2 and 8.9 respectively at 25° (Burrell, 1955). Since the solubility parameters for ortho- and meta-xylenes are 9.0 and 8.8

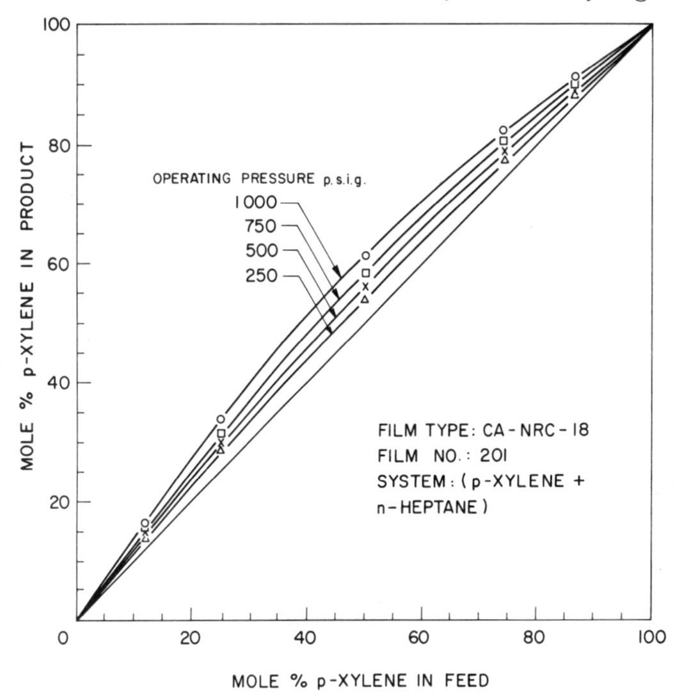

Figure 7-12. Separation data for the system p-xylene-n-heptane—effect of operating pressure. Data for Kopeček and Sourirajan (1969d).

respectively (Burrell, 1955), the reverse osmosis data given in Tables 7-4 and 7-5 for the systems p-xylene-benzene, p-xylene-toluene, p-xylene-o-xylene, and m-xylene-o-xylene would appear as exceptions to the above solubility parameter principle. It is probable that such exceptions arise more from uncertainties in the respective values of the solubility parameter than from the principle itself. Further, the contribution of the other intermolecular forces for the solubility of nonelectrolyte substances cannot be ignored. Consequently it seems reasonable to conclude that precise numerical expressions of the various intermolecular forces in nonelectrolyte substances may lead to some valid criteria of preferential sorption in reverse osmosis.

7. Significance of Results

The data presented in this chapter illustrate the applicability of the reverse osmosis technique for the separation of substances in nonaqueous solutions. Hydrocarbons tend to collapse the porous structure of the cellulose acetate membranes used; hence, for feed mixtures containing hydrocarbons, some other type of membranes will have to be developed for reverse osmosis application. For at least some organic feed solutions, such as alcohol-hydrocarbon mixtures containing more than about 25 mole% alcohol, CA-NRC-18 type porous cellulose acetate membranes appear sufficiently good for consideration for industrial applications. In view of the changing pore structure of the membrane in contact with organic feed solutions, the membrane-solution-operating systems discussed in this chapter could not be specified precisely in the manner done in Chapters 3 and 4 for aqueous solution systems and cellulose acetate membranes. Consequently, the performance data given

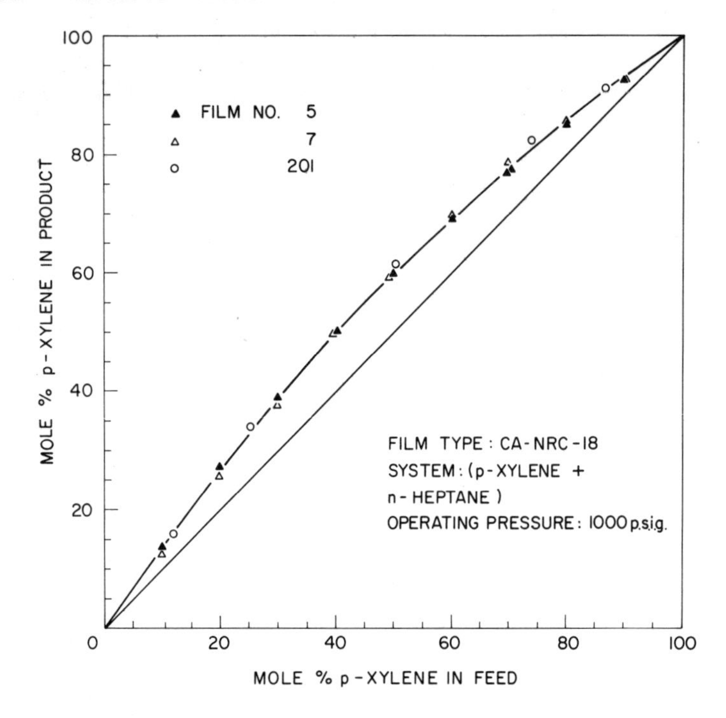

Figure 7-13. Separation data for the system p-xylene-n-heptane—effect of variation of initial porosity of membrane surface. Data of Kopeček and Sourirajan (1969d).

in this chapter have mainly relative significance, and, in particular, they do not represent limiting values obtainable in reverse osmosis systems involving organic liquid mixtures.

8. Summary

This chapter illustrates the applicability of the reverse osmosis technique for the separation of binary mixtures of alcohols and/or hydrocarbons including azeotropic and isomeric mixtures. Data are also presented to illustrate the existence of equisorptic compositions in reverse osmosis. The porous structures of the cellulose acetate membranes used were affected by the composition of the feed solution in contact with them; hydrocarbon liquids tended to collapse their porous structures on continued contact. For alcohol-hydrocarbon feed mixtures containing more than about 25 mole % alcohol, the CA-NRC-18 type porous cellulose acetate membranes appear suitable for practical reverse osmosis separations in industry. Hydrogen bonding and solubility parameter seem to offer valid criteria of preferential sorption in reverse osmosis for nonelectrolyte binary feed mixtures containing components whose solubilities are governed primarily either by polar or dispersion forces.

CHAPTER 8

Some Applications and Engineering Developments in the Reverse Osmosis Separation Process

In recent years, extensive work has been done, and considerable progress has been made, with respect to the application of the reverse osmosis process for specific practical problems of water treatment and solution concentration. Several reverse osmosis pilot plants have been designed, built, and operated, and considerable engineering experience is now available on the design, construction, operation, and performance of such plants. In this chapter some of the available data on the subject are reviewed from the points of view of practical applications and engineering developments.

1. Water Softening, Water Pollution Control, Water Renovation, and Waste Recovery

Hauck and Sourirajan (1969) studied the performance of a few typical Batch 18 type porous cellulose acetate membranes at the laboratory temperature, for the treatment of hard waters, polluted waters, and sewage waters, using the apparatus shown in Figures 1-23 and 1-24 (Chapter 1). The specifications of the films used by them are given in Table 8-1, and the generalised mass transfer coefficient correlation applicable to their experiments was given as $N_{Sh}/N_{Sc}^{0.33} = 124$. Thus their performance data expressed in terms of feed and product concentrations, and product rate have firm quantitative significance. In their experiments, calcium and magnesium ions were analysed by titration with EDTA, biochemical oxygen demand was determined using the azide modification of the iodometric method, alkyl benzene sulphonate was determined colorimetrically, and total dissolved solids were determined by evaporation following the procedures given for each in the Standard Methods for the Examination

Table 8-1. Membrane specifications

Data of Hauck and Sourirajan (1969)

Film no.	$A \times 10^6$ $\left(\frac{\text{g. mole } H_2O}{\text{cm}^2 \text{ sec. atm}}\right)$	$(D_{AM}/K\delta) \times 10^5$ (cm/sec)
H-1	1.618	0.679
H-2	1.765	6.990
H-3	1.914	9.232
H-4	0.886	0.972
H-5	1.695	3.457

Film type: CA-NRC-18
System: [NaCl-H_2O]
Operating pressure: 1000 psig

of Water and Waste Water (1965); phosphates and nitrates were analysed by Auto Analyzers, and very low concentrations of calcium and magnesium were determined by the atomic absorption technique. Other inorganic salts were analysed by the specific resistance measurements using a conductivity bridge.

Reverse Osmosis For Treatment of Hard Waters. Hardness is of special concern in municipal and industrial water supply. Hard water requires much soap before a lather is formed, and hard water deposits sludges or incrustations on surfaces with which it comes into contact and in vessels and boilers in which it is heated. The substances responsible are calcium and magnesium ions, and to a lesser extent (because of their normally smaller concentrations) those of iron, manganese, strontium, and aluminium. In the operation of boilers, foaming, priming, scale formation, caustic embrittlement, and corrosion increase with operating pressures. Foaming and priming entrain moisture and solids in steam. The solids carried over may then be deposited in steam lines, turbines, and other equipment. The tolerances for hardness (expressed as ppm $CaCO_3$) of water are 80, 40, 10, and 2 for boilers operating in the pressure ranges below 150, 150 to 250, 250 to 400, and over 400 psig respectively (Fair and Geyer, 1961b).

Therefore the treatment of hard waters to produce boiler feed waters of acceptable quality is an important industrial problem. The application of reverse osmosis for the treatment of hard waters was investigated by Hauck and Sourirajan, and some of their results are presented in Tables 8-2, 8-3, and 8-4, and Figure 8-1.

Table 8-2 gives the results obtained with three films. $CaCl_2$, $MgCl_2$, and a 1 : 1 mixture of $CaCl_2$ and $MgCl_2$ were used as solutes. Three different initial feed concentrations (~300, 500, and 800 ppm expressed as $CaCO_3$) were tested. The hardness and flux of product water were determined corresponding to 90 per cent volume fraction product recoveries. The results showed that product water hardness of 2 ppm or less were obtained with film H-1 with an average product rate of 38 gal/day/ft^2 at 1000 psig. Under the same conditions, the hardness of product water obtained ranged from 17 to 45 ppm for film H-2, and 22 to 64 ppm for film H-3 with corresponding average product rates of 43. 6 and 49. 2 gal/day ft^2.

Table 8-3 illustrates the performance of film H-4 for the separation of iron, manganese, strontium and aluminium ions present in low concentrations. The data show that the technique can be successfully applied for such separations.

Figure 8-1 shows the effect of pressure on the quality and flux of product water with a feed whose hardness was 500 ppm. Since solute separation generally decreases with decrease in operating pressure, data of this type determine the minimum operating pressure necessary for specific applications. The results showed that film H-1 could give product water hardness of 2 ppm or less in the entire pressure range 200 to 1000 psig tested.

Table 8-4 illustrates the performance of films H-1 and H-2 for softening some natural hard waters obtained from different sources.

The high solute separations and product rates obtained with the above films indicate the possibility that reverse osmosis can be successfully used for the economic treatment of industrial and natural hard waters to give product waters of acceptable quality for domestic use as well as for high pressure boilers.

Production of 'Ultrapure' Waters by Repeated Application of Reverse Osmosis. The development of new industries in the electronic, semiconductor,

Table 8-2. Separation of calcium and magnesium ions in aqueous solution

Data of Hauck and Sourirajan (1969)

Film no.	System	Feed water hardness (ppm $CaCO_3$)	Product water: Hardness (ppm $CaCO_3$)	Product water: Product rate (gal/day/ft^2)
H-1	$CaCl_2$-H_2O	302	2	38.6
	$CaCl_2$-H_2O	498	2	42.3
	$CaCl_2$-H_2O	782	2	41.4
	$MgCl_2$-H_2O	315	<1	41.1
	$MgCl_2$-H_2O	516	<1	37.4
	$MgCl_2$-H_2O	800	2	36.1
	($CaCl_2$ + $MgCl_2$)-H_2O	300	<1	34.2
	($CaCl_2$ + $MgCl_2$)-H_2O	495	<1	36.7
	($CaCl_2$ + $MgCl_2$)-H_2O	807	<1	35.1
H-2	$CaCl_2$-H_2O	302	17	42.8
	$CaCl_2$-H_2O	504	21	42.8
	$CaCl_2$-H_2O	806	26	42.6
	$MgCl_2$-H_2O	315	24	49.2
	$MgCl_2$-H_2O	509	36	43.7
	$MgCl_2$-H_2O	829	45	41.5
	($CaCl_2$ + $MgCl_2$)-H_2O	304	19	43.5
	($CaCl_2$ + $MgCl_2$)-H_2O	504	28	43.8
	($CaCl_2$ + $MgCl_2$)-H_2O	805	41	42.6
H-3	$CaCl_2$-H_2O	301	28	45.2
	$CaCl_2$-H_2O	508	42	50.4
	$CaCl_2$-H_2O	802	64	51.2
	$MgCl_2$-H_2O	302	29	51.6
	$MgCl_2$-H_2O	505	43	51.7
	$MgCl_2$-H_2O	806	60	47.3
	($CaCl_2$ + $MgCl_2$)-H_2O	289	22	51.9
	($CaCl_2$ + $MgCl_2$)-H_2O	498	36	50.0
	($CaCl_2$ + $MgCl_2$)-H_2O	812	52	43.5

Operating pressure: 1000 psig
Product recovery: 90%

and nuclear areas, and the expansion of many old industries such as the pharmaceutical, utility, and electrochemical fields have created a demand for large quantities of 'ultrapure' water. A leading contributor to the field of production of ultrapure water is the electrical power industry. The use of boilers operating at or close to critical pressure is on the increase. Typical specifications for feed water quality for subcritical boilers (operating at 1800 to 2400 psig), and supercritical boilers (operating at pressure >3200 psig) are respectively 0.5 and 0.05 ppm total dissolved solids (Calmon and Kingsbury, 1966). Production of such ultrapure waters can be accomplished by repeated application of the reverse osmosis process to the available feed water. This is illustrated by the data of Table 8-5 where the feed used was a composite

Table 8-3. Separation of iron, manganese, strontium and aluminium ions in aqueous solution

Data of Hauck and Sourirajan (1969)

System	Solute concentration in feed (ppm)	Product water	
		Solute concentration (ppm)	Product rate (gal/day/ft^2)
$FeCl_3$-H_2O	795	40	19.0
$MnSO_4$-H_2O	500	<1	25.7
$SrCl_2$-H_2O	485	17	23.7
$AlCl_3$-H_2O	503	11	26.9

Film No.: H-4
Operating pressure: 1000 psig
Product recovery: 90%

Table 8-4. Softening of natural hard waters

Data of Hauck and Sourirajan (1969)

Film no.	Source of feed water	Feed water hardness (ppm $CaCO_3$)	Product Water	
			Hardness (ppm $CaCO_3$)	Product rate (gal/day/ft^2)
H-1	Coalinga, Calif.	843	<1	26.5
H-1	Webster, S. D.	610	<1	28.6
H-1	Roswell, N. M.	641	4	29.6
H-1	San Diego, Calif.	340	2	30.4
H-1	Indianapolis, Ind.	247	4	30.0
H-2	Coalinga, Calif.	843	12	40.1
H-2	Webster, S. D.	610	5	43.2
H-2	Roswell, N. M.	641	14	42.9
H-2	San Diego, Calif.	340	7	45.1
H-2	Indianapolis, Ind.	247	11	45.0

Operating pressure: 1000 psig
Product recovery: 90%

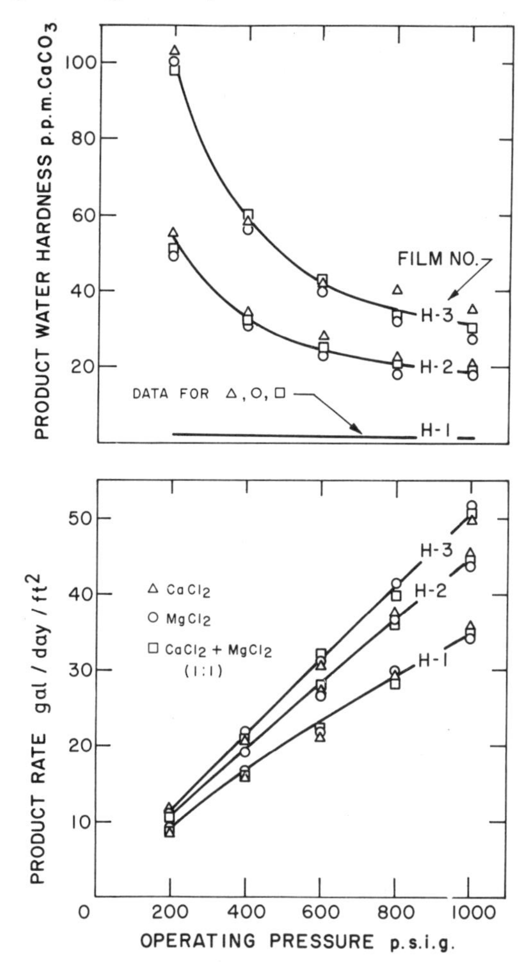

Figure 8-1. Effect of pressure on membrane performance for reverse osmosis water softening. Data of Hauck and Sourirajan (1969).

Film type: CA-NRC-18; feed concentration: 500 ppm; product recovery: 75%.

sample of the product waters obtained from the once processed hard waters (hardness = 300 to 800 ppm $CaCO_3$). The product water obtained by the twice operated reverse osmosis process (Table 8-5) is suitable as feed for sub-critical boilers. An additional reverse osmosis processing of the above waters can give waters suitable as feed for supercritical boilers. Since the data given in Tables 8-2 and 8-5 refer to 90 per cent product recovery and the product rates obtained are sufficiently high (~36 gal/day/ft^2 at 1000 psig), the technique of repeated operation of the reverse osmosis process might prove economical for the production of large quantities of ultrapure waters.

Some Common Water Pollutants. The presence of excessive amounts of nitrates, borates, fluorides, chlorides, phosphates, alkyl benzene sulphonate, and ammonium ions are usually regarded as pollutants in municipal and industrial water supply. Nitrates and phosphates serve as nutrients for the

Table 8-5. Results of repeated reverse osmosis operation

Data of Hauck and Sourirajan (1969)

System	Feed water* hardness (ppm $CaCO_3$)	Product water Recovery (%)	Hardness (ppm $CaCO_3$)	Product rate (gal/day/ft^2)
$CaCl_2$-H_2O	2.25	25	0.125	35.2
$CaCl_2$-H_2O	2.25	50	0.125	35.8
$CaCl_2$-H_2O	2.25	75	0.175	35.2
$CaCl_2$-H_2O	2.54	90	0.225	36.6
$MgCl_2$-H_2O	2.14	25	0.120	35.2
$MgCl_2$-H_2O	2.14	50	0.140	35.7
$MgCl_2$-H_2O	2.14	75	0.140	35.9
$MgCl_2$-H_2O	2.18	90	0.260	36.6

* Product obtained from a once processed hard water.

Film no.: H-1
Operating pressure: 1000 psig

growth of algae and other aquatic plants which render the water unfit for recreational and other uses; further, fish and other aquatic life are deprived of oxygen by the decomposing algae and plant life. While small concentrations of fluorides appear beneficial in reducing the prevalence of dental caries, excessive amounts of fluorides are definitely associated with the prevalence of mottled teeth in many communities. Excessive amounts of nitrates are held responsible for some infant illnesses (Fair and Geyer, 1961a), and excessive amounts of chlorides render the water unfit for drinking. The presence of alkyl benzene sulphonate even in concentrations of only 1 ppm will produce substantial frothing and, being non-biodegradable, it will accumulate in water, producing very undesirable foams. The pretreatment of water for the removal of these pollutants may be necessary in many water supply systems. The data given in Table 8-6 show that they can be effectively removed by the reverse osmosis process.

Pollutants From Plating Wastes. The waste effluents from metal finishing plants contain numerous toxic constituents (Promisel, 1960). The toxicity limits to fish life with respect to, for example, copper, lead, zinc, and chromium are 0.02, 0.1, 0.2, and 1.0 ppm respectively (Hawksley, 1967). Hence the treatment of plating wastes, especially the dilute wastes, is a serious problem in the metal finishing industry. Since profuse rinsing is a pre-requisite of a sound finish, very large quantities of water are involved; and further, some of the waste constituents may have some economic value (Lancy, 1955). Consequently, the problem of treatment of plating wastes is important from the points of view of water pollution control, water re-use, and waste recovery. The data presented in Table 8-7 illustrate the possible applicability of the reverse osmosis technique for the separation of substances present in

Table 8-6. Separation of some water pollutants

Data of Hauck and Sourirajan (1969)

System	Solute concentration in feed (ppm)	Product	
		Solute concentration (ppm)	Product rate (gal/day/ft^2)
$NaNO_3$-H_2O	492	87	27.3
$Na_2B_4O_7$-H_2O	524	16	26.1
NaF-H_2O	505	26	26.4
NaCl-H_2O	507	78	27.6
A.B.S.-H_2O	95	<1	20.9
A.B.S.-H_2O	300	<1	19.6
NH_4NO_3-H_2O	487	97	23.2
Na_3PO_4-H_2O	480	3	20.4

Film no.: H-4
Operating pressure: 1000 psig
Product recovery: 90%

Table 8-7. Separation of some salts present in plating wastes

Data of Hauck and Sourirajan (1969)

System	Solute concentration in feed (ppm)	Product	
		Solute concentration (ppm)	Product rate (gal/day/ft^2)
$ZnSO_4$-H_2O	535	48	20.7
$Pb(CH_3COO)_2$-H_2O	504	32	20.4
$CuSO_4$-H_2O	500	8	19.2
$NiCl_2$-H_2O	500	14	19.2
CrO_3-H_2O	512	22	21.5
$SnCl_2$-H_2O	500	49	20.8
$AgNO_3$-H_2O	500	135	22.6
$Fe(SO_4)_2(NH_4)_2$-H_2O	525	19	20.1
$Ni(SO_4)_2(NH_4)_2$-H_2O	515	22	20.9
$Cr(SO_4)_3$-H_2O	500	9	22.1
$HAuCl_4$-H_2O	500	109	19.1

Film no.: H-4
Operating pressure: 1000 psig
Product recovery: 90%

plating wastes. The data given are for 90 per cent recovery in a single-stage reverse osmosis process using ~500 ppm of solutes. The actual concentration of solutes in plating wastes is usually much less. Since essentially the same degree of separation can be expected at lower concentrations also, more than one reverse osmosis processing may not be necessary in many situations. In any case, by repeated operation of the process with 90 to 95 per cent recovery in each operation, product water of any desired quality can be obtained along with concentrated solutions suitable for waste recovery.

Sewage Water Treatment. The present primary and secondary sewage treatment facilities have as their main objectives the removal of biochemical oxygen demand and suspended solids. These treatments are not designed to remove nitrates, phosphates, or the nonbiodegradable surfactants. The removal of the latter would be the objective of tertiary sewage treatment facilities which are not extensive in use today. Reverse osmosis can effectively take the place of tertiary treatment, and sometimes both secondary and tertiary treatments, and offer an effective means of upgrading sewage water to a quality practically suitable for all water uses. Some pilot plant results of sewage water treatment by reverse osmosis have been reported (Bray et al., 1965; Sudak and Nusbaum, 1968) and they will be summarised later in this chapter.

Table 8-8 gives the results obtained with a typical film and a number of samples of raw sewage water obtained from the Ottawa City primary sewage treatment plant. Experiments were made at two operating pressures, 1000 and 500 psig, with particular reference to the removal of biochemical oxygen demand, nitrate, phosphate, alkyl benzene sulphonate, and total dissolved solids. The performance of the membrane was found to be very good with respect to the removal of all the above contaminants. The average biochemical oxygen demand removals were 85. 8 and 80. 8 per cent respectively at 1000 and 500 psig. Under the conditions of the experiments made, the average separations of nitrates, alkyl benzene sulphonate, and phosphates were 50. 3, 93, and $>$99 per cent respectively. The average product rates were 32. 7 and 18. 3 gal/day/ft^2 at 1000 and 500 psig respectively. The above results indicate that the reverse osmosis process using the Batch 18 type porous cellulose acetate membranes has the potentialities of becoming an economic means of renovation of waste waters.

The separation and product rate data given in Tables 8-2 to 8-8 do not represent the limits obtainable in the reverse osmosis process. By increasing the mass transfer coefficient on the high-pressure side of the membrane, significantly better performance can be obtained with a given membrane in most cases. These comments apply generally to all the experimental data presented in this chapter.

Table 8-9 (Ironside and Sourirajan, 1967), and Table 8-10 (Sourirajan and Sirianni, 1966) give some additional data on the performance of porous cellulose acetate membranes for sewage water treatment, water pollution control, and removal of some nonionic detergents from aqueous solutions.

2. Waste Waters From the Pulp and Paper Industry

The pulp and paper industry is a large user of water. An average size pulp mill producing 150 to 500 tons of cellulose pulp for making paper and allied products may commonly use from 5 to 50 million gallons of water daily. A large part of this water can be processed by reverse osmosis, and reused again; the concentrated wastes may then be disposed of more economically by conventional methods of evaporation and burning. Therefore the application

Table 8-8. Reverse osmosis for sewage water treatment

Data of Hauck and Sourirajan (1969).

Solute or Equivalent	Solute concentration in feed (ppm)	Operating pressure (psig)	Product: Solute concentration (ppm)	Product: Product rate (gal/day/ft^2)
Biochemical oxygen demand	24	1000	3	29.6
	46	1000	8	35.8
	37	1000	2	38.4
	25	1000	5	31.0
	36	1000	5	28.7
	44	500	15	14.6
	46	500	8	16.2
	24	500	5	19.6
	37	500	7	18.8
	21	500	4	18.4
NO_3^-	0.24	1000	0.1	34.2
	0.50	1000	0.25	34.0
	0.07	1000	0.03	28.7
PO_4^-	2.5	1000	0.01	34.2
	1.8	1000	0.01	34.0
	3.5	1000	0.02	29.7
Alkyl benzene sulphonate	0.7	1000	0.05	29.6
	0.8	1000	0.05	35.8
	0.4	1000	0.02	38.4
	1.2	1000	0.08	31.0
	1.5	1000	0.10	28.7
	1.8	500	0.20	14.6
	0.8	500	0.08	16.2
	1.2	500	0.01	19.6
	1.1	500	0.01	18.8
	1.3	500	0.01	18.4
Total dissolved solids	284	1000	32	29.6
	454	1000	9	35.8
	76	1000	0.1	38.4
	324	1000	9	31.0
	278	1000	7	28.7
	434	500	49	14.6
	385	500	19	16.2
	350	500	30	19.6
	265	500	23	18.8
	294	500	62	18.4

Film no.: H-5
Feed: Raw sewage water
Product recovery: 90%

Table 8-9. Performance of porous cellulose acetate membranes for water pollution control

Data of Ironside and Sourirajan (1967)

Film no.	Film shrinkage temp. (°C)	Feed solution	Analysis of Feed solution	Analysis of Product water	Product water rate, (gal/day/ft^2)
70	88	Sewage water, unchlorinated	B. Coli* average 100, 000/cm^3	Sterile	31.1
70	88	Sewage water, chlorinated	B. Coli* average 400/cm^3	Sterile	29.3
72	90	Aqueous solution of sodium dioctyl sulphosuccinate	370[a] ppm	<0.5[a] ppm	24.4
67	92	Aqueous solution of sodium alkyl-benzene sulphonate	512[a] ppm	0.6[a] ppm	24.0
75	88	Brown lignin solution (acidic)	10.4[a]%	0.035[a]	16.2

* Standard plate count

[a] Solute concentration

Film type: CA-NRC-18
Operating pressure: 1000 psig

Table 8-10. Removal of tritons from aqueous solutions using porous cellulose acetate membranes

Data of Sourirajan and Sirianni (1966)

Film no.	Film shrinkage temp. (°)	Solute-Triton type	Solute separation (%)	Product rate (gal/day/ft^2)
198	86	N-128	98.6	32.5
203	89	X-165	99.1	29.4
202	90	X-205	99.1	29.5
203	89	X-305	~100	26.2

Film type: CA-NRC-18
Feed concentration: 2 g Triton/100 g water
Feed rate: 120 cm^3/min
Operating pressure: 1500 psig

of reverse osmosis for the treatment of waste waters from the pulp and paper industry is an area of great practical importance from the points of view of waste water treatment, water renovation, and water pollution control. Some limited amount of work has been reported on the subject, a summary of which is given below.

A variety of different paper mill waste waters has been tested with reference to this application by Wiley et al. (1967) and Ammerlaan et al. (1968), using the commercially available reverse osmosis units supplied by Gulf General Atomic and by Havens Industries. The engineering details of these units are given later in this chapter. Some of their test results are summarised in Tables 8-11 and 8-12.

Table 8-11. Reverse osmosis for treatment of paper mill waste waters

Data of Wiley et al. (1967)

Sample	Flux (gal/day/ft^2)	mg/litre			pH	OD[3]
		Solids	C.O.D.[2]	Chloride		
Process feed[1]		1960	945	430	2.2	8.75
Product Water						
Membrane grade A*	6.1	14	100	36	2.6	0.194
Membrane grade C	7.0	72	132	251	2.2	0.261
Membrane grade B	10.9	822	516	400	2.3	3.63
Concentrated stream						
Membrane grade A*		12,120	6655	3260	2.2	80.4
Membrane grade C		6940	4366	1964	1.9	44.8
Membrane grade B		5160	3405	852	2.2	31.7

* A = Dense; C = Intermediate; B = Coarse

1 First stage sulphite bleach plant effluent

2 Chemical oxygen demand

3 Optical density at 281 millimicrons (a measure of the lignin content)

Reverse osmosis test unit: Gulf General Atomic A, B, and C Modules (Size 4)
Operating pressure: 450 psig
Feed rate: 1.8 gal/min
Product recovery: 80%

Table 8-11 gives the results obtained with the processing of first stage sulphite bleach plant effluent with the three Gulf General Atomic modules. The reductions obtained with the use of the 'dense' type-A membrane in terms of chemical oxygen demand (94 per cent), chlorides (92 per cent) and excellent colour removal between the process feed and product water, as well as the recovery of these components in the concentrate stream, are of significant interest to the pulp and paper industry.

The results of more extensive studies using the Havens unit have been reported by Ammerlaan et al. (1968) (Table 8-12). They found that the flux of product water depended on the type of waste water feed, concentration level of the feed, and the type of membrane used. In many cases a change in flux due to a change in operating temperature was inversely proportional to the change in viscosity of water over the same temperature range. The pH of the feed solution also had an effect on the flux; apparently a change in pH increased the amount of

Table 8-12. Reverse osmosis for treatment of paper mill waste waters

Date of Ammerlaan et al. (1968)

	Separation (%)							
	Feed A to F		Feed G					
			Membrane type 3			Membrane type 5		
			pH			pH		
Constituent	Membrane type 3	Membrane type 5	2.7	4.8	6.1	2.7	4.8	6.1
Solids	75-98	95-99	27	57	70	45	81	95
OD (colour)	98-99.8	99-99.8	66	55	48	79	74	80
B.O.D.**	80-95	90-99	26	45	58	42	62	82
C.O.D.***	85-98	90-99	19	43	57	37	62	81
Inorganics	60-95	94-99	—	--	—	—	—	—
Volatile acids			7	40	65	26	70	95

Reverse osmosis test unit: Havens Industries (membrane types 3 and 5*)
Operating pressure: 600 psig
Feed water: A. Diluted calcium base spent sulphite liquor
B. Calcium base pulp wash water
C. Kraft first stage bleach effluent
D. Kraft second stage bleach effluent
E. Sulphite first stage bleach effluent
F. Total effluent
G. Ammonia base liquor condensate
Product flux: Membrane type 3: 2 to 12 gal/day/ft^2
Membrane type 5: 2 to 5 gal/day/ft^2

* Membrane type 3: Intermediate
Membrane type 5: Very dense

** Biochemical oxygen demand

*** Chemical oxygen demand

insoluble material present in a colloidal, semi-flocculating or in some kind of dispersed state. They observed a coating of some slime material deposited on membrane surface during operation. The membrane surface however could be cleaned successfully by flushing a slightly oversized plastic foam ball through the membrane tubes. The test results given in Table 8-12 show that the reverse osmosis process can be successfully used for the treatment of paper mill waste waters.

Porous cellulose acetate membranes are used in the Gulf General Atomic, and Havens reverse osmosis units. Perona et al. (1967) showed that calcium base sulphite liquors formed dynamic membranes, when circulated under pressure over porous bodies, such as porous ceramic, and hence such dynamically formed membranes could be used for the treatment of pulp mill sulphite wastes.

In some of their pilot experiments, a membrane was formed with 1 per cent spent liquor on Type 02 and 03 porous ceramic tubes (average pore diameters 1.4 μ and 0.90 μ respectively) in series. After 15 hours of operation, typical solute separations were obtained with permeation rates of 7.5 and 9 gal/day/ft^2 for the 02 and 03 tubes respectively at 500 psig. The loop was drained and filled with the chlorination bleach waste. At 400 psig and 30° the type 03 tube gave 90 per cent solute separation based on optical density, and 22 per cent by chloride analysis; the corresponding product rate was 7 gal/day/ft^2. Solute separation for the 02 tube rose over 4.5 hours to 67 per cent by colour and 18 per cent of chloride with a flux of 9 gal/day/ft^2. The loop was again drained, rinsed with demineralised water, and filled with alkaline extraction bleach waste. Solute separation based on optical density for the 03 tube was 90 per cent (flux, 8 gal/day/ft^2), and for the 02 tube was 51 per cent (flux, 14 gal/day/ft^2). Chloride separations were 32 per cent and 16 per cent respectively.

Dynamic membranes were also formed on 0.1 μ Millipore supports with the chlorination bleach wastes. After about 5 hours of operation, solute separation based on optical density was 95 per cent and chloride separation was 34 per cent; and the product rate was 10 gal/day/ft^2 at 500 psig. In a similar test, second stage alkaline bleach waste formed a membrane separating 94 per cent of colour, and 43 per cent of chloride at 15 gal/day/ft^2. Even though the results obtained were still irreproducible, Perona et al. (1967) estimated that a production facility operating at 60° could attain product rates of about 30 gal/day/ft^2 and solute separations greater than 90 per cent at pressures less than 500 psig.

3. Other Applications

Extensive pilot plant test data are available on the application of the reverse osmosis process for the treatment of natural brackish waters, acid mine waters, and polluted waters; some of these data are presented later in this chapter.

Okey and Stavenger (1966) studied the performance of D-O membranes (non-ionic membranes produced by Dorr-Oliver, Inc.) with particular reference to their ability to separate viruses in the reverse osmosis treatment of waste waters. The upstream water was inoculated with E. Coli phage suspensions. The virus density in the upstream side of the membrane was 3.5×10^4 organisms per cm^3. The product water was collected at intervals, concentrated, and examined; no organisms were detected in the product water.

In view of the importance of the purification of waters contaminated with phenol, the possible separation of phenol present in low concentrations in water was investigated by Ironside and Sourirajan (1968) in reverse osmosis experiments using porous cellulose acetate membranes. It was found that phenol was transported through porous cellulose acetate membranes preferentially from aqueous solutions, i.e. the product water was enriched in phenol. Similar results were observed by Lonsdale et al. (1967). These results indicate the need for either changing the chemical nature of the membrane surface suitably, or converting the phenol into a compound which will not be preferentially transported through cellulose acetate membranes, for this application.

The development of reverse osmosis as a concentration technique has been discussed in Chapter 5, where some data on the concentration of natural maple sap are given. Further work has been done more recently by Peterson (1968) on the concentration of natural maple sap using Loeb-Sourirajan type porous cellulose acetate membranes in a bench scale reverse osmosis unit shown in Plate 8-1. This unit has a plate-and-frame flat membrane configuration and uses 13 plates (26 membranes) with a total effective membrane area of 6.18 ft^2. Baffles were provided on the membrane surface to increase fluid velocity. The experiments were carried out with Batch 47 type membranes at 1500 psig using natural maple sap feed containing 3.31 per cent equivalent sucrose. The usual problem encountered in the earlier experiments was the growth of bacteria on the membrane surfaces and the consequent decrease in permeability. In these runs the above problem was solved by operating the reverse osmosis process at low temperatures (2 to 6°) and subjecting the sap feed to ultraviolet radiation (by passing it though a Steroline ultraviolet tube) before entering the pump and membrane unit. The feed was also initially filtered through a 0.45 micron wire mesh filter cartridge. In a typical experiment, the membrane used had a pure water permeability of 49.3 gal/day/ft^2 at 38°. The maple sap concentration was done at 2 to 5°. The membrane permeate was essentially pure water. The initial permeation rate was 18.1 gal/day/ft^2 at 2°, and after 40 hours of continuous operation, the feed concentration was 10.84 per cent equivalent sucrose, and the corresponding membrane permeation rate was 6 gal/day/ft^2 at 5°. The feed rate used in the experiment was ~43 gal/hour.

4. Engineering of Reverse Osmosis Plants For Saline Water Conversion

General Considerations. The reverse osmosis process is currently in a stage of research and development and, in view of its practically unlimited potentialities, this situation will continue to prevail at all times with respect to some application or other. However, at least with respect to water treatment applications in general, and saline water conversion in particular, the reverse osmosis process has already emerged into the pilot plant stages, and is fast approaching the stage of large-scale industrial practice, primarily because of the chemical suitability, and the productive capability of the Loeb-Sourirajan type porous cellulose acetate membranes for such applications. Most of the accomplishments up to this time have been with particular reference to brackish water and sea water conversion. Consequently, unless otherwise stated, the discussions in the rest of this chapter are concerned only with this particular application. The details of process engineering, plant design, and parametric studies discussed below are however general enough to contribute significantly to the science and engineering of reverse osmosis in all its applications.

Saline water conversion plants. These are generally divided into two classes: brackish water conversion plants, and sea water conversion plants. The primary distinction lies in the operating pressure and the choice of the surface pore structure of the membrane. The brackish water conversion plants are generally designed to operate at about 600 psig or less, with a feed water containing 5000 ppm or less of NaCl. The reverse osmosis units designed for home use operate at 50 psig or less with feed waters containing 1500 ppm or less of total dissolved solids. Under such conditions, the osmotic pressure of the feed solution is low, and the degree of solute separation needed for obtaining potable water (<500 ppm total dissolved solids) is not high. Consequently a more porous membrane can be used; such a membrane also gives a high product rate. Further, since membranes which can be used for low pressure service have a more favorable A v. $(D_{AM}/K\delta)$ correlation (Figure 3-58) and are subject to less compaction effects (Table 2-56), brackish water conversion reverse osmosis units have the most favourable conditions for industrial success.

The conditions for sea water conversion on the other hand, are more severe. A higher level of solute separation is needed for obtaining potable water in a single step; and the osmotic pressure of the feed solution is high. Consequently high operating pressures are necessary. In addition, a high mass transfer coefficient on the high-pressure side of the membrane is necessary to maintain the quality of product water along with high product recoveries. At high operating pressures, the A v. $(D_{AM}/K\delta)$ correlation is less favourable (Figures 3-57 and 3-58), and compaction effects are generally severe (Table 2-56). For these reasons, sea water conversion is frequently considered in terms of a two-stage operation which would naturally increase the final cost of product water. There is however no doubt that single-stage sea water conversion plants will be successfully built in the future, and they may be expected to operate at pressures higher than 1000 psig.

Flow diagram. Figure 8-2 is the basic flow diagram for a single stage reverse osmosis plant as given by Bray (1966). Feed water, free of silt and gross debris, enters the feed sump. Chlorine is periodically added to the inlet feed line. From the feed sump, the solution is pumped through a filter assembly. A pH controller regulates an acid injection system which governs the pH of the feed between the feed sump and the main pump. The feed is de-aerated, though only if it is necessary, and then pumped under pressure into the membrane units. The product water is collected at the atmospheric pressure. The high-pressure brine is discharged into the brine hold-up tank through a recovery turbine.

Several modifications of the basic flow diagram can be used to suit particular circumstances. For example, de-aeration is not necessary for the reverse osmosis process, but it may be necessary as a method of corrosion prevention. Acid injection is not necessary if the feed pH is already in the desired

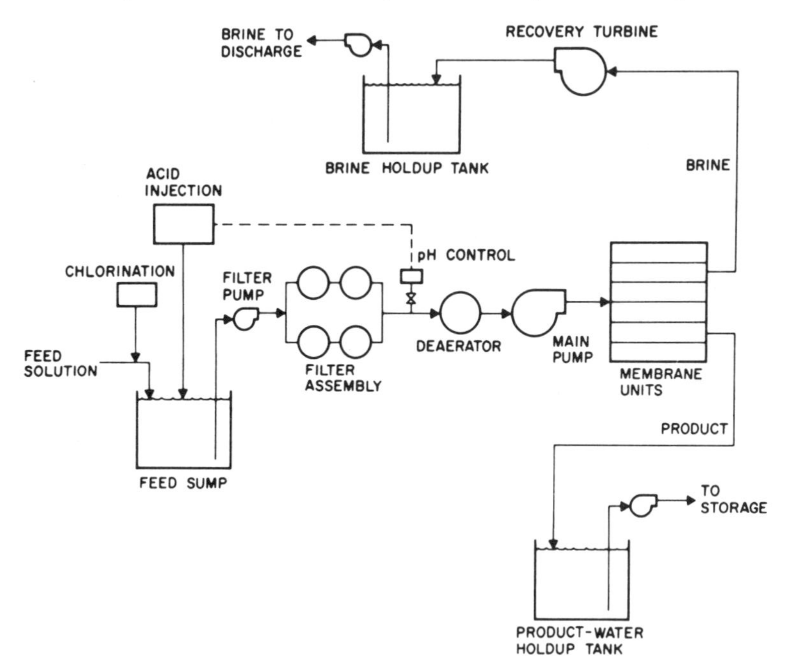

Figure 8-2. Basic flow diagram for a single-stage reverse osmosis plant (Bray, 1966).

range. A recovery turbine is necessary only if economic analysis justifies power recovery.

Materials of construction. The choice of materials is dictated by operating pressure, corrosion, and economic considerations. For the low-pressure side of the plant, the designer has the choice of carbon steel, copper, or plastic materials, the latter being the ones most preferred. Fibreglass cloth laminates and solid phenolics have been used as membrane supports. For sea water systems, Monel or 316 stainless steel may be used on smaller pipes and fittings on the high-pressure side of the unit, and clad carbon steels (with alloys of copper and nickel or stainless steels as the cladding material) may be used for larger pipes and tanks. For systems operating at <600 psig, fibreglass reinforced epoxy tubes are preferred. Organic coatings as protective surfacing materials for completely carbon-steel systems are often considered advantageous. Keilin and DeHaven (1965) have given some data on the behaviour of organic coatings in sea water. A Red Phenolic coating (Pittsburg Plate Glass No. H-13817A) is said to have been in salt water service for four years at 10,000 psig with no apparent coating deterioration (Keilin and DeHaven, 1965). A dominant current trend is to use plastic materials wherever possible in the entire reverse osmosis unit.

Feed Water Pretreatments. Filtration and acid injection are the two feed water pretreatments generally considered in the operation of reverse osmosis plants. The need for such pretreatment depends entirely on the nature of the feed and the design of the membrane units.

The object of filtration is to provide a clean feed, free from suspended matter, to the pump and membrane units, so that pump abrasion is minimised and the flow path and the membrane surface are not clogged with solid matter. Stainless steel wire mesh cartridges (25 μ), sand filters (10 to 25 μ) or diatomaceous earth filters (1 to 10 μ) may be used depending on the extent of filtration necessary. Further the feed may contain in solution ferrous and manganous salts which, on oxidation, might precipitate on the membrane surface and other parts of the plant. Conventionally they can be held back from the system by initial aeration and subsequent filtration of the feed water; or, they may be held in solution in the entire unit by preventing their oxidation and precipitation as accomplished by Loeb and Johnson (1967) by the addition of a few parts per million of catalysed sodium sulphite in the feed.

The object of acid injection to the feed is to prevent the precipitation of calcium carbonate and magnesium hydroxide (which are usually present in natural waters) during the concentration of the feed in the membrane unit, and to keep the pH of the feed solution slightly in the acid range, preferably between 5 and 6, so that the rate of hydrolysis of the cellulose acetate material is kept at its minimum. Hydrochloric or sulphuric acid (preferably the latter) is used for the purpose. Further periodic chlorination or other bactericide additions to the feed may also be necessary to reduce the growth of organisms within the plant. Feeds with 1 to 2 ppm of residual chlorine have been used for periods of several months with no apparent deterioration of membrane performance (Bray, 1966).

High-Pressure Pumps. The main pressure pumps are items of major capital cost in any reverse osmosis unit. Even though there is a large variety of pumps available in the market, the selection of suitable pumps for reverse osmosis application is still a problem. Feed waters saturated with air, in the pH range 3 to 7, and with dissolved solids in such concentrations as are present in sea water, are characteristics of reverse osmosis saline water conversion units. Very few data are available on the use of pumps of either piston or centrifugal types for medium- (400 to 1000 psig) and high-pressure (>1000 psig) service under these conditions. Intense development work on pumps specifically intended for reverse osmosis application may be expected.

Centrifugal pumps are generally preferred for large scale industrial practice because of their lower maintenance requirements; their use for reverse osmosis plants is probably restricted to plant capacities of 100, 000 to 200, 000 gal/day medium-pressure units, and ~500, 000 gal/day high-pressure units. Piston pumps are currently being used in the laboratory type, pilot plant, and demonstration reverse osmosis units. The leakage of brine through the packing glands is a usual problem with piston pumps. This is minimised by directing a spray of water on the part of the plunger outside of the packing gland. This prevents an accumulation of abrasive salt within the packing, and removes heat which tends to harden the packing and prevent it from sealing adequately.

Turbines For Power Recovery. A recent study (Bray and Menzel, 1966) indicates that recovery turbines are not economically justified for brine discharge volumes of less than about 500, 000 gal/day. Cheng et al. (1967) have developed the concept of a flow-work exchanger which may offer an economic means of power recovery even in comparatively smaller size reverse osmosis units.

A flow-work exchanger is a unified piece of equipment which simultaneously pressurises a condensed fluid stream and depressurises a substantially equivalent volume of another condensed fluid stream. It uses two displacement vessels to form closed loops with a high pressure processing system. It is conceivable that each of the displacement vessels may be filled alternately with the low pressure feed and the high pressure brine both pressurised and depressurised respectively, by substantially non-flow processes. The pressurised feed may be pushed into the membrane system by the high-pressure brine stream, while at the same time the depressurised brine may be pushed out of the displacement vessel. Two working units based on the above principle have been built and tested (Cheng and Fan, 1968). The first unit which has two floating-piston type displacement vessels is operable under 1500 psig and delivers 9 gal/minute. The second unit which has two bladder-type displacement vessels, is operable under 1500 psig and delivers 18 gal/minute. The study has been limited to the construction and hydraulic operation of the units without coupling them to a reverse osmosis unit. The hydraulic operation of these units has been generally satisfactory and the technical feasibility of such units has been demonstrated. The overall efficiencies of the units are as high as 90 to 95 per cent at the rated capacities. It would be interesting to watch further developments in this field.

Cheng and Fan (1968) have reviewed the conventional power recovery schemes which may be applicable for reverse osmosis systems. According to them, multistage centrifugal machines, impulse type hydraulic water turbines, or hydraulically driven reciprocating pumps may be used for power recovery; the estimated efficiencies are 60 per cent, 70 per cent, and 80 per cent respectively. Actual engineering experience in the use of turbines for power recovery in reverse osmosis systems seems limited.

Sumps and Storage Tanks. The sizes of the feed sump and product water storage tank are governed by the time needed to shut or start the pump involved on low-level or high-level signals. A sump tank of a few minutes' pumping capacity should be normally sufficient. Product water storage capacity should be consistent with the usage requirements. If the concentrated brine is to be used for sand filter back flush, a brine hold-up tank of sufficient capacity should be provided.

Plant Control and Instrumentation. The main control signals for the plant are the system pressure, operating temperature, product water quality, feed pH, pressure drop across membrane assembly and the fluid flow rates.

The main instruments used are pressure, temperature, and pH indicators and recorders, conductivity cells with recorder points to include the compositions of the feed and product waters, and the concentrated brine, flow meters, pressure differential indicators, hand-operated and automatic switches, bypass and relief valves, alarm signals, and automatic shutdown and startup circuits. The product water quality may be monitored for the unit as a whole or for different sections of the unit. On a signal indicating an excessive salt content, the affected section may be shut down or the product stream from the section may be diverted to the feed sump. Flow indicators may be installed at all strategic plant locations. Rotameters with pneumatic transmitters may be provided to record the flow of feed water, product water, and the concentrated brine. A variable stroke chemical feed pump, interlocked with feed water sump pumps, may be used for acid injection. The feed pump may be regulated pneumatically by a pH recorder controller on the control panel. Flow-through type conductivity cells may be used. The fouling of the cells is a common problem which can be avoided by daily cleaning of the electrodes with hydrochloric acid and water. The feed water sump tank input may be controlled by a local pneumatic level controller and a pneumatic diaphragm-actuated butterfly control valve. Conductance-type level probes may be provided to shut down pumps associated with feed water, brine, and product water tanks when tank levels become too low. An audible alarm with individual flashing lights may be located in the control panel to call attention to any abnormal plant conditions such as high conductivity feed water entering the product water tank, low feed pressure at the main pump suction, or too high or too low operating pressure in the membrane units. Provision should also be made to shut down the plant automatically under specified operating conditions.

Membrane Units—Design Concepts. There are essentially four different membrane configurations which have been successfully developed by different groups of workers. These are (i) flat membrane configuration, also called plate-and-frame design, (used in the early UCLA pilot plants, and the later Aerojet units), (ii) tubular membrane configuration (used in the UCLA-Coalinga pilot plant, and the Havens units), (iii) spiral-wound membrane configuration (used in the Gulf General Atomic units), and (iv) hollow fibre membrane configuration (used in the DuPont units). Each one of these design concepts is described in detail in the next section.

Briefly, the flat membrane configuration is just a closely spaced stack of porous or perforated support plates with the membrane mounted on either side of each plate. The feed flows through a baffled path in the narrow space between the membrane surfaces, and the product is withdrawn through the support plate openings.

In the tubular membrane configuration, the membrane is assembled into perforated metal support tubes, as in the UCLA-Coalinga design, or it is integrally lined in porous fibreglass support tubes, as in the Havens design. The feed water flows through the membrane tubes (~0.5 to 1 inch diameter) and the product water drips out through the perforations or pores in the support tubes.

In the spiral-wound membrane configuration, a pair of membranes is placed in contact with a porous flexible backing material, and the resulting three-layer sandwich is glued together around the edges. A perforated tube is included between the membranes, and the membrane assembly is rolled up on this tube, together with an appropriate spacer material which serves to separate adjacent membrane surfaces as the roll is made. The rolled assembly is then placed in a snug-fitting cylindrical pressure vessel. Feed water flows in the passage provided by the coarse brine side spacer material. Product water flows spirally inward to the centre tube which is integrally attached to the pressure vessel.

The hollow fibre membrane configuration was originally proposed by Mahon (1963) of the Dow Chemical Co. In this design, membranes in the form of hollow fibres (5 to 100μ outside diameter, with diameter to wall ratio of 3 to 5) are arranged in bundles with the ends sealed in headers with some material such as epoxy. The feed flow is inside the tubes and the product water flows out through the walls of the tubes. In the current DuPont design, the flow pattern is reverse; the feed flow is outside of the hollow fibres, and the product water permeates through the inside of the tubes.

While each one of the above configurations has its own advantages and disadvantages, no one design concept has yet been proved superior to the others. On ultimate analysis, the best design is one which produces potable water from a given feed at the lowest cost. Most cost estimates in the field tend to fall within narrow limits, and hence no clear choice in design concept has emerged.

The effect of concentration polarisation on the performance of membranes in reverse osmosis units is well recognised in general terms, and hence, in every design, attempt is made to keep the feed flow turbulent in the membrane unit. But very few experimental studies have been reported on the mass transfer coefficients obtained in the actual operation of the different membrane configurations. As pointed out in Chapter 4, the desired and the actually-obtainable $\lambda\theta$ values should be the most important considerations in the development of suitable design concepts for saline water and sea water conversion. From this point of view, the whole field of membrane configurations and design concepts seems still open.

5. Design, Construction, and Performance of Pilot Plants for Saline Water Conversion and Similar Applications

The term 'pilot plant' is used here in a rather general sense for all reverse osmosis units scaled up from the earliest laboratory sizes; the term is particularly appropriate because all reverse osmosis units currently available, or being built, for practical applications are still experimental and evolutionary in nature. In the conventional commercial sense, reverse osmosis engineering is growing strong rapidly as seen from the fact that, within the past 2 years, several organisations are advertising the availability of reverse osmosis units ranging in capacities from a few gallons per day (home units) to 1000 to 10,000 gallons per day. Details of their present engineering and performance will naturally contribute to their future development. Available information on the subject are hence summarised here. All plants (except those of DuPont) discussed below use the Loeb-Sourirajan type porous cellulose acetate membranes.

6. UCLA Pilot Plants

Four-inch desalination cell assembly. The first scaled-up reverse osmosis unit was designed, constructed, and operated by Loeb and Sourirajan (1961) in 1960. It resembled a plate-and-frame filter press with a close stacking arrangement of membranes as illustrated in Figures 8-3, 8-4 and 8-5. The basic membrane assembly consisted of a porous stainless steel collector plate with

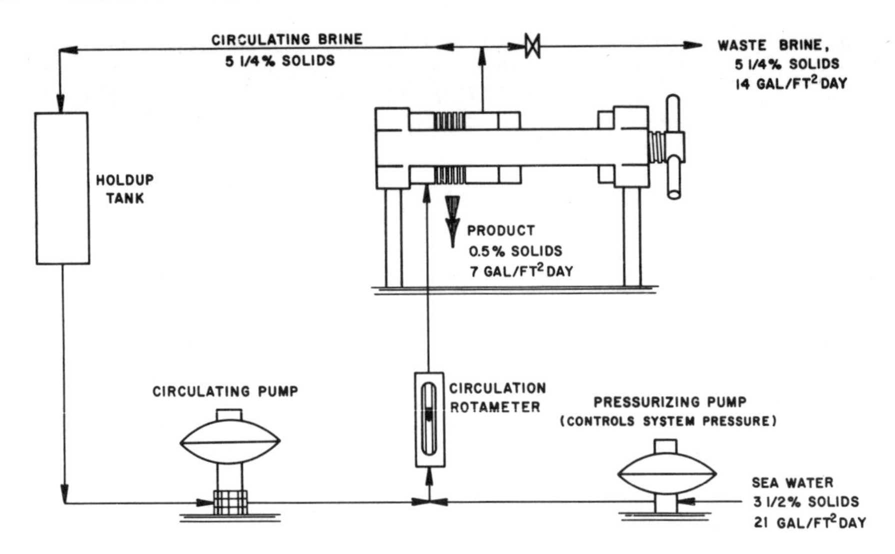

Figure 8-3. Flow diagram of early UCLA unit using 4-inch cells (Loeb and Sourirajan, 1961).

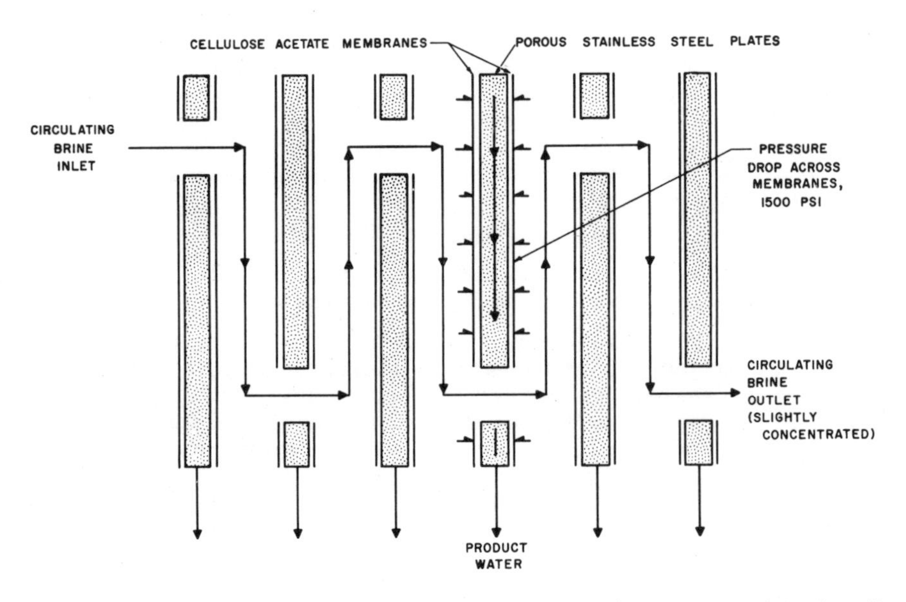

Figure 8-4. Feed flow arrangement in the early UCLA unit using 4-inch cells (Loeb and Sourirajan, 1961).

a film on each side, and a frame which provided the space for the brine to pass over the film. The frame backed up an O-ring which gave the system its external seal. The internal seal was maintained by means of an Epocast coating, a rubber gasket and a washer. The effective diameter of the membrane on each side was a little over 4 inches, and the cell was operated with as many as 12 films (six packages) at a time. This represented a 100-fold scale-up in the size of the reverse osmosis unit in terms of total membrane area compared

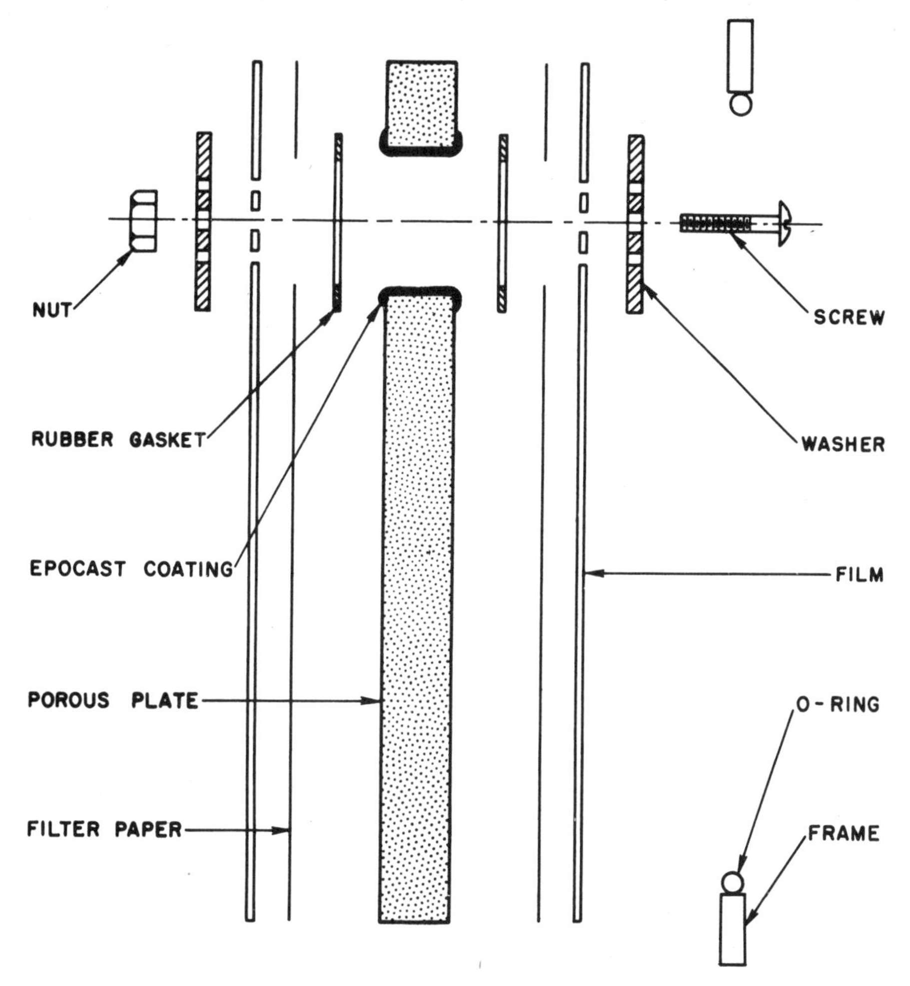

Figure 8-5. Cell assembly in the early UCLA unit using 4-inch cells (Loeb and Sourirajan, 1961).

to the laboratory test cell used up to that time. The operation of the cell proved entirely satisfactory, even though the performance of the larger size membranes made at that time was not reproducible. The experimental data obtained in the smaller test cells could be multiplied reliably in scaled up reverse osmosis units provided comparable flow conditions were maintained.

Sixteen-inch desalination cell assembly. The next step in the UCLA pilot plant programme was the design, construction and operation of a 500 gal/day reverse osmosis plant (Loeb and Milstein, 1962). This was also based on plate-and-frame design using circular plates 16 inches in diameter spaced 0.2 inch apart. The unit had 23 plates, and thus 46 membranes. Total effective membrane area was about 56 ft^2. A simplified version of the cell assembly is shown in Figure 8-6. The sea water or brine entered at one end of the space between two plates and flowed in a spiral path to the centre of the unit; here it passed through one plate to the next passage above, flowing spirally from the centre to the opening in the next plate above, and so on. The flow

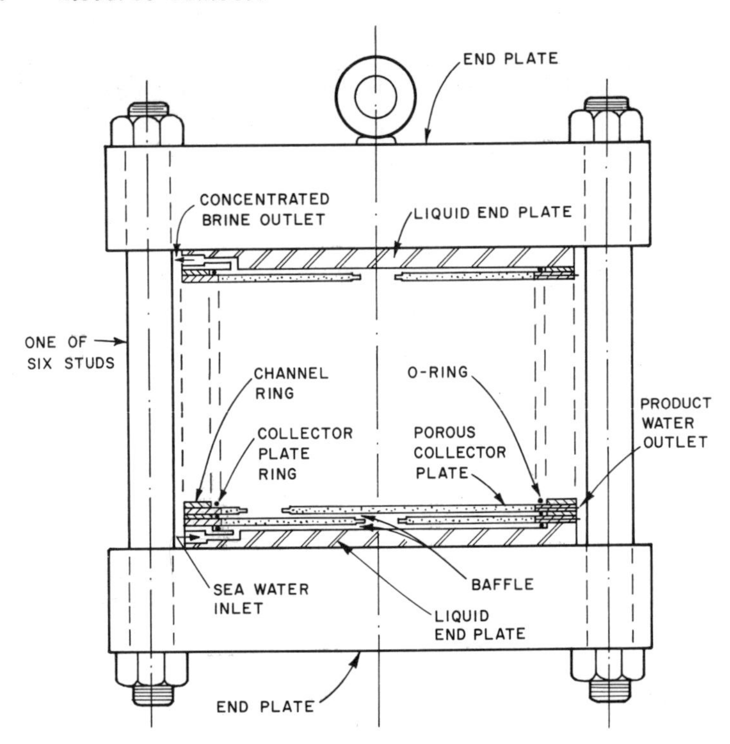

Figure 8-6. A simplified diagram of UCLA 16-inch desalination cell assembly (Loeb and Milstein, 1962).

arrangement was in series, and a baffle was provided on each side to induce spiral flow. The product water was manifolded from the ports at the edges of the collector plates.

A porous plastic (polyvinyl chloride) collector plate with an aluminium periphery solid plate to house the O-ring seals (Figure 8-7) was first employed, and found unsatisfactory because of problems experienced with the sealing of the plastic to the aluminium rings. Hence a solid collector plate was then developed, which relied on a layer of nylon parchment placed between the filter paper and the solid plate to carry the product water laterally to the collecting holes or slots in the solid plate (Figure 8-8). The number of product outlet openings required and the permissible width of the slots were important design considerations.

The pilot plant was equipped with a constant volume pump which supplied 1500 gal/day, and thus provided for a 2 : 1 brine-discard : product-water ratio at the design production rate of 500 gal/day. The feed water was passed through cartridge filters with a nominal 10 micron rating; no other feed water pretreatment was provided.

A 24-day continuous test run on sea water was conducted at Port Hueneme, California, with this pilot plant. Operation was at 1500 psig. Initially the product water contained 1600 ppm total dissolved solids, which increased gradually to 2000 ppm during the first two weeks of operation. At this point, it began to increase rapidly reaching a maximum of 9000 ppm before the unit was shut down and flushed out with distilled water. The flushing operation removed

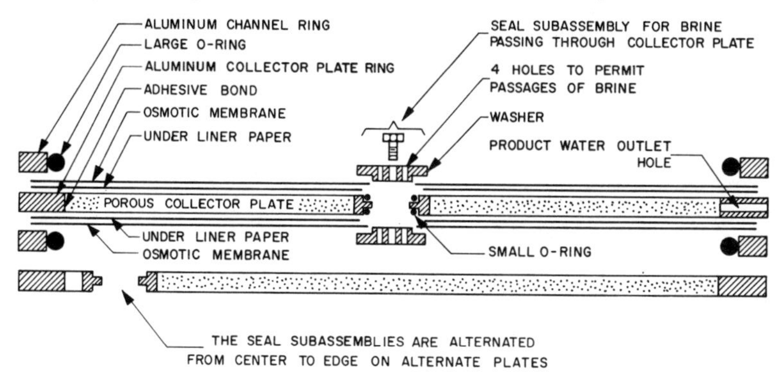

Figure 8-7. Porous collector plate assembly for the UCLA 16-inch desalination cell (Loeb and Milstein, 1962).

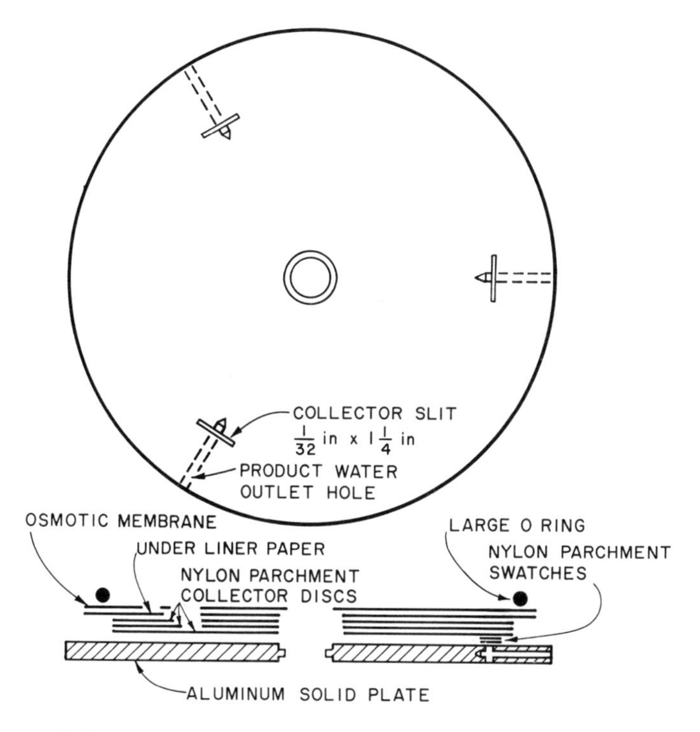

Figure 8-8. Solid collector plate assembly for the UCLA 16-inch desalination cell (Loeb and Milstein, 1962).

large amount of solid matter from the unit, and when operation was resumed, the total solids in product was down to less than 3000 ppm where it remained for the last few days of the experiment. Water production rates for the unit began at just under 700 gal/day, and remained in the 500 to 600 gal/day range during most of the run.

A comparison of the performance data of the pilot plant with those for individual samples of the membranes used, revealed that throughout the run the product salinities were higher and the water fluxes somewhat lower in the pilot plant than in smaller test cells indicating some deficiencies in the pilot plant design or the non-uniformity of the large-size membranes, or both. In any case, the unit showed reasonably satisfactory mechanical and hydraulic performance and this accomplishment was of inestimable value in demonstrating the feasibility of constructing larger plants.

Following the Port Hueneme experiments, the unit was used in two other pilot operations on synthetic brackish water (Loeb and Manjikian, 1963b). The general arrangement was similar to that used before, except that only five plates (10 membranes) were used in the first run, and seven plates (14 membranes) in the second. This reduction in membrane area (to about 12 ft^2 in the first case, and 17 ft^2 in the second) was made possible by the use of membranes with larger surface pores. The solid collector plates were again used, but with some of the aluminium plates replaced by plates of Plexiglas and polyvinyl chloride. The brine and product water were recombined and recycled.

Operating pressure in both runs was 600 psig. The first lasted for 16 days but was interrupted twice for a total of six days. About 300 gal/day of product water were produced during the operating periods, with a feed of 1200 gal/day of a 4300 ppm NaCl solution; the NaCl content in product water ranged from 400 to 800 ppm.

The second run lasted 42 days and was interrupted twice for a total of 4 days. The product rate varied between 450 and 300 gal/day during the course of the run. Part of the variability was attributable to temperature fluctuations and to periodic additions of a bactericidal agent which appeared to affect performance; the data also seemed to indicate the effect of membrane compaction. The feed solution contained 5000 ppm NaCl, and the average concentration in the product water was less than 500 ppm.

Coalinga Plant. This oustanding next step in the UCLA pilot plant programme was the result of the successful gravity drop technique for making tubular membranes and the composite tubular assembly developed by Loeb (Chapter 2). This development led to the authorisation of funds by the California State Legislature for the design and fabrication of a 5000 gal/day pilot plant using tubular membranes. The plant was built in Coalinga, California, and has been in operation since June 4, 1965. The plant has been providing a needed addition to the potable water supply at Coalinga since that date. The Coalinga plant is the first practical installation of its kind, whose operation has been large and long enough for its performance to be one of tremendous experience in reverse osmosis engineering (Loeb and Johnson, 1967; Loeb and Selover, 1967; Stevens and Loeb, 1967).

Figure 8-9 shows a simplified flow diagram of the Coalinga plant. It operates at 600 psig, and produces nominally 5000 gal/day of potable water from 10, 000 gal/day of a feed water containing 2500 ppm of dissolved salts. The raw feed water, supplied to the plant at 50 psig, first flows through a Cuno filter which minimises the entrance of particulate matter into the other parts of the plant. The feed can be treated by injection of a treating agent either upstream or downstream of the filter. The main plant pump is a Worthington triplex reciprocating pump. The feed flow rate can be adjusted as desired during normal operation by the U.S. Motors Varidrive. A Greer accumulator of one gallon capacity is mounted on the discharge side of the pump to smooth out flow and pressure cycling due to piston reciprocation. The discharge line has also an excess pressure relief valve and a low-pressure audible alarm (not shown in Figure 8-9).

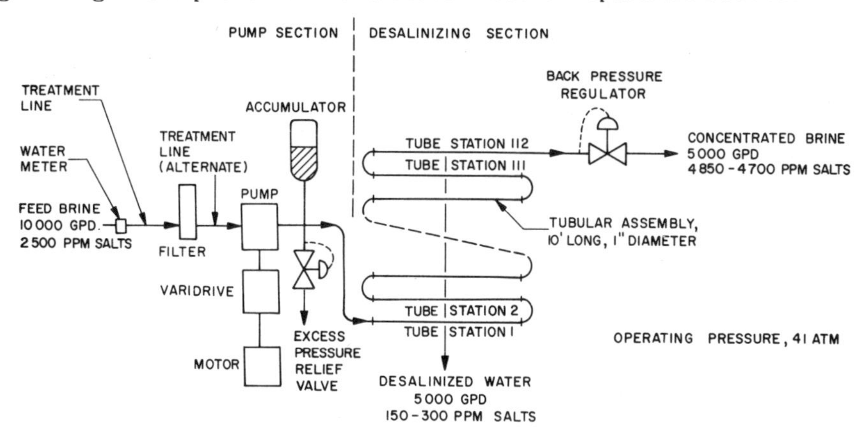

Figure 8-9. Flow diagram of Coalinga pilot plant (Loeb and Johnson, 1967).

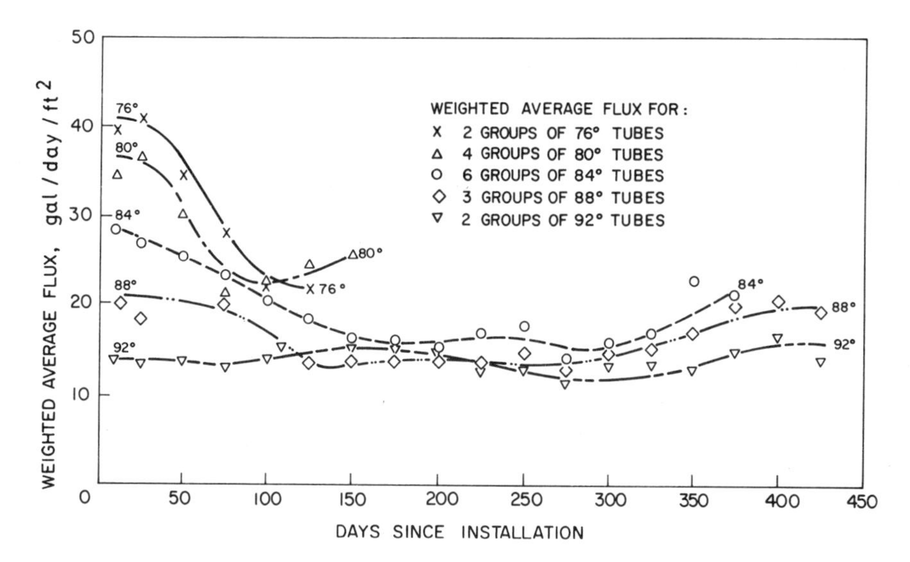

Figure 8-10. Membrane performance in Coalinga pilot plant (Loeb and Selover, 1967).

Details of the fabrication of the tubular membrane assembly are given in Chapter 2 (Figure 2-22). Each assembly is 10 ft long and nominally 1 inch in diameter, and the effective membrane area is 2.24 ft^2. The assembly consists of a tubular membrane around which are several porous wraps to provide a low-resistance path to the $1/16$ inch perforations in the copper support tube, which assumes the expansive stresses due to internal pressure everywhere except where the tube is perforated. In this region the porous wraps support the membrane and prevent it from extruding through the perforations. It is hence important to have adequate strength in the porous wraps. Most of the assemblies used in the early test period contained membrane tubes which were wrapped with one layer of smooth filter paper around which were $1\frac{1}{2}$ layers of nylon fabric (French Fabric Co. No. 627T), the latter to provide a low resistance path and strength. However, some of the later membrane tubes were wrapped only with nylon fabric, under which conditions fabrication was simpler and more strength was provided since three layers of nylon could then be included.

While in full operation, the plant contains 112 such tube assemblies spread in 4 tiers of 28 tubes in each. The tubes are connected in series through 1 inch U-bends. The valving permits either the upper or the lower two tiers to be operated independently so that half of the plant can be kept in operation during the 10- to 15-minute period when the other half is down for a tube assembly change. Simplified tube assembly replacement and continuity of plant operation during such changes are important features of the Coalinga plant.

There is a collection trough for each tier of 28 tubes, and each individual tube is enclosed in a plastic sleeve, with all production passing through a small side outlet to a representative station in the collection trough for that tier. By this means, production from each tube can be monitored separately. This relatively elaborate arrangement of troughs and plastic sleeves would be unnecessary for a plant designed for production only, but the arrangement is useful for experimental purposes, since it permits simultaneous performance testing of tubes fabricated in a number of different ways.

A small tank is used for collecting and monitoring salt content in the desalinised water from each of the four collection troughs. The desalinised water flows by gravity from the monitoring tank to a holding tank, from which it is periodically pumped into the Coalinga potable water mains through an integrating water meter.

The system pressure is controlled by setting the back-pressure regulator on the concentrated brine stream. A manual bypass valve is used during start-up and shut-down to relieve pressure without changing the setting on the back pressure regulator.

Overall performance of the Coalinga pilot plant. From 17 June 1965, when the desalinised water was accepted for use by the Fresno County authorities, until 30 June 1966, a total of 1, 744, 000 gallons of product water was pumped into the Coalinga potable water mains. This represents an average rate of 4630 gallons per day based on total time elapsed. The dissolved solids content of the desalinised water varied from 110 to 480 ppm. Table 8-13 shows a detailed analysis of the feed brine, the concentrated brine, and the desalinised water on 17 August 1965. The average production rate of the membrane was about 15 to 20 gal/day/ft^2 (Stevens and Loeb, 1967).

The plant was onstream 98.5 per cent of the total elapsed time from start-up to 30 June 1966. Of the 143 hours downtime, 104 hours were used for triplex pump maintenance (i.e. replacement of pistons and packing), 22½ hours for flushing the tubes with brine, ½ hour for replacing the back pressure regulator, 1 hour for power failure, and 15 hours for tube failures and changes.

Details of some problems and solutions relating to the operation and performance of the Coalinga pilot plant are given by Loeb and Selover (1967). Two steep declines in desalinised water production rate occurred in 1965 due primarily to membrane fouling. Fouling was caused primarily by deposition of a ferric hydroxide slime, probably formed in situ on the surface of the membranes by the reaction of ferrous ion and dissolved oxygen, and through the agency of iron bacteria. Chlorination minimised the decline but appeared to cause failures of the membrane and/or its nylon support wraps (Loeb and Johnson, 1967).

Production decline due to fouling could be controlled without membrane failures by feed pretreatment with catalysed sodium sulphite to eliminate the dissolved

Table 8-13. Coalinga pilot plant analyses, 17 August 1965

Data of Loeb and Johnson (1967)

	Feed brine	Concentrated brine	Desalinised water
pH	7.7	8.0	7.8
Iron (ppm)	0.018	0.06	0.012
Hydroxide (ppm)	nil	nil	nil
Boron (ppm)	2.75	3.50	2.25
Carbonates (CO_3) (ppm)	nil	nil	nil
Bicarbonates (HCO_3) (ppm)	161.7	375.2	24.4
Chlorides (Cl) (ppm)	262.4	517.7	74.5
Sulphates (SO_4) (ppm)	1260.5	2942.9	34.2
Phosphates (PO_4) (ppm)	nil	nil	nil
Silica (SiO_2) (ppm)	49.2	92.8	8.4
Iron and Alumina (R_2O_3) (ppm)	1.2	2.8	0.8
Calcium (Ca) (ppm)	128.7	299.4	6.9
Magnesium (Mg) (ppm)	89.0	199.1	1.0
Sodium (Na) (ppm)	521.0	1168.0	66.0
Total dissolved solids (ppm)	2477.6	5592.8	220.8
Total hardness (as $CaCO_3$) (ppm)	692.4	1577.7	25.2
Total incrustating solids (ppm)	742.8	1673.3	34.4
Free carbon dioxide (ppm)	5.0	6.0	1.0

oxygen (Loeb and Johnson, 1967). This technique was used effectively from November 1965 to July 1966 during which period membranes stayed clean whether or not the feed was initially filtered.

Since July 1966, membranes have been scrubbed in-place every 3 to 4 days by passage under hydraulic pressure of a polyurethane ball of about 2 in. uncompressed diameter, through the array. This entire operation takes about 15 minutes including the time for ball insertion and for ball and slime removal. This mechanical method adequately prevents production decline and requires no continuous expenditures for chemicals as did the sulphite dosing method.

More recent data on the performance of the Coalinga plant are given in Tables 8-14, 8-15, and 8-16 (McCutchan, 1968). The data show that the plant has been expanded, and its membrane section now contains 172 tubes, and the technique of ball flushing once in every three days is sufficiently effective, and the plant as a whole has maintained an outstanding performance record.

7. Reverse Osmosis Units of Havens Industries

Havens Industries, of San Diego, California, has pioneered the successful commercial development of reverse osmosis units using the Loeb-Sourirajan type

Table 8-14. Changes in the number of tubular assemblies in Coalinga pilot plant

Data of McCutchan (1968)

Date	Size of membrane unit	Effective membrane area (ft^2)
4 June 1965	56-1" tubes	125
11 Aug 1965	112-1" tubes	250
2 Oct 1966	134-1" tubes	300
12 Oct 1966	130-1" tubes + 4-2" tubes	309
16 Mar 1967	172-1" tubes + 4-2" tubes	403

Table 8-15. Antifouling methods used in Coalinga pilot plant

Data of McCutchan (1968)

Method*	Duration of test	Results
Chlorine (continuous dosing)	72 days	Flux decline somewhat reduced
Copper sulphate (continuous dosing)	14 days	No effect
Catalysed sodium sulphite (continuous dosing)	250 days	Flux decline noticeably lessened
Citric acid (once-through flushes)	5 times, ½ hr, each time	No effect
Rapid brine flushes	3 times, ½ hr, each time	No effect
Ball flushes (once-through every 3 days)	160 days	Same as Na_2SO_3 flushing
Ball flushes (once-through every 6 days)	85 days	Poor

* Prefiltration has always been used with filters, pore size 1μ to 75μ, normally 5μ.

porous cellulose acetate membranes integrally lined in porous fibreglass tubes. Several Havens units of over 1000 gal/day capacity are currently in operation in industrial and government establishments in different parts of the world. The applications are generally in the fields of water conservation, water purification, water pollution control, food concentration, and chemical processing.

In the Havens units, the feed water is pumped at high pressure through the fibreglass tubes which are grouped in series inside plastic pipes by means of

Table 8-16. Performance of Coalinga pilot plant
Data of McCutchan (1968)

Performance for first two years, 4 June 1965 to 4 June 1967	
Total production	3,863,000 gal.
Average product flux	20.0 gal/day/ft^2
Average total dissolved solids of feed	2450 ppm
Average total dissolved solids of product	226 ppm
Average solute separation	93.5%
Average desalination ratio	15.4
Plant availability	98.7%

end fittings. These groupings are called modules. The tubes, each of which has an integrally formed membrane layer inside, are imbedded in an inboard fitting at each end of the module. A shroud surrounds the tubes and collects the membrane permeated product water dripping through the porous walls of the tubes. The tubes are connected in series with U-turn connections at each end. Each end of the module has an outboard hexagonal fitting that supports the U-turn fittings, and provides a convenient method of connecting the individual modules together, and through which the product water is withdrawn (Figure 8-11).

Porous fibreglass tube development. The Havens reverse osmosis units utilise filament-wound porous fibreglass tubes of ½-inch inside diameter, in each of which the membrane forms an integral part of the tube by actual bonding to the internal surface of the tube. In this arrangement the membrane support tube is also the pressure vessel. It is flexible in designing systems around varying tubular configurations. High feed velocities can be maintained without excessive pressure drops. The membrane can be cleaned of precipitates, flocks, or organic deposits, through use of standard tube cleaning methods such as the use of polyurethane foam plugs. The manufacture and membrane coating of the tubes have been automated to a high degree even though all details of automation are not available. A major portion of the Havens development work has been on the high speed manufacture of these porous ½-inch diameter tubes with a wall thickness of 0.035 inch, withstanding 5000 psi static burst test. These tubes are produced at a cost of about five cents per foot (Havens and Guy, 1967). The porosity requires careful control in manufacture, some details of which are given below.

Fibreglass tube manufacture and assembly. Plate 8-2 illustrates one of Havens' continuous tube-making machines in operation. On the left side of the picture, the fibreglass longitudinal rovings are pulled off the supply spools into a positioning adhesive resin feeder. The longitudinals are then positioned around the periphery of the ½-inch diameter mandrel. The first of the counter-rotating wheel spiral wraps the inner circumferential layer. The second wheel applies the outer spiral wrap, followed by a heat curing cycle. The tube is then cooled sufficiently to hold its shape through the continuous puller. The tubes are then cut to length and stacked.

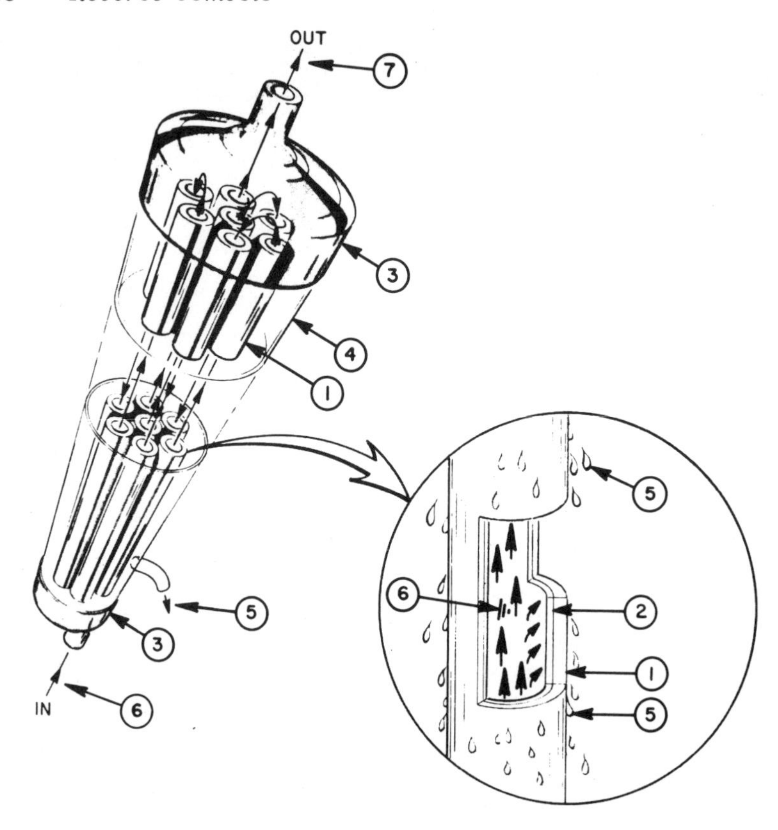

Figure 8-11. Schematic representation of Havens' module (Guy, 1968).

(1) fibreglass tube; (2) osmotic membrane; (3) end fitting; (4) PVC shroud to collect product water; (5) product water; (6) feed solution; (7) effluent.

Under the machine bed are the synchronised drive motors and feed back generators. The binary speed control computer is shown in front of the machine. The resin and heat control panel is on the front lip of the machine bed. Mounted on the top of the frame at the left is the mandrel load sensing strain gauge, and at the right, the infrared sensing camera. The mandrel design and resin control are most important in obtaining the smooth, glasslike surface needed to support the membrane.

After the tubing is processed off the winding machine, the membrane is cast on the tube interior and tested (the details of this casting procedure are not available). Then the multiple tube module assembly is made. The individual tubes, each about 8 ft long, are cemented into the inboard end fittings, and the sub-assembly is coated with the membrane. The turnarounds are then installed and the outboard fittings attached. After testing, the protective collector shroud is assembled. In some models, seven tubes are used in series giving fifty feet of tubing per module.

More recently the 'floating compact' eighteen-tube module has been developed. The term 'floating' denotes that the tubes are not cemented in place, as are those in the seven tube module. The term 'compact' signifies that the tubes are geometrically packed in the least area. In this module, the tube connectors are supported on the outside by the shroud flange and on the end by the module end fitting. The end fittings are then tied together with a longitudinal load carrying bar. The flow reversing and sealing tube connectors are structural members inserted into the tube ends. Feed solution flows from the module to the other through the moulded bosses on the end fitting. Product water is collected from the same bosses at the opposite end of the module, which are connected to the shroud area only.

In the 'floating compact' module, the 18 tubes are first assembled with the connectors in a flat pattern. The feed inlet is in the first tube, and the outlet is in the eighteenth tube. The tube set is pretested in the flat layout for leaks and tube performance. It is then rolled around the tie bar and inserted in the shroud. End caps are then assembled and the tie bar secured.

The module configuration has a maintenance advantage in that the tube set can be removed and any or all tubes cleaned or replaced. The tie bar carries longitudinal loads rather than the tubes. It also has a manufacturing cost advantage in excess of 50 per cent for large-scale manufacture.

All modules are tested for sealing leaks and faulty connections before they are racked in the system configuration required for the particular application.

General Performance of Membranes in Havens Units. The differences in the porous structure of the cellulose acetate membranes used in the Havens units are not adequately specified in terms of their reverse osmosis transport characteristics. For commercial purposes however, they are identified by numbers such as 2, 3, 3A etc. in terms of their performance in specific applications as illustrated in Table 8-17. Some additional data on the performance of their 3A, 4A, 5A, and 6A membranes for feed solutions such as brackish waters, sea water, sugar solutions, orange juice, whey solutions, and paper mill waste liquors are illustrated in Figure 8-12 where P/P_W is the ratio of product rate and pure water permeability through the film at constant temperature. An increase of 1° in the operating temperature was found to increase the pure water permeability of their films by about 2.8 per cent indicating the effect of the change in viscosity of water.

Havens also tested their membranes on sea water for 11 months, and on brackish water (pH = 7.8) for 18 months, and found no deterioration in performance.

Flow diagram. The flow diagram of a typical Havens two-stage 5000 gal/day sea water conversion reverse osmosis unit is shown in Figure 8-13. The unit consists of a heavy duty skid upon which are mounted two Moyno pumps, two motors, a control panel, a storage tank, two electrical blow down valves, three chemical injector pumps and associated components. The module package consists of 143 modules Type 5 (part number 10-101), and 66 modules Type 4 (part number 10-102), which are secured in a support frame. The Type 5 modules are used in the first stage and the Type 4 modules are used in the second stage. Each individual stage is connected together with its own separate associated plumbing. The product from the first stage is the feed for the second stage. Mounted on one end of the module package are two horizontal PVC pipes connected to the individual modules by means of 1/8-inch polyurethane tubings which collect the product water; each collector has its own sampling valve for testing the first and second stage product waters. A skid housing a product water tank and pump forms part of the unit. It has an automatic level device which will automatically start and stop the pump as set. The pump can also be operated manually.

Table 8-17. Havens Industries—identification and performance of membranes

Data of Guy (1968)

Membrane type	Part number	Remarks	Operating pressure (psig)	Product rate (gal/day/ft^2)	Rejection ratio
2	—	Will pass sugar, salts, and water while holding back pulp and large organic molecules (used for the concentration of algae, protein, etc.)	150	40-100	1
3	10-103	Will pass water and some monovalent salts, but will not pass bivalent salts and sugar	800	20	4
4	10-102	Used for brackish water conversion	800	13	15
5	10-101	Used for sea water conversion and food concentration	800 1100	8 9.5	25
6	10-107	Used for fruit juice concentration	800 1100	6 7.3	>40

Concentration of NaCl in feed water = 5000 ppm

$$\text{Rejection ratio} = \frac{\text{NaCl in feed}}{\text{NaCl in product}}$$

A Typical Havens Unit. Plate 8-3 shows a typical two-stage 6500 gal/day sea water conversion unit currently in commercial production at Havens. This unit uses 7-tube modules, and it can be converted to a 20,000 gal/day brackish water conversion unit. Total recovery of 38 per cent is accomplished on sea water, and 85 per cent on brackish water (5000 ppm NaCl). The first stage operates at 1100 psig, and the second stage at 800 psig. Operated on sea water, the second stage will produce from 100 to 200 ppm water depending upon the operating pressure and the corresponding water recovery rate. In one particular plant using sea water feed the operating pressures were 800 psig for the first stage, and 500 psig for the second stage; and the product rates were respectively 4 and 7 gal/day/ft^2 after six months of continuous operation; the product waters after the first and second stage purifications contained 1720 and 216 ppm of total dissolved solids respectively.

In the Havens' configuration, a closely packed stack of tubes (33 × 33 tubes approximately 20 × 16 in^2 in cross sectional area, 7½ ft in length) will produce an output in excess of 15,000 gal/day on a brackish water or 5000

Plate 8–1. A bench scale reverse osmosis unit (Peterson, 1968).

Plate 8–2. Havens' continuous tube making machine in operation (Havens and Guy, 1967).

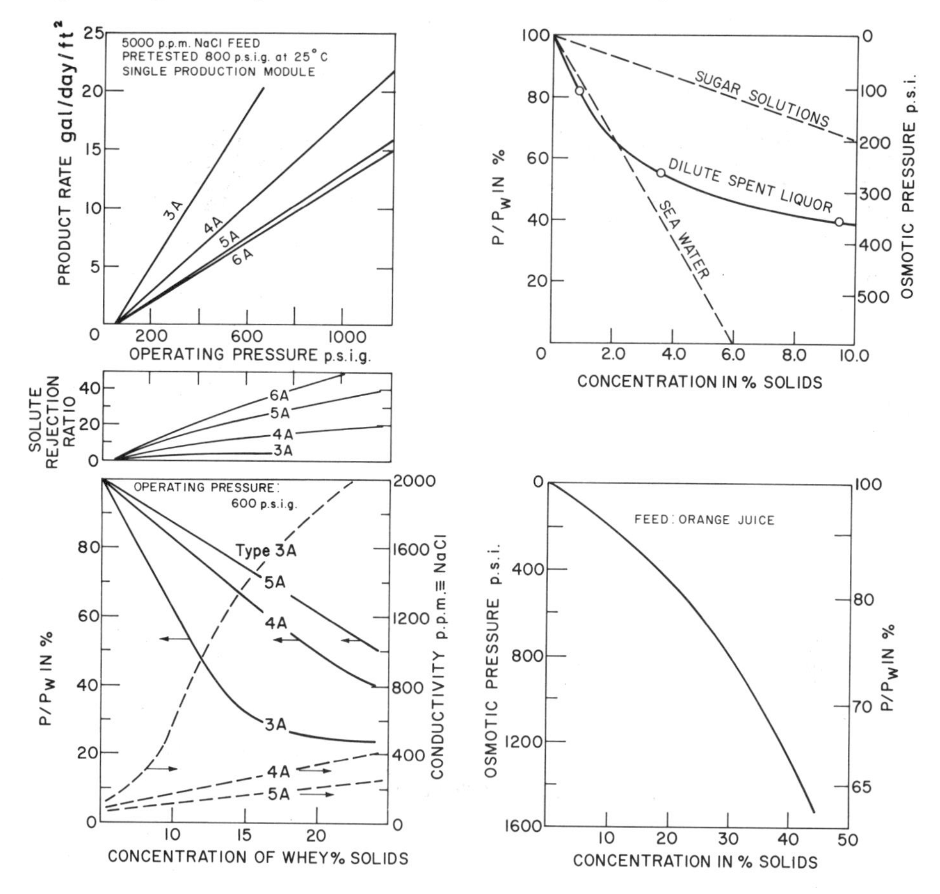

Figure 8-12. Performance of Havens' tubular membranes (Havens and Guy, 1968).

gal/day on a sea water feed solution utilising two stages or sets of modules. By multiplying the dimensions of this conceptual stack by 100, the stack will now produce 1.5 million gal/day on brackish, or 0.5 million gal/day on sea water. A desalting plant of this size would occupy 1700 ft^3 of space; the entire plant would occupy about twice this space to include the power plant, plumbing, work, and aisle areas.

Current problems and developments. Havens reports (Guy, 1968) that their worst current problem is in the reliability of auxiliary equipment, and particularly high-pressure pumps. They also have some way to go in obtaining the required quality control in their membrane-casting process. They are currently extending their plastics technology into the auxiliary equipment area, and are developing plastic pressure pumps, valves, accumulators, and back pressure regulators. They estimate that a million gal/day brackish water reverse osmosis unit with the 18-tube modules would require 53, 000 pounds of plastic materials.

Reverse Osmosis Engineering at Havens. Reverse osmosis units of sizes ranging from 5000 to 100, 000 gallons of product water per day made by Havens Industries are already in the market. They may soon start constructing

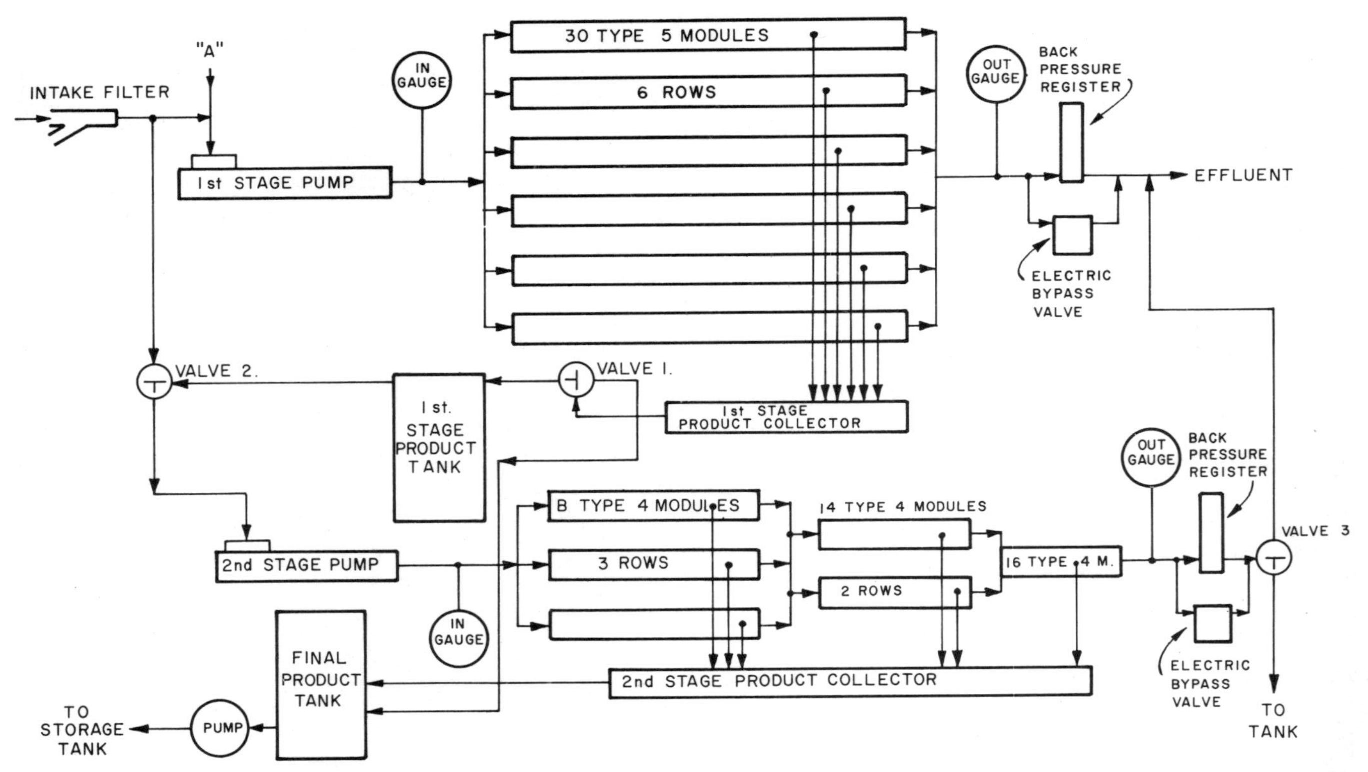

Figure 8-13. Flow diagram of a typical Havens' two-stage 5000 gal/day sea water conversion unit (Guy, 1968).

one-plus million gal/day plants for a variety of reverse osmosis applications. Recently (February 1968), they submitted a proposal to the U.S. Department of the Interior, for building a three million gal/day reverse osmosis plant (for Foss reservoir) for converting brackish water (containing about 1800 ppm total dissolved solids) to drinkable water (containing 200 ppm total dissolved solids, and 110 ppm total hardness). Even though this plant has not yet been built, the proposal itself is an indication of the progress of reverse osmosis engineering at Havens.

The main features of the proposed three million gal/day Havens 'OSMOTIK' brackish water conversion plant for Foss Reservoir are as follows (Guy, 1968).

Basic assumptions include (a) chemical costs* as of 1 February 1968; (b) building costs at \$10.00 per ft^2; (c) Foss Reservoir water available at 2 lb/in^2; (d) product water delivered at 50 lb/in^2; (e) 330 equivalent days annual plant operation; (f) electrical energy available at 7 mills per kWh; (g) 7 per cent annual amortisation; (h) 2.75 per cent annual interest charge; (i) brine disposal at \$0.05 per 1000 gallon concentrate; (j) membrane replacement at \$0.50 per ft^2; (k) membrane life, two years; and (l) sufficient power available at plant site.

Basic design parameters include (a) 100 per cent back-up capability for all pumps, drives, etc., i.e. failure of any one unit will not cause a decrease in plant capacity; (b) module piping designed so that failure of any one module will cause shutdown of no more than 1.25 per cent of total plant; (c) equipment included to adjust pH of feed water to 5.5 by H_2SO_4 injection in order to extend membrane life; (d) no additional chemical treatment or pretreatment, and (e) instrumentation permitting one-man operation.

Desalting equipments include (a) 11,545 Series 10-400, 18-tube OSMOTIK modules vertically mounted; (b) four Byron Jackson 4 × 6 × 9 D—DVM stainless steel 900 gal/min, 800 lb/in^2 main feed pumps (three operating and one spare), each with 700 h.p. motor and starter box; (c) piping and valving to ensure no more than $2\frac{1}{2}$ per cent shutdown of total system due to module failure; (d) four back-pressure valves programmed so that any three can handle total effluent; (e) four three-inch electric blowdown valves, three operating with electric switch to operate fourth valve in case of failure; (f) three chemical injection systems with feed rate of 0.15 gal/min sulphuric acid; (g) continuous-reading pH indicator; (h) four flow indicators; (i) four pneumatic controls for back-pressure valves; (j) two remote-reading conductivity meters with 200 reading points, Roto switches and high-conductivity shutdown; (k) six pressure gauges; (l) feed inlet temperature indicator; (m) 7400 gallons capacity chemical storage drum; and (n) accessory electrical equipment including wiring parts and labour.

Water treatment equipments include (a) four 24 ft × 24 ft × 9 ft sumps for feed, brine, and product water; (b) four sand filters with Byron Jackson 900 gal/min Sump Master Pump and 60 h.p. motor (three operating, one spare); and (c) chlorinator assembly.

The capital cost of the above plant is estimated as 48 cents per gallon-day, and the operating and maintenance costs are estimated as 20 cents per 1000 gallons of product water.

Performance of a Havens field test unit. A large mobile waste reclamation unit has been in operation in Appleton, Wisconsin, since October 1968. Presently (January 1969) it is recovering 90 per cent or more of the water content

---

* All cost figures in this book are in U.S. currency.

from paper pulp sulphite liquor, fed at a rate of 55, 000 gal/day. The system can be expanded to over 100, 000 gal/day intake merely by adding modules.

The system, wholly contained within a standard 40-ft trailer van, comprises 387 active 18-tube modules (6700 ft^2 of membrane area), arranged in 5 banks. A large triplex reciprocating pump, driven by a variable speed 50 h.p. motor, feeds the first bank, after which three centrifugal pumps can be arranged to boost pressure into following banks and/or recycle the liquor back through previous banks.

Starting with high molecular weight dissolved solids in the range ½ to 2 wt. %, the system concentrates the liquor to 10 to 20 wt. % solids. The normal operating pressure is 600 psig, although the system can be operated at pressures up to 1000 psig. The average permeate rate is about 8 gal/day/ft^2 on 1 per cent solids feed and 90 per cent water recovery (Leonard, 1969).

8. Reverse Osmosis Units of Universal Water Corporation

Universal Water Corporation, of Del Mar, California, also has developed its commercial systems for desalination of brackish and sea water using a tubular membrane configuration (Manjikian, 1968). Figure 8-14 presents assembly details of a single tubular unit. Modules are used which are made up of 18 high-strength plastic backing tubes each four feet long terminating in unitary end flanges and sealed end caps. Internal channels in the end caps provide connections for backing tube ends, one to another and to input and output piping. Each backing tube is lined with some porous material such as nylon cloth inside which is the tubular cellulose acetate membrane. The porous material

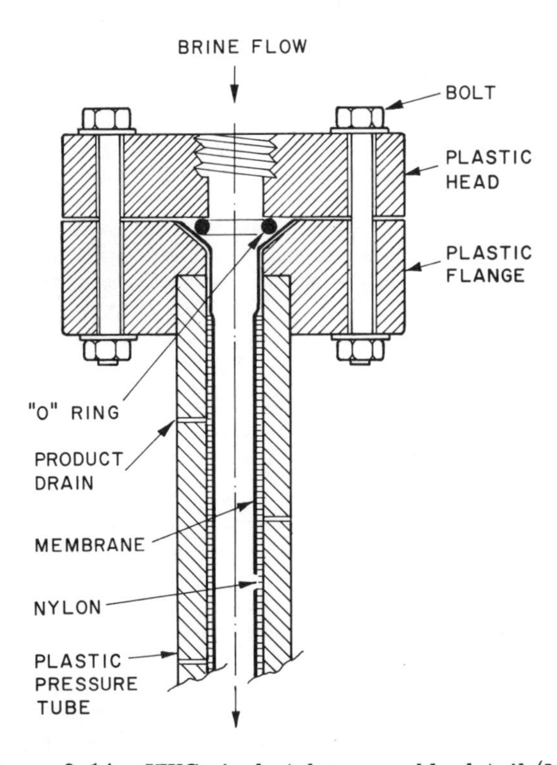

Figure 8-14. UWC single tube assembly detail (Manjikian, 1966).

provides lateral product water transfer to perforations drilled into the backing tube walls.

The tubular membranes are about 0.4 inch in diameter. Feed solution is circulated at Reynolds numbers of 4000 to 12,000.

Backing tubes are extruded of alkyl benzene sulphonate plastic of ample strength for use in brackish water systems operating at about 600 psig. The same tubes have been used successfully in sea water conversion units operating at 1250 psig. The end flanges and end caps are of filled epoxy composition combining strength, castability, and durability. Fastenings and small hardware are of 316 stainless steel for resistance to corrosion. Commercial use of alkyl benzene sulphonate backing tubes and epoxy cast ends has demonstrated this to be sound engineering practice. These plastic materials have adequate strength, good corrosion resistance, and are readily and inexpensively fabricated and assembled.

Modules are arranged in series, or series-parallel flow patterns according to unit requirements. In smaller size systems three modules in series will purify 500 gallons of brackish water per day. Larger units containing as many as 80 modules in multiple module banks have been produced.

The complete desalination unit includes conventionally a high-pressure pump with its accessories, and a regulator valve for the brine outflow. Filters, acid dosing for pH control, product water catch tanks or delivery pumps are some of the other common accessories. A service life of about one year is given for membranes used for brackish water purification under favorable conditions. Plate 8-4 is a photograph of a Universal Water Corporation reverse osmosis unit currently in the market.

9. Reverse Osmosis Units of American Radiator and Standard Sanitary Corporation

The design of the American Radiator and Standard Sanitary Corporation, of New Brunswick, New Jersey, utilises a $\frac{1}{2}$-inch porous fibreglass tube as a pressure vessel. In this design the membranes are cast as flat sheet and then placed within the tube. A small overlap of membrane provides the necessary seal within the tube. This arrangement reduces circumferential stress on the membranes and facilitates easy removal of the membranes. This procedure differs from other tubular designs in which the membranes are either cast directly on the inner walls of the tubes or prepared as separate cylinders and inserted into the pressure tubes (OSW, 1966).

10. Reverse Osmosis Units of Aerojet-General Corporation

Aerojet-General Corporation, of Azusa, California, has successfully built and operated several 1000 gal/day reverse osmosis units based on the flat membrane configuration (Aerojet, 1964, 1966). A photograph and process flow sheet of one of their units are shown in Plate 8-5 and Figure 8-15 respectively. This unit was trailer-mounted and consisted of 45 membrane support plates using 16-inch diameter membranes. The effective diameter was $14\frac{3}{8}$ inches, resulting in 1.12 ft^2 of effective area for each membrane. Hence, 45 plates with a membrane on either side provided 90 membranes with a total area of 100 sq. ft. The plates were grouped in three sections (Figure 8-16) so that product water could be collected separately from each section, and individual flow and other measurements could be made. The porous cellulose acetate membranes were bonded to the support plates with a strip of adhesive (93 per cent Pliobond, 5 per cent cellulose acetate, and 2 per cent acetone) $\frac{1}{8}$ inch wide at the peri-

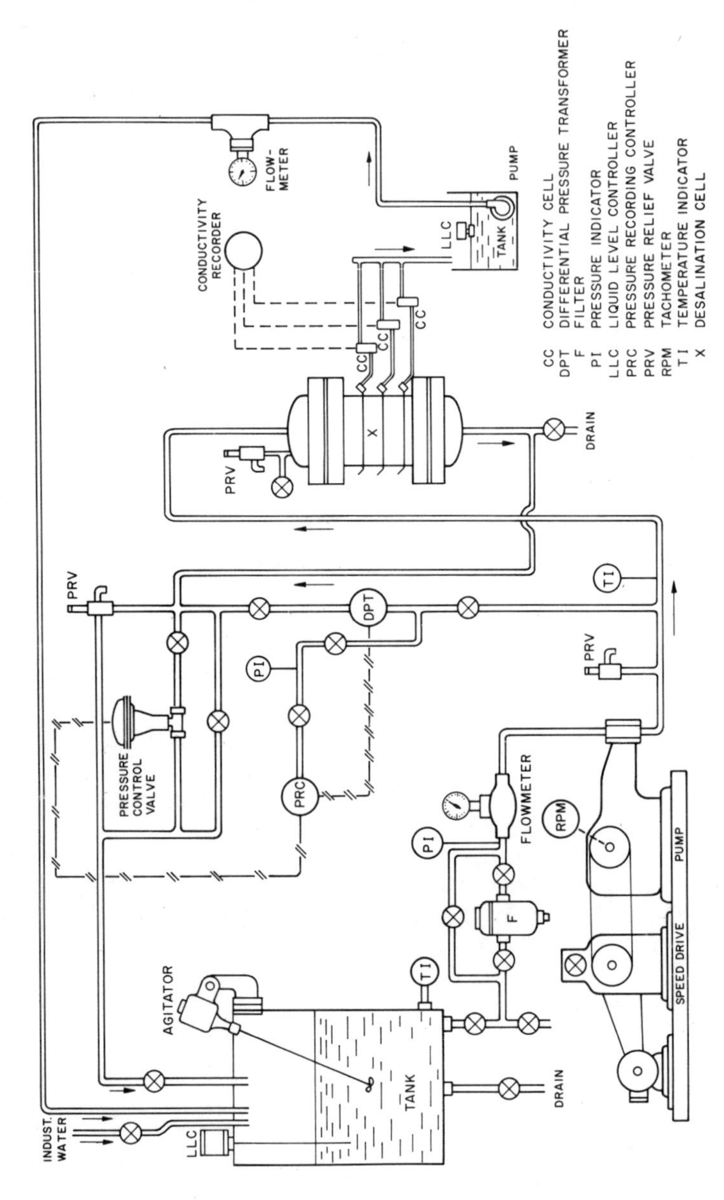

Figure 8-15. Flow diagram of Aerojet reverse osmosis unit (Aerojet, 1964).

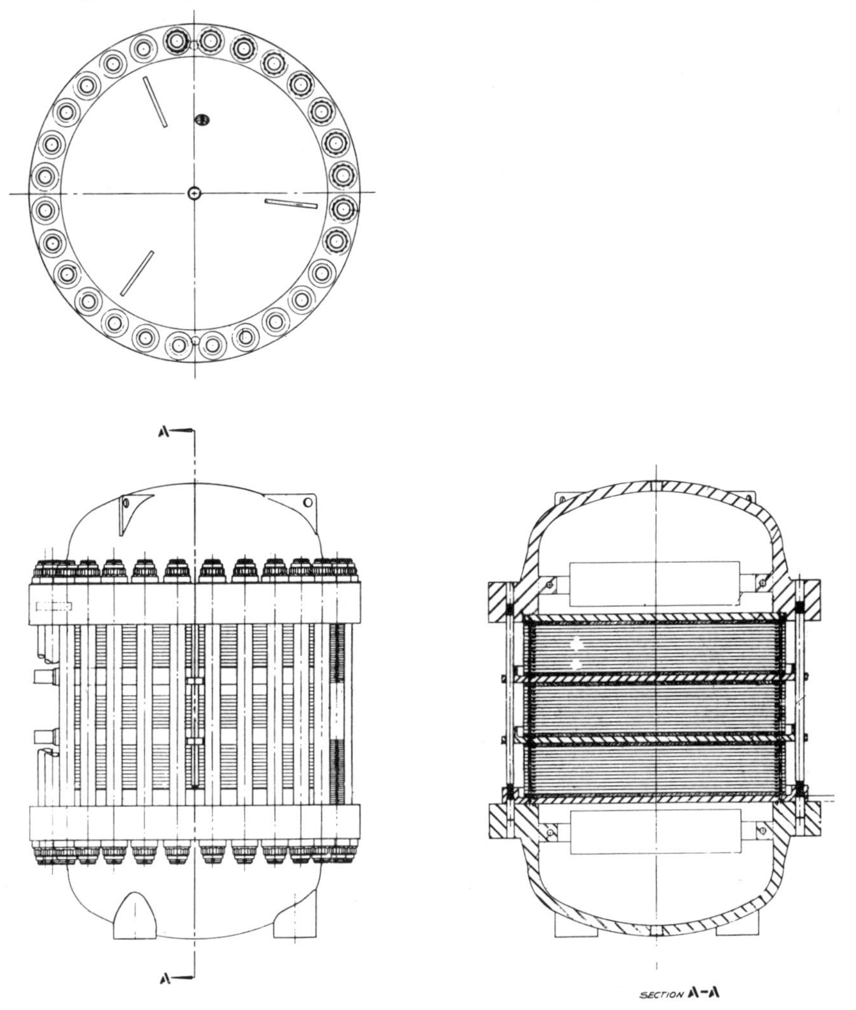

Figure 8-16. Membrane unit in Aerojet reverse osmosis unit (Aerojet, 1966).

phery of the membrane. Teflon tape, 1 inch wide, was used around the periphery of the membrane seal as reinforcement.

The unit was of tie-bolt construction with carbon steel vessel heads and was designed for pressures up to 1500 psig to permit operation on either sea or brackish waters. The nominal design capacity of 1000 gal/day with 100 ft^2 of membrane area assumed an average flux of 10 gal/day/ft^2, which in actual operation was expected to be higher or lower, depending on conditions. An 8 gal/min piston type pump was provided to pressurise the feed. This pump capacity was chosen to provide a brine channel flow of 1 ft/sec parallel to the membrane surface. The unit was provided with a 10 micron filter on the feed water line.

The seal between the plates was achieved with an O-ring. The saline feed flowed back and forth across the plates parallel to the membrane surfaces until it completed its path through the unit. Ports for transfer of the flow from plate to plate were provided near the periphery of the plates. The product was collected on the low-pressure side of the membrane flowing through the porous support plate to its periphery where holes were provided to manifold and collect the product. In actual assembly a layer of filter paper was placed between the membrane and the support plate.

The material chosen for the first support plates was sintered bronze. In view of their high cost, substitute materials were investigated, which resulted first in the development of a glass-fibre-epoxy-bonded porous support plate having both superior flow characteristics and lower cost. This was followed by a still lower-cost grooved phenolic plate having excellent hydraulic characteristics. This phenolic plate was made of solid material, instead of the previous porous materials. Test operations showed that the solid plate gave results as good as or better than those obtained with the porous epoxy plate. Test plates of the grooved design were made using various groove depths and groove spacing. Selection of groove width was guided in part by previous experience indicating that a membrane supported by filter paper will bridge a $1/32$-inch hole at 1500 psig. Dimensions found suitable in a test plate were grooves 0.010 in. wide and 0.030 in. deep, spaced 0.1 in. apart.

It was recognised that the grooved plate would present a variety of moulding and fabricating problems. Further development work produced the slotted phenolic plate as an alternative to the grooved plate. This plate could be moulded readily and was also improved hydraulically. In the final design, a $1/8$-inch thick plate was made up of two identical halves, each $1/16$-inch thick. Each plate contained /‾\-shaped parallel grooves or slots which were 10 mils wide at the top and 45 mils wide at the bottom. In assembling the plates, the two halves were joined with the wide grooves together, rotated so that the grooves in the two plates were at right angles, and permanently bonded together. The product water flowed through the interior grooves to the nearest product-water outlet.

Field experience at Aerojet, with the slotted phenolic plate in the 1000 gal/day units, has demonstrated the merits of the design and its serviceability. Along with the plate development, baffles were also designed for the 1000 gal/day units that had initially operated without baffles in the feed water channels. These baffles, shown in Figure 8-17, were made of vacuum-formed polystyrene sheet. The stream is split into two halves, above and below the baffles, and it is further divided into a right and left quarter. A flow path of from 30 to 40 feet was provided. Feed stream velocities were increased by a factor of about six over that of the unbaffled design.

Test Operations. Pilot plant test operations were conducted both with brackish and sea waters. The first brackish water tests were made at a well at the Irvine Ranch (Santa Ana, California), and later at Laguna Beach, California. Figure 8-18 shows the flux rates and product salt concentration for the 106-day period continuous test run at Irvine Ranch. The flux remained stable at 10 gal/day/ft^2 during the last 10 weeks of operation and the salt content of the product was nearly constant at 250 ppm. The unit was operated at 750 psig, and the only pretreatment received by the well water was filtration through a cartridge-type filter. A typical analysis of Irvine Ranch feed and product waters showed that the total dissolved solids were 4900 and 250 ppm, and total hardness 1600 and 65 ppm respectively (Sieveka, 1966).

Operations at Laguna Beach were a continuation of the Irvine Ranch run. A typical analysis of the feed and product streams at the Laguna Beach tests showed that the total dissolved solids were 4075 and 375 ppm, and total hardness 1650 and 80 ppm respectively (Sieveka, 1966). Difficulties were experi-

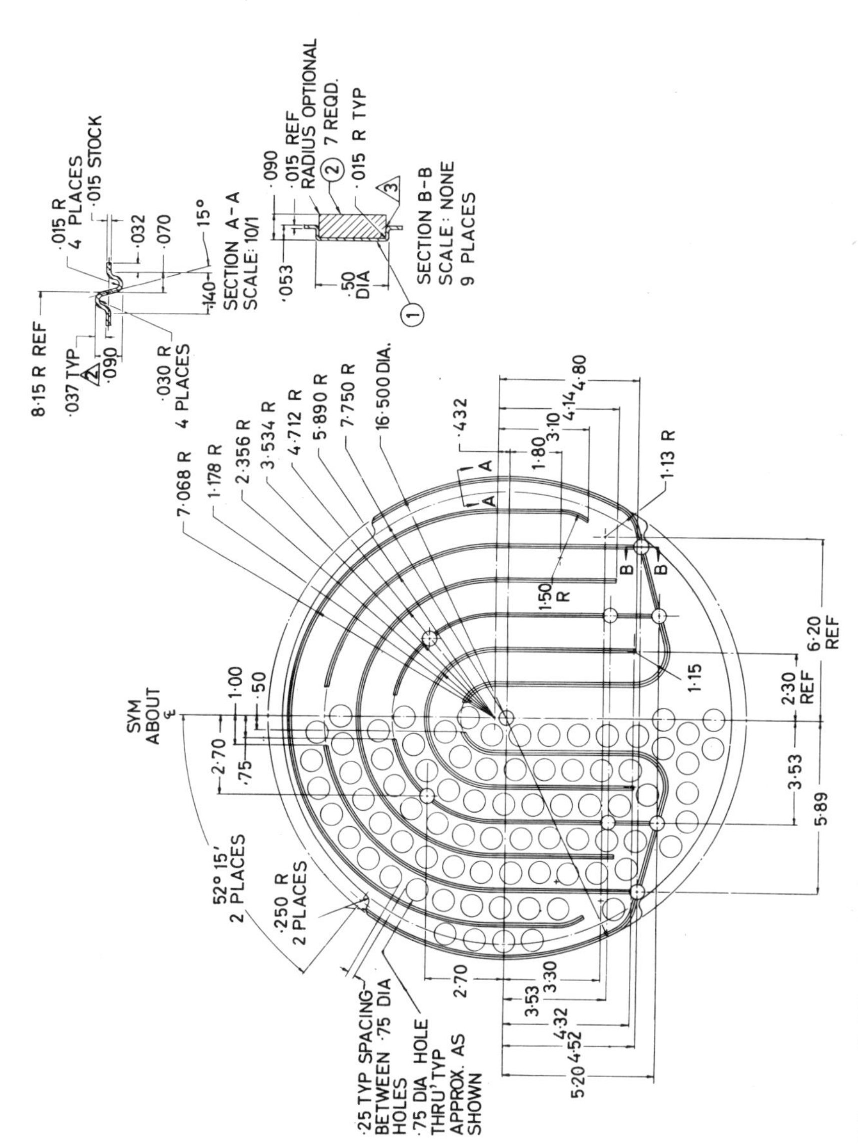

Figure 8-17. Baffle design in Aerojet reverse osmosis unit (Aerojet, 1966).

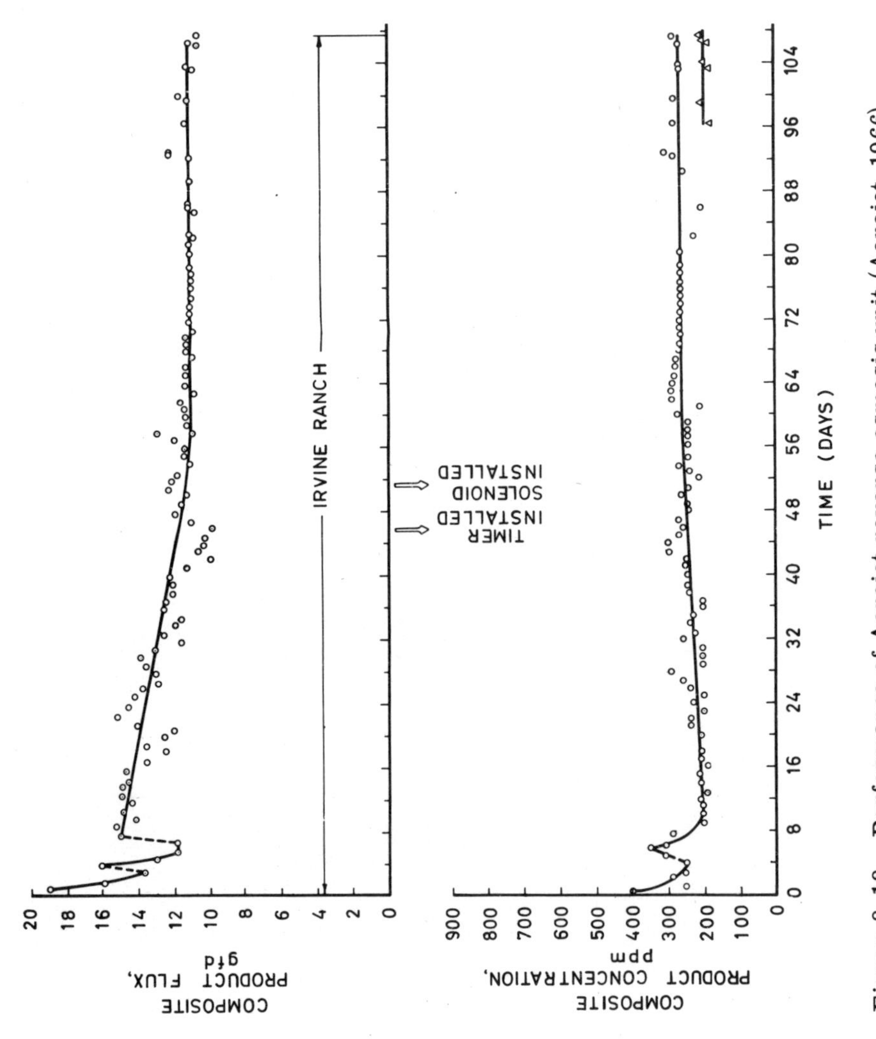

Figure 8-18. Performance of Aerojet reverse osmosis unit (Aerojet, 1966).

enced in the pilot plant operation on the Laguna brackish water due to iron hydroxide deposition on membrane surfaces. The feed water contained 4.7 ppm soluble ferrous iron which was readily oxidised to the insoluble ferric condition. Even these small amounts of iron in this state had a harmful effect on membrane flux. Successful cell operation was finally achieved by proper pretreatment of the feed to remove the iron. Oxidation of the iron was accomplished by thorough aeration of the feed stream. It was then filtered through a diatomaceous earth filter. When the membranes became fouled with iron, their flux rates decreased by as much as 80 per cent of their initial values. Membrane rejuvenation by flushing with water was not successful, but pickling with 2 per cent citric acid or a dilute mineral acid completely restored their performance (OSW, 1966).

The brackish water tests at Laguna produced two major technical advances during 1966. First, a method was demonstrated to prevent solids precipitation within the reverse osmosis cell during high product recovery operations. The Laguna water contained sufficient calcium salts for recovery of 42 per cent of the feed stream as product to saturate the remaining feed. It was shown that the addition of only 0.5 ppm of an organic polyelectrolyte to the feed stream effectively prevented precipitation within the cell while more than 80 per cent of the feed stream was being recovered as the product. This corresponds to a final feed stream concentration of calcium sulphate approximately four times the normal saturation limit. The second advance was the mere fact that the test was carried out continuously for a period of $8\frac{1}{2}$ months. Even after 7 months of continuous operation, the potable water flux was 22 gal/day/ft^2 at 750 psig.

Pilot plant operations on sea water demonstrated that membrane support plates and other components of the unit functioned satisfactorily at the operating pressure of 1500 psig. Likewise, the membranes themselves did not show mechanical or physical failure at the high pressure, nor was any difficulty encountered with the adhesives used to seal the membrane to the phenolic plate. The test results were characterised generally by runs that began with fluxes of about 10 gal/day/ft^2 and product salinities between 500 and 1000 ppm, but after as little as 2 weeks of operation, the fluxes declined to a few gal/day/ft^2 with product salinities of 1000 to 2000 ppm or more.

50,000 gal/day Pilot Plant Design. The United States Office of Saline Water awarded a contract to Aerojet in March 1966 to design and construct a trailer-mounted, brackish water, reverse osmosis pilot plant of the plate-and-frame design that will produce 50,000 gal/day of product water (OSW, 1966). Design criteria for the plant are based on a 5000 ppm (total dissolved solids) brackish feed water. Figure 8-19 shows a schematic layout of the trailer-mounted plant.

The proposed unit will contain approximately 2500 ft^2 of effective membrane area mounted on both sides of membrane support plates. Other details of the proposed design are illustrated in Figures 8-20 and 8-21. Each plate is an open centred disk 36 inches in diameter. Each surface has a network of grooves which are all connected to radial flow channels (Figure 8-20). The flow channels are open to the centre hole of the plate. A group of these plates are stacked together to form a module and are placed on a base plate through which a slotted product water channel projects. A divider plate will be placed on top of this module. Eight additional modules are assembled on the centre shaft forming the complete membrane assembly.

Figure 8-21 shows the complete stack. It has nine passes. The feed water enters through the open-centred top plate, down through the openings of the

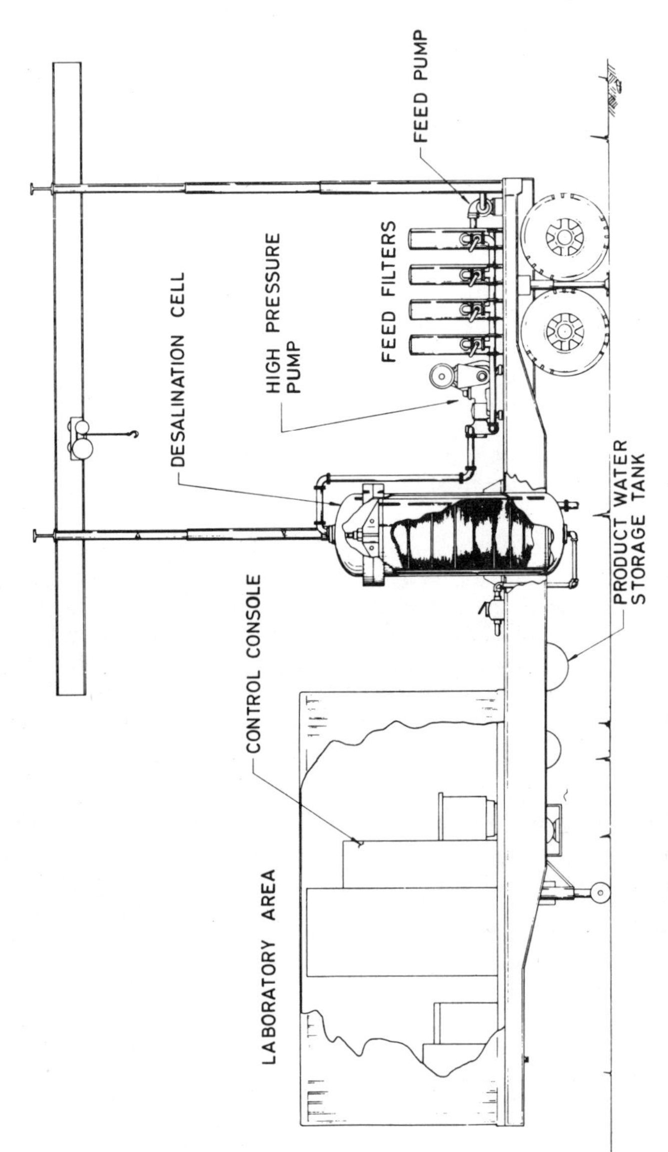

Figure 8-19. Schematic layout of proposed 50,000 gal/day Aerojet reverse osmosis unit (OSW, 1966).

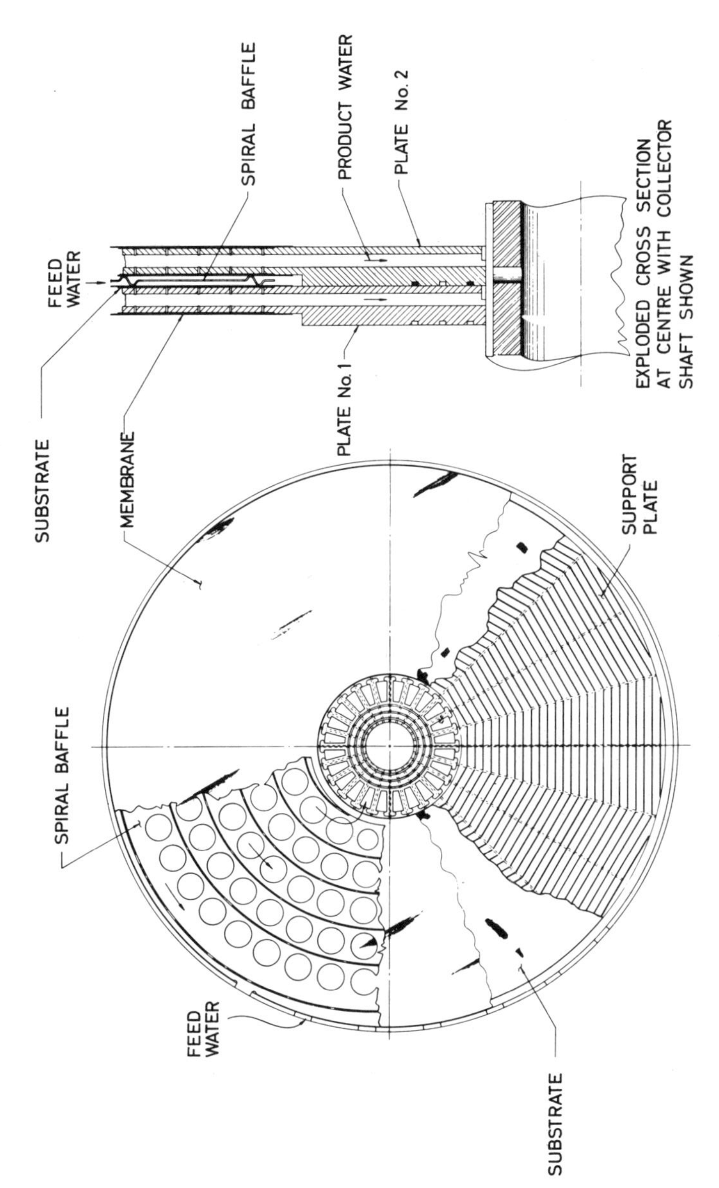

Figure 8-20. Design of membrane support plate for the proposed 50, 000 gal/day Aerojet reverse osmosis unit (OSW, 1966).

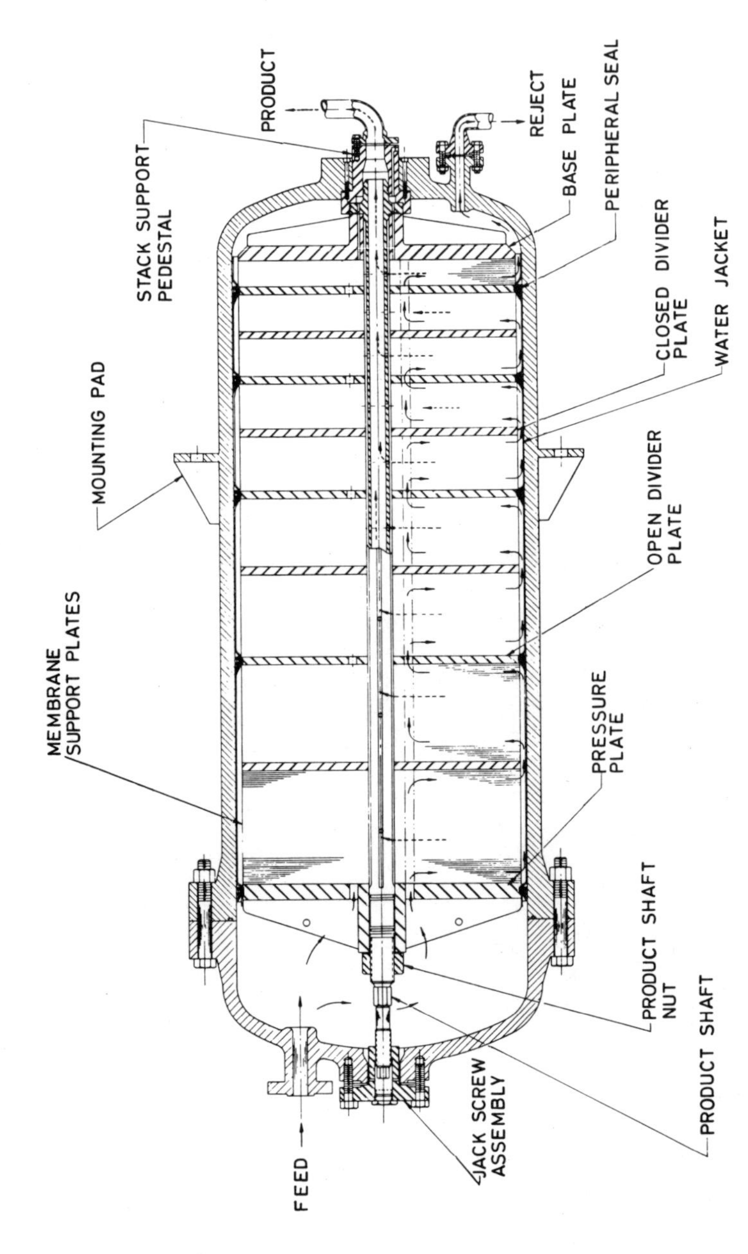

Figure 8-21. Design of desalination cell for the proposed 50, 000 gal/day Aerojet reverse osmosis plant (OSW, 1966).

plates in the first pass, and is stopped by the closed centre plate. The water is forced to flow between the membranes mounted on the plates and is directed to the outer periphery of the plate. At this point, it flows down around the closed centre divider plate and enters the second pass from the outer edges of the plates where it again flows between membranes towards the centre of the plates.

As the feed flows across the membrane under pressure, water passes through the membrane. The product water flows through the grooves of the plates, into the collecting channels, and then into the slotted product channel. Between each of the plates are O-rings which prevent the feed stream from flowing into the product channel.

An operating pressure of 750 psig and a water recovery of 80 per cent are assumed for brackish water operation. Hence the production of 50, 000 gal/day desalinated water would require 62, 500 gal/day of feed, and the power demand would be 27. 5 horsepower. The pressure vessels, pumps, and associated components are designed to operate on sea water as well as brackish feed waters. Hence it is required that the plant is designed for operation at 1500 psig. A 75-horsepower positive displacement quintuplex pump is to be provided to pressurise the feed.

11. Reverse Osmosis Units of Gulf General Atomic

In the reverse osmosis units of Gulf General Atomic Inc., of San Diego, California, the characteristic engineering feature is the design, construction and utilisation of spiral-wound membrane configuration (known as ROGA modules) one of which is illustrated in Figure 8-22. The module consists of one or more leaves wrapped around a product water take-off tube. These leaves consist of (i) the membrane, (ii) a porous product-water-side backing material, and (iii) a brine-side flow spacer. The membrane is bonded along the two sides, at the end, and around the product water tube, forming a sealed envelope that encloses the backing material except at the product water tube open end. The brine-side flow spacer is placed on the membrane, and the several layers are then wrapped around the product water tube to form a cylindrical module.

In operation, the module is placed in a snug-fitting cylindrical pressure vessel. Feed water flows in at one end of the vessel, axially through the module in the passage provided by the coarse brine-side spacer material, and out at the other end. Product water is collected in the product-water-side material, spirals inward to the central tube and is collected there. The central tube is attached to a penetration in the end wall of the pressure vessel, and product water escapes through this penetration.

Spiral-wound modules have been made in many sizes. In general, the module diameter varies from 2 to 4 in., and the length is 12 or 36 in. With the spiral design, it is expected that about 300 ft^2 of membrane area can ultimately be accommodated in 1 ft^3 of pressure vessel volume. In the units currently used, areas per unit volume range from 100 to 150 ft^2/ft^3. Details of the spiral-wound module design and development are given by Larson et al. (1968), and Riedinger et al. (1968).

The Spiral-Wound Module

Product-water-side considerations. The basic requirements of a satisfactory product-water-side backing material are that it (i) be able to support the operating pressure without collapsing or exhibiting excessive creep under long term service, (ii) has a fine enough structure so that the membrane is not

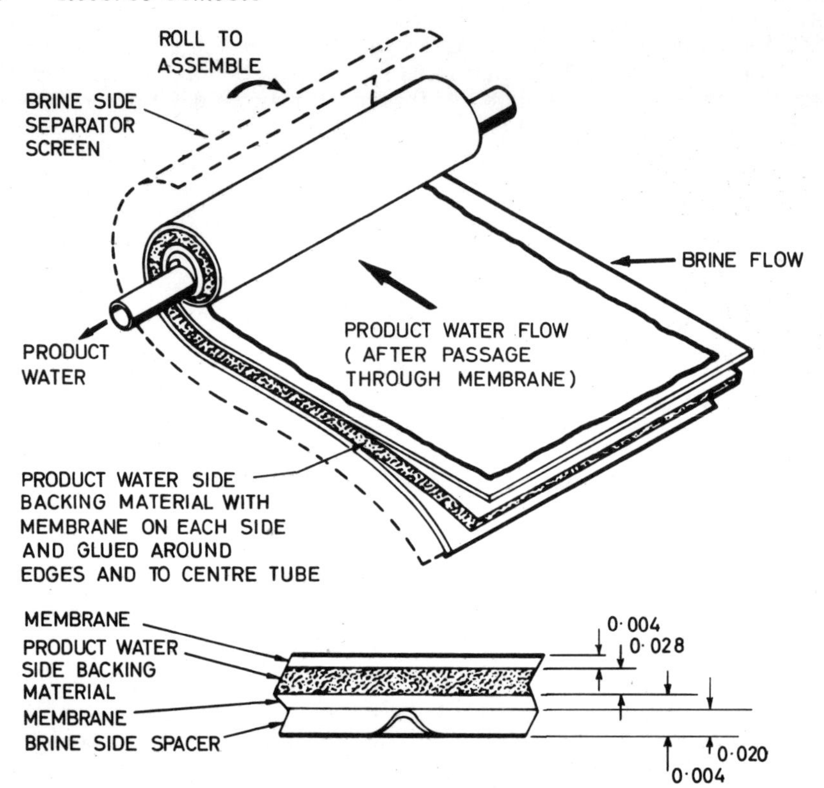

Figure 8-22. Gulf General Atomic spiral-wound module, unrolled (Schultz and Newby, 1966).

forced into the surface layers of the backing material, (iii) has sufficient porosity and large pore sizes so that the flow of product water through it does not cause an excessive pressure drop, (iv) be economical in cost, and (v) has a satisfactory life-time. Several materials were evaluated for the above purpose. Tests showed that the use of fibrous plastic materials was limited to operating pressures less than 200 psig. Because of the permeability decline caused by the creep of plastic backing materials, a major research effort has been the evaluation of less compressible backings including very fine metal screens, glass fibre felts, and glass cloths. One of the materials which appeared promising for operation at high pressures (1500 psig) consisted of uniform silica granules held in place on a layer of felt paper by a synthetic resin. However, several practical problems were experienced with the silica granule backing material. The sharp irregular edges and corners of the granules were found to cause numerous membrane punctures during high pressure operation. A major improvement was made by substituting glass microbeads for the silica granules. Pinholes in the membranes were greatly reduced, but not completely eliminated, in the operating pressure range 600 to 2000 psig.

Large bead size reduces pressure losses in the backing material. However, a definite limit has been imposed on the bead size by the need to limit the thickness of the product-water-side backing material to approximately 0.030 to 0.045 in., and the need for a smooth uniform membrane support. Such a support cannot be provided by the large (0.004 to 0.010 in.) diameter beads. A newly developed composite bead backing shows a good possibility of providing

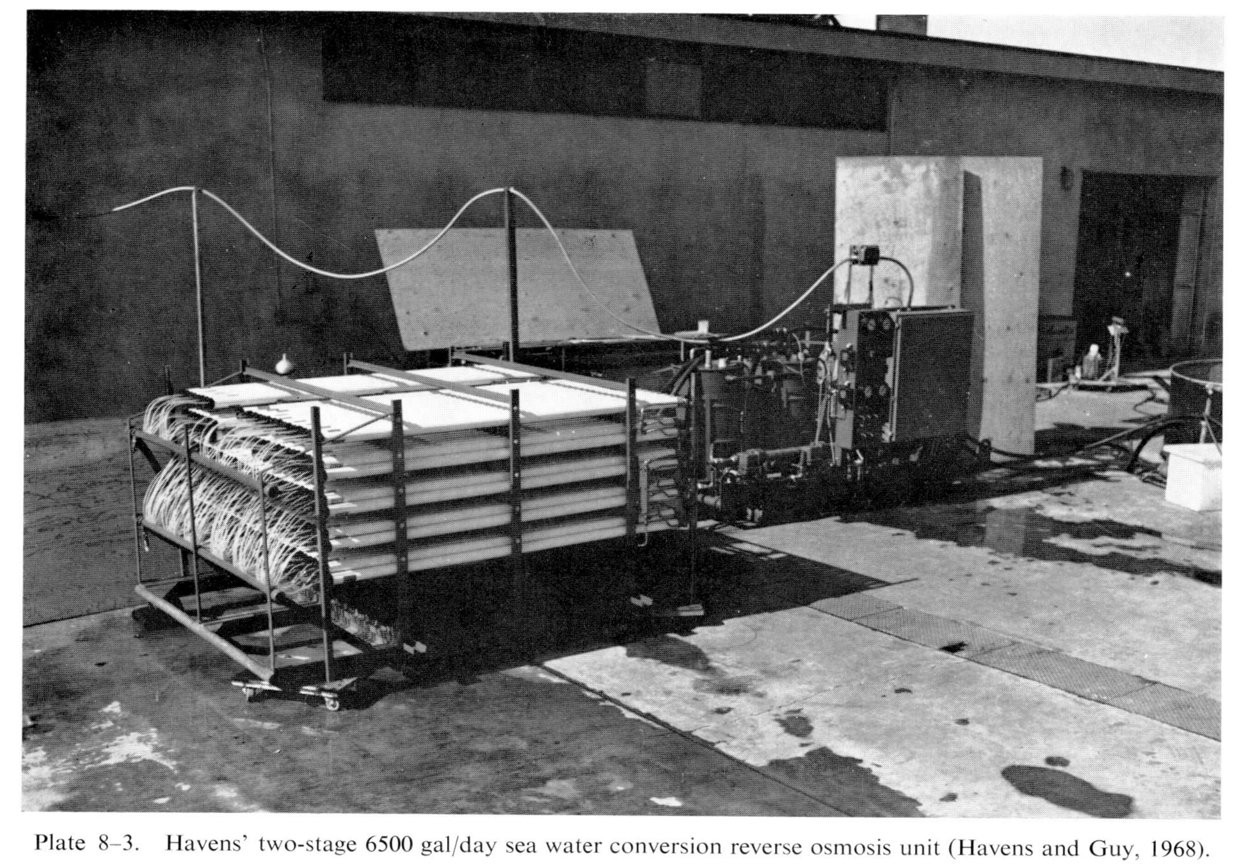

Plate 8–3. Havens' two-stage 6500 gal/day sea water conversion reverse osmosis unit (Havens and Guy, 1968).

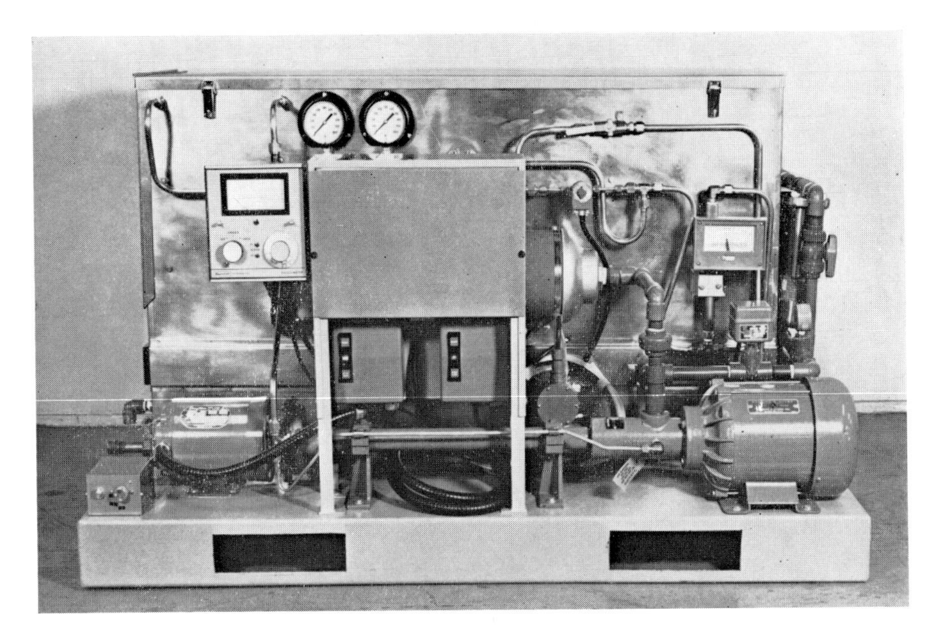

Plate 8–4. UWC 1500 gal/day brackish water conversion reverse osmosis unit (Manjikian, 1968).

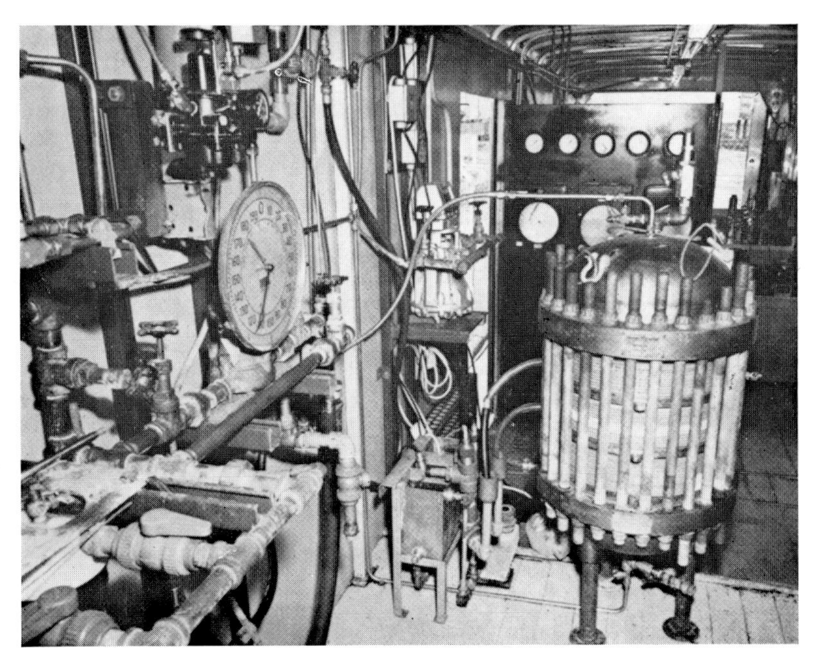

Plate 8–5. Interior of trailer-mounted Aerojet reverse osmosis unit (OSW, 1964).

the smooth support needed. This material consists of a layer of the large beads (0.004 to 0.008 in. diameter) sandwiched between two thin layers of very small beads (0.0005 to 0.0025 in. diameter). The large beads serve to provide a relatively porous, low resistance flow path for the product water; the small beads serve to smooth out the irregularities on the surface of the large bead matrix and hence present a somewhat smoother support plane to the felt backing material and the membrane.

A composite bead backing material consists of three basic elements: (i) the base material (felt or paper), (ii) beads (glass, plastic, or sand), and (iii) binder (a suitable water-insoluble adhesive). The properties of each of these components have a profound effect on the overall module performance. Efforts to develop the most suitable backing material are still continuing.

The base material acts as a physical barrier between the glass bead matrix and the membrane, to minimise the possibility of direct contact between the beads and the membrane, and to smooth out or moderate irregularities in the support matrix surface. In addition, the base material performs an important function in the fabrication process by providing a base support upon which the bead/latex matrix is formed. Experience indicates that the materials used in the fabrication of the felt base material and their geometry, on a microscopic scale, have a strong effect on the final properties of the module. Variables include fibre material, diameter, uniformity of size and distribution, binder and filler. The base material currently in use is a dacron polyester (Selmatex MU supplied by Kimberly Clark); other materials such as alpha cellulose (Patterson Parchment 532) and polyamides (Pellon Corp. 2504K, 2505K, and 2506K) appear promising. Since a good base material is essential for fabrication of a reliable bead backing composite, and since none of these materials is entirely satisfactory for long time service, the search for better base material continues.

The newly-developed composite bead backing configuration mentioned has been tested in pilot plant operations. After approximately 700 hours of operation, about one-half of which was at 1500 psig, the test data and module examination showed (i) there were no pinhole leaks and (ii) there were some erratic module performances due to glue line leaks, i.e. poor adhesion between the membrane and the backing material.

The binder latex provides the cementing agent which allows the backing material to act as an integral support structure. The latex must contain sufficiently high solids to allow convenient handling of the backing material without damage. Further, the solids content must be low enough so that it offers minimum interference to the flow of product water. The binder material which was used successfully was butyl rubber latex (Grade 8-21 supplied by Enjay Chemical Co.). Since the production of this material was discontinued, a survey to find a suitable replacement is still in progress. It was found by experience that a solids content of about 10 per cent for most latexes was required to produce a backing material matrix that was amenable to a reasonable production rate in module rolling. Tests indicate that the acrylic latex, Rhoplex AC-34 (supplied by Rohm and Haas Co.), is a promising substitute for the previously used butyl latex; the polyurethane and nitrile rubber latexes also appear worthy of investigation.

The adhesive used for forming the necessary seals in the spiral-wound module is a polyamide cured epoxy resin; it is one of the few materials which adheres to the moist cellulose acetate membrane. As more modules were fabricated and tested, however, it was observed that there was an uncomfortably high incidence of glue line failures. This was particularly true of modules operated at high pressures and subjected to frequent pressure cycles. This is being attributed to an adverse reaction between the glue and the membrane. The reference adhesive has a pH of 8 to 9 under which condition cellulose acetate material tends to hydrolyse. Consequently investigations are in progress to study the performance of neutral and acid adhesive materials.

Brine-side considerations in a spiral-wound module. The configuration of the brine-side spacer must be such as to permit good mixing of the brine without excessive pressure drop through the system. A variety of materials have been found satisfactory for this purpose, but the current material of choice is Vexar, a large mesh polyethylene screen supplied by DuPont. With this spacer in the module, pressure drops of the order of 1 lb/in^2 per linear foot of module provide acceptable mixing of brine on the high pressure side of the membrane.

As the brine flow increases, or the flow channels tend to foul, the pressure drop across the module increases. During operation of a module at an overall pressure drop of 5 lb/in^2 per module, it has been found that the membrane is displaced in the direction of brine flow, relative to the product water tube, unless the module is restrained. This effect is referred to as telescoping. Large diameter modules have been found to be more susceptible to telescoping than small diameter modules. Investigations are countinuing on the development of anti-telescoping devices.

Multileaf modules. Experience with spiral-wound modules showed that the leaf length was limited to about 18 in. because of the pressure drop characteristics of the product water backing material. For modules with a nominal 1 ft width, the effective membrane area per leaf was limited to about 3 ft^2. The options for increasing membrane area included adding leaves, widening the leaves, placing multiple product water take-off tubes along a long single leaf, or lengthening leaves with accompanying increase in pressure drop in the product water path, unless a better product water backing material was developed. Multileaf modules have been developed for use in large scale operations.

In a single leaf module, the composite leaf (consisting of the membrane, brine-side spacer, and the product water backing material) is wrapped around a central product water tube. In a 2-leaf module, two such composite leaves are wrapped in sequence around a product tube resulting in approximately 6 ft^2 of membrane area for a 1 ft wide module. The resulting module is about 2$^3/_8$ in. diameter. The technique of rolling 2-leaf modules was then extended to include two more leaves, thus producing a 4-leaf module with 12 ft^2 of membrane area; the product water tube in the latter case was a 14 in. long section of $^3/_8$ in. Schedule-80 PVC pipe.

The 'multileaf' rolling process was developed as a method for adding more than 4 leaves to a single product tube. Experience with the 2-leaf and 4-leaf rolling processes indicated that a maximum of 4 leaves could be attached without much difficulty using the method of gluing the leading edge of the product water backing material of each leaf to the product tube, and then rolling the leaves into a spiral cylinder. The multileaf process incorporated a larger diameter product tube which was slotted along its length. Each leaf was wrapped around a small plastic tube, called a key. The keys, with the leaf materials attached, were then inserted into the slots in the product tube and the module was rolled into the spiral form. Six-leaf modules rolled in the multileaf process had about 18 ft^2 of membrane area and a diameter of about 3$^3/_4$ in. The advantages of the multileaf process over the 4-leaf process are a reduction of leaf material and the capability of using up to six leaves. A multileaf rolling machine is illustrated in Figure 8-23.

One of the most promising module concept developed to date is the 3-ft module. The 3-ft module rolling process followed the same technique as the 2-leaf and 4-leaf processes except that the width of the leaf materials was increased from 12 in. to about 36 in. New fabricating techniques have been developed to produce 3 ft wide membrane and product water backing material. The initial 3-ft modules were made with four leaves, each 17 in. long. The module design was later changed to three leaves, each 26 in. long. The reduction of the

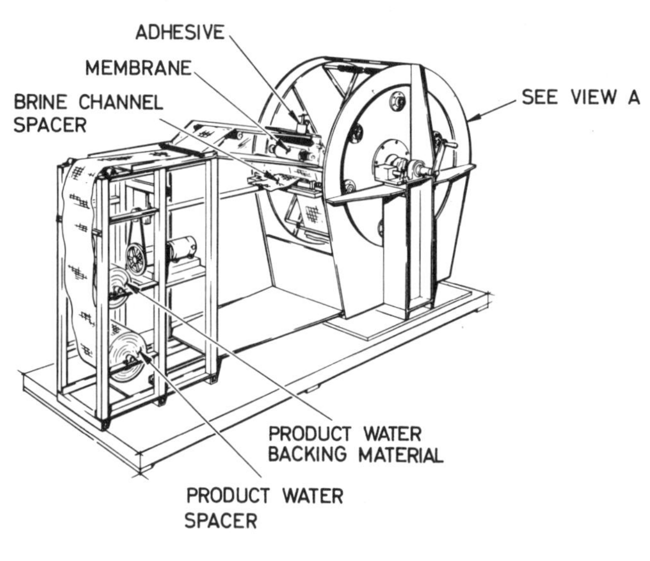

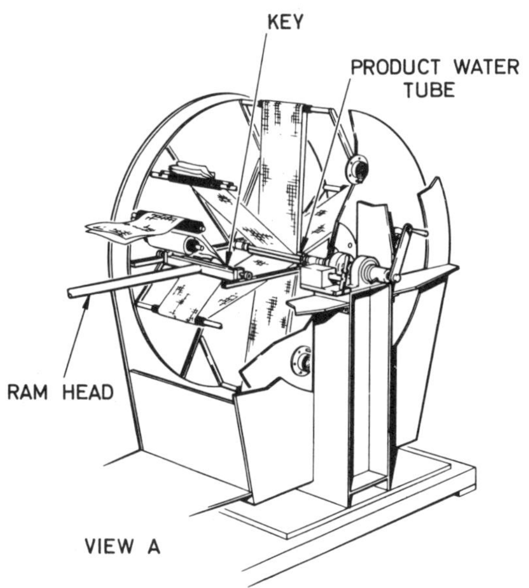

Figure 8-23. Gulf General Atomic multileaf rolling machine (Riedinger et al., 1968).

number of leaves improved fabrication time and resulted in an increased membrane area for a given module diameter. The increase in product water path length was, in this case, more than compensated for by the additional membrane area. Modules fabricated in this manner were chosen for use in the 10-100K (10,000 to 100,000 gal/day) test pilot plant.

Description of 3-ft modules. These modules are assembled in a spiral wound configuration as shown in Figure 8-24. The rectangular membrane, typically 0.004 in. thick and 35 in. wide, is folded over the brine channel spacer. This is then inserted between layers of product water backing material which has been previously attached to a central product water tube. An adhesive, applied during the rolling, seals the membrane to the backing material along the two sides, at the end, and around the product water take-off tube, thus forming an envelope which is sealed except for the open ends at the product water tube. The layers of material are then wrapped around the product water tube to form a cylindrical module. Materials currently being used in the 3-ft module include the following:

(i) Batch 18 type porous cellulose acetate membranes;

(ii) Product-water-side backing material: ~0.024 in. thick layer of 0.005 in. diameter glass beads attached with butyl latex to a 0.004 in. thick layer of Dacron felt; this composite is covered on both sides with a 0.006 in. thick layer of rayon felt;

(iii) Brine channel spacer: a 0.045 in. thick layer of polypropylene Vexar with ~0.025 in. fibres bonded to form a 12 mesh grid;

(iv) Reinforcing strip: 2 in. wide polypropylene tape attached to the product water side;

(v) Product water tube: $5/8$ in. o.d. PVC tubing.

The 3-ft module has three leaves, each with a 26-in. product water channel. The brine channel is 34 in. long, 31 in. of which is active membrane. The outside diameter of the module is $3\,5/8$ in. The module is designated as 36A2. Test results at 600 psig showed that the performance of the module was satisfactory under the operating conditions.

Module Interconnectors. Operating experience with systems using spiral-wound modules has disclosed problems associated with module interconnection leakage, brine seal leakage, and module telescoping as a result of high differential pressure. The interconnector design shown in Figure 8-25 was developed to solve the above problems.

The interconnector body is constructed of moulded polyvinyl chloride. The brine seal is a standard lip seal which fits between the interconnector body and the inside of the pressure vessel. The product water tube interconnection seals are effected by O-rings, which fit into grooves on the product water tubes, and seal against the bore of the interconnector body. Anti-telescoping restraint is achieved by taping the interconnector to the outside surface of the downstream module which keeps the outer wrap of the module in tension under flow conditions. Tests conducted with the interconnector showed that all three of the module problem areas had been corrected.

Because the number of interconnectors required for the 10-100K units was too small to justify the expense of a mould, a slightly different design was developed using a standard PVC pipe cap and a section of PVC pipe. With this design shown in Figure 8-26 a simplified fabrication process could be used for small production quantities.

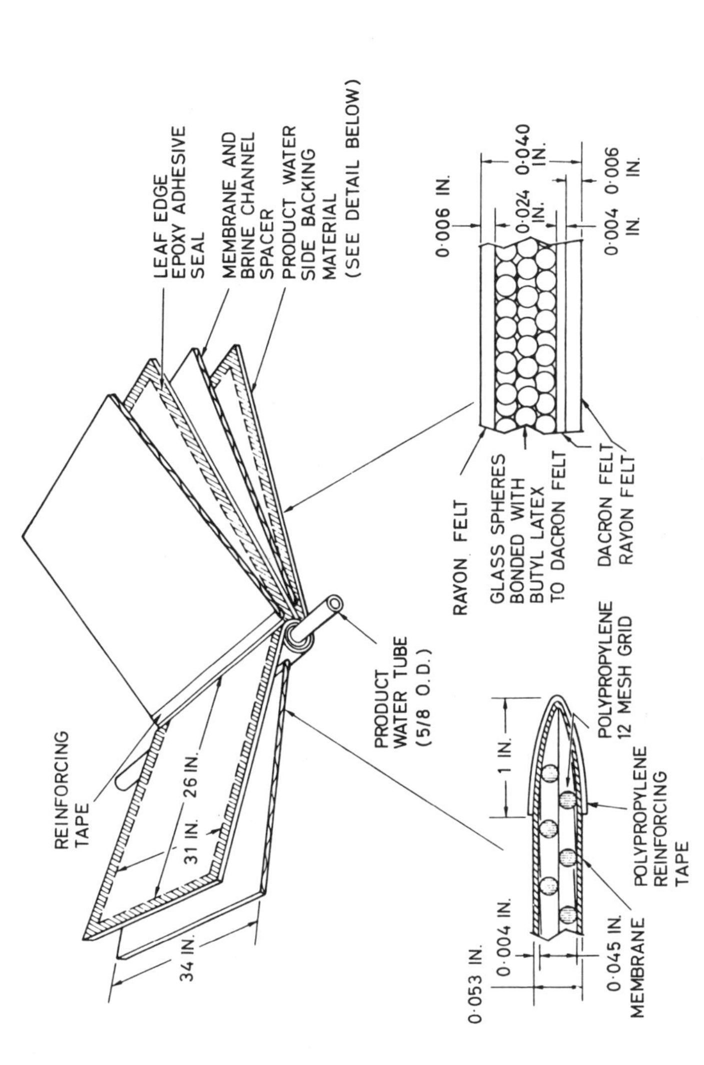

Figure 8-24. Gulf General Atomic 3 ft spiral-wound module, unrolled (Riedinger et al., 1968).

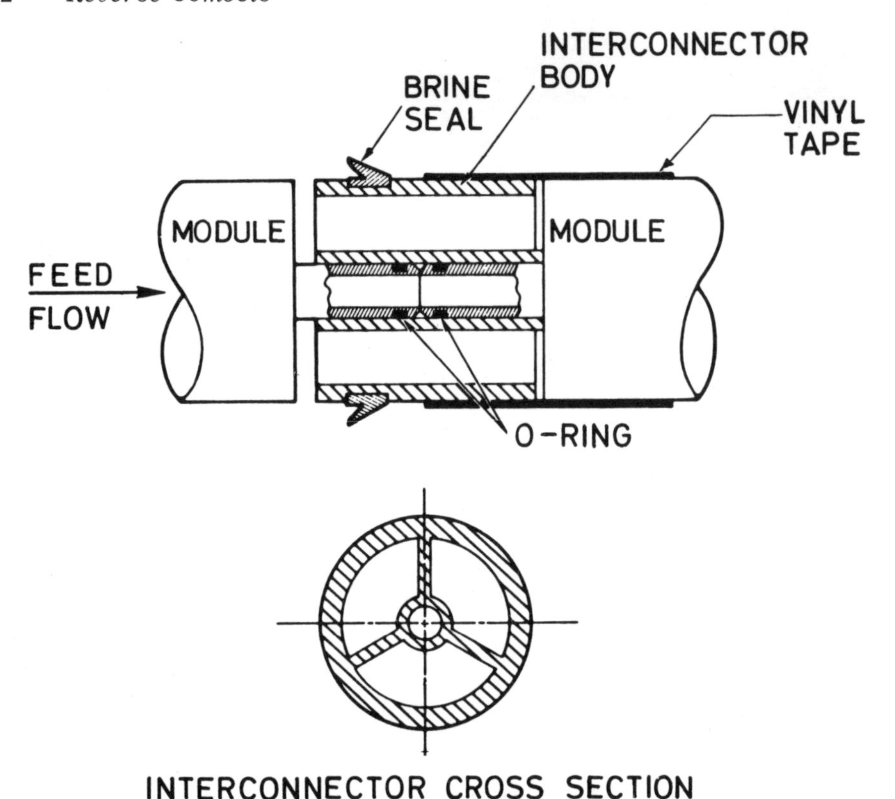

Figure 8-25. Gulf General Atomic moulded module interconnector (Riedinger et al., 1968).

10,000 to 100,000 gal/day Test Pilot Plant Design. A test pilot plant was designed and built to permit testing of large spiral-wound modules and investigation of engineering problems relating to the construction, operation and maintenance of large reverse osmosis systems. The pilot plant was designed to operate initially on brackish water (5000 ppm total dissolved solids with an osmotic pressure of 1.15 lb/in^2 per 100 ppm). The product water output was designed to be 10,000 gal/day (10K), to be readily expandable to a full capacity of 100,000 gal/day (100K). Selection of plant components was determined through optimisation studies based on a 100K system. Some details of this test pilot plant design are given below.

The capital and maintenance costs relating to pressure vessels and associated piping in the 10-100K unit represent a significant part of the total plant costs. Consequently, a study was conducted to optimise the design of pressure vessels from the standpoint of economics. This study compared the concept of clustering 7 or 19 module rows inside single large pressure-vessel shells with that of using individual pressure-vessels for each module row. The results of the study indicated, as shown in Table 8-18, that the individual tube pressure vessel concept shown in Figure 8-27 was the most advantageous in units of 100K capacity.

Following the above results, the existing individual pressure-vessel designs were reviewed for areas of possible improvement in cost and performance.

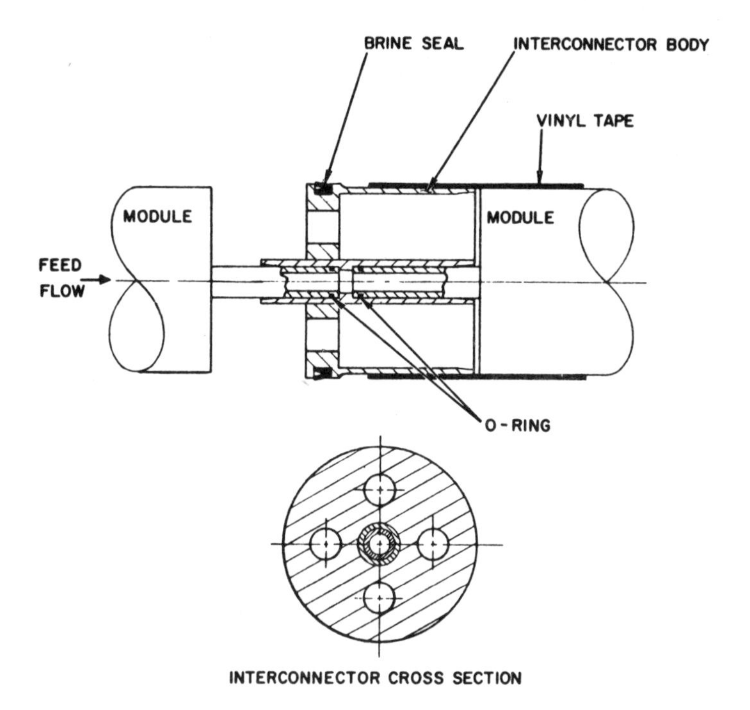

Figure 8-26. Cross section of Gulf General Atomic module interconnector (Riedinger et al., 1968).

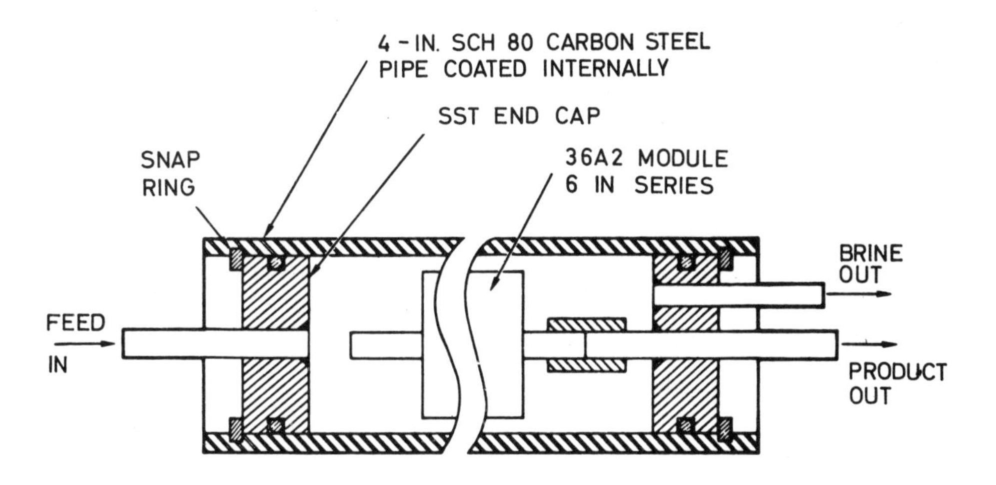

Figure 8-27. Gulf General Atomic individual tube pressure vessel design (Riedinger et al., 1968).

Table 8-18. Comparative factors for multitube and individual tube pressure vessel concepts in a 100, 000 gal/day reverse osmosis unit

(OSW, 1966)

Vessel concept	Number of vessels	Total vessel cost (dollars)	Total vessel wt. (lb)	Total vessel packing volume (ft^3)	Number of modules required
7-tube	10	34, 592	35, 250	893	420
19-tube	4	28, 783	30, 688	762	456
Individual-tube	70	21, 488	21, 490	216	420

Areas investigated included vessel body construction, internal corrosion protection, end closure design, and module interconnector and piping arrangements. Finally, the pressure-vessel design selected for the 10-100K unit is the one shown in Figure 8-28; this design incorporates a Schedule-40 carbon steel pipe body with upset ends and snap-ring closures. The vessel internals include a new module interconnector and a new concept for connection of product tube to end cap.

A number of methods and materials for internal corrosion protection of pressure vessels were surveyed, and Plasite 7122 was chosen as the most promising; this material is applied with a spray nozzle and then air-dried.

Because of the possible economic advantages offered by moulded end caps, a prototype PVC cap (Figure 8-29) was designed, fabricated and tested to determine its suitability in a simulated 10-100K environment. When the cap was pressurised with water to a maximum of 2100 psig, no leakage or significant deformations were observed. Further, the PVC cap proved to be much easier to insert and extract than stainless steel caps. Consequently this design (Figure 8-29) was chosen for the 10-100K pressure vessels. Although tests confirmed the structural integrity of the design, the effects of creep, which are characteristic of PVC material under load, remain to be checked on a long-term basis.

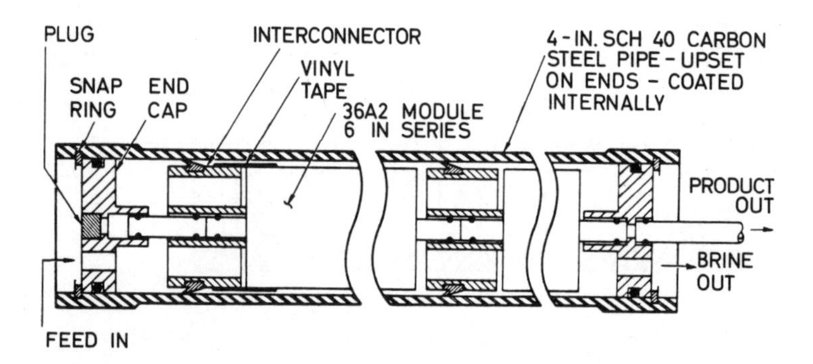

Figure 8-28. Pressure vessel assembly for the Gulf General Atomic 10-100K reverse osmosis unit (Riedinger et al., 1968).

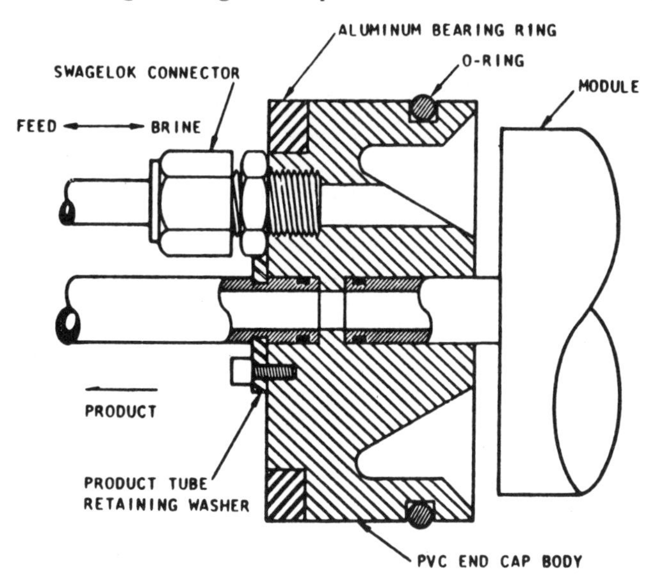

Figure 8-29. Gulf General Atomic prototype PVC end cap (Riedinger et al., 1968).

The pump requirements for the 10-100K unit were specified as 110 gal/min output at 800 psig for operation on sea water at a pH range of 4. 5 to 7. The results of a survey indicated that plunger-type pumps are the most suitable for the purpose. On the basis of cost, availability of parts, and field applications in similar service, the Gaso horizontal triplex plunger pump was selected for the 10-100K reverse osmosis unit. [The Gaso pump has been extensively used for pumping sea water at pressures up to 3600 psig and flows up to 300 gal/min in secondary oil recovery operations.] For the reverse osmosis unit operating at 10, 000 gal/day capacity, a 55 gal/min Gaso pump with its speed reduced to deliver 23 gal/min was chosen. The Gaso pump has an aluminium-bronze fluid end, delrin valves, Monel valve seats, ceramic plungers with stainless steel shafts and Teflon-impregnated packing. This combination of materials should result in a satisfactory service life for the pump.

The basic components of the 10-100K fluid system are water pretreatment equipment, a high pressure pump, pressure vessels, and the flow and pressure regulation equipments. The feed water enters the system through a PVC line and is treated for pH and filtered prior to pressurisation by the main pumps. The pH control is effected by a Uniloc pH controller operating in conjunction with an acid pump and an acid storage tank. After acid injection, the feed water passes through 25-micron filters which, in addition to providing the filtration requirements for the pump and modules, complete the mixing of acid with the feed water. To prevent pressure pulsations during operation, accumulators are positioned at the suction and discharge ends of the pump. Pneumatic controllers are used to control system pressure and fluid flow. The instrumentation for the system includes flow meters for measuring brine and product flows, a differential pressure gauge across the total pressure vessel feed brine path, a pressure gauge each at the feed inlet and at the brine outlet, a conductivity meter with probes located at the feed inlet, brine outlet, and product outlet, a pH-indicating meter, and a temperature recorder for indicating the product water temperature.

The electrical control system for the unit is designed either for hand or automatic operation and contains provisions for alarm and plant shutdown as functions of instrumentation settings.

The 10-100K unit is designed so that expansion to a product water output of 100, 000 gal/day can be achieved with a minimum of expense and complication. The basic additions required for expansion to 100K are a 55 gal/min pump, 60 pressure vessels with interconnection manifolding, a motor starter, a larger pressure regulating valve, a larger relief valve and changes in flow meter scales. The arrangement of the proposed 100K unit is illustrated in Figure 8-30.

The 10-100K test pilot plant was put into operation in March 1967, and since then the plant has been in service for several continuous periods. No functional problems were encountered, and the performance of the unit has been generally satisfactory.

Other Reverse Osmosis Units and Their Performance. Several commercial units are currently available from Gulf General Atomic. A single spiral-wound module is employed in two such units. One is a versatile laboratory unit that permits different modes of operation and can be used for testing a wide variety of reverse osmosis applications. The second is a home drinking-water unit that operates at normal tap water pressures and delivers 2 to 3 gal/day. An example of the performance of this unit on San Diego tap water is given in Table 8-19.

The major sub-units in each commercial reverse osmosis unit are the feed water system (including pretreatment facilities), the high-pressure pumps, the spiral-wound module assembly, and the facilities for storage or distribution of the product water and discharge of the concentrated brine. In large operating units a turbine also may be included for power recovery.

Figure 8-31 is a schematic diagram of a 100 gal/day unit (all specified capacities are nominally for brackish water conversion). It operates at 350 psig with a gasoline- or electrically-driven pump. Figure 8-32 represents a 1000 gal/day unit designed for operations at pressures up to 1500 psig. It has 6 pressure tubes which can be interconnected either in series or in parallel as

Table 8-19. Module performance in home drinking water unit

Data of Sudak and Nusbaum (1968)

	Concentration (ppm)	
Ion	Tap water	Product water
Calcium	84	1.2
Magnesium	30	0.7
Sodium	100	10.3
Chloride	110	18
Sulphate	300	4.1
pH	7.6	7.7

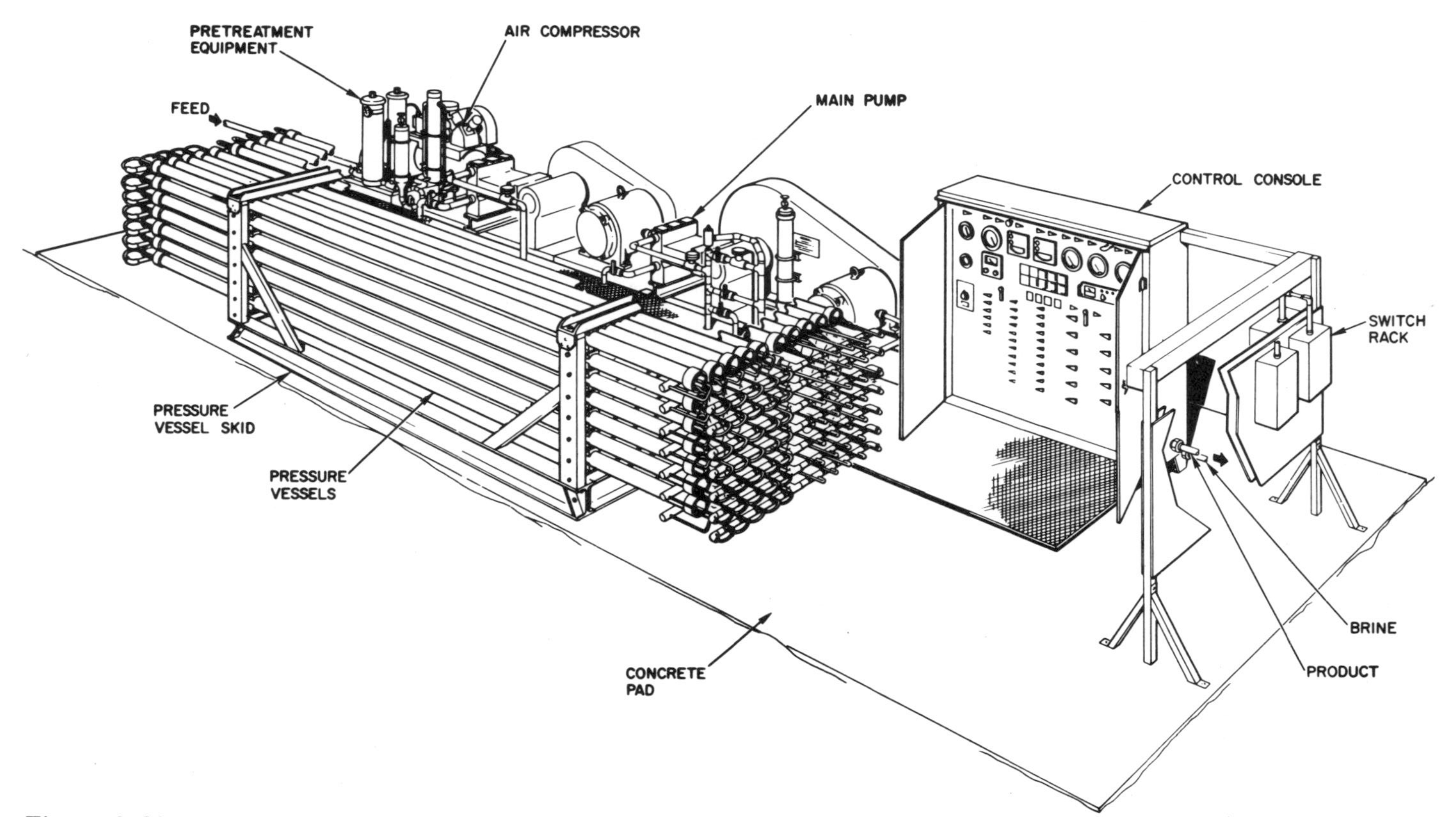

Figure 8-30. Schematic layout of Gulf General Atomic 100K reverse osmosis unit (Riedinger et al., 1968).

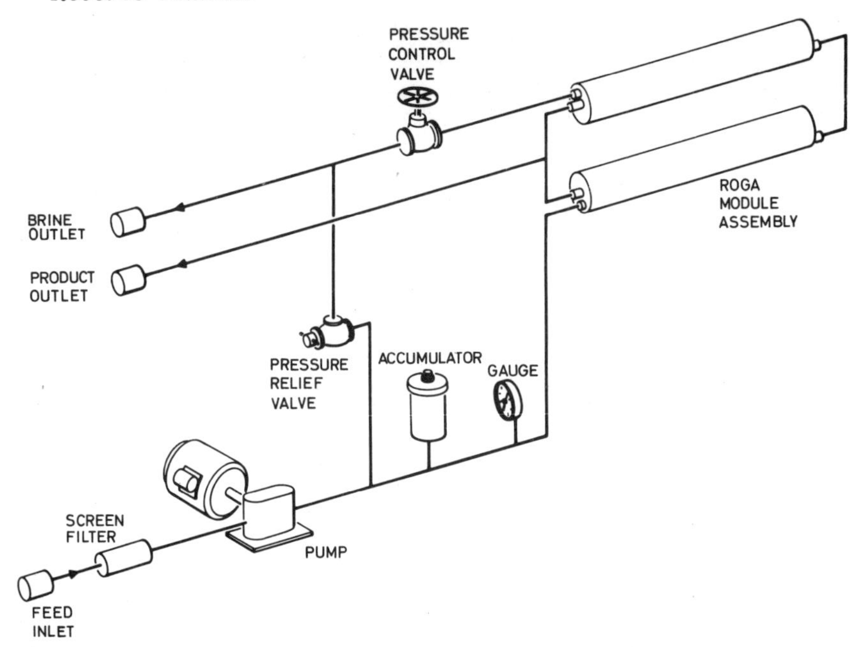

Figure 8-31. Schematic layout of a 100 gal/day Gulf General Atomic reverse osmosis unit.

required; each tube contains 6 modules, each of which has an effective area of 6 ft^2. The unit is installed on a conventional moving van trailer. Such a unit was first displayed at the First International Symposium on Water Desalination in Washington, D.C. in October 1965. The unit was later used for field tests which demonstrated successful performance in operation with Potomac River water in Washington, D.C. (OSW, 1965; Schultz, 1966), acid mine waters in Kittanning, Pennsylvania (OSW, 1965; Riedinger and Schultz, 1966), and three different brackish waters in Texas (Schultz et al., 1967). Some of the above test results are given in Tables 8-20, 8-21, and 8-22 respectively.

The performance of spiral-wound modules for the reverse osmosis treatment of sewage water has been under test at Pomona, California, since 1965. A schematic diagram of the test unit used originally is shown in Figure 8-33 (Bray et al., 1965). A 5000 gal/day unit is currently being tested for the above application (Sudak and Nusbaum, 1968). The results obtained over a recent extended test run with the 5000 gal/day unit are given in Table 8-23. The operating pressure was 400 psig and temperatures varied from 20° to 25°. During this run 80 to 85 per cent of the feed was recovered as product water. The system was operated on effluent from an activated carbon column. Table 8-23 is arranged to show the performance throughout the length of the flow path in the pilot plant. The data were obtained with 9 pressure vessels; eight vessels contained 10 modules and the remaining vessel contained 4 modules, giving a total of 84. Tubes 1A and 1B were in parallel; the remaining tubes were in series. The modules in the last four tubes were smaller in area and diameter to maintain satisfactory brine velocities. In general, solute separation decreased as the brine progressed through the tubes; apparent anomalies were attributed to leaks and differences in membrane performance.

A 10, 000 gal/day spiral-wound module pilot plant had been tested with a brackish well water at the River Valley Golf Course in San Diego, California.

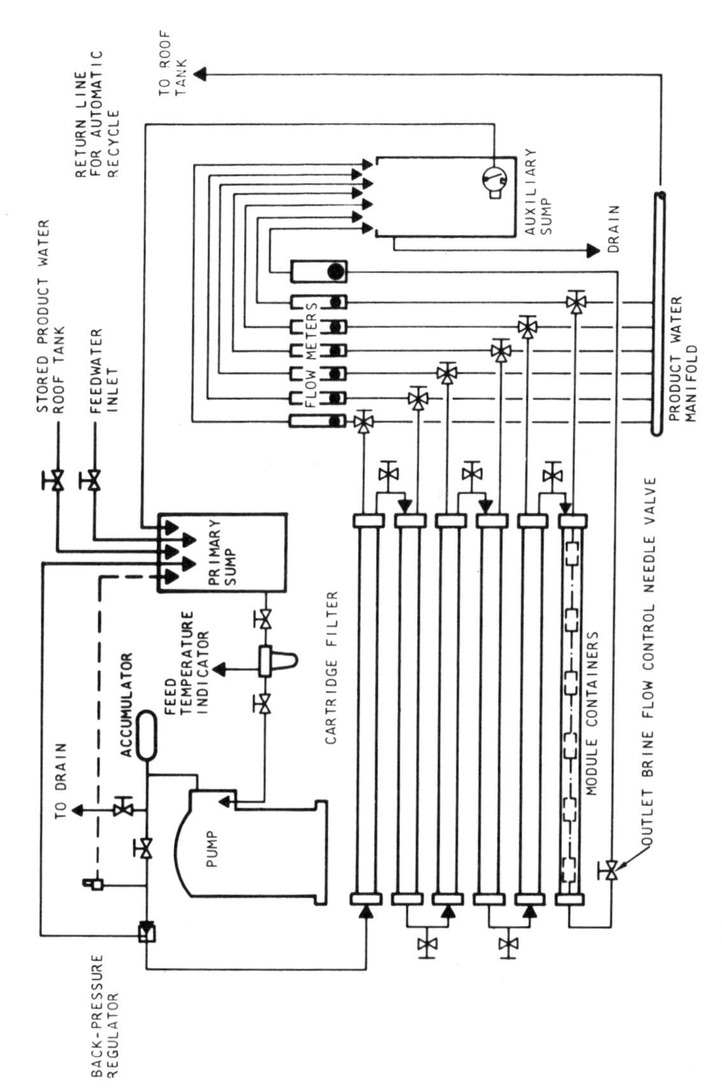

Figure 8-32. Flow diagram of 1000 gal/day Gulf General Atomic reverse osmosis unit (Larson, 1967).

Table 8-20. Typical chemical analysis of feed, product and brine in reverse osmosis test with Potomac River water

(OSW, 1965)

Sample	Total organic carbon (ppm)	Chemical oxygen demand (ppm)	Total dissolved solids (ppm)	Phosphate (ppm)	Chloride (ppm)	Nitrate (ppm)
Feed	3.5	8.7	366	0.05	25.7	0.16
Product	1.5	0.2	13.6	<0.01	< 1.0	0.02
Brine	5.0	12.8	448	0.07	37.6	0.21

A typical bacteriological test showed a plate-count reduction from $4200/cm^3$ in the river feed water to $7/cm^3$ in the product water. The coliforms were reduced from a most probable number of $4600/cm^3$ (positive) to less than $3/cm^3$ (negative) in the product. Alkyl benzene sulphonate and ammonia nitrogen were generally too low to measure.

The average water flux was 5 to 6 gal/day/ft^2 of membrane area at 600 psig (Schultz, 1966)

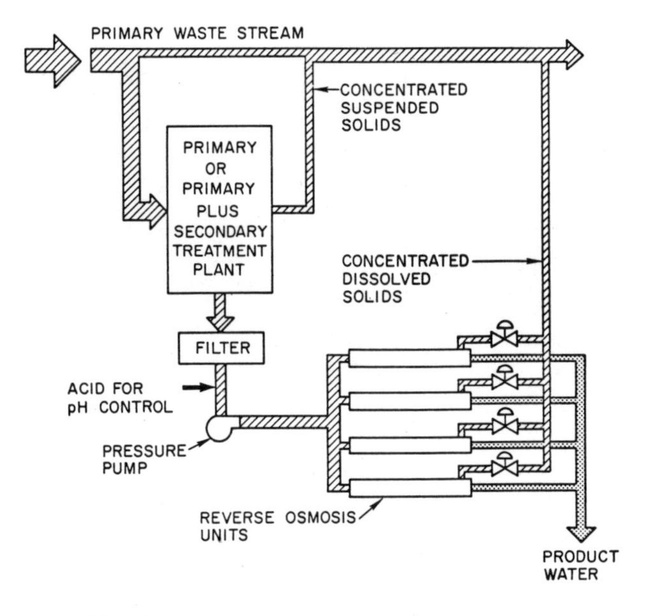

Figure 8-33. Flow diagram of a Gulf General Atomic reverse osmosis unit in waste water treatment system (Bray et al., 1965).

Table 8-21. Typical chemical analysis of feed, product, and brine in reverse osmosis test with acid mine water at coal mine, Decker No. 5, Kittanning, Pa.

(OSW, 1965)

Sample	pH	Acidity (as $CaCO_3$) (ppm)	Total dis-solved solids (ppm)	Sulphate (ppm)	Ca (as $CaCO_3$) (ppm)	Mg (as $CaCO_3$) (ppm)	Fe (as Fe_2O_3) (ppm)
Feed	2.35	695	1920	1326	389	261	279
Product	4.75	38	3.4	0.7	<1	<1	Trace
Brine	1.85	7270	19,364	12,650	1660	3687	2869

As of the end of January 1968, the modules had operated for a total of 1300 hours. The operating pressure was 600 psig, and the product recovery was 75 per cent. The plant had 36 modules installed in six pressure vessels. Each module was 36 in. long, and had an effective membrane area of 27.5 ft^2. At the 10,000 gal/day level, the modules were delivering 10.1 gal/day ft^2. A typical analysis of the well water feed, the brine and the product water is given in Table 8-24 (Sudak and Nusbaum, 1968).

12. Small Reverse Osmosis Units For Home Service

Small size, ~5 gal/day, reverse osmosis water purification units for home service, using the Loeb-Sourirajan type porous cellulose acetate membranes and operating at tap water pressures (20 to 50 psig), have come onto the market in recent years. The need for such units cannot be over-emphasised. When such units become available at low costs, they can be expected to become increasingly popular in all parts of the world. Ultimately, the total quantity of purified water obtained from all such units may well exceed that produced by all the big reverse osmosis plants put together.

Apart from commercial considerations, the possibility of developing such low-cost reverse osmosis home units offers an ideal project for students of engineering in academic institutions. Such a project will demand all the ingenuity of engineering design and construction, and can offer a worth-while sense of professional accomplishment, and possibly also a career of one's own making, to at least some of the participants. One such project which the author is aware of, was successfully carried out at the Thayer School of Engineering, Dartmouth College, Hanover, N.H., U.S.A., under the direction of Professor Myron Tribus. The accomplishments in the project are described in the Master of Engineering degree theses of Miller (1968) and Spatz (1968).

Miller (1968) has presented details of a machine, represented in Figure 8-34, for the continuous production of porous cellulose acetate reverse osmosis membrane on a polyester paper. Referring to Figure 8-34, the mixed cellulose acetate membrane casting solution is pressurised in the pressure vessel (2) by a regulated nitrogen supply (1). It flows into the extrusion jaws (4). The pressure in the extrusion jaws is regulated by a second regulated nitrogen source (3). Polyester filter paper from a 50-yard roll (6) is pulled at 5 ft/min into the casting zone by the application roller driven by the motor (5). In the

Table 8-22. Reverse osmosis test results with brackish well water at Midland, Texas

(Schultz et al., 1967)

	Concentration (ppm)					
			Product			
Component	Feed tube 1	Brine tube 6	Tube 1	Tube 2	Tube 3	Tube 6
Calcium	496	736	<0.02	<0.02	<0.02	19
Magnesium	243	253	0	0	0	2
Sodium	398	656	<0.2	<0.2	<0.2	34
Iron	0.07	0.05	0.03	<0.02	0.03	0.03
Bicarbonate	211	290	<0.5	<0.5	<0.5	18.5
Sulphate	1700	2450	3	3	3	73
Chloride	860	1140	17	20	25	80
Nitrate	25	44	14	12	13	18
Silica	73	—	3.5	3.5	3.5	—
Fluoride	4.3	7.5	0	0	0	0
Total dissolved solids	4010	6118	26.5	27.5	30	321
Alkalinity*	176	242	2	2	2	15.4
Hardness*	2240	2880	<1	<1	<1	96
pH	8.2	8.0	6.1	6.05	5.9	—

* Expressed as $CaCO_3$

Membrane area in each module: 6 ft^2
Number of modules in each tube: 6
Number of tubes in series: 6
Feed pressure: 600 psig
Outlet brine pressure: 550 psig
Feed temperature: 14.5°
Product recovery: 44.4%
Feed flow rate: 5360 cm^3/min
Product flow rate: 2380 cm^3/min
Brine flow rate: 2980 cm^3/min

casting zone the jaws (4) coat the filter paper with an even film of cellulose acetate. The 0.005 inch film backed with polyester filter paper is pulled through the drying zone by the take-up mechanism driven by the motor (5) in the take-up bath (9). From the drying zone the backed film travels into the first ice water bath (7) where it is flushed with ice water. In the storage bath (9) it is wound up and flushed again with ice water. The casting, evaporation, and gelation (in ice water bath) take place totally enclosed in a chest freezer (11) to facilitate the control of the environmental conditions of casting.

Table 8-23. Water quality data from the 5000 gal/day reverse osmosis unit used for sewage water treatment at Pomona, California

Data of Sudak and Nusbaum (1968)

	Average concentration (ppm)				
	Phosphate	Chemical oxygen demand	Ammonia as N	Nitrate as N	Total dissolved solids
Feed	30.9	10.8	9.2	2.4	623
Tube 1A	0.22	1.2	1.3	0.4	51
1B	0.16	1.1	1.2	0.5	36
2C	1.4	2.4	2.8	0.9	60
3C	1.1	1.6	1.8	0.7	70
4C	0.43	1.8	1.7	0.9	67
5C	0.48	2.0	1.7	0.9	53
6C	0.41	1.6	1.8	1.0	53
7C	0.37	2.3	2.6	1.0	67
8C	0.77	2.4	3.2	1.6	95
Product	0.57	1.7	1.7	0.8	73
Brine	177	43.8	94	7.5	3402
% Reduction	98.2	84	82	67	88

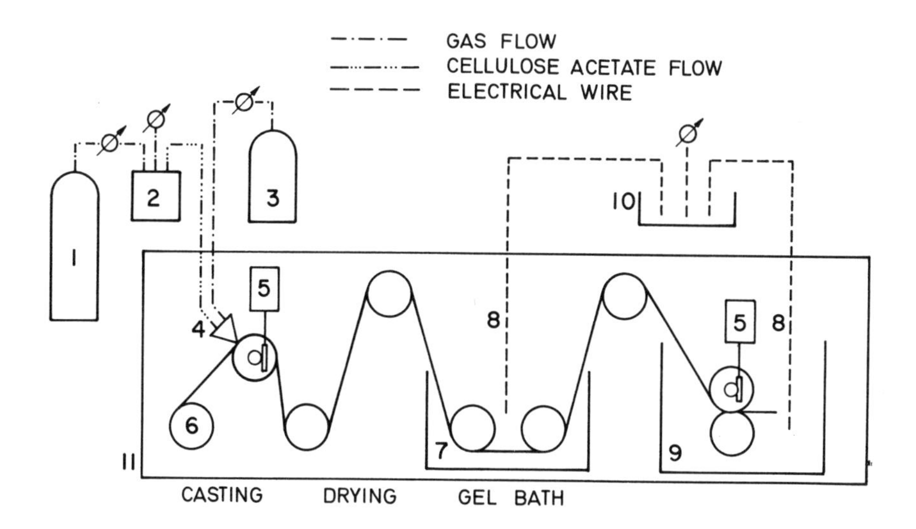

Figure 8-34. Flow diagram of continuous film casting machine (Miller, 1968).

(1) Nitrogen pressure source; (2) cellulose acetate supply; (3) device to sense pressure in extrusion jaws; (4) extrusion jaws; (5) drive motor; (6) roll of polyester filter paper; (7) initial ice bath; (8) thermocouple; (9) storage ice bath; (10) reference junction; (11) freezer.

Table 8-24. Typical analysis of feed, brine and product water from the 10,000 gal/day reverse osmosis pilot plant tested at the River Valley Golf Course, San Diego, California

Data of Sudak and Nusbaum (1968)

	Concentration (ppm)		
	Feed	Brine	Product
Cations:			
Calcium(Ca)	357	1500	5.6
Magnesium (Mg)	221	1020	3.4
Sodium (Na)	900	3700	106
Potassium (K)	26	103	3.8
Iron (Fe)	7	27	0.1
Manganese (Mn)	4.1	17	0.0
Anions:			
Bicarbonate (HCO_3)	61	18	1.8
Sulphate (SO_4)	612	2580	2.9
Chloride (Cl)	2260	8900	180
Nitrate (NO_3)	0.9	3	0.0
Fluoride (F)	1.1	3.6	0.1
Boron (B)	0.3	0.5	0.2
Silica (SiO_2)	35	58	6.4
Alkalinity ($CaCO_3$)	50	15	1.5
Hardness ($CaCO_3$)	1800	7940	28
Dissolved solids	4490	17,620	300
pH	6.0	4.6	6.0

According to Miller's estimation, a system such as that shown in Figure 8-34 could produce polyester filter-paper backed membranes for 15 cents per ft^2. Such a system offers a possibility for control over the casting variables not attainable with hand casting methods.

The test results from over 100 samples taken from 500 yards of membrane showed that the membranes produced by this casting machine gave product rates of 0.7 to 1.0 gal/day/ft^2 with 80 per cent solute separation from 0.025M NaCl-H_2O feed solutions at 50 psig, and they had reasonably uniform properties when taken from the same 10 ft section of cast film. Further, the production of such machine-made films contributed to the development of low cost reverse osmosis units for home service described by Spatz (1968).

Spatz (1968) presents details of the design, development, and field test data of several prototype 3 to 5 gal/day plate-and-frame reverse osmosis water purification units (Figure 8-35) operating at home water supply pressures (Figure 8-36). Extensive field test data from Guilford (Connecticut), Hanover (New Hampshire), Fairbanks (Alaska) and Lyme (New Hampshire) have demonstrated the successful performance of the units under conditions of actual operation for extended periods of time.

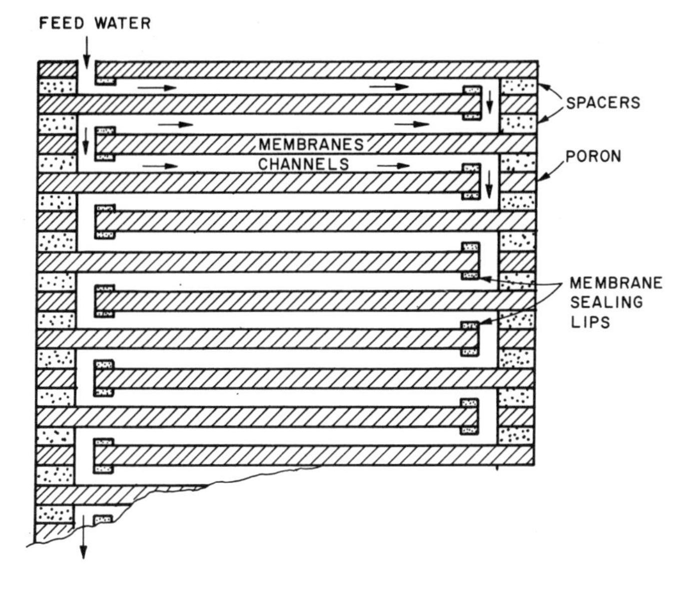

Figure 8-35. Film and spacer arrangement in the prototype home reverse osmosis unit (Spatz, 1968).

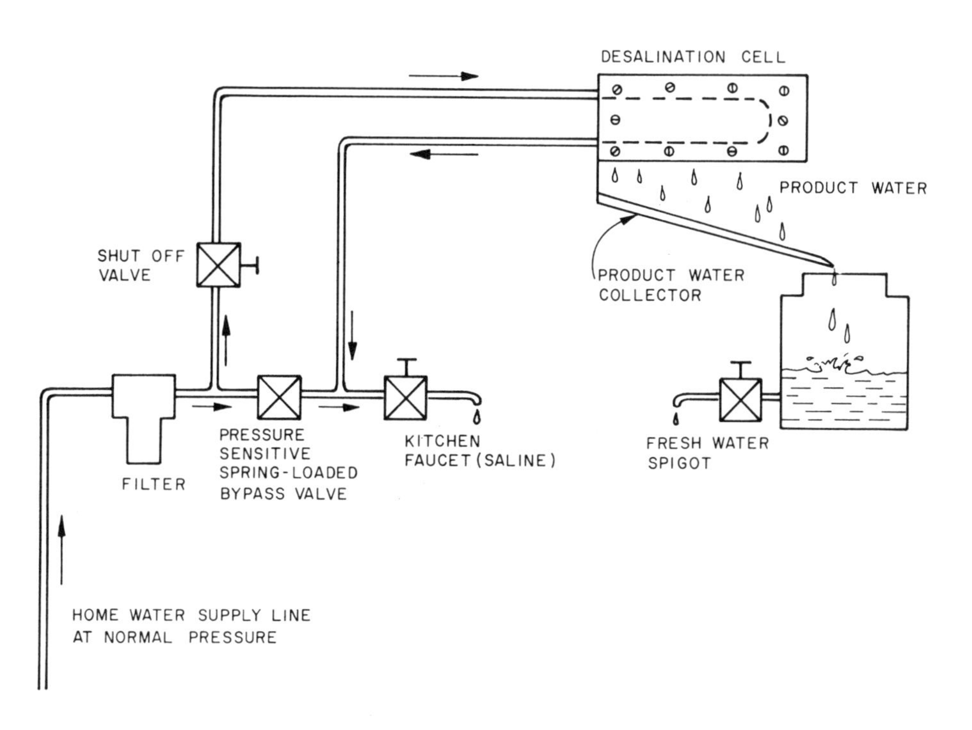

Figure 8-36. Reverse osmosis water purification unit in home service (Spatz, 1968).

The unit consists of a stack of membranes supported by porous Poron phenoxy resin (Figure 8-35). The membranes are placed on each side of the Poron backing. At one end of the backing a passage is machined through the Poron. The walls of the passage are impregnated with epoxy. Feed water is transported through this passage from one side of the backing to the other without allowing the saline feed water to contaminate the product water. Polyethylene spacers are used to maintain a channel through which the feed water can run over the membrane. The channel is 12 in. long and 2.5 in. wide. The spacer channel transports the feed from one end of the membrane to the other and then allows the feed to go through the Poron-epoxy impregnated passage into the next channel created by the next spacer. This chain of spacer plate channel, Poron passage to spacer plate channel, etc. continues until the feed runs over the last membrane in the cell and is discharged finally.

The spacer has dimensions of 0.125 in. thick, 5 in. wide and 15 in. long. The Poron has dimensions of 0.1 in. thick, 5 in. wide and 15 in. long. The 1.25 in. overlap on each side is used to seal the feed water from the product water. No O-ring seals are used.

One of the problems encountered in sealing was the seal between the membrane and the Poron at the entrance to the Poron-epoxy impregnated passage. This problem was overcome by leaving a small 0.05 in. strip of spacer material in front of the passage and applying a thin film of epoxy to the immediate area surrounding the passage. When the cell was fastened together the pressure exerted by the strip of spacer material in front of the passage coupled with the epoxy gave an exceptionally good seal.

Each prototype uses two half-cylindrical end caps made of type 304 stainless steel. Stainless steel 3/8-in. bolts are used to fasten the stack of membranes, membrane backing, and spacer plates together. Polyvinyl chloride was coated on the inside of the caps to prevent corrosion between the stainless steel and the silver solder used to put the prototypes together. In production, these end caps would be stamped into shape, eliminanting the need for the PVC coating. It was estimated that the plate-and-frame home unit could be produced for a materials and labour cost of $20.19.

Spatz (1968) has also presented the design and development of an axial flow spiral-wound reverse osmosis home unit using machine cast polyester backed membranes (Miller, 1968). Two membranes are placed back to back with the cellulose acetate coating facing outside. The polyester matting and the membrane are adhered at the edges using United Shoe Machinery's No. 4040 nitrile base contact type adhesive. The membranes are rolled around a 1/2-in. perforated PVC product water tube with Vexar spacers between the membrane surfaces. Feed water flows axially between Vexar and membrane. A 3-in. PVC tube is used as the pressure vessel. This unit produces 3 to 5 gal/day. It is estimated that the cost of materials and labour for this unit is $6.26.

13. DuPont Permasep Permeators

The reverse osmosis units of DuPont, called 'Permasep' permeators, use nylon hollow fibre membranes having outside diameters in the range 25 to 250μ with wall thicknesses of 5 to 50μ. They have a shell-and-tube configuration (Figure 8-37) similar to a single-end heat exchanger. The fibres are potted in a special epoxy resin which serves as the tube sheet. The feed enters the shell side near the head of the equipment and the concentrated brine leaves from the other end. Water permeates from the outside of the fibre to the inside through the walls. Product water inside the hollow fibres leaves the system countercurrent to the flow of feed in the shell.

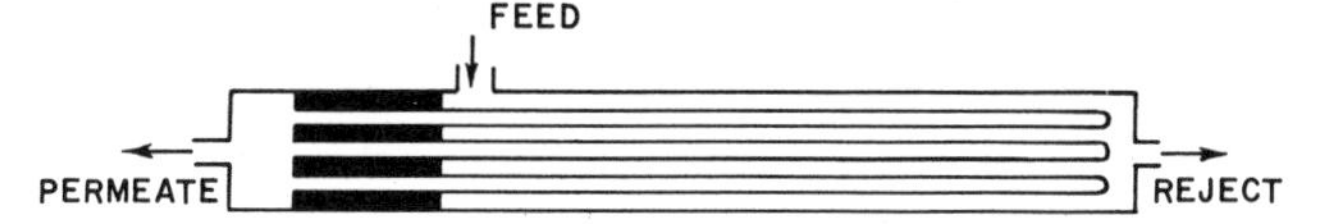

Figure 8-37. Configuration of DuPont 'Permasep' permeators.

The following reasons are given for this flow arrangement. The fibre wall withstands greater pressure under compression than under tension. There is less chance of plugging in the shell side and it is much easier to clean. Further, the outside-in flow is a fail-safe mechanism, since if a fibre should be weak, it would be closed rather than ruptured by pressure.

A permeation flux of 0.15 gal/day/ft^2 of membrane surface is currently obtainable with DuPont units. DuPont has built a variety of permeators for test purposes with diameters up to 6 in. and active fibre lengths up to 7 ft. Units of 12-in. diameter with 7 ft of active fibre length are currently under development. Such a unit will contain 15 to 30 million hollow fibres with surface areas of 50,000 to 80,000 ft^2.

14. Parametric Studies

Parametric studies and process optimisation together constitute the most important engineering requirement prior to the design and construction of any large-scale production plant. At least with respect to water desalination and similar applications, the reverse osmosis process is ready for large-scale operation. It is hence appropriate to conclude this chapter with a brief review of some of the parametric studies of the reverse osmosis process for saline water conversion. The ultimate object of all such parametric studies is the same, namely to find the optimum conditions for producing the required quality of product water from a given source at the minimum cost, and to identify the critical parameters of process design and the areas which require further study, research and development. However, the approaches employed in such studies have naturally been different with different workers depending on the choice of their own design variables. In the following discussion, the basis and the results obtained in each of these studies are briefly reviewed to illustrate the point of view that in the overall development of the subject, every approach serves a useful purpose.

15. Studies of Stevens and Loeb

The extensive experience of the Coalinga plant operation was the basis of the studies of Stevens and Loeb (1967). From the Coalinga experience it was concluded that (1) the composite tubular assembly was entirely adequate for extended field service and an on-stream time of at least 98 per cent could be maintained, (2) a membrane life of at least one year could be expected without any special feed water pretreatment other than simple filtration through 25 micron filters, and (3) an average product water flux of 15 to 20 gal/day/ft^2 could be maintained during the membrane life time by swabbing the membrane tubes with an oversized polyurethane foam ball passed through the tube series twice a week.

On the basis of the above conclusions, estimates were made on costs of fresh water produced from Coalinga type feed water plants employing the composite tubular assembly geometry. Four different types of such plants were considered. Their technical features and operating costs are given in Table 8-25 and 8-26 respectively. The total operating costs for Plants 1, 2, 3 and 4

Table 8-25. Technical features of the four plants studied by Stevens and Loeb (1967)

	5 kgal/day		50 kgal/day	
Features	Plant 1	Plant 2	Plant 3	Plant 4
Fraction product recovery	½	½	⅔	⅔
Operating pressure, (psig)	600	600	600	800
Average product water flux (gal/day/ft^2)	20	20	20	25
Composite tubular assembly characteristics:				
Membrane area (ft^2)	2.24	2.24	4.49	11.3
Nominal diameter (in.)	1	1	1	2
Overall length (ft)	10	10	20	20
End fitting type	Flare	Flare	Flare	Flange
Support tube and end fitting material	Copper	Copper	Copper	Fibreglass reinforced plastic
Tubular assembly arrangement	112 in series	112 in series	4 modules in parallel Each module 140 tubes in series	177 in series
Feed brine filter used	Cartridge	Cartridge	Sand	Sand

were estimated to be 1.94, 1.33, 0.50 and 0.37 dollars per 1000 gal of product water; the factors contributing to these costs are briefly as follows.

Plant 1. This represents the Coalinga pilot plant. The high cost of water production at this plant is largely due to amortisation of capital spent to build this plant both for reliable production and as a versatile experimental tool. The labour cost for membrane tube replacement, material cost for filter cartridge replacement, and the costs for pump and back pressure regulator maintenance are relatively high because of the small scale of the operation.

Plant 2. This represents a 5000 gal/day production plant. It would utilise a pump designed just for the pressure and flow needed, and incapable of flow variation. The tube rack would be for tube support only, and plastic sleeves around each tube would be omitted. Instrumentation costs would be minimised.

50,000 gal/day plants. In both Plants 3 and 4, the following design bases were utilised. (a) Two-thirds of the feed brine would be recovered as product water. (b) Tubes 20-ft long would be utilised instead of 10 ft tubes; this change effects two cost savings: first in amortisation costs, since only half as many end fit-

Table 8-26. Summary of operating costs in dollars per kilogallon

Data of Stevens and Loeb (1967)

Cost details	5 kgal/day		50 kgal/day	
	Plant 1	Plant 2	Plant 3	Plant 4
Electric power	0.098	0.087	0.070	0.091
Tube assembly replacement:				
(a) labour	0.237	0.237	0.119	0.038
(b) materials				
Membrane	0.003	0.003	0.003	0.002
Nylon	0.026	0.026	0.026	0.020
Supplies and maintenance				
(a) filter cartridge	0.076	0.076	—	—
(b) pump packing	0.042	0.042	0.004	0.004
(c) pressure regulator core	0.061	0.061	0.006	0.006
(d) gaskets	0.088	0.007	0.003	0.001
Operating labour:				
(a) replacements of tube assemblies	0.059	0.059	0.030	0.009
(b) tube-swabbing	0.019	0.019	0.018	0.006
(c) data-taking	0.175	0.175	0.018	0.018
Maintenance labour				
(a) for filters change cartridge	0.001	0.001	—	—
(b) pump packing replacement	0.017	0.017	0.002	0.002
(c) pressure regulator core replacement	0.008	0.008	0.001	0.001
Administration	0.084	0.084	0.021	0.011
Amortisation	0.734	0.331	0.141	0.125
Taxes and insurance	0.197	0.089	0.038	0.033
Interest on working capital	0.014	0.010	0.004	0.003
Total operating costs ($/kgal)	1.94	1.33	0.50	0.37

tings are required per ft^2 of membrane area; second, in labour costs necessary for the replacement of the membrane tube and nylon wrap; 20-ft long tubes can be fabricated about as quickly as 10 ft tubes, and hence, as with fittings, the labour cost per ft^2 of membrane area is cut in half. (c) A sand filter would be utilised, capable of removing all particulate matter coarser than 25 microns.

Plant 3. It was desired that this plant be immediately operational, i.e. that its design be based largely on components proven in the Coalinga runs. Thus copper tubes and flare-type fittings were chosen for the plant. The overall operational cost was reduced significantly because of the 10-fold increase in magnitude of operation and the use of longer tubes.

Plant 4. In this plant 2-in. fibreglass-reinforced plastic tubes having flanged end fittings would be used. A further modification in this plant is the use of

800 psig as the operating pressure, instead of 600 psig used in the other designs. By this increase in pressure, it was estimated that the average flux of desalinised water would be about 25 gal/day/ft^2. Additional savings in Plant 4 are due to decreases in operating labour costs for in-field tube assembly replacement and for cleaning tubes, because there are fewer tubes than in Plant 3.

16. Studies of Keilin and Co-workers

Keilin and DeHaven (1965) considered the design criteria of desalination plants built on the basis of flat membrane configuration for the membrane unit. Three possible methods for using flat plate membrane supports were evaluated; their main features are the following.

(a) The membrane unit is made up of top and bottom heads welded to flanges between which a stack of flat plates is interposed. Top and bottom heads are held together with tie-bolts.

(b) A solid vessel wall is substituted for the tie-bolts to hold the top and bottom heads together. The cylindrical vessel wall is not exposed to pressure, but serves only as a tension member to replace the tie-bolts. The membrane support plates serve to confine the feed stream.

(c) A complete pressure vessel is used to contain the feed water and channel it between plates. The plates would merely support the membrane and convey the product water, whereas the pressure vessel would contain the pressure.

Method (a) requires long tie-bolts, the full length of the vessel; methods (b) and (c) require only short bolts to join the two adjacent flanges. Method (b) requires a cylindrical shell about 60 per cent as thick as that needed for method (c). The studies showed that economically, the principal trade-off was between the incremental cost of the long tie-bolts in method (a), the cost of half thick shell in method (b), and that of the full thick shell in method (c). Based on such cost considerations, and other factors relating to seals on plates, seals in vessel, and plate materials and machining, it was concluded that method (c) was the most preferable.

Assuming a cylindrical pressure vessel containing flat plate membrane supports, two principal plate arrangements were considered, where plates were arranged with their planes either perpendicular or parallel to the vessel axis. It was concluded that the choice depended on the selection of vessel size, required length of feed flow path, and considerations of fluid dynamics.

It was proposed that in a large plant, the feed stream would be guided in a multilimb spiral baffle clamped between adjacent plates, all desalination plates would be in parallel flow relationship to the feed stream, and modules of plates would be preassembled outside the cell and then slipped into position in the pressure vessel. In a proposed design, allowing 0.125 in. for plate thickness, 0.100 in. for flow channel space, and 0.015 in. for two membranes and two filter papers, each plate and adjacent space would occupy 0.240 in., and approximately 50 plates would stack in 1 foot length. A pressure vessel 12 ft long would accommodate 5 modules (each containing 100 plates) with 2 ft allowance for module end plates and tube connections. The pressure vessel size was established as 40 in. outside diameter to accommodate 36 in. plates.

Based on these design details, Keilin and DeHaven estimated that a 10 mgd sea water conversion plant would cost $5.5 millions, and it would produce potable water at $0.44 per 1000 gal., whereas a brackish water conversion plant of the same capacity would cost $2.7 millions, and it would produce potable water at $0.21 per 1000 gal.

In a later report, Keilin (1966) has given another cost estimate for a 1 mgd plant; this is given in Table 8-27. In this estimate, all plants are assumed to have been constructed from a standard flat film configuration desalination cell, the number of cells and pumps varying. Each cell is assumed to contain approximately 1000 plates or 12, 700 ft^2 of membrane area.

17. Studies of Bray and Menzel

A parametric study of a mgd reverse osmosis plant operating on sea water feed has been made by the Gulf General Atomic, and the results have been reported by Bray and Menzel (1966) and OSW (1965). The study had as its primary objective the selection of system design pressure, water recovery ratio, and flow arrangements for the engineering design of a 50, 000 gal/day sea water feed prototype plant using the spiral-wound membrane modules.

The approach used was first to investigate a single-stage operation assuming a salt flux of five times 'theoretical'. A second stage which used as feed the product water of varying salt content from the first stage, was then considered. The first stage was designed for excess capacity so that the second stage could produce 1 mgd. An initial membrane flux of 1×10^{-5} g. $H_2O/cm^2/sec/atm$ was assumed, and the effect of decrease of this flux to 0. 5, 0. 7, and 0. 9 of initial values was then considered. System operating pressures of 720, 960, 1440, and 2160 psig were considered. These pressures were selected to correspond to the allowable pressures at which standard size carbon steel pipe could be used for the module pressure vessels.

The following economic ground rules were used for the study. For first-stage evaluation, all water produced had the same economic value, though most of the systems produced water with a salt content above the 500 ppm limit for potable water. The plant load factor was 0. 9. Power costs were 0. 7 mill/kWh. Plant operation was under the control of one operator with extra maintenance help for major equipment maintenance and for membrane replacement. The cost of the latter was an operating parameter. Annual capital charges were 9. 4 per cent.

Some of the results obtained in the study are given in Tables 8-28 and 8-29. It was found that for the first stage, optimum system pressure was about 1500 psig at water recoveries of over 50 per cent; 30 per cent of the capital cost was for the membrane units, and about 80 per cent of the plant cost was a function of plant size. As for the second stage, capital costs were relatively low, and not representative of brackish water systems, because no water pretreatment was need for this stage, the solute was essentially NaCl, and hence large concentration factors could be used without the problem of salt precipitation. Under the assumptions used in the study, an operating pressure of about 1000 psig, and product recoveries of 80 per cent appeared optimum for the second stage.

Similar parametric studies were carried out for a typical 1 mgd brackish water conversion unit using the Gulf General Atomic spiral-wound modules (OSW, 1966). Two systems were considered. Feed concentration was 5000 ppm of salt in both cases. In one system, highly selective 'Type A' membrane and an operating pressure of 960 psig were used; in the other system, a less selective 'Type C' membrane and an operating pressure of 720 psig were used. Type C membrane gave less separation but higher product rate than the Type A membrane; the former was adequate for brackish water conversion. It was estimated that the capital cost of the first plant would be approximately $700, 000, and it would produce potable water at a cost of about $0. 60/1000 gal; the capital cost of the second plant would be approximately $580, 000. and

Table 8-27. A typical cost estimate for saline water conversion by reverse osmosis

Data of Keilin (1966)

Description	Case A	Case B	Case C
Operating variables			
Total dissolved solids (ppm)	1500	5000	35,000
Flux, average (gal/day/ft^2)	40	20	10
Water recovery (%)	80	60	33
Reject stream (ppm)	6500	12,100	52,900
Membrane life (years)	1.0	1.0	0.5
Capital costs, dollars			
Desalination cells	124,000	248,000	496,000
Filters	35,000	46,000	83,000
Pumps	52,000	78,000	130,000
Aerator	10,000	14,000	25,000
Erection and assembly	66,000	116,000	220,000
Instruments	9000	15,000	29,000
Raw water supply	6000	8000	15,000
Product storage	10,000	10,000	10,000
Service facilities	22,000	39,000	73,000
Contingencies	33,000	57,000	108,000
Engineering	37,000	63,000	119,000
Interest on investment	16,000	28,000	52,000
Site preparation	3000	3000	3000
Working capital	17,000	24,000	44,000
Total	440,000	749,000	1,407,000
Operating costs, dollar/1000 gal			
Electric power ($0.007/kWh)	0.068	0.085	0.163
Membranes	0.008	0.015	0.060
Supplies	0.006	0.011	0.022
Maintenance and labour	0.060	0.060	0.180
Payroll extras	0.008	0.008	0.024
Administrative overheads	0.022	0.024	0.070
Amortisation	0.095	0.163	0.306
Taxes and insurance	0.025	0.043	0.082
Interest on capital	0.002	0.003	0.005
Total*	0.294	0.412	0.912

* The paper under reference (Keilin, 1966) gives the total costs as $0.288, 0.401, and 0.728 respectively.

Operating parameters:
- Plant capacity: 10^6 gal/day
- Total dissolved solids in product: 250 ppm
- Operating pressure: 750 psig
- Electrical efficiency: 80%

Table 8-28. Pumping power requirements, capital costs, and product water costs for single stage one million gallon per day plant

Data of Bray and Menzel (1966)

Operating pressure (psig)	Product water to brine ratio											
	0.5	1.0	1.5	2.0	0.5	1.0	1.5	2.0	0.5	1.0	1.5	2.0
	Power requirement (kWh/1000 gal)				Capital cost (dollars)*				Product water cost (cents/1000 gal)			
720	14.0	10.5	9.3	8.7	892,260	933,810	1,103,500	1,384,640	114.3	136.9	179.3	250.0
960	18.2	13.7	12.2	11.4	870,350	838,790	878,680	932,970	89.5	93.2	105.5	122.8
1440	26.5	20.0	17.8	16.7	868,130	790,780	772,030	761,920	81.1	75.5	76.4	79.5
2160	38.9	29.4	26.2	24.6	1,058,040	894,850	865,160	837,070	91.0	79.2	76.5	76.4

* The equipment is designed for a 25% excess flow capacity.

Feed: Sea water
Pump efficiency = 0.81
Recovery turbine efficiency: 0.66
Initial flux = 1×10^{-5} g. H_2O/cm/sec/atm
Compaction factor = 0.5
Membrane replacement cost = 50 cents/ft^2/year

Table 8-29. Capital costs and product water costs for second stage of one million gallon per day plant

Data of Bray and Menzel (1966)

	Salt concentration in feed water (ppm)							
Operating pressure (psig)	1000	3000	5000	7000	1000	3000	5000	7000
	Capital cost (dollars)				Product water cost (cents/1000 gal)			
720	206,200	214,400	224,600	239,000	34.2	36.8	40.1	44.2
960	234,400	242,600	250,800	261,800	33.3	35.0	37.0	39.4
1440	272,500	275,900	282,800	289,600	35.2	36.1	37.2	38.4
2160	350,200	355,300	360,300	365,400	42.4	43.1	43.8	44.5

Feed: Product from first stage
Product water to brine ratio = 4
Initial flux = 1×10^{-5} g. $H_2O/cm^2/sec/atm$
Compaction factor = 0.5
Membrane replacement cost = 50 cents/ft^2/year

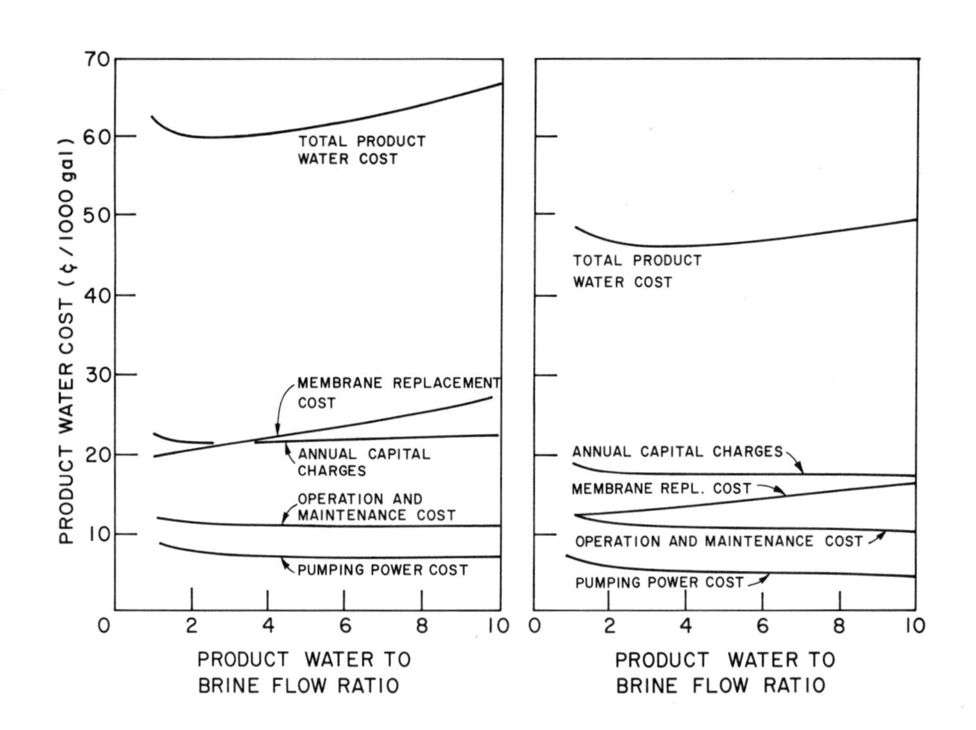

Figure 8-38. Results of Gulf General Atomic parametric studies for reverse osmosis brackish water conversion. Output: 10^6 gal/day. Feed: 5000 ppm NaCl. Membrane type A (left): Operating pressure, 960 psig. Membrane type C (right): Operating pressure, 720 psig (OSW, 1966)

it would produce potable water at a cost of about \$0.47/1000 gal. A breakdown of the respective product water costs is given in Figure 8-38.

18. Studies of Johnson et al.

Johnson et al. (1967) made some parametric studies on the design of a mgd sea water conversion plant operating at 1500 psig using flat or tubular membranes. They considered the possibilities of single-stage and two-stage operations, using single-step (constant cross-sectional area) or multi-step (variable cross-sectional area) channels for fluid flow. Their calculations took into account pressure drop, concentration polarisation, and the effect of pressure on solute separation. They concluded that (1) for large plants it would be advantageous to use a multi-step flow system in which the total cross-sectional area of flow channels was reduced stepwise as the brine approached its reject concentration, and (2) long flow channels would be appropriate, and tubes of 0.1 to 0.5-in. diameter or wide rectangular channels with a gap between membranes of 0.1 in. behaved similarly from the point of view of boundary layer and power requirement. They estimated that product water containing 500 ppm could be obtained from sea water in a two-stage process at a cost of 65 to 70 cents per 1000 gallons. Some of their other results are given in Tables 8-30 and 8-31.

19. Studies of Ohya and Sourirajan

The general equations for reverse osmosis process design derived in Chapter 4 offer an analytical means for parametric studies on water desalination systems. This is illustrated by the work of Ohya and Sourirajan (1969b). This work, which is described below, draws attention to some of the parameters of process design, which are important in practical desalination plants, and to the areas which require further experimental and theoretical studies. The approach used in the following discussion is from the point of view of reverse osmosis system specification and membrane performance, and the symbols used are the same as those given in Chapter 4.

Performance Data for Some Reverse Osmosis Systems. Using the general equations given in Chapter 4, Δ v. $\overline{C}_3$, τ (or X), and C_2 values have been calculated for several reverse osmosis systems. These are specified as follows:

$$\gamma = 0, 0.1, 0.3, \text{ or } 0.5$$

$$\theta = 0.001, 0.002, 0.003, 0.004, 0.005, \text{ or } 0.007$$

and

$$\lambda\theta = \infty, 5, 2, \text{ or } 1$$

These data are plotted in Figures 8-39, 8-40 and 8-41. These plots are general, and they describe the performance of any reverse osmosis system specified by the given values of γ, θ, and $\lambda\theta$. The data show that in the range of the θ values considered, the values of $\overline{C}_3$, C_2, and τ (or X) for given values of Δ are functions of γ and $\lambda\theta$, and, in addition, $\overline{C}_3$ is very sensitive to changes in the values of θ.

Application to Saline Water Conversion. The values of γ, θ, $\lambda\theta$ chosen above are of practical interest in saline water conversion. For purposes of illustration, consider the desalination of an aqueous solution of sodium chloride containing 3.5 wt. % NaCl, to give product water containing 500 or 300 ppm

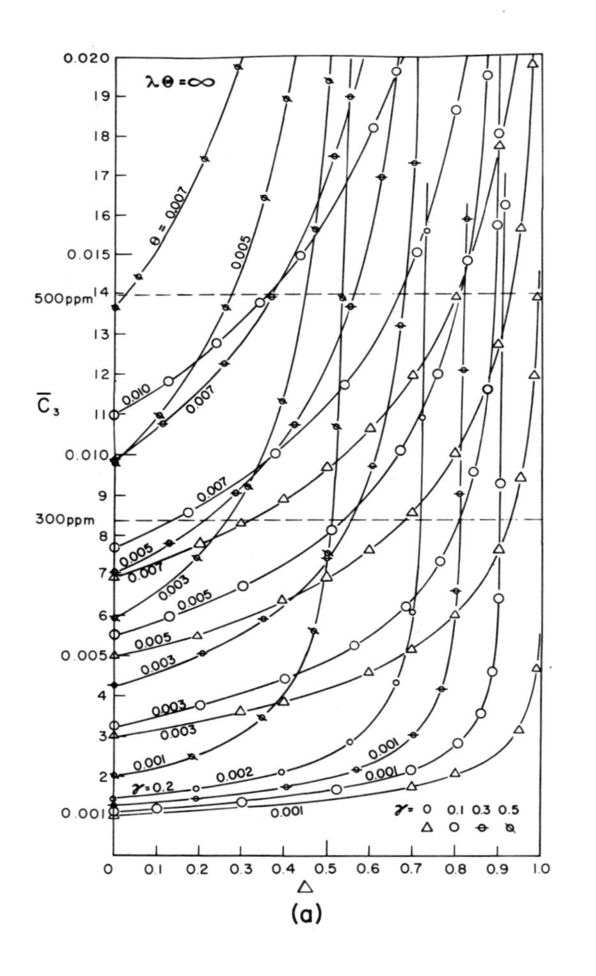
λΘ = ∞
θ = 0.007
500ppm
300ppm
$\bar{C}_3$
Δ
γ = 0.2
γ = 0 0.1 0.3 0.5
(a)

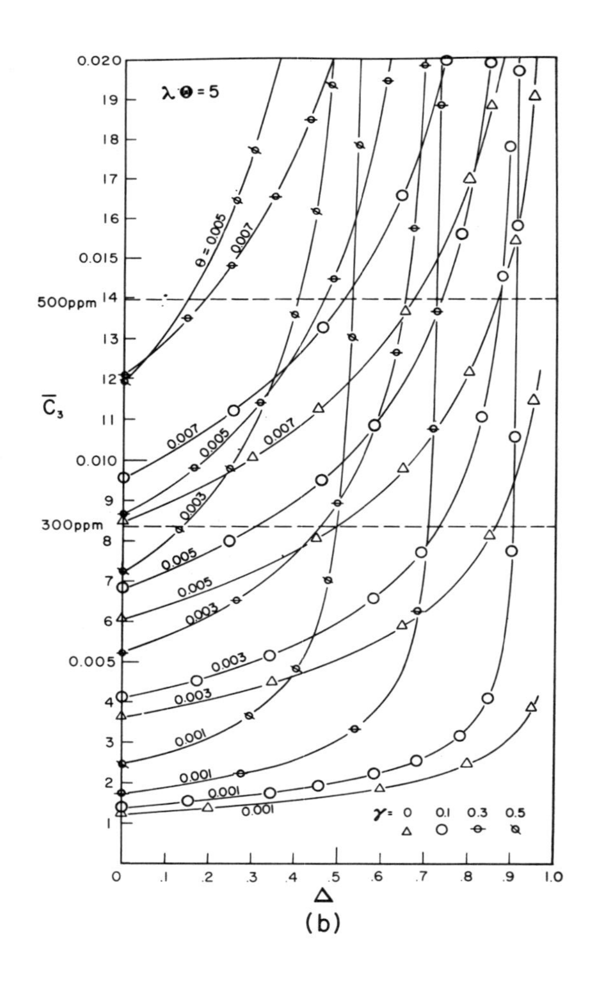
λΘ = 5
θ = 0.005
500ppm
300ppm
$\bar{C}_3$
Δ
γ = 0 0.1 0.3 0.5
(b)

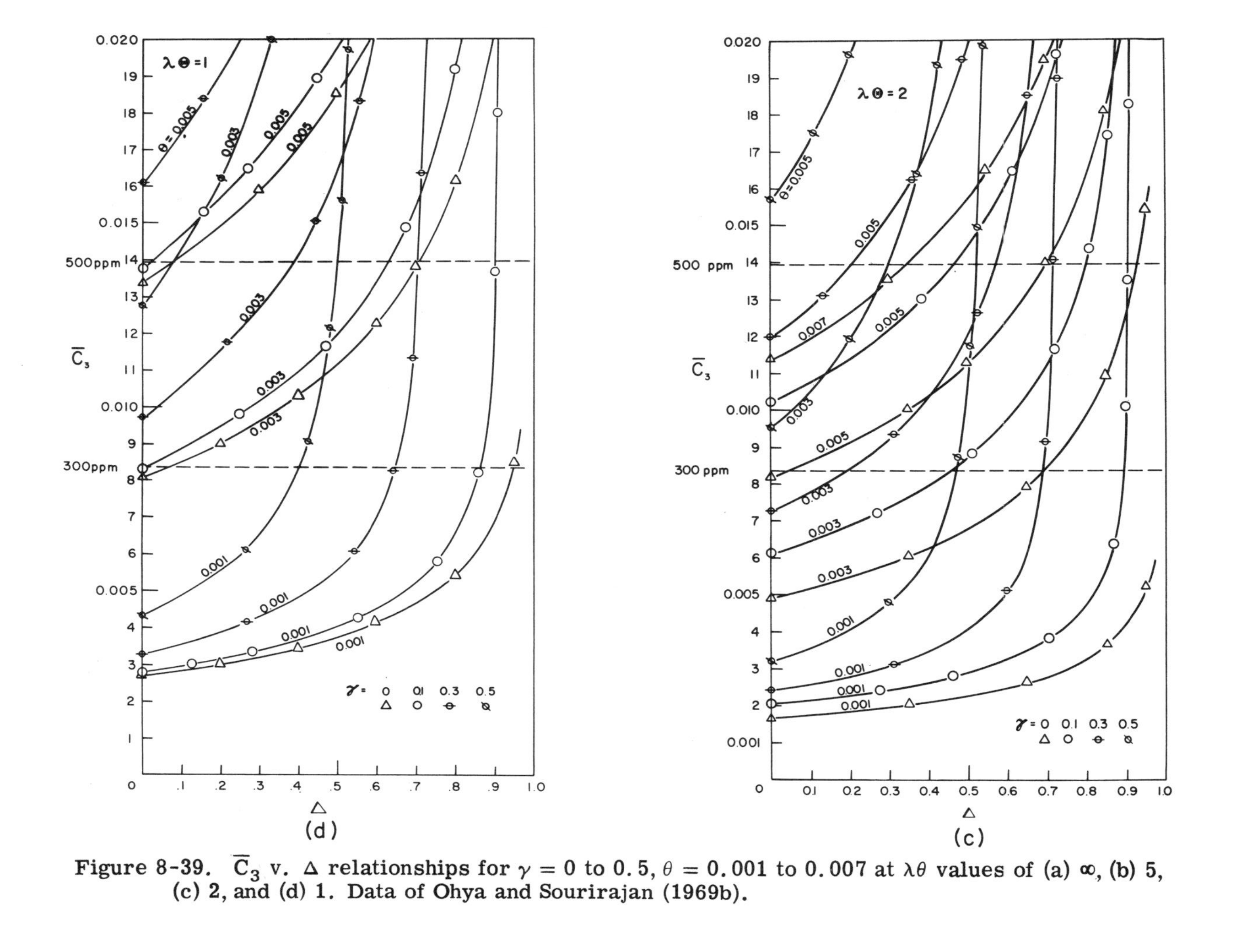

Figure 8-39. $\overline{C}_3$ v. Δ relationships for $\gamma = 0$ to 0.5, $\theta = 0.001$ to 0.007 at $\lambda\theta$ values of (a) ∞, (b) 5, (c) 2, and (d) 1. Data of Ohya and Sourirajan (1969b).

Table 8-30. Comparison of membrane areas in tubular assembly

Data of Johnson et al. (1967)

	Membrane area ($ft^2/10^6$ gal/day)				Feed velocity (ft/sec)			
Power MW	0.1 in. diam. tube		0.5 in. diam. tube		0.1 in. diam. tube		0.5. in diam. tube	
10^6 gal/day	a	b	a	b	a	b	a	b
Feed power								
5	42,700		43,000		2.9		5.3	
4.5	42,800	43,100	43,100	43,000	3.2	4.0	5.7	5.3
4.0	42,900	43,180	43,200	43,180	3.6	4.2	6.2	5.7
3.5	43,000	43,290	43,750		4.2	4.5	7.1	
3.0	43,400	43,480			5.0	4.8		
2.75	44,000							
Net power								
2.2			43,050				5.0	
2.0	42,900		43,150		3.0		5.6	
1.8	42,920		43,700		3.45		7.25	
1.6	43,070				4.2			
1.5	43,250				4.8			
1.45	43,700				5.25			

a Step terminal velocity = 0.6 × step feed velocity

b Step terminal velocity = 0.92 × step feed velocity

Operating pressure: 1500 psig
Feed: sea water
Product concentration: 5000 ppm

of salt. The mole fraction of salt in feed solution (X^0_{A1}) is 11.045×10^{-3}, and that in product solution ($\overline{X}_{A3}$) is 0.15403×10^{-3} or 0.09244×10^{-3} respectively. The corresponding $\overline{C}_3$ ($= \overline{X}_{A3}/X^0_{A1}$) are 0.0139 or 0.00834 respectively. Horizontal lines at the above values of $\overline{C}_3$ are drawn in Figure 8-39 which gives directly the γ, θ, and $\lambda\theta$ values corresponding to the given quality of product water along with the obtainable fraction product recovery; the latter of course is also limited by the maximum permissible value of $C_2 = 9$ which corresponds to a saturated solution of NaCl at the concentrated boundary layer at 25°.

θ v. Δ and τ (or X) correlations. Using Figures 8-39 and 8-40, the correlations of θ v. Δ and τ (or X) for various values of γ, and $\lambda\theta$ can be obtained for any specified value of $\overline{C}_3$. Such correlations are given in Figure 8-42 to 8-45 for $\overline{C}_3$ values corresponding to 500 or 300 ppm. Figures 8-42 to 8-45 are of practical interest in water desalination. They specify the reverse osmosis systems in terms of θ, γ, and λ which can given product water containing 500 ppm or less from 3.5 per cent NaCl-H_2O feed solutions. Figures 8-42 and 8-43 show that high recoveries of product water containing 500 or 300 ppm of

Table 8-31. Power and membrane area required rectangular channel 0.5 in. × 0.1 in.

Data of Johnson et al. (1967)

Power ($MW/10^6$ gal/day)	Membrane area ($ft^2/10^6$ gal/day)	Feed velocity (ft/sec)
Feed power		
5	42,800	3.7
4	43,050	4.3
3	43,800	5.7
Net power		
3	42,700	3.0
2.5	42,780	3.5
2.0	43,000	4.2
1.6	43,700	5.5
1.5	45,200	7.0

Operating pressure: 1500 psig
Feed: sea water
Product concentration: 5000 ppm
Step terminal velocity = 0.6 × feed velocity

salt can be obtained in a single stage from 3.5 per cent aqueous sodium chloride solutions by the proper choice of γ, θ, and $\lambda\theta$; from the τ values given in Figures 8-44 and 8-45, the membrane areas needed per unit quantity of product water can be calculated.

Performance of Some Real Membranes. Table 8-32 gives the specifications of two typical Batch 18 type porous cellulose acetate membranes for water desalinations. These specifications for the films were established from reverse osmosis experiments in the operating pressure range 40 to 102 atm; for the purpose of this illustration, these specifications are assumed valid for the operating pressure range up to 300 atm. Using Equations 3-78 and 3-79 (in Chapter 3) to express the effect of operating pressure on A and $(D_{AM}/K\delta)$, the values of γ, and θ for the reverse osmosis system using the above films can be calculated for the operating pressure range 50 to 300 atm. The results of such calculations are also given in Table 8-32.

The performance of the above films in reverse osmosis desalination can be expressed by plotting the γ and θ values given in Table 8-32 in Figures 8-42 and 8-43 giving rise to specific θ v. Δ relationships from which the corresponding θ v. τ (or X) relationships can also be obtained using Figure 8-40. The above membrane performance data are also plotted in Figures 8-42 to 8-45.

The effect of membrane compaction (A-factor) on solute separation and product rate has been discussed in Chapter 3. During long-time continuous reverse osmosis operation, A decreases due to membrane compaction but $(D_{AM}/K\delta)$ remains constant provided the surface pore structure of the membrane remains unchanged (Kimura and Sourirajan, 1968b). Consequently, as a result of membrane compaction, v_W^* and hence θ change; the new values of θ, and the values of γ given in Table 8-32 can also be plotted in Figures 8-42 to

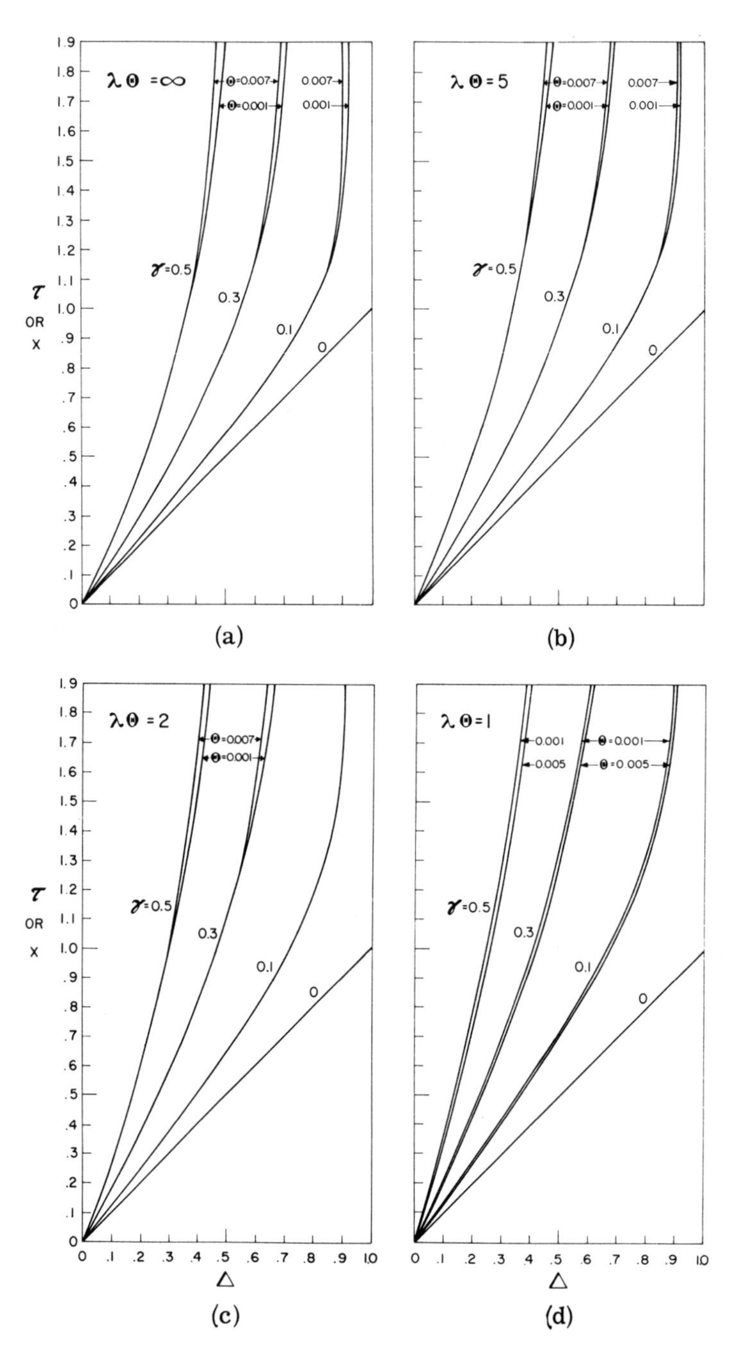

Figure 8-40. τ (or X) v. Δ relationships for $\gamma = 0$ to 0.5, $\theta = 0.001$ to 0.007, at $\lambda\theta$ values of (a) ∞, (b) 5, (c) 2, and (d) 1. Data of Ohya and Sourirajan (1969b).

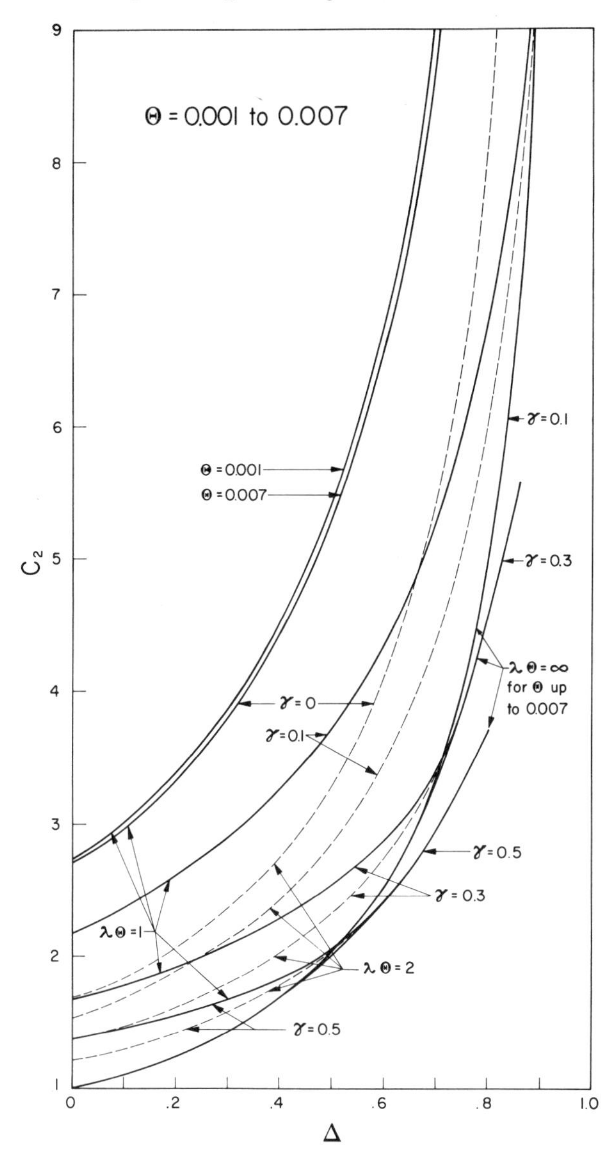

Figure 8-41. C_2 v. Δ relationships for $\gamma = 0$ to 0.5, $\theta = 0.001$ to 0.007 at $\lambda\theta$ values of ∞, 2, and 1. Data of Ohya and Sourirajan (1969b).

8-45 for different values of $\lambda\theta$ giving rise to new θ v. Δ, and hence θ v. τ (or X) relationships expressing the performance of compacted membrane. This is illustrated for film F3 in Figures 8-42 to 8-44 for A-factor = 0.5.

Table 8-33 gives a set of performance data for films F3 expressed in terms of fraction product recovery (Δ) of water containing 500 ppm of NaCl, and the corresponding τ values at different operating pressures and $\lambda\theta$ values. Similar data can be read from Figures 8-42 and 8-43 for film 1 also. The results of this analysis show that product water containing less than 500 ppm of NaCl can be obtained using films F3 or 1 from aqueous 3.5 per cent NaCl feed solutions at operating pressures of 102 atm with fraction product recoveries in

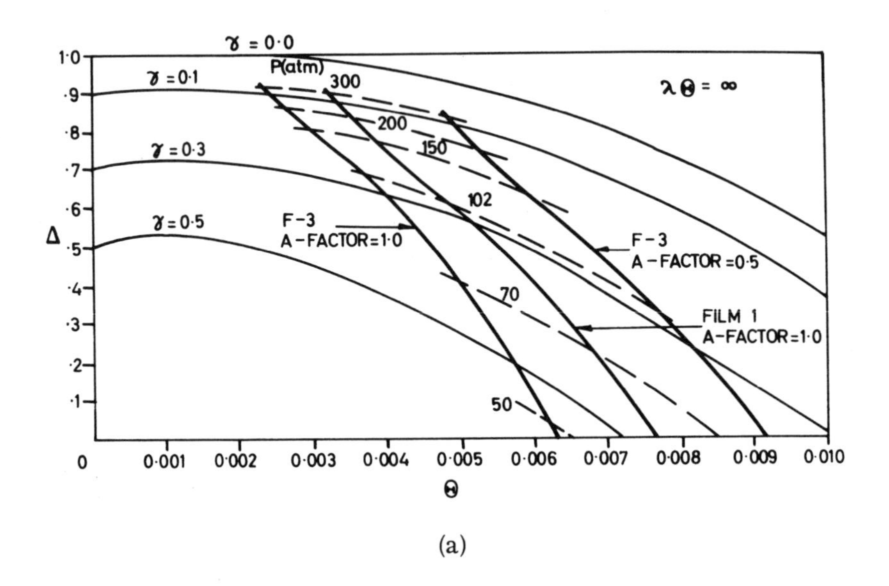

(a)

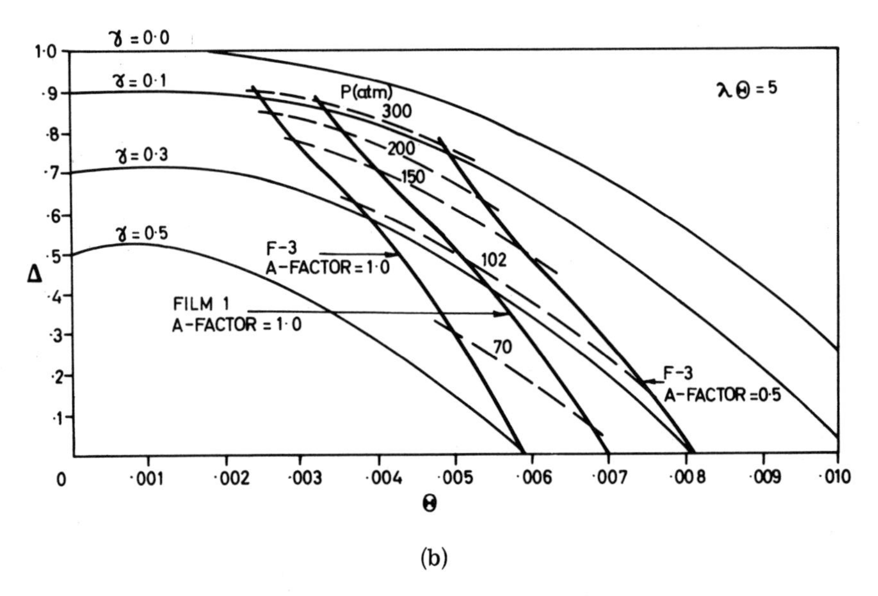

(b)

Figure 8-42. θ v. Δ relationships for product water containing 500 ppm of NaCl from 3.5% NaCl feed solutions, for $\gamma = 0$ to 0.5, at $\lambda\theta$ values of (a) ∞, (b) 5, (c) 2, and (d) 1. Data of Ohya and Sourirajan (1969b).

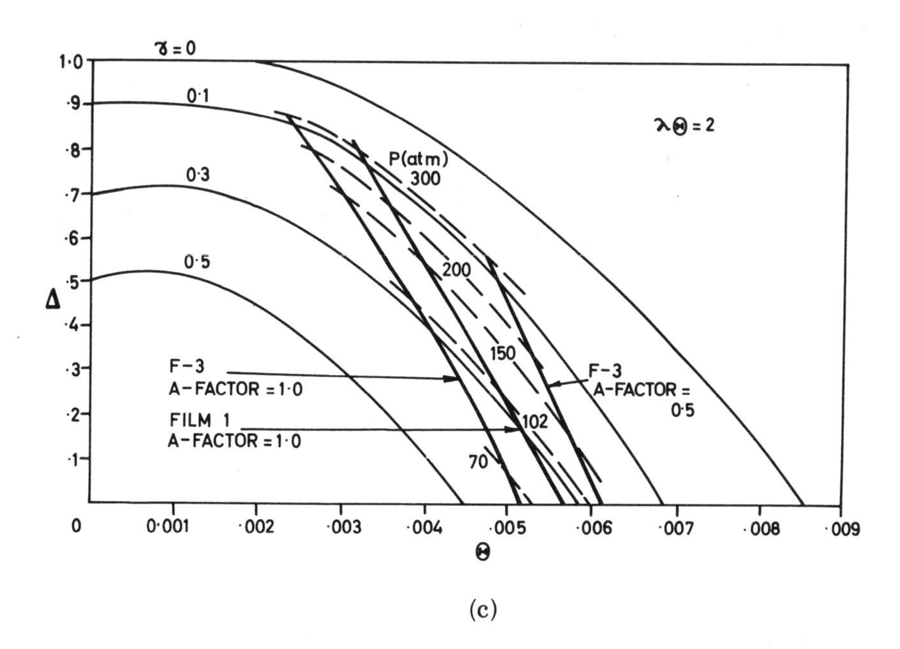

γ = 0
0·1
0·3
0·5
λΘ = 2
P(atm)
300
200
150
102
70
F-3
A-FACTOR = 1·0
FILM 1
A-FACTOR = 1·0
F-3
A-FACTOR =
0·5
Δ
Θ
1·0
·9
·8
·7
·6
·5
·4
·3
·2
·1
0
0·001
·002
·003
·004
·005
·006
·007
·008
·009

(c)

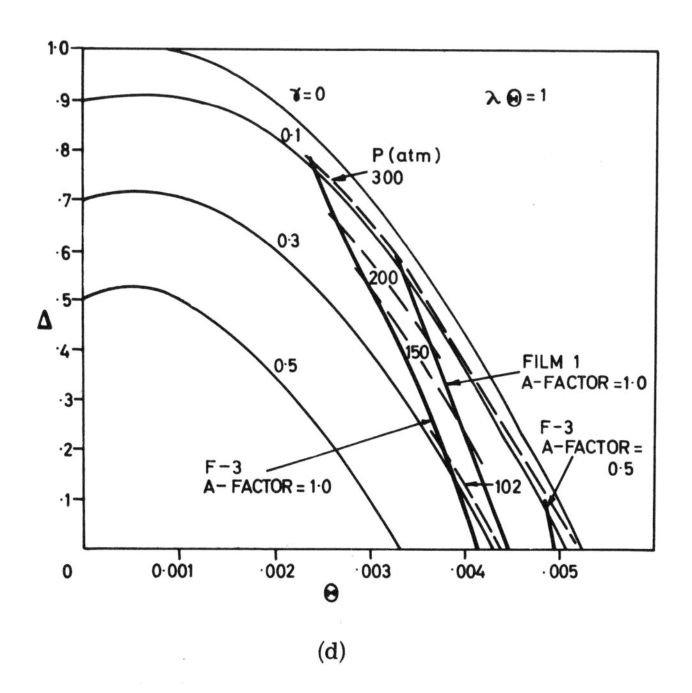

γ = 0
0·1
0·3
0·5
λΘ = 1
P (atm)
300
200
150
102
FILM 1
A-FACTOR = 1·0
F-3
A-FACTOR =
0·5
F-3
A- FACTOR = 1·0
Δ
Θ
1·0
·9
·8
·7
·6
·5
·4
·3
·2
·1
0
0·001
·002
·003
·004
·005

(d)

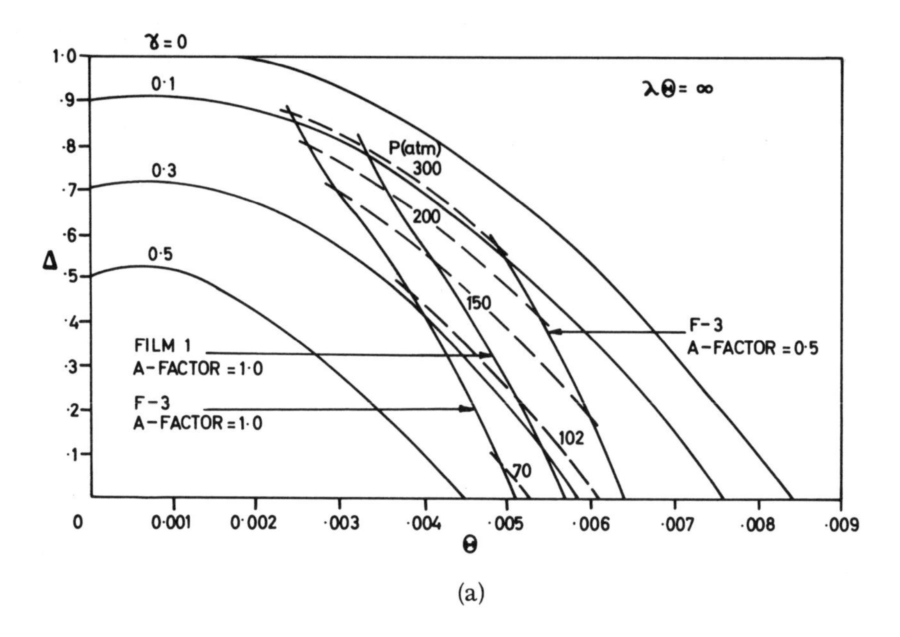

(a)

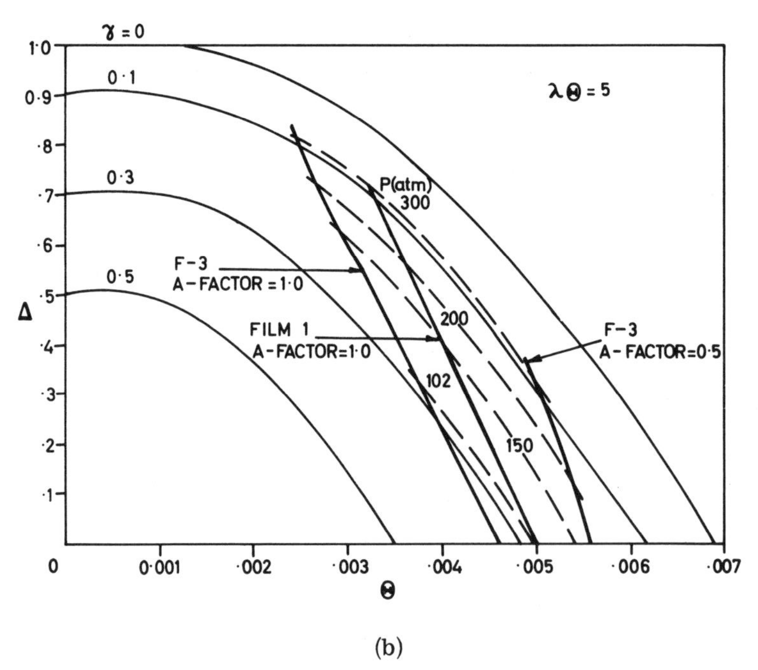

(b)

Figure 8-43. θ v. Δ relationships for product water containing 300 ppm of NaCl from 3.5% NaCl feed solutions, for $\gamma = 0$ to 0.5, at $\lambda\theta$ values of (a) ∞, (b) 5, (c) 2, and (d) 1. Data of Ohya and Sourirajan (1969b).

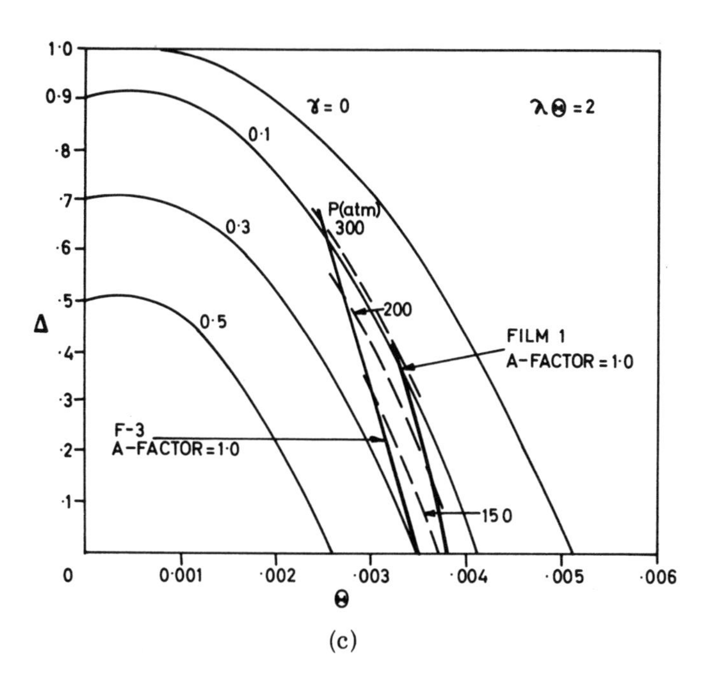
1·0
0·9
·8
·7
·6
·5
·4
·3
·2
·1
Δ
0
0·001
·002
·003
·004
·005
·006
Θ
γ = 0
λΘ = 2
0·1
0·3
0·5
P(atm)
300
200
150
FILM 1
A-FACTOR = 1·0
F-3
A-FACTOR = 1·0

(c)

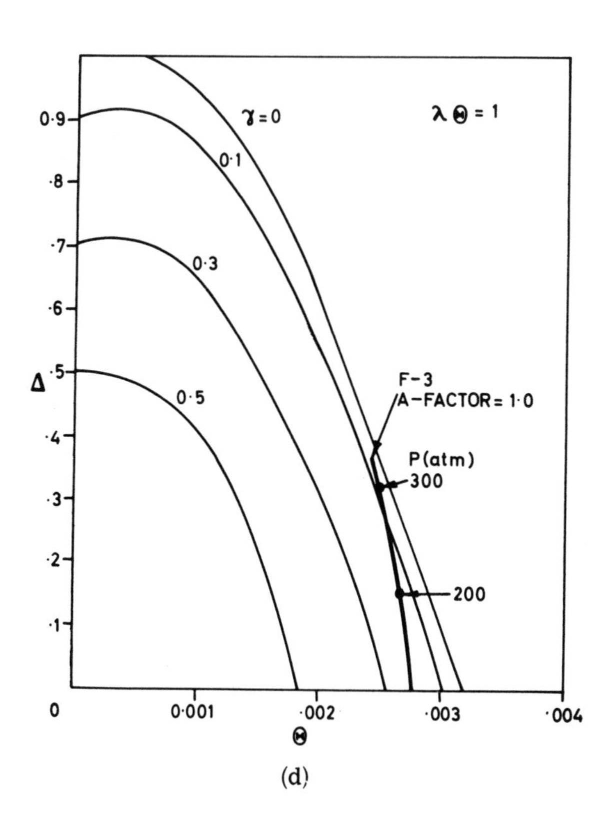
0·9
·8
·7
·6
·5
·4
·3
·2
·1
Δ
0
0·001
·002
·003
·004
Θ
γ = 0
λΘ = 1
0·1
0·3
0·5
F-3
A-FACTOR = 1·0
P(atm)
300
200

(d)

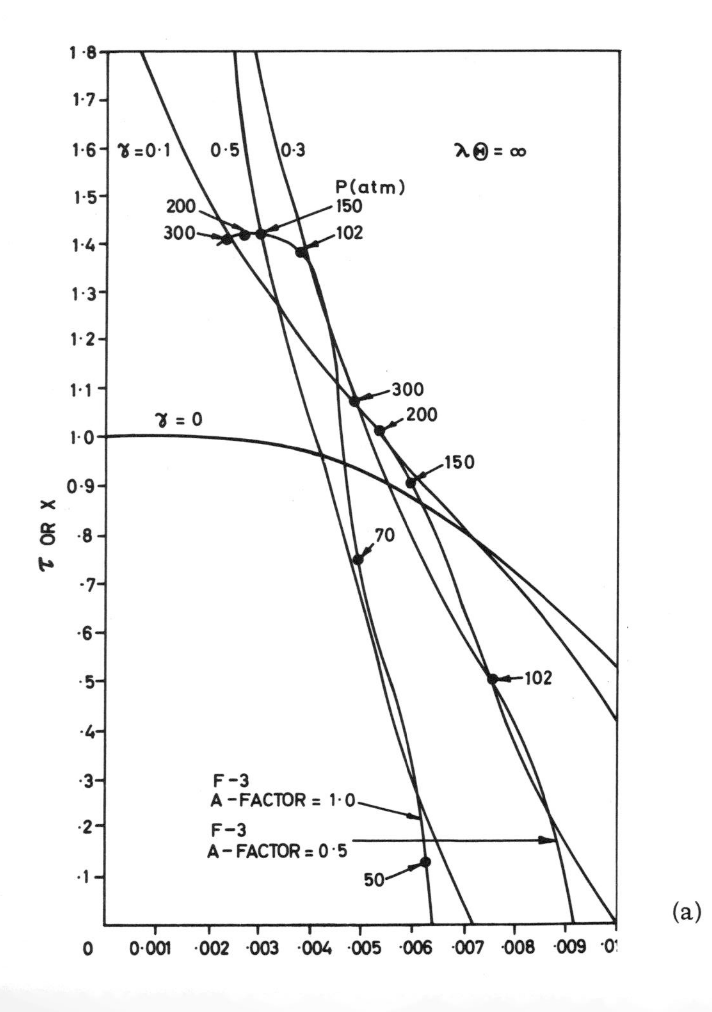

1·8
1·7
1·6
1·5
1·4
1·3
1·2
1·1
1·0
0·9
·8
·7
·6
·5
·4
·3
·2
·1
0
τ OR X
γ = 0·1
0·5
0·3
λ⊛ = ∞
P(atm)
200
300
150
102
300
200
150
γ = 0
70
102
F-3
A-FACTOR = 1·0
F-3
A-FACTOR = 0·5
50
0·001
·002
·003
·004
·005
·006
·007
·008
·009

(a)

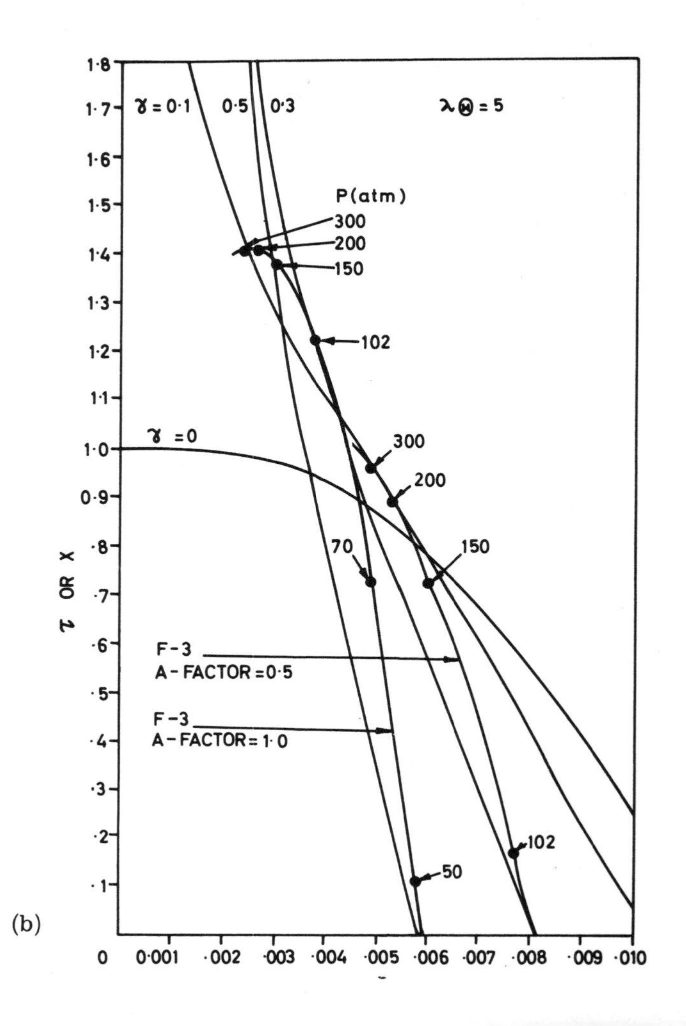

1·8
1·7
1·6
1·5
1·4
1·3
1·2
1·1
1·0
0·9
·8
·7
·6
·5
·4
·3
·2
·1
0
τ OR X
γ = 0·1
0·5
0·3
λ⊛ = 5
P(atm)
300
200
150
102
γ = 0
300
200
70
150
F-3
A-FACTOR = 0·5
F-3
A-FACTOR = 1·0
102
50
0·001
·002
·003
·004
·005
·006
·007
·008
·009
·010

(b)

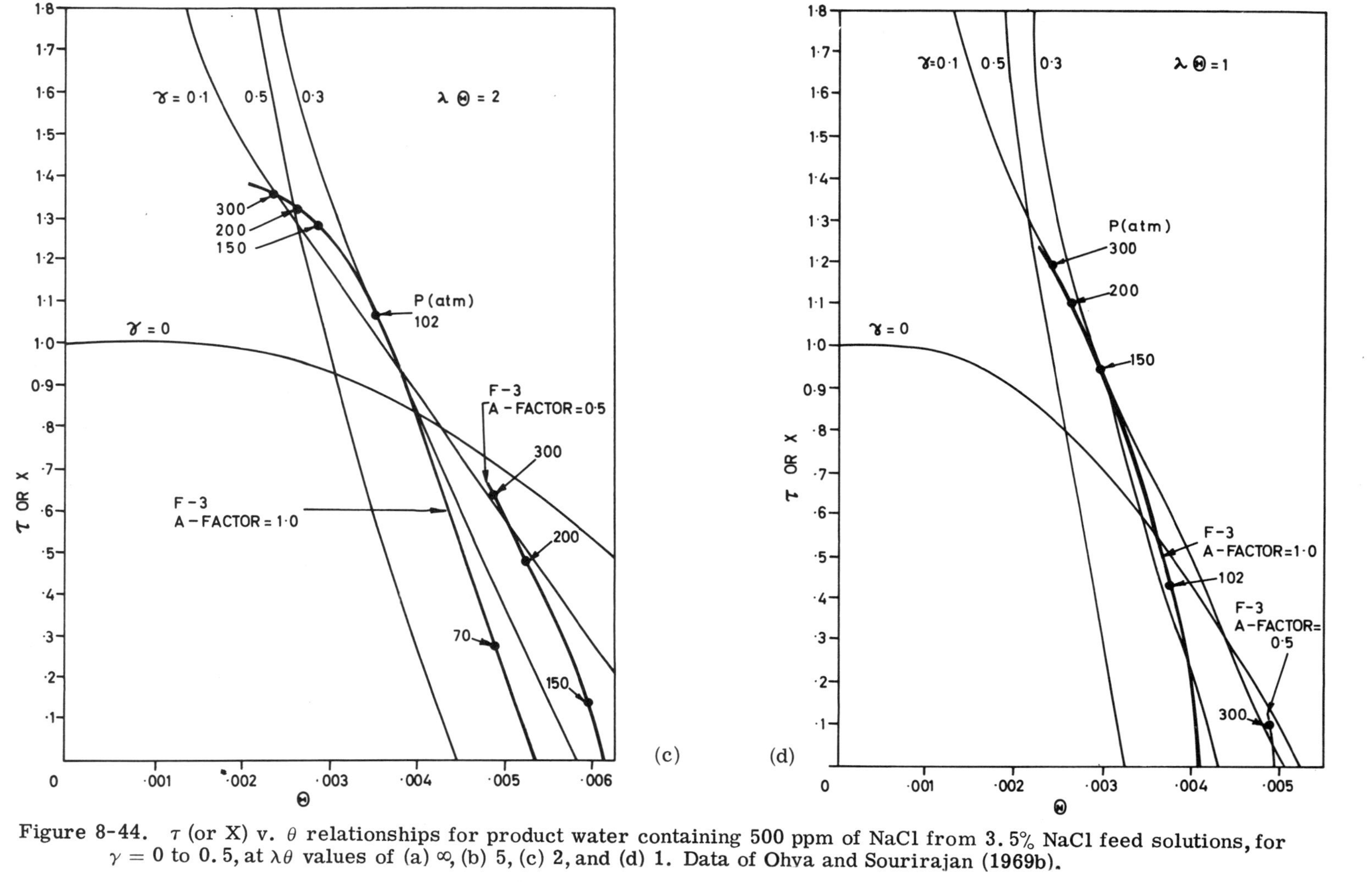

Figure 8-44. τ (or X) v. θ relationships for product water containing 500 ppm of NaCl from 3.5% NaCl feed solutions, for $\gamma = 0$ to 0.5, at $\lambda\theta$ values of (a) ∞, (b) 5, (c) 2, and (d) 1. Data of Ohya and Sourirajan (1969b).

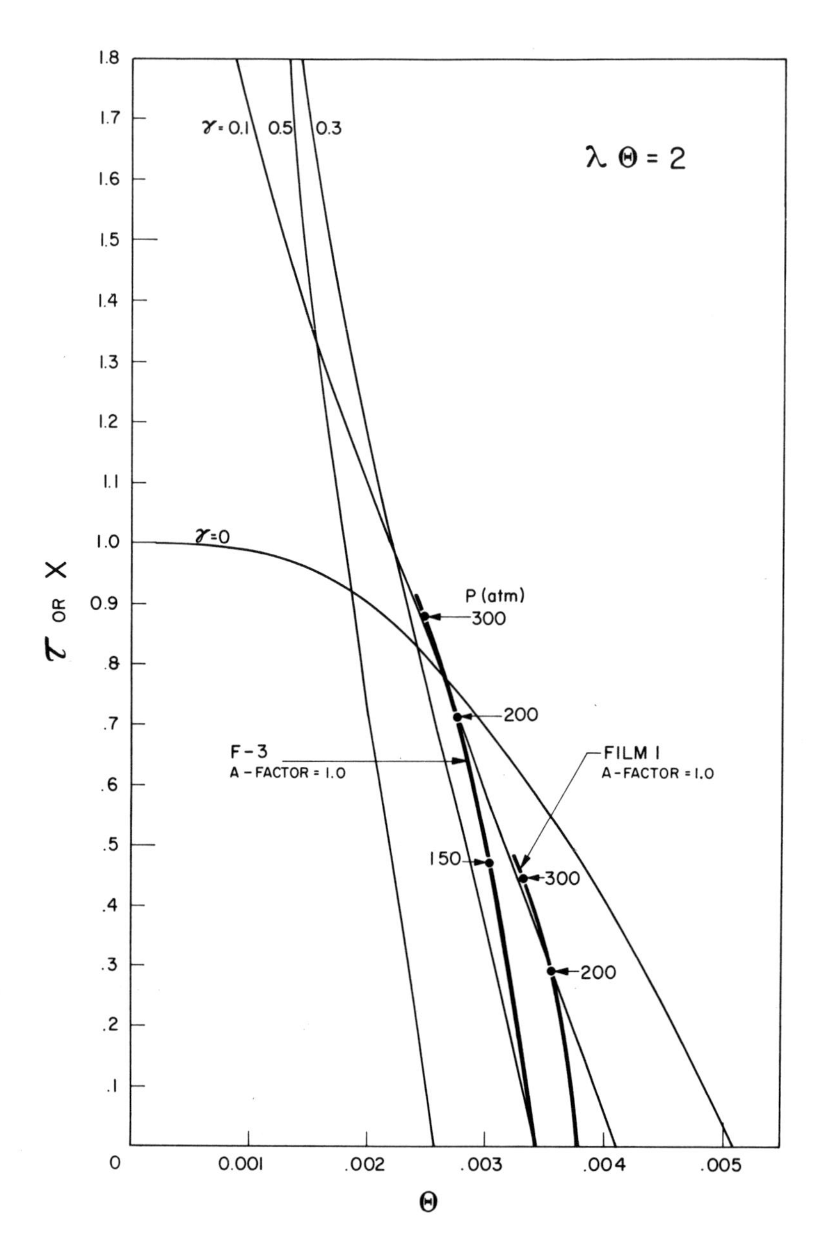

Figure 8-45. τ (or X) v. θ relationships for product water containing 300 ppm of NaCl from 3.5% NaCl feed solutions for $\gamma = 0$ to 0.5, at $\lambda\theta = 2$. Data of Ohya and Sourirajan (1969b).

Table 8-32. Specifications of two typical porous cellulose acetate membranes for reverse osmosis water desalination

Data of Ohya and Sourirajan (1969b)

Film specifications at 102 atm

	Film 1	Film F3	
A	0.97×10^{-6}	1.295×10^{-6}	$\frac{g.mole.H_2O}{cm^2\ sec.atm}$
$\left(\frac{D_{AM}}{K\delta}\right)_{NaCl}$	0.90×10^{-5}	0.891×10^{-5}	cm/sec
α	0.0033	0.0033	
β	0	0	

Feed solution: 3.5% $NaCl-H_2O$, $\Pi(X_{A1}^0) = 28.2$ atm

P atm		50	70	102	150	200	300
γ		0.565	0.402	0.276	0.188	0.141	0.094
θ	1	0.0087	0.00664	0.00505	0.00402	0.00356	0.00330
	F3	0.0062	0.00490	0.00376	0.00298	0.00264	0.00244

Film type: CA-NRC-18

the range 0.68 to 0.22 provided $\lambda\theta$ is very high approaching infinity; under the latter conditions, the reverse osmosis plant can be operated even at pressures as low as 50 atm with film F3 (A-factor = 1.0) and still produce water containing 500 ppm of NaCl at, of course, very low Δ values. At an operating pressure of 150 atm, product water containing <500 ppm of NaCl can be obtained using films F3 or 1 even when $\lambda\theta = 2$ and A-factor = 0.5. Provided the operating pressure is sufficiently high (~150 atm) and the mass transfer coefficient on the high pressure side of the membrane is also sufficiently high ($\lambda\theta > 2$) product water containing <500 ppm of NaCl can be obtained from aqueous 3.5 per cent NaCl solutions in a single-stage operation even with the presently available porous cellulose acetate membranes such as films F3 and 1. The above results illustrate the importance of the operating pressure and $\lambda\theta$ in reverse osmosis operation, and the necessity for reducing membrane compaction for obtaining good membrane performance.

Work requirements. Consider unit volume of feed solution at an operating pressure of P atm; let fraction product recovery be Δ, and fraction power recovery be η in a reverse osmosis unit.

$$\left.\begin{matrix}\text{Work required per unit}\\ \text{volume of product recovered}\end{matrix}\right\} = \frac{P\Delta + (1-\eta)P(1-\Delta)}{\Delta} \qquad (8\text{-}1)$$

Since (1 atm) (1 kilogallon) ≡ 0.1066 kWh, the work requirements in terms of kWh per kgal of product recovered can be calculated from Equation 8-1 using

Table 8-33. Performance of Film F3

Data of Ohya and Sourirajan (1969b)

A-Factor	1.0				0.5			
$\lambda\theta$	∞	5	2	1	∞	5	2	1
P atm	Δ Values							
50	0.032	—	—	—	—	—	—	—
70	0.418	0.310	0.098	—	—	—	—	—
102	0.688	0.622	0.462	0.197	0.340	0.135	—	—
150	0.800	0.775	0.702	0.533	0.625	0.486	0.075	—
200	0.860	0.845	0.792	0.660	0.740	0.640	0.335	—
300	0.910	0.895	0.858	0.755	0.825	0.755	0.495	0.065
P atm	τ Values							
50	0.080	—	—	—	—	—	—	—
70	0.800	0.730	0.280	—	—	—	—	—
102	1.377	1.200	0.955	0.435	0.500	0.220	—	—
150	1.420	1.385	1.260	0.945	0.905	0.727	0.135	—
200	1.415	1.405	1.320	1.100	1.010	0.890	0.485	—
300	1.410	1.401	1.360	1.180	1.065	0.965	0.640	0.100

Feed concentration = 3.5% NaCl
NaCl in product = 500 ppm

the appropriate P, Δ, and η values for any particular film and operating conditions. A set of such calculations has been made for film F3 for A-factor = 1.0, and 0.5, and $\eta = 0$ and 0.5 using the data given in Table 8-33. The results are presented in Figure 8-46. The results show the existence of an optimum operating pressure for the reverse osmosis unit depending on the values of $\lambda\theta$, η, and A-factor. Assuming that $\eta = 0.5$, and $\lambda\theta = 2$ are practical, the optimum operating pressures are 110 atm for A-factor = 1.0, and 220 atm for A-factor = 0.5; under these conditions, the work requirements are 16.4 and 40 kWh per kgal respectively; these values for the work requirement do not take into consideration friction and other losses. Hence the actual work requirements should be higher than those given above.

Membrane area requirements. From Equation 4-368 (in Chapter 4), the membrane area S is given by the relation

$$S = \frac{\tau V_1^0}{v_w^* t} \tag{8-2}$$

Since

$$(V_1^0/t) = \frac{\text{Product rate}}{\Delta} \tag{8-3}$$

and

$$v_w^* = \frac{AP}{c} = \frac{A_0 e^{-\alpha \bar{P}} P}{c} \tag{8-4}$$

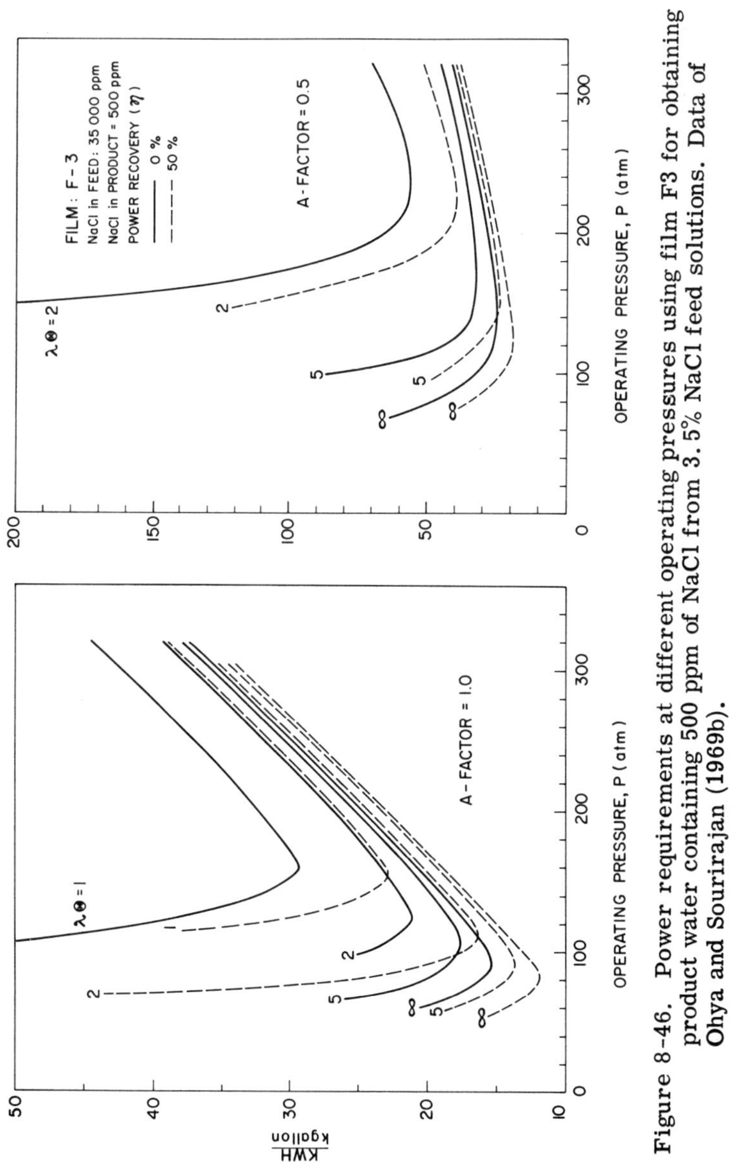

Figure 8-46. Power requirements at different operating pressures using film F3 for obtaining product water containing 500 ppm of NaCl from 3.5% NaCl feed solutions. Data of Ohya and Sourirajan (1969b).

and τ values are given in Figures 8-44 and 8-45, the total membrane area requirements per 10^6 gallons of product water per day using any film can be calculated using Equation 8-2. The results of such calculations for film F3 is given in Figure 8-47 for different values of P, $\lambda\theta$, and A-factor. The values of S decrease with increase in P and $\lambda\theta$; a $\lambda\theta$ value of 2 appears both necessary and sufficient to decrease membrane area requirements to a practical minimum under high operating pressures.

A Comparative Cost Analysis. A set of cost calculations has been made for a one million gallon per day plant operating for 330 days per year using the following assumed unit costs.

Cost of pump:	\$83/kW
Cost of power:	\$0.005/kWh
Cost of membrane (life 6 months):	\$0.27/ft^2
Cost of membrane assembly:	\$5.40/ft^2
Cost of energy recovery device:	\$47.30/kW
Annual charge based on capital cost:	15 per cent

The above unit costs are approximately the same as those given by Johnson et al. (1967).

A set of calculations is illustrated below for the cost of one kilogallon of product water. Let S be the membrane area required in ft^2.

Cost of membrane

$$= \$\,\frac{0.27 \times S \times 10^3}{10^6 \times 165}$$

$$= \$1.636 \times 10^{-6}\ \text{S/kgal}$$

Annual return per dollar of investment

$$= \$\,\frac{0.15 \times 10^3}{10^6 \times 330}$$

$$= \$\ 4.55 \times 10^{-7}/\text{kgal}$$

Example: Film F3

Operating pressure = 150 atm
$\lambda\theta = \infty$
$\eta = 0$
A-factor = 1

From Figure 8-46, power required

= 20 kwh./kgal.

From Figure 8-47, membrane area required

$= 2.8 \times 10^4\ \text{ft}^2/10^6\ \text{gal}$

Cost of membrane

$$= \$2.8 \times 10^4 \times 1.636 \times 10^{-6}$$

$$= \$0.046/\text{kgal}$$

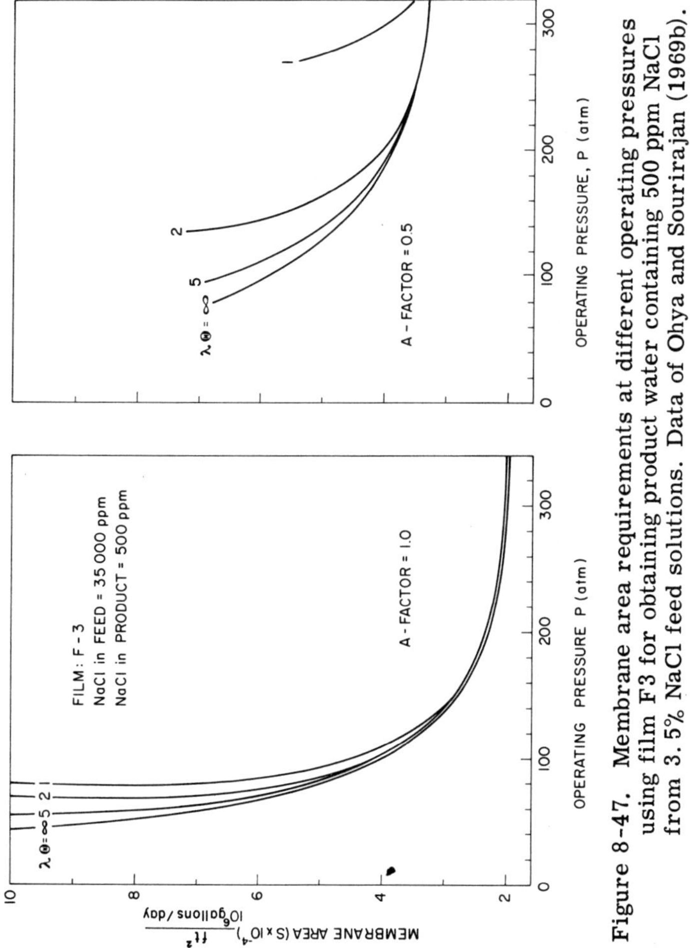

Figure 8-47. Membrane area requirements at different operating pressures using film F3 for obtaining product water containing 500 ppm NaCl from 3.5% NaCl feed solutions. Data of Ohya and Sourirajan (1969b).

Cost of power

$= \$ 20 \times 0.005$

$= \$ 0.100/\text{kgal}$

Investment on membrane assembly

$= \$5.4 \times 2.8 \times 10^4$

$= \$1.512 \times 10^5$

Annual return on above investment

$= \$1.512 \times 10^5 \times 4.55 \times 10^{-7}$

$= \$0.069/\text{kgal}$

Since 10^6 gal/day = 41.7 kgal/hour

pumping power

$= 20 \times 41.7$ kW

$= 834$ kW

Investment on pump

$= \$834 \times 83$

$= \$6.922 \times 10^4$

Annual return on above investment

$= \$6.922 \times 10^4 \times 4.55 \times 10^{-7}$

$= \$0.031/\text{kgal}$

Therefore total cost of water

$= \$(0.046 + 0.100 + 0.069 + 0.031)$

$= \$0.246/\text{kgal}$

Now assume power recovery $\eta = 0.5$ for the above example. From Figure 8-46, the power requirement is 18 kWH/kgal

Pumping power

$= 18 \times 41.7$ kW

$= 750$ kW

Therefore power recovered

$= 834 - 750 = 84$ kW

Investment on energy recovery device

$= \$84 \times 47.3$

$= \$3.973 \times 10^4$

Annual return on above investment

$= \$3.973 \times 10^4 \times 4.55 \times 10^{-7}$

$= \$0.002$

Therefore power cost

$= \$(18 \times 0.005) + \$0.002$

$= \$0.092$

Therefore total cost of water for the above case is $\$(0.046 + 0.092 + 0.069 + 0.031) = \$0.238$ per kgal

The results of a set of similar calculations are given in Figure 8-48. The results show the importance of reducing membrane compaction, operating at high pressures (~150 atm) with high values of $\lambda\theta$ (~5) for reducing the cost of product water. While the purpose of this cost analysis is only to isolate the parameters which require further studies, from the point of view of the economic conversion of saline waters, it also indicates that potable water at reasonable costs can be obtained even with the existing films provided sufficient improvement in the mechanical design of the reverse osmosis operating unit can be accomplished.

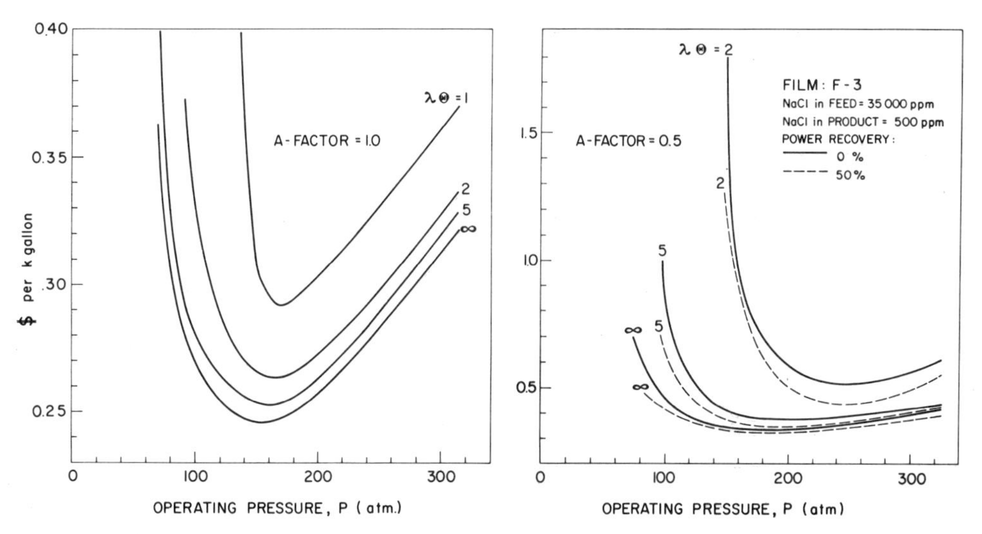

Figure 8-48. A comparative cost analysis for product water containing 500 ppm of NaCl from 3.5% NaCl feed solutions using film F3. Data of Ohya and Sourirajan (1969b).

Sea Water Conversion. According to Posnjak (1940), concentration of sea water by a factor of ~3.3 (~11.5 wt. per cent solids) is sufficient to cause calcium sulphate precipitation at 25°. This means that the maximum permissible value of C_2 in sea water conversion is only 3.3; this may have important consequences on the performance of a reverse osmosis plant for sea water conversion. However, the conclusions of the above analysis given for the system NaCl-H_2O seem generally applicable for sea water conversion also. Figure 8-41 shows that C_2 is little affected by θ in the range $\theta = 0.001$ to 0.007; and, for $\gamma = 0.188$, and $\lambda\theta = 5$, Δ value is about 0.65 by interpolation. Referring to Table 8-33, the Δ values obtainable for film F3 at 150 atm operating pressure are 0.775 for A-factor = 1.0, and 0.486 for A-factor = 0.5, for $\lambda\theta = 5$, and $\gamma = 0.188$; under practical operating conditions, Δ values will be intermediate between 0.775 and 0.486 and probably close to 0.6 which would correspond to a C_2 value less than 3.3. Consequently the general conclusion that operating pressures of about 150 atm, and $\lambda\theta$ values close to 5 are the conditions worth investigating for reducing the cost of desalinised water seems valid of sea water conversion also.

Significance of this work. The work just described illustrates a new approach to parametric studies of specific reverse osmosis applications. The results indicate that potable water can be obtained from sea water in a single-stage operation even with the presently available Loeb-Sourirajan type porous cellulose acetate membranes. Operating pressure, mass transfer coefficient on the high pressure side of the membrane, and A-factor (compaction effect) emerge as the important factors governing the economics of sea water conversion. Operating pressures in the range 150 to 200 atm, $\lambda\theta$ values of 5 or more, and A-factors close to 1.0 seem to be the most favourable conditions for sea water conversion. The studies reported here point out the need for extensive experimental investigations on the characteristics of the presently available porous cellulose acetate membranes in the operating pressure range 100 to 200 atm, on the mass transfer coefficients actually obtained on the high pressure side of the membrane with different membrane configurations and with the reverse osmosis pilot plants currently in operation, and on methods for reducing membrane compaction during continuous operation of the reverse osmosis process.

Effect of Longitudinal Diffusion in Reverse Osmosis. Though longitudinal diffusion has only a second order effect on membrane performance, it may be of practical significance in some cases. Ohya and Sourirajan (1969c) have developed equations and useful correlations describing the effect of longitudinal diffusion in reverse osmosis for both the general case and the limiting cases of no mixing and complete mixing. Their treatment of the subject, given below, compliments their earlier development of the general equations for reverse osmosis process design discussed in Chapter 4.

As before (Chapter 4), it is assumed that $(D_{AM}/K\delta)$ for the solute at a given operating pressure is independent of feed concentration and feed flow rate, the osmotic pressure of the feed solution is proportional to the mole fraction of solute, the molar density of the solution is constant, and the reverse osmosis system is specified in terms of γ, θ, and λ defined by Equations 4-11, 4-204, and 4-13 (in Chapter 4) respectively.

Basic Transport Equations. The following equations have been derived in Chapter 4; they are listed here for convenience. For all values of γ, θ, and λ,

$$V_W = v_W/v_W^* = 1 - \gamma(C_2 - C_3) = \theta/(\gamma C_3 + \theta) \quad (8\text{-}5)$$

$$C_1 = C_3 \left[1 + \frac{1}{(\gamma C_3 + \theta)} \exp\left\{-\frac{1}{\lambda(\gamma C_3 + \theta)}\right\}\right] \quad (8\text{-}6)$$

$$C_2 = \left[1 + \frac{1}{(\gamma C_3 + \theta)}\right] C_3 \quad (8\text{-}7)$$

$$C_2^0 = \left[1 + \frac{1}{(\gamma C_3^0 + \theta)}\right] C_3^0 \quad (8\text{-}8)$$

$$C_3 = \frac{\sqrt{[1 + \theta - \gamma C_2)^2 + 4\gamma\theta C_2]} - (1 + \theta - \gamma C_2)}{2\gamma} \quad (8\text{-}9)$$

$$C_3^0 = \frac{\sqrt{[(1 + \theta - \gamma C_2^0)^2 + 4\gamma\theta C_2^0]} - (1 + \theta - \gamma C_2^0)}{2\gamma} \tag{8-10}$$

$$C_1(1 - \Delta) + \overline{C}_3\Delta = 1 \tag{8-11}$$

where

$$\Delta = \text{fraction product recovery}$$

$$= 1 - \bar{u}/\bar{u}^0 \tag{8-12}$$

Equations 8-5 to 8-10 are applicable at any point in the reverse osmosis system and Equation 8-11 is simply a material balance equation for any section of the reverse osmosis unit. They are applicable for any reverse osmosis system whether or not longitudinal diffusion is negligible.

The performance data which are of basic interest in any reverse osmosis unit are $C_1, C_2, C_3, \overline{C}_3$, and Δ as functions of X. If any two of the five quantities C_1, $C_2, C_3, \overline{C}_3$, and Δ are available, the other three can be calculated simply by the use of Equations 8-5 to 8-11; the quantities C_3 and $\overline{C}_3$ are particularly convenient for this purpose. Hence the object of the following discussion is to present the C_3, $\overline{C}_3$, and X correlations in useful graphical forms for γ, θ, and λ values of practical interest for desalination and related applications. First the limiting cases of no longitudinal mixing and complete mixing are considered, and then the general case of longitudinal mixing is considered. For all cases, C_3 and $\overline{C}_3$ are expressed as $₵_3$ and $\overline{₵}_3$ defined as follows:

$$₵_3 = C_3/C_3^0 \tag{8-13}$$

$$\overline{₵}_3 = \overline{C}_3/C_3^0 \tag{8-14}$$

Case 1. No longitudinal mixing. For this case, the following equation has been derived in Chapter 4:

$$X = \int_{C_3^0}^{C_3} \left[\frac{\gamma}{\lambda(\gamma C_3 + \theta)} - \gamma + \frac{(\gamma C_3 + \theta)}{C_3}\left\{(\gamma C_3 + \theta)\exp\frac{1}{\lambda(\gamma C_3 + \theta)} + 1\right\}\right]\frac{\exp(-Z)}{\theta}dC_3 \tag{8-15}$$

where

$$Z = \int_1^{C_1} \frac{dC_1}{C_1 - C_3} \tag{8-16}$$

and

$$X = \frac{v_w^*}{\bar{u}^0}\frac{x}{h} \tag{8-17}$$

Computer solutions of the system of equations represented by Equations 8-5 to 8-17 can give the values of $C_1, C_2, C_3, \overline{C}_3$, and Δ for any given value of **X** for a reverse osmosis system specified by γ, θ, and λ. Such correlations for a number of systems have been illustrated (Chapter 4, and Appendix III). When the desired computer solutions are not available it is useful to have graphical correlations of X v. C_3 and $\overline{C}_3$ for values of γ, θ, and λ which are of interest.

A consideration of the condition $\gamma = 0$ leads to the form of correlations for calculating $₵_3$ and $\overline{₵}_3$ (and hence C_3 and $\overline{C}_3$) as functions of X for all values of γ, θ, and λ. When $\gamma = 0$, V_W is constant, and, as shown in Chapter 4,

$$₵_3 = [1 - X]^{-1/[1+\theta \exp(1/\lambda\theta)]} \tag{8-18}$$

Therefore

$$\overline{₵}_3 = \int_0^X V_w(1 - X)^{-1/[1+\theta \exp(1/\lambda\theta)]} dX \Big/ \int_0^X V_W dX \tag{8-19}$$

$$= \frac{1 + \theta \exp(1/\lambda\theta)}{\theta \exp(1/\lambda\theta)} \cdot \frac{\left[1 - (1 - X)^{\frac{\theta \exp(1/\lambda\theta)}{1+\theta \exp(1/\lambda\theta)}}\right]}{X} \tag{8-20}$$

For small values of θ,

$$\theta \exp\left(\frac{1}{\lambda\theta}\right) \ll 1 \tag{8-21}$$

and,

$$\left[(1 - X)^{\frac{\theta \exp(1/\lambda\theta)}{1+\theta \exp(1/\lambda\theta)}}\right] \approx 1 + \frac{\theta \exp(1/\lambda\theta)}{1 + \theta \exp(1/\lambda\theta)} \ln(1 - X) \tag{8-22}$$

Therefore for $\gamma = 0$,

$$\overline{₵}_3 = -\frac{\ln(1 - X)}{X} \tag{8-23}$$

Following the forms of Equations 8-18 and 8-23, $₵_3$ and $\overline{₵}_3$ can be expressed as follows for all values of γ, θ, and λ.

$$₵_3 = [1 - F(X)]^{-1/[1+\theta \exp(1/\lambda\theta)]} \tag{8-24}$$

$$\overline{₵}_3 = -\frac{\ln[1 - \overline{F}(X)]}{\overline{F}(X)} \tag{8-25}$$

From the values of $₵_3$ and $\overline{₵}_3$ obtained for different values of X from the computer solutions of the system of equations represented by Equations 8-5 to 8-17, the values of F(X) and $\overline{F}$(X) as functions of X can be calculated for specified reverse osmosis systems. Such X v. F(X) [which is the same as τ v. F(τ)] correlations have been given in Chapter 4 (Figures 4-46, 4-47, and 4-48). Figure 8-49 gives the corresponding X v. $\overline{F}$(X) correlations for $\gamma = 0$, 0.1, and 0.5, $\theta = 0.001$ and 0.1 and $\lambda = 1$, and ∞. The $\overline{F}$(X) values for other systems can be interpolated from Figure 8-49, or obtained and plotted in a similar manner. Figure 8-49, along with Figures 4-46, 4-47, and 4-48, gives the values of C_3 and $\overline{C}_3$ as functions of X for a wide range of γ, θ, and λ values of practical interest for the case where longitudinal mixing can be neglected.

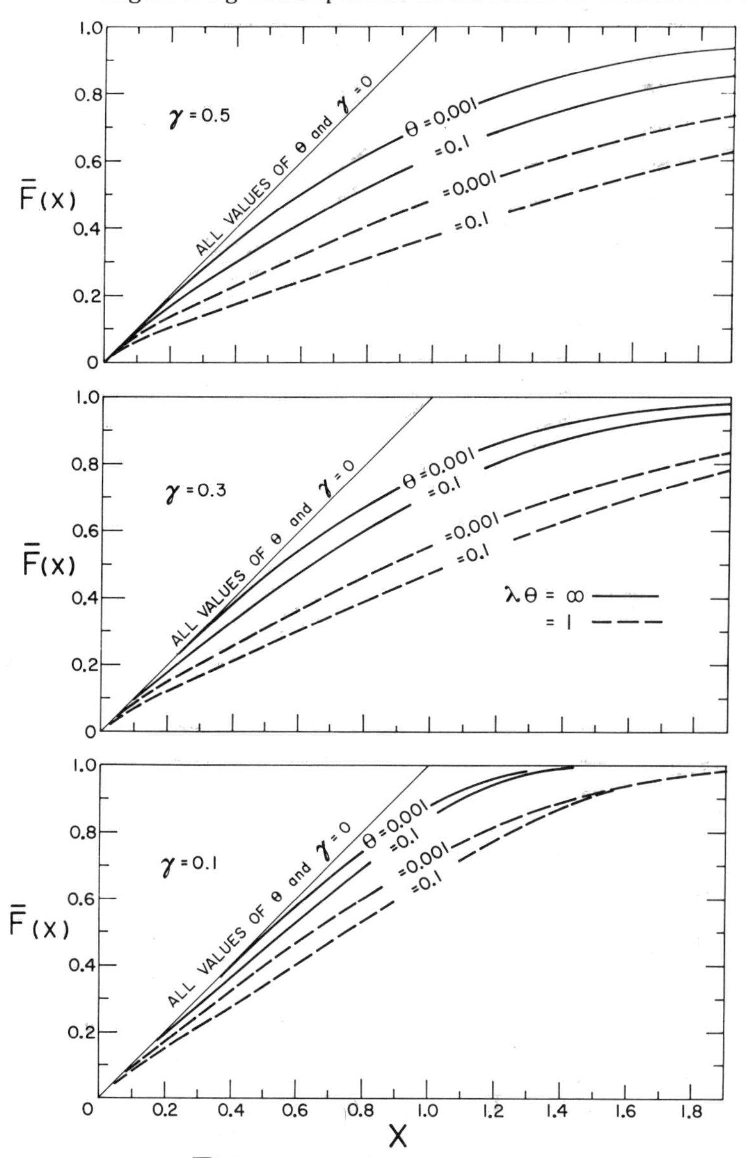

Figure 8-49. X v. $\bar{F}(X)$ for the case of no longitudinal mixing. Data of Ohya and Sourirajan.

Case 2. Complete longitudinal mixing. Consider a complete mixing reverse osmosis tank whose fluid volume is V, membrane area is (V/h), and to which the average feed rate is Q_0 as shown in Figure 8-50. Under operating conditions, the solute material balance can be written as

$$Q_0 C_1^0 = \frac{V}{h} v_W C_3 + \left(Q_0 - \frac{V}{h} v_W\right) C_1 \tag{8-26}$$

Let a new parameter M be defined as

$$M = \frac{h}{v_W^*} \cdot \frac{1}{(V/Q_0)} = \frac{h}{v_W^*} \frac{\bar{u}^0}{L} \tag{8-27}$$

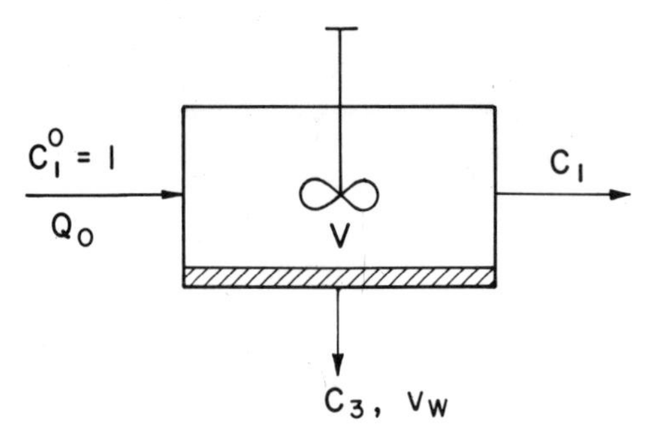

Figure 8-50. Complete mixing tank.

In Equation 8-27, (V/Q_0) or $(L/\bar{u}^0)$ expresses average residence time at inlet conditions, and L is the longitudinal length over which mixing is considered in the analysis. Therefore, in a practical case, by fixing L, M is defined. M is the important dimensionless parameter for parametric studies involving longitudinal mixing. Using Equation 8-5, and the fact that $C_1^0 = 1$ corresponding to Q_0, Equation 8-26 can be written as

$$\frac{C_3[1-\gamma(C_2-C_3)]}{M} + \left[1 - \frac{1-\gamma(C_2-C_3)}{M}\right]C_1 = 1 \tag{8-28}$$

and the fraction product recovery is given by

$$\Delta = \frac{(V/h)v_w}{Q_0} = \frac{\theta}{M(\gamma C_3 + \theta)} \tag{8-29}$$

For the complete mixing case under consideration,

$$C_3 = \overline{C}_3 \text{ or } \mathbb{C}_3 = \overline{\mathbb{C}}_3 \tag{8-30}$$

For the case $\gamma = 0$, an expression for $\mathbb{C}_3$ or $\overline{\mathbb{C}}_3$ can be obtained analytically as follows. When $\gamma = 0$, Equations 8-6, 8-28 and 8-29 become respectively

$$C_1 = C_3\left[1 + \frac{1}{\theta}\exp\left(-\frac{1}{\lambda\theta}\right)\right] \tag{8-31}$$

$$\frac{C_3}{M} + \left[1 - \frac{1}{M}\right]C_1 = 1 \tag{8-32}$$

and

$$\Delta = \frac{1}{M} \tag{8-33}$$

Combining Equations 8-31 and 8-32

$$C_3 = \frac{1}{1 + \left[1 - \frac{1}{M}\right]\frac{1}{\theta}\exp\left(-\frac{1}{\lambda\theta}\right)} \tag{8-34}$$

Since

$$C_3^0 = \frac{1}{1 + \frac{1}{\theta}\exp\left(-\frac{1}{\lambda\theta}\right)} \tag{8-35}$$

$$\bar{C}_3 = \bar{\bar{C}}_3 = \frac{1}{1 - \frac{1}{M}\left[\frac{1}{1 + \theta \exp\left(\frac{1}{\lambda\theta}\right)}\right]} \tag{8-36}$$

Equation 8-36 is applicable for the condition $\gamma = 0$.

Again following the form of Equation 8-23, $\bar{C}_3$ or $\bar{\bar{C}}_3$ for the complete mixing case may be represented by the relation

$$\bar{C}_3 = \bar{\bar{C}}_3 = -\frac{\ln\left[1 - \bar{F}_0\left(\frac{1}{M}\right)\right]}{\bar{F}_0\left(\frac{1}{M}\right)} \tag{8-37}$$

Using Equations 8-36 and 8-37, the values of $\bar{F}_0\left(\frac{1}{M}\right)$ as functions of $X\left(=\frac{1}{M}\right)$ can be obtained for $\gamma = 0$, and different values of θ and λ; a set of such correlations is given in Figure 8-51 which also includes similar correlations for γ values other than zero. The latter correlations have been obtained from the solutions of the general longitudinal diffusion equation given below with the appropriate boundary condition for complete mixing.

Case 3. Longitudinal mixing, general case. Let the average longitudinal coefficient be given as E, and let $\bar{u}$ be the average fluid velocity at any longitudinal position x. From material balance considerations, the following equations can be derived.

$$-E\frac{d^2C_1}{dx^2} + \bar{u}\frac{dC_1}{dx} = -(C_1 - C_3)\frac{d\bar{u}}{dx} \tag{8-38}$$

$$-d\bar{u}/dx = v_w/h = (v_w^*/h)[1 - \gamma(C_2 - C_3)] \tag{8-39}$$

Let the longitudinal diffusion and fluid velocity be expressed in dimensionless forms as follows:

$$PeB = \bar{u}^0 L/E \tag{8-40}$$

$$\bar{U} = \bar{u}/\bar{u}^0 \tag{8-41}$$

The limiting values of PeB = 0 and ∞ correspond respectively to complete mixing and zero longitudinal diffusion.

Changing the variables, Equations 8-38 and 8-39 can be written as follows:

$$-\frac{d^2C_1}{d(XM)^2} + PeB\,\bar{U}\frac{dC_1}{d(XM)} = -PeB\,(C_1 - C_3)\frac{d\bar{U}}{d(XM)} \tag{8-42}$$

$$-\frac{d\bar{U}}{d(XM)} = \frac{\theta}{M(\gamma C_3 + \theta)} \tag{8-43}$$

On integration, Equation 8-43 becomes

$$\bar{U} = 1 - \int_0^{XM} \frac{\theta}{(\gamma C_3 + \theta)}\frac{d(XM)}{M} \tag{8-44}$$

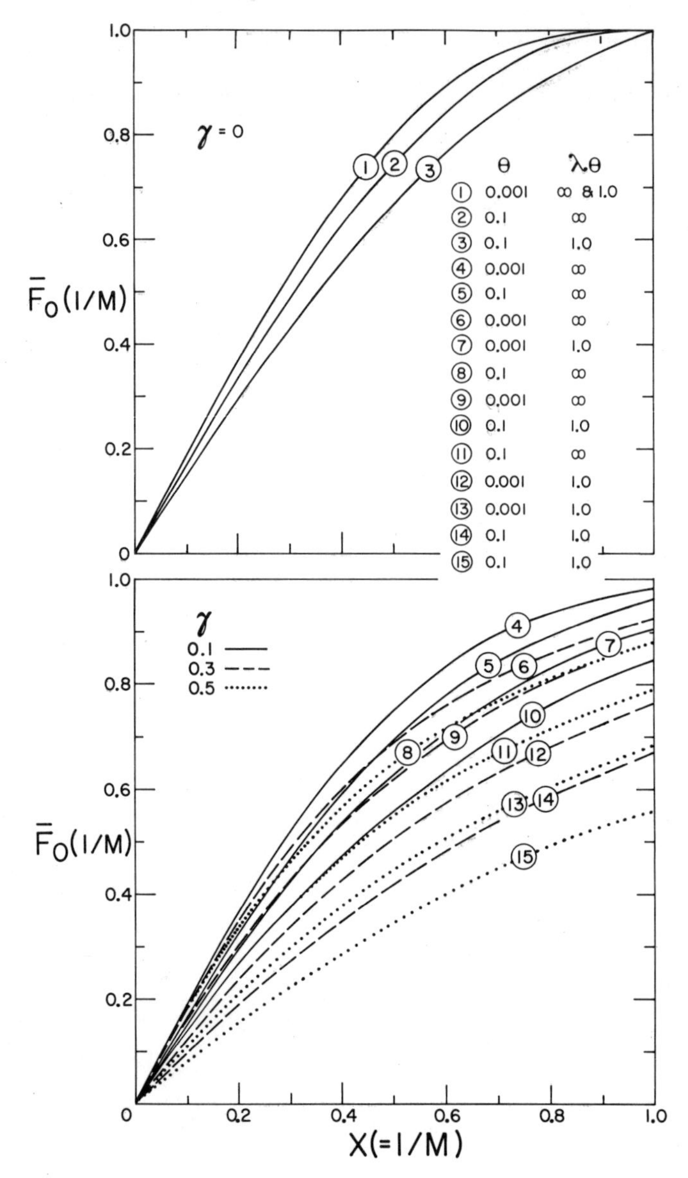

Figure 8-51. X(=1/M) v. $\overline{F}_0(1/M)$ for the case of complete mixing. Data of Ohya and Sourirajan.

The boundary conditions of Equation 8-42 are:

$$\text{at } MX = 0 \text{ (fluid inlet)}, -\frac{dC_1}{d(XM)} = PeB\ (1 - C_1) \tag{8-45}$$

$$\text{at } MX = 1 \text{ (fluid exit)}, -\frac{dC_1}{d(XM)} = 0 \tag{8-46}$$

Equation 8-42 is nonlinear, and it can be solved numerically on a computer by the Milne method of integration in which the four starting values are obtained by using up to the fourth derivative at initial state (Mickley et al., 1957). Thus the solution of Equation 8-42 gives the value of C_1 and (XM) for specified values of PeB and M for any reverse osmosis system; the corresponding Δ, C_3, and $\overline{C}_3$ values can then be obtained from Equations 8-44, 8-6, and 8-11 respectively. Examples of such calculations are given in Figures 8-51, 8-52, and 8-53.

The data for $X\left(=\frac{1}{M}\right)$ v. $\overline{F}_0\left(\frac{1}{M}\right)$ for various values of γ ($\neq 0$), θ, and λ (bottom of Figure 8-51) were calculated using Equations 8-37 and 8-6, and the values of C_1 and M obtained from the solutions of Equation 8-42 for the boundary condition PeB = 0 (complete mixing) and XM = 1 (exit condition).

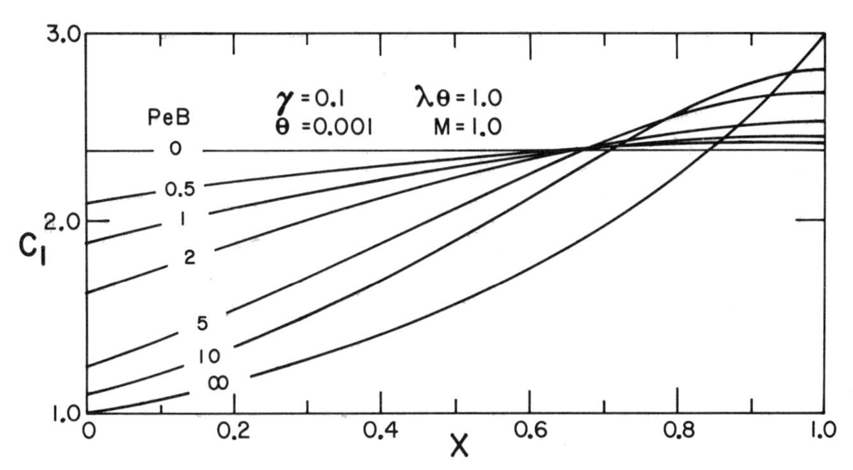

Figure 8-52. Effect of longitudinal diffusion on C_1 from inlet to exit of an operating unit. Data of Ohya and Sourirajan (1969c).

Figure 8-52 gives the X v. C_1 correlation for various values of PeB for the system $\gamma = 0.1$, $\theta = 0.001$, and $\lambda\theta = 1.0$ (chosen for illustration) for a fixed value of $M = 1$. This figure illustrates the effect of longitudinal diffusion on the values of C_1 from the inlet to the exit of the operating unit under consideration.

Figure 8-53 illustrates $X(=1/M)$ v. C_1, $\overline{C}_3$, and Δ correlations as functions of PeB for the same system $\gamma = 0.1$, $\theta = 0.001$, and $\lambda\theta = 1.0$. The data in this figure correspond to the exit condition (XM = 1) and not for a fixed value of M. These data show that up to $X(=1/M) \approx 0.5$, C_1, $\overline{C}_3$, and Δ values are not very much affected by longitudinal diffusion.

Following the form of Equation 8-37, $\overline{\text{¢}}_3$ for the general case of longitudinal mixing may be expressed as

$$\overline{\text{¢}}_3 = -\frac{\ln\left[1 - f_1 \cdot \overline{F}_0\left(\frac{1}{M}\right)\right]}{f_1 \cdot \overline{F}_0\left(\frac{1}{M}\right)} \tag{8-47}$$

where $\overline{F}_0\left(\frac{1}{M}\right)$ is the value obtained for the complete mixing case represented in Figure 8-51, and f_1 is the appropriate numerical factor to satisfy the equality of Equation 8-47 for the $\overline{\text{¢}}_3$ values obtained from the solutions of Equation 8-42. Values of f_1 have been calculated as function of PeB and $X(=1/M)$ (exit condition) for reverse osmosis systems specified by the parameters

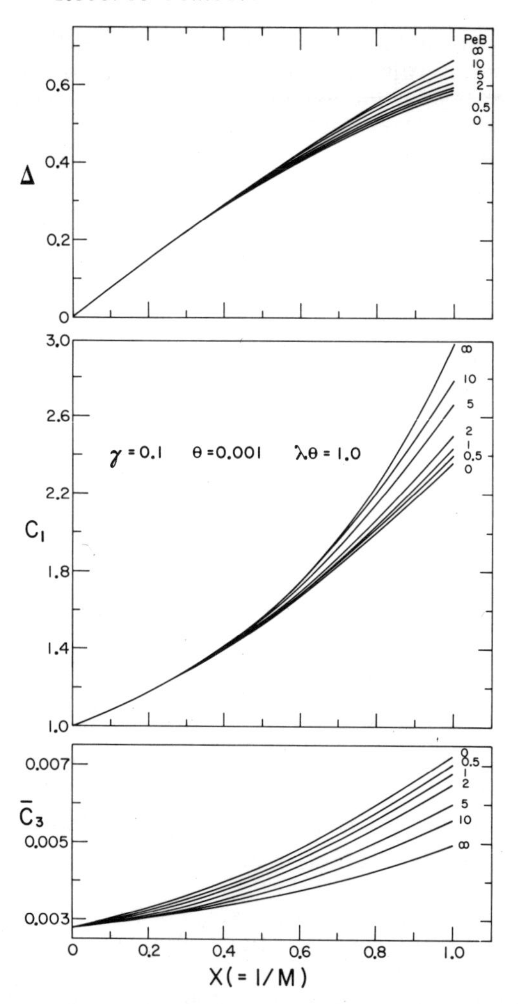

Figure 8-53. X(=1/M) v. $\overline{C}_3$, C_1, and Δ as functions of PeB for a reverse osmosis system. Data of Ohya and Sourirajan (1969c).

$\gamma = 0, 0.1, 0.3$, and 0.5, $\theta = 0.001$, and 0.1, and $\lambda\theta = 1$ and ∞; these data are given in Figure 8-54.

As for $₵_3$, it has been found that for the γ, θ, and λ values considered above, the differences between $₵_3$ (PeB $= \infty$) and $₵_3$ (PeB $\neq \infty$) are negligible at least up to X(=1/M) = 0.5; hence in this range, $₵_3$ for the case of longitudinal mixing may be considered essentially the same as that for the case of no longitudinal mixing. For values of X(=1/M) > 0.5, the values of $₵_3$ for any PeB, [$₵_3$(PeB)], may be represented by the relation

$$₵_3(\text{PeB}) = ₵_3(\text{PeB} = \infty) - f_2[₵_3(\text{PeB} = \infty) - ₵_3(\text{PeB} = 0)] \quad \text{(8-48)}$$

where $₵_3$(PeB $= \infty$), $₵_3$(PeB $= 0$) are the values of $₵_3$ for no mixing and complete mixing cases respectively, and f_2 is a numerical factor to satisfy the equality of Equation 8-48 for the $₵_3$ values obtained from the solutions of Equation 8-42. Values of f_2 have been calculated as functions of X(=1/M) and PeB for some reverse osmosis systems, and they are given in Figure 8-55.

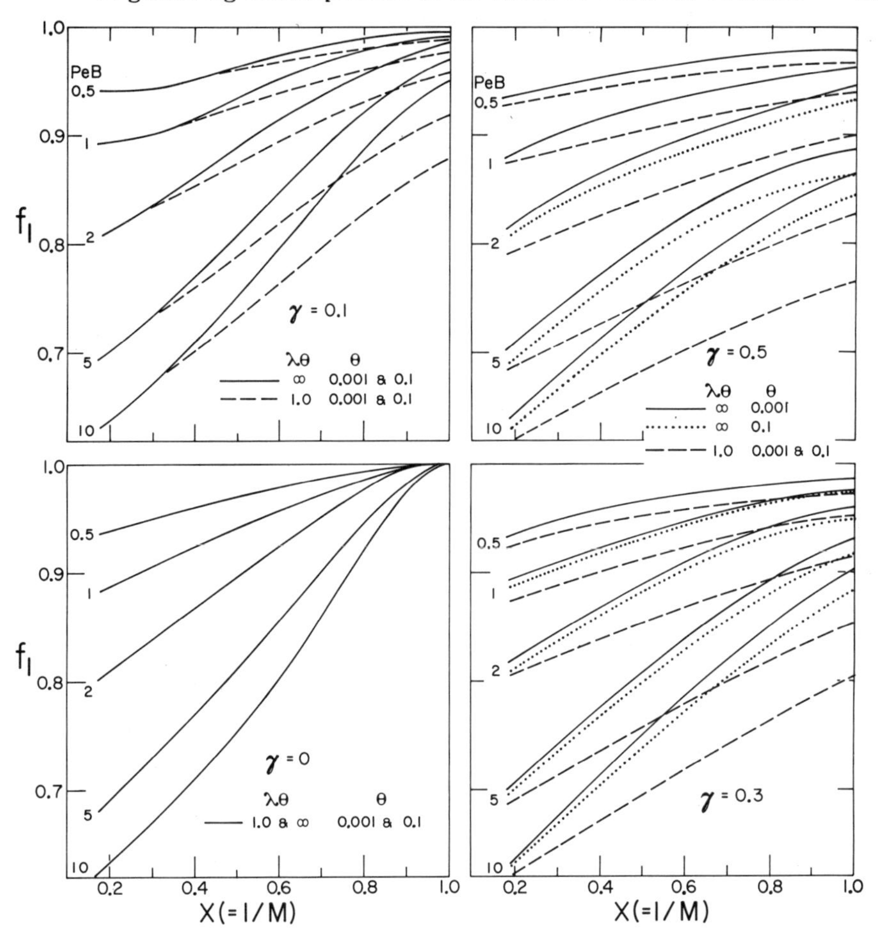

Figure 8-54. X(=1/M) v. f_1 as function of PeB, γ, θ, and λ. Data of Ohya and Sourirajan.

Referring to Figures 8-54 and 8-55, there are either 1, 2, or 3 different kinds of lines in each part of the graph corresponding to specified values of γ. In each part, the data for two θ values (0.001 to 0.1) and two $\lambda\theta$ values (1 and ∞) are plotted. When there is only one kind of line, the data correspond to all values of θ and $\lambda\theta$. When there are two kinds of lines, the data correspond to (i) $\lambda\theta = 1$, and (ii) $\lambda\theta = \infty$, as indicated, for both values of θ. When there are three different kinds of lines, the data correspond to (i) both values of θ, and $\lambda\theta = 1$, (ii) $\theta = 0.001$ and $\lambda\theta = \infty$, and (iii) $\theta = 0.1$ and $\lambda\theta = \infty$ as indicated. Thus Figures 8-54 and 8-55 give the correlations of $₵_3$ and $\bar{₵}_3$ as functions of M and PeB for the general case of longitudinal diffusion for the specified reverse osmosis systems.

Significance of these Studies. The foregoing analysis permits the determination of the overall behaviour of reverse osmosis systems affected by longitudinal dispersion of the fluid. This analysis, together with that given in Chapter 4 (Ohya and Sourirajan, 1969a, d), offers a unified approach to parametric study and system design of the reverse osmosis process as a general separation technique.

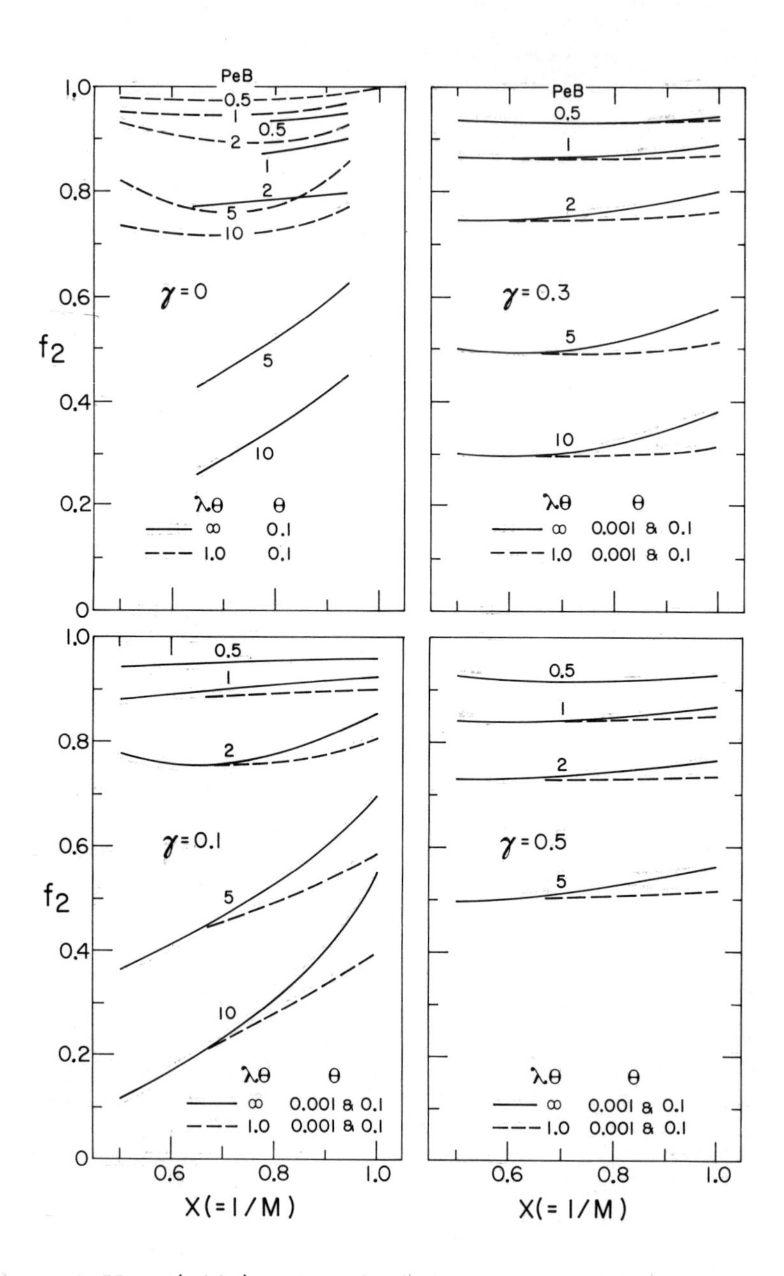

Figure 8-55. X(=1/M) v. f_2 as functions of PeB, γ, θ, and λ. Data of Ohya and Sourirajan.

20. Nomenclature for Chapter 8

A = pure water permeability constant, $\frac{\text{g. mole } H_2O}{\text{cm}^2\text{sec. atm}}$

A_0 = extrapolated value of A at P = 0

c = molar density of solution, g. mole/cm^3

c_A = solute concentration, g. mole/cm^3

$\bar{c}_A$ = average solute concentration, g. mole/cm^3

C = (c_A/c^0_{A1}) or (X_A/X^0_{A1})

$\overline{C}$ = $(\bar{c}_A/c^0_{A1})$ or $(\overline{X}_A/X^0_{A1})$

$¢_3$ = (C_3/C^0_3)

$\overline{¢}_3$ = $(\overline{C}_3/C^0_3)$

d = effective diameter of membrane surface, cm

D = diffusivity of solute, cm^2/sec

$(D_{AM}/K\delta)$ = solute transport parameter, cm/sec

E = longitudinal diffusion coefficient, cm^2/sec

f_1 = quantity defined by Equation 8-47

f_2 = quantity defined by Equation 8-48

$\overline{F}_0(1/M)$ = quantity defined by Equation 8-37

$F(X)$ = quantity defined by Equation 8-24

$\overline{F}(X)$ = quantity defined by Equation 8-25

$1/h$ = area of membrane surface per unit volume of fluid space (cm^{-1})

k = mass transfer coefficient on the high pressure side of membrane, cm/sec

L = longitudinal length over which mixing is considered, cm

M = longitudinal length parameter defined by Equation 8-27

N_{Sc} = ν/D

N_{Sh} = kd/D

P = operating pressure, atm

PeB = quantity defined by Equation 8-40

Q_0 = average feed rate, cm^3/sec

S = membrane surface area, cm^2

t = time, sec

$\bar{u}$ = average fluid velocity in the transverse length of the channel at a given longitudinal position, cm/sec

$\overline{U}$ = $(\bar{u}/\bar{u}^0)$

v_w = permeating velocity of solvent water through the membrane, cm/sec

v^*_w = (AP/c), cm/sec

V = volume of complete mixing tank, cm^3

V_1 = volume of solution on the high pressure side of membrane, cm^3

V_W = (v_W/v_W^*)

x = longitudinal distance from channel entrance, cm

X = quantity defined by Equation 8-17

X_A = mole fraction of solute

$\overline{X}_A$ = average mole fraction of solute

Z = quantity defined by Equation 8-16

Greek letters

α = constant

β = constant

γ = $\Pi(X_{A1}^0)/P$

Δ = fraction product recovery

η = fraction power recovery

θ = $\dfrac{(D_{AM}/K\delta)}{v_W^*}$

λ = $\dfrac{k}{(D_{AM}/K\delta)}$

ν = kinematic viscosity of feed solution, cm^2/sec

$\Pi(X_A)$ = osmotic pressure of solution corresponding to solute concentration X_A, atm

τ = $\dfrac{Sv_W^*t}{V_1^0}$

Subscripts

1 = bulk solution

2 = concentrated boundary solution

3 = membrane permeated product solution

Superscript

0 = initial condition, or condition at channel entrance.

21. Summary

This chapter deals with four major topics. First, the applications of the reverse osmosis process for water softening, water pollution control, water renovation, waste recovery and solute concentration are reviewed. Experimental data on the performance of a few typical Loeb-Sourirajan type porous cellulose acetate membranes for the treatment of hard waters, polluted waters, and sewage waters are given. Using feed waters containing 300 to 800 ppm hardness (expressed as $CaCO_3$), product waters containing 2 ppm or less have been obtained with 90 per cent product recovery and an average initial flux of 38 gal/day/ft^2 at 1000 psig. Further experimental data are given to illustrate the *possibilities of producing 'ultrapure' waters by repeated reverse osmosis processing, separating common water pollutants (such as nitrates, borates, fluorides, alkyl benzene sulphonate, ammonia, phosphates, and constituents of plating wastes), and treating sewage waters and paper mill waste waters from the point of view of water re-use. Some recent test results on the application of the reverse osmosis process for maple sap concentration are also given.*

Based on practical experience, some general aspects of the engineering of reverse osmosis plants are discussed next, with particular reference to saline water conversion. These aspects include basic differences between brackish water and sea water conversion units, general flow diagram of reverse osmosis plants, choice of materials of construction, feed water pretreatment practices, selection and operation of high pressure pumps, possible use of turbines for power recovery from high pressure brine, size of sumps and storage tanks, plant control signals and instrumentation, and the different membrane unit design concepts currently in practice.

Then follows a detailed engineering description of the design, construction, and performance of several reverse osmosis pilot plants which have been built and operated by different organisations. These plants include those built by UCLA (using flat membrane and tubular membrane configurations), Havens Industries, Universal Water Corporation, and American Radiator and Standard Sanitary Corp. (using tubular membrane configurations), Aerojet General Corporation (using flat membrane configuration), and the Gulf General Atomic Inc. (using spiral-wound membrane configuration). Some details of the home reverse osmosis units built and operated by the Thayer School of Engineering, Dartmouth College, Hanover, N.H. (using flat and spiral-wound membrane configurations), and of DuPont hollow fibre units are also given. Extensive data on field test results obtained with some of the above pilot plants are also presented. The UCLA-Coalinga pilot plant has maintained an outstanding record of continuous service for over 3 years.

There is currently sufficient engineering experience to design, construct, and operate successfully reverse osmosis brackish water conversion plants up to 100,000 gal/day capacity, and also to construct low cost 3 to 5 gal/day low pressure reverse osmosis units for home service.

The chapter concludes with a review of the different kinds of parametric studies made by different workers on the reverse osmosis process with particular reference to saline water conversion. Based on the UCLA-Coalinga pilot plant experience, Stevens and Loeb have given cost estimates of 5000 and 50,000 gallons per day brackish water conversion plants using tubular membranes. The studies of Keilin and co-workers are based on the different design criteria of plant construction using the flat membrane configuration for the membrane units. The studies of Bray and Menzel are directed toward the selection of system pressure, product recovery, and feed flow arrangements for the design of 1 million gallon per day single-stage and two-stage saline water conversion plants using the Gulf General Atomic spiral-wound membrane modules. Johnson and others have considered the possibilities of single-stage and two-stage operation of a 1 million gallon per day sea water conversion plant using single-step and multistep feed flow channels. Ohya and Sourirajan have presented a unified analytical approach to parametric study and system design of the reverse osmosis process as a general separation technique.

Conclusion

This book is an attempt to present the science and engineering of reverse osmosis from the point of view of the general applicability of the technique. There is now growing interest and activity in this field in many laboratories and institutions all over the world, especially in connection with water treatment and saline water conversion applications. In the United States of America, much of this work is being supported by the Office of Saline Water. Tremendous progress has already been made in recent years, and will continue to be made in the future.

From the author's point of view, there are several areas of work which have potentially far-reaching beneficial consequences to chemical industry. Studies on the physico-chemical criteria of preferential sorption and preferential repulsion at solid-solution interfaces, the mechanism of pore formation in the process of making asymmetric porous membranes, the scientific and engineering aspects of porous membrane technology, quality control of the Loeb-Sourirajan type porous cellulose acetate membranes in their large-scale manufacture and the utilisation of their fullest potentialities, integration of the science and engineering of reverse osmosis, development of different membranes for different applications, and the application of reverse osmosis for the separation of gaseous mixtures, organic liquid mixtures, and highly acid or alkaline aqueous solutions, are all areas of enormous practical interest.

Further work in this field can lead to the growth of reverse osmosis in all its applications, and contribute significantly to the economic prosperity and physical well-being of all mankind. Such is the scope of reverse osmosis.

Literature

Aerojet General Corporation, Design and Construction of a Desalination Pilot Plant (A Reverse Osmosis Process), U.S. Dept. Interior, Office of Saline Water, Research and Development Progress Report No. 86 (1964).

Aerojet General Corporation, Operation of a Reverse Osmosis Desalination Pilot Plant, U.S. Dept. Interior, Office of Saline Water, Research and Development Progress Report No. 213 (1966)

Agrawal, J. P., Sourirajan, S., J. Applied Polymer Sci. **13**, 1065 (1969a)

Agrawal, J. P., Sourirajan, S., Ind. Eng. Chem. Process Design Develop. **8**, 439 (1969b)

Agrawal, J. P., Sourirajan, S., Reverse Osmosis Separation of Some Inorganic Salts in Aqueous Solution Containing Mixed Solutes With a Common Ion, Ind. Eng. Chem. Process Design Develop. (in press) (1969c).

Ambard, L., Trautmann, S., Ultrafiltration, p. 3-14, Charles C. Thomas, Springfield, Ill. (1960a).

Ambard, L., Trautmann, S., Ultrafiltration, p. 23, Charles C. Thomas, Springfield, Ill. (1960b).

Ammerlaan, A. C. F., Lueck, B. F., Wiley, A. J., Membrane Processing of Dilute Pulping Wastes By Reverse Osmosis, Pulp Manufacturers Research League, Inc., Appleton, Wisc. (1968).

Andelman, J. B., Suess, M. J., Anal. Chem. **38**, 351 (1966).

Baldwin, W. H., Holcomb, D. L., Johnson, J. S., J. Polymer Sci. **A3**, 833 (1965).

Banks, W., Sharples, A., J. Applied Chem. **16**, 28 (1966a).

Banks, W., Sharples, A., J. Applied Chem. **16**, 94 (1966b).

Banks, W., Sharples, A., J. Applied Chem. **16**, 153 (1966c).

Barrer, R. M., Diffusion In and Through Solids, Chapters 9 and 10, Cambridge Univ. Press, London (1951).

Barrer, R. M., A. I. Ch. E.-I. Chem. E. Symp. Ser. **1**, 112 (1965).

Becher, P., J. Colloid Sci. **16**, 49 (1961).

Bell, R. P., Endeavour **17**, 31 (1958).

Benedict, M., Pigford, T. H., Nuclear Chemical Engineering, p. 384, McGraw-Hill, New York (1957a).

Benedict, M., Pigford, T. H., Nuclear Chemical Engineering, p. 385, McGraw-Hill, New York (1957b).

Benedict, M., Pigford, T. H., Nuclear Chemical Engineering, p. 386, McGraw-Hill, New York (1957c).

Bennion, D. N., Mass Transport of Binary Electrolyte Solutions in Membranes, Department of Engineering, University of California, Los Angeles, Report No. 66-17 (1966).

Bennion, D. N., Rhee, B. W., Ind. Eng. Chem. Fundamentals **8**, 36 (1969).

Berman, A. S., J. Appl. Phys. **24**, 1232 (1953).

Biget, A. M., Ann. Chim. [12], **5**, 66 (1950).

Bird, R. B., Stewart, W. E., Lightfoot, E. N., Transport Phenomena, p. 520, John Wiley, New York (1960).

Blatt, W. F., Feinberg, M. P., Hopfenberg, H. B., Saravis, C. A., Science **150**, 224 (1965).

Bloch, R., Kedem, O., Vofsi, D., Polymer Letters **3**, 965 (1965).

Bockris, J. O'M., Quart. Revs. (London) **3**, 173 (1949).

Bousfield, W. R., Trans. Faraday Soc. **13**, 401 (1918).

Bray, D. T., in Desalination by Reverse Osmosis, U. Merten, ed., p. 203, M.I.T. Press, Cambridge, Mass. (1966).

Bray, D. T., Menzel, H. F., Design Study of a Reverse Osmosis Plant for Sea Water Conversion, U.S. Dept. Interior, Office of Saline Water, Research and Development Progress Report No. 176 (1966).
Bray, D. T., Merten, U., Augustus, M., Reverse Osmosis For Water Reclamation, Gulf General Atomic Report No. GA-6337 (1965).
Breitenbach, J. W., Forster, E. L., Makromol. Chem. **8,** 140 (1952).
Breton, E. J., Jr., Water and Ion Flow Through Imperfect Osmotic Membranes, U.S. Dept. Interior, Office of Saline Water, Research and Development Progress Report No. 16 (1957)
Breton, E. J., Jr., Reid, C. E., A.I.Ch.E. Chem. Eng. Symp. Ser. **24,** 171 (1959).
Brian, P. L. T., Ind. Eng. Chem. Fundamentals **4,** 439 (1965a).
Brian, P. L. T., Concentration Polarization in Reverse Osmosis Deslaination with Variable Flux and Incomplete Salt Rejection, M.I.T. Desalination Research Laboratory, Report No. 295-7, DSR-4647 (1965b).
Brian, P. L. T., Proc. First International Symp. on Water Desalination, Washington, D.C. (1965) (Pub. U.S. Dept. Interior, Office of Saline Water, Washington, D.C.) Vol. 1, p. 349-370.
Brian, P. L. T., in Desalination by Reverse Osmosis, U. Merten, ed., Chapter 5, M.I.T. Press, Cambridge, Mass. (1966).
Brown, W. E., Tuwiner, S. B., Diffusion and Membrane Technology, Chapter **12,** Reinhold, New York (1962).
Brubaker, D. W., Kammermeyer, K., Anal. Chem. **25,** 424 (1953a).
Brubaker, D. W., Kammermeyer, K., Ind. Eng. Chem. **45,** 1148 (1953b).
Brubaker, D. W., Kammermeyer, K., Proc. Conf. Nuclear Engineering, F9-F28, Univ. California, Berkeley (1953c).
Brubaker, D. W., Kammermeyer, K., Ind. Eng. Chem. **46,** 733 (1954).
Burrell, H., Offic. Dig., Federation Paint Varnish Prod. Clubs, p. 726 (1955).
Calmon, C., Kingsbury, A. W., in Principles of Desalination, K. S. Spiegler, ed., p. 453, Academic Press, New York (1966).
Carnell, P. H., J. Applied Polymer Sci. **9,** 1863 (1965).
Carnell, P. H., Cassidy, H. G., J. Polymer Sci. **55,** 233 (1961).
Chemical Society of Japan, Tokyo, Chemical Handbook, p. 479 (1958).
Cheng, C., Cheng, S., Fan, L., A.I.Ch.E.J. **13,** 438 (1967).
Cheng, C., Fan, L., A Flow Work Exchanger For Desalination Processes, U.S. Dept. Interior, Office of Saline Water, Research and Development Progress Report No. 357 (1968).
Choo, C. Y., in Advances in Petroleum Chemistry, Vol. 6, Chapter 2, Interscience, New York (1962).
Clark, W. E., Science **138,** 148 (1963).
Clauss, A., Hofmann, U., Weiss, A., Z. Electrochem. **61,** 1284 (1957).
Dobry, A., Bull. Soc. Chim. [5], **3,** 312 (1936).
Dodge, B. F., Eshaya, A. M., Advan. Chem. Ser. **27,** 7 (1960).
Dresner, L., Boundary Layer Build-up in the Demineralization of Salt Water by Reverse Osmosis, Report ORNL-3621, Oak Ridge National Laboratory (1964).
Dresner, L., J. Phys. Chem. **69,** 2230 (1965).
Dresner, L., Kraus, K. A., J. Phys. Chem. **67,** 990 (1963).
Elford, W., Proc. Royal Soc. (London), **B106,** 216 (1940).
Erickson, D. L., Cellulose Acetate Membranes as a Means of Removing Scale Forming Ions of Natural Saline Waters, Department of Engineering, University of California, Los Angeles, Report No. 66-7 (1966).
Erickson, D. L., Glater, J., McCutchan, J. W., Ind. Eng. Chem. Prod. Res. Develop. **5,** 205 (1966).
Erschler, B., Kolloid-Z. **68,** 289 (1934).
Fair, G. M., Geyer, J. C., Water Supply and Waste Water Disposal, p. 12, Wiley, New York (1961a).

Fair, G. M., Geyer, J. C., Water Supply and Waste Water Disposal, p. 859, Wiley, New York (1961b).
Fisher, R. E., Sherwood, T. K., Brian, P. L. T., Salt Concentration at the Surface of Tubular Reverse Osmosis Membranes, U.S. Dept. Interior, Office of Saline Water, Research and Development Progress Report No. 141 (1965).
Flowers, L. C., Sestrich, D. E., Berg. D., Reverse Osmosis Membranes Containing Graphitic Oxide, U.S. Dept. Interior, Office of Saline Water, Research and Development Progress Report No. 224 (1966).
Francis, P. S., Cadotte, J. E., Second Report on Frabrication and Evaluation of New Ultrathin Reverse Osmosis Membranes, U.S. Dept. Interior, Office of Saline Water, Research and Development Progress Report No. 247 (1967).
Gill, W. N., Tien, C., Zeh, D. W., Ind. Eng. Chem. Fundamentals **4**, 433 (1965).
Gill, W. N., Tien, C., Zeh, D. W., Int. J. Heat Mass Transfer **9**, 907 (1966a).
Gill, W. N., Zeh, D. W., Tien, C., Ind. Eng. Chem. Fundamentals **5**, 367 (1966b).
Gill, W. N., Zeh, D. W., Tien, C., A.I.Ch.E.J. **12**, 1141 (1966c).
Glasstone, S., Text Book of Physical Chemistry, 2nd ed., p. 958, Van Nostrand, New York (1951a).
Glasstone, S., Text Book of Physical Chemistry, 2nd ed., p. 1254, Van Nostrand, New York (1951b).
Glueckauf, E., Proc. First International Symp. on Water Desalination, Washington, D.C. (1965) (Pub. U.S. Dept. Interior, Office of Saline Water, Washington, D.C.) Vol. 1, p. 143-156.
Goard, A. K., J. Chem. Soc. **127**, 2451 (1925).
Gosting, L. J., Akeley, D. F., J. Am. Chem. Soc. **74**, 2058 (1952).
Govindan, T. S., Sourirajan, S., Ind. Eng. Chem. Process Design Develop. **5**, 422 (1966).
Guy, D. B., (Havens Industries), private communication (1968).
Harkins, W. D., McLaughlin, H. M., J. Am. Chem. Soc. **47**, 2083 (1925).
Hasted, J. B., Ritson, D. M., Collie, C. H., J. Chem. Phys. **16**, 1 (1948).
Hauck, A. R., Sourirajan, S., Environmental Sci. Tech. **3**, 1269 (1969).
Havens, G. G., Guy, D. B., The Use of Porous Fiberglas Tubes for Reverse Osmosis. A.I.Ch.E. meeting, Salt Lake City, Utah (1967).
Havens, G. G., Guy, D. B., The Havens Osmotic Process (1968).
Havens Industries, Sea Water Conversion by Reverse Osmosis, Technical Brochure, San Diego, California (1964).
Haward, R. N., Trans. Faraday Soc. **38**, 394 (1942).
Hawksley, R. W., The Analyzer 8 (1), 13 (1967) (Pub. Beckman Instruments Inc.).
Helfferich, F., Ion Exchange, p. 339-420, McGraw-Hill, New York (1962).
Hildebrand, J., Chem. Rev. **44**, 37 (1949).
Hoffer, E., Kedem, O., Desalination **2**, 25 (1967).
Hook, A. V., Russell, H. D., J. Am. Chem. Soc. **67**, 370 (1945).
International Critical Tables, Vol. III, p. 79, McGraw-Hill, New York (1928a).
International Critical Tables, Vol. III, p. 89, McGraw-Hill, New York (1928b).
International Critical Tables, Vol. V, p. 15, McGraw-Hill, New York (1929a).
International Critical Tables, Vol. V, p. 70, McGraw-Hill, New York (1929b).
Ironside, R., Sourirajan, S., J. Water Research **1**, 179 (1967).
Ironside, R., Sourirajan, S., Unpublished work, 1968.
Jagur-Grodzinski, J., Kedem, O., Desalination **1**, 327 (1966).
Johnson, J. S., Bennion, D. N., Chem. Eng. Progress Symp. Series Vol. **64**, 270 (1968).
Johnson, J. S., Jr., Dresner, L., Kraus, K. A., in Principles of Desalination, K.S. Spiegler, ed., Chapter 8, Academic Press, New York (1966).
Johnson, J. S., Kraus, K. A., Hyperfiltration by Dynamically Formed Membranes, Paper presented before Division of Water, Air, and Waste Chemistry, American Chemical Society, San Francisco, March 31-April 5 (1968).
Johnson, K. D. B., Grover, J. R., Pepper, D., Desalination **2**, 40 (1967).

Kammermeyer, K., in Technique of Organic Chemistry, 3, Part 1, p. 37, Interscience, New York (1956).

Kammermeyer, K., Hagerbaumer, D. H., A.I.Ch.E.J. **1**, 215 (1955).

Kammermeyer, K., Wyrick, D. D., Ind. Eng. Chem. **50**, 1309 (1958).

Kedem, O., Katchalsky, A., Biochem. Biophys. Acta **27**, 229 (1958).

Kedem, O., Katchalsky, A., J. Gen. Physiol. **45**, 143 (1961).

Kedem, O., Katchalsky, A., Trans. Faraday Soc. **59**, 1918, 1931, 1941 (1963).

Keilin, B., The Mechanism of Desalination by Reverse Osmosis, U.S. Dept. Interior, Office of Saline Water, Research and Development Progress Report No. 84 (1963).

Keilin, B., in Membrane Processes for Industry, p. 80, Southern Research Institute, Birmingham, Alabama (1966).

Keilin, B., DeHaven, C. G., Proc. First International Symp. on Water Desalination, Washington, D.C. (1965) (Pub. U.S. Dept. Interior, Office of Saline Water, Washington, D.C.) Vol. 2, p. 367-380.

Keilin, B., Saltonstall, C. W., Higley, W. S., Kesting, R. E., Vincent, A. L., Structure and Properties of Reverse Osmosis Membranes, U.S. Dept. Interior, Office of Saline Water, Research and Development Progress Report No. 154 (1965).

Kesting, R. E., J. Applied Polymer Sci. **9**, 663 (1965).

Kesting, R. E., Barsh, M. K., Vincent, A. L., J. Applied Polymer Sci. **9**, 1873 (1965).

Kimura, S., Sourirajan, S., A.I.Ch.E.J. **13**, 497 (1967).

Kimura, S., Sourirajan, S., Ind. Eng. Chem. Process Design Develop. **7**, 41 (1968a).

Kimura, S., Sourirajan, S., Ind. Eng. Chem. Process Design Develop. **7**, 197 (1968b).

Kimura, S., Sourirajan, S., Ind. Eng. Chem. Process Design Develop. **7**, 539 (1968c).

Kimura, S., Sourirajan, S., Ind. Eng. Chem. Process Design Develop. **7**, 548 (1968d).

Kimura, S., Sourirajan, S., Ohya, H., Ind. Eng. Chem. Process Design Develop. **8**, 79 (1969).

Klemm, K., Friedman, L., J. Am. Chem. Soc. **54**, 2637 (1932).

Knudsen, J. G., Katz, D. L., Fluid Dynamics and Heat Transfer, p. 100, McGraw-Hill, New York (1958a).

Knudsen, J. G., Katz, D. L., Fluid Dynamics and Heat Transfer p. 167, McGraw-Hill, New York (1958b).

Knudsen, J. G., Katz, D. L., Fluid Dynamics and Heat Transfer, p. 171, McGraw-Hill, New York (1958c).

Knudsen, J. G., Katz, D. L., Fluid Dynamics and Heat Transfer, p. 207, McGraw-Hill, New York (1958d).

Knudsen, J. G., Katz, D. L., Fluid Dynamics and Heat Transfer, p. 364, McGraw-Hill, New York (1958e).

Knudsen, J. G., Katz, D. L., Fluid Dynamics and Heat Transfer, p. 366, McGraw-Hill, New York (1958f).

Kolthoff, I. M., Lingane, J. J., Polarography, 2nd ed., Vol. I, p. 41, Interscience, New York (1952a).

Kolthoff, I. M., Lingane, J. J., Polarography, 2nd ed., Vol. I, p. 51, Interscience, New York (1952b).

Kopeček, J., Sourirajan, S., J. Applied Polymer Sci. **13**, 637 (1969a).

Kopeček, J., Sourirajan, S., Ind. Eng. Chem. Prod. Res. Develop. 8, 274 (1969b).

Kopeček, J., Sourirajan, S., Canad. J. Chem: **47**, 3467 (1969c).

Kopeček, J., Sourirajan, S., Performance of Porous Cellulose Acetate Membranes for the Reverse Osmosis Separation of Mixtures of Organic Liquids (Paper submitted to Ind. Eng. Chem. Process Design Develop.) (1969d)

Kraus, K. A., Marcinkowsky, A. E., Johnson, J. S., Shor, A. J., Science **151**, 194 (1966a).
Kraus, K. A., Phillips, H. O., Marcinkowsky, A. E., Johnson, J. S., Shor, A. J., Desalination **1**, 225 (1966b).
Kraus, K. A., Shor, A. J., Johnson, J. S., Desalination **2**, 243 (1967).
Kuppers, J. R., Harrison, N., Johnson, J. S., Jr., J. Applied Polymer Sci. **10**, 969 (1966).
Kuppers, J. R., Marcinkowsky, A. E., Kraus, K. A., Johnson, J. S., Separation Science **2** (5), 617 (1967).
Lakshminarayanaiah, N., Chem. Rev. **65**, 504 (1965).
Lancy, L. E., The Integrated Treatment of Metallurgical Wastes, in Proc. Second Ontario Industrial Waste Conference, p. 41, Ontario Water Resources Commission, Toronto (1955).
Langmuir, I., J. Am. Chem. Soc. **39**, 1848 (1917).
Larson, T. J., Purification of Subsurface Waters by Reverse Osmosis (Pilot Plant Operations), Gulf General Atomic Report No. GA-7819 (1967).
Larson, T. J., Nusbaum, I., Riedinger, A. B., Astl, J., Reverse Osmosis Membrane Module (Spiral Wound Concept), U.S. Dept. Interior, Office of Saline Water, Research and Development Progress Report No. 338 (1968).
Leonard, F. B. (Havens Industries), private communication (1969).
Levich, V. G., Physicochemical Hydrodynamics, p. 390, Prentice-Hall, Englewood Cliffs, N.J. (1962).
Lewis, G. N., Randall, M., Thermodynamics, Revised by Pitzer, K. S., and Brewer, L., 2nd ed., p. 320, McGraw-Hill, New York (1961).
Lin, C. S., Denton, E. B., Gaskill, H. S., Putnam, G. L., Ind. Eng. Chem. **43**, 2136 (1951).
Linton, W. H., Jr., Sherwood, T. K., Chem. Eng. Progress **46**, 258 (1950).
Loeb, S., Sea Water Demineralization By Means of a Semipermeable Membrane, Department of Engineering, University of California, Los Angeles, Report No. 61-42 (1961).
Loeb, S., Sea Water Demineralization By Means of a Semipermeable Membrane, Department of Engineering, University of California, Los Angeles, Report No. 62-26 (1962).
Loeb, S., Sea Water Demineralization By Means of a Semipermeable Membrane, Department of Engineering, University of California, Los Angeles, Report No. 63-32 (1963).
Loeb, S., Desalination **1**, 35 (1966a).
Loeb, S., in Desalination by Reverse Osmosis, U. Merten, ed., Chapter 3, M.I.T. Press, Cambridge, Mass. (1966b).
Loeb, S., Johnson, J. S., Chem. Eng. Progress **63**, No. 1, 90 (1967).
Loeb, S., Manjikian, S., Brackish Water Desalination by an Osmotic Membrane, Department of Engineering, University of California, Los Angeles, Report No. 63-22 (1963a).
Loeb, S., Manjikian, S., Brackish Water Desalination by an Osmotic Membrane, Department of Engineering, University of California, Los Angeles, Report No. 63-37 (1963b).
Loeb, S., Manjikian, S., Field Tests on Osmotic Desalination Membranes, Department of Engineering, University of California, Los Angeles, Report No. 64-34 (1964).
Loeb, S., Manjikian, S., Ind. Eng. Chem. Process Design Develop. **4**, 207 (1965).
Loeb, S., McCutchan, J. W., Ind. Eng. Chem. Product Res. Develop. **4**, 114 (1965).
Loeb, S., Milstein, F., Dechema-Monographien **47**, 707 (1962).
Loeb, S., Nagaraj, G. R., U.S. Patent 3, 170, 867 dated Feb. 23, 1965.
Loeb, S., Selover, E., Desalination **2**, 75 (1967).
Loeb, S., Sourirajan, S., in Sea Water Research, Department of Engineering, University of California, Los Angeles, Report No. 59-3 (1958).

Loeb, S., Sourirajan, S., in Sea Water Research, Department of Engineering, University of California, Los Angeles, Report No. 59-28 (1959a).
Loeb, S., Sourirajan, S., in Sea Water Research, Department of Engineering, University of California, Los Angeles, Report No. 59-46 (1959b).
Loeb, S., Sourirajan, S., in Sea Water Research, Department of Engineering, University of California, Los Angeles, Report No. 60-5 (1960a).
Loeb, S., Sourirajan, S., in Sea Water Research, Department of Engineering, University of California, Los Angeles, Report No. 60-26 (1960b).
Loeb, S., Sourirajan, S., Sea Water Demineralization by Means of a Semipermeable Membrane, Department of Engineering, University of California, Los Angeles, Report No. 60-60 (1961).
Loeb, S., Sourirajan, S., Advan. Chem. Ser. No. **38**, 117 (1963).
Loeb, S., Sourirajan, S., U.S. Patent 3, 133, 132, dated May 12 (1964).
Loeb, S., Sourirajan, S., U.S. Patent 3, 223, 424, dated Dec. 14 (1965).
Loeb, S., Sourirajan, S., Weaver, D. E., U.S. Patent 3, 133, 137, dated May 12 (1964).
Lonsdale, H. K., in Desalination by Reverse Osmosis, U. Merten, ed., Chapter 4, M.I.T. Press, Cambridge, Mass. (1966).
Lonsdale, H. K., Merten, U., Riley, R. L., J. Applied Polymer Sci. **9**, 1341 (1965).
Lonsdale, H. K., Merten, U., Riley, R. L., Vos, K. D., Westmoreland, J. C., Reverse Osmosis for Water Desalination, U.S. Dept. Interior, Office of Saline Water, Research and Development Progress Report No. 111 (1964).
Lonsdale, H. K., Merten, U., Tagami, M., J. Applied Polymer Sci. **11**, 1807 (1967).
Lorimer, J. W., Boterenbrood, E. I., Hermans, J. J., Disc. Faraday Soc. **21**, 141 (1956).
Lowe, E., Durkee, E. L., Morgan, Jr., A. I., Food Technol. **22**, 915 (1968).
Mahon, H. I., Proc. Desalination Research Conference, Nat. Acad. Sci. Publication 942, 345 (1963).
Maier, K., Scheuermann, E., Kolloid-Z. **171**, 122 (1960).
Manjikian, S., Optimization of a Reverse Osmosis Sea Water Desalination System. Report No. CR67. 027 for U.S. Naval Civil Engineering Laboratory, Port Hueneme, California (1966).
Manjikian, S., Ind. Eng. Chem. Product Res. Develop. **6**, 23 (1967).
Manjikian, S., private communication, 1968.
Manjikian, S., Loeb, S., McCutchan, J. W., Proc. First International Symp. on Water Desalination, Washington, D.C. (1965) (Pub. U.S. Dept. Interior, Office of Saline Water, Washington, D.C.) Vol. 2, p. 159-173.
Marcinkowsky, A. E., Kraus, K. A., Phillips, H. O., Johnson, Jr., J. S., Shor, A. J., J. Am. Chem. Soc. **88**, 5744 (1966).
Markley, L. L., Cross, R. A., Bixler, H. J., Membranes for Desalination by Reverse Osmosis, U.S. Dept. Interior, Office of Saline Water, Research and Development Progress Report No. 281 (1967).
Marshall, P. G., Dunkley, W. L., Lowe, E., Food Technol. **22**, 969 (1968).
McBain, J. W., Colloid Science, p. 55, D. C. Heath & Co., Boston, Mass. (1950).
McBain, J. W., Dubois, R., J. Am. Chem. Soc. **51**, 3534 (1929).
McBain, J. W., Kistler, S. S., J. Gen. Physiol. **12**, 187 (1928).
McBain, J. W., Stuewer, R. F., J. Phys. Chem. **40**, 1157 (1936).
McCutchan, J. W., in Saline Water Research Progress Summary, p. 13, Department of Engineering, University of California, Los Angeles, Report No. 68-1 (1968).
McKelvey, J. G., Spiegler, K. S., Wyllie, M. R. J., Chem. Eng. Progress Symp. Ser. **55**, No. 24, 199 (1959).
Meares, P., European Polymer J. **2**, 241 (1966).
Merson, R. L., Morgan, Jr., A. I., Food Technol. **22**, 631 (1968).
Merten, U., Ind. Eng. Chem. Fundamentals **2**, 229 (1963).

Merten, U., in Desalination by Reverse Osmosis, U. Merten, ed., Chapter 2, M.I.T. Press, Cambridge, Mass. (1966a).
Merten, U., Desalination **1**, 297 (1966b).
Merten, U., Lonsdale, H. K., Riley, R. L., Ind. Eng. Chem. Fundamentals **3**, 210 (1964).
Merten, U., Lonsdale, H. K., Riley, R. L., Vos, K. D., Reverse Osmosis Membrane Research, U.S. Dept. Interior, Office of Saline Water, Research and Development Progress Report No. 265 (1967).
Michaeli, I., Kedem, O., Trans. Faraday Soc. **57**, 1185 (1961).
Michaels, A. S., Ind. Eng. Chem. **57**, No. 10, 32 (1965).
Michaels, A. S., Bixler, H. J., Hodges, R. M., Jr., J. Colloid Sci. **20**, 1034 (1965a).
Michaels, A. S., Bixler, H. J., Hausslein, R. W., Fleming, S. M., Polyelectrolyte complexes as Reverse Osmosis and Ion-Selective Membranes, U.S. Dept. Interior, Office of Saline Water, Research and Development Progress Report No. 149 (1965b).
Michaels, A. S., Miekka, R. G., J. Phys. Chem. **65**, 1765 (1961).
Mickley, H. S., Sherwood, T. K., Reed, C. E., Applied Mathematics in Chemical Engineering, p. 191, McGraw-Hill, New York (1957).
Miller, C. S., Design Construction and Operation of a Pilot System For the Production of Continuous Reverse Osmosis Membranes, M. Eng. Thesis, The Thayer School of Engineering, Dartmouth College, Hanover (1968).
Morgan, A. I., Jr., Lowe, E., Merson, R. L., Durkee, E. L., Food Technol. **19**, 1790 (1965).
Öholm, L. W., Finska Kemistsamfundets Meddelanden **43**, 55 (1934a).
Öholm, L. W., Finska Kemistsamfundets Meddelanden **43**, 121 (1934b).
Öholm, L. W., Finska Kemistsamfundets Meddelanden **45**, 71 (1936).
Öholm, L. W., Finska Kemistsamfundets Meddelanden **47**, 59 (1938a).
Öholm, L. W., Finska Kemistsamfundets Meddelanden **47**, 115 (1938b).
Ohya, H., Sourirajan, S., Ind. Eng. Chem. Process Design Develop. **8**, 131 (1969a).
Ohya, H., Sourirajan, S., Desalination **6**, 153 (1969b).
Ohya, H., Sourirajan, S., A.I.Ch.E.J. **15**, 780 (1969c).
Ohya, H., Sourirajan, S., A.I.Ch.E.J. **15**, 829 (1969d).
Okey, R. W., Stavenger, P. L., in Membrane Processes for Industry, p. 127, Southern Research Institute, Birmingham, Alabama (1966).
OSW, Saline Water Conversion Report, U.S. Dept. Interior, Office of Saline Water (1964).
OSW, Saline Water Conversion Report, U.S. Dept. Interior, Office of Saline Water (1965).
OSW, Saline Water Conversion Report, U.S. Dept. Interior, Office of Saline Water (1966).
Parks, G. A., Chem. Rev. **65**, 177 (1965).
Pauling, L., The Nature of Chemical Bond, 3rd ed., p. 321, Cornell University Press, Ithaca, N.Y. (1960).
Perona, J. J., Butt, F. H., Fleming, S. M., Mayr, S. T., Spitz, R. A., Brown, M. K., Cochran, H. D., Kraus, K. A., Johnson, Jr., J. S., Environmental Sci. Tech. **1**, 991 (1967).
Peterson, W. S., (Division of Applied Chemistry National Research Council of Canada, Ottawa), personal communication (1968).
Podall, H. E., Theoretical and Practical Aspects of the Evaluation of Candidate Reverse Osmosis Membranes for Desalination, U.S. Dept. Interior, Office of Saline Water, Research and Development Progress Report No. 267 (1967).
Podall, H. E., Transport Property Requirements of Membranes for Use in the Reverse Osmosis Process as Defined by the Method of Irreversible Thermodynamics, U.S. Dept. Interior, Office of Saline Water, Research and Development Progress Report No. 303 (1968a).

Podall, H. E., Transport Property Requirements of Membranes for Use in Pressure Dialysis Desalination, U.S. Dept. Interior, Office of Saline Water, Research and Development Progress Report No. 304 (1968b).
Posnjak, E., Am. J. Sci. **238,** 559 (1940).
Prideaux, E. B. R., Trans. Faraday Soc. **38,** 121 (1942).
Promisel, N. E., in Metal Finishing Guidebook, 28th ed., p. 658, Metals and Plastics Publications Inc., Westwood, N.J. (1960).
Reid, C. E., Breton, E. J., J. Applied Polymer Sci. **1,** 133 (1959).
Reid, C. E., Kuppers, J. R., J. Applied Polymer Sci. **2,** 264 (1959).
Reid, C. E., Spencer, H. G., J. Applied Polymer Sci. **4,** 354 (1960a).
Reid, C. E., Spencer, H. G., J. Phys. Chem. **64,** 1587 (1960b).
Reiss, L. P., Hanratty, T. J., A.I.Ch.E.J. **8,** 245 (1962).
Remy, H., Z. Physik. Chem. **89,** 467 (1915).
Riedinger, A. B., Laughlin, J. K. Sudak, R. G., Large Reverse Osmosis System Technology and Module Development, U.S. Dept. Interior, Office of Saline Water, Research and Development Progress Report No. 341 (1968).
Riedinger, A., Schultz, J., Acid Mine Water Reverse Osmosis Test at Kittanning, Pennsylvania, U.S. Dept. Interior, Office of Saline Water, Research and Development Progress Report No. 217 (1966).
Riley, R. L., Gardner, J. O., Merten, U., Science **143,** 801 (1964).
Riley, R. L., Lonsdale, H. K., Lyons, C. R., Merten, U., J. Applied Polymer Sci. **11,** 2143 (1967).
Riley, R. L., Merten, U., Gardner, J. O., Desalination **1,** 30 (1966).
Robinson, R. A., Stokes, R. H., Electrolyte Solutions, 2nd ed., p. 30, Butterworths, London (1959a).
Robinson, R. A., Stokes, R. H., Electrolyte Solutions, 2nd ed., p. 461, Butterworths, London (1959b).
Robinson, R. A., Stokes, R. H., Electrolyte Solutions, 2nd ed., p. 476-490, Butterworths, London (1959c).
Robinson, R. A., Stokes, R. H., Electrolyte Solutions, 2nd ed., p. 513-515, Butterworths, London (1959d).
Ruess, G. L., Monatsh. **76,** 381 (1947).
Rutz, L. O., Kammermeyer, K., AECU-3921, U.S. Atomic Energy Commission, Technical Information Service, Oak Ridge, Ten. (1958).
Rutz, L. O., Kammermeyer, K., AECU-4328, U.S. Atomic Energy Commission, Technical Information Service, Oak Ridge, Ten. (1959).
Scatchard, G., J. Phys. Chem. **68,** 1056 (1964).
Scatchard, G., Hamer, W. J., Wood, S. E., J. Am. Chem. Soc. **60,** 3061 (1938).
Schick, M. J., Atlas, S. M., Eirich, F. R., J. Phys. Chem. **66,** 1326 (1962).
Schmid, G., Schwarz, H., Z. Electrochem. **56, 35** (1952).
Schultz, J., Potomac River Reverse Osmosis Test, U.S. Dept. Interior, Office of Saline Water, Research and Development Progress Report No. 216 (1966).
Schultz, J., Newby, G. A., Desalination by Reverse Osmosis, Gulf General Atomic Report GA-7153 (1966).
Schultz, J., Riedinger, A., McCraken, H., Brackish Well Water Reverse Osmosis Tests at Midland, Fort Stockton, and Kermit, Texas, U.S. Dept. Interior, Office of Saline Water, Research and Development Progress Report No. 237 (1967).
Shaw, P. V., Reiss, L. P., Hanratty, T. J., A.I.Ch.E.J. **9,** 362 (1963).
Sherwood, T. K., Mass Transfer Between Phases, 33rd Annual Priestley Lectures, p. 38, Pennsylvania State Univ. (1959).
Sherwood, T. K., Brian, P. L. T., Fisher, R. E., Salt Concentration at Phase Boundaries in Desalination Processes, U.S. Dept. Interior, Office of Saline Water, Research and Development Progress Report No. 95 (1964).
Sherwood, T. K., Brian, P. L. T., Fisher, R. E., Dresner, L., Ind. Eng. Chem. Fundamentals **4,** 113 (1965).

Sherwood, T. K., Brian, P. L. T., Fisher, R. E., Ind. Eng. Chem. Fundamentals **6**, 2 (1967).
Sherwood, T. K., Pigford, R. L., Absorption and Extraction, p. 53, McGraw-Hill, New York (1952).
Shinoda, K., Nakagawa, T., Tamamushi, B.-I., Isemura, T., Colloidal Surfactants, p. 117, Academic Press, New York (1963).
Sieveka, E. H., in Desalination by Reverse Osmosis, U. Merten, ed., Chapter 7, M.I.T. Press, Cambridge, Mass. (1966).
Sinclair, D. A., J. Phys. Chem. **37**, 495 (1933).
Slavin, W., Atomic Absorption spectroscopy, Chapter IV, Wiley, Interscience, New York (1968).
Small, P. A., J. Applied Chem. **3**, 71 (1953).
Sollner, K., Z. Electrochem. **36**, 36 (1930).
Sollner, K., Svensk. Kem. Tidskr. **6-7**, 267 (1958).
Sollner, K., Dray, S., Grim, E., Neihof, R., Ion Transport Across Membranes, H. T. Clarke and D. Nachmanson, eds., p. 144-188, Academic Press, New York (1954).
Sourirajan, S., Ind. Eng. Chem. Fundamentals **2**, 51 (1963a).
Sourirajan, S., Nature **199**, 590 (1963b).
Sourirajan, S., Separation of Some Inorganic Salts in Aqueous Solution By Flow, Under Pressure, Through Porous Cellulose Acetate Membranes, National Research Council of Canada, N.R.C. No. 7498 (1963c).
Sourirajan, S., Ind. Eng. Chem. Fundamentals **3**, 206 (1964a).
Sourirajan, S., J. Applied Chem. **14**, 506 (1964b).
Sourirajan, S., Nature **203**, 1348 (1964c).
Sourirajan, S., Ind. Eng. Chem. Prod. Res. Develop. **4**, 201 (1965).
Sourirajan, S., Ind. Eng. Chem. Process Design Develop. **6**, 154 (1967a).
Sourirajan, S., in Water Resources of Canada, C. E. Dolman, ed., p. 154-182, University of Toronto Press, Toronto (1967b).
Sourirajan, S., The Canadian Scientist **1**, No. 3, 22 (1967c).
Sourirajan, S., Canadian Patent 797, 604 dated Oct. 29, 1968.
Sourirajan, S., Agrawal, J. P., Ind. Eng. Chem. **61** (11), 62 (1969).
Sourirajan, S., Govindan, T. S., Proc. First International Symp. on Water Desalination, Washington, D.C. (1965) (Pub. U.S. Dept. Interior, Office of Saline Water, Washington, D.C.) Vol. 1, p. 251-274.
Sourirajan, S., Kimura, S., Ind. Eng. Chem. Process Design Develop. **6**, 504 (1967).
Sourirajan, S., Loeb, S., in Sea Water Research, Department of Engineering, University of California, Los Angeles, Report No. 58-65 (1958).
Sourirajan, S., Sirianni, A.F., Ind. Eng. Chem. Prod. Res. Develop. **5**, 30 (1966).
Spatz, D. D., The Design, Development, and Field Testing of Home Reverse Osmosis Desalination Units, M. Eng. Thesis, The Thayer School of Engineering, Dartmouth College, Hanover (1968).
Spiegler, K. S., Trans. Faraday Soc. **54**, 1408 (1958).
Spiegler, K. S., Kedem, O., Desalination **1**, 311 (1966).
Srinivasan, S., Tien, C., Gill. W. N., Chem. Eng. Sci. **22**, 417 (1967).
Standard Methods for the Examination of Water and Waste Water, 12th ed., American Public Health Association, New York (1965).
Staverman, A. J., Trans. Faraday Soc. **48**, 176 (1952).
Stevens, D., Loeb, S., Desalination **2**, 56 (1967).
Stoughton, R. W., Lietzke, M. H., J. Chem. Eng. Data **10**, 254 (1965).
Sudak, R. G., Nusbaum, I., Pilot Plant Operation of Spiral Wound Reverse Osmosis Systems, Gulf General Atomic Report No. GA-8515 (1968).
The San Diego Union, July 30, 1967.
Tien, C., Gill, W. N., A.I.Ch.E.J. **12**, 722 (1966).

Trautman, S., Ambard, L., J. Chim. Phys. **49**, 220 (1952).
Treybal, R. E., Mass Transfer Operations, p. 38, McGraw-Hill, New York (1955).
Tribus, M., Asimow, R., Richardson, N., Gustaldo, C., Elliot, K., Chambers, J., Evans, R., Thermodynamic and Economic Considerations in the Preparation of Fresh Water from the Sea, Department of Engineering, University of California, Los Angeles, Report No. 59-34 (1960).
University of California, Office of Public Information Press Release, New Water Desalting Process Developed at UCLA, August 23, 1960.
Van Amerongen, G. J., J. Polymer Sci. **5**, 307 (1950).
Van Oss, C. J., Science **139**, 1123 (1963).
Vincent, A. L., Barsh, M. K., Kesting, R. E., J. Applied Polymer Sci. **9**, 2363 (1965).
Vos, K. D., Burris, Jr., F. O., Riley, R. L., J. Applied Polymer Sci. **10**, 825 (1966a).
Vos. K. D., Hatcher, A. P., Merten, U., Ind. Eng. Chem. Prod. Res. Develop. **5**, 211 (1966b).
Wasilewski, S., Microwave Investigation of Temperature Induced, Transient Changes in Cellulose Acetate Osmotic Membranes, Department of Engineering, University of California, Los Angeles, Report No. 65-10 (1965).
Weller, S., Steiner, W. A., J. Applied Phys. **21**, 279 (1950).
Wheaton, R. M., Bauman, W. C., Ind. Eng. Chem. **45**, 228 (1953).
Wiley, A. J., Ammerlaan, A. C. F., Dubey, G. A., Application of Reverse Osmosis to Processing of Spent Liquors from the Pulp and Paper Industry, Pulp Manufacturers Research League, Inc., Appleton, Wisc. (1967).
Wilson, J. R., ed., Demineralization by Electrodialysis, Butterworths, London (1960).
Willits, C. O., Underwood, J. C., Merten, U., Food Technol. **21**, 24 (1967).
Yuan, S. W., Finkelstein, A. B., Trans. ASME **78**, 719 (1956).
Yuster, S. T., personal communication, September 1956.
Yuster, S. T., Sourirajan, S., Bernstein, K., Sea Water Demineralization by the Surface-Skimming Process, Department of Engineering, University of California, Los Angeles, Report No. 58-26 (1958).

APPENDIX I

Text of Press Release dated 23 August 1960

Office of Public Information, University of California, Los Angeles
23 August 1960

New Water Desalting Process Developed at UCLA

A new and promising process for taking the salt out of salt water to get drinkable water is being developed in a UCLA engineering laboratory.

Research engineers Sidney Loeb and Dr. Srinivasa Sourirajan have successfully tested a special membrane, or film, with a large number of tiny pores, which in effect filters the salt water and separates the brine from the potable water.

Membranes have been used before to demineralize salt water, but the problem has always been to find a film which would keep back the salt but at the same time allow fresh water to filter through at a fast enough rate.

To make their film, the UCLA engineers mix cellulose acetate with aqueous magnesium perchlorate solution and acetone, and then cast the sticky mixture cold on a glass plate.

After putting the glass plate into cold water, a film four-thousandths of an inch thick is stripped from the plate and then shrunk slightly in hot water. The finished membrane, which can filter water a hundred times faster then previous commercial films, is mounted on each side of a porous steel disk.

Any number of disks are then stacked in a frame press, salt water is circulated through the unit at 1500 pounds per square inch pressure, and the desalted water flows out from the bottom of the unit.

With their present film, Loeb and Dr. Sourirajan have been able to get potable water at the rate of eight gallons a day per square foot of membrane area. Furthermore, the potable water is produced in a single filtering from a brine containing a 5.25 per cent salt concentration, which is much saltier than ordinary ocean water.

Loeb and Dr. Sourirajan believe that their process holds possibilities for future large-scale commercial use because the film is both cheap and durable, the equipment is simple and can run 24 hours a day without maintenance, and power requirements are as low as for any other desalinization method.

They are now working on a unit capable of producing 500 gallons a day in order to get cost and design information for a pilot plant which can yield 25,000 gallons a day.

The project, initiated by the late UCLA engineering professor S. T. Yuster, is supported by California state funds.

APPENDIX II

Osmotic Pressure, Molar Density, Kinematic Viscosity and Solute Diffusivity for Several Aqueous Solution Systems

The diffusivity data have been obtained from Gosting and Akeley (1952), Hook and Russell (1945), International Critical Tables (1929b), Öholm (1934a, 1934b, 1936, 1938a, 1938b), and Robinson and Stokes (1959d).

Table A-1. Data for the System [NH_4Cl-H_2O] at 25°

Molality	Mole fraction $\times 10^3$	Weight % solute	Osmotic pressure (lb/in^2)	Density of solution (g/cm^3)	Molar density ($mole/cm^3$ $\times 10^2$)	Kinematic viscosity (cm^2/sec $\times 10^2$)	Solute diffusivity (cm^2/sec $\times 10^5$)
0	0	0	0	0.9971	5.535	0.8963	1.994
0.1	1.798	0.5322	67	0.9987	5.524	0.8938	1.838
0.2	3.590	1.0587	131	1.0004	5.514	0.8911	1.836
0.3	5.375	1.5796	195	1.0020	5.504	0.8886	1.840
0.4	7.154	2.0952	259	1.0036	5.494	0.8861	1.850
0.5	8.927	2.6053	323	1.0051	5.483	0.8838	1.860
0.6	10.693	3.1102	386	1.0066	5.472	0.8820	1.870
0.7	12.453	3.6098	450	1.0081	5.462	0.8802	1.880
0.8	14.207	4.1043	514	1.0096	5.452	0.8784	1.892
0.9	15.955	4.5938	579	1.0111	5.442	0.8766	1.906
1.0	17.696	5.0783	644	1.0125	5.431	0.8748	1.917
1.2	21.160	6.0327	774	1.0154	5.411	0.8718	1.940
1.4	24.600	6.9681	905	1.0182	5.391	0.8689	1.964
1.6	28.016	7.8850	1038	1.0209	5.371	0.8661	1.986
1.8	31.408	8.7841	1172	1.0235	5.350	0.8633	2.009
2.0	34.777	9.6658	1307	1.0260	5.330	0.8606	2.030
2.5	43.096	11.7971	1652	1.0321	5.281	0.8577	2.085
3.0	51.273	13.8302	2001	1.0378	5.232	0.8551	2.134
3.5	59.312	15.7717	2361	1.0433	5.186	0.8545	2.170
4.0	67.216	17.6277	2727	1.0484	5.139	0.8542	2.199
4.5	74.987	19.4036	3096	1.0533	5.094	0.8565	2.222
5.0	82.631	21.1045	3459	1.0579	5.050	0.8592	2.243
5.5	90.149	22.7352	3828	1.0623	5.007	0.8628	2.258
6.0	97.545	24.2998	4204	1.0665	4.966	0.8665	2.264

Table A-2. Data for the System [$BaCl_2$-H_2O] at 25°

Molality	Mole fraction × 10^3	Weight % solute	Osmotic pressure (lb/in^2)	Density of solution (g/cm^3)	Molar density ($mole/cm^3 \times 10^2$)	Kinematic viscosity ($cm^2/sec \times 10^2$)	Solute diffusivity ($cm^2/sec \times 10^5$)
0	0	0	0	0.9971	5.535	0.8963	1.385
0.1	1.798	2.0402	91	1.0180	5.546	0.9002	1.159
0.2	3.590	3.9989	180	1.0320	5.519	0.9053	1.150
0.3	5.375	5.8808	272	1.0500	5.515	0.9115	1.151
0.4	7.154	7.6903	367	1.0680	5.512	0.9170	1.155
0.5	8.927	9.4315	465	1.0850	5.504	0.9248	1.160
0.6	10.693	11.1083	566	1.1010	5.491	0.9297	1.164
0.7	12.453	12.7241	671	1.1190	5.490	0.9391	1.168
0.8	14.207	14.2822	779	1.1340	5.474	0.9491	1.171
0.9	15.955	15.7857	889	1.1500	5.463	0.9604	1.175
1.0	17.696	17.2373	1005	1.1670	5.458	0.9675	1.177
1.2	21.160	19.9954	1249	1.1980	5.435	0.9978	1.179
1.4	24.600	22.5756	1510	1.2290	5.415	1.0245	1.180
1.6	28.016	24.9946	1781	1.2570	5.384	—	—
1.8	31.408	27.2670	2064	1.2840	5.352	—	—

Table A-3. Data for the System [$CaCl_2$-H_2O] at 25°

Molality	Mole fraction × 10^3	Weight % solute	Osmotic pressure (lb/in^2)	Density of solution (g/cm^3)	Molar density ($mole/cm^3 \times 10^2$)	Kinematic viscosity ($cm^2/sec \times 10^2$)	Solute diffusivity ($cm^2/sec \times 10^5$)
0	0	0	0	0.9971	5.535	0.8963	1.335
0.1	1.798	1.0977	92	1.0061	5.533	0.9167	1.285
0.2	3.590	2.1716	186	1.0149	5.531	0.9373	1.281
0.3	5.375	3.2224	283	1.0237	5.529	0.9562	1.292
0.4	7.154	4.2509	385	1.0323	5.526	0.9755	1.304
0.5	8.927	5.2577	494	1.0408	5.523	0.9959	1.318
0.6	10.693	6.2436	607	1.0492	5.519	1.0159	1.334
0.7	12.453	7.2092	726	1.0575	5.516	1.0355	1.350
0.8	14.207	8.1551	851	1.0657	5.512	1.0576	1.362
0.9	15.955	9.0819	985	1.0738	5.507	1.0800	1.376
1.0	17.696	9.9902	1126	1.0817	5.502	1.1028	1.389
1.2	21.160	11.7534	1430	1.0975	5.492	1.1513	1.414
1.4	24.600	13.4488	1765	1.1130	5.482	1.2052	1.441
1.6	28.016	15.0804	2132	1.1280	5.471	1.2589	1.465
1.8	31.408	16.6515	2531	1.1428	5.459	1.3253	1.483
2.0	34.777	18.1656	2967	1.1573	5.447	1.3894	1.501
2.5	43.096	21.7206	4240	1.1927	5.416	1.5259	1.512
3.0	51.273	24.9796	5803	1.2258	5.381	1.8485	1.486
3.5	59.312	27.9780	7579	1.2576	5.345	2.1317	1.436
4.0	67.216	30.7460	9584	1.2813	5.281	2.2621	—
4.5	74.987	33.3091	11813	1.3097	5.242	2.2783	—
5.0	82.631	35.6893	14248	1.3425	5.224	3.2315	—
5.5	90.149	37.9054	16768	1.3683	5.184	4.5573	—
6.0	97.545	39.9738	19348	1.3924	5.141	5.5843	—

Table A-4. Data for the System [$Ca(NO_3)_2$-H_2O] at 25°

Molality	Mole fraction × 10^3	Weight % solute	Osmotic pressure (lb/in^2)	Density of solution (g/cm^3)	Molar density (mole/cm^3 × 10^2)	Kinematic viscosity (cm^2/sec × 10^2)	Solute diffusivity (cm^2/sec × 10^5)
0	0	0	0	0.9971	5.535	0.8963	
0.1	1.798	1.6144	89	1.0075	5.512	0.8730	1.103
0.2	3.590	3.1775	176	1.0190	5.496	0.8636	1.086
0.3	5.375	4.6917	264	1.0300	5.479	0.8544	1.081
0.4	7.154	6.1593	354	1.0410	5.462	0.8597	1.065
0.5	8.927	7.5824	444	1.0530	5.451	0.8737	1.060
0.6	10.693	8.9630	537	1.0650	5.440	0.9015	1.043
0.7	12.453	10.3029	631	1.0755	5.423	0.9158	1.045
0.8	14.207	11.6039	726	1.0870	5.411	0.9475	1.033
0.9	15.955	12.8678	824	1.0970	5.392	0.9754	1.033
1.0	17.696	14.0960	925	1.1080	5.379	1.0018	1.033
1.2	21.160	16 4514	1136	1.1295	5.352	1.0624	1.010
1.4	24.600	18.6810	1353	1.1500	5.322	1.1304	1.010
1.6	28.016	20 7948	1579	1.1690	5.288	1.1976	0.998
1.8	31.408	22.8015	1810	1.1870	5.252	1.2721	0.980
2 0	34.777	24 7090	2052	1.2050	5.218	1.3610	0.975
2.2	38.122	26.5245	2337	1.2220	5.182	1.4320	0.981
2.4	41.444	28.2545	2596	1.2390	5.148	1.4850	1.007
2 6	44.743	29.9049	2870	1.2555	5.114	1.6009	0.989
2.8	48.019	31.4811	3152	1.2730	5.086	1.6810	0.996
3.0	51.273	32.9880	3445	1.2890	5.054	1.7688	1.002

Table A-5. Data for the System [$CuSO_4$-H_2O] at 25°

Molality	Mole fraction × 10^3	Weight % solute	Osmotic pressure (lb/in^2)	Density of solution (g/cm^3)	Molar density (mole/cm^3 × 10^2)	Kinematic viscosity (cm^2/sec × 10^2)	Solute diffusivity (cm^2/sec × 10^5)
0		0	0	0.9971	5 535	0.8963	0.627
0 1	1.798	1 5709	40	1.0132	5.546	0.9445	0.590
0.2	3.590	3.0933	74	1.0288	5.554	0.9914	0.578
0.3	5.375	4.5692	106	1.0444	5.562	1.0436	0.562
0.4	7.154	6.0009	137	1.0604	5.573	1.0967	0.544
0.5	8.927	7.3903	169	1.0761	5.582	1.1523	0.529
0.6	10.693	8.7391	200	1.0910	5.587	1.2099	0.517
0.7	12 453	10.0493	231	1.1056	5.590	1 2690	0.503
0.8	14 207	11.3224	263	1.1204	5.595	1.3308	0.494
0.9	15.955	12.5599	298	1.1351	5.599	1.3919	0.483
1.0	17.696	13.7634	334	1.1488	5.600	1.4580	0.474
1.2	21.160	16.0736	411	1.1772	5.603	1.6140	0.455
1.4	24.600	18.2632	498	1.2048	5.604	1.8011	0.438

Table A-6. Data for the System glycerol-water at 25°

Molality	Mole fraction × 10^3	Weight % solute	Osmotic pressure (lb/in^2)	Density of solution (g/cm^3)	Molar density ($mole/cm^3$ × 10^2)	Kinematic viscosity (cm^2/sec × 10^2)	Solute diffusivity (cm^2/sec × 10^5)
0	0	0	0	0.9971	5.535	0.8963	—
0.1	1.798	0.9126	36	0.9993	5.505	0.9106	0.9368
0.2	3.590	1.8087	72	1.0014	5.478	0.9287	0.9239
0.3	5.375	2.6887	108	1.0035	5.450	0.9467	0.9138
0.4	7.154	3.5531	144	1.0055	5.422	0.9647	0.9041
0.5	8.927	4.4023	181	1.0075	5.395	0.9876	0.8964
0.6	10.693	5.2366	217	1.0095	5.368	1.0054	0.8894
0.7	12.453	6.0565	253	1.0115	5.341	1.0282	0.8834
0.8	14.207	6.8624	290	1.0134	5.315	1.0480	0.8778
0.9	15.955	7.6545	326	1.0152	5.288	1.0688	0.8733
1.0	17.696	8.4333	363	1.0170	5.262	1.0865	0.8688
1.2	21.160	9.9521	436	1.0206	5.212	1.1268	0.8621
1.4	24.600	11.4213	510	1.0242	5.163	1.1668	0.8561
1.6	28.016	12.8434	584	1.0276	5.115	1.2116	0.8520
1.8	31.408	14.2205	658	1.0311	5.069	1.2559	0.8483
2.0	34.777	15.5548	732	1.0343	5.023	1.3004	0.8458
2.5	43.096	18.7157	919	1.0421	4.914	1.4202	0.8424
3.0	51.273	21.6485	1107	1.0494	4.811	1.5485	—
3.5	59.312	24.3771	1295	1.0564	4.714	1.6897	—
4.0	67.216	26.9220	1485	1.0628	4.622	1.8207	—
4.5	74.987	29.3011	1675	1.0689	4.535	1.9646	—
5.0	82.631	31.5303	1866	1.0748	4.453	2.1167	—
5.5	90.149	33.6232	2059	1.0803	4.375	2.2771	—
6.0	97.545	35.5919	2252	1.0855	4.300	2.4459	—

Table A-7. Data for the System [$LiCl-H_2O$] at 25°

Molality	Mole fraction × 10^3	Weight % solute	Osmotic pressure (lb/in^2)	Density of solution (g/cm^3)	Molar density ($mole/cm^3$ × 10^2)	Kinematic viscosity (cm^2/sec × 10^2)	Solute diffusivity (cm^2/sec × 10^5)
0	0	0	0	0.9971	5.535	0.8963	1.366
0.1	1.798	0.4222	67	0.9996	5.535	0.9066	1.269
0.2	3.590	0.8409	135	1.0020	5.535	0.9169	1.267
0.3	5.375	1.2560	204	1.0044	5.535	0.9270	1.269
0.4	7.154	1.6677	274	1.0068	5.535	0.9368	1.273
0.5	8.927	2.0760	346	1.0091	5.535	0.9468	1.277
0.6	10.693	2.4809	419	1.0115	5.535	0.9574	1.283
0.7	12.453	2.8824	495	1.0138	5.534	0.9681	1.288
0.8	14.207	3.2807	572	1.0161	5.534	0.9787	1.292
0.9	15.955	3.6757	650	1.0183	5.533	0.9895	1.296
1.0	17.696	4.0675	731	1.0206	5.533	1.0000	1.301
1.2	21.160	4.8417	898	1.0250	5.531	1.0236	1.312
1.4	24.600	5.6034	1073	1.0293	5.529	1.0471	1.323
1.6	28.016	6.3530	1254	1.0336	5.528	1.0704	1.334
1.8	31.408	7.0908	1444	1.0379	5.526	1.0936	1.346
2.0	34.777	7.8171	1643	1.0420	5.524	1.1167	1.358
2.5	43.096	9.5841	2180	1.0521	5.518	1.1812	1.389
3.0	51.273	11.2846	2776	1.0619	5.512	1.2447	1.419
3.5	59.312	12.9223	3442	1.0714	5.505	1.3146	1.448
4.0	67.216	14.5007	4174	1.0806	5.498	1.3837	—
4.5	74.987	16.0228	4970	1.0895	5.490	1.4634	—
5.0	82.631	17.4917	5834	1.0983	5.483	1.5420	—
5.5	90.149	18.9102	6760	1.1068	5.476	1.6352	—
6.0	97.545	20.2806	7750	1.1151	5.468	1.7271	—

Table A-8. Data for the System [$LiNO_3$-H_2O] at 25°

Molality	Mole fraction × 10^3	Weight % solute	Osmotic pressure (lb/in^2)	Density of solution (g/cm^3)	Molar density ($mole/cm^3$ × 10^2)	Kinematic viscosity (cm^2/sec × 10^2)	Solute diffusivity (cm^2/sec × 10^5)
0	0	0	0	0.9971	5.535	0.8963	1.336
0.1	1.798	0.6847	67	1.0000	5.523	0.9035	1.240
0.2	3.590	1.3600	134	1.0040	5.517	0.9097	1.243
0.3	5.375	2.0263	202	1.0075	5.509	0.9252	1.248
0.4	7.154	2.6836	272	1.0110	5.501	0.9211	1.254
0.5	8.927	3.3321	342	1.0150	5.496	0.9271	1.260
0.6	10.693	3.9721	414	1.0185	5.488	0.9313	1.267
0.7	12.453	4.6036	487	1.0220	5.480	0.9400	1.274
0.8	14.207	5.2269	562	1.0260	5.475	0.9450	1.280
0.9	15.955	5.8421	638	1.0295	5.468	0.9505	1.286
1.0	17.696	6.4494	716	1.0330	5.461	0.9603	1.293
1.2	21.160	7.6407	874	1.0400	5.447	0.9727	1.303
1.4	24.600	8.8020	1038	1.0480	5.439	0.9849	1.313
1.6	28.016	9.9346	1208	1.0550	5.427	0.9996	1.321
1.8	31.408	11.0393	1382	1.0620	5.414	1.0140	1.327
2.0	34.777	12.1173	1562	1.0700	5.408	1.0273	1.332
2.2	38.122	13.1694	1749	1.0760	5.392	1.0407	1.334
2.4	41.444	14.1967	1941	1.0835	5.384	1.0541	1.335
2.6	44.743	15.1999	2138	1.0900	5.371	1.0699	1.334
2.8	48.019	16.1799	2339	1.0970	5.362	1.0835	1.333
3.0	51.273	17.1376	2543	1.1030	5.348	1.0995	1.332
3.2	54.505	18.0736	2756	1.1100	5.339	1.1151	1.324
3.4	57.715	18.9887	2972	1.1160	5.326	1.1331	1.316
3.6	60.903	19.8836	3191	1.1225	5.316	1.1504	1.308
3.8	64.070	20.7589	3417	1.1290	5.306	1.1715	1.300
4.0	67.216	21.6153	3646	1.1350	5.294	1.1905	1.292
4.2	70.340	22.4534	3882	1.1405	5.281	1.2146	1.281
4.4	73.443	23.2738	4121	1.1470	5.272	1.2334	1.271
4.6	76.526	24.0770	4364	1.1530	5.262	1.2549	1.260
4.8	79.589	24.8635	4609	1.1580	5.247	1.2772	1.249
5.0	82.631	25.6339	4852	1.1640	5.238	1.2990	1.238
5.2	85.653	26.3887	5105	1.1700	5.229	1.3176	1.222
5.4	88.655	27.1283	5356	1.1750	5.215	1.3386	1.206
5.6	91.638	27.8532	5607	1.1800	5.202	1.3594	1.190
5.8	94.601	28.5638	5861	1.1855	5.192	1.4006	1.173
6.0	97.545	29.2606	6115	1.1910	5.182	1.4167	1.157

Table A-9. Data for the System [$MgCl_2$-H_2O] at 25°

Molality	Mole fraction × 10^3	Weight % solute	Osmotic pressure (lb/in^2)	Density of solution (g/cm^3)	Molar density ($mole/cm^3$ × 10^2)	Kinematic viscosity (cm^2/sec × 10^2)	Solute diffusivity (cm^2/sec × 10^5)
0	0	0	0	0.9971	5.535	0.8963	1.249
0.1	1.798	0.9434	93	1.0048	5.535	0.9197	1.074
0.2	3.590	1.8691	189	1.0123	5.534	0.9475	1.051
0.3	5.375	2.7777	290	1.0198	5.533	0.9766	1.041
0.4	7.154	3.6696	397	1.0271	5.532	1.0069	1.040
0.5	8.927	4.5453	511	1.0343	5.530	1.0368	1.039
0.6	10.693	5.4052	632	1.0414	5.527	1.0758	1.039
0.7	12.453	6.2497	759	1.0484	5.525	1.1143	1.039
0.8	14.207	7.0794	896	1.0554	5.522	1.1524	1.039
0.9	15.955	7.8944	1043	1.0622	5.519	1.1901	1.040
1.0	17.696	8.6953	1199	1.0690	5.516	1.2273	1.040
1.2	21.160	10.2560	1539	1.0823	5.508	1.3385	1.042
1.4	24.600	11.7643	1919	1.0954	5.501	1.4473	1.043
1.6	28.016	13.2226	2339	1.1080	5.491	1.5543	1.044
1.8	31.408	14.6336	2805	1.1205	5.482	1.6590	1.046
2.0	34.777	15.9994	3315	1.1326	5.471	1.7620	1.047
2.5	43.096	19.2301	4805	1.1619	5.444	2.1514	1.053
3.0	51.273	22.2215	6596	1.1901	5.416	2.5239	1.061
3.5	59.312	24.9992	8692	1.2167	5.385	—	—
4.0	67.216	27.5853	11,085	1.2428	5.356	—	—
4.5	74.987	29.9990	13,796	1.2671	5.323	—	—

Table A-10. Data for the System [$Mg(NO_3)_2$-H_2O] at 25°

Molality	Mole fraction × 10^3	Weight % solute	Osmotic pressure (lb/in^2)	Density of solution (g/cm^2)	Molar density ($mole/cm^3$ × 10^2)	Kinematic viscosity (cm^2/sec × 10^2)	Solute diffusivity (cm^2/sec × 10^5)
0	0	0	0	0.9971	5.535	0.8963	1.602
0.1	1.798	1.4616	92	1.0080	5.524	0.9120	1.047
0.2	3.590	2.8811	187	1.0185	5.511	0.935	1.032
0.3	5.375	4.2603	287	1.0290	5.498	0.964	1.029
0.4	7.154	5.6009	394	1.0385	5.481	0.992	1.028
0.5	8.927	6.9044	506	1.0490	5.470	1.025	1.028
0.6	10.693	8.1725	625	1.0570	5.446	1.065	1.029
0.7	12.453	9.4064	747	1.0670	5.433	1.105	1.031
0.8	14.207	10.6076	876	1.0770	5.421	1.150	1.033
0.9	15.955	11.7774	1013	1.0850	5.400	1.190	1.034
1.0	17.696	12.9170	1156	1.0950	5.389	1.230	1.035
1.2	21.160	15.1101	1465	1.1130	5.358	1.320	1.036
1.4	24.600	17.1954	1797	1.1310	5.330	1.420	1.037
1.6	28.016	19.1807	2155	1.1490	5.303	1.520	1.038
1.8	31.408	21.0730	2540	1.1650	5.270	1.640	1.039
2.0	34.777	22.8788	2954	1.1840	5.251	1.770	1.040
2.2	38.122	24.6037	3403	1.2000	5.221	1.910	1.040
2.4	41.444	26.2532	3881	1.2160	5.193	2.040	—
2.6	44.743	27.8321	4394	1.2330	5.171	—	—
2.8	48.019	29.3448	4913	1.2490	5.146	—	—
3.0	51.273	30.7953	5523	1.2640	5.118	—	—

Table A-11. Data for the System [$MgSO_4$-H_2O] at 25°

Molality	Mole fraction × 10³	Weight % solute	Osmotic pressure (lb/in²)	Density of solution (g/cm³)	Molar density (mole/cm³ × 10²)	Kinematic viscosity (cm²/sec × 10²)	Solute diffusivity (cm²/sec × 10⁵)
0	0	0	0	0.9971	5.535	0.8963	0.849
0.1	1.798	1.1896	43	1.0091	5.545	0.9335	0.602
0.2	3.590	2.3512	81	1.0209	5.554	0.9707	0.602
0.3	5.375	3.4858	116	1.0325	5.561	1.0107	0.586
0.4	7.154	4.5943	152	1.0440	5.569	1.0541	0.571
0.5	8.927	5.6777	187	1.0553	5.575	1.1005	0.556
0.6	10.693	6.7368	223	1.0665	5.581	1.1497	0.550
0.7	12.453	7.7723	260	1.0776	5.586	1.1991	0.543
0.8	14.207	8.7851	298	1.0885	5.591	1.2585	0.533
0.9	15.955	9.7759	336	1.0993	5.595	1.3192	0.519
1.0	17.696	10.7454	378	1.1100	5.599	1.3786	0.504
1.2	21.160	12.6232	470	1.1310	5.604	1.5326	0.493
1.4	24.600	14.4236	576	1.1515	5.608	1.6973	0.483
1.6	28.016	16.1513	686	1.1716	5.610	1.8878	0.473
1.8	31.408	17.8106	829	1.1913	5.611	2.1058	0.463
2.0	34.777	19.4055	977	1.2107	5.612	2.3700	0.453
2.5	43.096	23.1346	1438	1.2572	5.606	3.2349	—
3.0	51.273	26.5338	2046	1.3011	5.593	4.5428	—

Table A-12. Data for the System [KCl-H_2O] at 25°

Molality	Mole fraction × 10³	Weight % solute	Osmotic pressure (lb/in²)	Density of solution (g/cm³)	Molar density (mole/cm² × 10²)	Kinematic viscosity (cm²/sec × 10²)	Solute diffusivity (cm²/sec × 10⁵)
0	0	0	0	0.9971	5.535	0.8963	1.993
0.1	1.798	0.7400	67	1.0018	5.530	0.8912	1.844
0.2	3.590	1.4691	131	1.0064	5.524	0.8867	1.838
0.3	5.375	2.1876	195	1.0110	5.519	0.8822	1.838
0.4	7.154	2.8957	259	1.0155	5.513	0.8779	1.844
0.5	8.927	3.5936	323	1.0200	5.508	0.8735	1.849
0.6	10.693	4.2815	387	1.0244	5.502	0.8694	1.857
0.7	12.453	4.9597	452	1.0287	5.495	0.8655	1.865
0.8	14.207	5.6283	516	1.0330	5.489	0.8615	1.873
0.9	15.955	6.2876	581	1.0373	5.483	0.8576	1.881
1.0	17.696	6.9378	645	1.0415	5.477	0.8538	1.889
1.2	21.160	8.2114	776	1.0499	5.465	0.8482	1.907
1.4	24.600	9.4506	908	1.0580	5.452	0.8428	1.926
1.6	28.016	10.6569	1042	1.0660	5.439	0.8377	1.945
1.8	31.408	11.8314	1178	1.0739	5.426	0.8327	1.966
2.0	34.777	12.9754	1315	1.0817	5.414	0.8279	1.986
2.5	43.096	15.7096	1666	1.1004	5.381	0.8202	2.036
3.0	51.273	18.2773	2029	1.1184	5.348	0.8159	2.083
3.5	59.312	20.6931	2400	1.1357	5.315	0.8149	2.127
4.0	67.216	22.9703	2788	1.1524	5.283	0.8143	2.163
4.5	74.987	25.1203	3186	1.1684	5.250	—	2.193

Table A-13. Data for the System [KNO_3-H_2O] at 25°

Molality	Mole fraction × 10^3	Weight % solute	Osmotic pressure (lb/in^2)	Density of solution (g/cm^3)	Molar density ($mole/cm^3$ × 10^2)	Kinematic viscosity (cm^2/sec × 10^2)	Solute diffusivity (cm^2/sec × 10^5)
0	0	0	0	0.9971	5.535	0.8963	1.886
0.1	1.798	1.0010	65	1.0075	5.547	0.8905	1.831
0.2	3.590	1.9821	125	1.0112	5.522	0.8900	1.787
0.3	5.375	2.9440	183	1.0116	5.480	0.8906	1.760
0.4	7.154	3.8872	239	1.0222	5.478	0.8824	1.736
0.5	8.927	4.8122	293	1.0271	5.476	0.8782	1.718
0.6	10.693	5.7196	345	1.0325	5.462	0.8732	1.701
0.7	12.453	6.6099	397	1.0380	5.449	0.8670	1.689
0.8	14.207	7.4835	447	1.0436	5.437	0.8566	1.683
0.9	15.955	8.3409	497	1.0490	5.424	0.8446	1.679
1.0	17.696	9.1825	544	1.0550	5.414	0.8341	1.674
1.2	21.160	10.8203	636	1.0656	5.389	0.8262	1.654
1.4	24.600	12.4001	724	1.0768	5.368	0.8217	1.635
1.6	28.016	13.9249	806	1.0870	5.343	0.8229	1.613
1.8	31.408	15.3975	886	1.0972	5.320	0.8303	1.584
2.0	34.777	16.8205	963	1.1107	5.313	0.8463	1.536
2.2	38.122	18.1965	1034	1.1117	5.248		
2.4	41.444	19.5277	1102	1.1265	5.250		
2.6	44.743	20.8163	1180	1.1336	5.216		
2.8	48.019	22.0642	1242	1.1446	5.202		
3.0	51.273	23.2734	1305	1.1550	5.185		
3.2	54.505	24.4457	1367	1.1640	5.163		
3.4	57.715	25.5827	1428	1.1732	5.143		
3.6	60.903	26.6860	1491	1.1820	5.122		

Table A-14. Data for the System [K_2SO_4-H_2O] at 25°

Molality	Mole fraction × 10^3	Weight % solute	Osmotic pressure (lb/in^2)	Density of solution (g/cm^3)	Molar density (mole/cm^3 × 10^2)	Kinematic viscosity (cm^2/sec × 10^2)	Solute diffusivity (cm^2/sec × 10^5)
0	0	0	0	0.9971	5.535	0.8963	1.954
0.1	1.798	1.7128	84	1.0105	5.523	0.9067	1.301
0.2	3.590	3.3680	160	1.0225	5.504	0.9150	1.245
0.3	5.375	4.9683	234	1.0370	5.500	0.9214	1.198
0.4	7.154	6.5165	303	1.0490	5.483	0.9316	1.164
0.5	8.927	8.0151	373	1.0610	5.466	0.9388	1.141
0.6	10.693	9.4664	440	1.0735	5.453	—	
0.7	12.453	10.8726	506	1.0855	5.438	—	

Table A-15. Osmotic pressures of sea salt solutions at different temperatures (atm)

Data from Johnson et al. (1966)

Temp. (°)	Weight % Salts								
	1.0	2.0	3.45	5.0	7.5	10.0	15.0	20.0	25.0
25	7.1	14.3	25.1	37.5	59.3	84	145	230	350
40	7.4	14.9	26.3	39.3	62.4	89	153	240	360
60	7.8	15.7	27.7	41.5	65.9	94	162	250	380
80	8.1	16.4	28.9	43.3	68.8	98	168	260	390
100	8.4	16.9	29.9	44.7	71.1	101	173	270	400

Table A-16. Osmotic pressure of sea water in atmospheres at various temperatures and concentrations of dissolved solids

Data of Tribus et al. (1960)

% Solids by wt.	Temperature (°F)										
	32	40	60	80	100	120	140	160	180	200	212
2	12.50	13.21	11.62	12.32	14.85	15.91	15.77	14.16	12.01	9.70	9.10
4	24.34	25.24	25.79	26.59	26.71	30.64	30.00	28.17	26.42	24.68	24.09
6	37.51	38.08	40.81	42.48	43.85	48.02	48.37	46.02	41.50	39.59	39.23
8	51.81	53.19	59.11	61.15	62.61	67.50	68.28	65.11	62.21	59.37	59.07
10	70.63	68.00	76.51	82.24	85.54	89.38	88.04	83.91	82.44	82.22	81.70
12	88.48	86.27	97.79	105.83	110.28	112.95	111.87	110.11	109.35	107.41	106.33
14	107.58	111.11	120.13	131.40	137.67	140.40	139.49	138.54	139.84	135.24	133.46
16	130.95	129.15	144.15	159.11	168.74	169.30	170.87	171.63	170.83	167.97	164.95
18	149.98	156.55	160.70	192.50	184.10	205.10	205.34	206.92	205.57	204.02	201.65
20	177.90	185.49	206.30	225.60	236.10	237.80	237.51	242.26	245.11	244.25	241.77
22	210.00	234.00	243.00	263.30	274.06	276.00	276.95	284.22	290.12	291.83	291.16
24	248.30	260.70	286.90	305.90	316.88	318.00	324.37	332.38	359.28	340.93	340.58
26	344.80	302.90	333.90	351.30	362.41	363.00	371.46	382.44	397.63	398.19	399.96

Table A-17. Data for the System [NaCl — H_2O] at 25°

Molality	Mole fraction × 10^3	Weight % solute	Osmotic pressure (lb/in^2)	Density of solution (g/cm^3)	Molar density ($mole/cm^3$ × 10^2)	Kinematic viscosity (cm^2/sec × 10^2)	Solute diffusivity (cm^2/sec × 10^5)
0	0	0	0	0.9971	5.535	0.8963	1.610
0.1	1.798	0.5811	67	1.0011	5.535	0.9009	1.483
0.2	3.590	1.1555	133	1.0052	5.535	0.9054	1.475
0.3	5.375	1.7233	199	1.0091	5.535	0.9100	1.475
0.4	7.154	2.2846	264	1.0130	5.534	0.9147	1.475
0.5	8.927	2.8395	331	1.0169	5.534	0.9193	1.475
0.6	10.693	3.3882	398	1.0208	5.534	0.9242	1.475
0.7	12.453	3.9307	466	1.0248	5.534	0.9290	1.475
0.8	14.207	4.4671	534	1.0286	5.533	0.9338	1.477
0.9	15.955	4.9976	603	1.0322	5.532	0.9389	1.480
1.0	17.696	5.5222	673	1.0357	5.530	0.9440	1.483
1.2	21.160	6.5543	814	1.0427	5.526	0.9567	1.488
1.4	24.600	7.5640	959	1.0505	5.526	0.9685	1.492
1.6	28.016	8.5522	1109	1.0581	5.526	0.9802	1.497
1.8	31.408	9.5194	1262	1.0653	5.524	0.9923	1.505
2.0	34.777	10.4665	1419	1.0722	5.521	1.0044	1.513
2.2	38.122	11.3939	1580	1.0790	5.517	1.0206	1.521
2.4	41.444	12.3022	1745	1.0859	5.515	1.0365	1.530
2.6	44.743	13.1922	1915	1.0927	5.512	1.0523	1.539
2.8	48.019	14.0642	2089	1.0991	5.507	1.0683	1.548
3.0	51.273	14.9190	2270	1.1056	5.504	1.0840	1.556
3.2	54.505	15.7568	2453	1.1121	5.500	1.1047	1.565
3.4	57.715	16.5784	2651	1.1185	5.497	1.1252	1.570
3.6	60.903	17.3840	2834	1.1247	5.492	1.1457	1.575
3.8	64.070	18.1743	3034	1.1309	5.488	1.1660	1.580
4.0	67.216	18.9496	3238	1.1369	5.484	1.1862	1.585
4.2	70.340	19.7103	3446	1.1429	5.479	1.2108	1.589
4.4	73.443	20.4569	3661	1.1490	5.475	1.2350	1.594
4.6	76.526	21.1897	3879	1.1550	5.472	1.2591	1.593
4.8	79.589	21.9092	4104	1.1608	5.467	1.2832	1.593
5.0	82.631	22.6156	4333	1.1666	5.463	1.3070	1.592
5.2	85.653	23.3093	4568	1.1723	5.458		1.592
5.4	88.655	23.9908	4807	1.1778	5.453		1.591
5.6	91.638	24.6602	5054	1.1832	5.447		1.590
5.8	94.601	25.3179	5304	1.1887	5.443		
6.0	97.545	25.9643	5560	1.1941	5.438		

Table A-18. Effect of temperature on osmotic pressure and molar density for the system sodium chloride-water

	Osmotic Pressure (lb/in^2)			Molar density ($g.mole/cm^3 \times 10^2$)		
Molality	5°	15°	35°	5°	15°	35°
0	0	0	0	5.551	5.547	5.518
0.1	61	64	68	5.551	5.547	5.518
0.2	119	124	132	5.551	5.546	5.518
0.3	178	185	197	5.551	5.545	5.520
0.4	244	254	270	5.551	5.545	5.516
0.5	300	313	334	5.551	5.540	5.513
0.6	367	383	409	5.554	5.546	5.513
0.7	424	443	472	5.554	5.546	5.511
0.8	491	513	546	5.554	5.546	5.509
0.9	548	573	610	5.557	5.547	5.509
1.0	616	644	685	5.558	5.547	5.510
1.2	752	786	836	5.556	5.546	5.506
1.4	888	933	1000	5.558	5.545	5.505
1.6	1016	1068	1145	5.559	5.544	5.502
1.8	1152	1210	1294	5.559	5.541	5.499
2.0	1286	1344	1437	5.556	5.538	5.494
2.2	1437	1501	1610	5.556	5.538	5.494
2.4	1581	1662	1775	5.551	5.533	5.490
2.6	1737	1830	1957	5.549	5.531	5.486
2.8	1891	1992	2131	5.547	5.527	5.482
3.0	2060	2172	2319	5.543	5.523	5.478
3.2	2229	2352	2511	5.539	5.520	5.473
3.4	2410	2545	2713	5.538	5.516	5.470
3.6	2583	2729	2897	5.533	5.513	5.465

Table A-19. Osmotic pressures of aqueous sodium cloride solutions at different temperatures

Data from Johnson et al. (1966)

	Osmotic Pressure (atm) at			
Molality	25°	40°	60°	100°
0.001	0.05	0.05	0.05	0.06
0.01	0.47	0.49	0.52	0.57
0.05	2.31	2.41	2.53	2.75
0.10	4.56	4.76	5.00	5.42
0.20	9.04	9.44	9.93	10.74
0.40	18.02	18.84	19.83	21.45
0.60	27.12	28.40	29.92	32.35
0.80	36.37	38.14	40.22	43.48
1.00	45.80	48.08	50.76	54.87
2.00	96.2	101.3	107.3	115.9
3.00	153.2	161.6	171.0	184.2
4.00	218.9	230.5	243.3	260.8
5.00	295.2	309.4	325.2	346.5
6.00	384.1	400.2	418.0	442.2

Table A-20. Data for the System [$NaNO_3$—H_2O] at 25°

Molality	Mole fraction × 10^3	Weight % solute	Osmotic pressure (lb/in^2)	Density of solution (g/cm^3)	Molar density ($mole/cm^3$ × 10^2)	Kinematic viscosity (cm^2/sec × 10^2)	Solute diffusivity (cm^2/sec × 10^5)
0	0	0	0	0.9971	5.535	0.8963	1.568
0.1	1.798	0.8429	66	1.0027	5.529	0.8958	1.443
0.2	3.590	1.6718	130	1.0082	5.523	0.8950	1.427
0.3	5.375	2.4869	192	1.0137	5.517	0.8943	1.414
0.4	7.154	3.2886	253	1.0191	5.510	0.8937	1.407
0.5	8.927	4.0772	314	1.0245	5.504	0.8941	1.403
0.6	10.693	4.8531	374	1.0297	5.497	0.8960	1.399
0.7	12.453	5.6165	434	1.0351	5.492	0.8977	1.394
0.8	14.207	6.3677	494	1.0401	5.484	0.8997	1.389
0.9	15.955	7.1071	553	1.0453	5.477	0.9016	1.384
1.0	17.696	7.8350	613	1.0503	5.470	0.9036	1.379
1.2	21.160	9.2569	730	1.0603	5.456	0.9138	1.371
1.4	24.600	10.6356	846	1.0700	5.442	0.9241	1.362
1.6	28.016	11.9731	963	1.0796	5.427	0.9342	1.353
1.8	31.408	13.2711	1078	1.0892	5.414	0.9443	1.345
2.0	34.777	14.5314	1193	1.0984	5.399	0.9544	1.336
2.5	43.096	17.5275	1476	1.1210	5.363	0.9835	1.325
3.0	51.273	20.3206	1759	1.1405	5.317	1.0141	1.318
3.5	59.312	22.9308	2038	1.1635	5.291	1.0576	1.310
4.0	67.216	25.3754	2311	1.1836	5.256	1.1020	1.303
4.5	74.987	27.6696	2586	1.2027	5.220	1.1459	1.296
5.0	82.631	29.8270	2861	1.2210	5.185	1.1892	—
5.5	90.149	31.8594	3146	1.2390	5.151	—	—
6.0	97.545	33.7775	3439	1.2560	5.116	—	—

Table A-21. Data for the System [Na_2SO_4 — H_2O] at 25°

Molality	Mole fraction × 10^3	Weight % solute	Osmotic pressure (lb/in^2)	Density of solution (g/cm^2)	Molar density ($mole/cm^3$ × 10^2)	Kinematic viscosity (cm^2/sec × 10^2)	Solute diffusivity (cm^2/sec × 10^5)
0	0	0	0	0.9971	5.535	0.8963	1.230
0.1	1.798	1.4006	85	1.0097	5.536	0.9236	1.042
0.2	3.590	2.7625	162	1.0220	5.536	0.9511	1.008
0.3	5.375	4.0873	235	1.0340	5.535	0.9793	0.975
0.4	7.154	5.3765	304	1.0458	5.533	1.0101	0.941
0.5	8.927	6.6315	372	1.0574	5.530	1.0426	0.909
0.6	10.693	7.8536	439	1.0687	5.526	1.0767	0.889
0.7	12.453	9.0442	504	1.0800	5.522	1.1128	0.874
0.8	14.207	10.2043	568	1.0910	5.517	1.1502	0.861
0.9	15.955	11.3353	632	1.1019	5.511	1.1953	0.848
1.0	17.696	12.4382	694	1.1126	5.505	1.2423	0.836
1.2	21.160	14.5635	819	1.1335	5.492	1.3399	—
1.4	24.600	16.5881	949	1.1539	5.478	1.4367	—
1.6	28.016	18.5190	1080	1.1737	5.462	1.5461	—
1.8	31.408	20.3625	1217	1.1928	5.577	1.6849	—
2.0	34.777	22.1244	1359	1.2115	5.237	1.8317	—

Table A-22. Data for the System Sucrose-Water at 25°

Molality	Mole fraction $\times 10^3$	Weight % solute	Osmotic pressure (lb/in^2)	Density of solution (g/cm^3)	Molar density ($mole/cm^3$ $\times 10^2$)	Kinematic viscosity (cm^2/sec $\times 10^2$)	Solute diffusivity (cm^2/sec $\times 10^5$)
0	0	0	0	0.9971	5.535	0.8963	0.523
0.1	1.798	3.3097	36	1.0100	5.431	0.9615	0.509
0.2	3.590	6.4074	73	1.0222	5.330	1.0352	0.499
0.3	5.375	9.3127	110	1.0339	5.233	1.1151	0.490
0.4	7.154	12.0431	148	1.0453	5.140	1.2053	0.483
0.5	8.927	14.6138	186	1.0560	5.050	1.3033	0.477
0.6	10.693	17.0386	225	1.0665	4.965	1.4124	0.472
0.7	12.453	19.3295	265	1.0764	4.881	1.5330	0.467
0.8	14.207	21.4972	305	1.0862	4.802	1.6639	0.463
0.9	15.955	23.5515	345	1.0953	4.723	1.8083	0.459
1.0	17.696	25.5010	387	1.1042	4.649	1.9658	0.455
1.2	21.160	29.1162	470	1.1210	4.506	2.3270	0.448
1.4	24.600	32.3968	557	1.1367	4.373	2.7580	0.441
1.6	28.016	35.3872	645	1.1512	4.248	3.2701	0.434
1.8	31.408	38.1242	734	1.1649	4.131	3.8772	0.428
2.0	34.777	40.6387	826	1.1777	4.021	4.6023	0.421
2.5	43.096	46.1134	1069	1.2063	3.771	7.0584	0.404
3.0	51.273	50.6636	1324	1.2310	3.553	10.8171	0.387
3.5	59.312	54.5051	1592	1.2524	3.362	16.5067	0.370
4.0	67.216	57.7917	1866	1.2711	3.193	25.0529	—

Table A-23. Data for the System Urea-Water at 25°

Molality	Mole fraction × 10^3	Weight % solute	Osmotic pressure (lb/in^2)	Density of solution (g/cm^3)	Molar density ($mole/cm^3$ × 10^2)	Kinematic viscosity (cm^2/sec × 10^2)	Solute diffusivity (cm^2/sec × 10^5)
0	0	0	0	0.9971	5.535	0.8963	1.3817
0.1	1.798	0.5970	36	0.9986	5.520	0.8983	1.3739
0.2	3.590	1.1869	71	1.0002	5.506	0.8998	1.3663
0.3	5.375	1.7699	106	1.0017	5.492	0.9005	1.3591
0.4	7.154	2.3460	141	1.0033	5.478	0.9025	1.3520
0.5	8.927	2.9154	176	1.0048	5.464	0.9037	1.3453
0.6	10.693	3.4783	210	1.0063	5.450	0.9058	1.3380
0.7	12.453	4.0346	244	1.0078	5.435	0.9074	1.3311
0.8	14.207	4.5845	278	1.0092	5.422	0.9096	1.3243
0.9	15.955	5.1282	312	1.0107	5.409	0.9117	1.3180
1.0	17.696	5.6657	346	1.0121	5.395	0.9144	1.3110
1.2	21.160	6.7227	412	1.0150	5.369	0.9187	1.2987
1.4	24.600	7.7562	478	1.0178	5.343	0.9236	1.2874
1.6	28.016	8.7671	542	1.0206	5.318	0.9284	1.2763
1.8	31.408	9.7561	607	1.0232	5.292	0.9338	1.2703
2.0	34.777	10.7239	671	1.0259	5.267	0.9402	1.2544
2.5	43.096	13.0548	826	1.0322	5.206	0.9533	1.2301
3.0	51.273	15.2672	979	1.0381	5.147	0.9686	1.2085
3.5	59.312	17.3697	1128	1.0439	5.090	0.9848	1.1893
4.0	67.216	19.3704	1276	1.0495	5.036	1.0000	1.1709
4.5	74.987	21.2766	1421	1.0548	4.983	1.0154	1.1555
5.0	82.631	23.0947	1564	1.0599	4.932	1.0312	1.1413
5.5	90.149	24.8307	1704	1.0647	4.883	1.0468	1.1289
6.0	97.545	26.4900	1847	1.0694	4.835	1.0656	1.1174

APPENDIX III

Reverse Osmosis System Specification and Performance Data—An Illustration

System Specification

$\gamma = 0.3$, $\theta = 0.01$, and $\lambda = 100$

Δ	C_1	C_2	C_3	$\overline{C}_3$	τ or X
0	1.0000	1.6510	0.0315	0.0315	0
0.02	1.0197	1.6730	0.0323	0.0332	0.0391
0.04	1.0308	1.6852	0.0328	0.0330	0.0606
0.06	1.0617	1.7190	0.0341	0.0332	0.1189
0.08	1.0840	1.7431	0.0350	0.0335	0.1596
0.10	1.1075	1.7682	0.0361	0.0339	0.2010
0.12	1.1317	1.7938	0.0371	0.0344	0.2429
0.14	1.1573	1.8203	0.0383	0.0349	0.2856
0.16	1.1838	1.8476	0.0395	0.0354	0.3288
0.18	1.2116	1.8759	0.0408	0.0359	0.3728
0.20	1.2408	1.9053	0.0422	0.0365	0.4177
0.22	1.2715	1.9357	0.0438	0.0371	0.4634
0.24	1.3051	1.9681	0.0455	0.0377	0.5109
0.26	1.3393	2.0010	0.0473	0.0384	0.5588
0.28	1.3736	2.0339	0.0491	0.0391	0.6067
0.30	1.4126	2.0701	0.0513	0.0398	0.6574
0.32	1.4519	2.1065	0.0535	0.0406	0.7084
0.34	1.4944	2.1450	0.0560	0.0414	0.7613
0.36	1.5388	2.1847	0.0587	0.0423	0.8153
0.38	1.5868	2.2267	0.0618	0.0432	0.8716
0.40	1.6369	2.2700	0.0652	0.0442	0.9292
0.42	1.6910	2.3158	0.0690	0.0453	0.9893
0.44	1.7524	2.3661	0.0738	0.0466	1.0543
0.46	1.8152	2.4173	0.0789	0.0479	1.1203
0.48	1.8781	2.4684	0.0839	0.0492	1.1863
0.50	1.9513	2.5257	0.0908	0.0507	1.2597
0.52	2.0271	2.5846	0.0981	0.0524	1.3350
0.54	2.1112	2.6482	0.1073	0.0542	1.4167
0.56	2.2005	2.7149	0.1178	0.0562	1.5027
0.58	2.2986	2.7869	0.1308	0.0586	1.5964
0.60	2.4070	2.8652	0.1474	0.0613	1.7001
0.62	2.5303	2.9530	0.1708	0.0646	1.8203
0.64	2.6556	3.0421	0.1955	0.0680	1.9431
0.66	2.7998	3.1447	0.2319	0.0724	2.0911
0.68	2.9578	3.2586	0.2802	0.0777	2.2623
0.70	3.1343	3.3894	0.3482	0.0844	2.4700
0.72	3.3337	3.5448	0.4475	0.0933	2.7360
0.74	3.5493	3.7197	0.5722	0.1042	3.0528
0.76	3.7899	3.9258	0.7377	0.1184	3.4595
0.78	4.0571	4.1651	0.9456	0.1366	3.9767
0.80	4.3640	4.4516	1.2098	0.1611	4.6747
0.82	4.6916	4.7618	1.5011	0.1889	5.4692

Δ	C_1	C_2	C_3	$\overline{C}_3$	τ or X
0.84	5.0710	5.1287	1.8545	0.2243	6.4986
0.86	5.5009	5.5484	2.2636	0.2666	7.7444
0.88	6.0003	6.0397	2.7463	0.3170	9.2565
0.90	6.6274	6.6602	3.3602	0.3796	11.1769
0.92	7.3699	7.3970	4.0911	0.4516	13.4227
0.94	8.3284	8.3504	5.0393	0.5371	16.1273
0.96	9.6623	9.6797	6.3637	0.6414	19.4702
0.98	11.9506	11.9638	8.6433	0.7763	23.8425

The symbols are the same as those used in Chapter 4.

$$\Delta = 1 - V_1/V_1^0 = 1 - \bar{u}/\bar{u}^0 = \text{fraction product recovery}$$

$$C_1 = c_{A1}/c_{A1}^0;\ C_2 = c_{A2}/c_{A1}^0;\ C_3 = c_{A3}/c_{A1}^0;\ \overline{C}_3 = \bar{c}_{A3}/c_{A1}^0$$

$$\tau = \frac{S \cdot v_W^* \cdot t}{V_1^0};\ X = \frac{v_W^*}{\bar{u}^0}\frac{x}{h}$$

Author Index

Numbers underlined refer to pages where complete literature references are given.

Subject Index

Substances and solution systems mentioned in the text are indexed where appropriate under the following headings: Additive; Membrane; Plasticiser; Solution system; Solvent; and Surface active agent.